Student Survival
and
Solutions Manual

Karl J. Smith

Calculus

THIRD EDITION

Strauss ▪ Bradley ▪ Smith

Prentice
Hall

Upper Saddle River, NJ 07458

Acquisitions Editor: Eric Frank
Supplement Editor: Aja Shevelew
Assistant Managing Editor: John Matthews
Production Editor: Wendy A. Perez
Supplement Cover Manager: Paul Gourhan
Supplement Cover Designer: Joanne Alexandris
Manufacturing Buyer: Ilene Kahn

© 2002 by Pearson Education, Inc.
Pearson Education, Inc.
Upper Saddle River, NJ 07458

Printed in the United States of America

21 22 23 24 25 26 V036 15 14 13 12

ISBN 0-13-067245-9

Pearson Education Ltd., *London*
Pearson Education Australia Pty. Ltd., *Sydney*
Pearson Education Singapore, Pte. Ltd.
Pearson Education North Asia Ltd., *Hong Kong*
Pearson Education Canada, Inc., *Toronto*
Pearson Educacíon de Mexico, S.A. de C.V.
Pearson Education—Japan, *Tokyo*
Pearson Education Malaysia, Pte. Ltd.
Pearson Education, *Upper Saddle River, New Jersey*

PREFACE

This manual was designed to help you bridge the gap between the textbook and a working knowledge of calculus. It has been said that "Mathematics is not a Spectator Sport" and this means you cannot learn calculus by simply attending class, but instead you must build a body of information that will enable you to do problem solving in the real world. I decided to entitle this supplement, *Student Survival and Solutions Manual* because I want it to be more than a Student Solution's Manual. Thirty years of teaching experience have given me the ability to anticipate the types of errors and difficulties you may have while taking this course. Here I will show you some of the steps that are left out of the text, and most all of these steps in the included problems.

There are several things you must do if you wish to be successful with calculus:

- Attend every class.

- Read the book.
  Regardless of how clear and lucid your professor's lecture on a particular topic may be, do not attempt to do the problems without first reading the text and studying the examples. It will serve to reinforce and clarify the concepts and procedures.

- **Problems,** Problems, Problems, problems...
  You must work problems every day; work the assigned problems; work **WHAT DOES THIS SAY** problems. Look over the entire problem set (even those problems which are not assigned).

- Ask questions when you are stuck (and you will get stuck — that is part of the process).

- Keep asking questions until receiving answers that are understandable to you.

- Today's calculators and computers are good at obtaining answers, and if all that is desired is an answer, then you have relegated yourself to the level of a machine. Do not work problems to obtain answers. It is the *concepts* that are important. Even though a solutions manual is basically a "how to" document, always ask *why* a particular approach was used, and understand the concept the problem is illustrating.

For the most part, this manual includes complete solutions to the odd-numbered problems, all solutions to the Proficiency Examinations, but none of the solutions for the miscellaneous problems. Some hints are provided for **WHAT THIS SAYS**, **COUNTEREXAMPLE PROBLEMS**, and **EXPLORATION PROBLEMS**, but you should remember that you are expected to answer these problems in your own words. I provided all the solutions to the Proficiency Examinations to help you prepare for tests, but none of the solutions to the supplementary problems are included in this manual.

There are several places in this manual where we used computer software. Even though such software is, of course, optional, it can help us through much of the tedium. The output shown is from *Converge* 4.0 available from JEMware, The Kawaiahao Plaza Executive Center, 567 South King Street, Suite 178, Honolulu, Hawaii 96813. Phone: 808-523-9911.

CONTENTS

CHAPTER 1

Functions and Graphs

SURVIVAL HINT: *If your instructor does not begin with Chapter 1, you might wish to take some time looking over this chapter anyway. As you look through this first chapter you will notice that we cover calculator graphing, lines, absolute value, trigonometry, and inverse functions. Pay particular attention to the definition of a function in Section 1.3, as well as functional notation. In order to succeed in this course, you will need to be thoroughly familiar with the meaning and use of the notation $y = f(x)$. The name of this function is f and the value of f at a value x is denoted by $f(x)$. Even though all of the material of this chapter was covered in precalculus classes, you will notice that calculus is probably the first course that you take which* **actually assumes** *that you remember the content and ideas of previous courses.*

If you purchased the textbook new, you should have found a free CD that contains a Student Mathematics Handbook. This not only reviews geometry, algebra, trigonometry, curve sketching, and the conic sections, it also summarizes the ideas you will study in this course; namely limits, derivatives, integrals, and series. Finally, it contains a complete integration table. If you purchased the textbook used, you may find this CD is missing. You can purchase from your bookstore a printed copy of the Student Mathematics Handbook, or a replacement CD.

Find out from your instructor what is expected of you. You will probably need a copy of this textbook, engineering paper, a straight-edge, and a calculator. (By the way, the cover for your calculator makes a good straight-edge.) Put your name and phone number inside the cover of your calculator, so if you lose it, it is, at least, possible that it be returned.

As you begin on your calculus journey, Bon Voyage!

1.2 Preliminaries, Pages 10-13

SURVIVAL HINT: *Interval notation is very compact and we will use it frequently in the book. Problems 1-4 are designed to see if you are familiar with this notation.*

1. **a.** $-3 < x < 4$ means that x is between -3 and 4. In interval notation this is $(-3, 4)$.

    b. The interval [3, 5] consists of all numbers between 3 and 5, including the endpoints: $3 \leq x \leq 5$

    c. The interval $[-2, 1)$ consists of all numbers between -2 and 1, including the endpoint -2: $-2 \leq x < 1$

    d. $2 < x \leq 7$ means that x is between 2 and 7, and includes the endpoint $x = 7$. In interval notation this is $(2, 7]$

3. **a.** This interval is closed on the left and open on the right.

b. This interval is closed on both the right

and the left.

c. This interval is open on both the right

and the left.

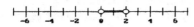

d. This interval is closed on the left and

open on the right.

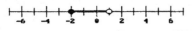

5. **a.**

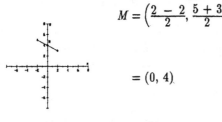

$$M = \left(\frac{2-2}{2}, \frac{5+3}{2}\right)$$

$$= (0, 4)$$

$$d = \sqrt{(-2-2)^2 + (5-3)^2}$$
$$= \sqrt{20}$$
$$= 2\sqrt{5}$$

b.

$$M = \left(\frac{-2+4}{2}, \frac{3+1}{2}\right)$$
$$= (1, 2)$$

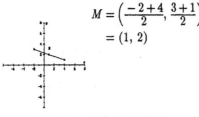

$$d = \sqrt{(4+2)^2 + (1-3)^2}$$
$$= \sqrt{40}$$
$$= 2\sqrt{10}$$

SURVIVAL HINT: *Solving quadratic equations is a routine skill that will be assumed throughout this course. Make sure you can carry out all of the steps in Problems 7-11. In this book, if the discriminant ($b^2 - 4ac$) is less than zero we usually do state the complex roots, because we assume we are working in the set of real numbers.*

7. $\quad x^2 - x = 0 \quad$ *This is a quadratic equation.*

$\quad x(x - 1) = 0 \quad$ *Factor, if possible.*

$\quad x = 0, 1 \quad$ *Set each factor equal to 0 (mentally) and solve.*

9. $\quad y^2 - 5y + 3 = 17 \quad$ *When working problems, copy the problem down first on your own paper.*

$\quad y^2 - 5y - 14 = 0 \quad$ *This is a quadratic equation, so begin by getting a 0 on one side.*

$\quad (y - 7)(y + 2) = 0 \quad$ *Factor, if possible.*

$\quad y = 7, -2$

Set each factor equal to 0 and solve.

11. $\quad 3x^2 - bx = c \quad$ *This is a quadratic equation.*

$\quad 3x^2 - bx - c = 0 \quad$ *This does not factor, so use the quadratic formula.*

$$x = \frac{-(-b) \pm \sqrt{b^2 - 4(3)(-c)}}{2(3)}$$

$$= \frac{b \pm \sqrt{b^2 + 12c}}{6}$$

SURVIVAL HINT: *Is $-a$ positive or negative? Without further information we do not know. It might be neither. Be comfortable with the definition of absolute value. Especially the fact that $-a$ is positive when $a < 0$.*

13. $\quad |2x + 4| = 16$ is an absolute value equation, and Property 7 from Table 1.1 tells us to write this as two equations, namely

$2x + 4 = \pm 16$. Thus,

$$2x + 4 = 16 \quad \text{or} \quad -(2x + 4) = 16$$
$$2x = 12 \quad \text{or} \quad -2x = 20$$
$$x = 6 \qquad\qquad x = -10$$

We summarize this by writing: $x = 6, -10$.

15. $|3 - 2w| = 7$ is an absolute value equation, and Property 7 from Table 1.1 tells us to write this as two equations, namely

$$3 - 2w = \pm 7.$$
$$3 - 2w = 7 \text{ or } -(3 - 2w) = 7$$
$$2w = -4 \quad \text{or} \quad 2w = 10$$
$$w = -2 \qquad\qquad w = 5$$

Thus, $w = -2, 5$.

17. $\emptyset$; (The empty set; an absolute value can never be equal to a negative number.)

Note: The TI-92 returns the answer: FALSE To see why this is correct, recall the three types of equations you studied in algebra, namely, *true, false,* and *open.* An equation such as $5 = 5$ is true, regardless of the values of the variable; an equation such as $5 = 4$ is false, no matter what the replacement for the variable, there is no value that will make $5 = 4$; finally, an equation such as $x = 5$ is true for the replacement of 5 and false for the replacement of 4.

SURVIVAL HINT: *Most students need some review of solving trigonometric equations. Notice the logo in the text that looks like*

$$\boxed{\text{s}^M\text{H}}$$

The techniques for solving trigonometric equations are reviewed in the handbook, and Problems 19-24 also review this skill.

19. $\sin x = -\frac{1}{2}$ on $[0, 2\pi)$.

The reference angle (in Quad I) is $\pi/6$. The sine is negative in Quad III and Quad IV, so $x = 7\pi/6, 11\pi/6$.

21. $(2\cos x + \sqrt{2})(2\cos x - 1) = 0$

Since there are two factors, use the factor theorem to find the solution by finding the values which make each of the factors zero.

$$2\cos x + \sqrt{2} = 0 \qquad\qquad 2\cos x - 1 = 0$$
$$\cos x = -\frac{\sqrt{2}}{2} \qquad\qquad \cos x = \frac{1}{2}$$
$$x = \frac{3\pi}{4}, \frac{5\pi}{4} \qquad\qquad x = \frac{\pi}{3}, \frac{5\pi}{3}$$

The reference angle (in Quad I) for the first equation is $\pi/4$. The cosine is negative in Quad II and III, so $x = 3\pi/4, 5\pi/4$. The reference angle for the second quadrant is $\pi/3$. The cosine is positive in Quad I and IV, so $x = \pi/3, 5\pi/3$. The entire solution is written as: $x = \frac{3\pi}{4}, \frac{5\pi}{4}, \frac{\pi}{3}, \frac{5\pi}{3}$

23.
$$\cot x + \sqrt{3} = \csc x$$
$$\cot^2 x + 2\sqrt{3}\cot x + 3 = \csc^2 x$$
$$\cot^2 x - \csc^2 x + 2\sqrt{3}\cot x + 3 = 0$$
$$-1 + 2\sqrt{3}\cot x + 3 = 0$$
$$2\sqrt{3}\cot x + 2 = 0$$
$$\cot x = -\frac{\sqrt{3}}{3}$$
$$x = \frac{2\pi}{3}, \frac{5\pi}{3}$$

A check is necessary because we squared both sides: $x = 5\pi/3$ is extraneous, so the solution is: $x = \frac{2\pi}{3}$.

SURVIVAL HINT: *Be careful to pay attention to the endpoints of intervals. Is the interval open, closed, or half-open? On most problems later in the text, the interval is not specified, but may be implied by the given function. For example, if*

$$f(x) = \sqrt{4 - x^2}$$

the endpoints are included, but if

$$f(x) = \frac{1}{\sqrt{4 - x^2}}$$

the endpoints are not included.

25. $3x + 7 < 2$

    $3x < -5$

    $x < -\frac{5}{3}$

    The endpoint is not included.

    Answer: $\left(-\infty, \ -\frac{5}{3}\right)$

27. $-5 < 3x < 0$ *This says that 3x is between -5 and 0.*

    $-\frac{5}{3} < x < 0$ *Divide by 3.*

    Answer: $\left(-\frac{5}{3}, 0\right)$

29. $3 \leq -y < 8$

    $-3 \geq y > -8$ *Reverse inequality since we multiplied by a negative number.*

    $-8 < y \leq -3$ *Restore proper order.*

    Note that the right endpoint is included.

    Answer: $(-8, \ -3]$

31. Problems 31 and 32 are quadratic inequalities.

    $t^2 - 2t \leq 3$

    $t^2 - 2t - 3 \leq 0$

$(t + 1)(t - 3) \leq 0$

Consider the factors, one at a time; plot the critical values. Determine where each factor is positive and where it is negative. We illustrate the procedure:

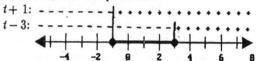

We wish the product to be nonpositive (≤ 0), so the part that is darkened is the part where the product of the factors is negative.

Answer: $[-1, 3]$.

33. Read this problem as a distance function: The distance between x and 8 is less than or equal to 0.001. The interval is $[7.999, 8.001]$.

SURVIVAL HINT: *Problems 35-38 require that you remember the equation of a circle. It is assumed that you have mastered the algebraic skill of completing the square. If not, do some review (see Section 2.6 of the Student Mathematics Handbook, for example).*

35. The equation of a circle with center (h, k) and radius r is: $(x - h)^2 + (y - k)^2 = r^2$. In this problem, we are given $h = -1$, $k = 2$, and $r = 3$, so we have:

    $$(x + 1)^2 + (y - 2)^2 = 9$$

37. The equation of a circle with center (h, k) and radius r is: $(x - h)^2 + (y - k)^2 = r^2$. In this problem, we are given $h = 0$, $k = 1.5 = \frac{3}{2}$, and $r = 0.25 = \frac{1}{4}$, so we have:

    $$x^2 + (y - 1.5)^2 = 0.0625$$

    or in fractional form,

$$x^2 + (y - \tfrac{3}{2})^2 = \tfrac{1}{16}$$

39. $x^2 - 2x + y^2 + 2y + 1 = 0$

We complete the square for both x and y:

$$(x^2 - 2x + 1) + (y^2 + 2y + 1) = -1 + 1 + 1$$
$$(x - 1)^2 + (y + 1)^2 = 1 \quad \textit{Factor}$$

Circle with center at $(1, -1)$ and $r = 1$

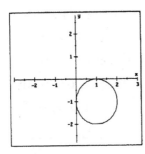

41. $x^2 + y^2 + 2x - 10y + 25 = 0$
$$(x^2 + 2x + 1) + (y^2 - 10y + 25) = 1$$
$$(x + 1)^2 + (y - 5)^2 = 1$$

Circle with center $(-1, 5)$ and $r = 1$

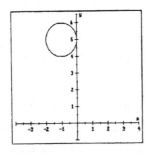

SURVIVAL HINT: *Problems 43-46 require that you remember the addition law identities from trigonometry:*

$$\cos(\alpha \pm \beta) = \cos \alpha \cos \beta \mp \sin \alpha \sin \beta$$

$$\sin(\alpha \pm \beta) = \sin \alpha \cos \beta \pm \cos \alpha \sin \beta$$

$$\tan(\alpha \pm \beta) = \frac{\tan \alpha \pm \tan \beta}{1 \mp \tan \alpha \tan \beta}$$

This, and other trigonometric identities are reviewed in Chapter 4 of the Student Mathematics Handbook.

43. $\sin(-\tfrac{\pi}{12}) = \sin(\tfrac{\pi}{4} - \tfrac{\pi}{3})$

$$= \sin \tfrac{\pi}{4} \cos \tfrac{\pi}{3} - \cos \tfrac{\pi}{4} \sin \tfrac{\pi}{3}$$

$$= \frac{\sqrt{2}}{2} \cdot \frac{1}{2} - \frac{\sqrt{2}}{2} \cdot \frac{\sqrt{3}}{2}$$

$$= \frac{\sqrt{2} - \sqrt{6}}{4}$$

$$\approx -0.2588$$

45. $\tan \tfrac{\pi}{12} = \tan\left(\tfrac{\pi}{4} - \tfrac{\pi}{6}\right)$

$$= \frac{\tan \tfrac{\pi}{4} - \tan \tfrac{\pi}{6}}{1 + \tan \tfrac{\pi}{4} \tan \tfrac{\pi}{6}}$$

$$= \frac{1 - \dfrac{\sqrt{3}}{3}}{1 + \dfrac{\sqrt{3}}{3}}$$

$$= 2 - \sqrt{3}$$

$$\approx 0.2679$$

47.   We will generally not give the results for **WHAT DOES THIS SAY?** problems, but this result is so critical to so many things in mathematics, we want to be sure you have this result memorized. You first obtain a 0 on one side, then try to factor. If it cannot be easily factored, use the *quadratic formula:* If $ax^2 + bx + c = 0$, $a \neq 0$, then

$$x = \frac{-b \pm \sqrt{b^2 - 4ac}}{2a}$$

49.   You should spend some time answering this question in your own words, but be sure you mention properties 8 and 9 from Table 1.1.

SURVIVAL HINT: *Section 4.4 of the Student Mathematics Handbook reviews graphing trigonometric functions.*

51. This problem reviews the shapes of the three trigonometric functions. You should remember the shapes of the graphs of each of these.

a. period 2π

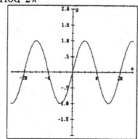

b. period 2π

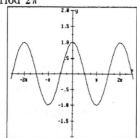

SURVIVAL HINT: *Note that the sine and cosine functions have the same shape, but are "out of phase" by $\pi/2$. Also note that calculus work is done in radian measure. Set your calculator to radian mode and change to degree mode only when you see the degree symbol.*

c. period π

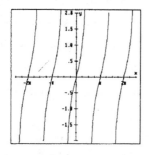

53. First rewrite the equation in the proper form.
$$y = \tan(2x - \tfrac{\pi}{2})$$
$$= \tan 2(x - \tfrac{\pi}{4})$$

Compare with the general form (see the *Student Mathematics Handbook*) Section 4.4:
$$y - k = a\tan b(x - h)$$

for a starting point of the from of (h, k) a period of $p = \frac{\pi}{b}$ and with a frame height of $2|a|$. For this problem, $a = 1$, $b = 2$, $p = \frac{\pi}{2}$, and $(h, k) = (\frac{\pi}{4}, 0)$.

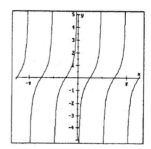

55. $y = 4\sin\left(\tfrac{1}{2}x + 2\right) - 1$

Write this in general form
$$y + k = a\sin b(x - h)$$
$$y + 1 = 4\sin\tfrac{1}{2}(x + 4)$$

so that we see $a = 4$ is the amplitude, $b = \frac{1}{2}$, so the period is $p = \frac{2\pi}{b} = \frac{2\pi}{1/2} = 4\pi$. The starting point for the frame is
$$(h, k) = (-4, -1).$$

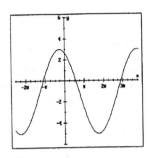

57. Begin by plotting the data points and then draw a smooth curve through those points.

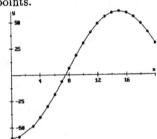

For the equation, consider

$$y - A = B \sin C(x - D)$$

Answers vary; from the point $(D, A) = (7.5, 0)$, it looks like $B = 60$ and the period is found by solving $30 = \frac{2\pi}{C}$ so that $C = \frac{\pi}{15}$. A possible equation is

$$y = 60 \sin \frac{\pi}{15}(x - 7.5).$$

$A = 0$, $B = 60$, $C = \frac{\pi}{15}$, and $D = 7.5$

59. sun curve: $y = \cos \frac{\pi}{6}x$;

moon curve: $y = 4 \cos \frac{\pi}{6}x$;

combined curve:

$$y = \cos \frac{\pi}{6}x + 4 \cos \frac{\pi}{6}x$$
$$= 5 \cos \frac{\pi}{6}x$$

61. We are given $s = \dfrac{3d\cos\theta}{\sqrt{7 + 9\cos^2\theta}}$.

a. For this part, $d = 5.0$ and $\theta = 37°$, which we substitute into the given equation.

$$s = \frac{3d\cos\theta}{\sqrt{7 + 9\cos^2\theta}}$$

$$= \frac{3(5.0)\cos 37°}{\sqrt{7 + 9\cos^2 37°}}$$

$$\approx 3.356208149 \quad \textit{by calculator}$$

The apparent depth is 3.4 m.

b. For this part, $s = 2.5$ and $d = 5.0$.

We need to solve for θ.

$$2.5 = \frac{3(5.0)\cos\theta}{\sqrt{7 + 9\cos^2\theta}}$$

$$\sqrt{7 + 9\cos^2\theta} = 6.0\cos\theta$$

$$7 + 9\cos^2\theta = 36\cos^2\theta$$

$$\cos^2\theta = \frac{7}{27}$$

$$\theta \approx 59.39110234°$$

By calculator; use the acute-angle result.

The angle of incidence is 59°.

63.

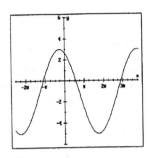

One revolution (360°) divided by 6 results in 6-60° central angles. Since the triangles are isosceles triangles, it follows (because of the 60° angle) that the triangles are also equilateral.

Thus, the radius of a circle can be applied exactly six times to its circumference as a chord.

65. If $a > 0$, the solution is positive if $b < 0$, 0 if $b = 0$, and negative if $b > 0$.

If $a < 0$, the solution is positive if $b > 0$, 0 if $b = 0$, and negative if $b < 0$.

67. If $a = 1$, the period is 2π; if a increases then the period decreases and vice-versa. The period is $2\pi/a$.

69. If $x \geq 0$, then $|x| = x$. Thus,

$$-|x| \leq 0 \leq x \leq |x|$$

If $x < 0$, then $|x| = -x$. Thus,

$$-|x| \leq x \leq 0 \leq |x|$$

Thus, $-|x| \leq x \leq |x|$.

71. $|a| < b$ and $b > 0$ implies $-b < a < b$.

If $a \geq 0$, then $|a| = a$ and $|a| < b$ means

$$a < b$$

If $a < 0$, then $|a| = -a$ and $|a| < b$ means

$$-a < b \text{ or } -b < a$$

Combining these results leads to the conclusion $-b < a < b$.

73. By Problem 72, if $|x| \geq |y|$, then

$$|x| \leq |x - y| + |y|$$
$$|x| - |y| \leq |x - y|$$

By Problem 72, if $|x| \leq |y|$, then

$$|y| \leq |y - x| + |x|$$
$$|y| - |x| \leq |x - y|$$

Thus,

$$\big||x| - |y|\big| \leq |x - y|$$

1.2 Lines in the Plane, Pages 17-19

SURVIVAL HINT: *Many mistakes are made by good students, when using concepts with which they are quite competent, because they try to do too much in their heads. Get into the habit NOW of showing work on all problems which could be handed in to your instructor. It is not a waste of your time to be neat and well organized and write a sufficient number of steps so that someone else could follow your work. In a job situation this will be required. Your boss will not want to see a page of scrap-work with an answer circled at the bottom!*

1. If the equation does not contain y, solve for x and draw the vertical line $x = c$. Otherwise, solve for $y = mx + b$. Plot the y-intercept $(0, b)$ and then count out the slope, m. Draw the line passing through the y-intercept and the slope point.

SURVIVAL HINT: *Generally we will not show answers to WHAT DOES THIS SAY? problems in this manual. The reason we show the answer to this one is to make sure you realize that plotting points is not the best way to be graphing lines.*

In Problems 2-15, you will need to remember that the standard form for the equation of a line is

$$Ax + By + C = 0$$

3. First, find the slope using $m = \dfrac{\Delta y}{\Delta x}$ where $\Delta y = y_2 - y_1$ and $\Delta x = x_2 - x_1$.

$$m = \frac{9 - 7}{-2 - (-1)} = \frac{2}{-1} = -2$$
$$y - 7 = -2(x + 1)$$
$$2x + y - 5 = 0$$

5. A line with slope 0 is a horizontal line. You should remember the special forms for horizontal and vertical lines. Since $(1, \frac{1}{2})$ is given we see, by inspection, that $k = \frac{1}{2}$, so that the equation is $y = \frac{1}{2}$ or, in standard form,

$$2y - 1 = 0$$

7. Since $(-2, -5)$ is given we see, by inspection, that $h = -2$, so that the equation is $x = -2$ or, in standard form, $x + 2 = 0$.

9. Since we are given the intercepts we see, by inspection, that for $\frac{x}{a} + \frac{y}{b} = 1$, $a = 7$ and $b = -8$ and we use the two-intercept form to write

$$\frac{x}{7} + \frac{y}{-8} = 1$$
$$8x - 7y = 56 \quad \textit{Multiply both sides by 56.}$$
$$8x - 7y - 56 = 0 \quad \textit{Subtract 56 from both sides.}$$

11. Any line parallel to $3x + y = 7$ has the same slope, namely -3. (To find this slope, write the equation in slope-intercept form, $y = -3x + 7$.) Since it passes through $(-1, 8)$ we see $(h, k) = (-1, 8)$ and then use the point-slope form of the equation.

$$y - k = m(x - h)$$
$$y - 8 = -3(x + 1)$$
$$3x + y - 5 = 0$$

13. To find the slope of the given line, write the equation in slope-intercept form:

$$4x - 3y + 2 = 0$$
$$3y = 4x + 2$$
$$y = \tfrac{4}{3}x + \tfrac{2}{3}$$

We see that it has slope $m = \frac{4}{3}$. Since

a perpendicular line has slope $-\frac{3}{4}$, we can find the desired line using the point-slope form for that line:

$$y - k = m(x - h) \quad \textit{Point-slope form}$$
$$y + 2 = -\tfrac{3}{4}(x - 3) \quad \textit{Given } (3, -2)$$
$$4y + 8 = -3x + 9 \quad \textit{Multiply by 4.}$$
$$3x + 4y - 1 = 0 \quad \textit{Write in standard form.}$$

SURVIVAL HINT: *You will frequently need to solve a system of equations, and the text (as does the following problem) assume you know how to solve a simple system of equations. You will find a review of of this topic in Chapter 3 of the Student Mathematics Handbook.*

15. $x - 4y + 5 = 0$
$$4y = x + 5$$
$$y = \tfrac{1}{4}x + \tfrac{5}{4}$$

so this line has $m = \frac{1}{4}$; the desired line must have $m = -4$. We now need to solve the system of equations

$$-2 \begin{cases} x - 4y = -5 \\ 2x + 3y = 1 \end{cases}$$

$$+ \begin{cases} -2x + 8y = 10 \\ 2x + 3y = 1 \end{cases}$$
$$\quad 11y = 11$$
$$\quad\quad y = 1$$

If $y = 1$, then from either of the given equations (we choose the first)

$$x - 4y = -5$$
$$x - 4(1) = -5$$
$$x = -1$$

We see their intersection at $(-1, 1)$. Now use the point and slope to find the line:

$$y - 1 = -4(x + 1)$$
$$4x + y + 3 = 0$$

17. By inspection, $m = -5/7$, so graph by first plotting the y-intercept $(0, 3)$ and then using the slope to count down 5 units and over 7.

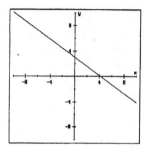

The x-intercept (where $y = 0$) is found by using the equation:

$$y = -\tfrac{5}{7}x + 3$$
$$0 = -\tfrac{5}{7}x + 3$$
$$\tfrac{5}{7}x = 3$$
$$x = \tfrac{21}{5}$$

We see the x-intercept is the point $(\tfrac{21}{5}, 0)$.

19. By inspection, $m = 6.001$, so graph this curve by plotting the given point $(3, 9)$ and from this point count up 6.0001 and over 1; draw the line as shown.

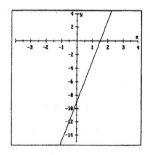

The x-intercept (where $y = 0$) is found by using the equation:

$$y - 9 = 6.001(x - 3)$$
$$0 - 9 = 6.001(x - 3)$$
$$\frac{-9}{6.001} = x - 3$$
$$x = \frac{-9}{6.001} + 3$$
$$\approx 1.50025$$

The y-intercept (where $x = 0$) is found by using the same equation:

$$y - 9 = 6.001(x - 3)$$
$$y - 9 = 6.001(0 - 3)$$
$$y = -18.003 + 9$$
$$= -9.003$$

The intercepts are: $(1.50025, 0)$, $(0, -9.003)$.

21. First, find the slope of the given line:

$$3x + 5y + 15 = 0$$
$$5y = -3x - 15$$
$$y = -\tfrac{3}{5}x - 3$$

We see $m = -3/5$, and we draw the line by first plotting the y-intercept $(0, -3)$, and then trace out the slope by going down 3 and over 5.

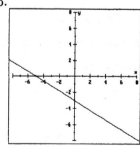

The x-intercept (where $y = 0$) is found by using the equation:

$$3x + 5y + 15 = 0$$
$$3x + 5(0) + 15 = 0$$
$$3x = -15$$
$$x = -5$$

The y-intercept (where $x = 0$) is found by using the same equation:

$$3x + 5y + 15 = 0$$
$$3(0) + 5y + 15 = 0$$
$$5y = -15$$
$$y = -3$$

The intercepts are $(-5, 0)$, $(0, -3)$.

23.
$$6x - 10y - 3 = 0$$
$$10y = 6x - 3$$
$$y = \frac{3}{5}x - \frac{3}{10}$$

Graph the line using $m = 3/5$ and intercept $y = -\frac{3}{10}$.

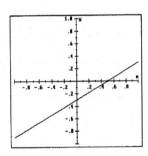

The x-intercept (where $y = 0$) is found by using the equation:

$$6x - 10y - 3 = 0$$
$$6x - 10(0) - 3 = 0$$

$$6x = 3$$
$$x = \frac{1}{2}$$

The y-intercept (where $x = 0$) is found by using the same equation:

$$6x - 10y - 3 = 0$$
$$6(0) - 10y - 3 = 0$$
$$10y = -3$$
$$y = -\frac{3}{10}$$

The intercepts are $(0.5, 0)$, $(0, -0.3)$.

25. Use the equation $\frac{x}{2} - \frac{y}{3} = 1$ to graph the line by plotting the intercepts (found by inspection): $(2, 0)$ and $(0, -3)$. For the slope, we can use these two points

$$m = \frac{\Delta y}{\Delta x}$$
$$= \frac{-3 - 0}{0 - 2}$$
$$= \frac{3}{2}$$

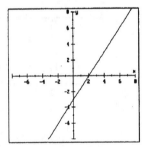

27. Write $x = 5y$ as $y = \frac{1}{5}x$ to find the slope $m = 1/5$. The y-intercept is 0, so the line passes through the origin, so the rise is 1 and run is 5, as shown in following figure.

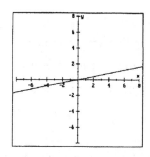

The x-and y-intercepts are the same, namely $(0, 0)$.

SURVIVAL HINT: *When working with lines, you need to distinguish between a line with zero slope (horizontal line) and a line with no slope (vertical line). When we say a line has no slope, we mean it is a vertical line; that is a line whose slope is undefined.*

29. Recognize this as a vertical line (no slope) passing through $(-3, 0)$; $x = -3$.

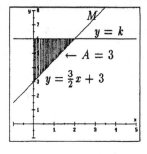

31. Begin by drawing the graph.

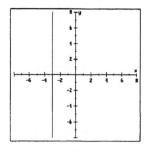

Let $y = k$ be the horizontal line. The height of the triangle at $y = k$ is $k - 3$, and since $x = \frac{2}{3}(y - 3)$, the base of the triangle is

$$x = \tfrac{2}{3}(k - 3)$$

The area of the triangle is

$$A = \tfrac{1}{2}(\text{BASE})(\text{HEIGHT})$$
$$= \tfrac{1}{2} \cdot \tfrac{2}{3}(k - 3)(k - 3) = \tfrac{1}{3}(k - 3)^2$$

So, when $A = 3$,

$$3 = \tfrac{1}{3}(k - 3)^2$$
$$9 = (k - 3)^2$$
$$k - 3 = \pm 3$$
$$k = 0, 6$$

There are two such horizontal lines:

$$y = 0 \text{ and } y = 6.$$

33. Let $A(1, 3)$, $B(3, -2)$, and $C(4, 11)$. Two parallelograms can be formed, as shown.

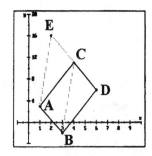

Label the unknown points $D(x_1, y_1)$ and

$E(x_2, y_2)$.

$\overline{AC}$ has slope $\dfrac{11 - 3}{4 - 1} = \dfrac{8}{3}$;

$\overline{BC}$ has slope $\dfrac{11 - (-2)}{4 - 3} = 13$;

$\overline{AB}$ has slope $\dfrac{-2 - 3}{3 - 1} = \dfrac{-5}{2}$.

Since the slopes of opposite sides of a parallelogram are equal, we have:

$$\frac{y_1 - (-2)}{x_1 - 3} = \frac{8}{3} \text{ so } \overline{BD} \parallel \overline{AC}$$

$$\frac{y_1 - 11}{x_1 - 4} = \frac{-5}{2} \text{ so } \overline{CD} \parallel \overline{AB}$$

Thus,

$$8x_1 - 3y_1 = 30$$

$$5x_1 + 2y_1 = 42$$

so $x_1 = 6$, $y_1 = 6$ and $D(6, 6)$.

Similarly, we find $E(2, 16)$.

SURVIVAL HINT: *Since simple systems of equations are very common in this book, it would be a good idea to know how to use a calculator to give approximate solutions. You can practice with Problems 34-45. We show a calculator solution for each of these problems.*

35.   no values

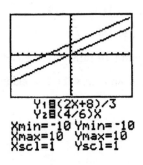

37. $(4, -1)$

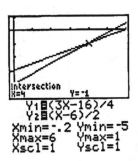

39. $\left(\dfrac{3}{4}, \dfrac{7}{2}\right)$

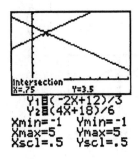

41. $\left(-\dfrac{64}{3}, \dfrac{100}{3}\right)$

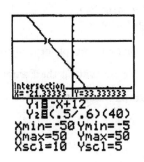

43. $(\sqrt{2}, \sqrt{2}), (-\sqrt{2}, -\sqrt{2})$

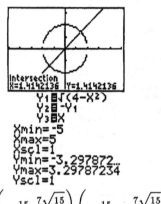

45. $\left(-\frac{15}{8}, \frac{7\sqrt{15}}{8}\right), \left(-\frac{15}{8}, -\frac{7\sqrt{15}}{8}\right)$

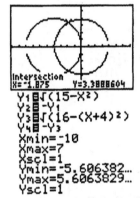

47. We use the slope-intercept form of the equation. We see that on the Fahrenheit scale, the graph passes through $(0, 32)$ so $b = 32$. Also,

$m = \dfrac{212 - 32}{100 - 0} = \dfrac{9}{5}$ so that $F = \frac{9}{5}C + 32$.

a. If $C = -39°$, then

$$F = \tfrac{9}{5}(-39) + 32$$
$$= -38.2°$$

b. If $F = 0°$, then

$$0 = \tfrac{9}{5}C + 32$$
$$-32 = \tfrac{9}{5}C$$
$$-160 = 9C$$
$$C = -\tfrac{160°}{9} \approx -17.8°$$

c. If $F = C$, then

$$F = \tfrac{9}{5}F + 32$$
$$-\tfrac{4}{5}F = 32$$
$$-4F = 160$$
$$F = -40$$
$$F = C = -40°$$

49. *The Spy problems weave a James Bond/Indiana Jones-type story where you, the reader, are asked to get the Spy out of trouble by solving a math problem. The story continues throughout the text, and even if you don't work all of these problems, you might enjoy reading the Spy problems when they appear in the book.*

For this problem, let x denote the time in hours the Spy has been traveling. Then $x - 2/3$ is the time the smugglers have been traveling. The distance the Spy travels is $72x$ km and the corresponding distance the smugglers travel is $168(x - 2/3)$ km. They reach the same point when:

$$72x = 168(x - \tfrac{2}{3})$$
$$72x = 168x - 112$$
$$x = \tfrac{112}{96} = \tfrac{7}{6}$$

This is 1 hr and 10 min. At the end of that time the Spy, and thus the smugglers, have traveled

$$72(\tfrac{7}{6}) = 84 \text{ km}$$

The Spy escapes, since freedom is reached at the border after only 83.8 km.

51. Let t denote the age in years of the machinery and V be a linear function of t. At the time of purchase, $t = 0$ and $V(0) = 200,000$. Ten years later, $t = 10$ and $V(10) = 10,000$. The slope of the line through $(0, 200\,000)$ and $(10, 10\,000)$ is

$$m = \frac{10,000 - 200,000}{10 - 0}$$
$$= -19,000$$

Thus, $V(t) = -19,000t + 200,000$.

In particular,

$$V(4) = -19,000(4) + 200,000$$
$$= 124,000$$

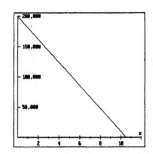

The value in 4 years is \$124,000.

53. a. The three possible parallelograms are $P_1P_2P_3A$, $P_1P_3P_2B$, $P_1P_3CP_2$ as shown:

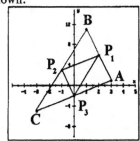

Note that P_1P_2, P_1P_3, P_2P_3 have slopes $m_{12} = 1$, $m_{13} = 4$, and $m_{23} = -5$, respectively. To find $A(a, b)$, use the fact that opposite sides of parallelogram $P_1P_2P_3A$ must have equal slopes. Thus,

$$\frac{b - (-2)}{a - 0} = 1 \text{ and } \frac{b - 6}{a - 2} = -5$$

so $a = 3$, $b = 1$. Similarly, the other two points are $B(1, 11)$ and $C(-3, -5)$.

b. The medians of $\triangle ABC$ are the lines $x + 2y = 5$, $-11x + 5y = 8$, and $13x - y = 2$, which intersect at the center $(\frac{1}{3}, \frac{7}{3})$. Similarly, the center of $\triangle P_1P_2P_3$ is also $(\frac{1}{3}, \frac{7}{3})$. Note a formula for obtaining these results is provided by Problem 65.

55. Let x be the number of days, and y the number of gallons. The slope between $(12, 200)$ and $(21, 164)$ is

$$m = \frac{164 - 200}{21 - 12} = -4$$

Thus,

$$y - 200 = -4(x - 12)$$
$$y = -4x + 248$$

In particular, on the 8th of the month,

$$y = -4(8) + 248 = 216$$

On the 8th of the month, the reservoir held 216 million gallons.

57. a.

Age:	0	10	20	30	40
Value:	3	6	12	24	48

At the end of 30 years the book is worth \$24 and at the end of 40 years it will be worth $48.

b. The slope between $(0, 3)$ and $(0, 6)$ is
$$\frac{6-3}{10-0} = \frac{3}{10}$$
and between $(10, 6)$ and $(20, 12)$ is
$$\frac{12-6}{20-10} = \frac{6}{10} = \frac{3}{5}$$
We see that the relationship is not linear because the slopes between pairs of points are not equal.

59. This is a system of equations, but it is not a linear system because the first equation is second degree.
$$\begin{cases} xy = 1 \\ x + y = a \end{cases}$$
Solve the first equation for y: $y = \frac{1}{x}$ and substitute this value into the second equation.
$$x + \frac{1}{x} = a$$
$$x^2 + 1 = ax$$
$$x^2 - ax + 1 = 0$$
$$x = \frac{a \pm \sqrt{a^2 - 4}}{2},$$
$$y = \frac{2}{a \pm \sqrt{a^2 - 4}}$$

61. $A(x_1, y_1)$, $B(x_2, y_2)$ are two points on a line whose equation is $y = mx + b$. Then,
$y_1 = mx_1 + b$ and $y_2 = mx_2 + b$.
$$mx_1 - y_1 = mx_2 - y_2$$
$$m(x_1 - x_2) = y_1 - y_2$$
$$m = \frac{y_2 - y_1}{x_2 - x_1}$$
The point $(0, b)$ is on the line since
$$b = m(0) + b$$

63. We know that $m_1 = \tan \alpha_1$ and $m_2 = \tan \alpha_2$, where $\phi = \alpha_2 - \alpha_1$, so
$$\tan(\alpha_2 - \alpha_1) = \frac{\tan \alpha_2 - \tan \alpha_1}{1 + \tan \alpha_2 \tan \alpha_1}$$
$$= \frac{m_2 - m_1}{1 + m_2 m_1}$$

65. Let $P(x_0, y_0)$ be a point on the line segment between vertex $A(x_1, y_1)$ and the midpoint
$$M\left(\frac{x_2 + x_3}{2}, \frac{y_2 + y_3}{2}\right)$$
For $P(x_0, y_0)$ to be the center of $\triangle ABC$, it must be located 2/3 the way from A to M, which means that $(x_0, 0)$ is 2/3 the way from $(x_1, 0)$ to $\left(\frac{x_2 + x_3}{2}, 0\right)$, and we have
$$x_0 - x_1 = \frac{2}{3}\left[\frac{x_2 + x_3}{2} - x_1\right]$$
$$x_0 = \frac{x_1 + x_2 + x_3}{3}$$
The formula for y_0 can be obtained similarly.

1.3 Functions and Graphs, Pages 31-33

Let D represent the domain in Problems 1-12.

1. $D = $ all reals or $D = \mathbf{R}$ or $D = (-\infty, \infty)$
$$f(-2) = 2(-2) + 3$$
$$= -1$$
$$f(1) = 2(1) + 3$$
$$= 5$$
$$f(0) = 2(0) + 3$$
$$= 3$$

3. $D = (-\infty, \infty)$

$$f(1) = 3(1)^2 + 5(1) - 2$$
$$= 6$$
$$f(0) = 3(0)^2 + 5(0) - 2$$
$$= -2$$
$$f(-2) = 3(-2)^2 + 5(-2) - 2$$
$$= 0$$

5. $D = (-\infty, -3) \cup (-3, \infty)$
$$f(2) = 2 - 2$$
$$= 0$$
$$f(0) = 0 - 2$$
$$= -2$$
$$f(-3) \text{ is undefined}$$

7. $D = (-\infty, -2] \cup [0, \infty)$
$$f(-1) = \sqrt{(-1)^2 + 2(-1)} \text{ is undefined}$$
$$f(\tfrac{1}{2}) = \sqrt{(\tfrac{1}{2})^2 + 2(\tfrac{1}{2})}$$
$$= \frac{\sqrt{5}}{2}$$
$$f(1) = \sqrt{(1)^2 + 2(1)}$$
$$= \sqrt{3}$$

9. $D = (-\infty, \infty)$
$$f(-1) = \sin[1 - 2(-1)]$$
$$= \sin 3$$
$$\approx 0.1411$$
$$f(\tfrac{1}{2}) = \sin[1 - 2(\tfrac{1}{2})]$$
$$= \sin 0$$
$$= 0$$
$$f(1) = \sin[1 - 2(1)]$$

$$= \sin(-1)$$
$$\approx -0.8415$$

11. $D = (-\infty, \infty)$
$$f(3) = 3 + 1$$
$$= 4$$
$$f(1) = -2(1) + 4$$
$$= 2$$
$$f(0) = -2(0) + 4$$
$$= 4$$

SURVIVAL HINT: *Problems 13-20 involve a calculation that is required when using the definition of derivative in Chapter 3.*

13.
$$\frac{f(x+h) - f(x)}{h} = \frac{[9(x+h) + 3] - (9x + 3)}{h}$$
$$= \frac{9h}{h}$$
$$= 9$$

15.
$$\frac{f(x+h) - f(x)}{h} = \frac{5(x+h)^2 - 5x^2}{h}$$
$$= \frac{5x^2 + 10xh + 5h^2 - 5x^2}{h}$$
$$= \frac{10xh + 5h^2}{h}$$
$$= 10x + 5h$$

17. $f(x) = |x| = -x$ since $x < 0$;
$$\frac{f(x+h) - f(x)}{h} = \frac{-x - h - (-x)}{h}$$
$$= \frac{-h}{h}$$
$$= -1$$

SURVIVAL HINT: *Spend some time with Problem 17. It illustrates important skills about how to deal with an absolute value. In Chapter 2 we will also calculate slopes of secant lines as a prelude to understanding the meaning of a line tangent to a curve at a given point.*

19.
$$\frac{f(x+h)-f(x)}{h} = \frac{\frac{1}{x+h}-\frac{1}{x}}{h}$$
$$= -\frac{h}{hx(x+h)}$$
$$= \frac{-1}{x(x+h)}$$

21.
$$f(x) = \frac{2x^2+x}{x}$$
$$= \frac{x(2x+1)}{x}$$
$$= 2x+1, \ x \neq 0$$

Domains of f and g are not the same, so they are not equal.

23.
$$f(x) = \frac{2x^2-x-6}{x-2}$$
$$= \frac{(2x+3)(x-2)}{x-2}$$
$$= 2x+3, \ x \neq 2$$

$f = g$ since they are equal and have the same domains.

25.
$$f(x) = \frac{3x^2-5x-2}{x-2}$$
$$= \frac{(3x+1)(x-2)}{x-2}$$
$$= 3x+1, \ x \neq 2$$

$f \neq g$ since the domains are not the same.

27. $f_1(x) = x^2+1;$
$$f_1(-x) = (-x)^2+1$$
$$= x^2+1$$
$$= f_1(x); \text{ even}$$

29. $f_3(x) = \dfrac{1}{3x^3-4}$
$$f_3(-x) = \frac{1}{3(-x)^3-4}$$
$$= \frac{1}{-3x^3-4}; \text{ neither}$$

31. $f_5(x) = \dfrac{1}{(x^3+3)^2}$
$$f_5(-x) = \frac{1}{[(-x)^3+3]^2}$$
$$= \frac{1}{(-x^3+3)^2}; \text{ neither}$$

33. $f_7(x) = |x|$
$$f_7(-x) = |-x|$$
$$= |x|$$
$$= f_7(x); \text{ even}$$

SURVIVAL HINT: *Most useful functions are compositions, and most in the text are compositions. If you are not really comfortable with Problems 35-50, spend a little extra time on them now and it will save you time later.*

35. $f(x) = x^2+1$ and $g(x) = 2x$
$$(f \circ g)(x) = f[g(x)]$$
$$= f(2x)$$
$$= (2x)^2+1$$
$$= 4x^2+1$$
$$(g \circ f)(x) = g[f(x)]$$
$$= g(x^2+1)$$
$$= 2x^2+2$$

37. $f(t) = \sqrt{t}$ and $g(t) = t^2$

$$(f \circ g)(t) = f[g(t)]$$
$$= f(t^2)$$
$$= \sqrt{t^2}$$
$$= |t|$$

$$(g \circ f)(t) = g[f(t)]$$
$$= g(\sqrt{t})$$
$$= t$$

39. $f(x) = \sin x$ and $g(x) = 2x + 3$

$$(f \circ g)(x) = f[g(x)]$$
$$= f(2x + 3)$$
$$= \sin(2x + 3)$$

$$(g \circ f)(x) = g[f(x)]$$
$$= g(\sin x)$$
$$= 2 \sin x + 3$$

41. $u(x) = 2x^2 - 1;\ g(u) = u^4$

43. $u(x) = 2x + 3;\ g(u) = |u|$

45. $u(x) = \tan x;\ g(u) = u^2$

47. $u(x) = \sqrt{x};\ g(u) = \sin u$

49. $u(x) = \dfrac{x+1}{2-x};\ g(u) = \sin u$

SURVIVAL HINT: *Don't underestimate the importance of Problems 51 and 52. It seems easy, but without the ability to name points as illustrated by the problem you will have a tough time as you progress through the course.*

51. By inspection, $P(5, f(5))$; $Q(x_0, f(x_0))$.

53.
$$3x^2 - 5x - 2 = 0 \qquad Given$$
$$(3x + 1)(x - 2) = 0 \qquad Factor$$
$$x = -\tfrac{1}{3},\ 2$$

55. $(x - 15)(2x + 25)(3x - 65)(4x + 1) = 0$

Set each factor to 0 and solve. You should be able to mentally solve this problem.

$$x = 15,\ -\tfrac{25}{2}, \tfrac{65}{3},\ -\tfrac{1}{4}$$

57.
$$5x^3 - 3x^2 + 2x = 0$$
$$x(5x^2 - 3x + 2) = 0$$
$$x = 0$$

(The other solutions are not real numbers.)

59.
$$\frac{x^2 - 1}{x^2 + 2} = 0$$
$$(x^2 - 1) = 0$$
$$x = \pm 1$$

61. $C(q) = q^3 - 30q^2 + 400q + 500$

a. $C(20) = 20^3 - 30(20)^2 + 400(20) + 500$
$$= 4,500$$

The cost is $4,500.

b. $C(20) - C(19) = 4,500 - 4,129$
$$= 371$$

The cost of the 20th unit is $371.

63. $I = K/s^2$; $s = 6t - t^2$

   a. $I = f(t)$

$$= \frac{K}{s^2}$$

$$= \frac{K}{(6t - t^2)^2}$$

$$= \frac{30}{t^2(6 - t)^2} \qquad K = 30 \text{ is given.}$$

   b. $I(1) = \dfrac{30}{1^2(6 - 1)^2}$

$$= \frac{6}{5} \text{ candles}$$

$$I(4) = \frac{30}{4^2(6 - 4)^2}$$

$$= \frac{15}{32} \text{ candles}$$

65. **a.** $C(25t) = (25t)^2 + (25t) + 900$

$$= 625t^2 + 25t + 900$$

   b. $C(75) = 75^2 + 75 + 900$

$$= \$6,600$$

   c. $11,000 = 625t^2 + 25t + 900$

$$0 = 625t^2 + 25t - 10,100$$

$$0 = 25t^2 + t - 404$$

$$0 = (25t + 101)(t - 4)$$

$$t = -\frac{101}{25}, 4$$

Negative values of time are not in the domain so $t = 4$ hours.

67. $V(T) = V_0\left(1 + \dfrac{T}{273}\right)$

   a.   Graph

$$V = 100\left(1 + \frac{T}{273}\right)$$

$$= \frac{100}{273}T + 100$$

The y-intercept is 100 and for the slope count out a rise of 100 and a run of 273.

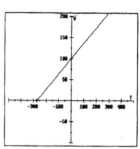

  b. $2V_0 = V_0\left(1 + \dfrac{T}{273}\right)$

$$2 = 1 + \frac{T}{273}$$

$$1 = \frac{T}{273}$$

$$273 = T$$

69. **a.**

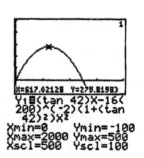

The maximum height of the cannonball is approximately 280 ft.

b. It hits the ground approximately 1,243 ft from the firing point.

c. The basic shape is that of a parabola (standard quadratic function).

71. a.
$$P(9) = 20 - \frac{6}{9+1}$$
$$= 19.4$$

This is about 19,400 people.

b.
$$P(9) - P(8) = \left(20 - \frac{3}{5}\right) - \left(20 - \frac{2}{3}\right)$$
$$= \frac{1}{15}$$

This is about 67 people.

c. The population will tend to 20,000 people in the long run. The denominator will have increased to the point of making $6/(t+1)$ negligible.

73. Pythagorean theorem: $\triangle ABC$ with sides a, b, and c is a right triangle if and only if $c^2 = a^2 + b^2$. Proofs vary.

1.4 Inverse Functions; Inverse Trigonometric Functions, Pages 41-42

1. Trigonometric functions are not one-to-one. Restrictions make them one-to-one so that the inverses can be defined.

3. We need to show that $f[g(x)] = g[f(x)] = x$.
$$f(x) = 5x + 3; \quad g(x) = \frac{x-3}{5}$$
$$f[g(x)] = 5\left(\frac{x-3}{5}\right) + 3$$

$$= x$$

$$g[f(x)] = \frac{(5x+3) - 3}{5}$$
$$= x$$

These are inverse functions.

5. $f(x) = \frac{4}{5}x + 4; \quad g(x) = \frac{5}{4}x + 3$
$$f[g(x)] = \frac{4}{5}\left(\frac{5}{4}x + 3\right) + 4$$
$$= x + \frac{12}{5} + 4$$
$$\neq x$$

These are not inverse functions.

7. $f(x) = x^2, \, x < 0; \quad g(x) = \sqrt{x}, \, x > 0$
$$f[g(x)] = (\sqrt{x})^2$$
$$= x$$
$$g[f(x)] = \sqrt{x^2}$$
$$= |x|$$
$$= -x \qquad \text{Since } x < 0.$$

These are not inverse functions.

9. To find the inverse interchange the domain and range values: $\{(5, 4), (3, 6), (1, 7), (4, 2)\}$.

11. Given $y = 2x + 3$;

The inverse is $x = 2y + 3$ or $y = \frac{1}{2}x - \frac{3}{2}$.

13. Given $y = x^2 - 5, \, x \geq 0$;

The inverse is $x = y^2 - 5, \, y \geq 0$ or

$y = \sqrt{x + 5}$ (positive value since $y \geq 0$).

15. Given $y = \sqrt{x} + 5$;

The inverse is $x = \sqrt{y} + 5$ or: $y = (x - 5)^2$.

17. Given $y = \dfrac{2x - 6}{3x + 3}$;

The inverse is $x = \dfrac{2y - 6}{3y + 3}$ or:

$$3xy + 3x = 2y - 6$$

$$(3x - 2)y = -3x - 6$$

$$y = \dfrac{3x + 6}{2 - 3x}$$

SURVIVAL HINT: *Table 1.2 in Section 1.1 provides what is known as the* **exact values** *of the trigonometric functions. You will find that you are expected to have this table memorized in much the same way that the times table from elementary school must be memorized. Problems 19-30 use this table in reverse along with the definition of the inverse trigonometric functions given in Table 1.4.*

19. a. $\dfrac{\pi}{3}$ b. $-\dfrac{\pi}{3}$

21. a. $-\dfrac{\pi}{4}$ b. $\dfrac{5\pi}{6}$

23. a. $-\dfrac{\pi}{3}$ b. π

25. Let $\theta = \sin^{-1}\frac{1}{2}$, so we know $\sin\theta = \frac{1}{2}$ and that θ is in Quad I because $\frac{1}{2}$ is positive.

$$\cos(\sin^{-1}\tfrac{1}{2}) = \cos\theta, \text{ and}$$

$$\cos\theta = \pm\sqrt{1 - \sin^2\theta}$$

$$= \sqrt{1 - \left(\tfrac{1}{2}\right)^2} \qquad \textit{Positive in Quad I.}$$

$$= \dfrac{\sqrt{3}}{2}$$

27. Let $\theta = \tan^{-1}\frac{1}{3}$, so we know $\tan\theta = \frac{1}{3}$ and

that θ is in Quad I because $\frac{1}{3}$ is positive.

$$\cot(\tan^{-1}\tfrac{1}{3}) = \cot\theta, \text{ and}$$

$$\cot\theta = \dfrac{1}{\tan\theta}$$

$$= \dfrac{1}{1/3}$$

$$= 3$$

29. Let $\alpha = \sin^{-1}\frac{1}{5}$ and $\beta = \cos^{-1}\frac{1}{5}$

Then $\sin\alpha = \frac{1}{5}$, $\cos\beta = \frac{1}{5}$ and using reference

triangles we find $\cos\alpha = \sin\beta = \dfrac{2\sqrt{6}}{5}$

$$\cos(\sin^{-1}\tfrac{1}{5} + 2\cos^{-1}\tfrac{1}{5}) = \cos(\alpha + 2\beta)$$

$$= \cos\alpha\cos 2\beta - \sin\alpha\sin 2\beta$$

$$= \cos\alpha(\cos^2\beta - \sin^2\beta) - \sin\alpha(2\cos\beta\sin\beta)$$

$$= \dfrac{2\sqrt{6}}{5}\left[\dfrac{1}{25} - \dfrac{24}{25}\right] - \dfrac{1}{5}\left[2\cdot\dfrac{1}{5}\cdot\dfrac{2\sqrt{6}}{5}\right]$$

$$= \dfrac{2\sqrt{6}}{5}\left[-\dfrac{23}{25} - \dfrac{2}{25}\right]$$

$$= -\dfrac{2\sqrt{6}}{5}$$

$$\approx -0.9798$$

31. Let the sides of the right triangle be $s^2 + t^2$ (hypotenuse), $s^2 - t^2$ (opposite α) and x (opposite β). Then (by the Pythagorean theorem)

$$x^2 + (s^2 - t^2)^2 = (s^2 + t^2)^2$$

$$x^2 + s^4 - 2s^2t^2 + t^4 = s^4 + 2s^2t^2 + t^4$$

$$x^2 = 4s^2t^2$$

$$x = 2st$$

By definition, $\tan \alpha = \dfrac{\frac{s^2 - t^2}{x}}{} = \dfrac{s^2 - t^2}{2st}$

so that

$$\alpha = \tan^{-1}\left(\frac{s^2 - t^2}{2st}\right).$$

33.

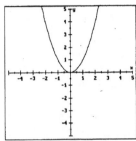

$f(x) = x^2$ for all x does not have an inverse, since the function is not one-to-one.

35.

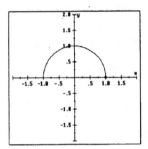

$f(x) = \sqrt{1 - x^2}$ does not have an inverse because it is not a one-to-one function.

37.

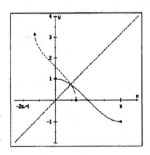

The inverse exists, since it passes the horizontal line test. $f^{-1}(x) = \cos^{-1}x$

39. If $\tan^{-1}x = \theta$, then $\tan\theta = x$, so (by a reference triangle)

$$\sin\theta = \frac{x}{\sqrt{x^2 + 1}} \text{ and } \cos\theta = \frac{1}{\sqrt{x^2 + 1}}.$$

$$\sin(2\tan^{-1}x) = \sin 2\theta$$

$$= 2\sin\theta\cos\theta$$

$$= 2\left(\frac{x}{\sqrt{x^2 + 1}}\right)\left(\frac{1}{\sqrt{x^2 + 1}}\right)$$

$$= \frac{2x}{x^2 + 1}$$

41. If $\cos^{-1}x = \theta$, then $\cos\theta = x$, then (by a reference triangle) $\tan\theta = \dfrac{\sqrt{1 - x^2}}{x}$.

43. In any right triangle, the sum of the acute angles is $\frac{\pi}{2}$, so for $|x| < 1$, we know

$$\sin^{-1}x + \cos^{-1}x = \frac{\pi}{2},$$

so

$$\sin(\sin^{-1}x + \cos^{-1}x) = \sin\frac{\pi}{2} = 1$$

45. Consider a reference triangle as shown.

a. $\cot\alpha = \frac{x}{1}$ so $\cot^{-1}x = \alpha$

$\tan\beta = \frac{x}{1}$ so $\tan^{-1}x = \beta$

Also, since the triangle is a right triangle,

$$\alpha + \beta = \tfrac{\pi}{2}$$

$$\cot^{-1} x + \tan^{-1} x = \tfrac{\pi}{2}$$

$$\cot^{-1} x = \tfrac{\pi}{2} - \tan^{-1} x$$

b. $\sin \alpha = \tfrac{1}{x}$, so $\alpha = \sin^{-1}\left(\tfrac{1}{x}\right)$

$\csc \alpha = x$, so $\alpha = \csc^{-1} x$

Thus, $\csc^{-1} x = \sin^{-1}\left(\tfrac{1}{x}\right)$

47. a. $\cot^{-1} 1.5 = \tfrac{\pi}{2} - \tan^{-1} 1.5 \approx 0.5880$

Use Theorem 1.3 (reciprocal identities for inverse trigonometric functions).

b. $\cot^{-1}(-1.5) = \tfrac{\pi}{2} - \tan^{-1}(-1.5)$

$$\approx 2.5536$$

c. $\sec^{-1}(-1.7) = \cos^{-1}\left(\tfrac{1}{-1.7}\right)$

$$\approx 2.1997$$

d. $\csc^{-1}(-1.84) = \sin^{-1}\left(\tfrac{1}{-1.84}\right)$

$$\approx -0.5746$$

49. Use Figure 1.35 in the text.

$$\tan \alpha = \frac{h}{|PC|} \text{ so } |PC| = \frac{h}{\tan \alpha};$$

$$\tan \beta = \frac{h}{|QP| + |PC|}$$

$$= \frac{h}{x + \dfrac{h}{\tan \alpha}}$$

$$= \frac{h}{\dfrac{x \tan \alpha + h}{\tan \alpha}}$$

$$= \frac{h \tan \alpha}{x \tan \alpha + h}$$

$$x \tan \beta \tan \alpha + h \tan \beta = h \tan \alpha$$

$$x \tan \beta \tan \alpha = h \tan \alpha - h \tan \beta$$

$$h = \frac{x \tan \beta \tan \alpha}{\tan \alpha - \tan \beta}$$

51. In $\triangle BCF$, $|BC| = |CF|$, so it is an isosceles triangle; $\angle F = \tfrac{\pi}{4}$. Thus, $\theta_1 = \tfrac{\pi}{4}$ so that $\tan \theta_1 = 1$ and $\theta_1 = \tan^{-1} 1$.

Since $\angle F = \tfrac{\pi}{4}$, and since $\triangle DEF$ is isosceles, $|EF| = |DE| = \sqrt{2}$. Also, $|BD| = \sqrt{1^2 + 3^2} = \sqrt{10}$, we see $|BE| = 2\sqrt{2}$. Thus, $\tan \theta_2 = \dfrac{2\sqrt{2}}{\sqrt{2}} = 2$ so that $\theta_2 = \tan^{-1} 2$. In $\triangle BCD$, $\tan \theta_3 = 3$, so $\theta_3 = \tan^{-1} 3$. Since $\theta_1 + \theta_2 + \theta_3 = \pi$,

$$\tan^{-1} 1 + \tan^{-1} 2 + \tan^{-1} 3 = \pi.$$

53. The identity is false. You should provide your own counterexample.

55.

$\sin \alpha = \tfrac{h}{c}$, by definition (in $\triangle ABD$) and

$\sin(\tfrac{\pi}{2} - \gamma) = \tfrac{h}{a}$ by definition (in $\triangle BDC$)

Thus, $h = c \sin \alpha$ (from the first equation)

and $h = a \sin(\frac{\pi}{2} - \gamma)$. Since

$\sin(\frac{\pi}{2} - \gamma) = \sin \gamma$, we have

$$c \sin \alpha = a \sin \gamma$$

$$\frac{\sin \alpha}{a} = \frac{\sin \gamma}{c}$$

CHAPTER 1 REVIEW

Proficiency Examination, Pages 42-43

SURVIVAL HINT: *The concept problems of each Proficiency Examination are designed to remind you what was covered in the chapter. It is a worthwhile activity to* **handwrite** *the answers to each of these questions onto your own paper. If you concentrate as you are doing this, some good things will happen. Here we show one possible answer for each question, but remember you will benefit only if you use the following list as a check after you answer the question.*

1. $\mathbf{N} = \{1, 2, 3, \cdots\}$;

    $\mathbf{W} = \{0, 1, 2, 3, \cdots\}$;

    $\mathbf{Z} = \{\cdots, -3, -2, -1, 0, 1, 2, 3, \cdots\}$;

    $\mathbf{Q} = \{p/q$ so that p is an integer and q is a

    nonzero integer$\}$;

    $\mathbf{Q}' = \{$nonrepeating or nonterminating

    decimals$\}$;

    $\mathbf{R} = \mathbf{Q} \cup \mathbf{Q}'$

2. $|a| = a$ if $a \geq 0$; $|a| = -a$ if $a < 0$

3. $|x + y| \leq |x| + |y|$

4. $d = \sqrt{(x_2 - x_1)^2 + (y_2 - y_1)^2}$

5. $m = \tan \theta$ where θ is the angle of inclination

6.   **a.** $Ax + By + C = 0$

    b. $y = mx + b$

    c. $y - k = m(x - h)$

    d. $y = k$

    e. $x = h$

7. Lines are parallel if they have the same slope, and perpendicular if their slopes are negative reciprocals of one another.

8. A function is a rule that assigns to each element x of the domain D a unique element of the range R.

9. $(f \circ g) = f[g(x)]$

10. The graph of a function f consists of all points (x, y) such that $y = f(x)$ for x in the domain of f.

11.   **a.**   **b.**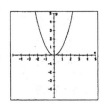

    c.   **d.**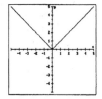

e. 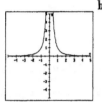 **f.**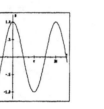

o. 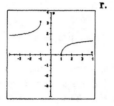 **p.**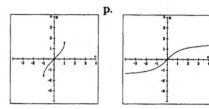

g. 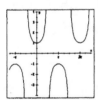 **h.**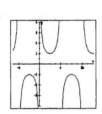

q. 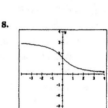 **r.**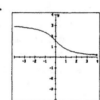

i. 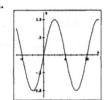 **j.**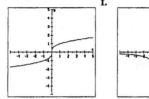

s.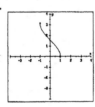

k. **l.**

m. **n.**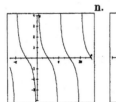

12. $f(x) = a_n x^n + a_{n-1} x^{n-1} + \cdots + a_2 x^2 + a_1 x + a_0$

13. $f(x) = \dfrac{P(x)}{D(x)}$, P and D are polynomials with

$D(x) \neq 0$

14. **a.** Let f be a function with domain D and range R. Then the function f^{-1} with domain R and range D is the inverse of f if

$f^{-1}[f(x)] = x$ for all x in D and

$f[f^{-1}(y)] = y$ for all y in R

b. Reflect the graph of $y = f(x)$ in the line $y = x$.

15. The horizontal line test says that a function f has an inverse f^{-1} if and only if no horizontal line meets the graph of f in more than one point (that is, f is a one-to-one function).

16. $\sin(\sin^{-1}x) = x$ for $-1 \le x \le 1$

$\sin^{-1}(\sin y) = y$ for $-\frac{\pi}{2} \le y \le \frac{\pi}{2}$

$\tan(\tan^{-1}x) = x$ for all x

$\tan^{-1}(\tan y) = y$ for $-\frac{\pi}{2} < y < \frac{\pi}{2}$

17. $\cot^{-1}x = \frac{\pi}{2} - \tan^{-1}x$

$\sec^{-1}x = \cos^{-1}\frac{1}{x}$ if $|x| \ge 1$

$\csc^{-1}x = \sin^{-1}\frac{1}{x}$ if $|x| \ge 1$

18. a. We use the point-slope form:

$$y - k = m(x - h)$$

$$y - 5 = -\frac{3}{4}\left[x - \left(-\frac{1}{2}\right)\right]$$

$$6x + 8y - 37 = 0$$

b. $m = \frac{2 - 5}{7 + 3} = \frac{-3}{10}$

We use the point-slope form.

$$y - k = m(x - h)$$

$$y - 5 = -\frac{3}{10}\left[x - (-3)\right]$$

$$3x + 10y - 41 = 0$$

c. $\frac{x}{a} + \frac{y}{b} = 1$ *intercept form*

$$\frac{x}{4} + \frac{y}{-\frac{3}{7}} = 1$$

$$3x - 28y - 12 = 0$$

d. Writing the given equation in slope-intercept form, $y = -\frac{2}{5}x + \frac{11}{5}$, we see that the slope is $-2/5$. A parallel line must have the same slope. Now use the point-slope form.

$$y - 5 = -\frac{2}{5}\left(x + \frac{1}{2}\right)$$

$$2x + 5y - 24 = 0$$

e. Find the slope of $\overline{PQ}$. A perpendicular line will have a slope which is the negative reciprocal of the given line. Find the midpoint of $\overline{PQ}$. Then use the point-slope form for the equation of the line. The slope of $\overline{PQ}$ is $-3/4$. The midpoint of $\overline{PQ}$ is $(1, 4)$.

$$y - k = m(x - h)$$

$$y - 4 = \frac{4}{3}(x - 1)$$

$$4x - 3y + 8 = 0$$

19. $y = -\frac{3}{2}x + 6$ **20.** $y - 3 = |x + 1|$

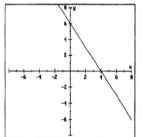
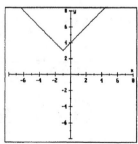

21. $y - 3 = -2(x-1)^2$

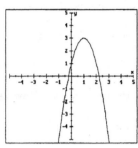

22. $y = (x-2)^2 - 14$

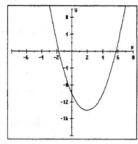

23. $y = 2\cos(x-1)$

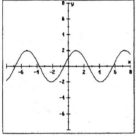

24. $y + 1 = \tan 2(x + \frac{3}{2})$

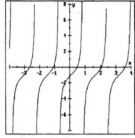

25. $y = \sin^{-1}(2x)$

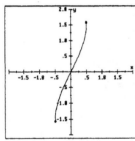

26. $y = \tan^{-1}x^2$

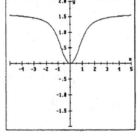

27. If $f(x) = \dfrac{1}{x+1}$, then

$$f\left(\frac{1}{x+1}\right) = \frac{1}{\dfrac{1}{x+1}+1} \text{ , and}$$

$$f\left(\frac{2x+1}{2x+4}\right) = \frac{1}{\dfrac{2x+1}{2x+4}+1}$$

So we need to find the values of x for which:

$$\frac{1}{\dfrac{1}{x+1}+1} = \frac{1}{\dfrac{2x+1}{2x+4}+1}$$

$$\frac{x+1}{1+x+1} = \frac{2x+4}{2x+1+2x+4}$$

$$4x^2 + 9x + 5 = 2x^2 + 8x + 8$$

$$2x^2 + x - 3 = 0$$

$$(2x+3)(x-1) = 0$$

$$x = -\frac{3}{2}, 1$$

28. The root of a quotient equals the quotient of the roots when they have the same domain. The domain of $f(x)$ is $(-\infty, 0] \cup (1, \infty)$. The domain of $g(x)$ is $(1, \infty)$. The two functions are not the same.

29. $f(x) = \sin x$; $g(x) = \sqrt{1 - x^2}$

$$(f \circ g)(x) = \sin\sqrt{1 - x^2}$$

$$(g \circ f)(x) = \sqrt{1 - \sin^2 x} = |\cos x|$$

30. Let the base of the box be x and the height z.

Then, $V = x^2 z$, so

sides: xz with a cost of $\$3xz$

bottom: x^2 with a cost of $\$8x^2$

Then $96 = 4(3xz) + 8x^2$

$$24 = 3xz + 2x^2$$

$$z = \frac{24 - 2x^2}{3x}$$

$$V = x^2\left(\frac{24 - 2x^2}{3x}\right) = \frac{2}{3}x(12 - x^2)$$

SURVIVAL HINT: *There is a large set of supplementary problems at the end of each chapter. You can use these problems for a variety of different uses. They are presented in random order, some routine, and some challenging. The answers to the odd numbered supplementary problems are given at the back of the textbook. You might try some of these problems to practice for examinations.*

CHAPTER 2

Limits and Continuity

SURVIVAL HINT: *If your instructor does not begin with Chapter 1, you might wish to take some time looking over that chapter anyway. Look at the survival hint on page 1 for some information about the **free** supplement that accompanies this book.*

 If your instructor does not assign the "What Does This Say?" problems, take a few minutes to think about these problems anyway. These problems involve the concepts of the section. Could you explain the concept to a classmate or another student? If not, it needs more work. Working with another student or a small group is highly recommended! Be bold and introduce yourself to a classmate.

2.1 The Limit of a Function, Pages 59-61

SURVIVAL HINT: *The limit is the basic concept underlying both the differential and integral calculus. We begin with developing your intuitive concept of limit. After you have learned about limits intuitively, you will find it essential to develop the ability to use the formal definition and learn some evaluation techniques. It is a spiral type concept, in that you will come back to it time and time again; each time at a higher level and with more understanding and greater perspective.*

1. Look at Figure 2.12.

   a. Find the graph named f, and look to see what happens as x approaches 3 from both the left and right. It looks like it approaches 0.

   b. Repeat part **a** as x approaches 2; it looks like it approaches 2.

   c. Repeat part **a** as x approaches 0; it looks like it approaches 6.

3. Look at Figure 2.12.

   a. Find the graph named t, and look to see what happens as x approaches 4 from both the left and right. It looks like it approaches 2.

   b. Repeat part **a** as x approaches -4; it look like it approaches 7.

   c. Notice the small "+" symbol next to -5; this means that we are only interested in approaching -5 from the right. It looks like it approaches 7.5.

5. Look at Figure 2.12.

   a. Find the graph named g, and look to see what happens as x approaches 3 from the left. It looks like y approaches 6.

   b. This time we see what happens as x approaches 3 from the right. This also looks like y approaches 6.

c. We need to see what happens as x approaches 3 from both the right and the left. It looks like it approaches 6.

7.

x	2	3	4	4.5	4.9	4.99
$f(x)$	3	7	11	13	14.6	14.96

$\lim\limits_{x \to 5^-} f(x)$ appears to be 15.

9.

x	1	1.9	1.99	1.999
$h(x)$	7	9.7	9.97	9.997

x	3	2.5	2.1	2.001
$h(x)$	13	11.5	10.3	10.003

$\lim\limits_{x \to 2} h(x)$ appears to be 10.

11. $\lim\limits_{x \to 2} f(x) = 8$ **13.** $\lim\limits_{x \to 4} h(x) = 2$

15. $\lim\limits_{x \to 3\pi/2} s(x) = -1$

17. Finding a limit means "getting closer and closer to" without being at a particular point. That is, the notation $\lim\limits_{x \to c} f(x) = L$ means that the functional values $f(x)$ can be made arbitrarily close to L by choosing x sufficiently close to c (but not equal to c).

SURVIVAL HINT: *In Problems 18-44, one of the most efficient methods is to graph the function using a graphing calculator. A graph (with TRACE) or a table of values using a graphing calculator works nicely for evaluating these limits.*

19. Construct a table of values approaching 0 from the right (1, 0.9, 0.5, 0.1, 0.01, $\cdots$, for example) using a calculator to find
$$\lim\limits_{x \to 0^+} \cos x = 1.00.$$

21. Construct a table of values approaching 3 from the left (2, 2.5, 2.9, 2.91, $\cdots$, for example) using a calculator to find
$$\lim\limits_{x \to 3^-} (x^2 - 4) = 5.00.$$

23. Construct a table of values approaching -3 from the right (-2, -2.5, -2.9, -2.99, for example) using a calculator to find
$$\lim\limits_{x \to -3^+} \frac{1}{x - 3} = -0.17.$$

25. Construct a table of values approaching $\frac{\pi}{2}$ from both the right and left; for example,

0, 1, 1.5, 1.57, 1.5701, 1.5705, $\cdots$

2, 1.7, 1.6, 1.58, 1.571, $\cdots$

Use your calculator to evaluate the function for these values and note that they seem to become infinitely large, so we say the limit does not exist.

27. a. Construct a table of values approaching 0 from both the right and the left and use your calculator to find
$$\lim\limits_{x \to 0} \frac{1 - \cos x}{x} = 0$$

b. Construct a table of values approaching π from both the right and the left and use your calculator to find
$$\lim\limits_{x \to \pi} \frac{1 - \cos x}{x} = 0.64$$

29. 8.00 (See the above problems for the procedure.)

31. -2.00 (Be sure to approach $\frac{\pi}{2}$ from both the right and the left.)

33. 0.17 (Be sure to approach 9 from both the right and the left.)

35. 0.25 (Be sure to approach 2 from both the right and the left.)

37. As x approaches 3 from the right the table of values is getting larger and larger, so we are led to the conclusion that the limit does not exist.

39. As x approaches 0 from both the right and the left, the calculator gives a table of values that looks like the function is approaching the value 2.00.

41. As x approaches 0 from both the right and the left, use your calculator to evaluate the function. Be careful with how you input the values for this function; that is,

$$\dfrac{1 + \dfrac{1}{x+1}}{x}$$

could be input (for a given value of x) as

【 ① + ① ÷ 〔 ⨂ + ① 〕 〕 ÷ ⨂ =

By looking at a table of value, it looks like the limit is 1.00.

43. As x approaches 0 from both the right and the left, use your calculator to evaluate the function to see that the values oscillate so limit does not exist.

45. a. $v(t) = \lim\limits_{x \to t} \dfrac{s(x) - s(t)}{x - t}$

$= \lim\limits_{x \to t} \dfrac{(-16x^2 + 40x + 24) - (-16t^2 + 40t + 24)}{x - t}$

$= \lim\limits_{x \to t} (-16x - 16t + 40)$

$= -32t + 40$

b. $v(0) = \lim\limits_{x \to 0} (-32x + 40)$
$= 40$ ft/s

c. $s(t) = 0$ if $-16t^2 + 40t + 24 = 0$ or if
$2t^2 - 5t - 3 = 0$
$(t - 3)(2t + 1) = 0$
$$t = 3, \ -\tfrac{1}{2}$$
Reject the negative solution. Impact velocity is
$$v(3) = -32(3) + 40 = -56 \text{ ft/s}$$

d. At the highest point on the trajectory, the ball has stopped moving upward and has not yet started on its downward fall. This occurs at $-32t + 40 = 0$ or $t = 1.25$ seconds.

47. Here is a table of values (found by calculator).

$$x \qquad f(x) = \dfrac{x^3 - 9x^2 - 45x - 91}{x - 13}$$

x	$f(x)$
14	259
13.5	243.25
13.1	231.01
13.01	228.3
13.001	228.03
⋮	⋮
12.999	227.97
12.99	227.7
12.9	225.01
12.5	213.25
12	199

It looks like the limit is 228.

SURVIVAL HINT: *Historical Quest problems are designed to get you thinking about some of the famous persons who contributed to calculus. With these problems, we have attempted to give you accessible problems from these mathematician's notebooks.*

49. The Cauchy statement (quoted in several sources) is: "When the successive values attributed to a variable approach indefinitely to a fixed value so as to end by differing from it by as little as one wishes, this last is called the limit of all the others. Thus, for example, an irrational number is the limit of diverse fractions which furnish more and more approximate values of it." The ϵ-δ definition was given by H. E. Heine (1821-1881) in his book *Elements* (1872). Heine was a student of Karl Weierstrass (1815-1897) and it is acknowledged that Heine's published definition came from Weierstrass' lectures. This definition, as stated in Carl B. Boyer's *History of Mathematics*, (© 1968 by John Wiley & Sons, Inc., p. 608) is: "If given any ϵ, there is a η_0 such that for $0 \leq \eta < \eta_0$ the difference $f(x_0 \pm \eta) - L$ is less in absolute value than ϵ, then L is the limit of $f(x)$ for $x = x_0$."

The use of the Greek letter ϵ in this context was originated by Cauchy and probably first used as an abbreviation for *error*. According to Judith V. Grabiner, *The Origins of Cauchy's Rigorous Calculus* (The MIT Press, Cambridge, 1981, p. 76), "the epsilon in a modern proof may be regarded as an inheritance from the days when inequalities belonged in approximations. The epsilon notation is a reminder that, paradoxically, the development of approximations and estimates of error brought forth many of the techniques necessary for the first exact and rigorous proofs about the concepts of calculus. Eighteenth-century mathematicians were

never more exact than when they were being approximate."

51.
$$\begin{aligned} |f(t) - L| &= |(3t - 1) - 0| \\ &< |3t - 1| \end{aligned}$$

The statement is false. Choose $\epsilon = 0.3$ and it is not possible to find a delta.

53.
$$\begin{aligned} |f(x) - L| &= |(2x - 5) + 3| \\ &= 2|x - 1| \\ &< 2\delta \end{aligned}$$

Choose $\delta = \frac{\epsilon}{2}$.

55. $\lim_{x \to 2} \frac{1}{x} = \frac{1}{2}$

Given $\epsilon > 0$, find $\delta > 0$ so that

$$|f(x) - L| < \epsilon \text{ whenever } |x - 2| < \delta$$

If $|x - 2| < \delta$, then

$$2 - \delta < x < 2 + \delta$$

$$\frac{1}{x} < \frac{1}{2 - \delta}$$

and

$$\frac{2 - x}{2x} < \frac{1}{2 - \delta} - \frac{1}{2} = \frac{\delta}{2(2 - \delta)}$$

Thus,

$$\begin{aligned} |f(x) - L| &= \left|\frac{1}{x} - \frac{1}{2}\right| \\ &= \frac{|x - 2|}{2|x|} \\ &< \frac{\delta}{2(2 - \delta)} \end{aligned}$$

We want $\frac{\delta}{2(2 - \delta)} < \epsilon$ which is equivalent to $\delta < \frac{4\epsilon}{1 + 2\epsilon}$. if we choose $\delta = \frac{4\epsilon}{1 + 2\epsilon}$ we have

$$|f(x) - L| < \frac{\delta}{2(2 - \delta)}$$

$$= \frac{\frac{4\epsilon}{1 + 2\epsilon}}{2\left[2 - \left(\frac{4\epsilon}{1 + 2\epsilon}\right)\right]}$$

$$= \epsilon$$

2.2 Algebraic Computation of Limits, Pages 68-69

SURVIVAL HINT: *When finding the limit of a polynomial (such as in Problem 1-3), you can evaluate the limits by substitution.*

1. $\lim\limits_{x \to -2} (x^2 + 3x - 7) = 4 - 6 - 7$

$$= -9$$

3. $\lim\limits_{x \to 3} (x + 5)(2x - 7) = (3 + 5)(2 \cdot 3 - 7)$

$$= -8$$

SURVIVAL HINT: *If the limit of a rational expression has the form 0/0 then it may have any value, L. But if it has the form a/0, where $a \neq 0$, then we note that as the denominator of a fraction decreases, the quotient gets very large positive or very large negative. We sometimes write that the limit is $+\infty$ or $-\infty$ to symbolize this idea.*

5. $\lim\limits_{z \to 1} \dfrac{z^2 + z - 3}{z + 1} = \dfrac{1 + 1 - 3}{1 + 1}$

$$= -\frac{1}{2}$$

SURVIVAL HINT: *If you are finding a limiting value of a trigonometric function which is defined at the limiting value, then you can evaluate the limit by substitution.*

7. $\lim\limits_{x \to \pi/3} \sec x = \sec \frac{\pi}{3}$

$$= 2$$

Theorem 2.2 (Limits of trigonometric functions)

9. $\lim\limits_{x \to 1/3} \dfrac{x \sin \pi x}{1 + \cos \pi x} = \dfrac{\frac{1}{3} \sin \frac{\pi}{3}}{1 + \cos \frac{\pi}{3}}$

$$= \frac{\sqrt{3}/6}{3/2}$$

$$= \frac{\sqrt{3}}{9}$$

SURVIVAL HINT: *When working limit problems, use proper form and notation to avoid errors and to make each statement value. For example, in Problem 11 write $\lim\limits_{x \to -2}$ for each transformed expression until you actually evaluate the limit. Do not be lazy in your notation.*

11. $\lim\limits_{u \to -2} \dfrac{4 - u^2}{2 + u} = \lim\limits_{u \to -2} \dfrac{(2 + u)(2 - u)}{(2 + u)}$

$$= \lim\limits_{u \to -2} (2 - u)$$

$$= 4$$

13. $\lim\limits_{x \to 1} \dfrac{\frac{1}{x} - 1}{x - 1} = \lim\limits_{x \to 1} \dfrac{\frac{1 - x}{x}}{x - 1}$

$$= \lim\limits_{x \to 1} \frac{-1}{x}$$

$$= -1$$

15. $\lim\limits_{x \to 1} \left(\dfrac{x^2 - 3x + 2}{x^2 + x - 2}\right)^2 = \lim\limits_{x \to 1} \left(\dfrac{(x - 2)(x - 1)}{(x + 2)(x - 1)}\right)^2$

$$= \lim\limits_{x \to 1} \dfrac{(x - 2)^2}{(x + 2)^2}$$

$$= \frac{1}{9}$$

SURVIVAL HINT: *Problem 17 illustrates a useful general procedure. The function is multiplied by 1 which is written in a form which rationalizes the numerator.*

17. $\lim\limits_{x\to 1} \dfrac{\sqrt{x}-1}{x-1} = \lim\limits_{x\to 1} \dfrac{(\sqrt{x}-1)(\sqrt{x}+1)}{(x-1)(\sqrt{x}+1)}$

$\qquad = \lim\limits_{x\to 1} \dfrac{x-1}{(x-1)(\sqrt{x}+1)}$

$\qquad = \lim\limits_{x\to 1} \dfrac{1}{\sqrt{x}+1}$

$\qquad = \dfrac{1}{2}$

19. $\lim\limits_{x\to 0} \dfrac{\sqrt{x+1}-1}{x}$

$\qquad = \lim\limits_{x\to 0} \dfrac{(\sqrt{x+1}-1)(\sqrt{x+1}+1)}{x(\sqrt{x+1}+1)}$

$\qquad = \lim\limits_{x\to 0} \dfrac{x+1-1}{x(\sqrt{x+1}+1)}$

$\qquad = \lim\limits_{x\to 0} \dfrac{x}{x(\sqrt{x+1}+1)}$

$\qquad = \lim\limits_{x\to 0} \dfrac{1}{\sqrt{x+1}+1}$

$\qquad = \dfrac{1}{\sqrt{0+1}+1}$

$\qquad = \dfrac{1}{2}$

21. $\lim\limits_{x\to 0^+} \dfrac{\sin x}{\sqrt{x}} = \lim\limits_{x\to 0^+} \dfrac{\sin x}{\sqrt{x}} \cdot \dfrac{\sqrt{x}}{\sqrt{x}}$

$\qquad = \lim\limits_{x\to 0^+} \left(\dfrac{\sin x}{x}\right)\sqrt{x}$

$\qquad = 1 \cdot 0$

$\qquad = 0$

SURVIVAL HINT: *Problem 23 illustrates a general technique which you will frequently use when evaluating limits of the type illustrated by this problem.*

23. $\lim\limits_{x\to 0} \dfrac{\sin 2x}{x} = \lim\limits_{x\to 0} 2\left(\dfrac{\sin 2x}{2x}\right)$

$\qquad = 2$

25. $\lim\limits_{t\to 0} \dfrac{\tan 5t}{\tan 2t} = \lim\limits_{t\to 0}\left(\dfrac{\sin 5t}{\cos 5t}\cdot\dfrac{\cos 2t}{\sin 2t}\right)$

$\qquad = \lim\limits_{t\to 0} \dfrac{\sin 5t}{5t} \lim\limits_{t\to 0} \dfrac{5}{\cos 5t} \lim\limits_{t\to 0} \dfrac{2t}{\sin 2t} \lim\limits_{t\to 0} \dfrac{\cos 2t}{2}$

$\qquad = \left(1\left(\dfrac{5}{1}\right)1\left(\dfrac{1}{2}\right)\right)$

$\qquad = \dfrac{5}{2}$

27. $\lim\limits_{x\to 0} \dfrac{1-\cos x}{\sin x} = \lim\limits_{x\to 0} \dfrac{\frac{1-\cos x}{x}}{\frac{\sin x}{x}}$

$\qquad\qquad = \dfrac{\lim\limits_{x\to 0} \frac{1-\cos x}{x}}{\lim\limits_{x\to 0} \frac{\sin x}{x}}$

$\qquad\qquad = \dfrac{0}{1}$

$\qquad\qquad = 0$

29. $\lim\limits_{x\to 0} \dfrac{\sin^2 x}{x^2} = \lim\limits_{x\to 0} \dfrac{\sin x}{x}\dfrac{\sin x}{x}$

$\qquad\qquad = (1)(1)$

$\qquad\qquad = 1$

31. $\lim\limits_{x\to 0} \dfrac{\sec x - 1}{x \sec x} = \lim\limits_{x\to 0} \dfrac{\frac{1}{\cos x}-1}{\frac{x}{\cos x}}$

$\qquad\qquad = \lim\limits_{x\to 0} \dfrac{\frac{1-\cos x}{\cos x}}{\frac{x}{\cos x}}$

$\qquad\qquad = \lim\limits_{x\to 0} \dfrac{1-\cos x}{x}$

$\qquad\qquad$ *Theorem 2.3 (Special limits)*

$\qquad\qquad = 0$

33. The limit of a polynomial can be found by substitution; that is $\lim\limits_{x\to a} f(x) = f(a)$.

35. $\lim\limits_{x\to 0} \dfrac{\sin ax}{x} = \lim\limits_{ax\to 0} a\left(\dfrac{\sin ax}{ax}\right)$

$\qquad\qquad = a(1)$

$\qquad\qquad = a$

37. In this problem, since the denominator is not

0 as x approaches 1 from the right, we can find the limit by direct substitution.

$$\lim_{x \to 1^+} \frac{\sqrt{x-1}+x}{1-2x} = \frac{0+1}{1-2}$$
$$= -1$$

39. For $\lim_{x \to 3} |3-x|$ we need to consider the left- and right-hand limits separately.

$$\lim_{x \to 3^-} |3-x| = \lim_{x \to 3^-} (3-x)$$
$$= 0$$

and

$$\lim_{x \to 3^+} |3-x| = \lim_{x \to 3^+} (x-3)$$
$$= 0$$

Thus

$$\lim_{x \to 3} |3-x| = 0$$

41. We need to consider the left- and right-hand limits separately.

$$\lim_{x \to -2^-} \frac{|x+2|}{x+2} = \lim_{x \to -2^-} \frac{-(x+2)}{x+2}$$
$$= -1$$

and

$$\lim_{x \to -2^+} \frac{|x+2|}{x+2} = \lim_{x \to -2^+} \frac{x+2}{x+2} = 1$$

The limit does not exist because the left- and right-hand limits are not equal.

43. $\lim_{s \to 1^+} \frac{s^2-s}{s-1} = \lim_{s \to 1^-} s = 1$ and

$$\lim_{s \to 1^-} \sqrt{1-s} = \sqrt{1-1} = 0$$

The limit does not exist because the left- and

right-hand limits are not equal.

45. As $x \to 2^+$, the denominator $\sqrt{x-2}$ is approaching 0, the result of dividing a constant (1 in this case) by a quantity approaching zero, becomes infinite.

47. $\lim_{x \to 3} \frac{x^2+4x+3}{x-3} = \lim_{x \to 3} \frac{(x+3)(x+1)}{x-3}$

As $x \to 3$, the numerator is approaching $(6)(4) = 24$ and the denominator $(x-3)$ is approaching 0, which when divided into a number close to 24, becomes infinite.

49. The limit does not exist because the limit from the left is 5 and the limit from the right is -1.

51. The limit does not exist because $\csc \pi x$ increases without bound as $x \to 1$ from the left and decreases without bound as $x \to 1$ from the right.

53. $\lim_{x \to 0} \left(\frac{1}{x} - \frac{1}{x^2} \right) = \lim_{x \to 0} \left(\frac{x-1}{x^2} \right)$

This fraction approaches $-\infty$ as $x \to 0^+$ and $+\infty$ as $x \to 0^-$, so the limit does not exist.

55. $\lim_{t \to 2} g(t) = 4$ since the left- and right-hand limits are both equal to 4.

57. $\lim_{x \to 3} f(x) = 8$ since the left- and right-hand limits are both equal to 8.

59. $\lim_{h \to 0} \frac{\cos h - 1}{h} = \lim_{h \to 0} \left(\frac{\cos h - 1}{h} \cdot \frac{\cos h + 1}{\cos h + 1} \right)$

$$= \lim_{h \to 0} \frac{\cos^2 h - 1}{h(\cos h + 1)}$$

$$= \lim_{h \to 0} \frac{-\sin^2 h}{h(\cos h + 1)}$$

$$= (-1) \lim_{h \to 0} \left[\left(\frac{\sin h}{h} \right) \left(\frac{\sin h}{\cos h + 1} \right) \right]$$

$$= (-1) \lim_{h \to 0} \left(\frac{\sin h}{h} \right) \lim_{h \to 0} \left(\frac{\sin h}{\cos h + 1} \right)$$

$$= (-1)(1)\left(\frac{0}{2} \right)$$

$$= 0$$

61. $f(x) = \frac{1}{x^2}$; if $|x| < \frac{1}{10\sqrt{L}}$, then

$f(x) = \frac{1}{x^2} > 100L$. As L grows, $|x|$ shrinks,

and $f(x)$ grows beyond all bounds. This

means $\lim_{x \to 0} f(x)$ does not exist.

63. To find $\lim_{x \to x_0} \cos x$, let

$$h = x - x_0 \text{ or } x = h + x_0$$

Note that $h \to 0$ as $x \to x_0$.

$$\lim_{x \to x_0} \cos x = \lim_{h \to 0} \cos(h + x_0)$$

$$= \lim_{h \to 0} [\cos h \cos x_0 - \sin h \sin x_0]$$

$$= 1(\cos x_0) - 0(\sin x_0)$$

$$= \cos x_0$$

2.3 Continuity, Pages 78–80

SURVIVAL HINT: *Once again, your intuitive notion of continuity will help. Think about all the possible situations that could make a function discontinuous. The formal definition takes care of all of these. Do not just memorize the definition,* **understand the concept.**

1. Temperature is continuous, so

TEMPERATURE $= f$(TIME) would be a

continuous function. The domain could be midnight to midnight say, $0 \le t < 24$.

3. The charges (range of the function) consist of rational numbers only (dollars and cents rounded to the nearest cent), so the function CHARGE $= f$(MILEAGE) would be a step function (that is, not continuous). The domain would consist of the mileage from the beginning of the trip to its end.

5. No suspicious points and no points of discontinuity with a polynomial.

7. The denominator factors to $x(x - 1)$, so suspicious points would be $x = 0, 1$. There will be a hole discontinuity at $x = 0$ and a pole discontinuity at $x = 1$.

9. $x = 0$ is suspicious and is a point of discontinuity, since the denominator vanishes. Points $x < 0$ are not in the domain since square roots of negative numbers are not defined in the set of real numbers.

11. $x = 1$ is a suspicious point; there are no points of discontinuity.

13. The sine and cosine are continuous on the reals, but the tangent is discontinuous at $x = \pi/2 + n\pi$, for any integer n. Each of these values will have a pole type discontinuity.

SURVIVAL HINT: *A common error is to apply a theorem when it really is not applicable. Theorems are usually if-then statement, and the conclusion is* *not justified unless the hypotheses are met. When learning a theorem, pay careful attention to the "if" part, and make sure the hypotheses are met.*

15. For continuity, $f(2)$ must equal

$$\lim_{x \to 2} f(x) = \lim_{x \to 2} \frac{x^2 - x - 2}{x - 2}$$

$$= \lim_{x \to 2} \frac{(x - 2)(x + 1)}{x - 2}$$

$$= \lim_{x \to 2} (x + 1)$$

$$= 3$$

17. For continuity, $f(2)$ must equal

$$\lim_{x \to 2} f(x) = \lim_{x \to 2} \frac{\sin(\pi x)}{x - 2}$$

$$= \pi$$

By table, graphing, or calculator.

19. The function is not defined at $x = 2$, and since $\lim_{x \to 2^-} f(x) = 11$ and $\lim_{x \to 2^+} f(x) = 9$ no value can be assigned to $f(2)$ to "tie together" the two pieces. This is sometimes called an "essential" discontinuity. Only hole type discontinuities are "removable".

21. **a.** No suspicious points on $[1, 2]$; continuous
    b. Suspicious point $x = 0$. The function is discontinuous on $[0, 1]$ since the pole $x = 0$ is in the domain. If the interval had been $(0, 1]$, the function would have been continuous on the interval.

23. Suspicious point $t = 0$; the limit from the left

is 15 while that from the right is 0. The function is discontinuous at $t = 0$.

25. Suspicious point $x = 0$. The function is not defined at $x = 0$, so it is discontinuous at $x = 0$.

27. $f(x) = \sqrt[3]{x} - x^2 - 2x + 1$ is continuous on $[0, 1]$ and $f(0) = 1$, $f(1) = -1$ so the hypotheses of the intermediate value theorem are met, and we are guaranteed that there is at least one number c on $(0, 1)$ such that $f(c) = 0$.

29. $f(x) = \sqrt[3]{x - 8} + 9x^{2/3} - 29$ is continuous on $[0, 8]$ and $f(0) = -31$, $f(8) = 7$, so the hypotheses of the intermediate value theorem are met, and we are guaranteed that there is at least one number c on $(0, 8)$ such that $f(c) = 0$.

31. $f(x) = \cos x - \sin x - x$ is continuous on the reals and $f(0) = 1$, $f(\pi/2) = -1 - \pi/2$. Since the hypotheses of the intermediate value theorem have been met, we are guaranteed that there exists at least one number c on $(0, \pi/2)$ such that $f(c) = 0$.

33. $\lim_{x \to 2^-} f(x) = 2 + 1 = 3 = f(2)$

Thus, f is continuous from the left.

$$\lim_{x \to 2^+} f(x) = 2^2 = 4 \neq f(2)$$

Thus, f is not continuous from the right.

35. At 12:00 the hands coincide. Let $f(x)$ denote the angle from the hour hand to the minute

hand, x minutes after the beginning of hour h (for $0 < h < 12$). Then, $f(0) = \frac{\pi}{6}(h) > 0$ and $f(60) = \frac{\pi}{6}(h + 1) - 2\pi \leq 0$. Since the sign of $f(x)$ changes between $x = 0$ and $x = 60$ and f is clearly continuous, there must be a time when $f(x) = 0$; that is, when the hands coincide.

37. a. $\lim\limits_{v \to v_w} \dfrac{Cv^k}{v - v_w}$ does not exist because there is a singularity at $v = v_w$. This means that the less variance between the rate at which the fish swim and the rate at which the river flows, the larger the expended energy (because the fish is not making any headway).

b. As $v \to +\infty$ the velocity of the fish gets large and the energy expended goes to infinity.

39. $f(5) = 8$; from the right, $5a + 3 = 8$, so $a = 1$. From the left, $25 + 5b + 1 = 8$, so $b = -18/5$.

41. We must have $a = 1$ or $\lim\limits_{x \to 1} f(x)$ does not exist. With $a = 1$,

$$\lim_{x \to 1} \frac{\sqrt{x} - 1}{x - 1} = \lim_{x \to 1} \frac{(\sqrt{x} - 1)(\sqrt{x} + 1)}{(x - 1)(\sqrt{x} + 1)}$$
$$= \lim_{x \to 1} \frac{1}{\sqrt{x} + 1}$$
$$= \frac{1}{2}$$

Since $f(1) = b$, we must have $b = \frac{1}{2}$.
So $a = 1$ and $b = \frac{1}{2}$.

43. $g(0) = 5$; if $x < 0$,

$$\lim_{x \to 0^-} \frac{\sin ax}{x} = a \lim_{x \to 0^-} \frac{\sin ax}{ax}$$
$$= a$$

Thus, $a = 5$.
Similarly,

$$\lim_{x \to 0^+} g(x) = 0 + b$$
$$= b$$

Thus, $b = 5$.

45. Cauchy first noted that "the value of $\sin \frac{1}{2}\alpha$ decreases indefinitely with that of α." In modern notation, we write

$$\lim_{\alpha \to 0} \sin(\tfrac{1}{2}\alpha) = 0$$

Next, consider

$$\lim_{\alpha \to 0} \sin(x + \alpha) - \sin x$$
$$= \lim_{\alpha \to 0} \left[2 \sin \tfrac{\alpha}{2} \cos(x + \tfrac{\alpha}{2}) \right]$$
$$= 2(0) \cos x = 0$$

Finally, if $s(\alpha) = \sin \alpha$, it follows that s is a continuous function of α.

47. a. Carry out the bisection method and verify with a calculator $x \approx 1.25872$.

b. Carry out the bisection method and verify with a calculator $x \approx 0.785398$.

49. Answers vary. The function

$$f(x) = \begin{cases} x \text{ if } x \text{ is rational} \\ 1 - x \text{ if } x \text{ is irrational} \end{cases}$$

continuous only at $x = \frac{1}{2}$.

51. If $f(x)$ is continuous at $x = c$, it must be continuous from the left and the right (where c is not an endpoint of an interval).

Conversely, suppose f is continuous from the right and the left at $x = c$. Then,

$$\lim_{x \to c^+} f(x) = f(c) = \lim_{x \to c^-} f(x)$$

This implies continuity at $x = c$.

53. $\lim_{x \to 0} f(u(x)) = \lim_{x \to 0} f(x) = 0$, since $f(x) = 0$

for $x \neq 0$, however

$$f\left[\lim_{x \to 0} u(x)\right] = f\left[\lim_{x \to 0} x\right] = f(0) = 1$$

2.4 Exponential and Logarithmic Functions, Pages 89-91

SURVIVAL HINT: *You need to be comfortable with logarithms as inverses of exponentials. The function defined by the equation $y = e^x$ has inverse $x = e^y$, where y is described as $\ln x$, i.e. the exponent of e that gives x. Logarithms are exponents. $\ln 5$ is the exponent of e that gives 5. $e^{\ln 3} = 3$ because the exponent on e that gives 3 is, of course, $\ln 3$. If you do not understand these concepts, it might be a good idea to review the sections on logarithms and inverses in a trigonometry or precalculus text. For applications involving the way the world is built, the two most important numbers are π and e.*

1.

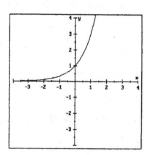

3.

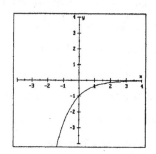

5. No calculator is necessary for this one:
$$32^{2/5} + 9^{3/2} = 2^2 + 3^3$$
$$= 31$$

SURVIVAL HINT: *The answers to Problems 7-21 are given here and if you do not obtain the same results with your calculator, it is important that you find out why. The proper use of a calculator throughout the text is essential.*

7. $e^3 e^{2.3} \approx 200.33681$

9. $5,000\left(1 + \frac{0.135}{12}\right)^{12(5)} \approx 9,783.225896$

11. $2,589 e^{0.45(6)} \approx 38,523.62544$

SURVIVAL HINT: *Problems 13-22 use properties of logarithms. Study the steps shown until you understand each one. Remember that exponential and natural logarithm are inverses.*

13. $\log_2 4 + \log_3 \frac{1}{9} = 2 + (-2)$
$$= 0$$

15. $5 \log_3 9 - 2 \log_2 16 = 5(2) - 2(4)$

$$= 2$$

17. $\left(3^{\log_7 1}\right)\left(\log_5 0.04\right) = 3^0 \left(\log_5 \frac{1}{25}\right)$

$$= -2$$

19. $\log_3 3^4 - \ln e^{0.5} = 4 - 0.5$

$$= 3.5$$

21. $\exp(\ln 3 - \ln 10) = \exp\left(\ln \frac{3}{10}\right)$

$$= \frac{3}{10}$$

23. $\log_x 16 = 2$

$x^2 = 16$ *Two is the exponent on an x that gives 16.*

$x = \pm 4$

Reject the negative value since bases of logarithms cannot be negative numbers. Thus, $x = 4$.

25. $e^{-3x} = 0.5$

$-3x = \ln 0.5$ *$-3x$ is the exponent on a base of e that gives 0.5.*

$x = -\frac{1}{3} \ln 0.5$

≈ 0.23104906

27. $7^{-x} = 15$

$-x = \log_7 15$ *$-x$ is the exponent on a base of 7 that gives 15.*

$x = -\log_7 15$

$= -\frac{\ln 15}{\ln 7}$

≈ -1.391662509

29. $\frac{1}{2} \log_3 x = \log_2 8$

$\log_3 x = 6$

$x = 3^6$

$$= 729$$

31. $3^{x^2 - x} = 9$

$3^{x^2 - x} = 3^2$

$x^2 - x = 2$

$(x - 2)(x + 1) = 0$

$x = 2, -1$

33. $2^x 5^{x+2} = 25{,}000$

$2^x 5^x 5^2 = 2^3 5^5$

$(2 \cdot 5)^x = 2^3 5^3$

$10^x = 10^3$

$x = 3$

35. $\left(\sqrt[3]{2}\right)^{x+10} = 2^{x^2}$

$2^{x/3 + 10/3} = 2^{x^2}$

$\frac{x}{3} + \frac{10}{3} = x^2$

$3x^2 - x - 10 = 0$

$(x - 2)(3x + 5) = 0$

$x = 2, -\frac{5}{3}$

37. $e^{2x+3} = 1$

$e^{2x+3} = e^0$

$2x + 3 = 0$

$x = -\frac{3}{2}$

39.
$$\log_3 x + \log_3(2x + 1) = 1$$
$$\log_3 x(2x + 1) = 1$$
$$x(2x + 1) = 3$$
$$2x^2 + x - 3 = 0$$
$$(x - 1)(2x + 3) = 0$$
$$x = 1, \ -\tfrac{3}{2}$$

Reject the negative value since logarithms of negative numbers are not defined. Thus, $x = 1$.

41. a. $\lim_{x \to 0^+} x^2 e^{-x} = 0$

b. $\lim_{x \to 1} x^2 e^{-x} = (1)^2 e^{-1}$
$$= e^{-1}$$

43. a. $\lim_{x \to 0^+} (1 + x)^{1/x} = \lim_{u \to \infty} \left(1 + \tfrac{1}{u}\right)^u$

Let $u = \tfrac{1}{x}$, so $u \to \infty$ as $x \to 0^+$.

$$= e \quad \textit{Definition of e.}$$

b. $\lim_{x \to 1} (1 + x)^{1/x} = (1 + 1)^{1/1}$
$$= 2$$

45. $\lim_{x \to e} x \ln x^2 = e \ln e^2$
$$= 2e$$

47. $\log_b 1{,}296 = 4$, so $b^4 = 1{,}296 = 6^4$, and $b = 6$.

Thus, $\left(\tfrac{3}{2} \cdot 6\right)^{3/2} = 27$.

49. Use the graph or solve utility on a graphing calculator to find $x \approx 0.4229976068447$.

Thus, to the nearest tenth, $x = 0.4$.

51. logarithmic

53. exponential

55.
$$I(x) = I_0 e^{kx}$$

$I(0) = I_0$; $I(2) = 0.05 I_0$ but also $I(2) = I_0 e^{2k}$ so that

$$0.05 I_0 = I_0 e^{2k}$$
$$e^{2k} = 0.05$$
$$2k = \ln 0.05$$
$$k = -1.497866$$

When does $I(x) = 0.01 I_0$?

$$0.01 I_0 = I_0 e^{-1.497866x}$$
$$-1.497866x = \ln 0.01$$
$$x \approx 3.07$$

It is approximately 3 meters below the surface.

57.
$$2P = P e^{rt}$$
$$2 = e^{rt}$$
$$\ln 2 = rt$$
$$t = \frac{\ln 2}{r}$$

It will take about $0.7/r$ years to double.

59.
$$2(\$8{,}500) = \$8{,}500 e^{10r}$$
$$2 = e^{10r}$$
$$10r = \ln 2$$

$$r = 0.1(\ln 2)$$

$$\approx 0.0693147181$$

The necessary rate is about 6.9%.

61. a. $$2P_0 = P_0(2^{60k})$$

$$60k = 1 \quad \text{so} \quad k = \tfrac{1}{60}$$

$$P(t) = P_0 2^{t/60}$$

Also, $P(20) = 1{,}000$ so

$$1{,}000 = P_0 2^{1/3}$$

$$P_0 \approx 793.7$$

b. $$5{,}000 = P_0(2^{t/60})$$

$$\tfrac{t}{60} = \log_2(5{,}000/P_0)$$

$$t = 60 \log_2(5{,}000/P_0)$$

$$t \approx 159.32$$

The time is about 2 hr and 40 min.

63. a. $$1.5M + 11.4 = \log E$$

$$E = 10^{1.5M+11.4}$$

b. $$\frac{E_1}{E_2} = 10^{1.5(M_1 - M_2)}$$

$$= 10^{1.5(8.5 - 6.5)}$$

$$= 10^3$$

$$= 1{,}000$$

1,000 times more energy is released

65. a.

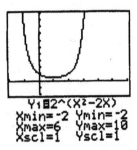

Y₁ = 2^(X²-2X)
Xmin=-2 Ymin=-2
Xmax=6 Ymax=10
Xscl=1 Yscl=1

b. It crosses the y-axis at $(0, 1)$.

Let $y = E(x)$; as $x \to +\infty$, $y \to +\infty$.

As $x \to -\infty$, $y \to +\infty$.

c. The smallest value of y is $E = 0.5$.

67. a. $$A = \$3{,}600\left(1 + \frac{0.15}{365}\right)^{7(365)}$$

$$\approx \$10{,}285.33$$

b. $$A = \$3{,}600\left(1 + \frac{0.15}{360}\right)^{7(360)}$$

$$\approx \$10{,}285.29$$

69. a. $F(0) = 70 - A = 35$, so $A = 35$;

$F(30) = 50$, so

$$50 = 70 - 35e^{-30k}$$

$$e^{-30k} = \tfrac{20}{35} = \tfrac{4}{7}$$

b. $$F(60) = 70 - 35e^{-60k}$$

$$= 70 - 35(e^{-30k})^2$$

$$= 70 - 35(\tfrac{4}{7})^2$$

$$\approx 58.57$$

The temperature is about 59°.

c. As $t \to +\infty$, $F(t) = 70 - 35(0) = 70$

$F(t)$ approaches $70°$.

71. a. $b^m b^n = \underbrace{(bbb \cdot \cdots \cdot b)}_{m} \underbrace{(bbb \cdot \cdots \cdot b)}_{n}$

$= \underbrace{bbb \cdot \cdots \cdot b}_{m+n} = b^{m+n}$

b. $\dfrac{b^m}{b^n} = \dfrac{\overbrace{bbb \cdot \cdots \cdot b}^{m}}{\underbrace{bbb \cdot \cdots \cdot b}_{n}}$

$= \underbrace{bbb \cdots \cdot b}_{m-n} = b^{m-n}$

73. Let $M = \log_b x$ and $N = \log_b y$. Then

$b^M = x$ and $b^N = y$.

a. $xy = b^M b^N = b^{M+N}$

By definition of logarithm,

$M + N = \log_b(xy)$

$\log_b x + \log_b y = \log_b(xy)$

b. $\dfrac{x}{y} = \dfrac{b^M}{b^N} = b^{M-N}$

By definition of logarithm,

$M - N = \log_b\left(\dfrac{x}{y}\right)$

$\log_b x - \log_b y = \log_b\left(\dfrac{x}{y}\right)$

75. a. Suppose $\log_a x = M$. Then,

$a^M = x$

$\log_b a^M = \log_b x$

$M \log_b a = \log_b x$

$M = \dfrac{\log_b x}{\log_b a}$

$\log_a x = \dfrac{\log_b x}{\log_b a}$

b. $(\log_a b)(\log_b a) = \left(\dfrac{\ln b}{\ln a}\right)\left(\dfrac{\ln a}{\ln b}\right) = 1$

CHAPTER 2 REVIEW

Proficiency Examination, Page 92

SURVIVAL HINT: *Since your class may have skipped Chapter 1, we will repeat the hint we gave in the proficiency examination for Chapter 1. The concept problems of each Proficiency Examination are designed to remind you what was covered in the chapter. It is a worthwhile activity to **handwrite** the answers to each of these questions onto your own paper. If you concentrate as you are doing this, some good things will happen.*

1. $\lim\limits_{x \to c} f(x) = L$ means that the function values

$f(x)$ can be made arbitrarily close to L by

choosing x sufficiently close to c.

2. $\lim\limits_{x \to c} f(x) = L$ means that for each $\epsilon > 0$ there

exists a number $\delta > 0$ such that

$|f(x) - L| < \epsilon$ whenever $0 < |x - c| < \delta$

3. a. $\lim\limits_{x \to c} k = k$ for any constant k

b. $\lim\limits_{x \to c}[sf(x)] = s \lim\limits_{x \to c} f(x)$

c. $\lim\limits_{x \to c}[f(x) + g(x)] = \lim\limits_{x \to c} f(x) + \lim\limits_{x \to c} g(x)$

d. $\lim\limits_{x \to c}[f(x) - g(x)] = \lim\limits_{x \to c} f(x) - \lim\limits_{x \to c} g(x)$

e. $\lim\limits_{x \to c}[f(x)g(x)] = [\lim\limits_{x \to c} f(x)][\lim\limits_{x \to c} g(x)]$

f. $\lim\limits_{x \to c} \dfrac{f(x)}{g(x)} = \dfrac{\lim\limits_{x \to c} f(x)}{\lim\limits_{x \to c} g(x)}$ if $\lim\limits_{x \to c} g(x) \neq 0$

g. $\lim\limits_{x \to c} [f(x)]^n = \left[\lim\limits_{x \to c} f(x) \right]^n$ n is a rational number and the limit on the right exists.

h. If P is a polynomial function, then
$$\lim_{x \to c} P(x) = P(c)$$

i. If Q is a rational function defined by
$$Q(x) = \frac{P(x)}{D(x)}, \text{ then } \lim_{x \to c} Q(x) = \frac{P(c)}{D(c)}$$
provided $\lim\limits_{x \to c} D(x) \neq 0$.

j. If T is a trigonometric, exponential, or a natural logarithmic function, defined at $x = c$, then
$$\lim_{x \to c} T(x) = T(c)$$

4.   If $g(x) \leq f(x) \leq h(x)$ for all x on an open interval containing c, and if
$$\lim_{x \to c} g(x) = \lim_{x \to c} h(x) = L,$$
then $\lim\limits_{x \to c} f(x) = L$.

5.   a. $\lim\limits_{x \to 0} \dfrac{\sin x}{x} = 1$ b. $\lim\limits_{x \to 0} \dfrac{\cos x - 1}{x} = 0$

6.   A function f is continuous at a point $x = c$ if
(1) $f(c)$ is defined
(2) $\lim\limits_{x \to c} f(x)$ exists
(3) $\lim\limits_{x \to c} f(x) = f(c)$

7.   If f is a polynomial, rational, power, trigonometric, logarithmic, exponential, or inverse trigonometric function, then f is continuous at any number $x = c$ for which $f(c)$ is defined.

8.   If $\lim\limits_{x \to c} g(x) = L$ and f is a function that is continuous at L, then $\lim\limits_{x \to c} f[g(x)] = f(L)$. That is,
$$\lim_{x \to c} f[g(x)] = f(L) = f(\lim_{x \to c} g(x))$$

9.   If f is a continuous function on the closed interval $[a, b]$ and L is some number strictly between $f(a)$ and $f(b)$, then there exists at least one number c on the open interval (a, b) such that $f(c) = L$.

10.   If f is continuous on the closed interval $[a, b]$ and if $f(a)$ and $f(b)$ have opposite algebraic signs (one positive and the other negative), then $f(c) = 0$ for at least one number c on the open interval (a, b).

11.   For any real number x, there exist rational numbers r_n such that
$$x = \lim_{n \to \infty} r_n$$
which means that for any number $\epsilon > 0$, there exists a number N such that $|x - r_n| < \epsilon$ whenever $n > N$.

12.   a. Let x be a real number, and let r_n be a sequence of rational numbers such that $x = \lim\limits_{n \to \infty} r_n$. Then, the exponential function with base $b > 0$ $(b \neq 1)$ is given by
$$b^x = \lim_{n \to \infty} b^{r_n}$$

   b. Exponential and logarithmic functions are inverse functions.

13. $e = \lim\limits_{n \to \infty} \left(1 + \dfrac{1}{n} \right)^n$

14.   a. If $b > 0$ and $b \neq 1$, the logarithm of x

to the base b is the function $y = \log_b x$ that satisfies $b^y = x$; that is

$$y = \log_b x \qquad \text{means} \qquad b^y = x$$

b. A common logarithm is a logarithm to the base 10, written $\log_{10} x = \log x$.

c. A natural logarithm is a logarithm to the base e, written $\log_e x = \ln x$.

15. a.

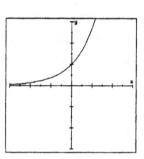

b.

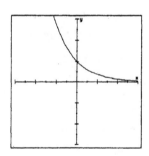

16. a.

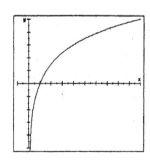

b.

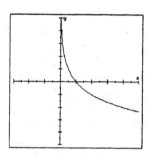

17. a. If $b \neq 1$, then $b^x = b^y$ if and only if

$x = y$.

b. If $x > y$ and $b > 1$, then $b^x > b^y$.

If $x > y$ and $0 < b < 1$, then $b^x < b^y$.

c. $b^x b^y = b^{x+y}$

d. $\dfrac{b^x}{b^y} = b^{x-y}$

e. $(b^x)^y = b^{xy}; \quad (ab)^x = a^x b^x; \quad \left(\dfrac{a}{b}\right)^x = \dfrac{a^x}{b^x}$

18. a. 0 **b.** 1 **c.** x **d.** y **e.** $e^{x \ln b}$

19. $\displaystyle\lim_{x \to 3} \frac{x^2 - 4x + 9}{x^2 + x - 8} = \frac{(3)^2 - 4(3) + 9}{3^2 + 3 - 8}$

$$= \frac{3}{2}$$

20. $\displaystyle\lim_{x \to 4} \frac{\sqrt{x} - 2}{x - 4} \cdot \frac{\sqrt{x} + 2}{\sqrt{x} + 2}$

$$= \lim_{x \to 4} \frac{x - 4}{(x - 4)(\sqrt{x} + 2)}$$

$$= \lim_{x \to 4} \frac{1}{\sqrt{x} + 2}$$

$$= \frac{1}{4}$$

21. $$\lim_{x \to 2} \frac{x^2 - 5x + 6}{x^2 - 4} = \lim_{x \to 2} \frac{(x-2)(x-3)}{(x+2)(x-2)}$$

$$= \lim_{x \to 2} \frac{x - 3}{x + 2}$$

$$= -\frac{1}{4}$$

22. $$\lim_{x \to 0} \frac{1 - \cos x}{2 \tan x} = \lim_{x \to 0} \frac{(1 - \cos x)(1 + \cos x)}{2 \frac{\sin x}{\cos x}(1 + \cos x)}$$

$$= \lim_{x \to 0} \frac{\sin^2 x}{2 \frac{\sin x}{\cos x}(1 + \cos x)}$$

$$= \lim_{x \to 0} \frac{\sin x \cos x}{2(1 + \cos x)}$$

$$= \frac{0}{4}$$

$$= 0$$

23. $$\lim_{x \to 0} \frac{\sin 9x}{\sin 5x} = \lim_{x \to 0} \frac{9x}{9x}(\sin 9x) \cdot \frac{5x}{5x}\left(\frac{1}{\sin 5x}\right)$$

$$= \lim_{x \to 0} \frac{9x}{5x}\left(\frac{\sin 9x}{9x}\right) \cdot \left(\frac{5x}{\sin 5x}\right)$$

$$= \frac{9}{5}(1)(1)$$

$$= \frac{9}{5}$$

24. $$\lim_{x \to \frac{1}{2}^-} \frac{|2x - 1|}{2x - 1} = \lim_{x \to \frac{1}{2}^-} \frac{-(2x - 1)}{2x - 1}$$

$$= -1$$

25. **a.** exponential **b.** exponential

 c. logarithmic **d.** logarithmic

26. We have suspicious points where the denominators are 0, at $t = 0, -1$. There are pole discontinuities at each of these points.

27. Suspicious points $x = -2$ (pole) and $x = 1$ (removable hole) are also points of discontinuity (since the denominator is 0).

28. **a.** $5,000 = 2,000\left(1 + \frac{0.08}{4}\right)^{4t}$

$$2.5 = 1.02^{4t}$$

$$4t = \log_{1.02} 2.5$$

$$t = \frac{1}{4} \log_{1.02} 2.5 \approx 11.56779247$$

 It will take 11 years, 3 quarters (since 11 years 2 quarters is insufficient).

 b. $5,000 = 2,000\left(1 + \frac{0.08}{12}\right)^{12t}$

$$t = \frac{1}{12} \log_{(1 + 0.08/12)} 2.5$$

$$\approx 11.49177065$$

 It will take 11 years, 6 months.

 c. $5,000 = 2,000 e^{0.08t}$

$$t = \frac{1}{0.08} \ln 2.5$$

$$\approx 11.45363415$$

 It will take 11 years, 166 days.

29. Polynomials are everywhere continuous, so the only problem is at $x = 1$. We need

$$\lim_{x \to 1^-} (Ax + 3) = 2, \text{ and } \lim_{x \to 1^+} (x^2 + B) = 2$$

$$A + 3 = 2, \quad A = -1 \text{ and } 1 + B = 2, \quad B = 1$$

30. Let $f(x) = x + \sin x - \dfrac{1}{\sqrt{x + 3}}$. This

function is continuous on $[0, \infty)$, and

$$f(0) = -\tfrac{1}{3}, \quad f(\pi) = \pi - \frac{1}{\sqrt{\pi + 3}} \approx 2.93$$

So by the root location theorem there must be

some value c on $(0, \pi)$ where $f(c) = 0$.

CHAPTER 3

Differentiation

SURVIVAL HINT: The derivative is one of the great ideas of calculus. Spending some extra time with this idea now will pay dividends throughout the book. It would be a good idea to memorize the definition of derivative given on page 100 of the text.

3.1 An Introduction to the Derivative: Tangents, Pages 107-110

SURVIVAL HINT: Recognize the concept of slope as the change in y divided by the change in x. In calculus, we symbolize this by $\Delta y/\Delta x$. Think of Δx as a single symbol for "change in x" and Δy as a single symbol for "change in y."

1. Some describe it as a five-step process:
    (1) Find $f(x)$

    (2) Find $f(x + \Delta x)$

    (3) Find $f(x + \Delta x) - f(x)$

    (4) Find $\dfrac{f(x + \Delta x) - f(x)}{\Delta x}$

    (5) Finally find $\displaystyle\lim_{\Delta x \to 0} \dfrac{f(x + \Delta x) - f(x)}{\Delta x}$

3. Answers vary; continuity does not imply differentiability, but differentiability implies continuity.

    SURVIVAL HINT: Look at the answer to Problem 3 even if your instructor did not assign this problem. This is an important principle that you should remember.

SURVIVAL HINT: For Problems 5-10, note that when the derivative is positive the graph of the function is increasing and when the derivative is negative, the graph of the function is decreasing. When the derivative is zero, the function may change from increasing to decreasing or from decreasing to increasing.

5.

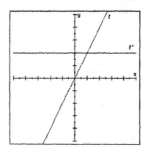

7.

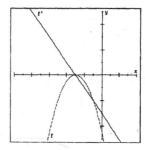

9.

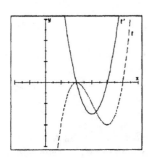

11. **a.** $\dfrac{f(-5 + \Delta x) - f(-5)}{\Delta x} = 0$

 b. $f'(-5) = \lim\limits_{\Delta x \to 0} 0$

 $= 0$

13. **a.** $\dfrac{f(1 + \Delta x) - f(1)}{\Delta x}$

 $= \dfrac{2 + 2\Delta x - 2}{\Delta x}$

 $= 2$

 b. $f'(1) = \lim\limits_{\Delta x \to 0} 2$

 $= 2$

15. **a.** $\dfrac{f(0 + \Delta x) - f(0)}{\Delta x}$

 $= \dfrac{2 - (\Delta x)^2 - 2}{\Delta x}$

 $= -\Delta x$

 b. $f'(0) = \lim\limits_{\Delta x \to 0} (-\Delta x)$

 $= 0$

17. $f'(x) = \lim\limits_{\Delta x \to 0} \dfrac{f(x + \Delta x) - f(x)}{\Delta x}$

 $= \lim\limits_{\Delta x \to 0} \dfrac{5 - 5}{\Delta x}$

 $= 0$

Differentiable for all x.

19. $f'(x) = \lim\limits_{\Delta x \to 0} \dfrac{f(x + \Delta x) - f(x)}{\Delta x}$

 $= \lim\limits_{\Delta x \to 0} \dfrac{[3(x + \Delta x) - 7] - [3x - 7]}{\Delta x}$

 $= \lim\limits_{\Delta x \to 0} 3$

 $= 3$

Differentiable for all x.

21. $g'(x) = \lim\limits_{\Delta x \to 0} \dfrac{g(x + \Delta x) - g(x)}{\Delta x}$

 $= \lim\limits_{\Delta x \to 0} \dfrac{[3(x + \Delta x)^2] - [3x^2]}{\Delta x}$

 $= \lim\limits_{\Delta x \to 0} (6x + 3\Delta x)$

 $= 6x$

Differentiable for all x.

23. $f'(x) = \lim\limits_{\Delta x \to 0} \dfrac{f(x + \Delta x) - f(x)}{\Delta x}$

 $= \lim\limits_{\Delta x \to 0} \dfrac{[(x + \Delta x)^2 - (x + \Delta x)] - [x^2 - x]}{\Delta x}$

 $= \lim\limits_{\Delta x \to 0} \dfrac{\Delta x(2x + \Delta x - 1)}{\Delta x}$

 $= \lim\limits_{\Delta x \to 0} (2x + \Delta x - 1)$

 $= 2x - 1$

Differentiable for all x.

25. $f'(s) = \lim_{\Delta s \to 0} \frac{f(s + \Delta s) - f(s)}{\Delta s}$

$= \lim_{\Delta s \to 0} \frac{[(s + \Delta s) - 1]^2 - (s - 1)^2}{\Delta s}$

$= \lim_{\Delta s \to 0} \frac{s^2 + 2s\Delta s + (\Delta s)^2 + 1 - 2s - 2\Delta s - s^2 + 2s - 1}{\Delta s}$

$= \lim_{\Delta s \to 0} \frac{2s\Delta s + (\Delta s)^2 - 2\Delta s}{\Delta s}$

$= \lim_{\Delta s \to 0} \frac{\Delta s(2s + \Delta s - 2)}{\Delta s}$

$= \lim_{\Delta s \to 0} (\Delta s + 2s - 2)$

$= 2s - 2$

Differentiable for all real s.

27. $f'(x) = \lim_{\Delta x \to 0} \frac{f(x + \Delta x) - f(x)}{\Delta x}$

$= \lim_{\Delta x \to 0} \frac{\sqrt{5(x + \Delta x)} - \sqrt{5x}}{\Delta x}$

$= \lim_{\Delta x \to 0} \frac{\sqrt{5x + 5\Delta x} - \sqrt{5x}}{\Delta x}\left[\frac{\sqrt{5x + 5\Delta x} + \sqrt{5x}}{\sqrt{5x + 5\Delta x} + \sqrt{5x}}\right]$

$= \lim_{\Delta x \to 0} \frac{5x + 5\Delta x - 5x}{\Delta x(\sqrt{5x + 5\Delta x} + \sqrt{5x})}$

$= \lim_{\Delta x \to 0} \frac{5}{\sqrt{5x + 5\Delta x} + \sqrt{5x}}$

$= \frac{5}{2\sqrt{5x}}$

$= \frac{\sqrt{5x}}{2x}$

Differentiable for $x > 0$.

29. $f'(x) = 3 = m$ (see Problem 19)

$y - 2 = 3(x - 3)$

$3x - y - 7 = 0$

31. $f'(s) = 3s^2$ (see Example 9)

$f'(-\tfrac{1}{2}) = 3(-\tfrac{1}{2})^2 = \tfrac{3}{4} = m$

$f(-\tfrac{1}{2}) = (-\tfrac{1}{2})^3 = -\tfrac{1}{8}$

$y - (-\tfrac{1}{8}) = \tfrac{3}{4}[s - (-\tfrac{1}{2})]$

$8y + 1 = 6s + 3$

$3s - 4y + 1 = 0$

33. $f'(x) = \lim_{\Delta x \to 0} \frac{f(x + \Delta x) - f(x)}{\Delta x}$

$= \lim_{\Delta x \to 0} \frac{1}{\Delta x}\left(\frac{1}{x + \Delta x + 3} - \frac{1}{x + 3}\right)$

$= \lim_{\Delta x \to 0} \frac{1}{\Delta x}\left[\frac{x + 3 - x - \Delta x - 3}{(x + 3)(x + \Delta x + 3)}\right]$

$= \lim_{\Delta x \to 0} \frac{-1}{(x + 3)(x + \Delta x + 3)}$

$= \frac{-1}{(x + 3)^2}$

$f'(2) = \frac{-1}{25} = m; \quad f(2) = \frac{1}{2 + 3} = \frac{1}{5}$

$y - \tfrac{1}{5} = \frac{-1}{25}(x - 2)$

$x + 25y - 7 = 0$

35. From Problem 19, $f'(3) = 3$; so the normal line has slope $m = -\frac{1}{3}$. Since it passes through $(3, 2)$, the equation is

$$y - 2 = -\tfrac{1}{3}(x - 3)$$

$$x + 3y - 9 = 0$$

37. From Problem 33, $f'(x) = \dfrac{-1}{(x + 3)^2}$, so

$f'(3) = -\tfrac{1}{36}$; the required line will have a

slope of 36 and must pass through the point

$(3, \tfrac{1}{6})$.

$$y - \tfrac{1}{6} = 36(x - 3)$$

$$216x - 6y - 647 = 0$$

39. $\dfrac{dy}{dx} = \lim\limits_{\Delta x \to 0} \dfrac{f(x + \Delta x) - f(x)}{\Delta x}$

$= \lim\limits_{\Delta x \to 0} \dfrac{2(x + \Delta x) - 2x}{\Delta x}$

$= \lim\limits_{\Delta x \to 0} 2$

$= 2$

$\left. \dfrac{dy}{dx} \right|_{x = -1} = 2$

41. $\dfrac{dy}{dx} = \lim\limits_{\Delta x \to 0} \dfrac{f(x + \Delta x) - f(x)}{\Delta x}$

$= \lim\limits_{\Delta x \to 0} \dfrac{[1 - (x + \Delta x)^2] - (1 - x^2)}{\Delta x}$

$= \lim\limits_{\Delta x \to 0} \dfrac{-2x\Delta x - (\Delta x)^2}{\Delta x}$

$= \lim\limits_{\Delta x \to 0} (-2x - \Delta x)$

$= -2x$

$\left. \dfrac{dy}{dx} \right|_{x = 0} = 0$

SURVIVAL HINT: *Do not confuse zero slope with no slope. Zero slope is a horizontal line, whereas no slope may mean the function is discontinuous at the given value, is continuous but has a cusp, or that the graph it is a vertical line.*

43. **a.** $f(-2) = 4$ and $f(-1.9) = 3.61$

$$m_{\text{sec}} = \frac{4 - 3.61}{-2 + 1.9} = -3.9$$

b. $f'(x) = \lim\limits_{\Delta x \to 0} \dfrac{(x + \Delta x)^2 - x^2}{\Delta x}$

$= \lim\limits_{\Delta x \to 0} \dfrac{x^2 + 2x\Delta x + (\Delta x)^2 - x^2}{\Delta x}$

$= \lim\limits_{\Delta x \to 0} (2x + \Delta x) = 2x$

$m_{\text{tan}} = f'(-2) = 2(-2) = -4$

$m_{\text{tan}} < m_{\text{sec}}$, but close

45. $f(x) = x^2 - x$

$f'(x) = \lim\limits_{\Delta x \to 0} \dfrac{f(x + \Delta x) - f(x)}{\Delta x}$

$= \lim\limits_{\Delta x \to 0} \dfrac{[(x + \Delta x)^2 - (x + \Delta x)] - [x^2 - x]}{\Delta x}$

$= \lim\limits_{\Delta x \to 0} \dfrac{x^2 + 2x\Delta x + (\Delta x)^2 - x - \Delta x - x^2 + x}{\Delta x}$

$= \lim\limits_{\Delta x \to 0} (2x + \Delta x - 1)$

$= 2x - 1$

The derivative is 0 when $x = \frac{1}{2}$; $f(\frac{1}{2}) = -\frac{1}{4}$, so the graph has a horizontal tangent at $(\frac{1}{2}, -\frac{1}{4})$.

47. a. $f(x) = 4 - 2x^2$

$$f'(x) = \lim_{\Delta x \to 0} \frac{f(x + \Delta x) - f(x)}{\Delta x}$$

$$= \lim_{\Delta x \to 0} \frac{[4 - 2(x + \Delta x)^2] - (4 - 2x^2)}{\Delta x}$$

$$= \lim_{\Delta x \to 0} \frac{-4x\Delta x - 2(\Delta x)^2}{\Delta x} = -4x$$

b. A horizontal line has $m = 0$, so

$f'(x) = 0$ when $-4x = 0$ or when

$x = 0$. At $x = 0$, $f(0) = 4$, so

$y - 4 = 0(x - 0)$

$y - 4 = 0$ is the equation of the

horizontal line.

c. $8x + 3y = 4$ has a slope of $-\frac{8}{3}$ so the

tangent line must have the same slope in

order to be parallel. At (x_0, y_0) on the

graph $f'(x_0) = -4x_0 = -\frac{8}{3}$, when x_0

$= \frac{2}{3}$ and $y_0 = f(\frac{2}{3}) = \frac{28}{9}$, so the point

at which the tangent is parallel to the

given line is $(\frac{2}{3}, \frac{28}{9})$.

49. $\lim_{\Delta x \to 0} \dfrac{f(x + \Delta x) - f(x)}{\Delta x}$

$$= \lim_{\Delta x \to 0} \frac{2|x + \Delta x + 1| - 2|x + 1|}{\Delta x}$$

$$= \lim_{\Delta x \to 0} \frac{2|2 + \Delta x| - 2(2)}{\Delta x} \quad \text{at } x = 1$$

$$= \lim_{\Delta x \to 0} \frac{4 + 2\Delta x - 4}{\Delta x}$$

$$= \lim_{\Delta x \to 0} \frac{2\Delta x}{\Delta x} = 2$$

Yes, f is differentiable at $x = 1$. By the way, f is not differentiable at $x = -1$.

51. a.

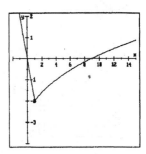

b. $f(1) = -2$ and since

$$\lim_{x \to 1^-}(-2x) = -2$$

and

$$\lim_{x \to 1^+}(\sqrt{x} - 3) = -2,$$

we see that the function is continuous at

$x = 1$.

To check differentiability, consider:

If $x < 1$,

$$\lim_{\Delta x \to 0} \frac{-2(x + \Delta x) + 2x}{\Delta x} = -2$$

If $x \geq 1$,

$$\lim_{\Delta x \to 0} \frac{\sqrt{x + \Delta x} - 3 - \sqrt{x} + 3}{\Delta x} = \frac{1}{2\sqrt{x}}$$

We see that at $x = 1$, the limits are not

the same, so f is not differentiable at

$x = 1$.

53. Let $\Delta x = 0.1$, $c = 1$, $f(x) = (2x - 1)^2$

$$f'(x) \approx \frac{f(x + \Delta x) - f(x)}{\Delta x}$$

$$= \frac{f(1.1) - f(1)}{0.1}$$

$$= \frac{1.44 - 1}{0.1}$$

$$= 4.4$$

Let $\Delta x = 0.01$

$$f'(x) \approx \frac{f(x + \Delta x) - f(x)}{\Delta x}$$

$$= \frac{f(1.01) - f(1)}{0.01}$$

$$= \frac{1.0404 - 1}{0.01}$$

$$= 4.04$$

It appears that $f'(x) = 4$.

55. We could proceed as shown in the solution for Problem 53, but here we present data which could be generated using a graphing calculator, spreadsheet, or a computer program.

$f(x) = \sin x$, $c = \frac{\pi}{3}$

Δx	$c + \Delta x$	$\frac{f(c + \Delta x) - f(c)}{\Delta x}$
0.5	1.5472	0.26739
0.125	1.1722	0.44464
0.03125	1.0784	0.48639
0.00781	1.0550	0.49661
0.00195	1.0491	0.49916
0.00049	1.0477	0.49979

We might guess that the derivative is $f'(\frac{\pi}{3}) = \frac{1}{2}$. Note: later we will find this to be $\cos \frac{\pi}{3} = \frac{1}{2}$.

57. We could proceed as shown in the solution for Problem 53, but here we present data which could be generated using a graphing calculator, spreadsheet, or a computer program.

$f(x) = \sqrt{x}$, $c = 4$

Δx	$c + \Delta x$	$\frac{f(c + \Delta x) - f(c)}{\Delta x}$
0.5	4.5	0.24264
0.125	4.125	0.24808
0.03125	4.0313	0.24951
0.00781	4.0078	0.24988
0.00195	4.0020	0.24997
0.00049	4.0005	0.24999

We might guess that the derivative is $f'(4) = \frac{1}{4} = 0.25$.

59. We find that $\frac{dy}{dx} = 2Ax$, $\frac{dy}{dx}\Big|_{x=c} = 2Ac$.

The point is $(c, f(c)) = (c, Ac^2)$

So the equation of the tangent line is:

$$y - Ac^2 = 2Ac(x - c).$$

The x-intercept occurs when $y = 0$:

$$-Ac^2 = 2Ac(x - c), \quad -c = 2(x - c), \quad x = \frac{c}{2}$$

The y-intercept occurs when $x = 0$:

$$y - Ac^2 = 2Ac(0 - c) \text{ or } y = -Ac^2$$

61. a. From Example 1, $f'(x) = 2x$

For $g(x) = x^2 - 3$, we have

$$g'(x) = \lim_{\Delta x \to 0} \frac{g(x + \Delta x) - g(x)}{\Delta x}$$

$$= \lim_{\Delta x \to 0} \frac{[(x + \Delta x)^2 - 3] - [x^2 - 3]}{\Delta x}$$

$$= \lim_{\Delta x \to 0} \frac{x^2 + 2x\Delta x + (\Delta x)^2 - 3 - (x^2 - 3)}{\Delta x}$$

$$= \lim_{\Delta x \to 0} (2x + \Delta x)$$

$$= 2x$$

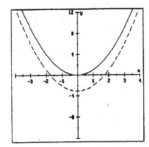

The graph of $y = x^2 - 3$ is the graph of $y = x^2$ lowered by 3 units. Thus, the tangents of the graphs have the same slopes for any value of x. This accounts geometrically for the fact their derivatives are identical.

b. The graph of $y = x^2 + 5$ is also parallel to the graph of $y = x^2$. Thus,

$$\frac{d}{dx}(x^2 + 5) = \frac{d}{dx}(x^2)$$

$$= 2x$$

63. For $x < 0$, the tangent lines have negative slope and for $x > 0$ the tangent lines have positive slope. There is a corner at $x = 0$, and there is no tangent at $x = 0$, and consequently no derivative at that point.

65. It is easy to show that f is continuous at $x = 0$ since $\lim_{\Delta x \to 0} \frac{\sin x}{x} = 1$. For the derivative we find

$$f'(x) = \lim_{\Delta x \to 0} \frac{f(x + \Delta x) - f(x)}{\Delta x}$$

$$= \lim_{\Delta x \to 0} \frac{\frac{\sin(x + \Delta x)}{x + \Delta x} - \frac{\sin x}{x}}{\Delta x}$$

At $x = 0$, we have $f'(0) = \lim_{\Delta x \to 0} \frac{\frac{\sin \Delta x}{\Delta x} - 1}{\Delta x}$

Using a table of values (or a graph) we guess the derivative of f at $x = 0$ is 0, so the equation of the tangent line, $y = 1$.

67. The equation of the circle is

$$(x - x_1)^2 + y^2 = (4 - x_1)^2 + 4^2$$

Substitute $y^2 = 4x$ into the equation of the circle.

$$(x - x_1)^2 + 4x = (4 - x_1)^2 + 16$$

Obtain a 0 on one side:

$$x^2 + (4 - 2x_1)x + 8(x_1 - 4) = 0$$

Since this is to be an equation with equal roots, Descartes would have compared this with $(x - r)^2 = 0$. Equating corresponding coefficients in

$$x^2 + (4 - 2x_1)x + 8(x_1 - 4) = x^2 - 2xr + r^2$$

This gives $4 - 2x_1 = -2r$ and $8(x_1 - 4) = r^2$

Thus, $(4 - 2x_1)^2 = 4r^2 = 32(x_1 - 4)$

$$(2 - x_1)^2 = 8(x_1 - 4)$$

or $x_1 = 6$. Thus, the point on the x-axis through which the normal to the parabola at $(4, 4)$ passes, is $(6, 0)$. The required tangent is then the line through $(4, 4)$ perpendicular to this normal.

69. For $f'(c) \neq 0, y - f(c) = -\dfrac{1}{f'(c)}(x - c)$

$$f'(c)y - f(c)f'(c) = -x + c$$

$$f'(c)y - f(c)f'(c) + x - c = 0$$

If $f'(c) = 0$, the normal line is vertical, with equation $x = c$.

3.2 Techniques of Differentiation, Pages 117-118

SURVIVAL HINT: *This section is a powerhouse section, and the power rule, product rule, and quotient rules must be thoroughly mastered. Since addition is commutative, the product rule can be used as*

$$(fg)' = fg' + gf$$

or $gf' + fg'$. To help you remember the product rule, the following rhyme might help:

> *f-g prime is one-two prime plus two prime one*
>
> *f-g prime is two prime one plus one-two prime*
>
> *You will remember this rule when you remember this rhyme.*

*On the other hand, since subtraction is **not** commutative, you must always begin with the denominator times the derivative of the numerator.*

The quotient rule is: $\left(\dfrac{f}{g}\right)' = \dfrac{gf' - fg'}{g^2}$

> *Quotient rule help rhyme:*
>
> *"To find the derivative of high-low prime, all you need is this stupid little rhyme,*
>
> *low dee high minus high dee low, over the square of the denominator must go."*

1. (11) $f'(x) = 0; f'(-5) = 0$

   (12) $f'(x) = 1; f'(2) = 1$

(13) $f'(x) = 2; f'(1) = 2$

(14) $f'(x) = 4x; f'(1) = 4$

(15) $f'(x) = -2x; f'(0) = 0$

(16) $f'(x) = -2x; f'(2) = -4$

3. (23) $f'(x) = 2x - 1$

   (24) $g'(t) = -2t$

   (25) $f(s) = (s - 1)^2$

$$= s^2 - 2s + 1;$$

$$f'(s) = 2s - 2$$

   (26) $f(x) = \frac{1}{2}x^{-1}$

$$f'(x) = -\frac{1}{2}x^{-2}$$

   (27) $f(x) = \sqrt{5}\,x^{1/2}$

$$f'(x) = \frac{\sqrt{5}}{2x^{1/2}} = \frac{\sqrt{5x}}{2x}$$

5. **a.** $f(x) = 3x^4 - 9$

$$f'(x) = 3(4)x^3 - 0$$

$$= 12x^3$$

   b. $g(x) = 3(9)^4 - x$

$$g'(x) = 0 - 1$$

$$= -1$$

SURVIVAL HINT: *Pay special attention to the differences between parts **a** and **b**. You do not apply the power rule to a constant. Look at Problem 7b; and remember that C^2 is a constant.*

7.　a.　$f(x) = x^3 + C$

$$f'(x) = 3x^2 + 0$$

$$= 3x^2$$

　　b.　$g(x) = C^2 + x$

$$g'(x) = 1$$

SURVIVAL HINT: *If you can write a quotient as a sum, you will often simply reduce the amount of work you need to do. In Problems 9-14 you will write each expression as a sum before you attempt to find the derivative.*

9.　$r(t) = t^2 - t^{-2} + 5t^{-4}$;

$$r'(t) = 2t + 2t^{-3} - 20t^{-5}$$

11.　$f(x) = 7x^{-2} + x^{2/3} + C$;

$$f'(x) = -14x^{-3} + \tfrac{2}{3}x^{-1/3}$$

13.　$f(x) = \dfrac{x^3 + x^2 + x - 7}{x^2}$　*Divide term-by-term*

$$= x + 1 + x^{-1} - 7x^{-2}$$

$$f'(x) = 1 - x^{-2} + 14x^{-3}$$

15.　$f(x) = (2x + 1)(1 - 4x^3)$

$$= 2x - 8x^4 + 1 - 4x^3$$

$$f'(x) = -32x^3 - 12x^2 + 2$$

SURVIVAL HINT: *Compare Problems 17-18 with Problems 9-14. You should use the quotient rule for these problems.*

17.　$f(x) = \dfrac{3x + 5}{x + 9}$

$$f'(x) = \frac{(x + 9)(3) - (3x + 5)(1)}{(x + 9)^2}$$

$$= \frac{22}{(x + 9)^2}$$

19.　$g(x) = x^2(x + 2)^2$

$$= x^2(x^2 + 4x + 4)$$

$$= x^4 + 4x^3 + 4x^2$$

$$g'(x) = 4x^3 + 12x^2 + 8x$$

21.　$f(x) = x^5 - 5x^3 + x + 12$

$$f'(x) = 5x^4 - 15x^2 + 1$$

$$f''(x) = 20x^3 - 30x$$

$$f'''(x) = 60x^2 - 30$$

$$f^{(4)}(x) = 120x$$

23.　$f(x) = -2x^{-2}$

$$f'(x) = 4x^{-3}$$

$$f''(x) = -12x^{-4}$$

$$f'''(x) = 48x^{-5}$$

$$f^{(4)}(x) = -240x^{-6}$$

25.　$y = 3x^3 - 7x^2 + 2x - 3$

$$\frac{dy}{dx} = 9x^2 - 14x + 2$$

$$\frac{d^2y}{dx^2} = 18x - 14$$

SURVIVAL HINT: *In Problems 27-32, remember the standard form equation of a line is*

$$Ax + By + C = 0$$

However, to get there you must first use the point-slope form of the equation of a line:

$$y - k = m(x - h)$$

27. $f(x) = x^2 - 3x - 5$

$f(-2) = 5$, so the point of tangency is $(-2, 5)$.

$f'(x) = 2x - 3$

and

$f'(-2) = -7 = m_{tan}$

Using the point-slope formula:

$$y - 5 = -7[x - (-2)]$$

$$y - 5 = -7x - 14$$

$$7x + y + 9 = 0$$

29. $f(x) = (x^2 + 1)(1 - x^3)$

$$= -x^5 - x^3 + x^2 + 1$$

$$f(1) = (1^2 + 1)(1 - 1^3)$$

$$= 0$$

The point of tangency is $(1, 0)$.

$$f'(x) = -5x^4 - 3x^2 + 2x$$

and

$$f'(1) = -6 = m_{tan}$$

Using the point-slope formula:

$$y - 0 = -6(x - 1)$$

$$6x + y - 6 = 0$$

31. $f(x) = \dfrac{x^2 + 5}{x + 5}$

$f(1) = 1$, so the point of tangency is $(1, 1)$.

$$f'(x) = \frac{(x + 5)(2x) - (x^2 + 5)(1)}{(x + 5)^2}$$

$$= \frac{x^2 + 10x - 5}{(x + 5)^2}$$

and

$$f'(1) = \tfrac{1}{6} = m_{tan}$$

Using the point-slope formula:

$$y - 1 = \tfrac{1}{6}(x - 1)$$

$$x - 6y + 5 = 0$$

33. $f(x) = 2x^3 - 7x^2 + 8x - 3$

$$f'(x) = 6x^2 - 14x + 8$$

Solve

$$6x^2 - 14x + 8 = 0$$

$$2(3x - 4)(x - 1) = 0$$

$$x = \tfrac{4}{3}, 1$$

$f(\tfrac{4}{3}) = -\tfrac{1}{27}$ and $f(1) = 0$

The points are $(1, 0)$ and $(\tfrac{4}{3}, -\tfrac{1}{27})$.

35. $g(x) = (3x - 5)(x - 8)$

$$= 3x^2 - 29x + 40$$

$$g'(x) = 6x - 29$$

$g'(x) = 0$ if $x = \frac{29}{6}$

$g(\frac{29}{6}) = -\frac{361}{12}$

The point is $(\frac{29}{6}, -\frac{361}{12})$.

37. $f(x) = \sqrt{x}(x - 3)$

$\qquad = x^{3/2} - 3x^{1/2}$

$f'(x) = \frac{3}{2}x^{1/2} - \frac{3}{2}x^{-1/2}$

Solve

$\frac{3}{2}x^{1/2} - \frac{3}{2}x^{-1/2} = 0$

$x - 1 = 0$

$x = 1$

$f(1) = \sqrt{1}(1 - 3) = -2$

The point is $(1, -2)$.

39. $h(x) = \dfrac{4x^2 + 12x + 9}{2x + 3}$

$\qquad = \dfrac{(2x + 3)^2}{2x + 3}$

$\qquad = 2x + 3$

$h'(x) = 2 \neq 0$, so there are no horizontal tangents.

41. **a.** $f(x) = \dfrac{2x - 3}{x^3}$

$f'(x) = \dfrac{x^3(2) - (2x - 3)(3x^2)}{x^6}$

$\qquad = \dfrac{-4x + 9}{x^4}$

b. $f(x) = x^{-3}(2x - 3)$

$f'(x) = x^{-3}(2) + (-3x^{-4})(2x - 3)$

$\qquad = 2x^{-3} - 6x^{-3} + 9x^{-4}$

$\qquad = -4x^{-3} + 9x^{-4}$

c. $f(x) = 2x^{-2} - 3x^{-3}$

$f'(x) = 2(-2)x^{-3} - 3(-3)x^{-4}$

$\qquad = -4x^{-3} + 9x^{-4}$

d. $\dfrac{-4x + 9}{x^4} = -4x^{-3} + 9x^{-4}$

43. The derivative of $y = x^4 - 2x + 1$ is $y' = 4x^3 - 2$. The given line has slope 2, so we want the point(s) of tangency (x_0, y_0) to satisfy

$4x_0^3 - 2 = 2$ or $x_0 = 1$

$y_0 = 1^4 - 2(1) + 1 = 0$. The equation of the tangent line is

$y - 0 = 2(x - 1)$

$2x - y - 2 = 0$

45. **a.** $f(x) = (x^3 - 2x^2)(x + 2)$

$\qquad = x^4 - 4x^2$

$f(1) = 1^4 - 4(1)^2 = -3$

$f'(x) = 4x^3 - 8x, \; f'(1) = -4 = m$

The equation of the tangent line is

$$y + 3 = -4(x - 1)$$

$$4x + y - 1 = 0$$

b. $f'(0) = 0$ so the slope is not defined and the normal line is vertical with equation $x = 0$ (that is, the y-axis).

47. We are looking for particular points (x_0, y_0) on the graph of $y = 4x^2$ which have a tangent at that point which will pass through the point $(2, 0)$.

$$y' = 8x, \quad f(x_0) = 4x_0{}^2, \quad f'(x_0) = 8x_0.$$

So we have a point $(x_0, 4x_0{}^2)$ and the slope of the line at that point: $8x_0$. We can now write the equation of the line:

$$y - 4x_0{}^2 = 8x_0(x - x_0)$$

This line must pass through the point $(2, 0)$, so

$$0 - 4x_0{}^2 = 8x_0(2 - x_0)$$
$$4x_0{}^2 - 16x_0 = 0$$
$$4x_0(x_0 - 4) = 0$$
$$x_0 = 0, 4$$

Therefore there are two points on the curve at which the tangent line will pass through $(2, 0)$; they are $(0, 0)$ and $(4, 64)$.

49. $f(x) = x^2 + 2x - 3$

$$f'(x) = 2x + 2; \ f''(x) = 2; \ f'''(x) = 0$$
$$y''' + y'' + y' = 2x + 2 + 2 + 0$$
$$= 2x + 4$$

The equation is not satisfied.

51. $f(x) = \frac{1}{2}x^2 + x$

$$f'(x) = x; \ f''(x) = 1; \ f'''(x) = 0$$
$$y''' + y'' + y' = 0 + 1 + x$$
$$= x + 1$$

This function satisfies the given equation.

53. $8(4 + 100) + 62,000 \approx 20,000\pi$

$$\pi \approx \frac{62,832}{20,000}$$

$$\approx 3.1416$$

This approximation is even better than the one commonly used approximation of 22/7 used in elementary school.

55. $F(x) = cf(x)$

$$F'(x) = \lim_{\Delta x \to 0} \frac{F(x + \Delta x) - F(x)}{\Delta x}$$

$$= \lim_{\Delta x \to 0} \frac{c[f(x + \Delta x) - f(x)]}{\Delta x}$$

$$= cf'(x)$$

57. $F(x) = [f(x)]^2$

$$F'(x) = \lim_{\Delta x \to 0} \frac{F(x + \Delta x) - F(x)}{\Delta x}$$

$$= \lim_{\Delta x \to 0} \frac{[f(x + \Delta x)]^2 - [f(x)]^2}{\Delta x}$$

$$= \lim_{\Delta x \to 0} \frac{[f(x + \Delta x) - f(x)][f(x + \Delta x) + f(x)]}{\Delta x}$$

$$= \lim_{\Delta x \to 0} [f(x + \Delta x) + f(x)] \lim_{\Delta x \to 0} \frac{f(x + \Delta x) - f(x)}{\Delta x}$$

$$= 2f(x)f'(x)$$

59. $\left(\dfrac{f}{g}\right)' = \lim\limits_{\Delta x \to 0} \dfrac{1}{\Delta x}\left[\dfrac{f(x + \Delta x)}{g(x + \Delta x)} - \dfrac{f(x)}{g(x)}\right]$

$= \lim\limits_{\Delta x \to 0} [f(x + \Delta x)g(x) - f(x)g(x)]$

$\qquad - f(x)g(x + \Delta x) + f(x)g(x)]/[\Delta x g(x)g(x + \Delta x)]$

$= \lim\limits_{\Delta x \to 0} \{g(x)[f(x + \Delta x) - f(x)]$

$\qquad - f(x)[g(x + \Delta x) - g(x)]\}/[\Delta x g(x)g(x + \Delta x)]$

$= g(x)\lim\limits_{\Delta x \to 0}\left[\dfrac{f(x + \Delta x) - f(x)}{\Delta x}\right]\left[\dfrac{1}{g(x)g(x + \Delta x)}\right]$

$\qquad - f(x)\lim\limits_{\Delta x \to 0}\left[\dfrac{g(x + \Delta x) - g(x)}{\Delta x}\right]\left[\dfrac{1}{g(x)g(x + \Delta x)}\right]$

$= g(x)f'(x)\left[\dfrac{1}{g^2(x)}\right] - f(x)g'(x)\left[\dfrac{1}{g^2(x)}\right]$

$= \dfrac{g(x)f'(x) - f(x)g'(x)}{[g(x)]^2}$

61. $(fgh)' = (fg)'h + (fg)h'$

$\qquad = (fg' + f'g)h + (fg)h'$

$\qquad = fg'h + f'gh + fgh'$

63. $y = Ax^3 + Bx + C$

Substitute $y' = 3Ax^2 + B$, $y'' = 6Ax$, and

$y''' = 6A$ into the given equation

$$y''' + 2y'' - 3y' + y = x$$

$6A + 2(6Ax) - 3(3Ax^2 + B) + Ax^3 + Bx + C = x$

$Ax^3 - 9Ax^2 + (12A + B - 1)x + (6A - 3B + C) = 0$

The only way for a polynomial to equal zero for all x is for the coefficients to be 0. Thus, $A = 0$, $-9A = 0$, $12A + B - 1 = 0$, and $6A - 3B + C = 0$. Solving, we obtain $A = 0$, $B = 1$, $C = 3$, so the required function is $y = x + 3$.

3.3 Derivatives of Trigonometric, Exponential, and Logarithmic Functions, Page 125

SURVIVAL HINT: *If your precalculus is rusty, it will save you time in the long run to do some serious review right now. About the second or third time you must look up a graph, exact value, or trigonometric identity, add it to your list of formulas, definitions, or facts to remember.*

1. $f(x) = \sin x + \cos x$

$\qquad f'(x) = \cos x - \sin x$

3. $g(t) = t^2 + \cos t + \cos \dfrac{\pi}{4}$

$\qquad g'(t) = 2t - \sin t$

5. Write $f(t) = (\sin t)(\sin t)$

$\qquad f'(x) = (\sin t)(\cos t) + (\cos t)(\sin t)$

$\qquad\qquad = 2 \sin t \cos t$

$\qquad\qquad = \sin 2t$

7. $f(x) = \sqrt{x} \cos x + x \cot x$

$\qquad f'(x) = \sqrt{x}(-\sin x) + (\cos x)\tfrac{1}{2}x^{-1/2}$

$\qquad\qquad + x(-\csc^2 x) + (\cot x)(1)$

$$= -\sqrt{x}\sin x + \tfrac{1}{2}x^{-1/2}\cos x$$

$$- x\csc^2 x + \cot x$$

9. $p(x) = x^2\cos x$

$$p'(x) = x^2(-\sin x) + 2x\cos x$$

$$= -x^2\sin x + 2x\cos x$$

11. $q(x) = \dfrac{\sin x}{x}$

$$q'(x) = \dfrac{x\cos x - \sin x}{x^2}$$

13. $h(t) = e^t\csc t$

$$h'(t) = e^t(-\csc t\cot t) + (\csc t)(e^t)$$

$$= e^t\csc t(1 - \cot t)$$

15. $f(x) = x^2\ln x$

$$f'(x) = x^2\Big(\dfrac{1}{x}\Big) + (\ln x)(2x)$$

$$= x + 2x\ln x$$

17. $h(x) = e^{2x}(\sin x - \cos x)$

$$h'(x) = e^{2x}(\cos x + \sin x) + 2e^{2x}(\sin x - \cos x)$$

$$= e^{2x}\cos x + e^{2x}\sin x + 2e^{2x}\sin x - 2e^{2x}\cos x$$

$$= 3e^{2x}\sin x - e^{2x}\cos x$$

19. $f(x) = \dfrac{\sin x}{e^x}$

$$f'(x) = \dfrac{e^x\cos x - \sin x(e^x)}{(e^x)^2}$$

$$= \dfrac{e^x(\cos x - \sin x)}{e^{2x}}$$

$$= e^{-x}(\cos x - \sin x)$$

21. $f(x) = \dfrac{\tan x}{1 - 2x}$

$$f'(x) = \dfrac{(1 - 2x)(\sec^2 x) - \tan x(-2)}{(1 - 2x)^2}$$

$$= \dfrac{\sec^2 x - 2x\sec^2 x + 2\tan x}{(1 - 2x)^2}$$

23. $f(t) = \dfrac{2 + \sin t}{t + 2}$

$$f'(t) = \dfrac{(t + 2)(\cos t) - (2 + \sin t)(1)}{(t + 2)^2}$$

$$= \dfrac{t\cos t + 2\cos t - 2 - \sin t}{(t + 2)^2}$$

25. $f(x) = \dfrac{\sin x}{1 - \cos x}$

$$f'(x) = \dfrac{(1 - \cos x)(\cos x) - \sin x(\sin x)}{(1 - \cos x)^2}$$

$$= \dfrac{\cos x - (\cos^2 x + \sin^2 x)}{(1 - \cos x)^2}$$

$$= \dfrac{\cos x - 1}{(1 - \cos x)^2}$$

$$= \dfrac{-1}{1 - \cos x}$$

$$= \dfrac{1}{\cos x - 1}$$

27. $f(x) = \dfrac{1 + \sin x}{2 - \cos x}$

$$f'(x) = \dfrac{(2 - \cos x)(\cos x) - (1 + \sin x)(\sin x)}{(2 - \cos x)^2}$$

$$= \dfrac{2\cos x - \cos^2 x - \sin x - \sin^2 x}{(2 - \cos x)^2}$$

$$= \dfrac{2\cos x - \sin x - 1}{(2 - \cos x)^2}$$

29. $f(x) = \dfrac{\sin x + \cos x}{\sin x - \cos x}$

$f'(x) = [(\sin x - \cos x)(\cos x - \sin x)$

$\qquad - (\sin x + \cos x)(\cos x + \sin x)]/(\sin x - \cos x)^2$

$\qquad = \dfrac{-2}{(\sin x - \cos x)^2}$

SURVIVAL HINT: *Sometimes you can simplify the given function before finding the derivative, as illustrated by the Problems 31 and 32.*

31. $g(x) = \sec^2 x - \tan^2 x + \cos x$

$\qquad = 1 + \cos x$

$\qquad g'(x) = -\sin x$

33. $f(\theta) = \sin \theta$

$\qquad f'(\theta) = \cos \theta;$

$\qquad f''(\theta) = -\sin \theta$

35. $f(\theta) = \tan \theta$

$\qquad f'(\theta) = \sec^2 \theta;$

$\qquad f''(\theta) = 2 \sec \theta \sec \theta \tan \theta$

$\qquad\qquad = 2 \sec^2 \theta \tan \theta$

37. $f(\theta) = \sec \theta$

$\qquad f'(\theta) = \sec \theta \tan \theta$

$\qquad f''(\theta) = \sec \theta (\sec^2 \theta) + \tan \theta (\sec \theta \tan \theta)$

$\qquad\qquad = \sec^3 \theta + \sec \theta \tan^2 \theta$

39. $f(x) = \sin x + \cos x$

$\qquad f'(x) = \cos x - \sin x$

$\qquad f''(x) = -\sin x - \cos x$

41. $f(x) = e^x \cos x$

$\qquad f'(x) = -e^x \sin x + e^x \cos x$

$\qquad f''(x) = -e^x \cos x + (-e^x)\sin x$

$\qquad\qquad\qquad + e^x(-\sin x) + \cos x(e^x)$

$\qquad\qquad = -2e^x \sin x$

43. $h(t) = \sqrt{t} \ln t$

$\qquad h'(t) = \sqrt{t}\left(\dfrac{1}{t}\right) + \tfrac{1}{2}t^{-1/2}\ln t$

$\qquad\qquad = \tfrac{1}{2}t^{-1/2}(2 + \ln t)$

$\qquad h''(t) = \tfrac{1}{2}t^{-1/2}\left(\dfrac{1}{t}\right) - \tfrac{1}{2}\left(\tfrac{1}{2}t^{-3/2}\right)(2 + \ln t)$

$\qquad\qquad = \tfrac{1}{4}t^{-3/2}(2 - 2 - \ln t)$

$\qquad\qquad = -\tfrac{1}{4}t^{-3/2}\ln t$

45. $f(\theta) = \tan \theta;\ f(\tfrac{\pi}{4}) = \tan \tfrac{\pi}{4} = 1$

$\qquad f'(\theta) = \sec^2 \theta;\ f'(\tfrac{\pi}{4}) = \sec^2 \tfrac{\pi}{4} = 2$

The equation of the tangent line is

$$y - 1 = 2(x - \tfrac{\pi}{4})$$

$\qquad 4x - 2y - \pi + 2 = 0$

47. $f(x) = \sin x;\ f(\tfrac{\pi}{6}) = \tfrac{1}{2}$

$\qquad f'(x) = \cos x;\ f'(\tfrac{\pi}{6}) = \dfrac{\sqrt{3}}{2}$

The equation of the tangent line is

$$y - \frac{1}{2} = \frac{\sqrt{3}}{2}\left(x - \frac{\pi}{6}\right)$$

$$\sqrt{3}x - 2y + \left(1 - \frac{\sqrt{3}\pi}{6}\right) = 0$$

49. $y = x + \sin x$; if $x = 0$, then $y = 0$;

$y' = 1 + \cos x$; if $x = 0$, then $y' = 2$.

The equation of the tangent line is

$$y - 0 = 2(x - 0)$$

$$2x - y = 0$$

51. $y = e^x \cos x$

$y' = e^x(-\sin x) + e^x \cos x$

If $x = 0$, then $y = 1$ and $y' = 1$.

The equation of the tangent line is

$$y - 1 = 1(x - 0)$$

$$x - y + 1 = 0$$

53. a. $y_1 = 2 \sin x + 3 \cos x$

$y_1' = 2 \cos x - 3 \sin x$

$y_1'' = -2 \sin x - 3 \cos x$

$y_1'' + y_1 = 0$; yes

b. $y_2 = 4 \sin x - \pi \cos x$

$y_2' = 4 \cos x + \pi \sin x$

$y_2'' = -4 \sin x + \pi \cos x$

$y_2'' + y_2 = 0$; yes

c. $y_3 = x \sin x$

$y_3' = x \cos x + \sin x$

$y_3'' = 2 \cos x - x \sin x$

$y_3'' + y_3 = 2 \cos x$; no

d. $y_4 = e^x \cos x$

$y_4' = e^x(-\sin x) + e^x \cos x$

$y_4'' = -e^x \cos x - e^x \sin x - e^x \sin x + e^x \cos x$

$\quad = e^x(-\cos x - \sin x - \sin x + \cos x)$

$\quad = -2e^x \sin x$

$y_4'' + y_4 = -2e^x \sin x + e^x \cos x$; no

55. $y = Ax \cos x + Bx \sin x$

$y' = (A \cos x - Ax \sin x) + (B \sin x + Bx \cos x)$

$\quad = (Bx + A)\cos x + (B - Ax)\sin x$

$y'' = (2B - Ax)\cos x - (2A + Bx)\sin x$

Since $y'' + y = -3 \cos x$ we have

$(2B - Ax)\cos x - (2A + Bx)\sin x$

$\quad + Ax \cos x + Bx \sin x = -3 \cos x$

$2B \cos x - 2A \sin x = -3 \cos x$

Thus, $2B = -3$ and $-2A = 0$ so that

$A = 0$, $B = -\frac{3}{2}$; $y = -\frac{3}{2}x \sin x$

57. $\dfrac{d}{dx}(\cot x) = \dfrac{d}{dx}\left(\dfrac{\cos x}{\sin x}\right)$

$\qquad = \dfrac{(\sin x)(-\sin x) - (\cos x)(\cos x)}{\sin^2 x}$

$\qquad = \dfrac{-\sin^2 x - \cos^2 x}{\sin^2 x}$

$\qquad = \dfrac{-1}{\sin^2 x} = -\csc^2 x$

59. $\dfrac{d}{dx}(\csc x) = \dfrac{d}{dx}\left(\dfrac{1}{\sin x}\right)$

$\qquad = \dfrac{(\sin x)(0) - (1)(\cos x)}{\sin^2 x}$

$\qquad = -\dfrac{\cos x}{\sin x}\left(\dfrac{1}{\sin x}\right)$

$\qquad = -\cot x \csc x$

61. Answers vary.

3.4 Rates of Change: Rectilinear Motion, Pages 134-137

SURVIVAL HINT: *Mathematical modeling is one of the most important processes you will learn in calculus. It is not just some terminology that is presented in this section, but rather it is a process for solving real world problems. Be sure you read the text description of this process and understand the illustration in Figure 3.25. Attempt to answer the questions in Problems 1 and 2 even if these questions are not assigned by your instructor.*

1. A mathematical model is a mathematical framework whose results approximate the real world situation. It involves abstractions, predictions, and then interpretations and comparisons with real world events. An excellent example of real world modeling from *Scientific American* (March 1991) is mentioned on page 129.

SURVIVAL HINT: *You should distinguish between the average rate of change and instantaneous rate of change. In the directions for Problems 3-16, you are simply asked for the "rate of change." This is the instantaneous rate of change and is given by $f'(x)$.*

3. $f(x) = x^2 - 3x + 5$

$f'(x) = 2x - 3$

$f'(2) = 1$

5. $f(x) = -2x^2 + x + 4$

$f'(x) = -4x + 1;$

$f'(1) = -4 + 1$

$\qquad = -3$

7. $f(x) = \dfrac{2x - 1}{3x + 5}$

$f'(x) = \dfrac{(3x + 5)(2) - (2x - 1)(3)}{(3x + 5)^2}$

$\qquad = \dfrac{13}{(3x + 5)^2}$

$f'(-1) = \dfrac{13}{4}$

9. $f(x) = x\cos x$

$f'(x) = -x \sin x + \cos x$

$f'(\pi) = -1$

11. $f(x) = \frac{1}{2}x \ln x$

$f'(x) = \frac{1}{2}\ln x + \frac{1}{2}x\left(\frac{1}{x}\right)$

$\quad = \ln\sqrt{x} + \frac{1}{2}$

$f'(1) = \frac{1}{2}$

13. $f(x) = \sin x \cos x$

$f'(x) = -\sin^2 x + \cos^2 x$

$f'(\frac{\pi}{2}) = -1$

15. Write $f(x) = \left(x - \frac{2}{x}\right)\left(x - \frac{2}{x}\right);$ then use the

product rule:

$f'(x) = \left(x - \frac{2}{x}\right)\left(1 + \frac{2}{x^2}\right) + \left(x - \frac{2}{x}\right)\left(1 + \frac{2}{x^2}\right)$

$\quad = 2\left(\frac{x^2 - 2}{x}\right)\left(\frac{x^2 + 2}{x^2}\right)$

$\quad = \frac{2(x^4 - 4)}{x^3};$

$f'(1) = -6$

17. $s(t) = t^2 - 2t + 6$

a. $s'(t) = v(t) = 2t - 2$

b. $s''(t) = a(t) = 2$

c. The object begins at $s(0) = 6$ and ends at $s(2) = 6;$ $s'(t) = 0$ when $2t - 2 = 0$ or when $t = 1.$ On $[0, 1)$ the object retreats to $s(1) = 5;$ on $(1, 2]$ the object advances. Distance covered:

$\left|s(2) - s(1)\right| + \left|s(1) - s(0)\right| = 2$

d. Because $a(t) > 0$, the object is always accelerating.

19. a. $s'(t) = v(t) = 3t^2 - 18t + 15$

b. $s''(t) = a(t) = 6t - 18$

c. The object begins at $s(0) = 25$ and ends at $s(6) = 7;$ $s'(t) = 0$ when

$$3t^2 - 18t + 15 = 0$$
$$3(t - 5)(t - 1) = 0$$
$$t = 1, 5$$

On $[0, 1)$ the object advances to $s(1) = 32;$ on $(1, 5)$ it retreats to $s(5) = 0;$ and on $(5, 6]$ it advances. Distance covered:

$\left|s(1) - s(0)\right| + \left|s(5) - s(1)\right| + \left|s(6) - s(5)\right|$
$\quad = |32 - 25| + |0 - 32| + |7 - 0|$
$\quad = 7 + 32 + 7$
$\quad = 46$

d. $s''(t) = 0$ when $6t - 18 = 0$ or $t = 3.$ On $[0, 3)$ the object is decelerating, and on $(3, 6]$ it is accelerating.

21. Write $s(t) = 2t^{-1} + t^{-2}$

a. $s'(t) = v(t) = -2t^{-2} - 2t^{-3}$

b. $s''(t) = a(t) = 4t^{-3} + 6t^{-4}$

c. The object begins at $s(1) = 3$ and ends at $s(3) = \frac{7}{9};$ $s'(t) = 0$ when

$$-2t^{-2} - 2t^{-3} = 0$$
$$-2t - 2 = 0$$

$$t = -1$$

-1 is not on the interval $[1, 3]$. On $[1, 3]$ the object retreats.

Distance covered:

$$\left| s(3) - s(1) \right| = \left| \frac{7}{9} - 3 \right|$$

$$= \frac{20}{9}$$

d. $s''(t) = 0$ when

$$4t^{-3} + 6t^{-4} = 0$$

$$4t + 6 = 0$$

$$t = -\frac{3}{2}$$

On $[1, 3]$, $s''(t) > 0$, so the object is accelerating.

23. $s(t) = 3\cos t$

a. $s'(t) = v(t) = -3\sin t$

b. $s''(t) = a(t) = -3\cos t$

c. Object begins at $s(0) = 3$ and ends at $s(2\pi) = 3$; $s'(t) = 0$ when

$$-3\sin t = 0$$

On $[0, 2\pi]$: $\qquad t = 0, \pi, 2\pi$

On $[0, \pi)$ the object retreats to $s(\pi) = -3$; on $(\pi, 2\pi]$ the object advances to $s(2\pi) = 3$.

Distance covered:

$$\left| s(\pi) - s(0) \right| + \left| s(2\pi) - s(\pi) \right|$$

$$= \left| -3 - 3 \right| + \left| 3 - (-3) \right|$$

$$= 6 + 6$$

$$= 12$$

d. $s''(t) = 0$ when $\qquad -3\cos t = 0$

On $[0, 2\pi]$: $\qquad t = \frac{\pi}{2}, \frac{3\pi}{2}$

On $[0, \frac{\pi}{2})$ the object is decelerating, on $(\frac{\pi}{2}, \frac{3\pi}{2})$ the object is accelerating, and on $(\frac{3\pi}{2}, 2\pi]$ it is decelerating again.

25. quadratic model

27. exponential model

29. quadratic model

31. cubic model

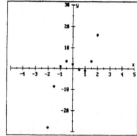

33. logarithmic model

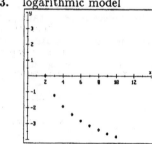

35. $f(x) = -6x + 1,082$

   a. $f'(x) = -6$

   b. The rate of change is negative, scores are declining. The rate of change is constant, the decline will be the same each year.

37. $x(t) = t^3 - 9t^2 + 24t + 20$

$$v(t) = x'(t)$$
$$= 3t^2 - 18t + 24$$

We need to find when $v(t) = 0$:

$$3t^2 - 18t + 24 = 0$$
$$t^2 - 6t + 8 = 0$$
$$(t - 2)(t - 4) = 0$$
$$t = 2, 4$$

That is, when $t = 2$ or 4, the velocity is 0; when v is positive it is advancing and when v is negative it is retreating, so we see that $x(t)$ is advancing on [0, 2) and (4, 8] and retreating on (2, 4). Thus, the total distance traveled is

$$\left| x(2) - x(0) \right| + \left| x(4) - x(2) \right| + \left| x(8) - x(4) \right|$$
$$= |40 - 20| + |36 - 40| + |148 - 36|$$
$$= 20 + 4 + 112 = 136 \text{ units}$$

39. $s(t) = 10t + te^{-t}$

   a.
$$v(t) = s'(t)$$
$$= 10 + e^{-t} - te^{-t}$$
$$v(4) = 10 - e^{-4} - 4e^{-4}$$
$$\approx 9.91 \text{ m/min}$$

   b. $v(t) > 0$ so the distance traveled during the 5th minute is

$$s(5) - s(4) \approx 50.034 - 40.073$$
$$\approx 9.961 \text{ m}$$

41. $s(t) = -16t^2 + v_0 t + s_0$

Consider earth's gravity and disregard air resistance and friction.

   a. The maximum height when

$$v(t) = -32t + v_0 = 0 \text{ or when } v_0 = 64$$

The initial velocity is 64 ft/sec.

b. The rock hits the ground when $s(t) = 0$.

The occurs when $t = 7$; so

$-16(7)^2 + 64(7) + s_0 = 0$ when $s_0 = 336$

The cliff is 336 ft high.

c. $s'(t) = -32t + 64$ ft/sec

d. $s'(7) = -32(7) + 64 = -160$ ft/sec

The negative sign indicates downward

motion, since upward is positive.

43. $s(t) = -16t^2 + v_0t + s_0$

Consider earth's gravity and disregard air

resistance and friction. The equation guiding

the path of the first rock is

$$s_1(t) = -16t^2 + s_1'(0)t + 90$$

and the second rock is

$$s_2(t) = -16(t - 1)^2 + s_2'(1)(t-1) + H$$

Time t is 0 when the first rock starts its

motion; $s'(0) = 0$ since the rocks are dropped.

The first rock hits the ground at time

$$-16t^2 + 90 = 0 \text{ or } t = \pm\frac{3\sqrt{10}}{4}$$

Reject the negative value, so $t \approx 2.3717$ sec.

The two rocks hit the ground simultaneously,

so $-16(\frac{3\sqrt{10}}{4} - 1)^2 + H = 0$ or $H \approx 30$ ft

45. $s(t) = -16t^2 + v_0t + s_0$

Assume earth's gravity and disregard air

resistance and friction.

$s(t) = -16t^2 + s_0$

The pavement, that is ground level, is reached

in 3 seconds, so $0 = -16(3^2) + s_0$ or

$s_0 = 144$. The height of the building is

144 ft.

47. On Mars, $g = 12$ ft/s^2; disregard air

resistance and friction. So

$$s(t) = -6t^2 + v_0t + s_0,$$

$$v(t) = -12t + v_0.$$

Since the rock goes up and back to cliff level

in 4 sec, it reaches its maximum height in 2

sec. At that time $v = 0$. So

$$0 = -12(2) + v_0 \text{ and } v_0 = 24 \text{ ft/s}$$

Therefore, the rock passes her on the way

down with $v = -24$ ft/s. The equation for

the rest of the rock's trip will be:

$$s(t) = -6t^2 + (-24)t + 0$$
$$s(3) = -6(3)^2 - 24(3)$$
$$= -126 \text{ ft}$$

The initial velocity is 24 ft/s, and the cliff is 126 ft. high. (The negative means the rock has traveled down 126 ft).

49. a. Because $C(t) = 100t^2 + 400t + 50t\ln t$ is the circulation t years from now, the rate of change of circulation t years from now is $C'(t) = 200t + 50\ln t + 450$ newspapers per year.

b. The rate of change of circulation 5 years from now is $C'(5) \approx 1,530$ newspapers per year.

c. The actual change in the circulation during the sixth year is

$$C(6) - C(5) = [100(6^2) + 400(6) + 50 \cdot 6 \cdot \ln 6]$$
$$- [100(5^2) + 400(5) + 50 \cdot 5 \cdot \ln 5]$$
$$\approx 1,635 \text{ newspapers}$$

51. $q(t) = 0.05t^2 + 0.1t + 3.4$

a. $q'(t) = 0.1t + 0.1$; $q'(1) = 0.2$ ppm/yr

b. Change in first year:

$$q(1) - q(0) = (0.05 + 0.1 + 3.4) - (3.4)$$
$$= 0.15 \text{ ppm}$$

c. Change in second year:

$$q(2) - q(1) = (0.2 + 0.2 + 3.4) - (3.55)$$
$$= 0.25 \text{ ppm}$$

53. $P(t) = P_0 + 61t + 3t^2$

$P'(t) = 61 + 6t$;

$P'(5) = 61 + 30$

$\quad = 91$ thousand/hr

55. $P(x) = 2x + 4x^{3/2} + 5,000$

a. $P'(x) = 2 + 6x^{1/2}$

$P'(9) = 2 + 18$

$\quad = 20$ persons per mo

b. $\dfrac{P'(9)}{P(9)}(100) = \dfrac{20}{5126}(100)$

$\quad \approx 0.39\%$ per mo

57. Let $t = 0$ for 1999 and $G(t)$ the GDP in billions of dollars. $G(0) = 125$, $G(2) = 155$, so the slope of the line is

$$m = \frac{155 - 125}{2} = 15 = G'(t)$$

$G(t) = 125 + 15t$; $G(5) = 125 + 75 = 200$.

The percentage rate of change is

$$\frac{100(15)}{200} = 7.5\% \text{ per year in 2000}$$

59. $P = \frac{4}{3}\pi N\left(\dfrac{\mu^2}{3kT}\right)$

$\quad = \dfrac{4\pi\mu^2 N}{9k} T^{-1}$

$\dfrac{dP}{dT} = -\dfrac{4\pi\mu^2 N}{9k} T^{-2}$

61. a. $s(t) = 7 \cos t$;

$$v(t) = s'(t) = -7 \sin t$$

$$a(t) = s''(t) = -7 \cos t$$

b. $s(0) = 7$, $s(2\pi) = 7$;

the period (one revolution) is 2π.

c. The highest point is reached at $t = \pi$ (downward is positive). $s(\pi) = -7$. The amplitude is 7.

63. This is our third encounter with the Spy.

$$s(t) = -\frac{g}{2}t^2 + v_0 t$$

It hits when $t = 5$, so

$$-\frac{g}{2}(5)^2 + v_0(5) = 0$$

$$v_0 = \frac{5}{2}g$$

The maximum height is reached when $v(t) = 0$:

$$v(t) = -gt + \frac{5}{2}g = 0$$

$$t = \frac{5}{2}$$

and

$$-\frac{1}{2}\left(\frac{5}{2}\right)^2 g + \frac{5}{2}\left(\frac{5}{2}\right)g = 37.5$$

$$g = \frac{(37.5)(8)}{25}$$

$$= 12$$

Our friendly Spy finds himself on Mars.

65. Let r be the radius of the sphere. Then

$$V(r) = \frac{4}{3}\pi r^3$$

$$V'(r) = 4\pi r^2 = S(r)$$

They are equal.

3.5 The Chain Rule, Pages 143-145

SURVIVAL HINT: *The chain rule is probably the most used of all the differentiation rules. Most interesting and/or useful functions are compositions, and their derivatives require the chain rule. When you are finding $f'[u(x)]$, identify the u function, and remember to include the du/dx. Write out intermediate steps; many mistakes are made by trying to do too much "in your head."*

1. The chain rule is the differentiation of a function of a function. If $y = f(u)$ and $u = g(x)$, then

$$\frac{dy}{dx} = \frac{dy}{du} \cdot \frac{du}{dx}$$

3. $\dfrac{dy}{dx} = \dfrac{dy}{du}\dfrac{du}{dx}$

$$= \frac{d}{du}(u^2 + 1)\frac{d}{dx}(3x - 2)$$

$$= 2u(3)$$

$$= 6(3x - 2)$$

5. $\dfrac{dy}{dx} = \dfrac{dy}{du}\dfrac{du}{dx}$

$$= \left(\frac{-4}{u^3}\right)(2x)$$

$$= \frac{-8x}{(x^2 - 9)^3}$$

7. $\dfrac{dy}{dx} = \dfrac{dy}{du} \dfrac{du}{dx}$

$= \left[\tan u + u \sec^2 u\right]\left[3 - \dfrac{6}{x^2}\right]$

where $u = 3x + \dfrac{6}{x}$.

9. **a.** $g'(u) = 3u^2$

b. $u'(x) = 2x$

c. $f'(x) = 3(x^2 + 1)^2(2x)$

$= 6x(x^2 + 1)^2$

11. **a.** $g'(u) = 7u^6$

b. $u'(x) = -8 - 24x$

c. $f'(x) = 7(5 - 8x - 12x^2)^6(-8 - 24x)$

$= -7(24x + 8)(12x^2 + 8x - 5)^6$

$= -56(3x + 1)(12x^2 + 8x - 5)^6$

13. $f(x) = (5x - 2)^5$

$f'(x) = 5(5x - 2)^4(5)$

$= 25(5x - 2)^4$

15. $f(x) = (3x^2 - 2x + 1)^4$

$f'(x) = 4(3x^2 - 2x + 1)^3(6x - 2)$

$= 8(3x - 1)(3x^2 - 2x + 1)^3$

17. $s(\theta) = \sin(4\theta + 2)$

$s'(\theta) = [\cos(4\theta + 2)](4)$

$= 4\cos(4\theta + 2)$

19. $f(x) = e^{-x^2 + 3z}$

$f'(x) = e^{-x^2 + 3x}(-2x + 3)$

21. $y = e^{\sec x}$

$y' = e^{\sec x}\sec x \tan x$

23. $f(t) = \exp(t^2 + t + 5)$

$f'(t) = (2t + 1)\exp(t^2 + t + 5)$

25. $g(x) = x\sin 5x$

$g'(x) = x(\cos 5x)(5) + \sin 5x$

$= 5x\cos 5x + \sin 5x$

27. $f(x) = (1 - 2x)^{-3}$

$f'(x) = -3(1 - 2x)^{-4}(-2)$

$= 6(1 - 2x)^{-4}$

29. $f(x) = xe^{1 - 2x}$

$f'(x) = e^{1 - 2x} + xe^{1 - 2x}(-2)$

$= (1 - 2x)e^{1 - 2x}$

31. $p(x) = \sin x^2 \cos x^2$

$p'(x) = (\sin x^2)\dfrac{d}{dx}(\cos x^2) + (\cos x^2)\dfrac{d}{dx}(\sin x^2)$

$= (\sin x^2)(-\sin x^2)(2x) + (\cos x^2)(\cos x^2)(2x)$

$= 2x(\cos^2 x^2 - \sin^2 x^2)$ or $2x\cos 2x^2$

33. $f(x) = x^3(2 - 3x)^2$

$f'(x) = x^3(2)(2 - 3x)(-3) + 3x^2(2 - 3x)^2$

$= 3x^2(2 - 3x)[-2x + (2 - 3x)]$

$= 3x^2(2 - 3x)(2 - 5x)$

35. $f(x) = \sqrt{\dfrac{x^2 + 3}{x^2 - 5}} = \left(\dfrac{x^2 + 3}{x^2 - 5}\right)^{1/2}$

$f'(x) = \dfrac{1}{2}\left(\dfrac{x^2 + 3}{x^2 - 5}\right)^{-1/2}\left[\dfrac{(x^2 - 5)(2x) - (x^2 + 3)(2x)}{(x^2 - 5)^2}\right]$

$\quad = \dfrac{1}{2}\left(\dfrac{x^2 + 3}{x^2 - 5}\right)^{-1/2}\left[\dfrac{-16x}{(x^2 - 5)^2}\right]$

37. $f(x) = \sqrt[3]{x + \sqrt{2x}} = (x + \sqrt{2x})^{1/3}$

$\quad f'(x) = \dfrac{1}{3}(x + \sqrt{2x})^{-2/3}\left(1 + \dfrac{1}{\sqrt{2x}}\right)$

39. $f(x) = \ln(\sin x + \cos x)$

$\quad f'(x) = \dfrac{\cos x - \sin x}{\sin x + \cos x}$

41. $f(x) = x\sqrt{1 - 3x} = x(1 - 3x)^{1/2}$

$\quad f'(x) = x(\tfrac{1}{2})(1 - 3x)^{-1/2}(-3) + \sqrt{1 - 3x}$

$\quad\quad = \dfrac{-3x + 2(1 - 3x)}{2(1 - 3x)^{1/2}}$

$\quad\quad = \dfrac{-9x + 2}{2(1 - 3x)^{1/2}}$

The tangent line is horizontal when
$f'(x) = 0$:

$\quad\quad \dfrac{-9x + 2}{2(1 - 3x)^{1/2}} = 0$

$\quad\quad\quad -9x + 2 = 0$

$\quad\quad\quad\quad x = \dfrac{2}{9}$

43. $q(x) = \dfrac{(x - 1)^2}{(x + 2)^3}$

$\quad q'(x) = \dfrac{(x + 2)^3 2(x - 1) - (x - 1)^2 3(x + 2)^2(1)}{(x + 2)^6}$

$\quad\quad = \dfrac{(x - 1)(-x + 7)}{(x + 2)^4}$

The tangent line is horizontal when
$q'(x) = 0$.

$\quad \dfrac{(x - 1)(-x + 7)}{(x + 2)^4} = 0$

$\quad (x - 1)(-x + 7) = 0$

$\quad\quad x = 1, 7$

45. $T(x) = x^2 e^{1 - 3x}$

$\quad T'(x) = x^2 e^{1 - 3x}(-3) + 2x e^{1 - 3x}$

$\quad\quad = -e^{1 - 3x}(3x^2 - 2x)$

The tangent line is horizontal when
$T'(x) = 0$.

$-e^{1 - 3x}(3x^2 - 2x) = 0$ *Note: $-e^{1 - 3x} \neq 0$, so divide both sides by this number.*

$\quad\quad x(3x - 2) = 0$

$\quad\quad\quad x = 0, \dfrac{2}{3}$

47. a. The graph indicates that $u \approx 5$ when $x = 2$. The slope of the tangent line to the curve $u = g(x)$ at $x = 2$ is about 1.

b. The graph indicates that $y \approx 3$ when $u = 5$. The slope of the tangent line is about $\dfrac{3}{2}$.

c. The slope of $y = f[g(x)]$ is about $(1)(\tfrac{3}{2}) = 1.5$

49. $h(x) = g[f(x)]$, so

$\quad\quad h'(x) = g'[f(x)] \cdot f'(x)$

a. From the graphs, $f(-1) \approx \frac{3}{4}$;

the slope of g at $\frac{3}{4}$ is $g'(\frac{3}{4}) \approx -\frac{3}{2}$

the slope of f at -1 is found as follows:

estimate the point to be $(-1, \frac{3}{4})$ and

draw a tangent line at this point. We

estimate the point of intersection of the

tangent line and the x-axis to be $(4, 0)$, so

that from $(-1, \frac{3}{4})$ we have a rise of $-\frac{3}{4}$

and a run of 5 for a slope of

$(-3/4)/5 = -\frac{3}{20}$; so $f'(-1) \approx -\frac{3}{20}$.

Thus,

$$h'(-1) = g'(\tfrac{3}{4}) \cdot f'(-1)$$

$$\approx -\tfrac{3}{2}(-\tfrac{3}{20})$$

$$= \tfrac{9}{40}$$

b. From the graphs, $f(1) \approx -\frac{1}{2}$;

the slope of g at $-\frac{1}{2}$ is $g'(-\frac{1}{2}) \approx \frac{3}{8}$

the slope of f at 1 is $f(1) \approx -1$.

Thus,

$$h'(1) = g'(-\tfrac{1}{2}) \cdot f'(1)$$

$$\approx \tfrac{3}{8} \cdot (-1)$$

$$= -\tfrac{3}{8}$$

c. From the graphs, $f(3) \approx -\frac{1}{2}$;

the slope of g at $-\frac{1}{2}$ is $g'(-\frac{1}{2}) \approx \frac{3}{8}$

the slope of f at 3 is $f'(3) \approx 1$.

Thus,

$$h'(3) = g'(-\tfrac{1}{2}) \cdot f'(3)$$

$$\approx \tfrac{3}{8}(1)$$

$$= \tfrac{3}{8}$$

51. From the chain rule,

$$h'(1) = f'[g(1)] \cdot g'(1)$$

$$= f'(2) \cdot g'(1)$$

We see that we need to find the values of $f'(2)$ and $g'(1)$. Use the given table of values to draw possible graphs for both f and g, along with the appropriate tangent lines:

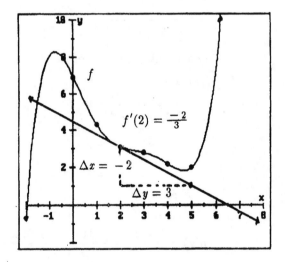

$$f'(2) \approx -\tfrac{2}{3}$$

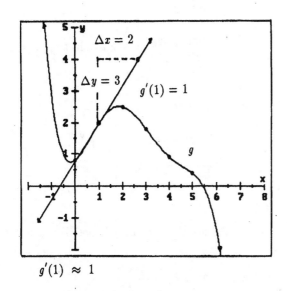

$g'(1) \approx 1$

Thus, $h'(1) = f'[g(1)] \cdot g'(1)$

$$\approx f'(2) \cdot g'(1)$$

$$\approx -\tfrac{2}{3} \cdot 1$$

$$= -\tfrac{2}{3}$$

53. Given

$$p(t) = 20 - \frac{6}{t+1}$$

and

$$L(p) = 0.5\sqrt{p^2 + p + 58}$$

We note

$$\frac{dp}{dt} = \frac{6}{(t+1)^2}$$

and

$$\frac{dL}{dp} = (0.5)(\tfrac{1}{2})(p^2 + p + 58)^{-1/2}(2p + 1)$$

Here we have a function of a function, so the chain rule is required.

$$\frac{dL}{dt} = \frac{dL}{dp} \cdot \frac{dp}{dt}$$

$$= (0.5)(\tfrac{1}{2})(p^2 + p + 58)^{-1/2}(2p + 1)\left[\frac{6}{(t+1)^2}\right]$$

$$= \frac{6(2p + 1)}{4(p^2 + p + 58)^{1/2}(t + 1)^2} \text{ ppm/year}$$

Two years from now, $t = 2$, $p(2) = 18$, and the rate of change is

$$\left.\frac{dL}{dt}\right|_{t=2,\ p=18} = \frac{6[2(18) + 1]}{4(18^2 + 18 + 58)^{1/2}(2 + 1)^2}$$

$$= \frac{37}{120}$$

$$\approx 0.31 \text{ ppm/year}$$

55. Given

$$D(p) = \frac{4{,}374}{p^2}$$

and

$$p(t) = 0.02t^2 + 0.1t + 6$$

We note

$$\frac{dD}{dp} = 4{,}374(-2p^{-3})$$

and

$$\frac{dp}{dt} = 0.04t + 0.1$$

Here we have a function of a function, so the chain rule is required.

$$\frac{dD}{dt} = \frac{dD}{dp} \cdot \frac{dp}{dt}$$

$$= 4{,}374(-2p^{-3})(0.04t + 0.1)$$

$$= -8{,}748p^{-3}(0.04t + 0.1)$$

When $t = 10$, $p(10) = 9$ and

$\left.\dfrac{dD}{dt}\right|_{t=10,\ p=9}$

$= -8{,}748(9)^{-3}[0.04(10) + 0.1]$

≈ -6 lb/wk

demand is decreasing.

57. $I = 20s^{-2}$ and $s(t) = 28 - t^2$

   a. $\dfrac{dI}{dt} = (20)(-2s^{-3})\dfrac{ds}{dt}$;

     $\dfrac{ds}{dt} = -2t$

     When $s(t) = 28 - t^2 = 19$, $t = 3$

       $\dfrac{dI}{dt} = -2(20)(19^{-3})(-2)(3)$

          ≈ 0.035

     Illuminance is increasing by 0.035 lux/s.

   b. Since $s(t) = 28 - t^2$, we have

     $t = \sqrt{28 - s}$ and the rate of change is

      $\dfrac{dI}{dt} = \dfrac{80t}{s^3}$

        $= \dfrac{80\sqrt{28 - s}}{s^3}$

     To solve $\dfrac{80\sqrt{28 - s}}{s^3} = 1$, we use a

calculator to obtain an approximate solution;

$s \approx 7.15$ m.

59. Given $\theta = \theta_M \sin kt$;

     $\dfrac{d\theta}{dt} = k\theta_M \cos kt$

     $\dfrac{d^2\theta}{dt^2} = -k^2\theta_M \sin kt$

Thus,

     $\dfrac{d^2\theta}{dt^2} + k^2\theta = -k^2\theta_M \sin kt + k^2(\theta_M \sin kt)$

          $= 0$

61. $T(t) = 58 + 17\sin\left(\dfrac{\pi t}{10} - \dfrac{5}{6}\right)$

    $T'(t) = 17\cos\left(\dfrac{\pi t}{10} - \dfrac{5}{6}\right)\left(\dfrac{\pi}{10}\right)$

    $T'(2) = 1.7\pi\cos\left(\dfrac{\pi}{5} - \dfrac{5}{6}\right)$

       ≈ 5.23

Since the derivative is positive, it is getting

hotter.

63. $I(\theta) = I_0\left(\dfrac{\sin \beta}{\beta}\right)^2$

$I'(\theta) = I_0(2)\left(\dfrac{\sin \beta}{\beta}\right)\left(\dfrac{\beta\cos \beta - \sin \beta}{\beta^2}\right)\left(\pi a \cos \dfrac{\theta}{\lambda}\right)\left(\dfrac{1}{\lambda}\right)$

    $= \dfrac{2a\pi I_0}{\lambda\beta^3}\sin \beta(\beta\cos \beta - \sin \beta)\cos \dfrac{\theta}{\lambda}$

65. $\dfrac{d}{dx}\cos u = \dfrac{d}{dx}[\sin(\tfrac{\pi}{2} - u)]$

        $= \cos(\tfrac{\pi}{2} - u)(-\dfrac{du}{dx})$ *Chain rule*

        $= -\sin u \dfrac{du}{dx}$

67. **a.** $f'(u) = \dfrac{1}{u^2 + 1}$

    Let $g(x) = f[u(x)]$ so

        $g'(x) = \dfrac{d}{du}f(u)\dfrac{du}{dx}$

    $u(x) = 3x - 1$;

        $g'(x) = \dfrac{3}{u^2 + 1}$

          $= \dfrac{3}{(3x - 1)^2 + 1}$

b. Let $h(x) = f[u(x)]$;

$$u(x) = x^{-1}$$

$$h'(x) = \frac{1}{x^{-2} + 1}(-x^{-2})$$

$$= \frac{-1}{x^2 + 1}$$

69. a. If $-\frac{\pi}{2} < x < \frac{\pi}{2}$, $\cos x > 0$, so

$$F'(x) = (\cos x)^{-1}(-\sin x)$$

$$= -\tan x$$

If $\frac{\pi}{2} < x < \frac{3\pi}{2}$, $\cos x < 0$, so

$$F'(x) = (-\cos x)^{-1}(\sin x)$$

$$= -\tan x$$

b. $$F'(x) = \frac{\sec x \tan x + \sec^2 x}{\sec x + \tan x}$$

$$= \frac{\sec x(\tan x + \sec x)}{\sec x + \tan x}$$

$$= \sec x$$

71. $$\frac{d}{dx}f'[f(x)] = f''[f(x)]f'(x)$$

and

$$\frac{d}{dx}f[f'(x)] = f'[f'(x)]f''(x)$$

3.6 Implicit Differentiation, Pages 155-157

SURVIVAL HINT: *Since y is assumed to be a function of x, implicit differentiation can be thought of as applying the chain rule with* $y = u(x)$.

1. $$x^2 + y^2 = 25$$

$$2x + 2y\frac{dy}{dx} = 0$$

$$\frac{dy}{dx} = -\frac{x}{y}$$

3. $$xy = 25$$

$$xy' + y = 0$$

$$y' = -\frac{y}{x}$$

5. $$x^2 + 3xy + y^2 = 15$$

$$2x + 3(x\frac{dy}{dx} + y) + 2y\frac{dy}{dx} = 0$$

$$(3x + 2y)\frac{dy}{dx} = -2x - 3y$$

$$\frac{dy}{dx} = \frac{-(2x + 3y)}{3x + 2y}$$

7. $$\frac{1}{y} + \frac{1}{x} = 1$$

$$-\frac{1}{y^2}\frac{dy}{dx} - \frac{1}{x^2} = 0$$

$$\frac{dy}{dx} = -\frac{y^2}{x^2}$$

9. $$\sin(x + y) = x - y$$

$$\cos(x + y)(1 + y') = 1 - y'$$

$$[\cos(x + y) + 1]y' = 1 - \cos(x + y)$$

$$y' = \frac{1 - \cos(x + y)}{\cos(x + y) + 1}$$

11. $$\cos xy = 1 - x^2$$

$$(-\sin xy)\frac{d}{dx}(xy) = -2x$$

$$(-\sin xy)(x\frac{dy}{dx} + y) = -2x$$

$$(-x \sin xy)\frac{dy}{dx} = y \sin xy - 2x$$

$$\frac{dy}{dx} = \frac{2x - y \sin xy}{x \sin xy}$$

13. Note that $xy \neq 0$, so we have

$$\ln(xy) = e^{2x}$$

$$\frac{1}{xy}(y + xy') = 2e^{2x}$$

$$\frac{1}{x} + \frac{y'}{y} = 2e^{2x}$$

$$y' = (2e^{2x} - x^{-1})y$$

15. a. $\qquad x^2 + y^3 = 12$

$$2x + 3y^2 \frac{dy}{dx} = 0$$

$$\frac{dy}{dx} = -\frac{2x}{3y^2}$$

b. $\quad x^2 + y^3 = 12$

$$y = (12 - x^2)^{1/3}$$

$$\frac{dy}{dx} = \frac{1}{3}(12 - x^2)^{-2/3}(-2x)$$

$$= \frac{-2x}{3(12 - x^2)^{2/3}}$$

$$= \frac{-2x}{3y^2}$$

17. a. $\qquad x + \frac{1}{y} = 5$

$$1 - \frac{1}{y^2}\frac{dy}{dx} = 0$$

$$\frac{dy}{dx} = y^2$$

b. $\quad xy + 1 = 5y$

$$(x - 5)y = -1$$

$$y = \frac{-1}{x - 5}$$

$$\frac{dy}{dx} = -\frac{-1}{(x - 5)^2}$$

$$= \frac{1}{(x - 5)^2} = y^2$$

SURVIVAL HINT: *The functions in Problems 19-32 are inverse functions, not reciprocal functions. Remember, $\sin^{-1}x$ is an inverse sine, whereas $(\sin x)^{-1}$ is the reciprocal of a sine function.*

19. $\qquad y = \sin^{-1}(2x + 1)$

$$\frac{dy}{dx} = \frac{1}{\sqrt{1 - (2x + 1)^2}}(2)$$

$$= \frac{2}{\sqrt{-4x^2 - 4x}}$$

$$= \frac{1}{\sqrt{-x^2 - x}}$$

21. $\qquad y = \tan^{-1}\sqrt{x^2 + 1}$

$$\frac{dy}{dx} = \frac{1}{1 + (x^2 + 1)}\frac{x}{\sqrt{x^2 + 1}}$$

$$= \frac{x}{(x^2 + 2)\sqrt{x^2 + 1}}$$

23. $\qquad y = (\sin^{-1}2x)^3$

$$\frac{dy}{dx} = 3(\sin^{-1}2x)^2\frac{1}{\sqrt{1 - (2x)^2}}(2)$$

$$= \frac{6(\sin^{-1}2x)^2}{\sqrt{1 - 4x^2}}$$

25. $\qquad y = \sec^{-1}(e^{-x})$

$$\frac{dy}{dx} = \frac{1}{|e^{-x}|\sqrt{e^{-2x} - 1}}(-e^{-x})$$

$$= \frac{-1}{\sqrt{e^{-2x} - 1}}$$

27. $\qquad y = \tan^{-1}\left(\frac{1}{x}\right)$

$$\frac{dy}{dx} = \frac{1}{1 + x^{-2}}(-x^{-2})$$

$$= \frac{-1}{x^2 + 1}$$

29. $\quad y = \sin^{-1}(\cos x)$

$$\frac{dy}{dx} = \frac{1}{\sqrt{1 - \cos^2 x}}(-\sin x)$$

$$= -\frac{\sin x}{|\sin x|}$$

$$= \pm 1$$

31. $\quad x\sin^{-1}y + y\tan^{-1}x = x$

$$x\frac{1}{\sqrt{1 - y^2}}y' + \sin^{-1}y + y\frac{1}{1 + x^2}$$
$$+ y'(\tan^{-1}x) = 1$$

$$y'\left(\frac{x}{\sqrt{1 - y^2}} + \tan^{-1}x\right)$$
$$= 1 - \sin^{-1}y - \frac{y}{1 + x^2}$$

$$y' = \frac{1 - \sin^{-1}y - \dfrac{y}{1 + x^2}}{\dfrac{x}{\sqrt{1 - y^2}} + \tan^{-1}x}$$

33. $\quad x^2 + y^2 = 13$

$$2x + 2yy' = 0$$

$$y' = -\frac{x}{y}$$

$$y'\Big|_{(-2,\,3)} = \frac{2}{3}$$

The equation of the tangent line is

$$y - 3 = \tfrac{2}{3}(x + 2)$$

$$2x - 3y + 13 = 0$$

35. $\quad \sin(x - y) = xy$

$$\cos(x - y)[1 - y'] = xy' + y$$

$$y' = \frac{\cos(x - y) - y}{\cos(x - y) + x}$$

$$y'\Big|_{(0,\,\pi)} = \frac{\cos(-\pi) - \pi}{\cos(-\pi)} = \pi + 1$$

The equation of the tangent line is

$$y - \pi = (\pi + 1)(x - 0)$$

$$(\pi + 1)x - y + \pi = 0$$

37. $\quad x\tan^{-1}y = x^2 + y$

$$\tan^{-1}y + x\left(\frac{1}{1 + y^2}\right)y' = 2x + y'$$

$$\left(\frac{x}{1 + y^2} - 1\right)y' = 2x - \tan^{-1}y$$

$$\left(\frac{x - 1 - y^2}{1 + y^2}\right)y' = 2x - \tan^{-1}y$$

$$y' = \frac{(2x - \tan^{-1}y)(1 + y^2)}{x - 1 - y^2}$$

At $(0, 0)$, $y' = 0$, so the tangent line is horizontal with equation $y = 0$.

39. $\quad (x^2 + y^2)^2 = 4x^2y$

$$2(x^2 + y^2)(2x + 2yy') = 4(2xy + x^2y')$$

At $(1, 1)$: $\qquad 2 + 2y' = 2 + y'$

$$y' = 0$$

41. $\quad x^3 + y^3 - \tfrac{9}{2}xy = 0$

$$3x^2 + 3y^2y' - \tfrac{9}{2}[xy' + y] = 0$$

$$y' = \frac{3y - 2x^2}{2y^2 - 3x}$$

At (2, 1): $y' = \frac{3-8}{2-6} = \frac{5}{4}$

43. $x^2 + 2xy = y^3$

$2x + 2xy' + 2y = 3y^2 y'$

$2x + 2xy' + 2y - 3y^2 y' = 0$

$$y' = \frac{2x + 2y}{3y^2 - 2x}$$

At (1, −1) $y' = 0$

The tangent line is horizontal so the normal line is a vertical line with equation $x - 1 = 0$.

45. $7x + 5y^2 = 1$

$7 + 10yy' = 0$

$$y' = -\frac{7}{10y}.$$

For y'' we again differentiate:

$$\frac{d}{dx}(y') = \frac{d}{dx}\left(-\frac{7}{10y}\right)$$

$$y'' = \frac{7}{10y^2}\, y'$$

$$= \left(\frac{7}{10y^2}\right)\left(-\frac{7}{10y}\right)$$

$$= -\frac{49}{100y^3}$$

47. a. Differentiate a variable base, constant exponent; $y' = 2x$

b. Differentiate a constant base ($\neq e$), variable exponent; $y' = 2^x \ln 2$

c. Differentiate a constant base e, variable exponent; $y' = e^x$

d. Differentiate a variable base, constant exponent of e; $y' = ex^{e-1}$

SURVIVAL HINT: *Logarithmic differentiation is advantageous when differentiating a complicated product or quotient. It is a technique for changing a product or a quotient to a sum or a difference, respectively. It is also used to change a power into a product.*

51. $y = \dfrac{(2x - 1)^5}{(x - 9)^{1/2}(x + 3)^2}$

$\ln y = 5 \ln(2x - 1) - \frac{1}{2}\ln(x - 9) - 2\ln(x + 3)$

$\frac{1}{y}\frac{dy}{dx} = 10(2x - 1)^{-1} - \frac{1}{2}(x - 9)^{-1} - 2(x + 3)^{-1}$

$\frac{dy}{dx} = y\left[\frac{10}{2x - 1} - \frac{1}{2(x - 9)} - \frac{2}{x + 3}\right]$

53. $y = \dfrac{e^{3x^2}}{(x^3 + 1)^2(4x - 7)^{-2}}$

$\ln y = 3x^2 - 2\ln(x^3 + 1) + 2\ln(4x - 7)$

$\frac{1}{y}\frac{dy}{dx} = 6x - 2(3x^2)(x^3 + 1)^{-1} + 8(4x - 7)^{-1}$

$\frac{dy}{dx} = y\left[6x - \frac{6x^2}{x^3 + 1} + \frac{8}{4x - 7}\right]$

55. $y = x^{\ln\sqrt{x}}$

$\ln y = \ln\sqrt{x}\,\ln x$

$\ln y = \frac{1}{2}\ln x \ln x$

$\ln y = \frac{1}{2}(\ln x)^2$

$\frac{1}{y}\frac{dy}{dx} = (\ln x)(\frac{1}{x})$

$\frac{dy}{dx} = \frac{y \ln x}{x}$

57. $2x^2 + 3xy + y^2 = -2$

$4x + 3xy' + 3y + 2yy' = 0$

$$y'\Big|_{(a,\,b)} = \frac{-(4a + 3b)}{3a + 2b}$$

Solve

$$\frac{-(4a + 3b)}{3a + 2b} = 0$$

$$4a + 3b = 0$$

$$b = -\frac{4a}{3}$$

Substitute $b = -\frac{4a}{3}$ into

$$2a^2 + 3ab + b^2 = -2$$

$$2a^2 + 3a\left(-\frac{4a}{3}\right) + \left(-\frac{4a}{3}\right)^2 = -2$$

$$18a^2 - 36a^2 + 16a^2 = -18$$

$$2a^2 = 18$$

$$a = \pm 3$$

Then $b = \mp 4$. Two such points are $(3, -4)$ and $(-3, 4)$.

59. a. Note $g^2(x)$ means $[g(x)]^2$.

$$x^2 + g^2(x) = 10$$

$$2x + 2g(x)[g'(x)] = 0$$

$$g'(x) = -\frac{x}{g(x)}$$

b. $g(x) = -\sqrt{10 - x^2}$ is a differentiable function of x for which $g(x) < 0$ by hypothesis. Using the chain rule:

$$g'(x) = -\frac{1}{2\sqrt{10 - x^2}}\frac{d}{dx}(10 - x^2)$$

$$= \frac{x}{\sqrt{10 - x^2}}$$

Verifying the result of (a):

$$\frac{x}{\sqrt{10 - x^2}} = \frac{-x}{-\sqrt{10 - x^2}} = \frac{-x}{g(x)}$$

61. a. $\quad x^2 + y^2 = 6y - 10$

$$2x + 2y\frac{dy}{dx} = 6\frac{dy}{dx}$$

$$(2y - 6)\frac{dy}{dx} = -2x$$

$$\frac{dy}{dx} = \frac{-2x}{2y - 6}$$

$$= \frac{x}{3 - y}$$

b. $\quad x^2 + y^2 = 6y - 10$

$$x^2 + (y - 3)^2 = -1$$

This is impossible since the sum of two squares is not negative (in the real number system).

c. The derivative does not exist.

63. $\quad x^2 + y^2 = \sqrt{x^2 + y^2} + x$

$$2x + 2yy' = \tfrac{1}{2}(x^2 + y^2)^{-1/2}(2x + 2yy') + 1$$

$$\left[2y - y(x^2 + y^2)^{-1/2}\right]y' = x(x^2 + y^2)^{-1/2} + 1 - 2x$$

$$y' = \frac{x(x^2 + y^2)^{-1/2} + 1 - 2x}{2y - y(x^2 + y^2)^{-1/2}}$$

$$= \frac{x + (1 - 2x)(x^2 + y^2)^{1/2}}{2y(x^2 + y^2)^{1/2} - y}$$

There will be a vertical tangent when the denominator is zero. That is,

$$y\left[2(x^2 + y^2)^{1/2} - 1\right] = 0$$

when $y = 0$ or $(x^2 + y^2)^{1/2} = \tfrac{1}{2}$.

If $y = 0$, the original equation is

$$x^2 = \sqrt{x^2} + x$$

$$x^2 = |x| + x$$

If $x \geq 0$, then $x^2 = 2x$, or $x = 0, 2$.

If $x < 0$, then $x^2 = 0$, which is impossible.

At $(0, 0)$, the numerator is also 0, so we exclude this point. Thus, there is a vertical tangent at $(2, 0)$.

If $(x^2 + y^2)^{1/2} = \frac{1}{2}$, we substitute into the original equation to find

$$\tfrac{1}{4} = \tfrac{1}{2} + x$$

$$x = -\tfrac{1}{4}$$

By substitution this gives the points

$$\left(-\tfrac{1}{4}, \frac{\sqrt{3}}{4} \right) \text{ and } \left(-\tfrac{1}{4}, -\frac{\sqrt{3}}{4} \right).$$

Hence, the tangent is vertical at $(2, 0)$,

$$\left(-\tfrac{1}{4}, \frac{\sqrt{3}}{4} \right) \text{ and } \left(-\tfrac{1}{4}, -\frac{\sqrt{3}}{4} \right).$$

65. Look at Figure 3.42 in the text.

$$\frac{dy}{dt} = 2 \text{ m/s}; \ \theta = \tan^{-1}\left(\frac{y}{4} \right)$$

$$\frac{d\theta}{dt} = \frac{\frac{1}{4}\frac{dy}{dt}}{1 + \frac{y^2}{16}}$$

$$= \frac{4}{16 + y^2}\frac{dy}{dt}$$

When $y = \frac{3}{2}$

$$\frac{d\theta}{dt} = \frac{4(2)}{16 + \frac{9}{4}}$$

$$= \frac{32}{73}$$

$$\approx 0.44 \text{ rad/s}$$

67.
$$\frac{x^2}{a^2} + \frac{y^2}{b^2} = 1$$

$$\frac{x}{a^2} + \frac{yy'}{b^2} = 0$$

$$y' = -\frac{b^2 x}{a^2 y}$$

At (x_0, y_0); $m = (-b^2 x_0)/(a^2 y_0)$, so the equation of the tangent line is

$$y - y_0 = -\frac{b^2 x_0}{a^2 y_0}(x - x_0)$$

$$a^2 y y_0 - a^2 y_0^2 + b^2 x_0 x - b^2 x_0^2 = 0$$

or

$$\frac{x_0 x}{a^2} + \frac{y_0 y}{b^2} = \frac{x_0^2}{a^2} + \frac{y_0^2}{b^2} = 1$$

since (x_0, y_0) lies on the curve and thus satisfies the equation of the curve.

69.
$$ax^2 + by^2 = c$$

$$2ax + 2by\frac{dy}{dx} = 0$$

$$\frac{dy}{dx} = -\frac{2ax}{2by}$$

$$= -\frac{ax}{by}$$

$$\frac{d^2y}{dx^2} = -\frac{by(a) - ax\left(b\dfrac{dy}{dx}\right)}{(by)^2}$$

$$= -\frac{aby - abx\dfrac{dy}{dx}}{b^2y^2}$$

Since $\dfrac{dy}{dx} = -\dfrac{ax}{by}$

$$\frac{d^2y}{dx^2} = -\frac{aby - abx\left(-\dfrac{ax}{by}\right)}{b^2y^2}$$

$$= -\frac{ab^2y^2 + a^2bx^2}{b^3y^3}$$

$$= -\frac{ab(by^2 + ax^2)}{b^3y^3}$$

$$= -\frac{a(by^2 + ax^2)}{b^2y^3}$$

From the original equation $by^2 + ax^2 = c$ so

$$\frac{d^2y}{dx^2} = -\frac{ac}{b^2y^3}$$

71. Let α be the angle between the tangent to C_1 and the positive x-axis. Let β be the angle between the tangent to C_2 and the positive x-axis. Then

$$\theta = \pi - [\alpha + (\pi - \beta)] = \beta - \alpha$$
$$\tan\theta = \tan(\beta - \alpha)$$
$$= \frac{\tan\beta - \tan\alpha}{1 + \tan\alpha\tan\beta}$$
$$= \frac{m_2 - m_1}{1 + m_2m_1} \text{ if } m_2 > m_1$$

Similarly, $\tan\theta = \dfrac{m_1 - m_2}{1 + m_1m_2}$ if $m_1 > m_2$;

Thus, $\tan\theta = \dfrac{|m_1 - m_2|}{1 + m_1m_2}$

73. Let $u = x^2$ so $x = \sqrt{u}$. Then By the chain rule,

$$\frac{d(y^2)}{d(x^2)} = \frac{d(y^2)}{du} = \frac{d(y^2)}{dx}\frac{dx}{dx^2}$$

$$= \left[2yy'\right]\left[\frac{1}{2}\frac{1}{\sqrt{u}}\right] = \frac{y}{\sqrt{u}}\frac{dy}{dx} = \frac{dy}{dx}\cdot\frac{y}{x}$$

The student was correct. Notice also that the answer does not depend on the original function. Suppose

$$y = x^3 - 3x^2 + \frac{7}{x}$$
$$\frac{d(y^2)}{d(x^2)} = \frac{dy}{dx}\frac{y}{x}$$
$$= \left(3x^2 - 6x - \frac{7}{x^2}\right)\left(\frac{y}{x}\right)$$
$$= y\left(3x - 6 - \frac{7}{x^3}\right)$$

75. Since $\sin^{-1}u + \cos^{-1}u = \dfrac{\pi}{2}$, we have $\cos^{-1}x = \dfrac{\pi}{2} - \sin^{-1}x$. Differentiating with respect to x, we have

$$\frac{d}{dx}(\cos^{-1}x) = \frac{d}{dx}\left(\frac{\pi}{2} - \sin^{-1}x\right)$$

$$= -\frac{1}{\sqrt{1 - x^2}}$$

where $|x| < 1$. By the chain rule,

$$\frac{d}{dx}(\cos^{-1}u) = -\frac{1}{\sqrt{1 - u^2}}\frac{du}{dx} \text{ where } |u| < 1$$

3.7 Related Rates and Applications, Pages 162-165

SURVIVAL HINT: *The common error when working*

related rate problems is to use constants in place of variables. For instance: if a 12 ft ladder is sliding down a wall and we wish to know how fast the top is descending when the bottom is 4 ft from the wall, **do not put the 4 ft on your figure.** The 12 ft is a constant and belongs on the figure. The vertical and horizontal distances are in motion, and should be given variable names. The 4 ft is a point at which you wish to evaluate your solved equation. Draw a careful figure, label constants and variables, then set up the **general situation.** Then, find a specific equation by substituting the given values to set up the **specific situation.**

1.
$$x^2 + y^2 = 25$$
$$2x \frac{dx}{dt} + 2y \frac{dy}{dt} = 0$$
When $x = 3$, $y = 4$, and $\frac{dx}{dt} = 4$
$$3(4) + 4 \frac{dy}{dt} = 0$$
$$\frac{dy}{dt} = -3$$

3.
$$5x^2 - y = 100$$
$$10x \frac{dx}{dt} - \frac{dy}{dt} = 0$$
When $x = 10$, $y = 400$, and $\frac{dx}{dt} = 10$
$$10(10)(10) - \frac{dy}{dt} = 0$$
$$\frac{dy}{dt} = 1,000$$

5.
$$y = 2\sqrt{x} - 9$$
$$\frac{dy}{dt} = 2(\tfrac{1}{2})x^{-1/2}\frac{dx}{dt}$$
When $x = 9$, $y = -3$, and $\frac{dy}{dt} = 5$
$$5 = (9)^{-1/2}\frac{dx}{dt}$$
$$\frac{dx}{dt} = 15$$

7.
$$xy = 10$$
$$x\frac{dy}{dt} + y\frac{dx}{dt} = 0$$
When $x = 5$, $y = 2$, and $\frac{dx}{dt} = -2$

$$5\left(\frac{dy}{dt}\right) + 2(-2) = 0$$
$$\frac{dy}{dt} = \frac{4}{5}$$

9.
$$x^2 + xy - y^2 = 11$$
$$2x \frac{dx}{dt} + x \frac{dy}{dt} + \frac{dx}{dt} y - 2y \frac{dy}{dt} = 0$$
When $x = 4$,
$$16 + 4y - y^2 = 11$$
$$y^2 - 4y - 5 = 0$$
$$(y + 1)(y - 5) = 0$$
$$y = -1 \text{ (discard), } 5$$
When $x = 4$, $y = 5$ and $\frac{dy}{dt} = 5$,
$$2(4) \frac{dx}{dt} + (4)(5) + \frac{dx}{dt}(5) - 2(5)(5) = 0$$
$$\frac{dx}{dt} = \frac{30}{13}$$

11.
$$F(x) = -12x$$
$$\frac{d}{dt}F(x) = -12 \frac{dx}{dt}$$
$$= -12(\tfrac{1}{4})$$
$$= -3$$

Notice that since F is a linear function of x, the change in F is a constant, and does not depend upon the value of x. (See Problems 10 and 11.)

13.
$$4x^2 + y^2 = 4$$
$$4x \frac{dx}{dt} + y \frac{dy}{dt} = 0$$
$$(2\sqrt{3})(5) + (1)\frac{dy}{dt} = 0$$
$$\frac{dy}{dt} = -10\sqrt{3} \text{ units/s}$$

15. The area of the ripple is $A = \pi r^2$ in.^{2}
$$\frac{dA}{dt} = 2\pi r \frac{dr}{dt} \text{ or } 4 = 2\pi(1)\frac{dr}{dt} \text{ so that}$$
$$\frac{dr}{dt} = \frac{2}{\pi} \approx 0.637 \text{ ft/s}$$

17. $R(x) = 0.5x^2 + 3x + 160$

$\dfrac{dR}{dt} = x\dfrac{dx}{dt} + 3\dfrac{dx}{dt}$

$x = x_0 + \left(\dfrac{dx}{dt}\right)t$, $x_0 = 10$

so when $t = 2$, $x = 14$, $\dfrac{dx}{dt} = 2$ and

$\dfrac{dR}{dt} = 14(2) + 3(2)$

$\quad\quad = 34$

In 2 years, $x = 14$, so the revenue will be increasing at \$34,000/yr.

19. Given $PV = C$, we are asked to find dP/dt for the specific values of $V = 30$, $P = 90$, and $dV/dt = 10$. Differentiating Boyle's law with respect to t:

$$P\dfrac{dV}{dt} + V\dfrac{dP}{dt} = 0\ .$$

Using the specified values:

$90(10) + 30\dfrac{dP}{dt} = 0$:

$\quad \dfrac{dP}{dt} = -30$ lb/in.^{2}/s. The

negative sign indicates the pressure is decreasing.

21. a. Draw a figure and assign variables.

b. Relate these variables through one or more equations (or formulas).

c. Differentiate.

d. Substitute numbers.

23. $S = 4\pi r^2$

$\dfrac{dS}{dt} = 8\pi r\dfrac{dr}{dt}$

When $r = 2$, $\dfrac{dS}{dt} = -3\pi$

$-3\pi = 8\pi(2)\dfrac{dr}{dt}$

$\dfrac{dr}{dt} = -\dfrac{3}{16} = -0.1875$ cm/s

Finally, since $V = \frac{4}{3}\pi r^3$

$\dfrac{dV}{dt} = 4\pi r^2\dfrac{dr}{dt}$

When $r = 2$, $\dfrac{dr}{dt} = -\dfrac{3}{16}$

$\dfrac{dV}{dt} = 4\pi(2^2)\left(-\dfrac{3}{16}\right) = -3\pi$ cm^3/s

25. Let x be the distance from the foot of the ladder to the wall, y the distance to the top of the ladder from the ground. We are asked to find dx/dt at the instant when $dy/dt = -3$ and $x = 5$. According to the Pythagorean theorem $x^2 + y^2 = 13^2$. So

$$2x\dfrac{dx}{dt} + 2y\dfrac{dy}{dt} = 0$$

Find y when $x = 5$:

$5^2 + y^2 = 13^2$; $\quad y = 12$. Now find dx/dt:

$2(5)\dfrac{dx}{dt} + 2(12)(-3) = 0$; $\quad \dfrac{dx}{dt} = 7.2$ ft/s

27. Let x be the horizontal distance from the boat to the pier and D the length of the rope; D is the hypotenuse of a right triangle with legs 12 and x.

$$12^2 + x^2 = D^2$$

$$0 + 2x\dfrac{dx}{dt} = (2D)\dfrac{dD}{dt}$$

We are given that $dD/dt = -6$ and find that $D = 20$ when $x = 16$. At this instant, we have

$$2(16)\frac{dx}{dt} = 2(20)(-6)$$

$$\frac{dx}{dt} = -\frac{240}{32} = -7.5$$

The negative value means the distance is decreasing at the rate of 7.5 ft/min.

29. $H(t) = -16t^2 + 160$ is the height of the ball at time t. The distance from the tip of the shadow to a point directly under the ball on the ground is x. $H(t)$ and x form a right triangle. The lamp is at 160 ft above the ground and at a horizontal distance $x + 10$ from the shadow of the ball. The line through the lamp, ball, and shadow is the hypotenuse of a right triangle similar to the one discussed above.

$$H = -16t^2 + 160 = \frac{160x}{x + 10}$$

Now,

$$-32t = 160\left[\frac{10}{(x + 10)^2}\right]\frac{dx}{dt}$$

When $t = 1$, $H = 144$ and

$$144 = \frac{160x}{x + 10} \text{ so that } x = 90$$

Thus,

$$-32(1) = 160\left[\frac{10}{(90 + 10)^2}\right]\frac{dx}{dt}$$

$$-200 = \frac{dx}{dt}$$

Thus, the shadow is moving toward the light at 200 ft/s.

31. *Note:* 20 m is $20/1,000 = 1/50$ km. Let x be the distance between the race car and the finish line, then $dx/dt = -200$ km/h.

$$x = \frac{1}{50}\tan\theta$$

$$\frac{dx}{dt} = \frac{1}{50}\sec^2\theta\,\frac{d\theta}{dt}$$

We are given that $\frac{dx}{dt} = -200$.

When $\theta = 0$, $\sec 0 = 1$:

$$-200 = \frac{1}{50}(1)^2\frac{d\theta}{dt}$$

$$-10,000 = \frac{d\theta}{dt} \qquad \textit{This is rad/hr.}$$

The angle of the line of sight is changing at the rate of 2.78 rad/s.

33. Assume the shape of the balloon is a sphere of radius r.

$$V = \frac{4}{3}\pi r^3$$

$$\frac{dV}{dt} = 4\pi r^2\,\frac{dr}{dt}$$

With $r = 4$, $\frac{dr}{dt} = 0.3$

$$\frac{dV}{dt} = 4\pi(4^2)(0.3) \approx 60.3 \text{ cm}^3/\text{min}$$

The volume is changing at the rate of 60.3 cm^3/min.

35. Let x be the radius of the top circle of the body of water and y its height. The radius of the top circle is 20, the height of the cone is 40 ft. By similar right triangles,

$$\frac{20}{40} = \frac{x}{y}$$

$$x = \frac{1}{2}y$$

The volume of the body of water is

$$V = \tfrac{1}{3}\pi r^2 h$$

$$= \tfrac{1}{3}\pi(\tfrac{1}{2}y)^2(y)$$

$$= \tfrac{1}{12}\pi y^3$$

Then,

$$\frac{dV}{dt} = \tfrac{1}{4}\pi y^2 \frac{dy}{dt}$$

We want the out volume, V_{out}, to satisfy

$$\frac{d}{dt}(80t - V_{out}) = \tfrac{\pi}{4}(12)^2(0.05)$$

$$\frac{d}{dt} V_{out} = 80 - \tfrac{\pi}{4}(12)^2(0.05)$$

$$\approx 74.35 \text{ ft}^3/\text{min}$$

37. At noon, the car is at the origin, while the truck is at $(250, 0)$. At time t, the truck is at position $(250 - x, 0)$, while the car is at $(0, y)$. Let H be the distance between them. Also, we are given $dx/dt = 25$, $dy/dt = 50$, $x = 25t$ and $y = 50t$.

$$H^2 = (250 - x)^2 + y^2$$

$$H^2 = (250 - 25t)^2 + 50^2t^2$$

$$H^2 = 3{,}125t^2 - 12{,}500t + 62{,}500$$

a. $$2H\frac{dH}{dt} = 6{,}250t - 12{,}500$$

$$\frac{dH}{dt} = \frac{3{,}125t - 6{,}250}{\sqrt{3{,}125t^2 - 12{,}500t + 62{,}500}}$$

b. $dH/dt = 0$ when $3{,}125t - 6{,}250 = 0$ or when $t = 2$.

c. $$H^2 = 3{,}125(2)^2 - 12{,}500(2) + 62{,}500$$

$$= 50{,}000$$

$$H = 100\sqrt{5} \approx 224 \text{ mi} \quad (\text{reject negative})$$

39. Let x denote the horizontal distance (in miles) between the plane and the observer. Let t denote the time (in hours), and draw a diagram representing the situation so that the angle is labeled θ and the distance is labeled D. It is given $dx/dt = -500$ mi/h. We are asked to find $d\theta/dt$ at the instant when $x = 4$ and $dx/dt = -500$.

$$\tan \theta = \tfrac{3}{x}$$

$$(\sec^2\theta) \frac{d\theta}{dt} = -\frac{3}{x^2}\frac{dx}{dt}$$

We note that $D = 5$ when $x = 4$ so we have

$$(\tfrac{5}{4})^2 \frac{d\theta}{dt} = -\frac{3}{4^2}(-500)$$

$$\frac{d\theta}{dt} = 60 \text{ rad/hr}$$

$$= 1 \text{ rad/min}$$

41. $d\theta/dt = 0.25$ rev/h $= \tfrac{\pi}{2}$ rad/h

$$x = 2\tan\theta$$

$$\frac{dx}{dt} = 2\sec^2\theta \frac{d\theta}{dt}$$

If $x = 1$, the hypotenuse is $\sqrt{2^2 + 1} = \sqrt{5}$ and $\sec\theta = \sqrt{5}/2$.

$$\frac{dx}{dt} = 2(\frac{\sqrt{5}}{2})^2(\tfrac{\pi}{2})$$

$$= \frac{5\pi}{4}$$

$$\approx 3.927 \text{ mi/h}$$

43. Draw a figure using the variables of A for the distance traveled by the first ship, B for the distance traveled by the second ship, D for the distance between them, and the constant angle of $60°$. We are asked to find dD/dt at $t = 2$ and $t = 5$. As the hint suggests, these variables are all generally related by the law of cosines:

$$D^2 = A^2 + B^2 - 2AB \cos \theta.$$

$$2D \frac{dD}{dt} = 2A \frac{dA}{dt} + 2B \frac{dB}{dt} - 2\left(A \frac{dB}{dt} + B \frac{dA}{dt}\right) \cos \theta$$

(Note that $\cos \theta = \frac{1}{2}$ is a constant.)

At $t = 2$ $A = 16$, $B = 12$, $\frac{dA}{dt} = 8$, $\frac{dB}{dt} = 12$,

$$D^2 = 16^2 + 12^2 - 2(16)(12)(\tfrac{1}{2})$$

$$D = \sqrt{208}$$

$$2\sqrt{208} \frac{dD}{dt} = 2(16)(8) + 2(12)(12)$$

$$- [16(12) + 12(8)]$$

$$\frac{dD}{dt} = \frac{128}{\sqrt{208}} = \frac{32\sqrt{13}}{13} \approx 8.875 \text{ knots}$$

At $t = 5$, $A = 40$, $B = 48$, $\frac{dA}{dt} = 8$, $\frac{dB}{dt} = 12$,

$$D^2 = 40^2 + 48^2 - 2(40)(48)(\tfrac{1}{2})$$

$$D = \sqrt{1984}$$

$$2\sqrt{1984} \frac{dD}{dt} = 2(40)(8) + 2(48)(12)$$

$$- [40(12) + 48(8)]$$

$$\frac{dD}{dt} = \frac{464}{\sqrt{1984}} = \frac{58\sqrt{31}}{31} \approx 10.417 \text{ knots}$$

45. Consider the figure. From similar triangles $\triangle AFG$ and $\triangle ABC$:

$$\frac{h_1}{1} = \frac{h_1 - 1}{\frac{3}{4}} \quad \textit{where } h_1 = |\,GA\,|$$

$$h_1 = 4$$

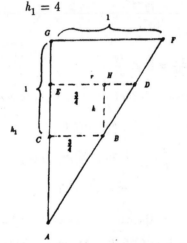

From $\triangle ADE$ and $\triangle AFG$, we have

$$r = \tfrac{1}{4}(3 + h)$$

The volume of water is that of a frustum with height h and radii r and 0.75

$$V = \frac{\pi h}{3}\left[r^2 + 0.75r + 0.75^2\right]$$

$$= \frac{\pi}{48}(h^3 + 9h^2 + 27h)$$

$$\frac{dV}{dt} = \frac{\pi}{48}(3h^2 + 18h + 27)\frac{dh}{dt}$$

When $h = \frac{1}{2}$, $dV/dt = 8$ in.^{3}/min or $1/216$ ft^3/min. Now,

$$\frac{1}{216} = \frac{\pi}{48}\left[3(\tfrac{1}{2})^2 + 18(\tfrac{1}{2}) + 27\right]\frac{dh}{dt}$$

$$\frac{dh}{dt} = \frac{8}{1{,}323\pi} \approx 0.001925 \text{ ft/min}$$

or $(0.001925)(12) \approx 0.02310$ in./min

47. **a.** $\frac{x}{150} = \cot\theta$ (Note: 50 yd = 150 ft)

$\theta = \cot^{-1}\frac{x}{150}$

$\frac{d\theta}{dt} = \frac{-150}{(150)^2 + x^2}(-40)$

b. If $x \to 0$, $\theta \to \frac{\pi}{2}$, and $\sin\theta \to 1$; thus,

$\lim_{\theta \to 0}\left|\frac{d\theta}{dt}\right| = \frac{4}{15}$

$d\theta/dt$ approaches 0.27 rad/s

c. $\frac{d\theta}{dt} = \frac{v}{150}\sin^2\theta \to \frac{v}{150}$ as $\theta \to \frac{\pi}{2}$; as v

increases so will $d\theta/dt$ and it becomes more

difficult to see the seals.

3.8 Linear Approximations and Differentials, Pages 173-176

SURVIVAL HINT: *The differential can be thought of as the change in f along the tangent. If Δx is "small" and the function is reasonably "well behaved," then the difference between the actual value of f and the corresponding value of the tangent will be relatively small. It is most useful when extrapolating data, that is, when we really do know how the function will behave beyond a given point, and our best bet is the tangent at that point.*

1. $d(2x^3) = 6x^2\,dx$

3. $d(2\sqrt{x}) = 2\cdot\frac{1}{2}x^{-1/2}\,dx = x^{-1/2}\,dx$

5. $d(x\cos x) = (\cos x - x\sin x)dx$

7. $d\left(\frac{\tan 3x}{2x}\right) = \frac{x(3\sec^2 3x) - (\tan 3x)(1)}{2x^2}\,dx$

$= \frac{3x\sec^2 3x - \tan 3x}{2x^2}\,dx$

9. $d(\ln|\sin x|) = \frac{1}{\sin x}\cdot\cos x\,dx$

$= \cot x\,dx$

11. $d(e^x\ln x) = \left[e^x\cdot\frac{1}{x} + e^x\ln x\right]dx$

$= \frac{e^x}{x}(1 + x\ln x)\,dx$

13. $d\left(\frac{x^2\sec x}{x - 3}\right)$

$= \frac{(x-3)(x^2\sec x\tan x + 2x\sec x) - (x^2\sec x)(1)}{(x - 3)^2}\,dx$

15. $d\left(\frac{x - 5}{\sqrt{x + 4}}\right)$

$= \frac{\sqrt{x + 4}(1) - (x - 5)(\frac{1}{2})(x + 4)^{-1/2}}{x + 4}\,dx$

$= \frac{\frac{1}{2}(x + 4)^{-1/2}[2(x + 4) - (x - 5)]}{x + 4}\,dx$

$= \frac{2x + 8 - x + 5}{2(x + 4)^{3/2}}\,dx$

$= \frac{x + 13}{2(x + 4)^{3/2}}\,dx$

17. $dx = \Delta x;\ dy = f'(x)dx$

19. Let $f(x) = \sqrt{x} = x^{1/2}$, then $f'(x) = \frac{1}{2\sqrt{x}}$;

$x_0 = 1.0;\ \Delta x = dx = -0.01$

$f(x_0 + \Delta x) \approx f(x_0) + f'(x_0)dx$, so

$\sqrt{0.99} = \sqrt{1 + (-0.01)}$

$\approx \sqrt{1} + \frac{1}{2\sqrt{1}}(-0.01)$

$= 0.995$

By calculator, $\sqrt{0.99} \approx 0.9949874371$

21. Let $f(x) = x^5 - 2x^3 + 3x^2 - 2$;

    $f'(x) = 5x^4 - 6x^2 + 6x$;

    $x_0 = 3$ and $\Delta x = dx = 0.01$.

    Now, $f(x_0 + \Delta x) \approx f(x_0) + f'(x_0)dx$, so

    $$f(3.01) \approx f(3) + f'(3)dx$$

    $$= [(3)^5 - 2(3)^3 + 3(3)^2 - 2]$$

    $$+ [5(3)^4 - 6(3)^2 + 6(3)](0.01)$$

    $$= 214 + 3.69$$

    $$= 217.69$$

    Comparing this to a calculator value of

    $f(3.01) = 217.7155882$ we see an error

    of approximately 0.03.

23. $A(r) = \pi r^2$; $A'(r) = 2\pi r$;

    $$\left| \frac{dA}{A} \right| = \left| \frac{2\pi r \, dr}{\pi r^2} \right|$$

    $$= 2 \left| \frac{dr}{r} \right|$$

    $$= 0.06 \text{ or } 6\%$$

25. $V = \frac{4}{3}\pi r^3$; $V' = 4\pi r^2$;

    $$\left| \frac{dV}{V} \right| = \left| \frac{3(4\pi r^2) \, dr}{4\pi r^3} \right|$$

    $$= 3 \left| \frac{dr}{r} \right|$$

    $$= 0.03 \text{ or } 3\%$$

27. Because the average level of carbon monoxide

    in the air t years from now will be

    $$Q(t) = 0.05t^2 + 0.1t + 3.4$$

    parts per million, the change in the carbon

    monoxide level during the next six months

    ($\Delta t = 0.5$) will be

    $$\Delta Q = Q(0.5) - Q(0) \approx Q'(0)(0.5)$$

    Since $Q'(t) = 0.1t + 0.1$ and $Q'(0) = 0.1$, it

    follows that

    $$\Delta Q \approx 0.1(0.5) = 0.05 \text{ parts per million}$$

29. $Q(L) = 60{,}000L^{1/3}$; $Q'(L) = 20{,}000L^{-2/3}$;

    $L_0 = 1{,}000$; $\Delta L = -60$

    $$\Delta Q \approx Q'(L_0)\Delta L$$
    $$= 20{,}000(1000)^{-2/3}(-60)$$
    $$= -12{,}000$$

    The output will be reduced by 12,000 units.

31. $V = \frac{4}{3}\pi r^3$ and $dV = 4\pi r^2 \, dr$

    When $r = 8.5/2$ and $\Delta r = dr = 1/8$, we have

    $$dV = 4\pi(4.25)^2(\tfrac{1}{8})$$

    $$\approx 28.37 \text{ in.}^3$$

33. Let $P(x)$ be the pulse in beats/min.

    $$P(x) = \frac{596}{\sqrt{x}}; \quad P'(x) = -\frac{298}{x^{3/2}}$$

    For $x_0 = 59$ and $\Delta x = 1$, we have

    $$\Delta P \approx P'(x_0) \, \Delta x$$

    $$= -\frac{298}{59^{3/2}} (1)$$

    $$\approx -0.658 \text{ beats/min}$$

A decrease of about 2 beats every 3 minutes.

35. Let $S(R)$ be the speed of the blood.

$$S(R) = cR^2$$
$$S'(R) = 2cR$$

We have

$$\frac{\Delta S}{S} \approx \frac{S'(R)\Delta R}{S} = \frac{2cR\Delta R}{cR^2} = 2\frac{\Delta R}{R}$$

A 1% error in R means $\Delta R/R = 0.01$ and the propagated error in S is

$$\frac{\Delta S}{S} \approx 2(0.01) = 0.02$$

The error in S is approximately $\pm 2\%$.

37. $S = 4\pi r^2$; $S' = 8\pi r$

$$\frac{\Delta S}{S} \approx \frac{S'(r)\Delta r}{S} = \frac{8\pi r\Delta r}{4\pi r^2} = 2\frac{\Delta r}{r}$$

A 1% increase in r means $\Delta r/r$ is 0.01, so

$$\frac{\Delta S}{S} \approx 0.02$$

or S increases by 2%. Since $V = \frac{4}{3}\pi r^3$;

$V' = 4\pi r^2$ and

$$\frac{\Delta V}{V} \approx \frac{V'(r)\Delta r}{V} = 3\frac{\Delta r}{r}$$

so if $\Delta r/r = 0.01$, $\Delta V/V = 0.03$ and the volume increases by 3%.

39. $\Delta L \approx L'(T)\Delta T$, so $\sigma = \frac{L'(T)}{L(T)} \approx \frac{\Delta L}{L(T)\Delta T}$
If $\sigma = 1.4 \times 10^{-5}$, $L = 75$, and
$\Delta T = 40 - (-10)$
$\quad = 50,$
then

$\Delta L \approx \sigma L(T)\Delta T$
$\quad = (1.4 \times 10^{-5})(75)(50)$
$\quad = 0.0525$

The length will change by about 0.0525 ft or 0.63 in.

41. Let N be the number of alpha particles falling on a unit area of the screen.

$$N(\theta) = \frac{1}{\sin^4(\frac{\theta}{2})}$$
$$N'(\theta) = -4[\sin^{-5}(\tfrac{\theta}{2})\cos(\tfrac{\theta}{2})]\tfrac{1}{2}$$
$$= -2\sin^{-5}(\tfrac{\theta}{2})\cos(\tfrac{\theta}{2});$$

With $\theta = 1$ and $\Delta\theta = 0.1$,

$\Delta N \approx N'(\theta)\Delta\theta$
$\quad = -2[\sin^{-5}(0.5)\cos(0.5)](0.1)$
$\quad \approx -6.93$

About 7 particles/unit area.

SURVIVAL HINT: *Anytime you see the adjective "marginal" in an economics application, you can translate it as derivative. It designates a rate of change.*

43. a. Since the cost is

$$C(q) = 0.1q^3 - 5q^2 + 500q + 200$$
$$C'(q) = 0.3q^2 - 10q + 500$$

The cost of producing the fourth unit is

$$C'(3) = 0.3(3)^2 - 10(3) + 500$$
$$= 472.7$$

b. The actual cost of manufacturing the fourth unit is

$$C(4) - C(3) = [0.1(4)^3 - 5(4)^2 + 500(4) + 200]$$
$$- [0.1(3)^3 - 5(3)^2 + 500(3) + 200]$$
$$= 468.70$$

45. Let $Q(L) = 360L^{1/3}$ be the daily output.
$$Q'(L) = 120L^{-2/3};$$
$$Q'(1,000) = \frac{120}{1,000^{2/3}}$$
$$= 1.2$$

$$\Delta Q \approx 1.2(1) = 1.2 \text{ units}$$

47. Let $x = -\sqrt{2}$; then $x^2 = 2$ or $x^2 - 2 = 0$.

We need to find a zero for $f(x) = x^2 - 2$;

$$f'(x) = 2x;$$
$$x_{n+1} = x_n - \frac{f(x_n)}{f'(x_n)} = x_n - \frac{x_n^2 - 2}{2x_n} = \frac{x_n^2 + 2}{2x_n}$$
$$x_0 = -1$$
$$x_1 = \frac{(-1)^2 + 2}{2(-1)} = -1.5$$
$$x_2 = \frac{(-1.5)^2 + 2}{2(-1.5)} \approx -1.416666667$$
$$x_3 = \frac{(-1.416666667)^2 + 2}{2(-1.416666667)} \approx -1.414215686$$
$$x_4 = \frac{(-1.414215686)^2 + 2}{2(-1.414215686)} \approx -1.414213562$$

The approximation (to four decimal places) is -1.4142.

49. Let $f(x) = x - e^{-x}$; $f'(x) = 1 + e^{-x}$
$$x_{n+1} = x_n - \frac{f(x_n)}{f'(x_n)}$$

$$= x_n - \frac{x_n - e^{-x_n}}{1 + e^{-x_n}}$$

$$x_0 = 1$$

$$x_1 = 1 - \frac{1 - e^{-1}}{1 + e^{-1}} \approx 0.5378828427$$

$$x_2 = x_1 - \frac{x_1 - e^{-x_1}}{1 + e^{-x_1}} \approx 0.5669869914$$

$$x_3 = x_2 - \frac{x_2 - e^{-x_2}}{1 + e^{-x_2}} \approx 0.567143286$$

$$x_4 = x_3 - \frac{x_3 - e^{-x_3}}{1 + e^{-x_3}} \approx 0.5671432904$$

The approximation (to four decimal places) is 0.5671.

51. $f(x) = -2x^4 + 3x^2 + \frac{11}{8}$

$$f'(x) = -8x^3 + 6x$$

a. $f(0) = \frac{11}{8}$ and $f(2) = -\frac{149}{8}$ so there is a root on $[0, 2]$. Also note that $f(x)$ is an even function and, therefore, symmetric about the y-axis. So it must have at least two roots.

b. $x_0 = 2$; $x_1 \approx 1.642$; $x_2 \approx 1.443$; $x_3 \approx 1.375$;

$x_4 \approx 1.367$; $x_5 \approx 1.367 \approx x$

c. $$x_{n+1} = x_n - \frac{f(x_n)}{f'(x_n)}$$

$$= x_n - \frac{-2x_n^4 + 3x_n^2 + \frac{11}{8}}{-8x_n^3 + 6x_n}$$

$$= \frac{-6x_n^4 + 3x_n^2 - \frac{11}{8}}{-8x_n^3 + 6x_n}$$

If $x_0 = \frac{1}{2}$ then $x_1 = -0.5$, $x_2 = 0.5$ and the values will continue to oscillate. This is due to the symmetry about the *y*-axis.

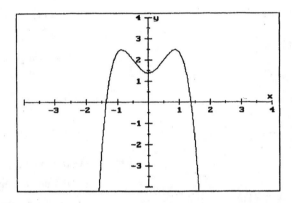

53. Since the ratio of the volume of the cap to the rest of the volume of the sphere is 1:2, the latter must be $\frac{2}{3}$ the total volume of the sphere. Thus (from Problem 52 with $H = x_C$ and $R = 1$),

$$\frac{2}{3}\left(\frac{4}{3}\pi(1)^3\right) = \frac{2}{3}\pi(1)^3 + \frac{\pi}{3}x_c^2(3 - x_c) \qquad \text{or} \qquad 3x_c^3 - 9x_c^2 + 2 = 0$$

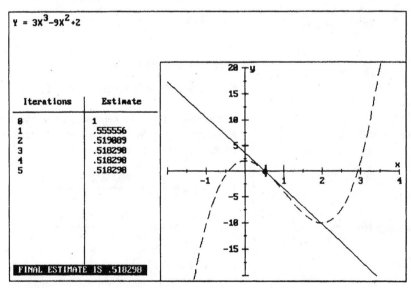

From the Newton-Raphson method, we find $x_c \approx 0.5183$. The other roots shown are not in the domain.

55. a. $f(x) = x^{1/2}$; $f'(x) = \frac{1}{2}x^{-1/2}$;

With $\Delta x = h$, $f'(1) = \frac{1}{2}$ so

$$f(1 + h) \approx f(1) + f'(1)h$$
$$= 1 + \frac{h}{2}$$

b. Let $f(x) = \frac{1}{x} = x^{-1}$

$$f'(x) = -x^{-2}$$

$$f(1 + h) \approx f(1) + f'(1)h$$

$$= 1 - h$$

57. Calculator check: $\sqrt{97} \approx 9.848857802$

$f(x) = \sqrt{x}$ and $f'(x) = \frac{1}{2x^{1/2}}$; $\Delta x = -3$

$$f(97) = f(100 - 3)$$
$$= f(100) + f'(100)(-3)$$
$$= 10 - \frac{3}{2\sqrt{100}}$$
$$= 9.85; \text{ Error} \approx 0.012\%$$

If $\Delta x = 16$,

$$f(97) = f(81 + 16)$$
$$= f(81) + f'(81)(16)$$
$$= 9 + \frac{16}{2\sqrt{81}}$$
$$\approx 9.89; \text{ Error} \approx 0.41\%$$

59. a. Let $P(x)$ be the profit when producing x units at a total cost of $C(x)$ and revenue of $R(x)$.

$$P(x) = R(x) - C(x)$$
$$P'(x) = R'(x) - C'(x)$$
$$P'(x) = 0 \text{ if } C'(x) = R'(x).$$

b. $A'(x) = \frac{xC'(x) - C(x)}{x^2} = \frac{1}{x}\left[C'(x) - \frac{C(x)}{x}\right]$

$$A'(x) = 0 \text{ if } C'(x) = A(x).$$

61. If $f(x_n) = 0$, x_n is a root. If $f'(x_n) \neq 0$, the tangent line through that root (x-intercept) is not horizontal. Successive tangent lines in the Newton-Raphson method fall on the same line and all points at which these tangent lines intersect the x-axis are the same as the root. Thus, $x_n = x_{n+1} = \cdots$

CHAPTER 3 REVIEW
Proficiency Examination, Page 177

SURVIVAL HINT: *To help you review the concepts of this chapter, **handwrite** the answers to each of these questions onto your own paper.*

1. $m_{\text{sec}} = \frac{\Delta y}{\Delta x} = \frac{f(x + \Delta x) - f(x)}{\Delta x}$

$m_{\text{tan}} = \lim\limits_{\Delta x \to 0} \frac{f(x + \Delta x) - f(x)}{\Delta x}$

2. If $y = f(x)$, then

$\frac{dy}{dx} = \lim\limits_{\Delta x \to 0} \frac{f(x + \Delta x) - f(x)}{\Delta x}$,

provided this limit exists.

3. A normal line is perpendicular to the tangent line at a point on the graph of a function.

4. If a function is differentiable at $x = c$, then it must be continuous at that point. The converse is not true: If a function is continuous at $x = c$, then it may or may not be differentiable at that point. Finally, if a function is discontinuous at $x = c$, then it cannot possibly have a derivative at that point.

5. Answers should include $f'(x)$, $\dfrac{dy}{dx}$ and y'.

6. a. $(cf)' = cf'$

    b. $(f + g)' = f' + g'$

    c. $(f - g)' = f' - g'$

    d. $(af + bg)' = af' + bg'$

    e. $(fg)' = fg' + f'g$

    f. $\left(\dfrac{f}{g}\right)' = \dfrac{gf' - fg'}{g^2}$ $(g \neq 0)$

    g. $d(cf) = c\,df$

    $d(f + g) = df + dg$

    $d(f - g) = df - dg$

    $d(af + bg) = a\,df + b\,dg$

    $d(fg) = f\,dg + g\,df$

    $d\left(\dfrac{f}{g}\right) = \dfrac{g\,df - f\,dg}{g^2}$ $(g \neq 0)$

7. a. $\dfrac{d}{dx}(k) = 0$

    b. $\dfrac{d}{dx}(x^n) = nx^{n-1}$

    c. $\dfrac{d}{dx}\sin x = \cos x;$

    $\dfrac{d}{dx}\cos x = -\sin x$

$\dfrac{d}{dx}\tan x = \sec^2 x;$

$\dfrac{d}{dx}\cot x = -\csc^2 x$

$\dfrac{d}{dx}\sec x = \sec x \tan x;$

$\dfrac{d}{dx}\csc x = -\csc x \cot x$

d. $\dfrac{d}{dx}e^x = e^x$

e. $\dfrac{d}{dx}\ln x = \dfrac{1}{x}$

f. $\dfrac{d}{dx}\left(\sin^{-1}x\right) = \dfrac{1}{\sqrt{1 - x^2}};$

$\dfrac{d}{dx}\left(\cos^{-1}x\right) = \dfrac{-1}{\sqrt{1 - x^2}};$

$\dfrac{d}{dx}\left(\tan^{-1}x\right) = \dfrac{1}{1 + x^2};$

$\dfrac{d}{dx}\left(\cot^{-1}x\right) = \dfrac{-1}{1 + x^2};$

$\dfrac{d}{dx}\left(\sec^{-1}x\right) = \dfrac{1}{|x|\sqrt{x^2 - 1}};$

$\dfrac{d}{dx}\left(\csc^{-1}x\right) = \dfrac{-1}{|x|\sqrt{x^2 - 1}}$

8. A higher-order derivative is a derivative of a derivative.

$$y''; \; y'''; \; y^{(4)}; \; \dots \; ; \; \frac{d^2y}{dx^2}; \; \frac{d^3y}{dx^3}; \; \frac{d^4y}{dx^4}; \; \dots$$

9. Rate of change refers to both average and instantaneous rates of change. The average rate of change for a function f is

$$\frac{f(x + \Delta x) - f(x)}{\Delta x}$$

The instantaneous rate of change is

$$\lim_{\Delta x \to 0} \frac{f(x + \Delta x) - f(x)}{\Delta x} = f'(x)$$

10. The relative rate of change of $y = f(x)$ with respect to x is given by the ratio $f'(x)/f(x)$.

11. $v(t) = s'(t); \; a(t) = v'(t) = s''(t)$

The speed is $|v(t)|$.

12. $\dfrac{dy}{dx} = \dfrac{dy}{du}\dfrac{du}{dx}$ or $[f(u(x))]' = f'[u(x)]u'(x)$

13. Logarithmic differentiation is a procedure in which logarithms are used to trade the task of differentiating products and quotients for that of differentiating sums and differences. It is especially valuable as a means for handling complicated product or quotient functions and power functions where variables appear in both the base and the exponent.

14. Apply all the rules of differentiation, treating y as a function of x and remembering the chain rule.

15. (1) Draw a figure.

(2) Relate the variables through a formula or equation.

(3) Differentiate the equation(s) (formulas).

(4) Substitute numerical values and solve algebraically for the required rate in terms of known rates.

16. $f(b) \approx f(a) + f'(a)(b - a)$

17. $dx = \Delta x; \; dy = f'(x)dx$

18. The propagated error is the difference between $f(x + \Delta x)$ and $f(x)$ and is defined by
$$\Delta f = f(x + \Delta x) - f(x)$$
The relative error is $\Delta f/f$, and the percentage error is $100|\Delta f/f|$.

19. Marginal analysis is the use of the derivative to approximate a change in a function produced by a unit change in the variable. It is especially useful in economics, where the function is cost, revenue, or profit.

20. The Newton-Raphson method approximates a root of a function by locating a point near a root, and then finding where the tangent line at this point intersects the x-axis. That is,
$$x_{n+1} = x_n - \frac{f(x_n)}{f'(x_n)}$$
Repetition of this technique usually closes in on the root.

21. $y = x^3 + x^{3/2} + \cos 2x$
$$\frac{dy}{dx} = 3x^2 + \frac{3}{2}x^{1/2} - 2\sin 2x$$

22. $y = \sqrt{3}\sqrt{x} + 3x^{-2}$
$$\frac{dy}{dx} = \tfrac{1}{2}(3x)^{-1/2}(3) + (-2)(3)x^{-3}$$
$$= \frac{\sqrt{3x}}{2x} - \frac{6}{x^3}$$

23. $\dfrac{dy}{dx} = \dfrac{1}{2}\left[\sin(3 - x^2)\right]^{-\frac{1}{2}}[\cos(3 - x^2)](-2x)$
$$= -x[\cos(3 - x^2)][\sin(3 - x^2)]^{-1/2}$$

24. $x\dfrac{dy}{dx} + y + 3y^2\dfrac{dy}{dx} = 0$
$$(x + 3y^2)\frac{dy}{dx} = -y$$
$$\frac{dy}{dx} = \frac{-y}{x + 3y^2}$$

25. $y' = x^2 \left(e^{-\sqrt{x}}\right)\left(-\frac{1}{2\sqrt{x}}\right) + e^{-\sqrt{x}}(2x)$

$\quad = -\frac{1}{2}x^{3/2}e^{-\sqrt{x}} + 2xe^{-\sqrt{x}}$

$\quad = \frac{1}{2}xe^{-\sqrt{x}}\left(4 - \sqrt{x}\right)$

26. $y' = \frac{(\ln 3x)(\frac{1}{x}) - (\ln 2x)(\frac{1}{x})}{(\ln 3x)^2}$

$\quad = \frac{\ln 1.5}{x(\ln 3x)^2}$

27. $y' = \frac{3}{\sqrt{1 - (3x + 2)^2}}$

28. $y' = \frac{1}{1 + (2x)^2}(2) = \frac{2}{1 + 4x^2}$

29. $\sin^2 a + \cos^2 a = 1.$ So $y = 1,$ $\frac{dy}{dx} = 0.$

30. $\ln y = \ln \frac{\ln(x^2 - 1)}{\sqrt[3]{x}(1 - 3x)^3}$

$\quad = \ln[\ln(x^2 - 1)] - \frac{1}{3}\ln x - 3\ln(1 - 3x)$

$\frac{1}{y}y' = \frac{1}{\ln(x^2 - 1)}\frac{1}{x^2 - 1}(2x) - \frac{1}{3x} - \frac{3}{1 - 3x}(-3)$

$\quad = \frac{2x}{(x^2 - 1)\ln(x^2 - 1)} - \frac{1}{3x} - \frac{9}{3x - 1}$

$y' = y\left[\frac{2x}{(x^2 - 1)\ln(x^2 - 1)} - \frac{1}{3x} - \frac{9}{3x - 1}\right]$

31. $y' = x^2(3)(2x - 3)^2(2) + 2x(2x - 3)^3$

$\quad = 2x(2x - 3)^2(5x - 3)$

$y'' = [2(2x - 3)^2(5x^2 - 3x)]'$

$\quad = 2(2x - 3)^2(10x - 3)$

$\quad\quad + 2(2)(2x - 3)(2)(5x^2 - 3x)$

$\quad = 2(2x - 3)[(2x - 3)(10x - 3)$

$\quad\quad + 2(2)(5x^2 - 3x)]$

$\quad = 2(2x - 3)[20x^2 - 36x + 9$

$\quad\quad + 20x^2 - 12x]$

$\quad = 2(2x - 3)(40x^2 - 48x + 9)$

32. $f(x) = x - 3x^2;$

$f(x + \Delta x) = (x + \Delta x) - 3(x + \Delta x)^2$

$\quad = x + \Delta x - 3x^2 - 6x\Delta x - 3\Delta x^2$

$\frac{dy}{dx} = \lim_{\Delta x \to 0} \frac{f(x + \Delta x) - f(x)}{\Delta x}$

$\quad = \lim_{\Delta x \to 0} \frac{(x + \Delta x - 3x^2 - 6x\Delta x - 3\Delta x^2) - (x - 3x^2)}{\Delta x}$

$\quad = \lim_{\Delta x \to 0} \frac{\Delta x - 6x\Delta x - 3\Delta x^2}{\Delta x}$

$\quad = \lim_{\Delta x \to 0} (1 - 6x - 3\Delta x)$

$\quad = 1 - 6x$

33. Given

$y = (x^2 + 3x - 2)(7 - 3x)$

When $x = 1,$ $y = 8,$ so the point is $(1, 8);$

$\frac{dy}{dx} = (x^2 + 3x - 2)(-3) + (7 - 3x)(2x + 3)$

At $x = 1,$ $\frac{dy}{dx} = 14$

$\quad\quad y - 8 = 14(x - 1)$

$\quad 14x - y - 6 = 0$

34. Given

$$y = f(x) = \sin^2 \frac{\pi x}{4}$$

$y = f(1) = \frac{1}{2}$, so the point is $(1, \frac{1}{2})$

$$\frac{dy}{dx} = 2 \sin(\tfrac{\pi x}{4}) \cos(\tfrac{\pi x}{4}) \left(\tfrac{\pi}{4}\right)$$

$$= \tfrac{\pi}{4} \sin\left(\tfrac{\pi}{2} x\right)$$

At $x = 1$, $\dfrac{dy}{dx} = 2 \left(\dfrac{\sqrt{2}}{2}\right)\left(\dfrac{\sqrt{2}}{2}\right)\left(\tfrac{\pi}{4}\right) = \tfrac{\pi}{4}$

The equation of the tangent line:

$$y - \tfrac{1}{2} = \tfrac{\pi}{4}(x - 1)$$

The equation of the normal line at that same point must have a slope that is the negative reciprocal of the slope of the tangent,

$m = -\frac{4}{\pi}$. The equation of the normal:

$$y - \tfrac{1}{2} = -\tfrac{4}{\pi}(x - 1)$$

35. $A = \pi r^2$ so $\dfrac{dA}{dt} = 2\pi r \dfrac{dr}{dt}$

$\dfrac{dr}{dt} = 0.5$ when $r = 2$ and

$$\frac{dA}{dt} = 2\pi(2)(0.5) = 2\pi \ \text{ft}^2/\text{s}$$

CHAPTER 4

Additional Applications of the Derivative

SURVIVAL HINT: *It is likely that you have had a test covering Chapter 3. The concepts of limit and derivative, and skill in finding derivatives accurately and quickly, are essential to success in the remainder of the course. It is essential that your examination not be filed away until you have mastered the material of Chapter 3! Analyze your mistakes. They are usually one of three types: you did not understand the concept, you made a computational or algebraic error, or you did not have time to finish the exam. For the first type, ask the instructor, a tutor, or another student to explain the concept to you. It is a good idea to form a small group of 3 to 5 students to do a "post mortem" on each examination. For the second type of error, attempt to organize your work better. Put more steps on paper and do less in your head. Stay composed during the examination, and do not rush. It is usually better to do fewer problems and get them right, than to turn in more completed problems full of errors. The solution for the third type of difficulty, assuming most other students completed the examination, is to practice. The more problems you solve, the faster you become. Do all of the chapter review problems, and some of the supplementary problems before the next examination. This takes time, but the standard rule-of-thumb is two to three hours of time for every hour of classroom time.*

4.1 Extreme Values of a Continuous Function, Pages 193-195

SURVIVAL HINT: *Remember that the candidates for extrema are the endpoints, and points where the derivative is either zero or does not exist.*

1. $f(x) = 5 + 10x - x^2$ on $[-3, 3]$

$f(-3) = -34; f(3) = 26;$

$f'(x) = 10 - 2x$

$\qquad 10 - 2x = 0$

$\qquad\qquad x = 5 \quad$ Not in the interval

Maximum value is 26 and the minimum value is -34.

3. $f(x) = x^3 - 3x^2$ on $[-1, 3]$

$f(-1) = -4; f(3) = 0;$

$f'(x) = 3x^2 - 6x$

$\qquad 3x^2 - 6x = 0$

$\qquad 3x(x - 2) = 0$

$\qquad\qquad x = 0, 2$

$f(0) = 0; f(2) = -4;$ maximum value is 0 and the minimum value is -4.

5. $f(x) = x^3$ on $[-\frac{1}{2}, 1]$

$f(-\frac{1}{2}) = -\frac{1}{8}; f(1) = 1;$

$f'(x) = 3x^2$

$$3x^2 = 0$$

$$x = 0$$

$f(0) = 0$; maximum value is 1 and the minimum value is $-\frac{1}{8}$.

7. $f(x) = x^5 - x^4$ on $[-1, 1]$

$$f(-1) = -2; \quad f(1) = 0;$$

$$f'(x) = 5x^4 - 4x^3$$

$$5x^4 - 4x^3 = 0$$

$$x^3(5x - 4) = 0$$

$$x = 0, \tfrac{4}{5}$$

$f(0) = 0$; $f(\tfrac{4}{5}) \approx -0.0819$; maximum value is 0 and the minimum value is -2.

9. $h(t) = te^{-t}$ on $[0, 2]$

$$h(0) = 0; \quad h(2) = 2e^{-2}$$

$$h'(t) = e^{-t}(1 - t)$$

$$e^{-t}(1 - t) = 0$$

$$t = 1$$

$h(1) = e^{-1}$; maximum value is e^{-1} and minimum value is 0.

11. $f(x) = |x|$ on $[-1, 1]$

$$f(-1) = 1; \quad f(1) = 1;$$

$f'(x)$ is not defined at $x = 0$; $f(0) = 0$

Maximum value is 1 and the minimum value is 0.

13. $f(u) = \sin^2 u + \cos u$ on $[0, 2]$

$f(0) = 1$; $f(2) = \sin^2 2 + \cos 2 \approx 0.41067$;

$$f'(u) = 2 \sin u \cos u - \sin u$$

$$2 \sin u \cos u - \sin u = 0$$

$$\sin u(2 \cos u - 1) = 0$$

$$\sin u = 0 \qquad 2 \cos u - 1 = 0$$

$$u = 0 \qquad\qquad \cos u = \tfrac{1}{2}$$

$$u = \tfrac{\pi}{3}$$

$f(0) = 1$; $f(\tfrac{\pi}{3}) = \tfrac{5}{4}$; maximum value is 1.25 and the minimum value is approximately 0.41067.

15.   a.   Find the value of the function at the endpoints of an interval.

   b.   Find the critical points, that is, points at which the derivative of the function is zero or undefined.

   c   Find the value of the function at each critical point.

   d.   State the absolute extrema.

17.   The calculator does not seem to take the derivative at $x = 0$ and $x < 0$ into account. The derivative is not defined at $x = 0$, but it certainly is defined for $x < 0$. If you enter $(x^2)\hat{\ }(1/3)(5 - 2x)$ you will obtain the correct graph.

19. $g(t) = (50 + t)^{2/3}$ on $[-50, 14]$

$g(-50) = 0$; $g(14) = 16$;

$g'(t) = \frac{2}{3}(50 + t)^{-1/3}$ which is not defined at

$t = -50$. The maximum value is 16 and the

minimum value is 0.

21. $g(x) = x^3 + 3x^2 - 24x - 4$ on $[-4, 4]$

$g(-4) = 76$; $g(4) = 12$

$g'(x) = 3x^2 + 6x - 24$

$3x^2 + 6x - 24 = 0$

$3(x + 4)(x - 2) = 0$

$x = -4, 2$

$g(2) = -32$

The maximum value is 76 and the minimum

value is -32.

23. $f(x) = \frac{1}{6}(x^3 - 6x^2 + 9x + 1)$ on $[0, 2]$

$f(0) = \frac{1}{6}$; $f(2) = \frac{1}{2}$

$f'(x) = \frac{1}{6}(3x^2 - 12x + 9)$

$3x^2 - 12x + 9 = 0$

$3(x - 1)(x - 3) = 0$

$x = 1, 3$

but 3 is not in the interval, so check $x = 1$:

$f(1) = \frac{5}{6}$.

The maximum value is $\frac{5}{6}$ and the minimum

value is $\frac{1}{6}$.

25. $s(t) = t \cos t - \sin t$ on $[0, 2\pi]$

$s'(t) = t(-\sin t) + \cos t - \cos t$

$= -t \sin t$

$s'(t) = 0$ when $t = 0, \pi, 2\pi$

$s(0) = 0$; $s(\pi) = -\pi$; $s(2\pi) = 2\pi$

The maximum value is 2π and the minimum

value is $-\pi$.

27. $g(x) = \cot^{-1}\left(\frac{x}{9}\right) - \cot^{-1}\left(\frac{x}{5}\right)$ on $[0, 10]$

$g(0) = 0$; $g(10) \approx 0.2692$

$g'(x) = \frac{4(45 - x^2)}{(x^2 + 25)(x^2 + 81)}$

$\frac{4(45 - x^2)}{(x^2 + 25)(x^2 + 81)} = 0$

$45 - x^2 = 0$

$x = \pm 3\sqrt{5}$

$g(3\sqrt{5}) = \cot^{-1}\frac{\sqrt{5}}{3} - \cot^{-1}\frac{3\sqrt{5}}{5}$

≈ 0.2898

$g(-3\sqrt{5}) = \cot^{-1}\left(-\frac{\sqrt{5}}{3}\right) - \cot^{-1}\left(-\frac{3\sqrt{5}}{5}\right)$

≈ -0.2898

The maximum value is

$\cot^{-1}\frac{\sqrt{5}}{3} - \cot^{-1}\frac{3\sqrt{5}}{5} \approx 0.2898$ and the minimum value is

$$\cot^{-1}\left(-\frac{\sqrt{5}}{3}\right) - \cot^{-1}\left(-\frac{3\sqrt{5}}{5}\right) \approx -0.2898$$

29. $f(x) = \begin{cases} 8 - 3x \text{ if } x < 2 \\ -x^2 + 3x \text{ if } x \geq 2 \end{cases}$ on $[-1, 4]$

$f(-1) = 11$; $f(4) = -4$; $f(2) = 2$

$f'(x) = -3$ if $x < 2$; $f'(x) = -2x + 3$

for $x \geq 2$; $f'(x) = 0$ when $x = \frac{3}{2}$, but $\frac{3}{2} < 2$.

$f'(x)$ is not defined at $x = 2$; $f(2) = 2$

The maximum value is 11 and the minimum value is -4.

31. $g(x) = \frac{9}{x} + x - 3$ on $[1, 9]$

$g(1) = 7$; $g(9) = 7$; $g'(x) = -9x^{-2} + 1$

which is equal to 0 when $x = 3$ (-3 is not in the

domain). $g(3) = 3$. The smallest value is 3.

33. $f(t) = \begin{cases} -t^2 - t + 2 \text{ if } t < 1 \\ 3 - t \quad \text{ if } t \geq 1 \end{cases}$ on $[-2, 3]$

$f(-2) = 0$; $f(3) = 0$; $f(1) = 2$

The function f is not continuous at $t = 1$.

$f'(t) = -2t - 1$ if $t < 1$;

$f'(t) = -1$ if $t \geq 1$

$f'(t) = 0$ when $t = -\frac{1}{2}$; $f(-\frac{1}{2}) = \frac{9}{4}$

The largest value is $\frac{9}{4}$.

35. $g(x) = \frac{\ln x}{\cos x}$ on $[2, 3]$

$g(2) = \frac{\ln 2}{\cos 2} \approx -1.6656$;

$g(3) = \frac{\ln 3}{\cos 3} \approx -1.1097$

$g'(x) = \dfrac{\cos x\left(\frac{1}{x}\right) - \ln x(-\sin x)}{\cos^2 x}$

$\dfrac{\cos x + x(\ln x)(\sin x)}{x\cos^2 x} = 0$

$x\ln x = -\cot x$

$x \approx 2.8098445$

$g(2.8098445) \approx -1.09271$; the largest value of g on $[2, 3]$ is approximately -1.1, which occurs at $x \approx 2.8$.

37. $g(\theta) = \theta \sin \theta$ on $[-2, 2]$

$g(-2) = -2 \sin(-2) \approx 1.819$;

$g(2) = 2 \sin 2 \approx 1.819$;

$g'(\theta) = \theta \cos \theta + \sin \theta$

$\theta \cos \theta + \sin \theta = 0$

$\tan \theta = -\theta$

Graphically, we see the only solution on $[-2, 2]$ is $\theta = 0$; $g(0) = 0$. The maximum value is approximately 1.819 and the minimum value is 0.

39. $g(u) = 98u^3 - 4u^2 + 72u$ on $[0, 4]$

$g(0) = 0$; $g(4) = 6,496$;

$g'(u) = 294u^2 - 8u + 72$

$$294u^2 - 8u + 72 = 0$$

$$147u^2 - 4u + 36 = 0$$

There are no real roots for this equation. The maximum value is 6,496 and the minimum value is 0.

41. $h(x) = x^{1/3}(x-3)^{2/3}$ on $[-1, 4]$

$$h(-1) = -2 \cdot 2^{1/3}$$

$$\approx -2.52$$

$$h(4) = 2^{2/3}$$

$$\approx 1.59$$

$$h'(x) = \tfrac{1}{3}x^{-2/3}(x-3)^{2/3} + \tfrac{2}{3}x^{1/3}(x-3)^{-1/3}$$

$$= \tfrac{1}{3}x^{-2/3}(x-3)^{-1/3}[(x-3)+2x]$$

$$= \tfrac{1}{3}x^{-2/3}(x-3)^{-1/3}(3x-3)$$

$$= x^{-2/3}(x-3)^{-1/3}(x-1)$$

$h'(x) = 0$ when $x = 1$;

does not exist at $x = 0, 3$.

$$h(1) = (-2)^{2/3} \approx 1.59; \; h(0) = h(3) = 0$$

The maximum value is 1.59 and the minimum value is -2.52.

43. $f(x) = e^{-x}(\cos x + \sin x)$ on $[0, 2\pi]$

$$f(0) = 1$$

$$f(2\pi) = e^{-2\pi}$$

$$\approx 0.001867$$

$$f'(x) = -e^{-x}(\cos x) + e^{-x}(-\sin x)$$

$$- e^{-x}(\sin x) + e^{-x}(\cos x)$$

$$= -2e^{-x}(\sin x)$$

$$-2e^{-x}(\sin x) = 0$$

$$\sin x = 0$$

$$x = 0, \pi, 2\pi$$

$$f(\pi) = -e^{-\pi} \approx -0.0432$$

The maximum value is 1 and the minimum value is $-e^{-\pi}$.

Answers to Problems 44-49 may vary.

45. a. $f(x) = \begin{cases} x^2 & \text{for } -0.5 \le x \le 1 \\ -x + 3 & \text{for } 1 < x \le 1.5 \end{cases}$

The minimum is 0, but there is no maximum.

b. $f(x) = \begin{cases} -x^2 & \text{for } -1 < x \le 1 \\ x - 3 & \text{for } 1 < x \le 1.5 \end{cases}$

The maximum is 0, but there is no minimum.

c. $f(x) = \begin{cases} \sin x & \text{for } -\pi \le x \le \frac{\pi}{2} \\ 0.5 & \text{for } \frac{\pi}{2} < x \le 3 \end{cases}$

The maximum is 1 and the minimum is -1.

d. $f(x) = \begin{cases} (1+x)/2 & -1 \le x < 1 \\ 0 & 1 \le x \le 2 \\ (x-4)/2 & 2 < x \le 4 \end{cases}$

There is no maximum and there is no minimum.

47. **a.** No such function can be found because of the extreme value theorem (Theorem 4.1).

b. No such function can be found because of the extreme value theorem.

c. $f(x) = \sin x$ for $[0, 2\pi]$.

d. No such function can be found because of the extreme value theorem.

49. $f(x) = x^{-1}$ on $(0, 1)$ has no extremum.

51. $v(t) = s'(t) = 4t^3 - 6t^2 - 24t + 60$

$v(0) = 60;\ v(3) = 42$

$v'(t) = 12t^2 - 12t - 24$

$\qquad = 12(t - 2)(t + 1)$

$v'(t) = 0$ when $t = -1, 2$

Reject $t = -1$ since it is not in the domain.

$v(2) = 20$; The maximum velocity is 60 when $t = 0$.

53. Let x and y be the numbers we are seeking on $[0, 6]$. Then $2x + y = 12$ or $y = 12 - 2x$, so

$P = x(12 - 2x)$.

$P'(x) = 12 - 4x = 0$ when $x = 3$

$P(0) = P(6) = 0;\ P(3) = 18$.

The largest product occurs when $x = 3$ and $y = 6$.

55. $P = xy = x(126 - 3x)$ on $[0, 42]$.

$P(0) = P(42) = 0$

$\quad P'(x) = -6x + 126 = 0$ when $x = 21$

$P(21) = 1{,}323$

The largest product occurs when $x = 21$ and $y = 63$.

57. Let x and y be the sides of a rectangle. Then the perimeter is $P = 2x + 2y$ or $y = \dfrac{P - 2x}{2}$.

$$A(x) = x\left(\frac{P - 2x}{2}\right) = \tfrac{1}{2}(xP - 2x^2)$$

$$A'(x) = \tfrac{1}{2}(P - 4x)$$

$$= 0 \text{ when } x = \frac{P}{4}$$

$$A(0) = A\left(\frac{P}{2}\right) = 0$$

$$A\left(\frac{P}{4}\right) = \frac{P^2}{16}$$

The largest area occurs when $x = P/4$.

$y = \dfrac{P - 2(P/4)}{2} = \dfrac{P}{4}$. Thus, $x = y$.

59. **a.** $\qquad x \geq x^2$

$\qquad x - x^2 \geq 0$

$\qquad x(1 - x) \geq 0$

Solution is in $[0, 1]$ since it is impossible for $x < 0$ and $1 - x < 0$ simultaneously.

Let $P(x) = x - x^2$

$P'(x) = 1 - 2x$

$= 0$ at $x = \frac{1}{2}$.

$P(0) = P(1) = 0$

$P(\frac{1}{2}) = \frac{1}{4}$

The greatest difference occurs at $x = \frac{1}{2}$.

b. $\qquad x \geq x^3$

$x - x^3 \geq 0$

$x(1 - x^2) \geq 0$

Solution for nonnegative x is in $[0, 1]$

Let $P(x) = x - x^3$

$P'(x) = 1 - 3x^2$

$= 0$ at $x = \frac{1}{\sqrt{3}}$

(disregard negative value)

$P(0) = P(1) = 0$

$P\left(\frac{1}{\sqrt{3}}\right) = \frac{2}{3\sqrt{3}}$

The greatest difference occurs at $x = \frac{1}{\sqrt{3}}$.

c. $\qquad x \geq x^n$

$x - x^n \geq 0$

$x(1 - x^{n-1}) \geq 0$

Solution for nonnegative x is in $[0, 1]$

Let $P(x) = x - x^n$; $P'(x) = 1 - nx^{n-1} = 0$

at $x = (\frac{1}{n})^{1/(n-1)}$.

$P(0) = P(1) = 0$

$P[(\frac{1}{n})^{1/(n-1)}]$

$= n^{-1/(n-1)} - n^{-n/(n-1)}$

The greatest difference occurs at

$x = (\frac{1}{n})^{1/(n-1)}$.

61. Since $f(x)$ has a relative minimum at $x = c$,

$$f(c) \leq f(c + \Delta x)$$

$f(c) - f(c + \Delta x) \leq 0$

$f(c + \Delta x) - f(c) \geq 0$

If $\Delta x > 0$, then $\dfrac{f(c + \Delta x) - f(c)}{\Delta x} \geq 0$

$\displaystyle \lim_{\Delta x \to 0} \dfrac{f(c + \Delta x) - f(c)}{\Delta x} \geq 0$

$f'(c) \geq 0$

If $\Delta x < 0$, then $\dfrac{f(c + \Delta x) - f(c)}{\Delta x} \leq 0$

$\displaystyle \lim_{\Delta x \to 0} \dfrac{f(c + \Delta x) - f(c)}{\Delta x} \leq 0$

$f'(c) \leq 0$

Since $f'(c) \geq 0$ and $f'(c) \leq 0$, it follows that $f'(c) = 0$ and c is a critical number.

4.2 The Mean Value Theorem, Pages 199-201

SURVIVAL HINT: *Never use a theorem until you have verified that the hypotheses are met.*

1. If f is a continuous function on $[a, b]$ and

Student Survival Manual for *Calculus, Third Edition* by Strauss/Bradley/Smith

differentiable on (a, b) with $f(a) = f(b)$, there

is at least one number c on (a, b) such that

$f'(c) = 0$. The tangent line is horizontal at

$(c, f(c))$. The importance of this theorem is

in its use in proving the mean value theorem.

3. $f(x) = 2x^2 + 1$; polynomials are everywhere

continuous and differentiable, so the

hypotheses of the MVT are met.

$f'(x) = 4x$, so there exists a c on the

interval $[0, 2]$ such that

$$f'(c) = \frac{f(2) - f(0)}{2 - 0}$$

$$4c = \frac{9 - 1}{2}$$

$$8c = 8$$

$$c = 1$$

5. $f(x) = x^3 + x$; polynomials are everywhere

continuous and differentiable, so the

hypotheses of the MVT are met.

$f'(x) = 3x^2 + 1$, so there exists a c on the

interval $[1, 2]$ such that

$$f'(c) = \frac{f(2) - f(1)}{2 - 1}$$

$$3c^2 + 1 = \frac{10 - 2}{1}$$

$$3c^2 = 7$$

$$c = \pm\sqrt{\frac{7}{3}}$$

The number $c = \sqrt{\frac{7}{3}} \approx 1.5275$ is on the

interval $[1, 2]$.

7. $f(x) = x^4 + 2$; polynomials are everywhere

continuous and differentiable, so the

hypotheses of the MVT are met.

$f'(x) = 4x^3$, so there exists a c on the

interval $[-1, 2]$ such that

$$f'(c) = \frac{f(2) - f(-1)}{2 - (-1)}$$

$$4c^3 = \frac{18 - 3}{3}$$

$$4c^3 = 5$$

$$c = \sqrt[3]{\frac{5}{4}}$$

The number $c = \sqrt[3]{\frac{5}{4}} \approx 1.0772$ is in the

interval $[1, 2]$.

9. $f(x) = \sqrt{x}$ is continuous and differentiable

everywhere on $[1, 4]$, so the hypotheses of the

MVT are met. $f'(x) = \frac{1}{2\sqrt{x}}$, so there exists

a c on the interval $[1, 4]$ such that

$$f'(c) = \frac{f(4) - f(1)}{4 - 1}$$

$$\frac{1}{2\sqrt{c}} = \frac{2 - 1}{3}$$

$$\sqrt{c} = \frac{3}{2}$$

$$c = \frac{9}{4}$$

The number $c = \frac{9}{4}$ is in the interval $[1, 4]$.

11. $f(x) = \frac{1}{x + 1}$ is continuous and differentiable everywhere on $[0, 2]$, so the hypotheses of the MVT are met. $f'(x) = -\frac{1}{(x + 1)^2}$, so there exists a c on the interval $[0, 2]$ such that

$$f'(c) = \frac{f(2) - f(0)}{2 - 0}$$

$$-\frac{1}{(c + 1)^2} = \frac{\frac{1}{3} - 1}{2},$$

$$\frac{2}{3}(c + 1)^2 = 2$$

$$c + 1 = \pm \sqrt{3}$$

$$c = -1 \pm \sqrt{3}$$

Note that only $c = -1 + \sqrt{3} \approx 0.73$ lies in the interval $[0, 2]$.

13. $f(x) = \cos x$ is continuous and differentiable everywhere on $[0, \pi/2]$, so the hypotheses of the MVT are met. $f'(x) = -\sin x$, so there exists a c on the interval $[0, \pi/2]$ such that

$$f'(c) = \frac{f(\frac{\pi}{2}) - f(0)}{\frac{\pi}{2} - 0}$$

$$-\sin c = \frac{0 - 1}{\frac{\pi}{2} - 0}$$

$$\sin c = \frac{2}{\pi}$$

The desired solution $c = \sin^{-1}(\frac{2}{\pi}) \approx 0.6901$.

15. $f(x) = e^x$ is continuous and differentiable everywhere on $[0, 1]$, so the hypotheses of the MVT are met.

$f'(x) = e^x$ so there exists a c on $[0, 1]$ so that

$$f'(c) = \frac{f(1) - f(0)}{1 - 0}$$

$$e^c = e - 1$$

$c = \ln(e - 1) \approx 0.54$, which is on $[0, 1]$.

17. $f(x) = \ln x$ is continuous and differentiable everywhere on $[\frac{1}{2}, 2]$, so the hypotheses of the MVT are met.

$f'(x) = \frac{1}{x}$, so that

$$f'(c) = \frac{f(2) - f(\frac{1}{2})}{2 - \frac{1}{2}}$$

$$\frac{1}{c} = \frac{2}{3}[\ln 2 - \ln \frac{1}{2}]$$

$$c = \frac{3}{4 \ln 2} \approx 1.082$$

is on the interval $[\frac{1}{2}, 2]$.

19. $f(x) = \tan^{-1} x$ is continuous and differentiable everywhere on $[0, 1]$, so the

hypotheses of the MVT are met.

$$f'(x) = \frac{1}{1 + x^2}, \text{ so that}$$

$$f'(c) = \frac{f(1) - f(0)}{1 - 0}$$

$$= \tan^{-1}1$$

$$= \frac{\pi}{4}$$

$$\frac{1}{1 + c^2} = \frac{\pi}{4}$$

$$c = \frac{\pm\sqrt{4 - \pi}}{\sqrt{\pi}}$$

$$\approx \pm 0.5227$$

We see that 0.5227 is on $[0, 1]$.

21. $f(x) = |x - 2|$ is continuous on the closed interval, but is not differentiable at $x = 2$, so Rolle's theorem does not apply.

23. Rolle's theorem is applicable since $f(x) = \sin x$ is continuous on $[0, 2\pi]$ and differentiable on $(0, 2\pi)$, $f(0) = f(2\pi) = 0$.

25. Rolle's theorem is not applicable since $f(x) = \sqrt[3]{x} - 1$ is continuous on $[-8, 8]$ but $f'(0) = \frac{1}{3}0^{-2/3}$ does not exist.

27. Rolle's theorem is not applicable on $[1, 2]$ since $f(x) = \frac{1}{x - 2}$ is not continuous at $x = 2$.

29. $f(x) = \sin^2 x$ is continuous everywhere, and $f'(x) = 2\sin x \cos x = \sin 2x$ so

$f(x)$ is differentiable everywhere, in particular on $[-\frac{\pi}{2}, \frac{\pi}{2}]$. $f(-\frac{\pi}{2}) = f(\frac{\pi}{2}) = 1$. Rolle's theorem applies.

31. $g(x) = 8x^3 - 6x + 8$; by the constant difference theorem $f(x) = 8x^3 - 6x + 8 + K$, for some constant K.

$$f(1) = 8 - 6 + 8 + K = 12$$

$$10 + K = 12$$

$$K = 2$$

Thus, $f(x) = 8x^3 - 6x + 10$.

33. $f(x) - g(x) = \frac{x + 4}{5 - x} - \frac{-9}{x - 5} = -1$

The constant difference theorem does not apply because f and g are not continuous at $x = 5$. If the interval does not contain 5, then the theorem applies.

$$f'(x) = \frac{9}{(x - 5)^2}$$

$$g'(x) = \frac{9}{(x - 5)^2};$$

$f'(x) = g'(x)$, $x \neq 5$.

35. $f(x) = (x - 1)^3$ and $g(x) = (x^2 + 3)(x - 3)$. f and g are polynomials so they are both continuous and differentiable, so the constant difference theorem applies.

$f'(x) = 3(x - 1)^2(1) = 3x^2 - 6x + 3$

$g'(x) = (x^2 + 3)(1) + 2x(x - 3)$

$= 3x^2 - 6x + 3$

We see $f'(x) = g'(x)$. To complete the task, we show $f(x) - g(x) = 8$, so we see the functions differ by a constant.

37. It looks like $c = 2.5$ and $c = 6.25$ satisfy the conditions of the mean value theorem. On $[4, 8]$, it looks like $f(4) = f(8)$, so Rolle's theorem (and consequently also the MVT) applies.

39. $f(x) = \tan x$, $f'(x) = \sec^2 x$, so that

$$\left| \frac{\tan u - \tan v}{u - v} \right| = \sec^2 c \geq 1$$

$$|\tan u - \tan v| \geq |u - v|$$

41. $g(x) = |x|$ is continuous everywhere, but is not differentiable at $x = 0$, so the mean value theorem does not apply.

43. The zero derivative theorem does not apply because $f(x)$ is not continuous at $x = 0$.

45. a. Let $f(x) = \cos x - 1$ on $[0, x]$. The hypotheses of the MVT apply, so there exists a w on the interval such that

$$f'(w) = \frac{f(x) - f(0)}{x - 0}$$

$$-\sin w = \frac{(\cos x - 1) - 0}{x}$$

b. Since w is on the interval $[0, x]$, as x approaches 0, w must approach 0 also (by the squeeze theorem). So

$$\lim_{x \to 0} \frac{\cos x - 1}{x} = \lim_{w \to 0} \frac{\cos w - 1}{w}$$

$$= \lim_{w \to 0} (-\sin w) = 0$$

47. $f(x) = 1 + \frac{1}{x}$; $f'(x) = -\frac{1}{x^2} < 0$ for all x

$f(b) > 1$ and $f(a) < 1$. Then, $f(b) - f(a) > 0$ and $\frac{f(b) - f(a)}{b - a} > 0$. As a result,

$\frac{f(b) - f(a)}{b - a} = f'(w)$ is impossible if $a < w < b$ as $f'(x) < 0$ for all x. The MVT does not apply because $f(x)$ is not continuous at $x = 0$.

49. $v(t_1) = v(t_2)$ implies $v'(c) = 0$ for some c between t_1 and t_2 by Rolle's theorem. Since the acceleration is $v'(t)$, this is what was to have been shown.

51. Let f and g represent the positions at time t of the two race cars on the time interval $[a, b]$. Let $F(t) = f(t) - g(t)$. Since, they begin and finish the race at the same time, $f(a) = g(a)$

and $f(b) = g(b)$; $F(a) = F(b) = 0$. By Rolle's theorem, there is a number $a \leq c \leq b$ such that $F'(c) = 0$ so $f'(c) - g'(c) = F'(c) = 0$. Thus, $f'(c) = g'(c)$ so that the race cars must be traveling at the same speed at some time c during the race.

53. If $x > 15$ then $f(x) = \sqrt{1 + x}$ is continuous and differentiable on $[15, x]$ and the hypotheses of the MVT are met. By the MVT,

$$\frac{f(x) - f(15)}{x - 15} = \frac{1}{2\sqrt{1 + c}}$$

$$\frac{\sqrt{1 + x} - 4}{x - 15} = \frac{1}{2\sqrt{1 + c}}$$

$$< \frac{1}{2} \cdot \frac{1}{4}$$

(Since $\sqrt{1 + c} > \sqrt{1 + 15} = 4$.)

Thus, $\sqrt{1 + x} - 4 < \frac{1}{8}(x - 15)$

55. $f(x) = \tan x$; $f'(x) = \sec^2 x$ and $\sec^2 x \neq 0$ on $0 < x < \pi$. The hypotheses of the MVT are not satisfied. $f(x) = \tan x$ is not continuous at $x = \pi/2$.

57. $f(x) = x^3 + ax - 1 = 0$; $f'(x) = 3x^2 + a$.

$f(0) = -1$; $f(2) = 7 + 2a > 0$. By the intermediate value theorem, there is at least one real root. Assume that there are two roots, x_1 and x_2. Then $f(x_1) = f(x_2)$, and by Rolle's theorem $3c^2 + a = 0$, which is impossible since $a > 0$.

59. Since $f''(x) = 0$, we have $f'(x) = A$ by the zero derivative theorem. Let $g(x) = Ax$, then $g'(x) = A$, and $f(x) = g(x) + B = Ax + B$ by the zero derivative theorem.

4.3 Using Derivatives to Sketch the Graph of a Function, Pages 214-217

SURVIVAL HINT: *Curve sketching is one of the major topics of beginning calculus. Don't think you can rely on software or a graphing calculator to help you graph curves. Usually, in fact, calculus is used, along with a graphing calculator, to sketch the graph of a complicated function.*

1. The first derivative test is a test to determine where a function is increasing or decreasing, leading to relative extrema. Specifically, the first-derivative test is used to:

(1) find all critical numbers of f.

(2) Classify each critical point $(c, f(c))$ as

follows:

    a. relative maximum if $f'(x) > 0$ to the

    left of c and $f'(x) < 0$ to the right of c;

    b. relative minimum if $f'(x) < 0$ to the

    left of c and $f'(x) > 0$ to the right of c;

    c. not an extremum if the derivative has

    the same sign on both sides of c.

3.   The second derivative test uses concavity to

classify critical points as relative maxima or

relative minima. Let f be a function such

that $f'(c) = 0$ and the second derivative

exists on an open interval containing c. If

$f''(c) > 0$, there is a relative minimum at

$x = c$. If $f''(x) < 0$, there is a relative

maximum at $x = c$. If $f''(c) = 0$, then the

second-derivative test fails and gives no

information.

5.   Let $P(t)$ be the price function in terms of

time. Then $dP/dt \geq 0$ and $d^2P/dt^2 < 0$.

7.   The black curve is the function and the blue

curve is the derivative.

9.

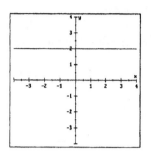

11.

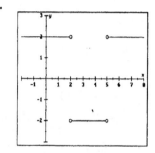

SURVIVAL HINT: $f''(x) = 0$ *gives **candidates** for inflection points. Even though the second derivative is zero, there may not be a change in concavity. These points must be tested.*

13. $f(x) = x^3 + 3x^2 + 1$

    a. $f'(x) = 3x^2 + 6x = 3x(x + 2)$

       critical numbers: $x = 0$, $x = -2$

    b. increasing on $(-\infty, -2) \cup (0, +\infty)$

       decreasing on $(-2, 0)$

    c. critical points:

       $(0, 1)$, relative minimum;

       $(-2, 5)$, relative maximum

d. $f''(x) = 6x + 6$;

second-order critical number, $x = -1$;

concave down on $(-\infty, -1)$ and concave

up on $(-1, \infty)$

e.

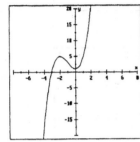

15. $f(x) = x^3 + 35x^2 - 125x - 9{,}375$

a. $f'(x) = 3x^2 + 70x - 125$

critical numbers: $x = \frac{5}{3}$, $x = -25$

b. increasing on $(-\infty, -25) \cup (\frac{5}{3}, +\infty)$

decreasing on $(-25, \frac{5}{3})$

c. critical points:

$(\frac{5}{3}, -9{,}481)$, relative minimum;

$(-25, 0)$, relative maximum

d. $f''(x) = 6x + 70$, second-order critical

number is $x = -\frac{35}{3}$;

concave down on $(-\infty, -\frac{35}{3})$ and

concave up on $(-\frac{35}{3}, \infty)$

e.

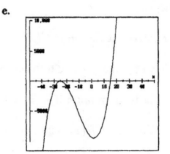

17. $f(x) = \dfrac{x - 1}{x^2 + 3}$

a. $f'(x) = \dfrac{(x^2 + 3)(1) - (x - 1)(2x)}{(x^2 + 3)^2}$

$= \dfrac{-x^2 + 2x + 3}{(x^2 + 3)^2}$

critical numbers: $x = -1$, $x = 3$

b. increasing on $(-1, 3)$

decreasing on $(-\infty, -1) \cup (3, +\infty)$

c. critical points:

$(-1, -\frac{1}{2})$, relative minimum;

$(3, \frac{1}{6})$, relative maximum

d. $f''(x) = [(x^2 + 3)^2(-2x + 2)$

$- (-x^2 + 2x + 3)2(x^2 + 3)(2x)](x^2 + 3)^{-4}$

Approximate second-order critical

numbers are $x \approx -2.0642$, $x \approx 0.3054$,

$x \approx 4.7588$; concave down on $(-\infty, -2.1)$

and (0.3, 4.8); concave up on $(-2.1, 0.3)$

and $(4.8, \infty)$.

e.

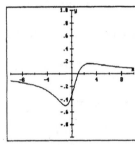

19. $f(t) = (t + 1)^2(t - 5)$

   a. $f'(t) = (t + 1)^2 + 2(t + 1)(t - 5)$

$$= (t + 1)(t + 1 + 2t - 10)$$

$$= (t + 1)(3t - 9)$$

$$= 3(t + 1)(t - 3)$$

   critical numbers: $t = -1$, $t = 3$

   b. increasing on $(-\infty, -1) \cup (3, +\infty)$

   decreasing on $(-1, 3)$

   c. critical points:

   $(3, -32)$, relative minimum;

   $(-1, 0)$, relative maximum

   d. $f''(t) = 6t - 6$;

   second-order critical number, $t = 1$;

   concave down on $(-\infty, 1)$ and concave

   up on $(1, \infty)$

e.

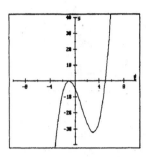

21. $f(x) = 1 + 2x + 18x^{-1}$

$$f'(x) = 2 - 18x^{-2}$$

$$f''(x) = 36x^{-3}$$

critical points: $(-3, -11)$, relative

    maximum; $(3, 13)$, relative minimum;

increasing on $(-\infty, -3) \cup (3, +\infty)$

decreasing on $(-3, 0) \cup (0, 3)$;

concave up on $(0, +\infty)$;

concave down on $(-\infty, 0)$

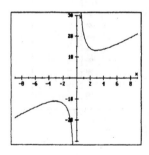

23. $g(u) = u^4 + 6u^3 - 24u^2 + 26$

$g'(u) = 4u^3 + 18u^2 - 48u$

$\quad = 2u(2u^2 + 9u - 24)$

$g''(u) = 12u^2 + 36u - 48$

$\quad = 12(u + 4)(u - 1)$

critical points:

$\quad$ (0, 26), relative maximum;

$\quad$ (−6.38, −852.22), relative minimum;

$\quad$ (1.88, −6.47), relative minimum;

$\quad$ inflection points (−4, −486), (1, 9);

increasing on $(-6.38, 0) \cup (1.88, +\infty)$;

decreasing on $(-\infty, -6.38) \cup (0, 1.88)$

concave up on $(-\infty, -4) \cup (1, +\infty)$;

concave down on $(-4, 1)$

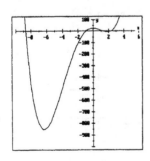

25. $g(t) = (t^3 + t)^2$

$g'(t) = 2(t^3 + t)(3t^2 + 1)$

$\quad = 2t(t^2 + 1)(3t^2 + 1)$

$g''(t) = 2(15t^4 + 12t^2 + 1)$

critical point:

$\quad$ (0, 0), relative minimum;

increasing on $(0, +\infty)$; decreasing on

$\quad (-\infty, 0)$; concave up on $(-\infty, +\infty)$

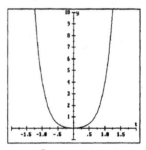

27. $f(x) = \dfrac{x}{x^2 + 1}$

$f'(x) = \dfrac{x^2 + 1 - x(2x)}{(x^2 + 1)^2}$

$\quad = \dfrac{1 - x^2}{(x^2 + 1)^2}$

$f''(x) = \dfrac{(x^2+1)^2(-2x) - (1 - x^2)(2)(x^2+1)(2x)}{(x^2 + 1)^4}$

$\quad = \dfrac{2x(x^2 - 3)}{(x^2 + 1)^3}$

critical points:

$\quad (-1, -\frac{1}{2})$, relative minimum;

$\quad (1, \frac{1}{2})$, relative maximum;

$\quad$ (0, 0) is a point of inflection

additional points of inflection:

$(\sqrt{3}, \frac{\sqrt{3}}{4}), (-\sqrt{3}, -\frac{\sqrt{3}}{4})$;

increasing on $(-1, 1)$;

decreasing on $(-\infty, -1) \cup (1, +\infty)$;

concave up on $(-\sqrt{3}, 0) \cup (\sqrt{3}, +\infty)$;

concave down on $(-\infty, -\sqrt{3}) \cup (0, \sqrt{3})$

Note that $\lim\limits_{|x| \mapsto \infty} \dfrac{x}{x^2 + 1} = 0$.

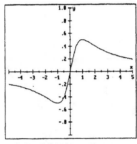

29. $f(t) = t^2 e^{-3t}$

$f'(t) = t^2(-3)e^{-3t} + 2te^{-3t}$

$\quad = (2 - 3t)te^{-3t}$

$f''(t) = (9t^2 - 12t + 2)e^{-3t}$

critical points:

$(0, 0)$, relative minimum;

$(0.67, 0.06)$, relative maximum;

$(0.195, 0.021)$, point of inflection;

$(1.138, 0.043)$, point of inflection

decreasing on $(-\infty, 0)$, $(0.67, +\infty)$

increasing on $(0, 0.67)$

concave up on $(-\infty, 0.195) \cup (1.138, +\infty)$

concave down on $(0.195, 1.138)$

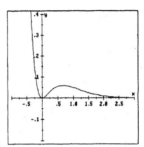

31. $f(x) = (\ln x)^2$

$f'(x) = 2(\ln x)x^{-1}$;

$f''(x) = 2[(\ln x)(-x^{-2}) + x^{-2}]$

$\quad = 2(x^{-2})(1 - \ln x)$

critical points:

$(1, 0)$, relative minimum;

$(e, 1)$, point of inflection

decreasing on $(0, 1)$

increasing on $(1, +\infty)$

concave up for $(0, e)$

concave down for $(e, +\infty)$

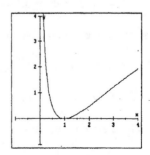

33. $t(\theta) = \theta + \cos 2\theta$

$t'(\theta) = 1 - 2\sin 2\theta$

$t''(\theta) = -4\cos 2\theta$

critical points:

$\left(\frac{\pi}{12},\ 1.13\right)$, relative maximum;

$\left(\frac{5\pi}{12},\ 0.44\right)$, relative minimum;

inflection points: $\left(\frac{\pi}{4},\ 0.79\right)$, $\left(\frac{3\pi}{4},\ 2.36\right)$;

increasing on $\left(0,\ \frac{\pi}{12}\right) \cup \left(\frac{5\pi}{12},\ \pi\right)$;

decreasing on $\left(\frac{\pi}{12},\ \frac{5\pi}{12}\right)$;

concave up on $\left(\frac{\pi}{4},\ \frac{3\pi}{4}\right)$;

concave down on $\left(0,\ \frac{\pi}{4}\right) \cup \left(\frac{3\pi}{4},\ \pi\right)$

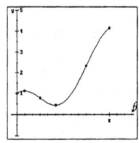

35. $f(x) = x^3 + \sin x$

$f'(x) = 3x^2 + \cos x > 0$ for all x because

both $3x^2$ and $\cos x$ are positive in Quadrants I

and IV and $3x^2 > \cos x$ for $\left(\frac{\pi}{2},\ \pi\right)$ and

$\left(-\pi,\ -\frac{\pi}{2}\right)$; $f''(x) = 6x - \sin x = 0$ when

$x = 0$. There are no critical points; the

function is increasing for all x. Inflection

point $(0, 0)$; concave down $\left(-\frac{\pi}{2},\ 0\right)$; concave

up $\left(0,\ \frac{\pi}{2}\right)$

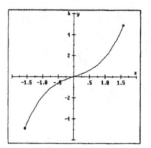

37. $f(x) = \dfrac{e^{-x^2}}{3 - 2x}$

$f'(x) = \dfrac{(3 - 2x)(-2xe^{-x^2}) - e^{-x^2}(-2)}{(3 - 2x)^2}$

$\qquad = \dfrac{2e^{-x^2}(2x - 1)(x - 1)}{(2x - 3)^2}$

At $x = \frac{1}{2}$ (check left and right sides), relative

maximum; at $x = 1$, relative minimum

39. $f(x) = \sqrt[3]{x^3 - 48x}$

$f'(x) = \frac{1}{3}(x^3 - 48x)^{-2/3}(3x^2 - 48)$

At $x = 4$ (check left and right sides),

relative minimum

41. $f(x) = \dfrac{x^2 - x + 5}{x + 4}$

$f'(x) = \dfrac{x^2 + 8x - 9}{(x + 4)^2}$

$f''(x) = \dfrac{50}{(x + 4)^3}$

$f''(1) = \frac{2}{5} > 0$; relative minimum

$f''(-9) = -\frac{2}{5} < 0$; relative maximum

43. $f(x) = \sin x + \frac{1}{2}\cos 2x$

$f'(x) = \cos x - \sin 2x$

$f''(x) = -2\cos 2x - \sin x$

$f''(\frac{\pi}{2}) = 1 > 0$; relative minimum

$f''(\frac{\pi}{6}) = -\frac{3}{2} < 0$; relative maximum

SURVIVAL HINT: *Exploration Problems 44-46 make excellent test questions to determine if you understand the concepts. It helps if you are able to visualize different graphical possibilities based on the signs of the function, its derivative, and its second derivative.*

Problems 45-49, answers may vary.

45.

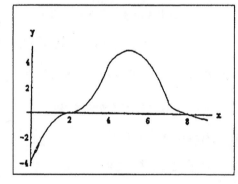

47.

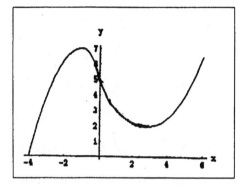

49.

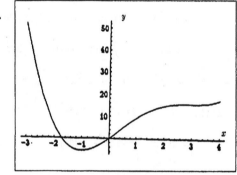

51. $v(T) = v_0\left(1 + \frac{1}{273}T\right)^{1/2}$

$v'(T) = v_0 \frac{1}{2}\left(1 + \frac{1}{273}T\right)^{-1/2}\left(\frac{1}{273}\right).$

For $T > 0$ this is always positive, so the

function is monotonically increasing.

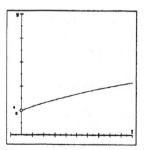

53. a. $-(2)^3 + 6(2)^2 + 13(2) = 42$

$-\frac{1}{3}(2)^3 + \frac{1}{2}(2)^2 + 25(2) = 49\frac{1}{3}$

b. $N(x) = -x^3 + 6x^2 + 13x - \frac{1}{3}(4 - x)^3$

$+ \frac{1}{2}(4 - x)^2 + 25(4 - x)$

c. $N'(x) = -2x^2 + 5x = 0$ when $x = 2.5$, so

the optimum time for the break is 10:30

A.M. The optimum time is half the value

for the diminishing return; $N''(2.5) < 0$.

55. $D(x) = \frac{9}{4}x^4 - 7\ell x^3 + 5\ell^2 x^2$ on $[0, \ell]$

$D'(x) = 9x^3 - 21\ell x^2 + 10\ell^2 x$

$= x(9x^2 - 21\ell x + 10\ell^2)$

$= x(3x - 5\ell)(3x - 2\ell)$

$D'(x) = 0$ when $x = 0, \frac{2}{3}\ell$

$(\frac{5}{3}\ell$ is not in the domain$)$

$D''(x) = 27x^2 - 42\ell x + 10\ell^2$

$D''(\frac{2}{3}\ell) < 0$, so $\frac{2}{3}\ell$ is a maximum;

$D(\frac{2}{3}\ell) = \frac{16}{27}\ell^4$.

Check the endpoints $(x = 0$ and $x = \ell)$:

$D(0) = 0$; $D(\ell) = \frac{\ell^4}{4}$

Maximum deflection at $x = \frac{2\ell}{3}$

57. Let $f(x) = \frac{1}{\sin x} - \frac{1}{x}$

To show $f(x) > 0$ for $0 < x \leq \frac{\pi}{2}$ note

$x > \sin x$ or

$\frac{1}{x} < \frac{1}{\sin x}$

so that $\frac{1}{\sin x} - \frac{1}{x} > 0$.

To show that f is strictly increasing, find

$f'(x) = \frac{-\cos x}{\sin^2 x} + \frac{1}{x^2}$

$= \frac{-x^2\cos x + \sin^2 x}{x^2 \sin^2 x} > 0$

because $-x^2 \cos x + \sin^2 x = 0$ has solution

$x = 0$ (found using CAS), which is not in the

domain $(0, \frac{\pi}{2}]$.

59. Let f and g be two functions concave up on

$[a, b]$. Then, for all c on the interval,

$f''(c) > 0$ and $g''(c) > 0$. Then,

$(f'' + g'')(c) = f''(c) + g''(c) > 0$,

since $f''(c) > 0$ and $g''(c) > 0$.

Thus, the sum is concave up.

61. $f(x) = Ax^3 + Bx^2 + C$

$f'(x) = 3Ax^2 + 2Bx$

$f''(x) = 6Ax + 2B$

$E(2, 11)$ is an extremum, so

$f'(2) = 12A + 4B = 0$ or $B = -3A$

$I(1, 5)$ is a point of inflection, so

$f''(1) = 6A + 2B = 0$ or $B = -3A$

Since the points E and I are on the curve, their coordinates must satisfy its equation. Thus,

$$f(2) = 8A + 4B + C$$

$$= 8A - 12A + C$$

$$= 11$$

$$f(1) = A + B + C$$

$$= A - 3A + C$$

$$= 5$$

Solving these equations simultaneously, we find $A = -3$, $B = 9$, and $C = -1$. Thus,

$$f(x) = -3x^3 + 9x^2 - 1$$

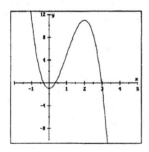

63. $y = x^3 + bx^2 + cx + d$

$y' = 3x^2 + 2bx + c = 0$ when

$$x = \frac{-b \pm \sqrt{b^2 - 3c}}{3}$$

If $b^2 - 3c > 0$, there is both a relative maximum and a relative minimum. The distance between the extrema changes as b varies. If $b^2 - 3c = 0$ there is a horizontal tangent but no relative extremum. If $b^2 - 3c < 0$ the are no horizontal tangents.

4.4 Curve Sketching with Asymptotes: Limits Involving Infinity, Pages 227-228

1. In general, to sketch a curve, check extent, find intercepts, look for symmetry, set $f'(x) = 0$, rough out intervals of curve increase and decrease, set $f''(x) = 0$, rough out intervals of curve concavity, look for horizontal and vertical asymptotes. Details are in Table 4.1 on page 226.

SURVIVAL HINT: *Do not define an asymptote as a line that is approached, but not reached. A function f may cross the asymptotes. For instance, $y = \sin x/x$ has $y = 0$ as a horizontal asymptote, and it is crossed an infinite number of times.*

3. Concavity indicates the shape of the curve as cupped up or cupped down; a point of inflection indicates where the curve changes concavity.

5. $\lim\limits_{x \to +\infty} \dfrac{2,000}{x + 1} = 0$

7. $\lim\limits_{x \to +\infty} \dfrac{3x + 5}{x - 2} = \lim\limits_{x \to +\infty} \dfrac{3 + \frac{5}{x}}{1 - \frac{2}{x}}$

$$= \frac{3 + 0}{1 - 0}$$

$$= 3$$

9. $\lim\limits_{t \to +\infty} \dfrac{9t^5 + 50t^2 + 800}{t^5 - 1{,}000} = \lim\limits_{t \to +\infty} \dfrac{9 + \frac{50}{t^3} + \frac{800}{t^5}}{1 - \frac{1{,}000}{t^5}}$

$$= \frac{9}{1}$$

$$= 9$$

11. $\lim\limits_{x \to +\infty} \dfrac{x}{\sqrt{x^2 + 1{,}000}} = \lim\limits_{x \to +\infty} \dfrac{1}{\sqrt{1 + \frac{1{,}000}{x^2}}}$

$$= \frac{1}{\sqrt{1 + 0}}$$

$$= 1$$

13. $\lim\limits_{x \to +\infty} \dfrac{x^{5.916} + 1}{x^{\sqrt{35}}}$

Note: $\sqrt{35} > 5.916$, so the denominator overpowers the numerator, and

$$\lim\limits_{x \to +\infty} f(x) = 0.$$

15. $\lim\limits_{x \to 1^-} \dfrac{x - 1}{|x^2 - 1|} = \lim\limits_{x \to 1^-} \dfrac{x - 1}{|x + 1||x - 1|}$

$$= \lim\limits_{x \to 1^-} \dfrac{-1}{x + 1}$$

$$= -\frac{1}{2}$$

17. $\lim\limits_{x \to 0^+} \dfrac{x^2 - x + 1}{x - \sin x}$

Since $\sin x < x$ for $x > 0$, the denominator will

approach 0 through positive values.

Meanwhile the numerator is approaching 1.

The quotient approaches $+\infty$.

19. $\lim\limits_{x \to +\infty} x \sin \frac{1}{x} = \lim\limits_{u \to 0^+} \dfrac{\sin u}{u}$ Let $u = \frac{1}{x}$.

$$= 1$$

21. $\lim\limits_{x \to 0^+} \dfrac{\ln \sqrt[3]{x}}{\sin x} = \frac{1}{3} \lim\limits_{x \to 0^+} \dfrac{\ln x}{\sin x}$

$$= -\infty$$

Since $|\sin x| \le 1$ (bounded) and $\ln x \to -\infty$

as $x \to 0^+$.

23. $\lim\limits_{x \to -\infty} e^x \sin x = \lim\limits_{x \to -\infty} \dfrac{\sin x}{e^{-x}}$

$$= 0$$

Since $|\sin x| \le 1$ (bounded) and $e^{-x} \to +\infty$

as $x \to -\infty$)

SURVIVAL HINT: *Values for which the denominator of a rational expression are zero are **candidates** for vertical asymptotes. If the numerator is also zero for that value, then you have a hole in the graph, and not an asymptote. The steps listed in Table 4.1 are a powerful set of analytic tools. You need not use every tool in your toolbox on each problem. Skill in knowing which tool to use comes with practice. Often you have some idea about the general shape of the graph, and calculators that help with the process are common.*

25. $f(x) = \dfrac{3x + 5}{7 - x}$

$$f'(x) = \frac{(7 - x)(3) - (3x + 5)(-1)}{(7 - x)^2}$$

$$= \frac{21 - 3x + 3x + 5}{(7 - x)^2}$$

$$= \frac{26}{(7 - x)^2}$$

$$f''(x) = 26(-2)(7 - x)^{-3}(-1)$$

$$= \frac{52}{(7 - x)^3}$$

asymptotes: $x = 7$, $y = -3$;

graph rising on $(-\infty, 7) \cup (7, +\infty)$;

concave up on $(-\infty, 7)$;

concave down on $(7, +\infty)$;

no critical points;

no points of inflection

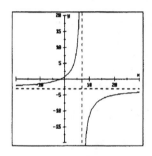

27. $f(x) = 4 + \dfrac{2x}{x - 3}$

$$f'(x) = \frac{(x - 3)(2) - 2x(1)}{(x - 3)^2}$$

$$= -\frac{6}{(x - 3)^2}$$

$$f''(x) = (-6)(-2)(x - 3)^{-3}(1)$$

$$= \frac{12}{(x - 3)^3}$$

asymptotes: $x = 3$, $y = 6$;

graph falling on $(-\infty, 3) \cup (3, +\infty)$;

concave up on $(3, +\infty)$;

concave down on $(-\infty, 3)$;

no critical points

no points of inflection

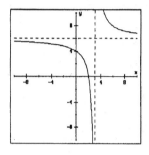

29. $f(x) = \dfrac{x^3 + 1}{x^3 - 8}$

$$\frac{x^3 + 1}{x^3 - 8} = 0$$

$$x^3 + 1 = 0$$

$$x = -1$$

The graph passes through the x-axis at $(-1, 0)$.

$$f'(x) = \frac{(x^3 - 8)(3x^2) - (x^3 + 1)(3x^2)}{(x^3 - 8)^2}$$

$$= -\frac{27x^2}{(x^3 - 8)^2}$$

$$\frac{-27x^2}{(x^3 - 8)^2} = 0$$

$$-27x^2 = 0$$

$x = 0$; undefined when $x^3 - 8 = 0$ or

when $x = 2$

$$f''(x) = \frac{(x^3 - 8)^2(-2)(27x) - 2(x^3 - 8)(3x^2)(-27x^2)}{(x^3 - 8)^4}$$

$$= \frac{-54x(x^3 - 8)[(x^3 - 8) - 3x^3]}{(x^3 - 8)^4}$$

$$= \frac{-54x(-2x^3 - 8)}{(x^3 - 8)^3}$$

$$= \frac{108x(x^3 + 4)}{(x^3 - 8)^3}$$

$$\frac{108x(x^3 + 4)}{(x^3 - 8)^3} = 0$$

$$108x(x^3 + 4) = 0$$

$$x^3 = 0, -4$$

$$x = 0, -\sqrt[3]{4}$$

asymptotes: $x = 2,\ y = 1$

graph falling on $(-\infty, 2) \cup (2, +\infty)$

concave up on $(-\sqrt[3]{4}, 0)$ or $(2, +\infty)$

concave down on $(-\infty, -\sqrt[3]{4})$ or $(0, 2)$

critical point is $(0, -\frac{1}{8})$

points of inflection $(0, -\frac{1}{8})$, $(-\sqrt[3]{4}, \frac{1}{4})$

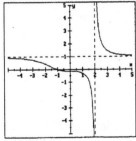

31. $g(x) = \dfrac{8}{x - 1} + \dfrac{27}{x + 4}$

$$= 8(x - 1)^{-1} + 27(x + 4)^{-1}$$

$$= \frac{8(x + 4) + 27(x - 1)}{(x - 1)(x + 4)}$$

$$= \frac{35x + 5}{(x - 1)(x + 4)}$$

$$g'(x) = -8(x - 1)^{-2} - 27(x + 4)^{-2}$$

$$= -\frac{8(x + 4)^2 + 27(x - 1)^2}{(x - 1)^2(x + 4)^2}$$

Undefined at $x = 1,\ x = -4$;

$g'(x) = 0$ when

$$8(x + 4)^2 + 27(x - 1)^2 = 0$$

$$35x^2 + 10x + 155 = 0$$

$$7x^2 + 2x + 31 = 0$$

No real roots.

$$g''(x) = 16(x - 1)^{-3} + 54(x + 4)^{-3}$$

$$= \frac{10(x + 1)(7x^2 - 4x + 97)}{(x - 1)^3(x + 4)^3}$$

Undefined at $x = 1,\ x = -4$;

$g''(0) = 0$ when

$$(x + 1)(7x^2 - 4x + 97) = 0$$

$$x = -1$$

asymptotes: $x = -4,\ x = 1,\ y = 0$

graph falling on

$(-\infty,\, -4) \cup (-4,\, 1) \cup (1,\, +\infty)$

concave up on $(-4,\, -1)$ or $(1,\, +\infty)$

concave down on $(-\infty,\, -4)$ or $(-1,\, 1)$

no critical points

point of inflection is $(-1,\, 5)$

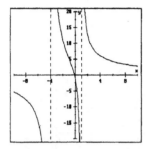

33. $g(t) = (t^3 + t)^2$

$g'(t) = 2t(t^2 + 1)(3t^2 + 1)$

$\quad\quad = (2t^3 + 2t)(3t^2 + 1)$

$\quad\quad = 6t^5 + 8t^3 + 2t$

$g''(t) = 30t^4 + 24t^2 + 2$

$\quad\quad = 2(15t^4 + 12t^2 + 1)$

no asymptotes; graph rising on $(0,\, +\infty)$;

graph falling on $(-\infty,\, 0)$; concave up on

$(-\infty,\, +\infty)$; critical point $(0,\, 0)$ is a relative

minimum; no points of inflection

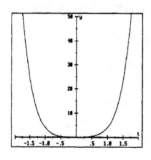

35. $f(x) = (x^2 - 9)^2$

$f'(x) = 2(x^2 - 9)(2x)$

$\quad\quad = 4x(x - 3)(x + 3)$

$\quad\quad = 4x^3 - 36x$

$f''(x) = 12x^2 - 36$

$\quad\quad = 12(x^2 - 3)$

no asymptotes

graph rising on $(-3,\, 0) \cup (3,\, +\infty)$

graph falling on $(-\infty,\, -3) \cup (0,\, 3)$

concave up on $(-\infty,\, -\sqrt{3}) \cup (\sqrt{3},\, +\infty)$

concave down on $(-\sqrt{3},\, \sqrt{3})$

critical points are $(-3,\, 0)$, relative minimum;

$(0,\, 81)$, relative maximum; $(3,\, 0)$, relative

minimum

points of inflection are $(-\sqrt{3},\, 36)$, $(\sqrt{3},\, 36)$

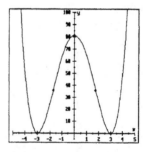

37. $f(x) = x^{1/3}(x - 4)$

$\quad\quad = x^{4/3} - 4x^{1/3}$

$f'(x) = \frac{4}{3}x^{1/3} - \frac{4}{3}x^{-2/3}$

$\quad\quad = \frac{4(x - 1)}{3x^{2/3}}$

$f''(x) = \frac{4}{9}x^{-2/3} + \frac{8}{9}x^{-5/3}$

$= \frac{4(x + 2)}{9x^{5/3}}$

no asymptotes

graph rising on $(1, +\infty)$

graph falling on $(-\infty, 1)$

concave up on $(-\infty, -2) \cup (0, +\infty)$

concave down on $(-2, 0)$

critical points:

   $(1, -3)$, relative minimum;

   $(0, 0)$, vertical tangent

   $(0, 0)$ and $(-2, 6\sqrt[3]{2})$ are points of

      inflection

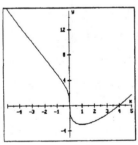

39. $f(x) = \tan^{-1}x^2$

$\lim_{x \to \infty} \tan^{-1}x^2 = \frac{\pi}{2}$;

horizontal asymptote at $y = \frac{\pi}{2}$

$f'(x) = \frac{2x}{1 + x^4}$

critical value at $x = 0$

$f''(x) = \frac{-2(3x^4 - 1)}{(1 + x^4)^2}$ and $f''(0) > 0$, so

there is a relative minimum at $x = 0$

The second-order critical number is at $x^4 = \frac{1}{3}$

or when $x \approx \pm 0.76$.

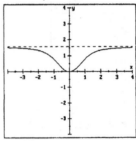

41. $t(\theta) = \sin\theta - \cos\theta$

$t'(\theta) = \cos\theta + \sin\theta$

critical numbers at $\theta = \frac{3\pi}{4}, \frac{7\pi}{4}$

$t''(\theta) = -\sin\theta + \cos\theta$

second-order critical numbers at $\theta = \frac{\pi}{4}, \frac{5\pi}{4}$

no asymptotes

graph rising on $(0, \frac{3\pi}{4}) \cup (\frac{7\pi}{4}, 2\pi)$

graph falling on $(\frac{3\pi}{4}, \frac{7\pi}{4})$

concave up on $(0, \frac{\pi}{4}) \cup (\frac{5\pi}{4}, 2\pi)$

concave down on $(\frac{\pi}{4}, \frac{5\pi}{4})$

critical points are $(\frac{3\pi}{4}, \sqrt{2})$, relative

maximum; $(\frac{7\pi}{4}, -\sqrt{2})$, relative minimum

points of inflection: $(\frac{\pi}{4}, 0)$, $(\frac{5\pi}{4}, 0)$

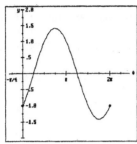

43. $f(x) = \sin^2 x - 2\sin x + 1$

$f'(x) = 2\sin x \cos x - 2\cos x$

$\quad = 2\cos x(\sin x - 1)$

critical number at $x = \frac{\pi}{2}$

$f''(x) = -4\sin^2 x + 2\sin x + 2$

$\quad = -2(\sin x - 1)(2\sin x + 1)$

critical number at $x = \frac{\pi}{2}$

no asymptotes

graph rising on $(\frac{\pi}{2}, \pi)$

graph falling on $(0, \frac{\pi}{2})$

concave up on $(0, \pi)$

critical points are $(\frac{\pi}{2}, 0)$, relative minimum;

$(0, 1)$, and $(\pi, 1)$ relative maxima;

no points of inflection

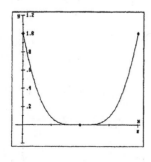

45. $v = \sqrt{gr}\,\tan^{1/2}\theta$

$v' = \frac{\sqrt{gr}}{2}\left(\frac{\sec^2\theta}{\sqrt{\tan\theta}}\right)$

This is not defined at $\theta = 0$ and $\frac{\pi}{2}$

$v'' = \frac{\sqrt{gr}}{2}\left(\frac{\sqrt{\tan\theta}(2)\sec^2\theta\,\tan\theta}{\tan\theta}\right)$

$\quad - \frac{\sqrt{gr}}{2}\left(\frac{\sec^2\theta(\frac{1}{2})\tan^{-1/2}\theta\,\sec^2\theta}{\tan\theta}\right)$

$\quad = \frac{\sqrt{gr}}{4}\left[\frac{\sec^2\theta(4\tan^2\theta - \sec^2\theta)}{(\tan\theta)^{3/2}}\right]$

In terms of sines and cosines,

$v'' = \frac{\sqrt{gr}}{4}\left[\frac{4\sin^2\theta - 1}{\sin^{3/2}\theta\,\cos^{5/2}\theta}\right]$

$v'' = 0$ if $\quad 4\sin^2\theta - 1 = 0$

$$\sin\theta = \pm\frac{1}{2}$$

On $[0, \frac{\pi}{2}]$ the solution is $\theta = \frac{\pi}{6}$

There is an asymptote at $x = \frac{\pi}{2}$

Point of inflection at $(0.52, 0.76\sqrt{gr})$

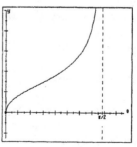

47. Answers may vary.

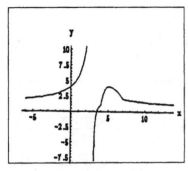

49. Frank is not correct.

$$\lim_{x \to 0^+} \left(\frac{1}{x^2} - \frac{1}{x} \right) = \lim_{x \to 0^+} \frac{1 - x}{x^2}$$

$$= +\infty$$

51. a. $+\infty$ **b.** $+\infty$ **c.** $-\infty$ **d.** $-\infty$ **e.** $+\infty$

53. In order to have a vertical asymptote, it is necessary that $3 - 5b = 0$, or $b = 3/5$. Thus,

$$\lim_{x \to +\infty} \frac{ax + 5}{3 - \frac{3x}{5}} = \lim_{x \to +\infty} \frac{5ax + 25}{15 - 3x}$$

$$= -\frac{5a}{3}$$

Since this limit is to be equal to -3, we see $a = 9/5$.

55. $x^2 = 1 + \frac{2}{3} y^2$ implies $y = \pm \frac{1}{2} \sqrt{6(x^2 - 1)}$

but since x and y are nonnegative, we have

$y = f(x) = \frac{1}{2} \sqrt{6(x^2 - 1)}$.

$\lim_{x \to +\infty} \frac{1}{2} \sqrt{6(x^2 - 1)} = +\infty$

57. True; f and g are concave up by hypotheses

(so $f'' > 0$ and $g'' > 0$). Let

$F(x) = f(x) + g(x)$; then

$$F''(x) = f''(x) + g''(x) > 0$$

so F is concave up.

59. False; find a counterexample; for example, let

$f(x) = x^2 - 4$; show that this is concave up

on $[-2, 2]$ but $f^2(x)$ is concave down on part

of this interval.

61. a. If $m > n$

$$\lim_{x \to +\infty} \frac{a_n x^n + \cdots + a_1 x + a_0}{b_m x^m + \cdots + b_1 x + b_0}$$

$$= \lim_{x \to +\infty} \frac{a_n + \cdots + a_1 x^{1 - n} + a_0 x^{-n}}{b_m x^{m - n} + \cdots + b_1 x^{1 - n} + b_0 x^{-n}}$$

$$= 0$$

b. If $m = n$, $\lim_{x \to +\infty} f(x) = \frac{a_n}{b_m}$

c. If $m < n$, $\lim_{x \to +\infty} f(x)$ will not have a

finite limit L.

63. By hypotheses, $\lim_{x \to +\infty} f(x) = L$ and

$\lim_{x \to +\infty} g(x) = M$. Thus for each $\epsilon > 0$ there

exists $N_1 > 0$ and $N_2 > 0$ such that

$$\left| f(x) - L \right| < \frac{\epsilon}{2} \text{ and } \left| g(x) - M \right| < \frac{\epsilon}{2}$$

Then,

$$\left| [f(x) + g(x)] - (L + M) \right|$$

$$= \left| [f(x) - L] + [g(x) - M] \right|$$

$$\leq \left| f(x) - L \right| + \left| g(x) - M \right|$$

$< \frac{\epsilon}{2} + \frac{\epsilon}{2} = \epsilon$ for $x > N$ where $N = \max(N_1, N_2)$

This means that

$$\lim_{x \to +\infty}[f(x) + g(x)] = M + L$$

$$= \lim_{x \to +\infty} f(x) + \lim_{x \to +\infty} g(x)$$

4.5 l'Hôpital's Rule, Pages 236-237

SURVIVAL HINT: *Do not attempt to apply l'Hôpital's rule unless you have verified that the limit has the form 0/0 or ∞/∞. It is good practice to indicate which form you have and then evaluate the limit.*

1. **a.** The limit is not an indeterminate form. In fact,

$$\lim_{x \to \pi} \frac{1 - \cos x}{x} = \frac{1 - (-1)}{\pi} = \frac{2}{\pi}$$

    b. The limit is not an indeterminate form.

$$\lim_{x \to \pi/2} \frac{\sin x}{x} = \frac{1}{\pi/2} = \frac{2}{\pi}$$

3. $\lim_{x \to 1} \frac{x^3 - 1}{x^2 - 1} = \lim_{x \to 1} \frac{3x^2}{2x}$ *Form* $\frac{0}{0}$

$$= \frac{3}{2}$$

5. $\lim_{x \to 1} \frac{x^{10} - 1}{x - 1} = \lim_{x \to 1} \frac{10x^9}{1}$ *Form* $\frac{0}{0}$

$$= 10$$

7. $\lim_{x \to 0} \frac{1 - \cos^2 x}{\sin^3 x}$ *Form* $\frac{0}{0}$

$$= \lim_{x \to 0} \frac{-2\cos x(-\sin x)}{3\sin^2 x \cos x}$$

$$= \frac{2}{3} \lim_{x \to 0} \frac{1}{\sin x} = \infty$$

9. $\lim_{x \to \pi} \frac{\cos \frac{x}{2}}{\pi - x}$ *Form* $\frac{0}{0}$

$$= \lim_{x \to \pi} \frac{-\frac{1}{2}\sin \frac{x}{2}}{-1}$$

$$= \frac{1}{2}$$

11. The limit is not an indeterminate form.

$$\lim_{x \to 0} \frac{\sin ax}{\cos bx} = \frac{\sin 0}{\cos 0}$$

$$= 0$$

13. $\lim_{x \to 0} \frac{x - \sin x}{\tan x - x}$ *Form* $\frac{0}{0}$

$$= \lim_{x \to 0} \frac{1 - \cos x}{\sec^2 x - 1}$$

$$= \lim_{x \to 0} \frac{\sin x}{2\sec^2 x \tan x}$$

$$= \lim_{x \to 0} \frac{1}{2\sec^3 x}$$

$$= \frac{1}{2}$$

15. $\lim_{x \to \pi/2} \frac{3\sec x}{2 + \tan x}$ *Form* $\frac{\infty}{\infty}$

$$= 3 \lim_{x \to \pi/2} \frac{\sec x \tan x}{\sec^2 x}$$

$$= 3 \lim_{x \to \pi/2} \sin x$$

$$= 3$$

17. $\lim_{x \to 0} \frac{\sin 3x \sin 2x}{x \sin 4x}$ *Form* $\frac{0}{0}$

$$= \lim_{x \to 0} \frac{6\left(\frac{\sin 3x}{3x}\right)\left(\frac{\sin 2x}{2x}\right)}{4\left(\frac{\sin 4x}{4x}\right)}$$

$$= \frac{6(1)(1)}{4}$$

$$= \frac{3}{2}$$

19. $\displaystyle\lim_{x\to 0} \frac{x^2 + \sin x^2}{x^2 + x^3} = \lim_{x\to 0} \frac{2x + 2x\,\cos x^2}{2x + 3x^2}$

$$= \lim_{x\to 0} \frac{2 + 2\cos x^2}{2 + 3x}$$

$$= 2$$

21. $\displaystyle\lim_{x\to +\infty} x^{3/2} \sin \frac{1}{x};$ Let $u = \frac{1}{x};$

$\displaystyle\lim_{u\to 0^+} \frac{\sin u}{u} \cdot u^{-1/2} = \lim_{u\to 0^+} \frac{\sin u}{u} \lim_{u\to 0^+} \frac{1}{\sqrt{u}}$

$$= (1)(+\infty)$$

$$= +\infty$$

23. $\displaystyle\lim_{x\to 1} \frac{(x-1)\sin(x-1)}{1 - \cos(x-1)};$ Let $u = x - 1$

$\displaystyle\lim_{u\to 0} \frac{u \sin u}{1 - \cos u} = \lim_{u\to 0} \frac{u \cos u + \sin u}{\sin u}$

$$= \lim_{u\to 0} \frac{-u \sin u + 2 \cos u}{\cos u}$$

$$= 2$$

25. $\displaystyle\lim_{x\to 0} \frac{x + \sin(x^2 + x)}{3x + \sin x}$

$$= \lim_{x\to 0} \frac{1 + (2x+1)\cos(x^2 + x)}{3 + \cos x}$$

$$= \frac{1 + 1}{3 + 1}$$

$$= \frac{1}{2}$$

27. $\displaystyle\lim_{x\to 0} \left(\frac{1}{\sin 2x} - \frac{1}{2x}\right)$ is not in the proper form

to apply l'Hôpital's rule;

$\displaystyle\lim_{x\to 0} \frac{2x - \sin 2x}{2x \sin 2x} = \lim_{x\to 0} \frac{2 - 2\cos 2x}{2[2x \cos 2x + \sin 2x]}$

$$= \lim_{x\to 0} \frac{4 \sin 2x}{8 \cos 2x - 8x \sin 2x}$$

$$= 0$$

29. $\displaystyle\lim_{x\to 0^+} x^{-5} \ln x = -\infty$ from Theorem 4.10.

31. $\displaystyle\lim_{x\to +\infty} \frac{\ln(\ln x)}{x} = \lim_{x\to +\infty} \frac{(\ln x)^{-1} x^{-1}}{1}$

$$= \lim_{x\to +\infty} (x \ln x)^{-1}$$

$$= 0$$

33. Let $u = 2x$, so $x = \frac{u}{2}$

$\displaystyle\lim_{x\to +\infty}\left(1 + \frac{1}{2x}\right)^{3x} = \lim_{u\to +\infty}\left(1 + \frac{1}{u}\right)^{3(u/2)}$

$$= \lim_{x\to +\infty}\left[\left(1 + \frac{1}{u}\right)^u\right]^{3/2}$$

$$= e^{3/2}$$

35. Let $L = \displaystyle\lim_{x\to 0^+} (e^x + x)^{1/x};$

$\ln L = \displaystyle\lim_{x\to 0^+} \ln(e^x + x)^{1/x}$

$$= \lim_{x\to 0^+} \frac{1}{x} \ln (e^x + x)$$

$$= \lim_{x\to 0^+} \frac{e^x + 1}{e^x + x}$$

$$= 2$$

Thus, $L = e^2$.

37. $\lim\limits_{x \to (\pi/2)^-}\left(\dfrac{1}{\pi - 2x} + \tan x\right)$ Let $u = \pi - 2x$,

then $\lim\limits_{u \to 0^+}\left[\dfrac{1}{u} + \tan\left(\dfrac{\pi - u}{2}\right)\right]$

$= \lim\limits_{u \to 0^+}\left[\dfrac{1}{u} + \cot\left(\dfrac{u}{2}\right)\right]$

$= +\infty$

39. $\lim\limits_{x \to +\infty}[x - \ln(x^3 - 1)]$

$= \lim\limits_{x \to +\infty}[\ln e^x - \ln(x^3 - 1)]$

$= \lim\limits_{x \to +\infty}\ln\left(\dfrac{e^x}{x^3 - 1}\right)$

$= \ln\lim\limits_{x \to +\infty}\left(\dfrac{e^x}{x^3 - 1}\right)$ $Form\ \dfrac{\infty}{\infty}$

$= \ln\lim\limits_{x \to +\infty}\left(\dfrac{e^x}{3x^2}\right)$

$= \ln\lim\limits_{x \to +\infty}\left(\dfrac{e^x}{6x}\right)$

$= \ln\lim\limits_{x \to +\infty}\left(\dfrac{e^x}{6}\right)$

$= +\infty$

41. $\lim\limits_{x \to 0^+}\left(\dfrac{1}{x^2} - \ln\sqrt{x}\right)$ is not the proper form for l'Hôpital's rule.

$\lim\limits_{x \to 0^+}\left(\dfrac{1}{x^2} - \ln\sqrt{x}\right) = \lim\limits_{x \to 0^+}\left(\dfrac{1 - x^2\ln\sqrt{x}}{x^2}\right)$

This still is not the proper form, so l'Hôpital's rule does not apply. Since

$\lim\limits_{x \to 0^+}\dfrac{1}{x^2} = \infty$ and $\lim\limits_{x \to 0^+}(-\ln\sqrt{x}) = \infty$

We conclude $\lim\limits_{x \to 0^+}\left(\dfrac{1}{x^2} - \ln\sqrt{x}\right) = +\infty$.

43. $\lim\limits_{x \to 0^+}(\ln x)(\cot x) = -\infty$

45. $\lim\limits_{x \to +\infty}\dfrac{x + \sin 3x}{x} = \lim\limits_{x \to +\infty}\left(1 + \dfrac{\sin 3x}{x}\right)$

$= \lim\limits_{x \to +\infty}1 + \lim\limits_{x \to +\infty}\dfrac{\sin 3x}{x}$

$= 1 + 0$

$= 1$

47. $\lim\limits_{x \to 0^+}\left(\dfrac{2\cos x}{\sin 2x} - \dfrac{1}{x}\right)$

$= \lim\limits_{x \to 0^+}\left(\dfrac{2\cos x}{2\sin x\cos x} - \dfrac{1}{x}\right)$

$= \lim\limits_{x \to 0^+}\left(\dfrac{1}{\sin x} - \dfrac{1}{x}\right)$

$= \lim\limits_{x \to 0^+}\dfrac{x - \sin x}{x\sin x}$ $Form\ \dfrac{0}{0}$

$= \lim\limits_{x \to 0^+}\dfrac{1 - \cos x}{\sin x + x\cos x}$ $Form\ \dfrac{0}{0}$

$= \lim\limits_{x \to 0^+}\dfrac{\sin x}{\cos x + \cos x - x\sin x} = 0$

49. l'Hôpital's rule does not apply.

$\lim\limits_{x \to 0^+}\dfrac{(2 - x)(e^x - x - 2)}{x^3} = -\infty$

$\lim\limits_{x \to 0^-}\dfrac{(2 - x)(e^x - x - 2)}{x^3} = +\infty$

Limit does not exist.

51. Let $u = \dfrac{1}{x}$

$\lim\limits_{x \to +\infty}x^5\left[\sin\left(\dfrac{1}{x}\right) - \dfrac{1}{x} + \dfrac{1}{6x^3}\right]$

$= \lim\limits_{u \to 0}\dfrac{\sin u - u + \frac{1}{6}u^3}{u^5}$ $Form\ \dfrac{0}{0}$

$= \lim\limits_{u \to 0}\dfrac{\cos u - 1 + \frac{1}{2}u^2}{5u^4}$ $Form\ \dfrac{0}{0}$

$$= \lim_{u \to 0} \frac{-\sin u + u}{20u^3} \qquad Form \ \frac{0}{0}$$

$$= \lim_{u \to 0} \frac{-\cos u + 1}{60u^2} \qquad Form \ \frac{0}{0}$$

$$= \lim_{u \to 0} \frac{\sin u}{120u}$$

$$= \frac{1}{120}$$

53. $\lim_{x \to -\infty} f(x) = -\infty$

$$\lim_{x \to +\infty} f(x) = 0$$

Horizontal asymptote, $y = 0$.

55. $\lim_{x \to +\infty} f(x) = L$

$$\ln L = \lim_{x \to \infty} \frac{2\ln(\ln \sqrt{x})}{x}$$

$$= \lim_{x \to \infty} \frac{\frac{2}{\ln \sqrt{x}}\left(\frac{1}{2x}\right)}{1}$$

$$= 0$$

$L = e^0 = 1$; horizontal asymptote, $y = 1$.

61. a.

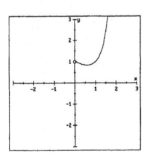

It looks like the limit is 1

b. Let $L = \lim_{x \to 0^+} x^{x^x}$, so

$$\ln L = \lim_{x \to 0^+} \ln x^{x^x}$$

$$= \lim_{x \to 0^+} x \ln(x^x) \qquad \text{Let } u = \frac{1}{x}$$

$$= \lim_{u \to +\infty} \frac{\ln(\frac{1}{u})^{1/u}}{u}$$

$$= \lim_{u \to +\infty} \frac{\frac{1}{u}\ln(\frac{1}{u})}{u}$$

$$= \lim_{u \to +\infty} \frac{\ln(\frac{1}{u})}{u^2} \qquad Form \ \frac{\infty}{\infty}$$

$$= \lim_{u \to +\infty} \frac{-\frac{1}{u}}{2u}$$

$$= \lim_{u \to +\infty} -\frac{1}{2u^2} = 0$$

Thus, $L = e^0 = 1$.

c. Answers vary. Some calculators will interpret the expression x^{x^x} as $x^{\left(x^x\right)}$ to obtain the answer 0.

63. $\lim_{x \to 0}\left(\frac{\sin 2x}{x^3} + \frac{a}{x^2} + b\right)$

$$= \lim_{x \to 0} \frac{\sin 2x + ax + bx^3}{x^3}$$

$$= \lim_{x \to 0} \frac{2\cos 2x + a + 3bx^2}{3x^2}$$

$$= \frac{0}{0} \text{ if } a = -2;$$

$$\lim_{x \to 0} \frac{-4\sin 2x + 6bx}{6x}$$

$$= \lim_{x \to 0} \frac{-8\cos 2x + 6b}{6}$$

$$= \frac{-8 + 6b}{6} = 1 \text{ if } -8 + 6b = 6 \text{ or}$$

$$b = \frac{7}{3}$$

65. Note that $(x^4 + 5x^3 + 3)^C$ behaves like $(x^4)^C$ when $x \to +\infty$; thus,

$$\lim_{x \to +\infty}[(x^4 + 5x^3 + 3)^C - x] = +\infty \text{ if } C > \tfrac{1}{4}$$

and $-\infty$ if $C < \tfrac{1}{4}$. Evaluate

$$\lim_{x \to +\infty}[(x^4 + 5x^3 + 3)^{1/4} - x] \quad \text{Let } u = \tfrac{1}{x}$$

$$= \lim_{u \to 0^+}\left[\left(\frac{1}{u^4} + \frac{5}{u^3} + 3\right)^{1/4} - \frac{1}{u}\right]$$

$$= \lim_{u \to 0^+}\frac{(1 + 5u + 3u^4)^{1/4} - 1}{u}$$

$$= \lim_{u \to 0^+}\frac{\tfrac{1}{4}(1 + 5u + 3u^4)^{-3/4}(5 + 12u^3)}{1}$$

$$= \frac{5}{4}$$

67. *Note*: treat β as variable and α as fixed.

$$\lim_{\beta \to \alpha}\left[\frac{C}{\beta^2 - \alpha^2}(\sin \alpha t - \sin \beta t)\right]$$

$$= C\lim_{\beta \to \alpha}\frac{\sin \alpha t - \sin \beta t}{\beta^2 - \alpha^2} \quad \text{Form } \tfrac{0}{0}$$

$$= C\lim_{\beta \to \alpha}\frac{0 - t\cos \beta t}{2\beta} \quad \text{l'Hôpital's rule}$$

$$= \frac{-Ct}{2\alpha}\cos \alpha t$$

4.6 Optimization in the Physical Sciences and Engineering, Pages 245–250

SURVIVAL HINT: *If the hypotheses of the extreme value theorem apply to a particular problem, you can sometimes avoid testing a candidate to see if it is a maximum or a minimum. Suppose you have three candidates and two are minima — then the third* **must** *be a maximum.*

1. (1) Draw a figure (if appropriate) and label all quantities relevant to the problem.

(2) Find a formula for the quantity to be maximized or minimized.

(3) Use conditions in the problem to eliminate variables in order to express the quantity to be maximized or minimized in terms of a single variable.

(4) Find the practical domain for the variables in Step 3; that is, the interval of possible values determined from the physical restrictions in the problem.

(5) If possible, use the methods of calculus to obtain the required optimum value.

3. Let x and y be the dimensions of the rectangular plot. The fencing (perimeter) is $P = 2(x + y)$ and $A = xy = 64$; domain, $x > 0$. Minimize P, so write

$$P = 2\left(x + \frac{64}{x}\right)$$

$$P' = 2(1 - 64x^{-2})$$

$P' = 0$ if $x = 8$ and $y = 8$. Since $P''(8) > 0$, $x = 8$ is a minimum. The dimensions of the garden should be 8 ft by 8 ft.

5. The size of the sheet (measured to the nearest centimeter) is 28 cm by 22 cm. Let x be the side of the square that is cut out $(0 \le x \le 11 \text{ cm})$. We want to maximize the volume, V,

$$V = \ell w h$$

$$= (28 - 2x)(22 - 2x)x$$

$$= 4x^3 - 100x^2 + 616x$$

$$V' = 12x^2 - 200x + 616$$

$V' = 0$ when

$$x = \frac{200 \pm \sqrt{(-200)^2 - 4(12)(616)}}{2(12)}$$

≈ 12.59 (not in domain), 4.08

$$V_{max} \approx (28 - 2 \cdot 4.08)(22 - 2 \cdot 4.08)(4.08)$$

$$\approx 1{,}120 \text{ cm}^3$$

Since a liter is $1{,}000$ cm^3, we see that the maximum volume is about 1 liter.

7. Let $2x$ and y be the lengths of the sides of the rectangle and R be the radius of the semicircle. Then $y = \sqrt{R^2 - x^2}$ for $0 \leq x \leq R$. Then

$$A = 2xy = 2x\sqrt{R^2 - x^2}$$

$$A' = 2(R^2 - x^2)^{1/2} + 2x(\tfrac{1}{2})(R^2 - x^2)^{-1/2}(-2x)$$

$$= \frac{2(-2x^2 + R^2)}{\sqrt{R^2 - x^2}}$$

$A' = 0$ if $x_c = \dfrac{R}{\sqrt{2}}$ and $y_c = \sqrt{R^2 - \dfrac{R^2}{2}} = \dfrac{R}{\sqrt{2}}$

$A' > 0$ for $x < x_c$ and $A' < 0$ for $x > x_c$. $A(0) = A(R) = 0$. The largest rectangle has sides of length $R/\sqrt{2}$ and $\sqrt{2}\,R$.

9. Let R be the radius of the sphere and r the radius of the cylinder with h so that $0 \leq h \leq R$. Since $r^2 = R^2 - h^2$,

$$V(h) = 2\pi r^2 h$$

$$= 2\pi h(R^2 - h^2)$$

$$= 2\pi(R^2 h - h^3)$$

$$V'(h) = 2\pi(R^2 - 3h^2)$$

$V'(h) = 0$ when $h = R/\sqrt{3}$. Since $h = 0$ and $h = R$ give minima, we see by the extreme value theorem that $h = R/\sqrt{3}$ must be a maximum. The dimensions are $h = \frac{R}{3}\sqrt{3}$ and $r = \frac{R}{3}\sqrt{6}$.

11. A vertex of the inner square subdivides a side of the outer square into line segments x and y ($0 \leq x \leq L$). The given outer square has side $x + y$. Thus, $x^2 + y^2 = L^2$. The area of the outer square is

$$A(x) = (x + y)^2 = (x + \sqrt{L^2 - x^2})^2$$

$$A'(x) = 2(x + \sqrt{L^2 - x^2})\left(1 + \frac{-2x}{2\sqrt{L^2 - x^2}}\right)$$

$A'(x) = 0$ if $x = \sqrt{L^2 - x^2}$ or $x_c = \dfrac{L}{\sqrt{2}} = y_c$

Since $x = 0$ or $x = L$ give minima, we see that the other critical value must be a maximum. Thus, a circumscribed square of side

$$s = \frac{L}{\sqrt{2}} + \frac{L}{\sqrt{2}} = \sqrt{2}\,L$$

yields a maximum area.

13. Draw a figure with the car at the origin of a Cartesian coordinate system and the truck at $(250, 0)$. At time t, the truck is at position

$(250 - x, 0)$, while the car is at $(0, y)$. Let D be the distance that separates them.

$$\frac{dx}{dt} = 60 \text{ and } \frac{dy}{dt} = 80,$$

so that $x = 60t$ and $y = 80t$.

$$D^2 = (250 - x)^2 + y^2$$

$$D^2 = (250 - 60t)^2 + (80t)^2$$

$$= 2,500(4t^2 - 12t + 25)$$

$$(D^2)' = 10,000(2t - 3)$$

The derivative of the distance squared is 0 when $t = 1.5$ hr. Substituting this into the equation for D^2 produces the shortest distance: $x = 60(1.5) = 90$, $y = 80(1.5) = 120$

$$D^2 = (250 - 90)^2 + 120^2 = 1,600(25)$$

$$D = 40(5) = 200$$

Since there is no maximum distance, we see the minimum distance is 200 mi.

15. Let x and y denote the dimensions of a rectangle with fixed area A and perimeter P. $A = xy$ or $y = A/x$. The perimeter is to be minimized:

$$P = 2x + 2y = 2x + \frac{2A}{x}$$

$$P' = 2 - \frac{2A}{x^2}$$

$P' = 0$ when $x = \pm\sqrt{A}$. For $x > 0$ and with $P'' = 4A/x^3 > 0$, the minimal perimeter is attained when $x = \sqrt{A}$ and $y = A/x = \sqrt{A}$. This rectangle is a square.

17. Let x be the dimension of one side of the square base and y the length. The volume is $V = x^2y$. The cross sectional perimeter plus the length is $4x + y = 108$ (maximum). Thus, $y = 4(27 - x)$ and

$$V = 4x^2(27 - x)$$

$$= 4(27x^2 - x^3)$$

$$V' = 4(54x - 3x^2)$$

$$= 12x(18 - x)$$

$V' = 0$ when $x = 18$ and $y = 4(27 - 18) = 36$. $V''(18) < 0$, so $x = 18$ is a maximum. The largest possible volume is

$$V = 18^2(36) = 11,664 \text{ in.}^3 \text{ or } 6.75 \text{ ft}^3$$

19. Refer to the figure; Missy rows to a point P which is x mi from point B. The rowing distance is $H_r = \sqrt{36 + x^2}$ and the corresponding time is $t_r = \frac{\sqrt{36 + x^2}}{6}$. The time taken to walk the $H_w = s - x$ distance is $t_w = \frac{s - x}{10}$. The total time, t, is

$$t = \frac{\sqrt{36 + x^2}}{6} + \frac{s - x}{10}$$

$$t' = \frac{2x}{12\sqrt{36 + x^2}} - \frac{1}{10}$$

$t' = 0$ when $x = 4.5$ km.

a. When $s = 4$, Missy should row all the way.

b. When $s = 6$, she should land at a point 4.5 km from point B and run the rest of the way (1.5 km).

21. Draw a figure. Let x and y be the dimensions of the rectangular printed matter. Then, $xy = 108$ or $y = 108/x$. The area of the poster to be minimized is:

$$A = (x + 4)(y + 12)$$

$$= xy + 4y + 12x + 48$$

$$= 108 + 432x^{-1} + 12x + 48$$

$$A' = 12 - 432x^{-2}$$

$A' = 0$ when $x^2 = 36$ or $x = 6$. Then, $y = 108/6 = 18$. The cost of the poster (in cents) is

$$C(6) = 20(6 + 4)(18 + 12) = 6,000$$

cents or $60.00.

23. We once again meet the Spy! Draw a figure; pick a point x km down the road (toward the power plant) from the nearest point on the paved road and have the jeep head toward that point. The distance traveled by the jeep on the sand is $\sqrt{x^2 + 32^2}$ and the distance on the paved road is $16 - x$ (assuming $x \le 16$). The total time traveled (in hours) is

$$t = \frac{\sqrt{32^2 + x^2}}{48} + \frac{16 - x}{80}$$

$$t' = \frac{1}{48}\left(\frac{1}{2}\right)(32^2 + x^2)^{-1/2}(2x) - \frac{1}{80}$$

$t' = 0$ when

$$\frac{1}{48}\left(\frac{1}{2}\right)(32^2 + x^2)^{-1/2}(2x) - \frac{1}{80} = 0$$

$$\frac{x}{48(32^2 + x^2)^{1/2}} = \frac{1}{80}$$

$$80x = 48(32^2 + x^2)^{1/2}$$

$$5x = 3(32^2 + x^2)^{1/2}$$

$$25x^2 = 9(32^2 + x^2)$$

$$16x^2 = 9 \cdot 32^2$$

$$x = \frac{3 \cdot 32}{4} = 24$$

This distance is further than the power plant. The minimum time corresponds to heading for the power plant on the sand. The time is

$$t = \frac{\sqrt{32^2 + 16^2}}{48} + 0 = \frac{\sqrt{5}}{3} \approx 0.745$$

This is 44 minutes 43 seconds; so he has about 5 minutes 17 seconds to defuse the bomb.

25. The amount of material is $M = 2(\pi r^2) + 2\pi rh$ and the volume is $V = \pi r^2 h = 355$. Solving for h we obtain $h = (355/\pi)r^{-2}$ so that

$$M = 2\pi r^2 + 2\pi r\left(\frac{355}{\pi}r^{-2}\right)$$

$$= 2\pi r^2 + 710r^{-1}$$

$$M' = 4\pi r - 710r^{-2}$$

$M' = 0$ when $r \approx 3.84$ and $h \approx 7.67$ cm.

The actual dimensions of a Coke can are $h \approx 12$ cm (larger than necessary) and

$r \approx 3.5$ cm (a bit smaller than necessary). They differ from the optimal dimensions for historical and marketing reasons.

27. **a.** Using similar triangles, $\triangle BAO$ and $\triangle ORS$, since the vertical angles are equal,

$$\frac{|RS|}{|AB|} = \frac{q}{p}$$

Also, $\triangle COF$ and $\triangle FRS$ are similar since the vertical angles are equal,

$$\frac{|RS|}{|OC|} = \frac{q - f}{f}$$

Since $|AB| = |OC|$,

$$\frac{q}{p} = \frac{q - f}{f}$$

$$fq = pq - pf$$

$$fq + pf = pq$$

$$\frac{fq}{fpq} + \frac{pf}{fpq} = \frac{pq}{fpq}$$

$$\frac{1}{p} + \frac{1}{q} = \frac{1}{f}$$

b. $q = 24 - p$ so that

$$\frac{p(24 - p)}{24} = f$$

$$\frac{df}{dp} = \frac{24 - 2p}{24}$$

$df/dp = 0$ if $p = 12$.

When $p = 12$, the largest value of f is 6.

29. $E' = -Mg + \dfrac{2mgx}{\sqrt{x^2 + d^2}} = 0$ when

$$Mg\sqrt{x^2 + d^2} = 2mgx,$$

$$\sqrt{x^2 + d^2} = \frac{2mx}{M}$$

$$x^2 + d^2 = \frac{4m^2 x^2}{M^2}$$

$$\left(\frac{4m^2}{M^2} - 1\right)x^2 = d^2$$

$$x^2 = \frac{d^2 M^2}{4m^2 - M^2}$$

$$x = \frac{Md}{\sqrt{4m^2 - M^2}}.$$

31. $T(x) = N(k + \frac{c}{x})p^{-x}$

a. We use logarithmic differentiation.

$$\ln T(x) = \ln N(k + \tfrac{c}{x}) - x \ln p$$

$$\frac{1}{T}T'(x) = \frac{-cx^{-2}}{k + c/x} - \ln p$$

$$T'(x) = -T\left[\frac{c}{(kx + c)x} + \ln p\right]$$

Note: if you use technology to find this derivative, you may find a different form. For example, you might obtain

$$T'(x) = \frac{-Np^{-x}[(kx^2 + cx)\ln p + c]}{x^2}$$

You can show these forms are equivalent.

b. Solve

$$-T\left[\frac{c}{(kx + c)x} + \ln p\right] = 0$$

$$(kx^2 + cx)\ln p + c = 0$$

$$x = \frac{-c\ln p + \sqrt{c^2(\ln p)^2 - 4kc\ln p}}{2k\ln p}$$

(Reject the negative root since x must be positive.)

c.

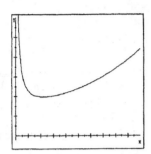

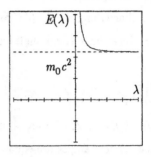

33. **a.** Hold m fixed, and maximize y as a function of x.

$$y = -16(1 + m^2)(x/v)^2 + mx$$

$$y' = \frac{-32x(m^2 + 1)}{v^2} + m$$

The maximum height occurs when

$$y' = 0 \text{ or when } x = \frac{mv^2}{32(m^2 + 1)}$$

b. Hold x fixed at x_0, so

$$y = -16(1 + m^2)\left(\frac{x_0}{v}\right)^2 + mx_0$$

$$\frac{dy}{dm} = -16(2m)\left(\frac{x_0}{v}\right)^2 + x_0$$

$$y' = \frac{x_0(v^2 - 32mx_0)}{v^2}$$

The maximum height occurs when

$$y' = 0 \text{ or when } m = v^2/(32x_0).$$

35. $E(\lambda) = \sqrt{\left(\frac{hc}{\lambda}\right)^2 + m_0{}^2 c^4}$

$$E'(\lambda) = \frac{h^2}{\lambda}\left|\frac{c}{\lambda}\right|(c^2\lambda^2 m_0{}^2 + h^2)^{-1/2} \neq 0,$$

so there are no critical values.

As $\lambda \to +\infty$, $E(\lambda) \to m_0 c^2$

37. Refer to Figure 4.69 in the text.

Note: $\cos \theta = \frac{x}{L}$ and

$$\cos(\pi - 2\theta) = (20 - x)/x \text{ then}$$

$$\cos 2\theta = (x - 20)/x; \text{ where } 0 \le x \le 15$$

$$\frac{x - 20}{x} = \cos 2\theta$$

$$= 2\cos^2\theta - 1$$

$$= 2\left(\frac{x}{L}\right)^2 - 1$$

$$L^2(x - 20) = 2x^3 - xL^2$$

$$L^2(x - 20) + xL^2 = 2x^3$$

$$L^2 = \frac{2x^3}{2x - 20}$$

$$L = \frac{x^{3/2}}{(x - 10)^{1/2}}$$

$$L' = \frac{(x-10)^{1/2}(\frac{3}{2})x^{1/2} - x^{3/2}(\frac{1}{2})(x-10)^{-1/2}}{x - 10}$$

$$= \frac{3(x - 10)x^{1/2} - x^{3/2}}{2(x - 10)^{3/2}}$$

$$= \frac{x^{1/2}(2x - 30)}{2(x - 10)^{3/2}}$$

$$L' = 0 \text{ if } x = 0 \text{ or if } x = 15$$

(undefined if $x = 10$)

$$L(15) = \sqrt{\frac{15^3}{5}} = 15\sqrt{3} \approx 26 \text{ cm}$$

By the first derivative test, we see this is a minimum.

39. $V(T) = 1 - 6.42 \times 10^{-5}T + 8.51 \times 10^{-6}T^2$
$$- 6.79 \times 10^{-8}T^3$$

$V'(T) = -6.42(10^{-5}) + 17.02(10^{-6})T$
$$- 20.37(10^{-8})T^2$$

$V'(T) = 0$ when

$$T = \frac{0.851(10^{-5}) \pm \sqrt{0.724(10^{-10}) - 0.1308(10^{-10})}}{2.037(10^{-7})}$$

$$\approx 4, 80$$

Reject 80 °C since the mathematical model is only valid around the freezing temperature.
$T = 4°$ is a minimum since V' is falling on $(0, 4)$ and rising on $(4, 80)$.

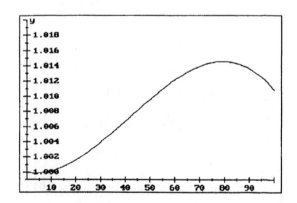

Liquid water and solid ice can coexist at 0 °C. The water below the surface is denser since the trapped temperature is a little higher than 0. As the temperature above the surface drops, the change in temperature is passed to the next higher level of water and transforms to ice. The minimum value occurs when $T \approx 4$ °C.

41. Refer to Figure 4.72 in the text. Let s be the side of the square base with $0 \leq s \leq 10\sqrt{2}$. Consider a plane that passes through the vertex of the pyramid (is perpendicular to the base) and is parallel to two sides in the base. The slanted sides of the pyramid intersect this plane in a length ℓ. If one-half of the pyramid is unfolded, we obtain half of our original sheet of paper. The hypotenuse is

$$20\sqrt{2} = 2\ell + s$$
$$\ell = \frac{20\sqrt{2} - s}{2}$$

Return to the pyramid and the vertical cutting plane. Half of this intersection consists of a right triangle with altitude h (the height of the pyramid), base $s/2$, and hypotenuse ℓ.

$$h = \sqrt{\ell^2 - \frac{s^2}{4}}$$
$$= \sqrt{\frac{(20\sqrt{2} - s)^2 - s^2}{4}}$$
$$= \frac{1}{2}\sqrt{800 - 40\sqrt{2}\,s}$$
$$= \sqrt{200 - 10\sqrt{2}\,s}$$

The volume of the pyramid is

$$V = \tfrac{1}{3}s^2 h = \tfrac{1}{3}s^2 \sqrt{200 - 10\sqrt{2}\, s}$$

with $0 \le s \le 10\sqrt{2}$.

$$V' = \frac{1}{3}\left[s^2 \frac{-10\sqrt{2}}{2\sqrt{200 - 10\sqrt{2}\, s}} + 2s\sqrt{200 - 10\sqrt{2}\, s} \right]$$

$$= \frac{1}{3}\left[\frac{-5\sqrt{2}\, s^2 + 2s(200 - 10\sqrt{2}\, s)}{\sqrt{200 - 10\sqrt{2}\, s}} \right]$$

$$= \frac{1}{3} \frac{-25\sqrt{2}\, s^2 + 400s}{\sqrt{200 - 10\sqrt{2}\, s}}$$

$V' = 0$ if

$$-25\sqrt{2}\, s^2 + 400s = 0$$

$$-25s(\sqrt{2}\, s - 16) = 0$$

$$s = 0,\ 8\sqrt{2}$$

If $s = 0$ we obtain a minimum, and if $s = 8\sqrt{2}$ we obtain a maximum (from the first derivative test).

$$V = \tfrac{1}{3}s^2 \sqrt{200 - 10\sqrt{2}\, s}$$

$$= \tfrac{128}{3}\sqrt{200 - 10\sqrt{2}(8\sqrt{2})}$$

$$= \tfrac{128}{3}\sqrt{200 - 160}$$

$$= \tfrac{256}{3}\sqrt{10}$$

$$\approx 270 \text{ cm}^3$$

43. **a.** At A, the incoming beam of light is deflected through $(\alpha - \beta)$; at B, the

deflection is $(\pi - 2\beta)$; and at C, it is again $(\alpha - \beta)$. Thus, the total deflection is

$$D = (\alpha - \beta) + (\pi - 2\beta) + (\alpha - \beta)$$

$$= \pi + 2\alpha - 4\beta$$

b. $\dfrac{dD}{d\alpha} = 2 - 4\dfrac{d\beta}{d\alpha}$

Differentiating $\sin \alpha = 1.33 \sin \beta$ implicitly with respect to α, we obtain

$$\cos \alpha = 1.33 \cos \beta \frac{d\beta}{d\alpha}, \text{ so}$$

$$D'(\alpha) = 2 - 4\left[\frac{\cos \alpha}{1.33 \cos \beta} \right]$$

c. Solving $D'(\alpha) = 0$, we obtain

$$2 - 4\left[\frac{\cos \alpha}{1.33 \cos \beta} \right] = 0$$

$$\cos \beta = 1.504 \cos \alpha$$

$$\cos^2 \beta = 2.261 \cos^2 \alpha$$

$$1 - \sin^2 \beta = 2.261(1 - \sin^2 \alpha)$$

and since $\sin \alpha = 1.33 \sin \beta$,

$$1 - \left(\frac{\sin \alpha}{1.33} \right)^2 = 2.261(1 - \sin^2 \alpha)$$

$$\sin \alpha = 0.8624$$

so the critical number is $\alpha_0 = 1.04$ (this is $59.6°$) and since $\sin \alpha_0 = 1.33 \sin \beta_0$, we have $\beta_0 = 0.7055$ (which is $40.4°$). To show this is a minimum, we find $D''(\alpha_0)$. First, note that since

$$\cos \alpha = 1.33 \cos \beta \, \frac{d\beta}{d\alpha},$$

we have

$$-\sin \alpha = -1.33 \left[\sin \beta \, \frac{d\beta}{d\alpha} - \cos \beta \, \frac{d^2\beta}{d\alpha^2} \right]$$

so

$$\frac{d^2\beta}{d\alpha^2} = \frac{\sin \beta \cos \alpha - \sin \alpha \cos \beta}{1.33 \cos^2 \beta}$$

$$= \frac{\sin(\beta - \alpha)}{1.33 \cos^2 \beta}$$

Since $D'(\beta) = 2 - 4 \dfrac{d\beta}{d\alpha}$

$$D''(\beta) = -4 \frac{d^2\beta}{d\alpha^2} = \frac{-4 \sin(\beta - \alpha)}{1.33 \cos^2 \beta}$$

Thus,

$$D''(\alpha_0) = \frac{-4 \sin(\beta_0 - \alpha_0)}{1.33 \cos^2 \beta_0} \geq 0$$

since $0 \leq \alpha_0 - \beta_0 \leq \frac{\pi}{2}$ and since α_0 is the only critical number between 0 and $\frac{\pi}{2}$, it must correspond to the minimum value of $D(\alpha)$ on the interval.

Finally, we compute the rainbow angle. The minimum deflection is

$$D(\alpha_0) = \pi + 2\alpha_0 - 4\beta_0$$

$$= \pi + 2(1.04) - 4(0.7055)$$

$$\approx 2.39959$$

so the rainbow angle is

$$\theta = \pi - D(\alpha_0) \approx 0.742$$

(or about 42.5°).

4.7 Optimization in Business, Economics, and the Life Sciences, Pages 259-263

1. The profit is maximized when the marginal revenue equals the marginal cost.

$$C(x) = \tfrac{1}{8}x^2 + 5x + 98 \text{ and } p(x) = \tfrac{1}{2}(75 - x)$$

$$C'(x) = \tfrac{1}{8}(2x) + 5$$

$$= \tfrac{1}{4}x + 5$$

$$R(x) = xp$$

$$= \tfrac{1}{2}x(75 - x)$$

$$= \tfrac{1}{2}(75x - x^2)$$

$$R'(x) = \tfrac{1}{2}(75 - 2x)$$

$$R'(x) = C'(x)$$

$$\tfrac{1}{2}(75 - 2x) = \tfrac{1}{4}x + 5$$

$$150 - 4x = x + 20$$

$$5x = 130$$

$$x = 26$$

3. The profit is maximized when the marginal revenue equals the marginal cost.

$$C(x) = \tfrac{1}{5}(x + 30) \text{ and } p(x) = \frac{70 - x}{x + 30}$$

$$C'(x) = \tfrac{1}{5}$$

$$R(x) = xp$$

$$= \frac{70x - x^2}{x + 30}$$

$$R'(x) = \frac{(x+30)(70-2x)-(70x-x^2)}{(x+30)^2}$$

$$R'(x) = C'(x)$$

$$\frac{(x+30)(70-2x)-(70x-x^2)}{(x+30)^2} = \frac{1}{5}$$

$$(x+30)(70-2x)-(70x-x^2) = \tfrac{1}{5}(x+30)^2$$

$$5(-x^2-60x+2{,}100) = x^2+60x+900$$

$$6x^2+360x-9{,}600 = 0$$

$$x^2+60x-1{,}600 = 0$$

$$(x+80)(x-20) = 0$$

$x = 20$ (-80 is not in the domain). To maximize profit produce 20 items.

5. $C(x) = 3x^2 + x + 48$

$$A(x) = \frac{C(x)}{x} = 3x + 1 + 48x^{-1}$$

$$A'(x) = 3 - 48x^{-2} = 0 \text{ at } x = 4$$

$A''(4) > 0$, so $x = 4$ determines the minimum average cost of $A(4) = 12 + 1 + 12 = 25$.

7. a. Total cost, $C(x)$, is the average cost, $A(x) = 5 + \frac{x}{50}$ times the number of items produced, x.

$$C(x) = x\left(5 + \frac{x}{50}\right)$$

$$= 5x + \frac{x^2}{50}$$

$$R(x) = x\left(\frac{380-x}{20}\right)$$

$$P(x) = R(x) - C(x)$$

$$= \frac{380x-x^2}{20} - \frac{250x+x^2}{50}$$

$$= \frac{-7x^2+1{,}400x}{100}$$

$$= -0.07x^2 + 14x$$

b. The maximum profit occurs when

$$R'(x) = C'(x).$$

$$19 - \frac{x}{10} = 5 + \frac{x}{25}$$

$$14 = \frac{7}{50}x$$

$$x = 100$$

This gives a price per item,

$$p = \frac{380-x}{20} = \$14 \text{ per item.}$$

The maximum profit is

$$P(x) = \frac{(-7x+1{,}400)x}{100}$$

$$= \frac{(-700+1{,}400)100}{100}$$

$$= \$700$$

9. Number sold is $s(x) = 1{,}000e^{-0.1x}$ where x is the price per pair; the cost per pair is $50, so the profit is

$$P(x) = R(x) - C(x)$$

$$= x(1{,}000e^{-0.1x}) - 50(1{,}000e^{-0.1x})$$

$$= 1{,}000e^{-0.1x}(x - 50)$$

$P'(x)$

$= 1,000[e^{-0.1x} + e^{-0.1x}(-0.1)(x - 50)]$

$= 1,000e^{-0.1x}(6 - 0.1x)$

$P'(x) = 0$ when $x = 60$. That is, profit is maximized when $x = 60$.

11. **a.** $P(t) = 50e^{0.02t}$;

$P'(t) = e^{0.02t}$;

$P'(10) = e^{0.2}$

≈ 1.22 million people/yr

b. The percentage rate is $\dfrac{100e^{0.02t}}{50e^{0.02t}} = 2$; that is, 2%.

13. We want to find the largest value of the population rate $R(t) = p'(t)$.

$p(t) = 160(1 + 8e^{-0.01t})^{-1}$

$p'(t) = -160(1+8e^{-0.01t})^{-2}(-0.01)(8e^{-0.01t})$

$= 12.8e^{-0.01t}(1 + 8e^{-0.01t})^{-2}$

Using the product rule, we find that

$R'(t) = p''(t)$

$= 12.8(-0.01)e^{-0.01t}(1 + 8e^{-0.01t})^{-2}$

$+(-2)(12.8e^{-0.01t})(1+ 8e^{-0.01t})^{-3}(8)(-0.01)e^{-0.01t}$

$= 12.8(-0.01)e^{-0.01t}(1 + 8e^{-0.01t})^{-3}[1$

$+ 8e^{-0.01t} - 16e^{-0.01t}]$

$= 12.8(-0.01)e^{-0.01t}(1 + 8e^{-0.01t})^{-3}[1$

$- 8e^{-0.01t}]$

This expression is 0 when $R'(t) = 0$ or when

$1 - 8e^{-0.01t} = 0$ or when

$e^{0.01t} = 8$, $t = \dfrac{\ln 8}{0.01} \approx 208$ years from now.

15. Let x denote the number of cases of connectors in each shipment, and $C(x)$ the corresponding (variable) cost. Then,

$C(x) = \text{STORAGE COST} + \text{ORDERING COST}$

$= \dfrac{4.5x}{2} + 20\left(\dfrac{18,000}{x}\right)$

$= 2.25x + 360,000x^{-1}$ on $[0, 18,000]$

$C'(x) = 2.25 - 360,000x^{-2}$

$C'(x) = 0$ when $x = 400$. Since this is the only critical point in the interval, and since $C''(x) = 720,000x^{-3} > 0$ when $x = 400$, C is minimized when $x = 400$. The number of shipments should be $18,000/400 = 45$ times per year.

17. $R(x) = -2x^2 + 68x - 128$ is the total revenue.

a. Average revenue per unit is total revenue divided by number of units:

$$A(x) = -2x + 68 - \frac{128}{x}$$

Marginal revenue is $R'(x) = -4x + 68$.

They are equal when:

$$-2x + 68 - \frac{128}{x} = -4x + 68$$

$$2x = \frac{128}{x}$$

$$x^2 = 64$$

$$x = 8 \quad (\text{reject } x = -8)$$

b. $A'(x) = -2 + \frac{128}{x^2}$;

$A'(8) = 0; \quad A'(8^-) > 0, \quad A'(8^+) < 0$

So there is a relative maximum at $(8, 36)$.
The function is increasing on $[0, 8)$ and
decreasing on $(8, +\infty)$.

c.

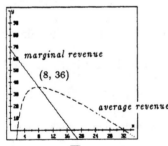

19. $P(t) = 300e^{\sqrt{3t}}e^{-0.08t}$

$$= 300e^{\sqrt{3t}-0.08t}$$

$P'(t) = 300\exp(\sqrt{3t} - 0.08t)[\frac{1}{2}(3t)^{-1/2}(3) - 0.08]$

$P'(t) = 0$ if $t = 117.19$

Since the optimum solution is over 100 years,

you should will the book to your heirs so they
can sell it in 117.19 years. (Some things —
like rare books — take time!)

21. $\quad P(x) = R(x) - C(x)$

Let $x =$ the increase in price,

$40 + x =$ cost per board,

$45 - 3x =$ number of boards sold.

$$P(x) = (40 + x)(45 - 3x) - 29(45 - 3x)$$

$$= -3x^2 + 12x + 495$$

$P'(x) = -6x + 12 = 0$ when $x = 2$.

therefore, raise the price \$2, sell the boards
at a price of \$42, sell 39 per month, and have
the maximum profit of \$507.

23. Let x be the number of people above the 100
level $(0 \le x \le 50)$. Then, the number of
travelers is $100 + x$. The fare per traveler is
$2,000 - 10x$. The revenue is

$$R(x) = (100 + x)(2,000 - 10x)$$

and the profit is

$$P(x) = (100 + x)(2,000 - 10x)$$

$$- 125,000 - 500(x + 100)$$

$$= -10x^2 + 500x + 25,000$$

$P'(x) = -20x + 500 = 0$ at $x = 25$

Thus, lower the fare by $10(25) = \$250$.

25. We want to maximize the yield, $Y(x)$. Let x be the additional number of trees to be planted.

$$\text{NUMBER OF TREES} = 60 + x$$

$$\text{AVERAGE YIELD} = 400 - 4x$$

$$Y(x) = (60 + x)(400 - 4x)$$

$$= -4x^2 + 160x + 24,000$$

$Y'(x) = -8x + 160 = 0$ when $x = 20$

Plant 80 total trees, have an average yield per tree of 320 oranges, and a maximum total crop of 25,600 oranges.

27. Let x denote the number of additional grapevines to be planted and $N(x)$ the corresponding yield. Since there are 50 grapevines to begin with and x additional grapevines are planted, the total number is $50 + x$. For each additional grapevine, the average yield is decreasing by 2 lb, so that the yield is $150 - 2x$ lb. Thus,

$N(x) = $ (LBS OF GRAPES PER VINE)(NO. OF GRAPEVINES)

$$= (150 - 2x)(50 + x)$$

$$= 2(75 - x)(50 + x)$$

$$= 2(3,750 + 25x - x^2) \text{ on } [0, 20]$$

$N'(x) = 2(25 - 2x) = 0$ when $x = 12.5$

$N(0) = 7,500$; $N(12) = 7,812$; $N(13) = 7,812$; $N(20) = 7,700$. The greatest possible yield is 7,812 pounds of grapes, which is generated by planting 12 or 13 (choose 12 for practical reasons) additional vines; that is, 62 vines are planted.

29. a. Let x be the number of days and $R(x)$ the corresponding revenue. Over the period of 80 days, 24,000 pounds of glass have been collected at a rate of 300 pounds per day, and so for each day over 80, an additional 300 pounds will be collected. Thus, the total number of pounds collected and sold is $24,000 + 300x$. Currently the recycling center pays 1¢ per pound. For each additional day, it reduces the price it pays by 1¢ per 100

pounds; that is, by 0.01¢/lb. Hence, after

x additional days, the price per pound

will be $1 - 0.01x$ cents.

$R(x) = $ (NO. OF LBS OF GLASS)(PRICE/LB)

$\qquad = (24,000 + 300x)(1 - 0.01x)$

$\qquad = 24,000 + 60x - 3x^2$ on $[0, 100]$

$R'(x) = 60 - 6x = 0$ when $x = 10$

$\qquad R(0) = 24,000;\ R(100) = 0;$

$\qquad R(10) = 24,300$

The most profitable time to conclude the

project is 10 days from now.

b. Assume R is continuous over $[0, 10]$.

31. $p(x) = \dfrac{b - x}{a}$ is the demand function.

a. $\qquad R(x) = xp(x)$

$\qquad\qquad = \dfrac{bx - x^2}{a}$ on $[0, b]$

$\qquad R'(x) = \dfrac{1}{a}(b - 2x)$

$\qquad R'(x) = 0$ if $x = \dfrac{b}{2}$.

R is increasing on $\left(0, \dfrac{b}{2}\right)$ and deceasing

on $\left(\dfrac{b}{2}, b\right)$.

b.

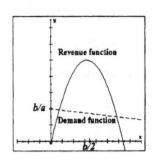

33. $t = \dfrac{s}{v - v_1};$

$\qquad E = cv^k t = cs\left(\dfrac{v^k}{v - v_1}\right)$

$\qquad \dfrac{dE}{dv} = cs\left[\dfrac{(v - v_1)kv^{k-1} - v^k}{(v - v_1)^2}\right]$

$dE/dv = 0$ if $\dfrac{csv^{k-1}}{(v - v_1)^2}(kv - kv_1 - v) = 0$

$\qquad v = \dfrac{kv_1}{k - 1}$

35. **a.** $P(x) = Ax^s e^{-sx/r}$

$\qquad P'(x) = Asx^{s-1}e^{-sx/r} + Ax^s\left(-\dfrac{s}{r}\right)e^{-sx/r}$

$\qquad\qquad = \dfrac{As}{r}x^{s-1}e^{-sx/r}(r - x)$

$\qquad P'(x) = 0$ if $x = r$

$P''(x) = \dfrac{As}{r^2}x^{s-2}e^{-sx/r}(sx^2 - 2rsx + r^2s - r^2)$

$P''(r) = -Ase^{-s}r^{s-2}$

which is negative so $x = r$ is a maximum.

b. $P''(x) = 0$ when $x = \dfrac{r}{\sqrt{s}}(\sqrt{s} \pm 1)$

$\qquad = \dfrac{r}{s}(s \pm \sqrt{s})$ which are inflection points,

so if $s > 1$, there are two inflection points.

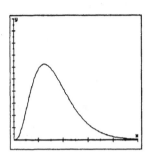

The production rate $P'(x)$ is increasing

for $0 < x < \frac{r}{s}(s - \sqrt{s})$ and for

$x > \frac{r}{s}(s + \sqrt{s})$. The rate is decreasing for

$\frac{r}{s}(s - \sqrt{s}) < x < \frac{r}{s}(s + \sqrt{s})$.

c. $s = 0$ means $P = A$, a horizontal line. If
$0 < s < 1$ one of the inflection points is
negative, so only one appears in $x \geq 0$.

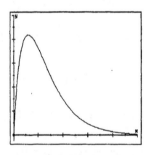

The production rate $P'(x)$ is decreasing
for $0 < x < \frac{r}{s}(s + \sqrt{s})$ and increasing for

$x > \frac{r}{s}(s + \sqrt{s})$.

37. $E = \frac{1}{v}[a(v - b)^2 + c]$

$E' = \frac{1}{v}[2a(v - b)] - \frac{1}{v^2}[a(v-b)^2 + c]$

$= \frac{1}{v^2}(av^2 - ab^2 - c)$

$E'(v) = 0$ when

$$v^2 = \frac{ab^2 + c}{a}$$

When $a = 0.04$, $b = 36$, and $c = 9$, then

$$v^2 = \frac{(0.04)(36)^2 + 9}{0.04} = 1{,}521$$

Thus, $v = 39$ units of length per unit of
time.

39. a.

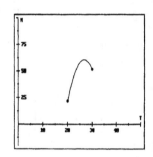

$N(t) = -0.85T^2 + 45.45T - 547$

$N'(t) = -1.7T + 45.45$

$N'(T) = 0$ when $T \approx 26.74$

The largest survival percentage is 60.56%

for $T \approx 26.74$ and the smallest survival

percentage is 22% for T = 20.

b. $S(T) = (-0.03T^2 + 1.67T - 13.67)^{-1}$

$S'(T) = -(-0.06T + 1.67)(-0.03T^2$

$+ 1.67T - 13.67)^{-2}$

$S'(T) = 0$ when

$-(-0.06T + 1.67) = 0$

$$T = \frac{1.67}{0.06}$$

$$\approx 27.83$$

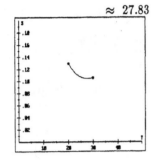

c.

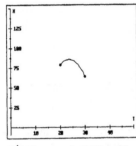

$H'(T) = -1.06T + 25$

$H'(T) = 0$ when $T \approx 23.58$

The largest hatching occurs when $T \approx 23.58$ and the smallest when $T = 30$.

41. a. Let x be number of machines to be set up. The set-up cost is Sx, the number of units produced per hour is nx, the number of hours of operation is $pQ/(nx)$, and the

total cost is

$$C = Sx + \frac{pQ}{nx}$$

$$C' = S - \frac{pQ}{nx^2}$$

$C' = 0$ when $x^2 = \dfrac{pQ}{nS}$ or $x = \sqrt{\dfrac{pQ}{nS}}$

b. The set-up cost equals the operating cost

if

$$x = \sqrt{\frac{pQ}{nS}}$$

43. The epidemic is spreading most rapidly when the rate of change $R(t)$ is a maximum, that is, when $R'(t) = f''(t) = 0$.

$$f(t) = \frac{A}{1 + Ce^{-kt}}$$

$$= A(1 + Ce^{-kt})^{-1}$$

$$f'(t) = A(-1)(1 + Ce^{-kt})^{-2}(-Cke^{-kt})$$

$$= kACe^{-kt}(1 + Ce^{-kt})^{-2}$$

$$f''(t) = kAC(1 + Ce^{-kt})^{-4}[(1 + Ce^{-kt})^2(-ke^{-kt})$$

$$- (e^{-kt})(2)(1 + Ce^{-kt})(-Cke^{-kt})]$$

$$= \frac{k^2ACe^{-kt}(Ce^{-kt} - 1)}{(1 + Ce^{-kt})^3}$$

$$= 0 \text{ if } Ce^{-kt} = 1$$

or $t = k^{-1}\ln C$. Substituting in the original

equation leads to

$$f(k^{-1}\ln C) = A(1 + Ce^{-\ln C})^{-1}$$

$$= \tfrac{1}{2}A$$

This means half of the total number of

susceptible residents. $R'(t) < 0$ when

$t > (\ln C)k^{-1}$ (the graph of the curve is

decreasing) and $R'(t) > 0$ when $t < (\ln C)k^{-1}$

(the graph of the curve is increasing).

45.   COST = STORAGE COST + ORDERING COST

$$C(x) = \tfrac{tx}{2} + \tfrac{QS}{x}$$

$$C'(x) = \tfrac{t}{2} - \tfrac{QS}{x^2}$$

$$C'(x) = 0 \text{ if } x^2 = \tfrac{2QS}{t} \text{ or } x = \sqrt{\tfrac{2QS}{t}}$$

We now substitute this value into the cost

equations:

STORAGE COST $= \tfrac{tx}{2} = \tfrac{t}{2}\sqrt{\tfrac{2QS}{t}} = \tfrac{1}{\sqrt{2}}\sqrt{QSt}$

ORDERING COST $= \tfrac{QS}{x} = QS\sqrt{\tfrac{t}{2QS}} = \tfrac{1}{\sqrt{2}}\sqrt{QSt}$

Thus, the total cost is minimized when the
storage cost is equal to the ordering cost.

47.   The revenue is $R(x) = px$ and its derivative

$$\frac{dR}{dx} = p + x\frac{dp}{dx}$$

$$= p\Big[1 + \tfrac{x}{p}\tfrac{dp}{dx}\Big]$$

$$= \tfrac{xp}{x}\Big[1 + \tfrac{x}{p}\tfrac{dp}{dx}\Big]$$

$$= \tfrac{R(x)}{x}\Big[1 + \tfrac{1}{E(x)}\Big]$$

49. $C(t) = \dfrac{k}{b-a}(e^{-at} - e^{-bt})$

$$C'(t) = \tfrac{k}{b-a}(-ae^{-at} + be^{-bt}) \qquad a < b$$

$$C'(t) = 0 \text{ when } t = t_c = \tfrac{1}{b-a}\ln(b/a)$$

$$C''(t) = \tfrac{k}{b-a}(a^2e^{-at} - b^2e^{-bt})$$

Since $ae^{-at_c} = be^{-bt_c}$ and $a < b$,

we have

$$a^2e^{-at_c} = abe^{-bt_c} < b^2e^{-bt_c}$$

so

$$C''(t_c) = \tfrac{k}{b-a}\Big[a^2e^{-at_c} - b^2e^{-bt_c}\Big] < 0$$

and it follows that a relative maximum occurs
where $t = t_c$.

51. $S(\theta) = 6sh + 1.5s^2(-\cot\theta + \sqrt{3}\csc\theta)$

$$S'(\theta) = 1.5s^2(\csc^2\theta - \sqrt{3}\csc\theta\cot\theta)$$

$$= \tfrac{1.5s^2}{\sin^2\theta}(1 - \sqrt{3}\cos\theta)$$

$S'(\theta) = 0$ when $\theta \approx 0.9553$; this is about $55°$.

Chapter 4 Review

Proficiency Examination, Page 263

SURVIVAL HINT: *To help you review the concepts of this chapter, handwrite the answers to each of these questions onto your own paper.*

1. A relative extremum is an extremum only in the neighborhood of the point of interest. An absolute extremum is the largest (or smallest) extremum for all values in the domain.

2. A continuous function f on a closed interval $[a, b]$ has an absolute maximum and an absolute minimum.

3. A critical number of a function is a value of the independent variable at which the derivative of the function is zero or is not defined. A critical point $(c, f(c))$ is the point on $y = f(x)$ that corresponds to the critical number c.

4. Find critical numbers, evaluate the function at these values and at the endpoints of the closed interval. Finally, determine the absolute extrema by selecting the largest and smallest of the evaluated functional values.

5. If f is continuous on the closed interval $[a, b]$ and differentiable on the open interval (a, b), then there exists at least one number c such that

    $$\frac{f(b) - f(a)}{b - a} = f'(c) \text{ for } a < c < b$$

    Rolle's theorem is a special case where

    $f(a) = f(b)$ (so $f'(c) = 0$).

6. Suppose f is a continuous function on the closed interval $[a, b]$ and is differentiable on the open interval (a, b), with $f'(x) = 0$ for all x on (a, b). Then the function f is constant on $[a, b]$.

7. Use the second-derivative test first:

    Given $f(x)$ and c such that $f'(c) = 0$. If $f''(x) > 0$ there is a relative minimum at $x = c$ (concave up). If $f''(c) < 0$, there is a relative maximum at $x = c$ (concave down). If this test fails, *i.e.* $f''(c) = 0$, $f''(c)$ is difficult to evaluate, or $f''(c)$ does not exist, then use the first derivative test:

    Given $f(x)$, find c such that $f'(c) = 0$ or $f'(c)$ is not defined (that is, c is a critical number). If $f'(x) < 0$ for $x < c$ and $f'(x) > 0$ for $x > c$ (what we have been calling the $\downarrow\uparrow$ pattern), there is a relative minimum at $x = c$. If $f'(x) > 0$ for $x < c$ and $f'(x) < 0$ for $x > c$ (the $\uparrow\downarrow$ pattern), there is a relative maximum at $x = c$.

8. For a plane curve, an asymptote is a line which has the property that the distance from a point P on the curve to the line approaches zero as the distance from P to the origin increases without bound and P is on a suitable portion of the curve.

9. Suppose the function f is continuous at the point $P(c, f(c))$. Then, the graph of f has a cusp at P if $\lim_{x \to c^-} f'(x)$ and $\lim_{x \to c^+} f'(x)$ have opposite signs. The graph has a vertical tangent at P if $\lim_{x \to c^-} f'(x)$ and $\lim_{x \to c^+} f'(x)$ are either both positive or both negative;

10. Let f and g be functions that are differentiable on an open interval containing c (except possibly at c itself). If $\lim_{x \to c} \frac{f(x)}{g(x)}$ produces an indeterminate form $\frac{0}{0}$

or $\frac{\infty}{\infty}$, and $\lim_{x \to c} \frac{f'(x)}{g'(x)}$ exists, then
$$\lim_{x \to c} \frac{f(x)}{g(x)} = \lim_{x \to c} \frac{f'(x)}{g'(x)}.$$

11. $\lim_{x \to +\infty} f(x) = L$, given an $\epsilon > 0$, there exists a number N_1 such that $\left| f(x) - L \right| < \epsilon$ whenever $x > N_1$ for x in the domain of f.

$\lim_{x \to c} f(x) = +\infty$ if for any number $N > 0$, it is possible to find a number $\delta > 0$ such that $f(x) > N$ whenever $0 < |x - c| < \delta$.

12. Find the domain and range of a function, locate intercepts, if any. Investigate extent, symmetry, asymptotes, and find extrema and/or points of inflection, if any. Determine where the graph is rising, where it is falling, and determine the concavity. (See Table 4.1).

13. An optimization problem involves finding the largest, or smallest, value of a function.
    Procedure:

    (1) Draw a figure (if appropriate) and label all quantities relevant to the problem.

    (2) Find a formula for the quantity to be maximized or minimized.

(3) Use conditions in the problem to eliminate variables in order to express the quantity to be maximized or minimized in terms of a single variable.

(4) Find the practical domain for the variables in Step 3; that is, the interval of possible values determined from the physical restrictions in the problem.

(5) If possible, use the methods of calculus to obtain the required optimum value.

14. Light travels between two points in such a way as to minimize the time of transit.

15. If a beam of light strikes the boundary between two media with an angle of incidence α and is refracted through an angle β, then

$$\frac{\sin \alpha}{\sin \beta} = \frac{v_1}{v_2}$$

where v_1 and v_2 are the rates at which light travels through the first and second medium, respectively. The constant ratio

$$n = \frac{\sin \alpha}{\sin \beta}$$

is called the relative index of refraction of the two media.

16. Profit is maximized when marginal revenue equals marginal cost. Average cost is minimized at the level of production where the marginal cost equals the average cost.

17. The resistance to the flow of blood in an artery is directly proportional to the artery's length and inversely proportional to the fourth power of its radius.

18. $\lim\limits_{x \to \pi/2} \dfrac{\sin 2x}{\cos x} = \lim\limits_{x \to \pi/2} \dfrac{2 \cos 2x}{-\sin x}$

$= \dfrac{-2}{-1}$

$= 2$

19. $\lim\limits_{x \to 1} \dfrac{1 - \sqrt{x}}{x - 1} = \lim\limits_{x \to 1} \dfrac{-\frac{1}{2} x^{-1/2}}{1}$

$= -\dfrac{1}{2}$

20. $\lim\limits_{x \to +\infty} \left(\dfrac{1}{x} - \dfrac{1}{\sqrt{x}} \right) = \lim\limits_{x \to +\infty} \dfrac{1 - \sqrt{x}}{x}$

$= \lim\limits_{x \to +\infty} \dfrac{-\frac{1}{2} x^{-1/2}}{1}$

$= 0$

21. $\lim\limits_{x \to +\infty} \left(1 + \dfrac{2}{x} \right)^{3x}$ Let $\dfrac{1}{u} = \dfrac{2}{x}$:

$\lim\limits_{u \to +\infty} \left(1 + \dfrac{1}{u} \right)^{3(2u)} = \lim\limits_{u \to +\infty} \left[\left(1 + \dfrac{1}{u} \right)^{u} \right]^{6}$

$= e^6$

22. $f(x) = x^3 + 3x^2 - 9x + 2,$

$f'(x) = 3x^2 + 6x - 9$

$\quad = 3(x + 3)(x - 1) = 0$ when $x = -3, 1$

$f''(x) = 6x + 6 = 0$ when $x = -1$

Relative maximum at $(-3, 29)$;

relative minimum at $(1, -3)$;

inflection point at $(-1, 13)$

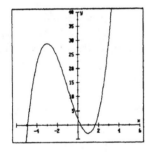

23. $f(x) = 27x^{1/3} - x^{4/3}$

$f'(x) = 9x^{-2/3} - \frac{4}{3}x^{1/3} = 0$ when

$\dfrac{9}{x^{2/3}} = \dfrac{4x^{1/3}}{3}$

$\quad x = \dfrac{27}{4}$ (not defined at $x = 0$)

$f''(x) = -6x^{-5/3} - \frac{4}{9}x^{-2/3}$

$\quad = 0$

when

$-6x^{-5/3} - \frac{4}{9}x^{-2/3} = 0$

$-\frac{2}{9}x^{-5/3}(27 + 2x) = 0$

$27 + 2x = 0$

$x = -\frac{27}{2}$

$f''(-\frac{27}{2}^-) > 0, \; f''(-\frac{27}{2}^+) < 0,$

as well as $f''(0^-) > 0, \; f''(0^+) < 0$ so there

are inflection points at approximately

$(-\frac{27}{2}, -96.43)$ and $(0, 0)$.

Relative maximum at $(\frac{27}{4}, 38.27)$

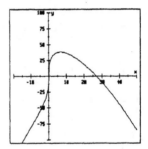

24. $f(x) = \dfrac{x^2 - 1}{x^2 - 4}$

Checking for horizontal asymptotes:

$\lim\limits_{x \to +\infty} f(x) = 1; \; \lim\limits_{x \to -\infty} f(x) = 1$

So $y = 1$ is a horizontal asymptote.

Factoring the denominator we see that there

are two candidates for vertical asymptotes; 2

and -2. Testing these:

$\lim\limits_{x \to -2^-} f(x) = +\infty; \; \lim\limits_{x \to -2^+} f(x) = -\infty$

$\lim\limits_{x \to 2^-} f(x) = -\infty; \; \lim\limits_{x \to 2^+} f(x) = +\infty$

So there are two vertical asymptotes.

$$f'(x) = \frac{(x^2 - 4)2x - (x^2 - 1)2x}{(x^2 - 4)^2} = \frac{-6x}{(x^2 - 4)^2}$$

This is equal to 0 when $x = 0$, and it is easy

to see that the slope is positive to the left of

0, and negative to the right of 0. So there is

a relative maximum at $(0, \frac{1}{4})$. This should

be sufficient information to sketch the graph:

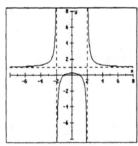

25. $y = f(x) = (x^2 - 3)e^{-x} = 0$ when

$x = \pm \sqrt{3}$,

When $x = 0$, $y = -3$, which is the y-

intercept.

$$y' = (x^2 - 3)e^{-x}(-1) + e^{-x}(2x)$$

$$= - e^{-x}(x^2 - 2x - 3)$$

Solve $y'(x) = 0$:

$$- e^{-x}(x^2 - 2x - 3) = 0$$

$$(x - 3)(x + 1) = 0$$

$x = -1, 3$

$$y'' = - e^{-x}(2x - 2) + (x^2 - 2x - 3)e^{-x}$$

$$= e^{-x}(x^2 - 4x - 1)$$

Solve $y''(x) = 0$:

$$e^{-x}(x^2 - 4x - 1) = 0$$

$$x^2 - 4x - 1 = 0$$

$$x = \frac{4 \pm \sqrt{4^2 - 4(1)(-1)}}{2(1)}$$

$$x = 2 \pm \sqrt{5}$$

$y''(-1) = 4e > 0$, so there is a relative

minimum at $(-1, -2e)$;

$y''(3) = - \frac{4}{e^3} < 0$, so there is a relative

maximum at $(3, 6e^{-3})$.

$$\lim_{x \to +\infty} f(x) = \lim_{x \to +\infty} \frac{x^2 - 3}{e^x}$$

$$= \lim_{x \to +\infty} \frac{2x}{e^x}$$

$$= \lim_{x \to +\infty} \frac{2}{e^x} = 0$$

Also

$$\lim_{x \to -\infty} f(x) = \lim_{x \to -\infty} \frac{x^2 - 3}{e^x}$$

$$= \frac{+\infty}{0}$$

$$= +\infty$$

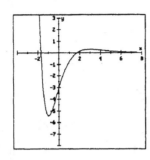

26. $f(x) = x + \tan^{-1} x$

$f'(x) = 1 + (1 + x^2)^{-1} > 0$; the curve is

rising for all x.

$f''(x) = -2x(1 + x^2)^{-2} = 0$ when $x = 0$;

$(0, 0)$ is a point of infection. The graph is

concave up for $x < 0$ and down for $x > 0$.

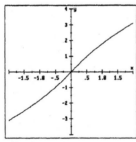

27. $f(x) = \sin^2 x - 2\cos x$

$f'(x) = 2 \sin x \cos x + 2 \sin x$

$\qquad = 2 \sin x(\cos x + 1) \ = \ 0$ when

$\qquad\qquad x \ = \ 0, \ \pi, \ 2\pi \ \text{ on } [0, 2\pi].$

$f''(x) = -2 \sin^2 x + 2 \cos^2 x + 2 \cos x$

$\qquad = 4 \cos^2 x + 2 \cos x - 2 = 0$ when

$\qquad\qquad x = \frac{\pi}{3}, \ \pi, \ \frac{5\pi}{3}$

The graph is rising for $0 < x < \pi$ and is falling

for $\pi < x < 2\pi$.

The graph is concave up for $0 < x < \frac{\pi}{3}$ and for

$\frac{5\pi}{3} < x < 2\pi$ and is concave down for

$\frac{\pi}{3} < x < \frac{5\pi}{3}$.

There is a relative maximum where $x = \pi$ and

inflection points at $x = \frac{\pi}{3}$ and $x = \frac{5\pi}{3}$.

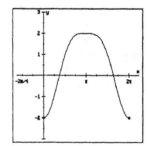

28. This is a continuous function on a closed inter-

val, so the extreme value theorem guarantees

an absolute maximum and an absolute

minimum. The candidates are the critical

points and the endpoints.

$f(x) = x^4 - 2x^5 + 5$

$f'(x) = 4x^3 - 10x^4 = 2x^3(2 - 5x)$ which is 0

at $x \ = \ 0, \frac{2}{5} = 0.4$. Testing our candidates:

$f(0) = 5$, $f(0.4) = 5.00512$, $f(1) = 4$. The absolute maximum is at $(0.4, 5.005)$ and the absolute minimum at $(1, 4)$.

29. We are asked to minimize the amount of material, $S = x^2 + 4xh$, with the restriction that $V = x^2 h = 2$. We can use the volume formula to find $h = 2/x^2$ to express S as a function of x.

$$S(x) = x^2 + 4x\left(\frac{2}{x^2}\right)$$
$$= x^2 + \frac{8}{x}$$
$$S'(x) = 2x - \frac{8}{x^2}$$

Solve $S'(x) = 0$

$$2x - \frac{8}{x^2} = 0$$
$$2x^3 - 8 = 0$$
$$x^3 = 4$$
$$x = \sqrt[3]{4}$$

Then, $h = \frac{2}{x^2} = \frac{2}{\sqrt[3]{4^2}}$

The dimensions of the box are approximately 1.587 by 1.587 by 0.794 ft, or to the nearest inch: 19 in. × 19 in. × 10 in.

30. $R(N) = -3N^4 + 50N^3 - 261N^2 + 540N$

$R'(N) = -12N^3 + 150N^2 - 522N + 540$

$$= -6(N - 2)(N - 3)(2N - 15)$$

$R'(N) = 0$ when $N = 2, 3, 7.5$;

$R(0) = 0$, $R(9) = 486$,

$R(2) = 388$, $R(3) = 378$, $R(7.5) = 970.3125$

We cannot hire 7.5 people, so we look at $N = 7$ and $N = 8$: $R(7) = 938$ and $R(8) = 928$, so we should hire 7 people.

CHAPTER 5

Integration

SURVIVAL HINT: *In this chapter you will find the third of the three main ideas of beginning calculus, namely the integral. (Remember the first two? Of course: the limit and the derivative!) Using the concept of an integral, you will be introduced to a significant type of mathematics equation, namely the differential equation. As you study this chapter, keep in mind that the operations of differentiation and integration are opposite operations, and that the derivative and integral of a function are themselves functions. After you finish this chapter you will have been introduced to the basic ideas of calculus.*

5.1 Antidifferentiation, Pages 281-282

SURVIVAL HINT: *Antiderivatives constitute a parametric family of curves. There is an infinite set of "parallel" curves that all have the same slope at any given value of x. Any curve that is an antiderivative can be translated vertically by C units and still be a solution. Always remember the $+C$.*

1. $\int 2\, dx = 2x + C$

3. $\int (2x + 3)\, dx = x^2 + 3x + C$

5. $\int (4t^3 + 3t^2)\, dt = t^4 + t^3 + C$

7. $\int \frac{dx}{2x} = \frac{1}{2}\ln|x| + C$

9. $\int (6u^2 - 3\cos u)\, du = 2u^3 - 3\sin u + C$

11. $\int \sec^2\theta\, d\theta = \tan\theta + C$

13. $\int 2\sin\theta\, d\theta = -2\cos\theta + C$

15. $\int \frac{5\, dy}{\sqrt{1-y^2}} = 5\sin^{-1}y + C$

17. $\int (u^{3/2} - u^{1/2} + u^{-10})\, du$

$$= \frac{u^{5/2}}{\frac{5}{2}} - \frac{u^{3/2}}{\frac{3}{2}} + \frac{u^{-9}}{-9} + C$$

$$= \frac{2}{5}u^{5/2} - \frac{2}{3}u^{3/2} - \frac{1}{9}u^{-9} + C$$

19. $\int x(x + \sqrt{x})\, dx = \int (x^2 + x^{3/2})\, dx$

$$= \frac{1}{3}x^3 + \frac{2}{5}x^{5/2} + C$$

21. $\int \left(\frac{1}{t^2} - \frac{1}{t^3} + \frac{1}{t^4}\right) dt = \int (t^{-2} - t^{-3} + t^{-4})\, dt$

$$= -t^{-1} + \frac{1}{2}t^{-2} - \frac{1}{3}t^{-3} + C$$

23. $\int (2x^2 + 5)^2 \, dx = \int (4x^4 + 20x^2 + 25) \, dx$

$$= \tfrac{4}{5}x^5 + \tfrac{20}{3}x^3 + 25x + C$$

25. $\int \left(\dfrac{x^2 + 3x - 1}{x^4} \right) dx = \int (x^{-2} + 3x^{-3} - x^{-4}) \, dx$

$$= -x^{-1} - \tfrac{3}{2}x^{-2} + \tfrac{1}{3}x^{-3} + C$$

27. $\int \dfrac{x^2 + x - 2}{x^2} \, dx = \int (1 + x^{-1} - 2x^{-2}) \, dx$

$$= x + \ln|x| + 2x^{-1} + C$$

29. $\int \dfrac{\sqrt{1 - x^2} - 1}{\sqrt{1 - x^2}} \, dx = \int \left[1 - \dfrac{1}{\sqrt{1 - x^2}} \right] dx$

$$= x - \sin^{-1} x + C$$

31. $F(x) = \int (x^2 + 3x) \, dx = \tfrac{1}{3}x^3 + \tfrac{3}{2}x^2 + C$

$$F(0) = \tfrac{1}{3}(0) + \tfrac{3}{2}(0) + C = 0 \text{ so } C = 0;$$

$$F(x) = \tfrac{1}{3}x^3 + \tfrac{3}{2}x^2$$

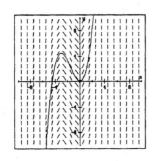

33. $F(x) = \int (\sqrt{x} + 3)^2 \, dx$

$$= \int (x + 6x^{1/2} + 9) \, dx$$

$$= \tfrac{1}{2}x^2 + 4x^{3/2} + 9x + C;$$

$$F(4) = \tfrac{1}{2}(4)^2 + 4(4)^{3/2} + 9(4) + C$$

$$= 36$$

so $C = -40$

$$F(x) = \tfrac{1}{2}x^2 + 4x^{3/2} + 9x - 40$$

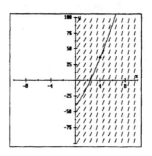

35. $F(x) = \int \dfrac{x + 1}{x^2} \, dx$

$$= \int (x^{-1} + x^{-2}) \, dx$$

$$= \ln|x| - x^{-1} + C$$

$$F(1) = 0 - 1 + C$$

$$= -2$$

so $C = -1;$

$$F(x) = \ln|x| - x^{-1} - 1$$

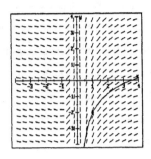

37. $F(x) = \int (x + e^x)\, dx$

$$= \tfrac{1}{2}x^2 + e^x + C$$

$F(0) = 1 + C$

$$= 2$$

so $C = 1$

$F(x) = \tfrac{1}{2}x^2 + e^x + 1$

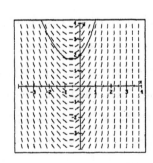

39. **a.** $F(x) = \int (x^{-1/2} - 4)\, dx$

$$= 2x^{1/2} - 4x + C$$

$F(1) = 2 - 4 + C = 0$, so $C = 2$;

$$F(x) = 2\sqrt{x} - 4x + 2$$

b.

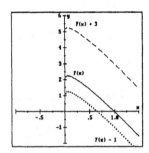

c. $G(x) = F(x) + C_0$;

$$G'(x) = F'(x) = \frac{1}{\sqrt{x}} - 4 = 0$$

At $x = \tfrac{1}{16}$ (note the original function),

$G(\tfrac{1}{16}) = 2(\tfrac{1}{4}) - 4(\tfrac{1}{16}) + 2 + C_0 = 0$, so

$$C_0 = -\tfrac{9}{4}$$

41. $C(x) = \int C'(x)\, dx$

$$= \int (6x^2 - 2x + 5)\, dx$$

$$= 2x^3 - x^2 + 5x + C$$

$C(1) = 2(1)^3 - (1)^2 + 5(1) + C = 5$, so

$C = -1$.

$C(x) = 2x^3 - x^2 + 5x - 1$, and

$C(5) = 250 - 25 + 25 - 1 = \$249$.

43. Let $P(t)$ be the population in the town at time t in months.

$$P'(t) = 4 + 5t^{2/3}$$

$$P(t) = \int (4 + 5t^{2/3}) \, dt$$

$$= 4t + 3t^{5/3} + C$$

$P(0) = 10,000$, so $C = 10,000$ and

$$P(t) = 4t + 3t^{5/3} + 10,000$$

The population in 8 months will be

$$P(8) = 4(8) + 3(8)^{5/3} + 10,000$$

$$= 10,128$$

45. $a(t) = k$,

$$v(t) = \int a(t) \, dt$$

$$= \int k \, dt$$

$$= kt + C$$

But $v(0) = 0$, so $C = 0$

$$s(t) = \int v(t) \, dt$$

$$= \int kt \, dt$$

$$= \frac{kt^2}{2} + C$$

We know that $s(0) = C$ and $s(6) = 18k + C$
and are given

$$s(6) - s(0) = 18k = 360$$

Thus, $k = 20$ ft/s^2.

47. With $a(t) = k$, $v(t) = kt + C_1$; $C_1 = 0$
because the plane starts from rest ($v_0 = 0$)
and the distance $s(t) = kt^2/2 + C_2$. For
convenience, measure the distance from the
point where the plane begins moving so
$s(0) = 0$ and $C_2 = 0$. Let t_1 be the time
required for liftoff. Since

$$s(t_1) = \frac{kt_1^2}{2} = 900 \text{ and } v(t_1) = kt_1 = 88,$$

we find (by dividing these equations)

$$\frac{kt_1^2}{kt_1} = \frac{1,800}{88}$$

$$t_1 \approx 20.4545 \text{ seconds}$$

From the velocity equation,

$$k = \frac{88}{t_1} = \frac{88}{20.4545} \approx 4.3$$

The acceleration of the plane at liftoff is 4.3
ft/s^2.

49. With $a(t) = k$, $v(t) = kt + C_1$; the initial
speed of the car is $v_0 = 88$ so $v(0) = C_1 = 88$
and the distance is $s(t) = kt^2/2 + 88t + C_2$.
For convenience, measure the distance from
the point where the car begins moving so
$s(0) = 0$ so $C_2 = 0$. Let t_1 be the time
required for stopping.

$$s(t_1) = \frac{kt_1^2}{2} + 88t_1 = 121 \text{ and } v(t_1) = 0$$

or $t_1 = -\frac{88}{k}$. Substituting into the distance

formula leads to

$$\frac{88^2 k}{2k^2} - \frac{88^2}{k} - 121 = 0, \text{ or}$$

$$k = -\frac{88^2}{242} = -32 \text{ ft/s}^2$$

SURVIVAL HINT: *This probably does not qualify as a "survival hint." We use this device to mention the Spy problem (whose solution is not given because it is an even problem). You need to be creative in your solution to this Spy problem. If you work this problem, you will find that the camel is toast! However, if you want to have some fun with this solution, you might argue that if the camel is standing so that the car is positioned between the camel's front and rear legs, and if the hood of the car is more than 0.9 ft ≈ 10.8 in. in length, the camel will escape undamaged. Here, of course, it is assumed that the camel's stomach is above the car's hood ornament.*

51. $s(t) = \int (t^{-1} + t)\, dt$

$$= \ln|t| + \tfrac{1}{2}t^2 + C$$

Since $0 = s(1) = \ln 1 + \tfrac{1}{2} + C, \; C = -\tfrac{1}{2}$

$$s(e^2) = \ln e^2 + \tfrac{1}{4}e^4 + C$$

$$= 2 + \tfrac{1}{4}e^4 - \tfrac{1}{2}$$

$$\approx 28.8$$

53. $F(x) = \int x^2\, dx$

$$= \frac{x^3}{3} + C$$

Since $F(1) = 0, \; \frac{1^3}{3} + C = 0,$ so $C = -\tfrac{1}{3}$

$$F(4) = \frac{4^3}{3} + C = \frac{64}{3} - \frac{1}{3} = 21$$

55. $F(x) = \int (e^x - x)\, dx$

$$= e^x - \tfrac{1}{2}x^2 + C$$

Since $F(0) = 0, \; e^0 - \tfrac{1}{2}(0)^2 + C = 0,$

so $C = -1.$

$$F(2) = e^2 - \tfrac{1}{2}(2)^2 + C = e^2 - 2 - 1$$

$$= e^2 - 3$$

57. $F(x) = \int \cos x\, dx$

$$= \sin x + C$$

Since $F(0) = 0, \; \sin 0 + C, \; C = 0$

$$F(\tfrac{\pi}{2}) = \sin \tfrac{\pi}{2} + C = 1 + 0 = 1$$

59. a. $\displaystyle\int \frac{dy}{2y\sqrt{y^2 - 1}} = \tfrac{1}{2}\sec^{-1} y + C$

b. $\tfrac{1}{2}\tan^{-1}\sqrt{y^2 - 1}$

c. Technology does not use "$+ C$," and to reconcile the inverse trigonometric functions, use a reference triangle where $\theta = \sec^{-1} y$, so $y = \sec \theta$. Then $\tan \theta = \sqrt{y^2 - 1}$ so $\tan^{-1}(y^2 - 1) = \sec^{-1} y.$

61.

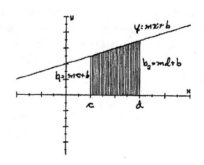

Using calculus,

$$F(x) = \int (mx + b)\, dx$$

$$= \frac{mx^2}{2} + bx + C$$

$F(0) = 0 + 0 + C = 0$, so $C = 0$

$F(x) = \frac{m}{2}x^2 + bx$; the area is

$$A = F(d) - F(c)$$

$$= \left(\frac{m}{2}d^2 + bd\right) - \left(\frac{m}{2}c^2 + bc\right)$$

$$= \frac{m}{2}(d^2 - c^2) + b(d - c)$$

$$= \tfrac{1}{2}(d - c)[m(d + c) + 2b]$$

Using geometry,

$$A = \tfrac{1}{2}h(b_1 + b_2)$$

$$= \tfrac{1}{2}\underbrace{(d - c)}_{h}[\underbrace{(mc + b)}_{b_1} + \underbrace{(md + b)}_{b_2}]$$

$$= \tfrac{1}{2}(d - c)[m(d + c) + 2b]$$

5.2 Area As the Limit of a Sum, Pages 288-290

SURVIVAL HINT: *In order to evaluate sums, you will need to use the summation formulas on page 287. If the sum does not look like one of those four formulas, then you should rewrite the sum to match one of those forms.*

1. $\displaystyle\sum_{k=1}^{6} 1 = 6(1) = 6$

3. $\displaystyle\sum_{k=1}^{15} k = \frac{(15)(16)}{2} = 120$

5. $\displaystyle\sum_{k=1}^{5} k^3 = \frac{(5^2)(6^2)}{4} = 225$

7. $\displaystyle\sum_{k=1}^{100}(2k - 3) = 2\sum_{k=1}^{100} k - 3\sum_{k=1}^{100} 1$

$$= 2\left[\frac{(100)(101)}{2}\right] - 300$$

$$= 9{,}800$$

9. $\displaystyle\lim_{n\to+\infty}\sum_{k=1}^{n}\frac{k}{n^2} = \lim_{n\to+\infty}\frac{1}{n^2}\sum_{k=1}^{n} k$

$$= \lim_{n\to+\infty}\frac{1}{n^2}\left[\frac{n(n+1)}{2}\right]$$

$$= \lim_{n\to+\infty}(1)\left(\tfrac{1}{2}\right)\left(1 + \tfrac{1}{n}\right)$$

$$= \tfrac{1}{2}$$

11. $\displaystyle\lim_{n\to+\infty}\sum_{k=1}^{n}\left(1 + \tfrac{k}{n}\right)\left(\tfrac{2}{n}\right)$

$$= \lim_{n\to+\infty}\sum_{k=1}^{n}\left(\tfrac{2}{n} + \tfrac{2k}{n^2}\right)$$

$$= \lim_{n\to+\infty}\left[\tfrac{2}{n}(n) + \tfrac{2}{n^2}\left(\frac{n(n+1)}{2}\right)\right]$$

$$= \lim_{n\to+\infty}\left[2 + \left(1 + \tfrac{1}{n}\right)\right]$$

$$= 3$$

13. $f(x) = 4x + 1$

    a. $n = 4$, $\Delta x = 0.25$, $f(a + k\Delta x) = k + 1$; $\;S = 3.5$

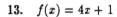
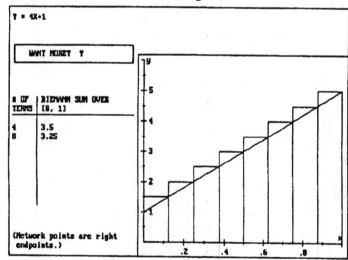

    b. $n = 8$, $\Delta x = 0.125$, $f(a + k\Delta x) = \frac{k}{2} + 1$; $S = 3.25$

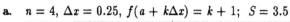
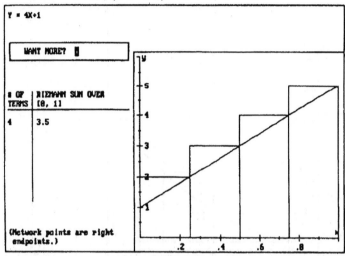
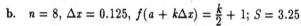

15. $f(x) = x^2$

    a. $n = 4$, $\Delta x = 0.25$, $f(a + k\Delta x) = \left(1 + \frac{k}{4}\right)^2$; $S = 2.71875$

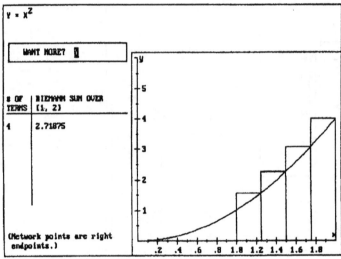

    b. $n = 6$, $\Delta x = \frac{1}{6}$, $f(a + k\Delta x) = \left(1 + \frac{k}{6}\right)^2$; $S \approx 2.58796$

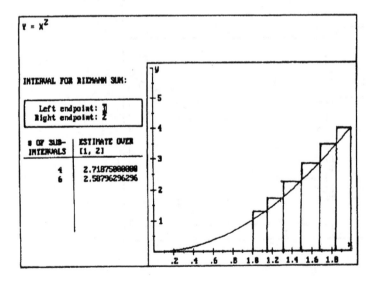

17. $f(x) = x + \sin x$

    $n = 3$, $\Delta x = \frac{\pi}{12}$, $f(a + k\Delta x) = \left(\frac{\pi}{12}k\right) + \sin\left(\frac{\pi}{12}k\right)$; $S \approx 0.795$

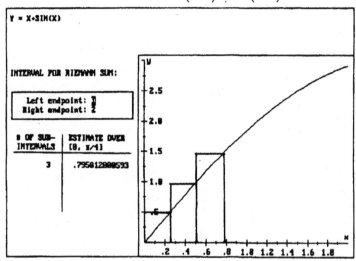

19. $f(x) = \frac{2}{x}$

    $n = 4$, $\Delta x = 0.25$, $f(a + k\Delta x) = 2\left(1 + \frac{k}{4}\right)^{-1}$; $S = 1.269$

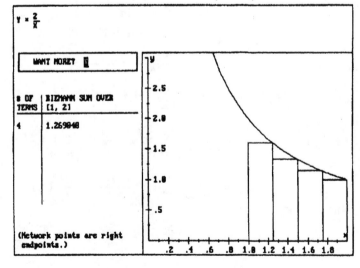

21. $f(x) = \sqrt{1 + x^2}$

$n = 4$, $\Delta x = 0.25$, $f(a + k\Delta x) = \sqrt{1 + \left(\frac{k}{4}\right)^2}$; $S \approx 1.2033$

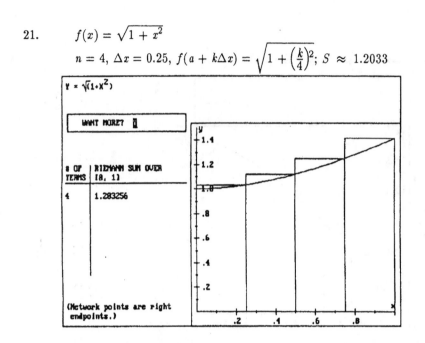

23. $\Delta x = \dfrac{b - a}{n} = \dfrac{2 - 1}{n} = \dfrac{1}{n}$

$A = \lim\limits_{n \to +\infty} \sum\limits_{k=1}^{n} f(1 + k\Delta x)\Delta x$

$= \lim\limits_{n \to +\infty} \sum\limits_{k=1}^{n} f(1 + \frac{k}{n})\Delta x$

$= \lim\limits_{n \to +\infty} \sum\limits_{k=1}^{n} \left[4(1 + \frac{k}{n})^3 + 2(1 + \frac{k}{n})\right]\left(\frac{1}{n}\right)$

$= \lim\limits_{n \to +\infty} \sum\limits_{k=1}^{n} \left[4\left(1 + \frac{3k}{n} + \frac{3k^2}{n^2} + \frac{k^3}{n^3}\right) + 2 + \frac{2k}{n}\right]\left(\frac{1}{n}\right)$

$= \lim\limits_{n \to +\infty} \sum\limits_{k=1}^{n} \left[6 + \frac{14k}{n} + \frac{12k^2}{n^2} + \frac{4k^3}{n^3}\right]\left(\frac{1}{n}\right)$

$= \lim\limits_{n \to +\infty} \left\{6n + \dfrac{14n(n + 1)}{2n}\right.$

$\left. + \dfrac{12n(n + 1)(2n + 1)}{6n^2} + \dfrac{4n^2(n + 1)^2}{4n^3}\right\}\dfrac{1}{n}$

$= \lim\limits_{n \to +\infty} \left\{6 + \dfrac{14n^2 + 14n}{2n^2} + \dfrac{24n^3 + 36n^2 + 12n}{6n^3}\right.$

$\left. + \dfrac{4n^4 + 8n^3 + 4n^2}{4n^4}\right\}$

$= (6 + 7 + 4 + 1) = 18$

25. $\Delta x = \dfrac{b - a}{n} = \dfrac{3 - 1}{n} = \dfrac{2}{n}$

$A = \lim\limits_{n \to +\infty} \sum\limits_{k=1}^{n} f(1 + \frac{2k}{n})\Delta x$

$= \lim\limits_{n \to +\infty} \sum\limits_{k=1}^{n} \left[6\left(1 + \frac{2k}{n}\right)^2 + 2\left(1 + \frac{2k}{n}\right) + 4\right]\left(\frac{2}{n}\right)$

$= \lim\limits_{n \to +\infty} \sum\limits_{k=1}^{n} \left(12 + \frac{28k}{n} + \frac{24k^2}{n^2}\right)\left(\frac{2}{n}\right)$

$= \lim\limits_{n \to +\infty} \left[\dfrac{24}{n} \sum\limits_{k=1}^{n} 1 + \dfrac{56}{n^2} \sum\limits_{k=1}^{n} k + \dfrac{48}{n^3} \sum\limits_{k=1}^{n} k^2\right]$

$$= \lim_{n \to +\infty} \left[24 + \frac{56}{n^2} \frac{n(n+1)}{2} + \frac{48}{n^3} \frac{n(n+1)(2n+1)}{6} \right]$$

$$= 24 + 28 + 16 = 68$$

27. $\Delta x = \frac{b-a}{n} = \frac{1-0}{n} = \frac{1}{n}$

$$A = \lim_{n \to +\infty} \sum_{k=1}^{n} f\left(\frac{k}{n}\right) \Delta x$$

$$= \lim_{n \to +\infty} \sum_{k=1}^{n} \left[4\left(\frac{k}{n}\right)^3 + 3\left(\frac{k}{n}\right)^2 \right]\left(\frac{1}{n}\right)$$

$$= \lim_{n \to +\infty} \left[\frac{4}{n^4} \sum_{k=1}^{n} k^3 + \frac{3}{n^3} \sum_{k=1}^{n} k^2 \right]$$

$$= \lim_{n \to +\infty} \left[\frac{4}{n^4} \frac{n^2(n+1)^2}{4} + \frac{3}{n^3} \frac{n(n+1)(2n+1)}{6} \right]$$

$$= 1 + 1$$

$$= 2$$

29. This statement is false. We are dealing with a trapezoid of base $(b - a)$ and parallel sides of length aC and bC. The area is

$$A = \tfrac{1}{2}(b - a)[aC + bC] = \tfrac{1}{2}C(b^2 - a^2)$$

31. The statement is true. $y = \sqrt{1 - x^2}$ or $x^2 + y^2 = 1$ (for $y \geq 0$) represents the equation of a semicircle. The area is $A = \pi/2$.

33. The statement is true. The graph of f is symmetric with respect to the y-axis. The area on the right of the x-axis is equal to that on the left.

35. $f(x) = C$ on $[a, b]$; $\Delta x = \frac{b-a}{n}$;

$$f(a + k\Delta x) = C$$

$$A = \lim_{n \to +\infty} \frac{b-a}{n} \sum_{k=1}^{n} C$$

$$= \frac{b-a}{n} nC$$

$$= (b - a)C$$

where $b - a$ is the length of the rectangle of width C.

37. a. $f(x) = 2x^2; \ \Delta x = \frac{1}{n}$;

$$f(a + k\Delta x) = 2\left(1 + \frac{k}{n}\right)^2$$

$$A = \lim_{n \to +\infty} \sum_{k=1}^{n} \left[2\left(1 + \frac{k}{n}\right)^2 \right]\left(\frac{1}{n}\right)$$

$$= \lim_{n \to +\infty} \left[\frac{2n}{n} + \frac{4}{n^2} \sum_{k=1}^{n} k + \frac{2}{n^3} \sum_{k=1}^{n} k^2 \right]$$

$$= \lim_{n \to +\infty} \left[2 + \frac{4n(n+1)}{2n^2} + \frac{2n(n+1)(2n+1)}{6n^3} \right]$$

$$= 2 + 2 + \tfrac{2}{3}$$

$$= \frac{14}{3}$$

b. If $g(x) = \frac{2}{3}x^3; \ g'(x) = 2x^2;$

$$g(2) - g(1) = \frac{16}{3} - \frac{2}{3}$$

$$= \frac{14}{3}$$

$$= A$$

c. If $h(x) = \frac{2}{3}x^3 + C; \ h'(x) = 2x^2;$

$$h(2) - h(1) = \frac{16}{3} + C - \frac{2}{3} - C$$

$$= \frac{14}{3}$$

$$= A$$

regardless of C. The statement is true.

39. The area seems to be 21.36459.

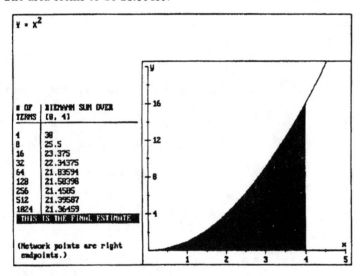

41. The area seems to be 0.6018907.

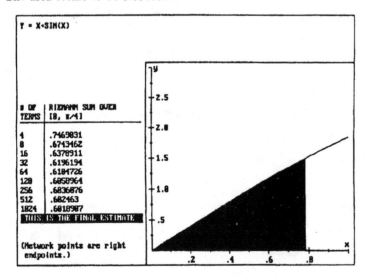

43. The area seems to be 0.5038795.

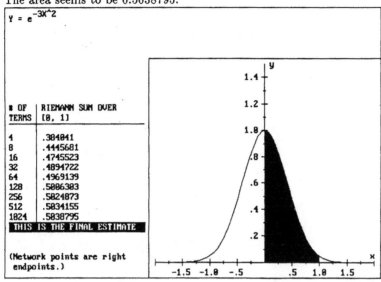

45. **a.** The area seems to be 2.

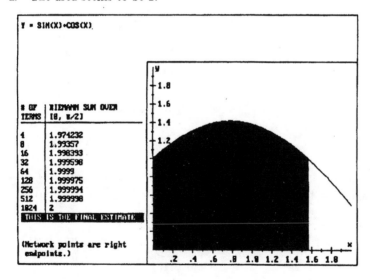

b. With $g(x) = -\cos x + \sin x$; $g'(x) = \sin x + \cos x = f(x)$

$$g(\tfrac{\pi}{2}) = -0 + 1$$

$$= 1$$

$$g(0) = -1 + 0$$

$$= -1,$$

and

$$g(\tfrac{\pi}{2}) - g(0) = 1 - (-1)$$

$$= 2$$

$$= A$$

c. $h(x) = -\cos x + \sin x + C$

$h'(x) = \sin x + \cos x = f(x)$

$h(\tfrac{\pi}{2}) = -0 + 1 + C$

$$= 1 + C$$

$h(0) = -1 + 0 + C$

$$= -1 + C,$$

and

$h(\tfrac{\pi}{2}) - h(0) = 1 + C - (-1) - C$

$$= 2$$

$$= A$$

The statement is true.

47. a. $k^3 = a[k^4 - (k - 1)^4] + bk^2 + ck + d$

$= a[k^4 - (k^4 - 4k^3 + 6k^2 - 4k + 1)]$

$+ bk^2 + ck + d$

$= 4ak^3 - 6ak^2 + 4ak - a$

$+ bk^2 + ck + d$

$= 4ak^3 + (b - 6a)k^2 + (4a + c)k$

$$+ (-a + d)$$

Since $4a = 1$; $b - 6a = 0$; $4a + c = 0$;

$-a + d = 0$; we find $a = \tfrac{1}{4}$, $b = \tfrac{3}{2}$,

$c = -1$, and $d = \tfrac{1}{4}$.

b. $\displaystyle\sum_{k=1}^{n} [k^3 - (k - 1)^3]$

$= (1^3 - 0^2) + (2^3 - 1^2) + \cdots$

$+ [n^3 - (n - 1)^3] = n^3$

Thus,

$n^3 = \displaystyle\sum_{k=1}^{n} [k^3 - (k - 1)^3]$

$= 3 \displaystyle\sum_{k=1}^{n} k^2 - 3 \displaystyle\sum_{k=1}^{n} k + \displaystyle\sum_{k=1}^{n} 1$

$= 3 \displaystyle\sum_{k=1}^{n} k^2 - 3 \dfrac{n(n + 1)}{2} + n$

From Problem 46.

$= 3 \displaystyle\sum_{k=1}^{n} k^2 - \dfrac{3n^2 + n}{2}$

Solve for the summation to find:

$3 \displaystyle\sum_{k=1}^{n} k^2 = n^3 + \dfrac{3n^2 + n}{2}$

$= \dfrac{n(2n^2 + 3n + 1)}{2}$

Consequently,

$\displaystyle\sum_{k=1}^{n} k^2 = \dfrac{n(n + 1)(2n + 1)}{6}$

49. $\Delta x = \dfrac{b - a}{n}$

$x_0 = a + \dfrac{\Delta x}{2}$

$$= a + \frac{b-a}{2n}$$

$$x_1 = a + \frac{3\Delta x}{2}$$

$$= a + \frac{3(b-a)}{2n}$$

$$x_2 = a + \frac{5\Delta x}{2}$$

$$= a + \frac{5(b-a)}{2n}$$

$$\vdots$$

$$x_k = a + \frac{(2k+1)\Delta x}{2}$$

$$= a + \frac{(2k+1)(b-a)}{2n}$$

$$A = \lim_{n \to +\infty} \left[\sum_{k=1}^{n} f\left(a + \frac{(2k+1)(b-a)}{2n}\right)\left(\frac{b-a}{n}\right) \right]$$

5.3 Riemann Sums and the Definite Integral, Pages 301-302

SURVIVAL HINT: *Thoroughly understand the significance of each symbol in the limit of a Riemann sum as given on page 291. This concept of an infinite sum of infinitely small parts will be used repeatedly throughout the text. Any successful calculus student should be able to state and explain the definitions of limit, derivative, and definite integral.*

1. $f(x) = 2x + 1$; $a = 0$; $\Delta x = \frac{1}{4}$

$$f(a + k\Delta x) = f(\tfrac{k}{4}) = \tfrac{k}{2} + 1$$

$$\int_0^1 (2x + 1)\, dx \approx \sum_{k=1}^{4} \left(\tfrac{k}{2} + 1\right)\left(\tfrac{1}{4}\right)$$

$$= \left[\left(\tfrac{1}{2} + 1\right) + \left(\tfrac{2}{2} + 1\right) + \left(\tfrac{3}{2} + 1\right) + \left(\tfrac{4}{2} + 1\right) \right]\left(\tfrac{1}{4}\right)$$

$$= \left[9\left(\tfrac{1}{4}\right)\right.$$

$$= 2.25$$

3. $f(x) = x^2$; $a = 1$; $\Delta x = \frac{1}{2}$

$$f(a + k\Delta x) = f(1 + \tfrac{k}{2}) = \left(1 + \tfrac{k}{2}\right)^2$$

$$\int_1^3 x^2\, dx \approx \sum_{k=1}^{4} \left(1 + \tfrac{k}{2}\right)^2\left(\tfrac{1}{2}\right)$$

$$= \left[\left(1 + \tfrac{1}{2}\right)^2 + \left(1 + \tfrac{2}{2}\right)^2 + \left(1 + \tfrac{3}{2}\right)^2 + \left(1 + \tfrac{4}{2}\right)^2\right]\left(\tfrac{1}{2}\right)$$

$$= \left[\tfrac{9}{4} + 4 + \tfrac{25}{4} + 9\right]\left(\tfrac{1}{2}\right)$$

$$= \left[\tfrac{43}{2}\right]\left(\tfrac{1}{2}\right)$$

$$= \frac{43}{4}$$

5. $f(x) = 1 - 3x$; $a = 0$; $\Delta x = \frac{1}{4}$

$$f(a + k\Delta x) = f(\tfrac{k}{4}) = 1 - \tfrac{3k}{4}$$

$$\int_0^1 (1 - 3x)\, dx \approx \sum_{k=1}^{4} \left(1 - \tfrac{3k}{4}\right)\left(\tfrac{1}{4}\right)$$

$$= \left[\left(1 - \tfrac{3}{4}\right) + \left(1 - \tfrac{6}{4}\right) + \left(1 - \tfrac{9}{4}\right) + \left(1 - \tfrac{12}{4}\right)\right]\left(\tfrac{1}{4}\right)$$

$$= \left[-\tfrac{7}{2}\right]\left(\tfrac{1}{4}\right)$$

$$= -\frac{7}{8}$$

7. $f(x) = \cos x$; $a = -\frac{\pi}{2}$; $\Delta x = \frac{\pi}{8}$

$$f(a + k\Delta x) = f(-\tfrac{\pi}{2} + \tfrac{k\pi}{8}) = \cos\left(-\tfrac{\pi}{2} + \tfrac{k\pi}{8}\right)$$

$$\int\limits_{-\pi/2}^{0} \cos x \, dx \approx \sum_{k=1}^{4} \cos\left(-\frac{\pi}{2} + \frac{k\pi}{8}\right)\left(\frac{\pi}{8}\right)$$

$$\approx 1.18$$

9. $f(x) = e^x; \ a = 0; \ \Delta x = \frac{1}{4}$

$$f(a + k\Delta x) = f\left(\frac{k}{4}\right) = e^{k/4}$$

$$\int\limits_{0}^{1} e^x \, dx \approx \sum_{k=1}^{4} (e^{k/4})\left(\frac{1}{4}\right)$$

$$\approx 1.94$$

11. $v(t) = 3t + 1; \ a = 1; \ \Delta t = \frac{3}{4};$

$$v(a + k\Delta t) = v\left(1 + \frac{3k}{4}\right) = 3\left(1 + \frac{3k}{4}\right) + 1$$

$$S_4 = \sum_{k=1}^{4} \left(4 + \frac{9k}{4}\right)\left(\frac{3}{4}\right)$$

$$= \frac{231}{8}$$

$$= 28.875$$

13. $v(t) = \sin t; \ a = 0; \ \Delta t = \frac{\pi}{4}$

$$v(a + k\Delta t) = v\left(\frac{k\pi}{4}\right) = \sin\frac{k\pi}{4}$$

$$S_4 = \sum_{k=1}^{4} \sin\frac{k\pi}{4}\left(\frac{\pi}{4}\right)$$

$$= \frac{\pi}{4}\left[\sqrt{2} + 1\right]$$

$$\approx 1.896$$

15. $v(t) = e^{-t}; \ a = 0; \ \Delta x = \frac{1}{4}$

$$v(a + k\Delta t) = v\left(\frac{k}{4}\right) = e^{-k/4}$$

$$S_4 = \sum_{k=1}^{4} e^{-k/4}\left(\frac{1}{4}\right)$$

$$\approx 0.556$$

17. $$\int\limits_{0}^{-1} x^2 \, dx = -\int\limits_{-1}^{0} x^2 \, dx$$

$$= -\frac{1}{3}$$

19. $$\int\limits_{-1}^{2} (2x^2 - 3x) \, dx = 2\int\limits_{-1}^{2} x^2 \, dx - 3\int\limits_{-1}^{2} x \, dx$$

$$= 2(3) - 3\left(\frac{3}{2}\right)$$

$$= \frac{3}{2}$$

21. $$\int\limits_{-1}^{0} x \, dx = \int\limits_{-1}^{2} x \, dx + \int\limits_{2}^{0} x \, dx$$

$$= \int\limits_{-1}^{2} x \, dx - \int\limits_{0}^{2} x \, dx$$

$$= \frac{3}{2} - 2$$

$$= -\frac{1}{2}$$

23. On $[0, 1]$ $x^3 \leq x$, so $\int\limits_{0}^{1} x^3 \, dx \leq \int\limits_{0}^{1} x \, dx$

Now $\int\limits_{0}^{1} x \, dx$ is the same as the area of a right

triangle with length 1 and height 1, so

$$\int\limits_{0}^{1} x \, dx = \frac{1}{2}(1)(1) = \frac{1}{2}$$

25. Let $F = \int\limits_{-2}^{4} f(x) \, dx$ and $G = \int\limits_{-2}^{4} g(x) \, dx$.

Then $\displaystyle\int_{-2}^{4} [5f(x) + 2g(x)]\,dx = 5F + 2G = 7,$

and $\displaystyle\int_{-2}^{4} [3f(x) + g(x)]\,dx = 3F + G = 4$

Subtracting the first from twice the second leads to $F = 1$ and $G = 1$.

27. By the subdivision and opposite properties,

$$\int_{-1}^{2} f(x)\,dx = \int_{-1}^{1} f(x)\,dx + \int_{1}^{3} f(x)\,dx + \int_{3}^{2} f(x)\,dx$$

$$= \int_{-1}^{1} f(x)\,dx + \int_{1}^{3} f(x)\,dx - \int_{2}^{3} f(x)\,dx$$

$$= 3 + 5 - (-2)$$

$$= 10$$

29.

For continuity, suspicious points are $x = -1$ and $x = 2$. By considering the left and right hand limits at these points, it is easy to show f is continuous on $[-3, 5]$.

$$\int_{-3}^{5} f(x)\,dx = \int_{-3}^{-1} 5\,dx + \int_{-1}^{2} (4 - x)\,dx + \int_{2}^{5} (2x - 2)\,dx$$

$$= 10 + \frac{21}{2} + 15$$

$$= \frac{71}{2}$$

31. $\displaystyle\int_{a}^{b} f(x)\,dx = \int_{a}^{c} f(x)\,dx + \int_{c}^{b} f(x)\,dx$

$$= \int_{a}^{c} f(x)\,dx + \left[\int_{c}^{d} f(x)\,dx + \int_{d}^{b} f(x)\,dx\right]$$

33. $f(x) = x^2$ and $\Delta x = \dfrac{b - a}{n}$; $x_k = a + \dfrac{b - a}{n}k$;

$$S_n = \sum_{k=1}^{n}\left[a + \frac{b-a}{n}k\right]^2\left(\frac{b-a}{n}\right)$$

$$= \sum_{k=1}^{n} \frac{b-a}{n}\left[a^2 + \frac{2(b-a)}{n}ak + \frac{(b-a)^2}{n^2}k^2\right]$$

$$= \frac{b-a}{n}\left(a^2\sum_{k=1}^{n}1 + \frac{2(b-a)}{n}a\sum_{k=1}^{n}k\right.$$
$$\left. + \frac{(b-a)^2}{n^2}\sum_{k=1}^{n}k^2\right)$$

$$= \frac{b-a}{n}\left(a^2 n + \frac{2(b-a)}{n}\frac{an(n+1)}{2}\right.$$
$$\left. + \frac{(b-a)^2}{n^2}\frac{n(n+1)(2n+1)}{6}\right)$$

$$\int_{a}^{b} x^2\,dx = \lim_{n\to+\infty} S_n$$

$$= \lim_{n\to+\infty} \frac{b-a}{n}\left(a^2 n + \frac{2(b-a)}{n}\frac{an(n+1)}{2}\right.$$
$$\left. + \frac{(b-a)^2}{n^2}\frac{n(n+1)(2n+1)}{6}\right)$$

$$= \left(a^2 + ab - a^2 + \frac{(b-a)^2}{3}\right)(b - a)$$

$$= \left(\frac{b-a}{3}\right)(3a^2 + 3ab - 3a^2 + b^2 - 2ab + a^2)$$

$$= \tfrac{1}{3}(b - a)(b^2 + ab + a^2) = \tfrac{1}{3}(b^3 - a^3)$$

35. $f(x) = 4 - 5x$

k:	1	2	3	4	5
x_k^*:	-0.5	0.8	1	1.3	1.8
$f(x_k^*)$:	6.5	0	-1	-2.5	-5
Δx_k:	0.8	1.1	0.4	0.4	0.3

$$R_5 = \sum_1^5 f(x_k)\Delta x_k$$

$$= 6.5(0.8) + 0 + (-1)(0.4)$$

$$+ (-2.5)(0.4) + (-5)(0.3)$$

$$= 2.3$$

$$\| P \| = 1.1$$

37.
$$\int_a^b f(x)\,dx = \lim_{n \to +\infty} \sum_{k=1}^n f(x_k^*)\Delta x_k$$

$$\le \lim_{n \to +\infty} \sum_{k=1}^n g(x_n^*)\Delta x_k$$

$$= \int_a^b g(x)\,dx$$

because $f(x) \le g(x)$ implies

$$a_k = f(x_k^*) \le b_k = g(x_k^*)$$

39. It is true; for example, consider the function

$y = x$ on $[-1, 1]$. The value of the integral

$$\int_{-1}^1 x\,dx = A_1 - A_2$$

where A_1 is the area of the triangle to the

right of the origin and A_2 is the area of the

triangle to the left of the origin. That is,

$$A_1 = \tfrac{1}{2}(1)(1) = \tfrac{1}{2} \quad \text{and} \quad A_2 = \tfrac{1}{2}(1)(1) = \tfrac{1}{2}$$

Thus,

$$\int_{-1}^1 x\,dx = \tfrac{1}{2} - \tfrac{1}{2} = 0$$

However, the area enclosed by the two

triangles is

$$\tfrac{1}{2} + \tfrac{1}{2} = 1$$

41. Since $m \le f(x) \le M$ on $[a, b]$, we have

$$\int_a^b m\,dx \le \int_a^b f(x)\,dx \le \int_a^b M\,dx$$

so

$$m(b - a) \le \int_a^b f(x)\,dx \le M(b - a)$$

5.4 The Fundamental Theorems of Calculus;
Pages 308-309

SURVIVAL HINT: *The fundamental theorem of calculus sounds really important — and it is. However, most calculus students seem not to appreciate its significance upon first exposure. It will become more meaningful as you encounter it in more advanced contexts, when your have a greater perspective. For now, see that it relates the two major concepts of elementary calculus — the derivative and the Riemann sum. Try to see that the increase in the area under a curve, as x is incremented, depends upon the slope of the curve.*

1. $\displaystyle\int_{-10}^{10} 7\,dx = 7x\Big|_{-10}^{10}$

$\qquad\qquad = 7[10 - (-10)]$

$\qquad\qquad = 140$

3. $\displaystyle\int_{-3}^{5} (2x + a)\,dx = (x^2 + ax)\Big|_{-3}^{5}$

$\qquad\qquad = (5)^2 + 5a - [(-3)^2 + (-3)a]$

$\qquad\qquad = 16 + 8a$

5. $\displaystyle\int_{-1}^{2} ax^3\,dx = \tfrac{1}{4}ax^4\Big|_{-1}^{2}$

$\qquad\qquad = \tfrac{1}{4}a[16 - 1]$

$\qquad\qquad = \dfrac{15}{4}a$

7. $\displaystyle c\int_{1}^{2} x^{-3}\,dx = \dfrac{cx^{-2}}{-2}\Big|_{1}^{2}$

$\qquad\qquad = -\tfrac{1}{2}c[\tfrac{1}{4} - 1]$

$\qquad\qquad = \dfrac{3c}{8}$

9. $\displaystyle\int_{0}^{9} x^{1/2}\,dx = \tfrac{2}{3}x^{3/2}\Big|_{0}^{9}$

$\qquad\qquad = \tfrac{2}{3}(27 - 0)$

$\qquad\qquad = 18$

11. $\displaystyle\int_{0}^{1} (5u^7 + \pi^2)\,du = \left(\tfrac{5}{8}u^8 + \pi^2 u\right)\Big|_{0}^{1}$

$\qquad\qquad = \tfrac{5}{8} + \pi^2$

13. $\displaystyle\int_{1}^{2} x^{2a}\,dx = \dfrac{x^{2a+1}}{2a+1}\Big|_{1}^{2}$

$\qquad\qquad = \dfrac{2^{2a+1}}{2a+1} - \dfrac{1}{2a+1}$

$\qquad\qquad = \dfrac{2^{2a+1} - 1}{2a+1}$

15. $\displaystyle 5\int_{\ln 2}^{\ln 5} e^x\,dx = 5e^x\Big|_{\ln 2}^{\ln 5}$

$\qquad\qquad = 25 - 10$

$\qquad\qquad = 15$

17. $\displaystyle\int_{0}^{4} \sqrt{x}(x + 1)\,dx = \int_{0}^{4} (x^{3/2} + x^{1/2})\,dx$

$\qquad\qquad = \left[\tfrac{2}{5}x^{5/2} + \tfrac{2}{3}x^{3/2}\right]\Big|_{0}^{4}$

$\qquad\qquad = \dfrac{272}{15}$

19. $\displaystyle\int_{1}^{2} \dfrac{x^3 + 1}{x^2}\,dx = \int_{1}^{2} (x + x^{-2})\,dx$

$\qquad\qquad = \left[\tfrac{1}{2}x^2 - x^{-1}\right]\Big|_{1}^{2}$

$\qquad\qquad = 2$

21. $\displaystyle 6a\int_{1}^{\sqrt{3}} \dfrac{dx}{1 + x^2} = 6a\tan^{-1}x\Big|_{1}^{\sqrt{3}}$

$\qquad\qquad = 6a(\tfrac{\pi}{3} - \tfrac{\pi}{4})$

$\qquad\qquad = \dfrac{a\pi}{2}$

23. **SURVIVAL HINT:** *Note the use of a trigonometric identity **before** integration.*

Simplify expressions whenever possible.

$$\int_{-2}^{3} (\sin^2 x + \cos^2 x)\, dx = \int_{-2}^{3} dx$$

$$= x\Big|_{-2}^{3}$$

$$= 5$$

25. $\displaystyle\int_{0}^{1} (1 - e^t)\, dt = (t - e^t)\Big|_{0}^{1}$

$$= 2 - e$$

27. $\displaystyle\int_{0}^{1} \frac{x^2 - 4}{x - 2}\, dx = \int_{0}^{1} \frac{(x - 2)(x + 2)}{x - 2}\, dx$

$$= \int_{0}^{1} (x + 2)\, dx$$

$$= \left(\frac{x^2}{2} + 2x\right)\Big|_{0}^{1}$$

$$= \frac{5}{2}$$

29. Recall $|x| = x$ if $x \geq 0$ and $|x| = -x$ if $x < 0$.

$$\int_{-1}^{2} (x + |x|)\, dx = \int_{-1}^{0} (x - x)\, dx + \int_{0}^{2} (x + x)\, dx$$

$$= 0 + \frac{2x^2}{2}\Big|_{0}^{2}$$

$$= 4$$

31. $f(x) = x^2 + 1$ is continuous on $[-1, 1]$ and $f(x) \geq 0$ on the interval, we have

$$\int_{-1}^{1} (x^2 + 1)\, dx = \left(\frac{x^3}{3} + x\right)\Big|_{-1}^{1} = \frac{8}{3}$$

33. $f(x) = \sec^2 x$ is continuous on $[0, \frac{\pi}{4}]$ and $f(x) \geq 0$ on the interval, we have

$$\int_{0}^{\pi/4} \sec^2 x\, dx = \tan x\Big|_{0}^{\pi/4}$$

$$= \tan \frac{\pi}{4} - \tan 0$$

$$= 1$$

35. $f(t) = e^t - t$ is continuous on $[0, 1]$ and $f(t) \geq 0$ on the interval, we have

$$\int_{0}^{1} (e^t - t)\, dt = (e^t - \tfrac{1}{2}t^2)\Big|_{0}^{1}$$

$$= e - \frac{3}{2}$$

37. $f(x) = \dfrac{x^2 - 2x + 3}{x} = x - 2 + 3x^{-1}$ is continuous on $[1, 2]$ and $f(x) \geq 0$ on the interval, we have

$$\int_{1}^{2} (x - 2 + 3x^{-1})\, dx = \left(\frac{x^2}{2} - 2x + 3\ln|x|\right)\Big|_{1}^{2}$$

$$= 3 \ln 2 - \frac{1}{2}$$

39. $F'(x) = \dfrac{x^2 - 1}{\sqrt{x + 1}}$ or $(x - 1)\sqrt{x + 1}$

41. $F'(t) = \dfrac{\sin t}{t}$

43. $F'(x) = \dfrac{-1}{\sqrt{1 + 3x^2}}$

45. $\dfrac{d}{du}\left(\dfrac{u}{2} + \dfrac{\sin 2au}{4a} + C\right) = \dfrac{1}{2} + \dfrac{2a}{4a}\cos 2au$

$$= \frac{1}{2} + \frac{1}{2} \cos 2au$$

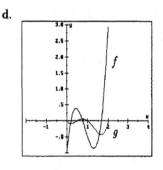

$$= \tfrac{1}{2} + \tfrac{1}{2}(2\cos^2 au - 1)$$

$$= \cos^2 au$$

47. $\dfrac{d}{du}\left(\dfrac{1}{\sqrt{a^2 - u^2}} + C \right) = -\tfrac{1}{2}(a^2 - u^2)^{-3/2}(-2u)$

$$= \dfrac{u}{(a^2 - u^2)^{3/2}}$$

49. $\dfrac{d}{du}[-(a^2 - u^2)^{1/2} + C]$

$$= -\tfrac{1}{2}(a^2 - u^2)^{-1/2}(-2u)$$

$$= \dfrac{u}{\sqrt{a^2 - u^2}}$$

51. If $f(x) \geq 0$ on $[a, b]$, the integral $\displaystyle\int_a^b f(x)\,dx$ is the same as the area under $y = f(x)$ and above the x-axis on $[a, b]$. If $f(x) \leq 0$ on $[a, b]$, the integral is the opposite of the area above $y = f(x)$ and below the x-axis on $[a, b]$. If $f(x) < 0$ over part of the interval of integration, the integral and the area are not the same, and will be discussed in Section 6.1. Thus, area is an integral, but an integral is not necessarily an area.

53. $\displaystyle\int_{-1}^{1} \dfrac{dx}{x^2}$ does not make sense since $\dfrac{1}{x^2}$ is not defined at $x = 0$.

55. $\displaystyle\int_0^{\pi} f(x)\,dx = \int_0^{\pi/2} f(x)\,dx + \int_{\pi/2}^{\pi} f(x)\,dx$

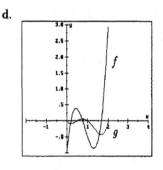

$$= \int_0^{\pi/2} \cos x\,dx + \int_{\pi/2}^{\pi} x\,dx$$

$$= \sin x \Big|_0^{\pi/2} + \tfrac{1}{2}x^2 \Big|_{\pi/2}^{\pi}$$

$$= 1 - 0 + \dfrac{\pi^2}{2} - \dfrac{\pi^2}{8}$$

$$= \tfrac{1}{8}(8 + 3\pi^2)$$

$$\approx 4.7011$$

57. a. It is a relative minimum because the derivative (function f) shows the $\downarrow\uparrow$ pattern. Also, $g(a) < 0$, g' is rising.

b. Obviously it is concave down at $x = 0.5$ and concave up at $x = 1.25$, so there must be an inflection point somewhere between those points, say at $x = 1$. We estimate $g(1) = 0$.

c. The function f crosses at x-axis 3 times, at $x = a$, $x = b$, and $x = c$. Now $g(a)$ is a relative minimum, and at $x = c$, $g(c)$ is also a relative minimum (function f has the $\uparrow\downarrow$ pattern). Thus, g has a relative maximum at $x = b$ (f has the $\downarrow\uparrow$ pattern), and we estimate this to be at $x = 0.75$.

d.

59. a. For our purpose with these counterexamples, we let all the constants of integration be zero.

$$\int x\sqrt{x}\ dx = \int x^{3/2}\ dx = \tfrac{2}{5}x^{5/2}$$

$$\int x\ dx = \tfrac{1}{2}x^2; \quad \int \sqrt{x}\ dx = \tfrac{2}{3}x^{3/2}$$

Since $\tfrac{2}{5}x^{5/2} \neq \left(\tfrac{1}{2}x^2\right)\left(\tfrac{2}{3}x^{3/2}\right)$

the result follows.

b. $\int \dfrac{\sqrt{x}}{x}\ dx = \int x^{-1/2}\ dx = 2\sqrt{x}$

Since $2\sqrt{x} \neq \dfrac{\tfrac{2}{3}x^{3/2}}{\tfrac{1}{2}x^2}$, the result follows.

61. $\dfrac{d}{dx}\displaystyle\int_0^{x^4}(2t\ -\ 3)\ dt = \dfrac{d}{du}\left[\displaystyle\int_0^u(2t\ -\ 3)\ dt\right]\dfrac{du}{dx}$

$$= (2u\ -\ 3)(4x^3) = (2x^4\ -\ 3)(4x^3)$$

$$= 8x^7\ -\ 12x^3$$

Check: $\displaystyle\int_0^{x^4}(2t\ -\ 3)\ dt = t^2\ -\ 3t\ \Big|_0^{x^4}$

$$= x^8\ -\ 3x^4$$

Thus, $\dfrac{d}{dx}(x^8\ -\ 3x^4) = 8x^7\ -\ 12x^3$

63. Let $G(x) = \displaystyle\int_1^x\dfrac{2t+1}{t+2}\ dt.$

Then, $F(x) = G(u(x))$ where $u(x) = \sqrt{x}$ and

$$F'(x) = G'(u)u'(x)$$

$$= \left[\dfrac{2u+1}{u+2}\right]\left(\dfrac{1}{2}\ \dfrac{1}{\sqrt{x}}\right)$$

$$= \dfrac{2\sqrt{x}+1}{2\sqrt{x}(\sqrt{x}+2)}$$

$F(1) = 0$ and $F'(1) = \tfrac{1}{2}$, so the equation of the tangent line at P is

$$y = \tfrac{1}{2}(x\ -\ 1)$$

65. $\displaystyle\int_u^v f(t)\ dt = \int_u^a f(t)\ dt + \int_a^v f(t)\ dt$

$$= -\int_a^u f(t)\ dt + \int_a^v f(t)\ dt$$

Differentiate both sides:

$$\dfrac{d}{dx}\left[\int_u^v f(t)\ dt\right] = \dfrac{d}{dx}\left[-\int_a^u f(t)\ dt\right] + \dfrac{d}{dx}\left[\int_a^v f(t)\ dt\right]$$

$$= -f(u)\dfrac{du}{dx} + f(v)\dfrac{dv}{dx}$$

Thus, $F'(x) = f(v)\dfrac{dv}{dx}\ -\ f(u)\dfrac{du}{dx}$

5.5 Integration by Substitution, Pages 315–316

1. a. $\displaystyle\int_0^4(2t+4)\ dt = (t^2+4t)\Big|_0^4$

$$= 32$$

b. $\displaystyle\int_0^4(2t+4)^{-1/2}\ dt = \int_4^{12} u^{-1/2}(\tfrac{1}{2}\ du)$

$$\boxed{\begin{array}{l} u = 2t+4;\ du = 2\ dt; \\ \text{if } t = 0,\ u = 4;\ \text{if } t = 4,\ u = 12 \end{array}}$$

$$= u^{1/2}\Big|_4^{12}$$

$$= 2\sqrt{3} - 2$$

3. a.
$$\int_0^\pi \cos t \, dt = \sin t \Big|_0^\pi$$

$$= 0$$

b.
$$\int_0^{\sqrt{\pi}} t \cos t^2 dt = \frac{1}{2}\int_0^\pi \cos u \, du$$

$$u = t^2;\ du = 2t\ dt;$$

$$\text{if } t = 0,\ u = 0;\ \text{if } t = \sqrt{\pi},\ u = \pi$$

$$= \frac{1}{2}\sin u \Big|_0^\pi$$

$$= 0$$

5. a.
$$\int_0^{16} \sqrt[4]{x}\ dx = \frac{4}{5}x^{5/4}\Big|_0^{16}$$

$$= \frac{128}{5}$$

b.
$$\int_{-16}^0 \sqrt[4]{-x}\ dx = -\int_{16}^0 u^{1/4} du$$

$$u = -x;\ du = -dx;$$

$$\text{if } x = -16,\ u = 16;\ x = 0,\ u = 0$$

$$= -\frac{4}{5}u^{5/4}\Big|_{16}^0$$

$$= \frac{128}{5}$$

7. a.
$$\int x^2 \sqrt{2x^3}\ dx = \sqrt{2}\int x^{7/2}\ dx$$

$$= \frac{2}{9}\sqrt{2}\,x^{9/2} + C$$

b.
$$\int x^2 \sqrt{2x^3 - 5}\ dx = \frac{1}{6}\int u^{1/2}\ du$$

$$u = 2x^3 - 5;\ du = 6x^2\ dx$$

$$= \frac{1}{9}(2x^3 - 5)^{3/2} + C$$

SURVIVAL HINT: *Do not try to do any but the simplest u substitutions in your head. You will have fewer errors, and save time in the long run, if for each problem you write down:* $u = \ldots$ and $du = \ldots$. *Most errors in u substitutions come from a failure to properly introduce constants to "set-up the du."*

9.
$$\text{Let } u = 2x + 3;\ du = dx$$

$$\int (2x + 3)^4\ dx = \frac{1}{2}\int u^4\ du$$

$$= \frac{1}{2}\cdot\frac{1}{5}u^5 + C$$

$$= \frac{1}{10}(2x + 3)^5 + C$$

11.
$$\text{Let } u = x^2 + 5x + 3;\ du = (2x + 5)\ dx$$

$$\int \tan(x^2 + 5x + 3)(2x + 5)\ dx$$

$$= \int \tan u\ du$$

$$= -\ln\big|\cos(x^2 + 5x + 3)\big| + C$$

Alternatively, $\ln\big|\sec(x^2 + 5x + 3)\big| + C$.

13. For the second part of the integral, let

$$\text{Let } u = 3x;\ du = 3\ dx$$

$$\int (x^2 - \cos 3x)\ dx = \frac{1}{3}x^3 - \frac{1}{3}\int \cos u \cdot du$$

$$= \frac{1}{3}x^3 - \frac{1}{3}\sin 3x + C$$

15. Let $u = 4 - x,\ du = -\,dx$

$$\int \sin(4 - x)\,dx = -\int \sin u\,du$$

$$= \cos(4 - x) + C$$

17. Let $u = t^{3/2} + 5;\ du = \tfrac{3}{2}t^{1/2}\,dt$

$$\int \sqrt{t}(t^{3/2}+5)^3\,dt = \tfrac{2}{3}\int u^3\,du$$

$$= \tfrac{1}{6}(t^{3/2} + 5)^4 + C$$

19. Let $u = 3 + x^2;\ du = 2x\,dx$

$$\int x\sin(3 + x^2)\,dx = \tfrac{1}{2}\int \sin u\,du$$

$$= -\tfrac{1}{2}\cos(3 + x^2) + C$$

21. Let $u = 2x^2 + 3;\ du = 4x\,dx$

$$\int \frac{x\,dx}{2x^2 + 3} = \tfrac{1}{4}\int \frac{du}{u}$$

$$= \tfrac{1}{4}\ln|u| + C$$

$$= \tfrac{1}{4}\ln(2x^2 + 3) + C$$

23. Let $u = 2x^2 + 1;\ du = 4x\,dx$

$$\int x\sqrt{2x^2 + 1}\,dx = \tfrac{1}{4}\int u^{1/2}\,du$$

$$= \tfrac{1}{4}(\tfrac{2}{3})u^{3/2} + C$$

$$= \tfrac{1}{6}(2x^2 + 1)^{3/2} + C$$

25. Let $u = x\sqrt{x} = x^{3/2};\ du = \tfrac{3}{2}x^{1/2}\,dx$

$$\int \sqrt{x}\,e^{x\sqrt{x}}\,dx = \tfrac{2}{3}\int e^u\,du$$

$$= \tfrac{2}{3}e^{x^{3/2}} + C$$

27. Let $u = x^2 + 4;\ du = 2x\,dx$

$$\int x(x^2 + 4)^{1/2}\,dx = \tfrac{1}{2}\int u^{1/2}\,du$$

$$= \tfrac{1}{2}\cdot\tfrac{2}{3}u^{3/2} + C$$

$$= \tfrac{1}{3}(x^2 + 4)^{3/2} + C$$

29. Let $u = \ln x;\ du = \tfrac{1}{x}\,dx$

$$\int \frac{\ln x}{x}\,dx = \int u\,du$$

$$= \tfrac{1}{2}u^2 + C$$

$$= \tfrac{1}{2}(\ln x)^2 + C$$

31. $u = x^{1/2} + 7;\ du = \tfrac{1}{2}x^{-1/2}\,dx$

$$\int \frac{dx}{\sqrt{x}(\sqrt{x} + 7)} = 2\int u^{-1}\,du$$

$$= 2\ln|u| + C$$

$$= 2\ln(\sqrt{x} + 7) + C$$

33. $u = e^t + 1;\ du = e^t\,dt$

$$\int \frac{e^t\,dt}{e^t + 1} = \int u^{-1}\,du$$

$$= \ln|u| + C$$

$$= \ln(e^t + 1) + C$$

35. $$\int_0^1 \frac{5x^2\,dx}{2x^3 + 1} = \tfrac{5}{6}\int_0^1 \frac{6x^2\,dx}{2x^3 + 1}$$

Let $u = 2x^3 + 1;\ du = 6x^2\,dx$

$x = 0,\ u = 1;\ x = 1,\ u = 3$

$$= \frac{5}{6} \int_1^3 \frac{du}{u}$$

$$= \frac{5}{6}(\ln 3 - \ln 1)$$

$$= \frac{5}{6}\ln 3$$

37. $\displaystyle\int_{-\ln 2}^{\ln 2} \frac{1}{2}(e^x - e^{-x})dx$

$$= \frac{1}{2}\int_{-\ln 2}^{\ln 2} e^x\,dx + \frac{1}{2}\int_{-\ln 2}^{\ln 2} e^{-x}(-dx)$$

$$= \frac{1}{2}(e^{\ln 2} - e^{-\ln 2} + e^{-\ln 2} - e^{\ln 2})$$

$$= 0$$

39. $\displaystyle\int_1^2 \frac{e^{1/x}\,dx}{x^2} = -\int_1^2 e^{1/x}\left(-\frac{dx}{x^2}\right)$

> Let $u = \frac{1}{x};\ du = -\frac{dx}{x^2};$ if $x = 1,\ u = 1;$
> if $x = 2,\ u = \frac{1}{2}.$

$$= -\int_1^{1/2} e^u\,du$$

$$= -e^u\Big|_1^{1/2}$$

$$= e - e^{1/2}$$

41. $\displaystyle\int_0^{\pi/6} \tan 2x\,dx = -\frac{1}{2}\int_0^{1/2} \frac{du}{u}$

> $u = \cos 2x;\ du = -2\sin 2x\,dx;$
> $x = 0,\ u = 1;\ x = \pi/6,\ u = 1/2$

$$= -\frac{1}{2}\ln|u|\Big|_1^{1/2}$$

$$= \frac{1}{2}\ln 2$$

43. $\displaystyle\int_0^5 \frac{0.58}{1 + e^{-0.2x}}\,dx = -\frac{0.58}{0.2}\int_0^5 \frac{-0.2\,dx}{1 + e^{-0.2x}}$

> $u = -0.2x;\ du = -0.2\,dx;$
> $x = 0,\ u = 0;\ x = 5,\ u = -1$

$$= -2.9\int_0^{-1} \frac{du}{1 + e^u}$$

$$= = 2.9\int_0^{-1} \frac{-e^{-u}\,du}{e^{-u} + 1}$$

$$= 2.9\ln|1 + e^{-u}|\Big|_0^{-1}$$

$$= 2.9[\ln(1 + e) - \ln 2]$$

$$= 2.9\ \ln\left(\frac{e}{2} + \frac{1}{2}\right) \approx 1.80$$

45. a. We take 1 Frdor as the variable so the note from the students reads, "Because of illness I cannot lecture between Easter and Michaelmas."

b. The Dirichlet function is defined as a function f so that $f(x)$ equals a determined constant c (usually 1) when the variable x takes a rational value, and equals another constant d (usually 0) when this variable is irrational. This famous function is one which is discontinuous everywhere.

47. $f(t) = \frac{1}{t^2}\sqrt{5 - \frac{1}{t}}$ is continuous and positive on $[\frac{1}{5}, 1]$.

$$\int_{1/5}^{1} \frac{1}{t^2}\sqrt{5 - \frac{1}{t}}\, dt = \int_{0}^{4} u^{1/2}\, du$$

$u = 5 - t^{-1};\ du = t^{-2}\, dt;\ t = 1/5,$
then $u = 0;\ t = 1,$ then $u = 4$

$$= \frac{2}{3}u^{3/2}\Big|_{0}^{4}$$

$$= \frac{16}{3}$$

49. $f(x) = \dfrac{x}{\sqrt{x^2 + 1}}$ is continuous and positive on

$[1, 3]$.

$$\int_{1}^{3} \frac{x}{\sqrt{x^2 + 1}}\, dx = \int_{2}^{10} \frac{1}{2}\frac{1}{\sqrt{u}}\, du$$

$u = x^2 + 1;\ du = 2x\, dx;\ x = 1,$
then $u = 2,\ x = 3,$ then $u = 10$

$$= \frac{1}{2}\int_{2}^{10} u^{-1/2}\, du$$

$$= \frac{1}{2}\cdot\frac{2}{1}u^{1/2}\Big|_{2}^{10}$$

$$= \sqrt{10} - \sqrt{2}$$

51. $\displaystyle\int_{-\pi}^{\pi} \sin x\, dx = 0$ since $\sin x$ is odd.

53. $\displaystyle\int_{-3}^{3} x\sqrt{x^4 + 1}\, dx = 0$ since $x\sqrt{x^4 + 1}$ is odd.

55. **a.** $f(x) = 7x^{1001} + 14x^{99}$ is odd, so

$$\int_{-175}^{175} (7x^{1001} + 14x^{99})\, dx = 0;\ \text{true}$$

b. $\displaystyle\int_{0}^{\pi} (\sin^2 x - \cos^2 x)\, dx = -\int_{0}^{\pi} \cos 2x\, dx$

$$= -\frac{\sin 2x}{2}\Big|_{0}^{\pi}$$

$$= 0$$

so $\displaystyle\int_{0}^{\pi} \sin^2 x\, dx = \int_{0}^{\pi} \cos^2 x\, dx$ is true

c. $\displaystyle\int_{-\pi/2}^{\pi/2} \cos x\, dx = \sin x\Big|_{-\pi/2}^{\pi/2} = 2$

$$\int_{-\pi}^{0} \sin x\, dx = -\cos x\Big|_{-\pi}^{0} = -2$$

The given statement is false.

57. $\dfrac{dy}{dx} = \dfrac{2x}{1 - 3x^2}$

$$F(x) = \int \frac{2x}{1 - 3x^2}\, dx$$

$u = 1 - 3x^2;\ du = -6x\, dx$

$$= -\frac{1}{3}\int \frac{du}{u}$$

$$= -\frac{1}{3}\ln|u| + C$$

$$= -\frac{1}{3}\ln|1 - 3x^2| + C$$

$F(0) = -\frac{1}{3}\ln|1| + C = 5$ implies $C = 5$, so
$F(x) = -\frac{1}{3}\ln|1 - 3x^2| + 5$

59. Water flows into the tank at the rate of

$$R(t) = V'(t) = t(3t^2 + 1)^{-1/2} \text{ ft}^3/\text{s}$$

The volume at time t is

$$V(t) = \int t(3t^2 + 10)^{-1/2} \, dt$$

$$u = 3t^2 + 10; \quad du = 6t \, dt;$$

$$= \frac{1}{6} \int u^{-1/2} \, du$$

$$= \frac{1}{3} u^{1/2}$$

$$= \frac{1}{3}\sqrt{3t^2 + 1} + C$$

The tank is empty to start, so

$$V(0) = 0 = \frac{1}{3} + C, \text{ so } C = -\frac{1}{3}$$

Thus,

$$V(t) = \frac{1}{3}(\sqrt{3t^2 + 1} - 1)$$

$$V(4) = \frac{1}{3}(\sqrt{49} - 1) = 2 \text{ ft}^3$$

The amount of water at 4 seconds is 2 ft^3.

The height h is given by the equation

$$100h = 2, \text{ so } h = \frac{1}{50} \text{ ft or } \frac{12}{50} = 0.24 \text{ in.}$$

The depth at that time is about $\frac{1}{4}$ in.

61. a.

$$L(t) = \int \frac{0.24 - 0.03t}{\sqrt{36 + 16t - t^2}} \, dt$$

$$= \int \frac{0.015(16 - 2t) \, dt}{\sqrt{36 + 16t - t^2}}$$

Let $u = 36 + 16t - t^2; \quad du = (16 - 2t) \, dt$

$$= 0.015 \int \frac{du}{\sqrt{u}}$$

$$= 0.015(2)u^{1/2} + C$$

$$= 0.03\sqrt{36 + 16t - t^2} + C$$

If $t = 0$, then $4 = 0.18 + C$ or $C = 3.82$

$$L(t) = 0.03\sqrt{36 + 16t - t^2} + 3.82$$

b.

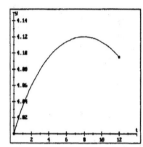

The highest level occurs when $L'(t) = 0$ or when $0.24 - 0.03t = 0$; $t = 8$ (at 3 P.M.). The highest level is 4.12 ppm.

c.

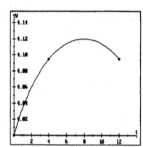

It is the same as 11:00 A.M. ($t = 4$) at 7:00 P.M. (when $t = 12$).

5.6 Introduction to Differential Equations, Pages 325-328

SURVIVAL HINT: *The solution to a differential equation, like the process of antidifferentiation, has an infinite set of functions (all translated vertically). Technology, if available, will draw a direction field of the solution, and can be used to sketch a particular member of that infinite set. Unless you are given initial values that will specify one particular function from this set, you must remember to include "+ C" in the general solution.*

1. $x^2 + y^2 = 7$

    $2x + 2y\dfrac{dy}{dx} = 0$

    $\dfrac{dy}{dx} = -\dfrac{x}{y}$

3. $xy = C$

    $x\dfrac{dy}{dx} + y = 0$

    $\dfrac{dy}{dx} = -\dfrac{y}{x}$

5. $\dfrac{dy}{dx} = A\cos(Ax + B)$

    $\dfrac{d^2y}{dx^2} = -A^2\sin(Ax + B)$

    Thus,

    $\dfrac{d^2y}{dx^2} + A^2y = -A^2\sin(Ax + B) + A^2\sin(Ax + B)$

    $\qquad\qquad = 0$

7. $y = 2e^{-x} + 3e^{2x}$

    $y' = -2e^{-x} + 6e^{2x}$

    $y'' = 2e^{-x} + 12e^{2x}$

    $y'' - y' - 2y$

    $= (2e^{-x} + 12e^{2x}) - (-2e^{-x} + 6e^{2x})$

    $\quad - 2(2e^{-x} + 3e^{2x})$

    $= 0$

9. $\dfrac{dy}{dx} = -\dfrac{x}{y}$

    $y\,dy = -x\,dx$

    $\displaystyle\int y\,dy = -\int x\,dx$

$\dfrac{y^2}{2} = -\dfrac{x^2}{2} + C_1$

$x^2 + y^2 = C$

Passes through $(2, 2)$, so $4 + 4 = C$;

$x^2 + y^2 = 8$

11. $\dfrac{dy}{dx} - y^2 = 1$

    $\dfrac{dy}{dx} = y^2 + 1$

    $\displaystyle\int\dfrac{dy}{y^2 + 1} = \int dx$

    $\tan^{-1}y = x + C$

    $y = \tan(x + C)$

    Note: $y = \tan(x + C) \neq \tan x + C$, so be careful here; in other words, we know that the constant must be supplied immediately after the last integration. Since the curve passes through $(\pi, 1)$,

    $\dfrac{\pi}{4} = \pi + C$

    $-\dfrac{3\pi}{4} = C$

    Thus, $y = \tan(x - \dfrac{3\pi}{4})$.

13. $\dfrac{dy}{dx} = \sqrt{\dfrac{x}{y}}$

    $y^{1/2}\,dy = x^{1/2}\,dx$

    $\displaystyle\int y^{1/2}\,dy = \int x^{1/2}\,dx$

    $\dfrac{2}{3}y^{3/2} = \dfrac{2}{3}x^{3/2} + C_1$

$$x^{3/2} - y^{3/2} = C$$

Since the curve passes through $(4, 1)$,

$$8 - 1 = C$$

$$C = 7$$

Thus, $x^{3/2} - y^{3/2} = 7$

15. First plot the given point, namely $(0, 1)$. Next, sketch the curve (using a pencil) from point to point using adjacent slope marks to move from point to point.

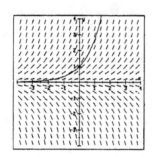

17. First plot the given point, namely $(0, 0)$. Next, sketch the curve (using a pencil) from point to point using adjacent slope marks to move from point to point.

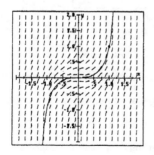

19. First plot the given point, namely $(1, 0)$. Next, sketch the curve (using a pencil) from point to point using adjacent slope marks to move from point to point.

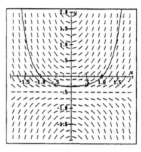

21.
$$\frac{dy}{dx} = 3xy$$

$$y^{-1} \, dy = 3x \, dx$$

$$\int y^{-1} \, dy = \int 3x \, dx$$

$$\ln|y| = \tfrac{3}{2}x^2 + C$$

$$y = \exp[(\tfrac{3}{2})x^2 + C]$$

$$= Be^{(3/2)x^2} \quad \text{for a constant } B$$

23.
$$\frac{dy}{dx} = \frac{x}{y}\sqrt{1 - x^2}$$

$$y \, dy = x\sqrt{1 - x^2} \, dx$$

$$\int y \, dy = \int x\sqrt{1 - x^2} \, dx$$

$$\int y \, dy = \int (-\tfrac{1}{2})u^{1/2} du$$

> Let $u = (1 - x^2)$; $\quad du = -2x \, dx$

$$\frac{y^2}{2} = -\frac{1}{2}\frac{(1-x^2)^{3/2}}{\frac{3}{2}} + C_1$$

$$\frac{(1-x^2)^{3/2}}{3} + \frac{y^2}{2} = C_1$$

$$2(1-x^2)^{3/2} + 3y^2 = C$$

25. $xy\ dx + \sqrt{xy}\ dy = 0$

$$\sqrt{xy}\ dx = -\,dy \quad (\text{if } xy \neq 0)$$

$$x^{1/2}\ dx = -y^{-1/2}\ dy$$

$$\int x^{1/2}\ dx = -\int y^{-1/2}\ dy$$

$$\frac{2}{3}x^{3/2} = -2y^{1/2} + C_1$$

$$x^{3/2} + 3y^{1/2} = C$$

27. $\frac{dy}{dx} = \frac{\sin x}{\cos y}$

$$\cos y\ dy = \sin x\ dx$$

$$\int \cos y\ dy = \int \sin x\ dx$$

$$\sin y = -\cos x + C$$

$$\cos x + \sin y = C$$

29. $xy\frac{dy}{dx} = \frac{\ln x}{\sqrt{1-y^2}}$

$$\int y\sqrt{1-y^2}\ dy = \int \frac{\ln x}{x}\ dx$$

$u = 1 - y^2$	$v = \ln x$
$du = -2y\ dy$	$dv = \frac{1}{x}\ dx$

$$\int \sqrt{u}\left(-\frac{1}{2}\ du\right) = \int v\ dv$$

$$-\frac{1}{2}\frac{u^{3/2}}{\frac{3}{2}} = \frac{v^2}{2} + C$$

$$-\frac{1}{3}(1-y^2)^{3/2} = \frac{1}{2}(\ln x)^2 + C$$

31. $x\ dy + y\ dx = 0$

$$d(xy) = 0, \text{ so } xy = C$$

33. Write $y\ dx = x\ dy$ as $\frac{x\ dy - y\ dx}{x^2} = 0;$

$$\frac{d}{dx}\left(\frac{y}{x}\right) = 0, \text{ so } \frac{y}{x} = C \text{ or } y = Cx$$

35. Family of curves: $2x - 3y = C;$ differentiating with respect to x leads to the slope of the tangent lines $dy/dx = 2/3$. For the orthogonal trajectories, the slope is the negative reciprocal, or $dY/dX = -3/2$. Integrating leads to the orthogonal trajectories: $2Y + 3X = K$

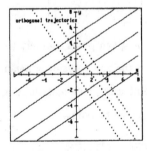

37. Family of curves: $y = x^3 + C;$ differentiating with respect to x leads to the slope of the tangent lines $dy/dx = 3x^2$. For the orthogonal trajectories, the slope is the negative reciprocal, or $dY/dX = -X^{-2}/3$. Integrating leads to the orthogonal

trajectories: $Y = \frac{1}{3}X^{-1} + K$

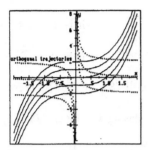

Note the differences in scale between the x- and y-axis. This makes the orthogonal curves "look like" they do not meet at right angles.

39. Family of curves: $xy^2 = C$; differentiating implicitly with respect to x leads to the slope of tangent lines to $dy/dx = -y/2x$. For the orthogonal trajectories, the slope is the negative reciprocal, or $dY/dX = 2X/Y$. Integrating leads to the orthogonal trajectories:

$$Y^2 - 2X^2 = K_1, \text{ or } X^2 - \frac{Y^2}{2} = K$$

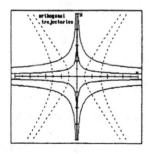

41. $$x^2 + y^2 = r^2$$

$$2x + 2yy' = 0$$

$$y' = -\frac{x}{y}$$

The slope of the orthogonal trajectory is the negative reciprocal:

$$\frac{dY}{dX} = \frac{Y}{X}$$

$$\int Y^{-1}\, dY = \int X^{-1}\, dX$$

$$\ln|Y| = \ln|X| + K$$

$$\left|\frac{Y}{X}\right| = e^K = C$$

$$Y = CX$$

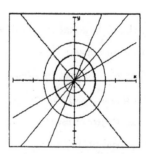

43. Let Q denote the number of bacteria. Then, dQ/dt is the rate of change of Q, and since this rate of change is proportional to Q, it follows that

$$\frac{dQ}{dt} = kQ$$

where k is a positive constant of proportionality.

45. Let T be temperature, t be time, T_m be temperature of the surrounding medium, and c be the constant of proportionality. Then:

$$\frac{dT}{dt} = c(T - T_m)$$

47. Let t denote time, Q the number of residents who have been infected, and B the total number of susceptible residents. The differentiable equation describing the spread of the epidemic is

$$\frac{dQ}{dt} = kQ(B - Q)$$

where k is the positive constant of proportionality.

49. Conjecture: the orthogonal trajectories are circles. Family of curves: $x^2 - y^2 = C$; differentiating with respect to x leads to the slope of the tangent lines $dy/dx = x/y$. For the orthogonal trajectories, the slope is the negative reciprocal of $dY/dX = -Y/X$. Integrating leads to $Y = C/X$.

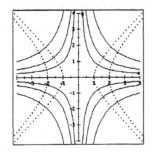

51.
$$Q = Q_0 e^{kt}$$

$$\frac{Q}{Q_0} = e^{k(2,047)} \qquad \text{where } t = 1947 + 100$$

$$\approx 0.780655 \qquad \text{where } k = \frac{\ln(1/2)}{5,730}$$

There was still 78% of the ^{14}C present.

53. For carbon dating $k = \ln(1/2)/5,730$;

$$\frac{Q}{Q_0} = e^{kt}$$

$$0.923 = e^{kt} \quad k \text{ is the known value given above}$$

$$t = \frac{\ln 0.923}{k} \approx 662$$

This dates the Shroud at around $1988 - 663 \approx 1326$ AD.

55. By Torricelli's Law, $\frac{dV}{dt} = -4.8 A_0 \sqrt{h}$.

Given a square hole with $s = 1.5$ in. $= \frac{1}{8}$ ft,

$A_0 = (\frac{1}{8})^2 = \frac{1}{64}$. So $\frac{dV}{dt} = -4.8(\frac{1}{64})\sqrt{h}$.

$\frac{dV}{dt} = 9\pi \frac{dh}{dt}$. Substituting, we have:

$$-(4.8)\frac{\sqrt{h}}{64} = 9\pi \frac{dh}{dt}$$

$$\int -\frac{4.8}{9\pi(64)} \, dt = \int h^{-1/2} \, dh$$

$$-\frac{1}{120\pi} t = 2h^{1/2} + C$$

$$t = -240\pi\sqrt{h} + C$$

When the tank is full, $t = 0$, $h = 5$.

$$0 = -240\pi\sqrt{5} + C, \quad C = 240\pi\sqrt{5}$$

We have the formula relating time and the

height of the water:

$$t = -240\pi\sqrt{h} + 240\pi\sqrt{5}$$

The tank will be empty when $h = 0$. The time required for this is: $t = 240\pi\sqrt{5}$ sec ≈ 28 min.

57. **a.** $v^2 = \dfrac{2gR^2}{s} + v_0{}^2 - 2gR$

$$= \frac{2(32)(3,956)^2(5,280)^2}{(3,956)(5,280) + 200}$$

$$+ 150^2 - 2(32)(3,956)(5,280)$$

$$= 9,700 \text{ or } v \approx 98.5 \text{ ft/s}$$

b. The velocity at the maximum height is 0:

$$0 = \frac{2gR^2}{s} + v_0{}^2 - 2gR \text{ so}$$

$R/s = 1 - v_0{}^2/(2gR) \approx 0.9999831689$; thus,

$s \approx 3,956.067$. We now find

$h \approx 0.067$ mi ≈ 352 ft.

59. $\dfrac{dP}{dt} = 1,500 \; t^{-1/2}$ so

$$P(t) = \int 1,500 \; t^{-1/2}dt$$

$$= 3,000 \sqrt{t} + C.$$

1994 is four years after 1990, so

$$P(4) = P(4)$$

$$= 39,000$$

and

$$39,000 = 3000\sqrt{4} + C,$$

$$C = 33,000.$$

$$P(t) = 3,000\sqrt{t} + 33,000$$

a. $P(0) = 33,000$

b. $P(9) = 3,000\sqrt{9} + 33,000 = 42,000$

61. $$\frac{dP}{dt} = k\sqrt{P}$$

$$\int P^{-1/2} \; dP = \int k \; dt.$$

$$2\sqrt{P} = kt + C$$

$P(0) = 9,000$, so $C = 60\sqrt{10}$, and $P(-10) = 4,000$, so

$$2\sqrt{4,000} = -10k + 60\sqrt{10}$$

$$k = 2\sqrt{10}$$

We now have the equation:

$$2\sqrt{P} = 2\sqrt{10}t + 60\sqrt{10}$$

$$\sqrt{P} = \sqrt{10}t + 30\sqrt{10}$$

To find t for $P = 16,000$,

$$\sqrt{16,000} = \sqrt{10}t + 30\sqrt{10}$$

$$t = 10$$

The population will be 16,000 10 years from now.

63. Differentiating $V = 9\pi h^3$ leads to

$$\frac{dV}{dt} = 9\pi\left(3h^2 \frac{dh}{dt}\right) = -4.8A_0\sqrt{h}$$

from Torricelli's law. Since the area of the

hole (in ft^2) is $A_0 = \pi\left(\frac{1}{12}\right)^2 = \frac{\pi}{144}$

$$27\pi h^2 \frac{dh}{dt} = (-4.8)(\frac{\pi}{144})\sqrt{h}$$

$$\int h^{3/2}\, dh = -\int \frac{4.8\pi}{144(27\pi)}\, dt$$

$$\frac{2}{5} h^{5/2} = -\frac{1}{810}t + C$$

If $t = 0$, then $h = 4$, so that $C = \frac{64}{5}$. The height is zero when

$$0 = -\frac{1}{810}t + \frac{64}{5}$$

$$10{,}368 = t$$

10,368 sec is about 173 min or 2 hr and 53 min.

65. $Q(t) = Q_0 e^{kt}$

For neptunium-139:

$0.7336 = e^{24k}$, so $k = \dfrac{\ln 0.7336}{24}$

$$\approx -0.0129079733$$

For 43% we have:

$0.43 = e^{kt}$

$$t = \frac{\ln 0.43}{k} \approx 65.38$$

It will take about 65 h for 43% to be left.

The half-life is

$$t = \frac{\ln 0.5}{k} \approx 53.70$$

Neptunium-139 has a half-life of 53.70 hours.

67. $Q(t) = Q_0 e^{kt}$

$$0.50 = e^{46.5k}$$

$$k = \frac{\ln 0.5}{46.5}$$

$$\approx -0.014906391$$

$$Q(30) - Q(35) = 100 e^{k(30)} - 100 e^{k(35)}$$

$$= 4.59246357$$

The isotope loses

$$100\left(\frac{4.59246357}{63.94213007}\right) \approx 7.182\%$$

of its original volume over the 5 hour period. For any other 5 hour period, beginning at time t_1, the percentage lost is

$$100\left[\frac{100(e^{kt_1} - e^{k(t_1+5)})}{100 e^{kt_1}}\right] = 100(1 - e^{5k})$$

$$\approx 7.182\%$$

The percentage lost is always the same.

69. On the surface of the planet, $s = R$ and $a = -g$ Since $F = ma$

$$F = ma = \frac{mk}{R^2} \text{ so } k = -gR^2.$$

In general, at a distance of s,

$$F = ma = \frac{mk}{s^2}$$

$$a = \frac{k}{s^2} = -\frac{gR^2}{s^2}$$

5.7 The Mean Value Theorem for Integrals; Average Value, Pages 332-333

1. $f(x)$ is continuous on $[1, 2]$, so the MVT guarantees the existence of c such that:

$$\int_1^2 4x^3 \, dx = f(c)(2 - 1)$$

$$15 = f(c)$$

$$15 = 4c^3$$

$$c^3 = \frac{15}{4}$$

$$c = \frac{\sqrt[3]{30}}{2}$$

$$\approx 1.55$$

1.55 is in the interval $[1, 2]$

3. $$\int_1^5 15x^{-2} \, dx = f(c)(5 - 1)$$

$$-15x^{-1}\Big|_1^5 = 15c^{-2}(4)$$

$$12 = 60c^{-2}$$

$$c^2 = 5$$

$$c = \pm\sqrt{5}$$

$\sqrt{5} \approx 2.24$ is in the interval.

SURVIVAL HINT: *Remember, that the MVT is a theorem and as such, has hypotheses that must be verified before the theorem is valid. Know the hypotheses, and understand why they are necessary, for every theorem you learn.*

5. The mean value theorem does not apply because $f(x) = \csc x$ is discontinuous at $x = 0$.

7. $$\int_{-1/2}^{1/2} e^{2x} \, dx = f(c)[\tfrac{1}{2} - (-\tfrac{1}{2})]$$

$$\tfrac{1}{2}e^{2x}\Big|_{-1/2}^{1/2} = e^{2c}$$

$$\tfrac{1}{2}(e - e^{-1}) = e^{2c}$$

$$c \approx 0.0807$$

This value is in the interval.

9. $$\int_{-1}^1 \frac{x + 1}{1 + x^2} \, dx = f(c)(1 + 1)$$

$$\int_{-1}^1 \left[\frac{x}{1 + x^2} + \frac{1}{1 + x^2}\right] dx = \frac{2(c + 1)}{1 + c^2}$$

$$\left[\tfrac{1}{2}\ln(x^2 + 1) + \tan^{-1}x\right]\Big|_{-1}^1 = \frac{2(c + 1)}{1 + c^2}$$

$$\frac{\pi}{2} = \frac{2(c + 1)}{1 + c^2}$$

$$\tfrac{\pi}{4}c^2 - c + \tfrac{\pi}{4} - 1 = 0$$

$$c \approx 1.46, \, -0.187$$

-0.187 is in the interval

11. $$A = \int_0^{10} \frac{x}{2} \, dx$$

$$= \frac{x^2}{4}\Big|_0^{10}$$

$$= 25$$

$$f(c)(b - a) = (\tfrac{c}{2})(10)$$

$$= 25$$

so $c = 5$ and $f(5) = 2.5$

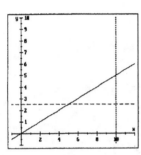

13. $\int\limits_{0}^{2} (x^2 + 2x + 3) \, dx = (\frac{1}{3}x^3 + x^2 + 3x)\Big|_{0}^{2}$

$$= \frac{38}{3}$$

$f(c)(b - a) = (c^2 + 2c + 3)(2)$

$$= \frac{38}{3}$$

so $3c^2 + 6c - 10 = 0$;

$c \approx 1.08$ (negative root, $c \approx -3.08$, is

not in the domain)

$f(1.08) \approx 6.33$

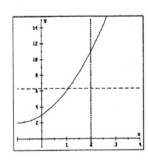

15. $A = \int\limits_{-1}^{1.5} \cos x \, dx$

$= \sin x \Big|_{-1}^{1.5}$

$= \sin 1.5 - \sin (-1)$

≈ 1.83897

$f(c)(b - a) = (\cos c)(2.5) \approx 1.83897$

so $c \approx 0.744264$ or $- 0.744264$ in $[-1, 1.5]$.

$f(c) \approx 0.735586$

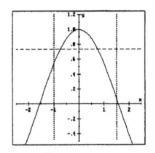

17. $\dfrac{1}{2 - (-1)} \int\limits_{-1}^{2} (x^2 - x + 1) \, dx$

$= \frac{1}{3}(\frac{1}{3}x^3 - \frac{1}{2}x^2 + x)\Big|_{-1}^{2}$

$= \frac{3}{2}$

19. $\dfrac{1}{1 - (-1)} \int\limits_{-1}^{1} (e^x - e^{-x}) \, dx$

$= \frac{1}{2}(e^x + e^{-x})\Big|_{-1}^{1}$

$= 0$

21. $\dfrac{1}{\frac{\pi}{4} - 0} \int\limits_{0}^{\pi/4} \sin x \, dx = -\frac{4}{\pi} \cos x \Big|_{0}^{\pi/4}$

$$= -\frac{4}{\pi}\left(\cos\frac{\pi}{4} - 1\right)$$

$$\approx 0.3729$$

23. $\displaystyle\frac{1}{4-0}\int_{0}^{4}\sqrt{4-x}\,dx = \frac{1}{4}\left(-\frac{2}{3}\right)(4-x)^{3/2}\Big|_{0}^{4}$

$$= \frac{4}{3}$$

25. $\displaystyle\frac{1}{1-0}\int_{0}^{1}(2x-3)^3\,dx = \frac{1}{2}\frac{(2x-3)^4}{4}\Big|_{0}^{1}$

$$= -10$$

27. $\displaystyle\frac{1}{1-(-2)}\int_{-2}^{1}x(x^2+1)^3\,dx = \frac{1}{6}\frac{(x^2+1)^4}{4}\Big|_{-2}^{1}$

$$= -\frac{609}{24}$$

$$= -25.375$$

29. $\displaystyle\frac{1}{3-(-3)}\int_{-3}^{3}\sqrt{9-x^2}\,dx$

$$= \frac{1}{6}(\text{AREA OF A SEMI-CIRCLE WITH } r = 3)$$

$$= \frac{1}{6}\left(\frac{9\pi}{2}\right)$$

$$= \frac{3\pi}{4}$$

$$\approx 2.36$$

31. Average height $= \displaystyle\frac{1}{t_1-t_0}\int_{t_0}^{t_1}\left(-\frac{1}{2}gt^2 + v_0 t\right)dt$

$$= \frac{1}{t_1-t_0}\left(-\frac{1}{6}gt^3 + v_0\frac{t^2}{2}\right)\Big|_{t_0}^{t_1}$$

$$= \frac{1}{t_1-t_0}\left[-\frac{1}{6}gt_1^3 + v_0\frac{t_1^2}{2} - \left(-\frac{1}{6}gt_0^3 + v_0\frac{t_0^2}{2}\right)\right]$$

$$= -\frac{g}{6}(t_1^2 + t_1 t_0 + t_0^2) + \frac{1}{2}v_0(t_1 + t_0)$$

33. Avg temp $= \displaystyle\frac{1}{12-9}\int_{9}^{12}(-0.1t^2 + t + 50)\,dt$

$$= \frac{1}{3}\left[-\frac{1}{30}t^3 + \frac{1}{2}t^2 + 50t\right]\Big|_{9}^{12}$$

$$= 49.4\ ^\circ\text{F}$$

35. a. $\displaystyle\frac{1}{20-10}\int_{10}^{20}\frac{2,000}{1+0.3\,e^{-0.276t}}\,dt$

$$= \frac{2,000}{10}\int_{10}^{20}\frac{e^{0.276t}}{e^{0.276t}+0.3}\,dt$$

> Let $u = e^{0.276t} + 0.3$
>
> $du = 0.276\,e^{0.276t}\,dt$

$$= \frac{200}{0.276}\ln(e^{0.276t}+0.3)\Big|_{10}^{20}$$

$$\approx 1,987.24$$

b. $\displaystyle\frac{2,000}{1+0.3e^{-0.276t}} = 1,987.24$

$$\frac{2,000}{1,987.24} = 1 + 0.3e^{-0.276t}$$

$$e^{-0.276t} \approx 0.0214032192$$

$$-0.276t \approx \ln(0.0214032192)$$

$$t \approx 13.93$$

The average population is reached at just under 14 minutes.

37. $\displaystyle\frac{1}{x+1}\int_{-1}^{x}f(t)\,dt = \sin x;$

$$\frac{d}{dx}\int_{-1}^{x} f(t)\, dt = [(x+1)\sin x]'$$

$$f(x) = (x+1)\cos x + \sin x$$

5.8 Numerical Integration: The Trapezoidal Rule and Simpson's Rule, Pages 339-342

SURVIVAL HINT: *Successful use of the trapezoidal rule and Simpson's rule is highly dependent upon your careful organization of the data. To simultaneously compute and keep a running total of values in your calculator takes considerable skill. Make a table of x_n values, $f(x_n)$, and $f(x_n)$ times the appropriate multiplier. It will save you time in the long run. Graphing calculators or computer software can often help with this process.*

1. $\displaystyle\int_{1}^{2} x^2\, dx$

$$\Delta x = \frac{2-1}{4} = \frac{1}{4}$$

$x_0 = 1$	$f(x_0) = 1$
$x_1 = 1.25$	$f(x_1) = 1.5625$
$x_2 = 1.5$	$f(x_2) = 2.25$
$x_3 = 1.75$	$f(x_3) = 3.0625$
$x_4 = 2$	$f(x_4) = 4$

Trapezoidal rule:

$$A \approx \tfrac{1}{2}[1 + 2(1.5625) + 2(2.25) + 2(3.0625) + 4](\tfrac{1}{4})$$
$$= \tfrac{1}{8}(18.75) = 2.34375$$

Simpson's rule:

$$A \approx \tfrac{1}{3}[1 + 4(1.5625) + 2(2.25) + 4(3.0625) + 4](\tfrac{1}{4})$$
$$= \tfrac{1}{12}(28) \approx 2.33333$$

Check with exact value:

$$A = \int_{1}^{2} x^2\, dx = \frac{x^3}{3}\Big|_1^2 = \frac{8}{3} - \frac{1}{3} = \frac{7}{3}$$

3. $\displaystyle\int_{0}^{1} \frac{dx}{1+x^2}$

$$\Delta x = \frac{1-0}{4} = \frac{1}{4}$$

$x_0 = 0$	$f(x_0) = 1.00$
$x_1 = 0.25$	$f(x_1) = 0.94$
$x_2 = 0.5$	$f(x_2) = 0.80$
$x_3 = 0.75$	$f(x_3) = 0.64$
$x_4 = 1$	$f(x_4) = 0.50$

a. Trapezoidal rule:

$$A \approx \tfrac{1}{2}[1(1) + 2(0.94) + 2(0.80) + 2(0.64)$$
$$+ 1(0.50)](\tfrac{1}{4})$$
$$= \tfrac{1}{8}(6.26)$$
$$= 0.7825$$

b. Simpson's rule:

$$A \approx \tfrac{1}{3}[1(1) + 4(0.94) + 2(0.80) + 4(0.64)$$
$$+ 1(0.50)](\tfrac{1}{4})$$
$$= \tfrac{1}{12}(9.42)$$
$$= 0.785$$

5. $\displaystyle\int_{2}^{4} \sqrt{1 + \sin x}\, dx$

$$\Delta x = \frac{4-2}{4} = \frac{1}{2}$$

$x_0 = 2$	$f(x_0) = 1.381773$
$x_1 = 2.5$	$f(x_1) = 1.264307$
$x_2 = 3$	$f(x_2) = 1.068232$
$x_3 = 3.5$	$f(x_3) = 0.805740$
$x_4 = 4$	$f(x_4) = 0.493151$

a. Trapezoidal rule:

$$A \approx \tfrac{1}{2}\left[1.381773 + 2(1.264307) + 2(1.068232) \right.$$

$$\left. + 2(0.805740) + 0.493151\right](\tfrac{1}{2})$$

$$= \tfrac{1}{4}(8.151482)$$

$$\approx 2.037871$$

b. Simpson's rule:

$$A \approx \tfrac{1}{3}\left[1.381773 + 4(1.264307) + 2(1.068232) \right.$$

$$\left. + 4(0.805740) + 0.493151\right](\tfrac{1}{2})$$

$$= \tfrac{1}{6}(12.291576)$$

$$\approx 2.048596$$

7. $\displaystyle\int_0^2 xe^{-x}\,dx$

$$\Delta x = \frac{2 - 0}{6} = \frac{1}{3} \approx 0.3333$$

$x_0 = 0.0000$		$f(x_0) = 0.000$
$x_1 = 0.3333$		$f(x_1) = 0.239$
$x_2 = 0.6667$		$f(x_2) = 0.342$
$x_3 = 1.0000$		$f(x_3) = 0.368$
$x_4 = 1.3333$		$f(x_4) = 0.351$
$x_5 = 1.6667$		$f(x_5) = 0.315$
$x_6 = 2.0000$		$f(x_6) = 0.271$

a. Trapezoidal rule:

$$A \approx \tfrac{1}{2}\left[1(0.000) + 2(0.239) + 2(0.342) \right.$$

$$+ 2(0.368) + 2(0.351) + 2(0.315)$$

$$\left. + 1(0.271)\right](\tfrac{1}{3})$$

$$\approx \tfrac{1}{6}(3.501)$$

$$\approx 0.584$$

b. Simpson's rule:

$$A \approx \tfrac{1}{3}\left[1(0.000) + 4(0.239) + 2(0.342) \right.$$

$$+ 4(0.368) + 2(0.351) + 4(0.315)$$

$$\left. + 1(0.271)\right](\tfrac{1}{3})$$

$$\approx \tfrac{1}{9}(5.345)$$

$$\approx 0.594$$

9. Approximate the area under a curve (evaluate an integral) by taking the sum of areas of trapezoids whose upper line segments join two consecutive points on an arc of the curve.

11. For the trapezoidal rule, $\displaystyle |E_n| \leq \frac{(b-a)^3}{12n^2}M,$ where M is the maximum value of $\left|f''(x)\right|$ on $[a,\ b]$.

$$f(x) = \frac{1}{x^2 + 1}$$

$$f'(x) = \frac{-2x}{(x^2 + 1)^2}$$

$$f''(x) = \frac{6x^2 - 2}{(x^2 + 1)^3}$$

The maximum of $\left|f''(x)\right|$ on $[0,\ 1]$ is 2 (at $x = 0$), so we need

$$\frac{1}{12n^2}(2) \leq 0.05$$

$$\frac{1}{n^2} \leq 0.30$$

$$n^2 \geq \frac{10}{3}$$

$$n \geq 2$$

$$A \approx \frac{1}{2}[f(0) + 2f(0.5) + f(1)](\frac{1}{2})$$

$$= \frac{1}{4}[1 + 2(0.8) + 0.5]$$

$$= 0.775$$

The exact answer is between $0.775 - 0.05$ and $0.775 + 0.05$.

13. $f(x) = \cos 2x$

$$f'(x) = -2\sin 2x$$

$$f''(x) = -4\cos 2x$$

$$f'''(x) = 8\sin 2x$$

$$f^{(4)}(x) = 16\cos 2x$$

The maximum value of $\left| f^{(4)}(x) \right|$ on $[0, 1]$ is 16 on $[0, 1]$. For Simpson's rule

$$\frac{1^5(16)}{180n^4} < 0.0005$$

$$n^4 > 178$$

$$n > 3.65$$

from which we will pick $n = 4$ (n must be even).

$$\Delta x = \frac{1 - 0}{4} = \frac{1}{4}$$

$x_0 = 0.00$	$f(x_0) = 1.000$
$x_1 = 0.25$	$f(x_1) = 0.878$
$x_2 = 0.50$	$f(x_2) = 0.540$
$x_3 = 0.75$	$f(x_3) = 0.071$
$x_4 = 1.00$	$f(x_4) = -0.416$

$A \approx 0.455$; the actual answer is between $0.455 - 0.0005$ and $0.455 + 0.0005$.

15. $f(x) = x(4 - x)^{1/2}$

$$f'(x) = \frac{8 - 3x}{2(4 - x)^{1/2}};$$

$$f''(x) = \frac{3x - 16}{4(4 - x)^{3/2}}$$

The maximum value of $\left| f''(x) \right|$ on $[0, 2]$ occurs at the endpoint $x = 2$, since it is an increasing function. $\left| f''(2) \right| \approx 0.88388$.

$$\frac{2^3}{12n^2}(0.88388) \leq 0.01$$

$$n^2 \geq 58.92533$$

$$n \geq 7.68$$

so we pick $n = 8$ terms. The trapezoidal approximation gives $A \approx 3.25$; the exact answer is between $3.25 + 0.01$ and $3.25 - 0.01$.

17. $f(x) = \tan^{-1} x$

$$f'(x) = \frac{1}{x^2 + 1}$$

$$f''(x) = \frac{-2x}{(x^2 + 1)^2}$$

$$f'''(x) = \frac{2(3x^2 - 1)}{(x^2 + 1)^3}$$

$$f^{(4)}(x) = \frac{-24x(x^2 - 1)}{(x^2 + 1)^4}$$

By calculator, the maximum value of $\left| f^{(4)}(x) \right|$ on $[0, 1]$ is approximately 4.669 at

$x \approx 0.325$. For Simpson's rule,

$$\frac{(1-0)^5}{180n^4}(4.669) < 0.01$$

$$n > 1.27$$

Pick $n = 2$. Use Simpson's rule to find $A \approx 0.44$; the exact answer is between $0.44 - 0.01$ and $0.44 + 0.01$.

19. $f(x) = x^{-1}$

$f'(x) = -x^{-2}$

$f''(x) = 2x^{-3}$

$f'''(x) = -6x^{-4}$

$f^{(4)}(x) = 24x^{-5}$

a. $\dfrac{2^3(2)}{12n^2} \le 0.00005$ or $n \approx 163.3$;

pick $n = 164$

b. $\dfrac{2^5(24)}{180n^4} \le 0.00005$ or $n \approx 17.09$

pick $n = 18$

21. $f(x) = \dfrac{1}{\sqrt{x}} = x^{-1/2}$

$f'(x) = -\frac{1}{2}x^{-3/2}$

$f''(x) = \frac{3}{4}x^{-5/2}$

$f'''(x) = -\frac{15}{8}x^{-7/2}$

$f^{(4)}(x) = \frac{105}{16}x^{-9/2}$

a. $\dfrac{3^3}{12n^2}\left(\frac{3}{4}\right) \le 0.00005$, or $n \approx 183.71$

pick $n = 184$

b. $\dfrac{3^5}{180n^4}\left(\dfrac{105}{16}\right) \le 0.00005$, or $n \approx 20.52$

pick $n = 22$ (n must be even)

23. $f(x) = e^{-2x}$

$f'(x) = -2e^{-2x}$

$f''(x) = 4e^{-2x}$

$f'''(x) = -8e^{-2x}$

$f^{(4)}(x) = 16e^{-2x}$

a. $\dfrac{(1)^3}{12n^2}(4) \le 0.00005$; pick $n = 82$

b. $\dfrac{(1)^5}{180n^4}(16) \le 0.00005$; pick $n = 8$

25. $f(x) = (1 - x^2)^{1/2}$; the second derivative is unbounded on $[0, 1]$, so the number of intervals needed to guarantee a certain accuracy cannot be predicted.

a. If $n = 8$,

$T_8 \approx (0.5)(12.347)(0.125)$

$\approx 0.772.$

Thus, $\pi \approx 4(0.772) \approx 3.09$; to one decimal place this is 3.1.

b. If $n = 4$,

$S_4 \approx (0.3333)(9.2508)(0.25)$

$\approx 0.771.$

Thus, $\pi \approx 4(0.771) \approx 3.08$; to one decimal place this is 3.1.

27. 5 m regions, so $\Delta x = 5$

$$A \approx \tfrac{1}{2}[(0) + 2(9) + 2(15) + 2(20) + 2(27) + (30)](5)$$

$$= 430$$

29. The area of the spill is approximately

$$A \approx \tfrac{1}{3}[0 + 4(7.3) + 2(9.1) + 4(10.3) + \cdots$$

$$+ 2(9.7) + 4(7.3) + 0](5)$$

$$\approx 613 \text{ ft}^2$$

31. $\Delta x = 0.5$

$$A \approx \tfrac{1}{3}[10 + 4(9.75) + 2(10) + 4(10.75) + 2(12)$$

$$+ 4(13.75) + 2(16) + 4(18.75)$$

$$+ 2(22) + 4(25.75) + 30](0.5)$$

$$= \tfrac{1}{3}(475)(0.5)$$

$$\approx 79.17$$

33. a. $\displaystyle\int_0^{\pi} (9x - x^3)\,dx = \left[\tfrac{9}{2}x^2 - \tfrac{1}{4}x^4\right]\Big|_0^{\pi}$

$$= \tfrac{1}{4}(18\pi^2 - \pi^4) \approx 20.06094705$$

TYPE OF ESTIMATE	# OF SUB-INTERVALS	ESTIMATE OVER [0, π]
Right endpt	10	19.3882917476
Right endpt	20	19.7855000789
Right endpt	40	19.9384437331
Right endpt	80	20.0035004324
Trapezoid	10	19.8174243188
Trapezoid	20	20.0000663645
Trapezoid	40	20.0457268759
Trapezoid	80	20.0571420838
Simpson	10	20.0609470464
Simpson	20	20.0609470464
Simpson	40	20.0609470464
Simpson	80	20.0609470464

For Simpson's rule the same values occur. The error term involves $f^{(4)}(x)$ which is 0 for cubics; that is, the Simpson error is 0.

b. The problem is that $f^{(4)}(x)$ is unbounded near $x = 2$.

35. Let d be the diameter of the larger interior circle (as shown in the figure).

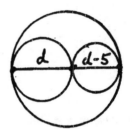

Then, $(d - 5)$ is the diameter of the smaller interior circle, and

$$\left[\frac{d + d - 5}{2}\right]^2 \pi - \left[\pi\left(\frac{d}{2}\right)^2 + 2\left(\frac{d - 5}{2}\right)^2 \pi\right] = 120$$

Use a solve utility to find (positive solution only) $d \approx 13.33$. So the inscribed circles have diameters $13\tfrac{1}{3}$ and $8\tfrac{1}{3}$.

37. $\displaystyle\int_a^b p(x)\,dx = \int_a^b (a_3 x^3 + a_2 x^2 + a_1 x + a_0)\,dx$

$$= \left[\frac{a_3}{4} x^4 + \frac{a_2}{3} x^3 + \frac{a_1}{2} x^2 + a_0 x\right]\Big|_a^b$$

$$= \frac{a_3}{4}(b^4 - a^4) + \frac{a_2}{3}(b^3 - a^3)$$

$$+ \frac{a_1}{2}(b^2 - a^2) + a_0(b - a)$$

$$= \tfrac{1}{12}(b - a)[3a_3b^3 + 3a_3ab^2 + 3a_3a^2b$$

$$+ 3a_3a^3 + 4a_2b^2 + 4a_2ab$$

$$+ 4a_2a^2 + 6a_1b + 6a_1a + 12a_0]$$

Now,

$$\frac{(b - a)}{6}\left[p(a) + 4p\left(\frac{a + b}{2}\right) + p(b)\right]$$

$$= \frac{(b - a)}{6}[\tfrac{3}{2}a_3b^3 + \tfrac{3}{2}a_3ab^2 + \tfrac{3}{2}a_3a^2b$$

$$+ \tfrac{3}{2}a_3a^3 + 2a_2b^2 + 2a_2ab + 2a_2a^2$$

$$+ 3a_1b + 3a_1a + 6a_0]$$

Thus,

$$\int_a^b p(x)dx = \frac{(b - a)}{6}\left[p(a) + 4p\left(\frac{a + b}{2}\right) + p(b)\right]$$

39. $a = -1$, $b = 3$, $b - a = 4$,

$$\frac{b - a}{6} = \frac{2}{3}, \text{ and } \frac{a + b}{2} = 1$$

$$\int_{-1}^3 (x^3 + 2x^2 - 7)\,dx = \frac{32}{3}$$

41. Using the error formula for Simpson's rule, $f^{(4)}(x)$ for a third degree polynomial will be zero.

5.9 An Alternative Approach: The Logarithm as an Integral, Pages 345-346

SURVIVAL HINT

The material of this section may seem confusing because you already "know" too much about logarithms and exponents. In precalculus you were introduced to $y = b^x$ without any proof that it was continuous for irrational values. The approach here is really better because the area function used is continuous to begin with, so the other functions derived from it, e^x, b^x, and $\log_b x$ will also be continuous. Try to read the section as if you were seeing $\ln x$ for the first time.

1. Let $L(x) = \displaystyle\int_1^x \frac{dt}{t}$. Then we have

$$L(xy) = L(x) + L(y) \text{ and } L(x^r) = rL(x)$$

in particular,

$$L(2^{-N}) = -NL(2) < 0$$

since

$$L(2) = \int_1^2 \frac{dt}{t} > 0$$

As $N \to +\infty$, $2^{-N} \to 0$ and $L(2^{-N}) \to -\infty$. If x is very small and positive, then $0 < x < 2^{-N}$ for some large N so

$$L(x) < L(2^{-N})$$

and for x decreasing to 0,

$$L(x) \to -\infty$$

3. $f'(t) = -t^{-2}$; $f''(t) = 2t^{-3}$;

$$f'''(t) = -6t^{-4}; f^{(4)}(t) = 24t^{-5}$$

Thus, $K = 24$. The error estimate for 8 subintervals is

$$E = \frac{2^5(24)}{180 \cdot 8^4} \approx 0.00104$$

For an error of 0.00005, the number of subintervals should be

$$0.00005 \geq \frac{2^5(24)}{180n^4}$$

$$n^4 \geq 85{,}333$$

$n \approx 17.09$; the number of subintervals should be 18.

5. **a.** $F(x) = \ln x^p$,

$$F'(x) = \frac{px^{p-1}}{x^p}$$

$$= \frac{p}{x}$$

$$G(x) = p\ln x,$$

$$G'(x) = \frac{p}{x}$$

$$= F'(x).$$

Therefore, $F(x) = G(x) + C$.

b. If $x = 1$, then $x^p = 1$ and $\displaystyle\int_1^1 \frac{dt}{t} = 0$.

$0 = p \cdot 0 + C$, so $C = 0$

Thus, $\qquad F(x) = G(x)$

$$\ln x^p = p \ln x$$

7. **a.** $f(xy) = f(x) + f(y)$ leads to

$f(1) = f(1) + f(1)$ when $x = y = 1$.

Thus, $f(1) = 2f(1)$ holds only when $f(1) = 0$.

b. $f(1) = f(-1) + f(-1)$ when $x = y = -1$. Thus, $0 = f(1) = 2f(-1)$ holds when $f(-1) = 0$.

c. $f(-x) = f(-1) + f(x)$ so $f(-x) = 0 + f(x)$ or $f(-x) = f(x)$

d. Hold x fixed in the equation

$$f(xy) = f(x) + f(y)$$

and differentiate with respect to y

by the chain rule.

$$xf'(xy) = 0 + f'(y)$$

In particular, when $y = 1$, $x > 0$,

$$f'(x) = \frac{f'(1)}{x}$$

so

$$f(x) = \int_1^x \frac{f'(1)}{t}\, dt = f'(1)\int_1^x \frac{dt}{t}$$

e. From part d, it can be seen that f' is continuous and hence integrable on any closed interval $[a,\, b]$ not including the origin. By the fundamental theorem of calculus,

$$f(x) - f(c) = \int_c^x f'(t)\, dt = f'(1)\int_c^x \frac{dt}{t}$$

for $x > 0$ if $c > 0$ and $x < 0$ if $c < 0$. Since $f(1) = 0$, we can use $c = 1$ to obtain

$$f(x) = f'(1)\int_1^x \frac{dt}{t} \qquad \text{if } x > 0$$

If $x < 0$, then $-x > 0$ and since $f(x) = f(-x)$, we obtain

$$f(x) = f'(1)\int_1^{-x} \frac{dt}{t} \qquad \text{if } x < 0$$

Combining these two formulas:

$$f(x) = f'(1)\int_1^{|x|} \frac{dt}{t} \qquad \text{if } x \neq 0$$

Finally, if $f'(1) \neq 0$ (that is, f is not

identically zero), we can let

$$F(x) = \frac{f(x)}{f'(1)} = \int\limits_{1}^{|x|} \frac{dt}{t}$$

It is easy to show that if

$$f(xy) = f(x) + f(y), \text{ then}$$

$$F(xy) = F(x) + F(y)$$

All solutions of $f(xy) = f(x) + f(y)$ can be obtained as multiples of $F(x)$

9.　a.　$\ln 2 = \int\limits_{1}^{2} \frac{dt}{t}$ is the area under $y = \frac{1}{t}$ on $[1, 2]$. Since $y = t^{-1}$ is always concave up and decreasing, this area is less than the area of a trapezoid with height $2 - 1 = 1$, and parallel bases $1/1$ and $1/2$. Thus,

$$\ln 2 = \int\limits_{1}^{2} \frac{1}{t}\, dt \; < \; \tfrac{1}{2}(1)\left(\tfrac{1}{1} + \tfrac{1}{2}\right) = \tfrac{3}{4} < 1$$

　　b.　Since the graph of $y = t^{-1}$ is always concave up, it will be above the tangent line at any point — say at $t_0 = 2$. The equation of the tangent line at $t_0 = 2$ is

$$y = \tfrac{1}{2} + \left(-\tfrac{1}{2^2}\right)(t - 2) \quad \textit{Since } f'(t) = -t^{-2}$$
$$= 1 - \tfrac{1}{4}t$$

Thus we have,

$$\ln 3 = \int\limits_{1}^{3} \frac{dt}{t} > \int\limits_{1}^{3} (1 - \tfrac{1}{4}t)\, dt = 1$$

We know that $\ln e = 1$ and we have just shown that $\ln 2 < 1 < \ln 3$, so

$$\ln 2 < \ln e < \ln 3$$

and it follows that

$$2 < e < 3$$

since $f(x) = \ln x$ is an increasing function.

11.　$\ln[E(x)]^p = p \ln[E(x)]$　　*Property of logs*

$$= px \qquad\qquad \textit{Inversion formula}$$

$$= \ln E(px) \qquad \textit{Inversion formula}$$

Because $\ln x$ is a one-to-one function, we conclude that

$$[E(x)]^p = E(px)$$

CHAPTER 5 REVIEW

Proficiency Examination, Pages 355-356

SURVIVAL HINT: *To help you review the concepts of this chapter, handwrite the answers to each of these questions onto your own paper.*

1.　An antiderivative of a function f is a function F that satisfies $F' = f$.

2.　$\displaystyle \int u^n\, du = \begin{cases} \dfrac{u^{n+1}}{n+1} + C; & n \neq -1 \\[2mm] \ln|u| + C; & n = -1 \end{cases}$

3.　$\displaystyle \int e^u\, du = e^u + C$

4. $\displaystyle\int \sin u\ du = -\cos u + C$

$\displaystyle\int \cos u\ du = \sin u + C$

$\displaystyle\int \tan u\ du = -\ln|\cos u| + C$

$\qquad\qquad = \ln|\sec u| + C$

$\displaystyle\int \cot u\ du = \ln|\sin u| + C$

$\displaystyle\int \sec^2 u\ du = \tan u + C$

$\displaystyle\int \sec u \tan u\ du = \sec u + C$

$\displaystyle\int \csc u \cot u\ du = -\csc u + C$

$\displaystyle\int \csc^2 u\ du = -\cot u + C$

5. $\displaystyle\int \frac{du}{\sqrt{1 - u^2}} = \sin^{-1} u + C$

$\displaystyle\int \frac{du}{1 + u^2} = \tan^{-1} u + C$

$\displaystyle\int \frac{du}{|u|\sqrt{u^2 - 1}} = \sec^{-1} u + C$

6. The *area function*, $A(t)$, as the area of the region bounded by the curve $y = f(x)$, the x-axis, and the vertical lines $x = a$, $x = t$. When we say that the area function can be written as an antiderivative, we mean that the area function can be written as a definite integral and by the fundamental theorem of calculus, the definite integral can be written as an antiderivative. The conditions that are necessary for an integral to represent an area

can be stated as restrictions on the function: f must be a continuous or a piecewise continuous function such that $f(x) \geq 0$ for all x on the closed interval $[a, b]$.

7. Suppose f is continuous or piecewise continuous and $f(x) \geq 0$ throughout the interval $[a, b]$. Then the *area* of the region under the curve $y = f(x)$ over this interval is given by

$$A = \lim_{n \to +\infty} \sum_{k=1}^{n} f(a + k\Delta x)\Delta x$$

where $\Delta x = \dfrac{b - a}{n}$.

8. Let $f(x)$ be a continuous function on a closed interval $[a, b]$. Given a partition,

$$a \leq x_1^* \leq x_2^* \leq \cdots \leq x_n^* \leq b$$

a Riemann sum is

$$f(x_1^*)\Delta x_1 + f(\bar{x}_2^*)\Delta x_2 + \ldots + f(\bar{x}_n^*)\Delta x_n$$
$$= \sum_{k=1}^{n} f(\bar{x}_k^*)\Delta x_k$$

9. If f is defined on the closed interval $[a, b]$ we say f is integrable on $[a, b]$ if

$$I = \lim_{\|P\| \to 0} \sum_{k=1}^{n} f(\bar{x}_k^*)\Delta x_k$$

exists. This limit is called the definite integral of f from a to b. The definite integral is denoted by

$$I = \int_a^b f(x)\ dx \qquad \text{or} \qquad I = \int_{x=a}^{x=b} f(x)\ dx$$

10. a. $$\int_a^a f(x)\,dx = 0$$

    b. $$\int_a^b f(x)\,dx = -\int_b^a f(x)\,dx$$

11. The total distance traveled by an object with continuous velocity $v(t)$ along a straight line from time $t = a$ to $t = b$ is

    $$S = \int_a^b |v(t)|\,dt$$

12. If f is continuous on the interval $[a, b]$ and F is any function that satisfies $F'(x) = f(x)$ throughout this interval, then

    $$\int_a^b f(x)\,dx = F(b) - F(a)$$

13. Let $f(t)$ be continuous on the interval $[a, b]$ and define the function G by the integral equation

    $$G(x) = \int_a^x f(t)\,dt$$

    for $a \le x \le b$. Then G is an antiderivative of f on $[a, b]$; that is,

    $$G'(x) = \frac{d}{dx}\left[\int_a^x f(t)\,dt\right] = f(x)$$

    on $[a, b]$.

14. Define a new variable of integration, $u = g(x)$. Find dx as a function of du and transform the limits. Make sure that the new integrand involves only the new variables and be careful if there are limits of integration.

15. A differential equation is an equation that contains derivatives.

16. A separable differential equation can be rewritten with only one variable on the left side of the equation and another variable on the right side. Each side of the equation is then integrated (if possible).

17. The growth/decay equation is
    $$Q(t) = Q_0 e^{kt}$$
    where $Q(t)$ is the amount of the substance present at time t, Q_0 is the initial amount of the substance, and k is a constant. The sign of k depends on the substance: growth if $k > 0$ and decay if $k < 0$. For carbon dating, $k = \ln 0.5/5{,}730$.

18. An orthogonal trajectory of a given family of curves is any curve that cuts all curves in the family at right angles.

19. If f is continuous on the interval $[a, b]$, there is at least one number c between a and b such that

    $$\int_a^b f(x)\,dx = f(c)(b - a)$$

20. If f is continuous on the interval $[a, b]$, the average value of f on this interval is given by the integral

$$\frac{1}{b-a}\int_a^b f(x)\,dx$$

21. **a.** Divide the interval $[a, b]$ into n subintervals, each of width $\Delta x = \frac{b-a}{n}$, and let $\overset{*}{x}_k$ denote the right endpoint (just for convenience) of the kth subinterval. The base of the kth rectangle is the kth subinterval, and its height is $f(\overset{*}{x}_k)$. Hence, the area of the kth rectangle is $f(\overset{*}{x}_k)\Delta x$. The sum of the areas of all n rectangles is an approximation for the area under the curve and hence an approximation for the corresponding definite integral. Thus,

$$\int_a^b f(x)\,dx \approx \sum_{k=1}^{n} f(\overset{*}{x}_k)\Delta x$$

b. Let f be continuous on $[a, b]$. The trapezoidal rule is

$$\int_a^b f(x)\,dx \approx \tfrac{1}{2}[f(x_0) + 2f(x_1)$$
$$+ 2f(x_2) + \cdots + 2f(x_{n-1}) + f(x_n)]\Delta x$$

where $\Delta x = \frac{b-a}{n}$ and, for the kth subinterval, $x_k = a + k\Delta x$.

c. Let f be continuous on $[a, b]$. Simpson's rule is

$$\int_a^b f(x)\,dx \approx \tfrac{1}{3}[f(x_0) + 4f(x_1) + 2f(x_2)$$
$$+ \cdots + 4f(x_{n-1}) + f(x_n)]\Delta x$$

where $\Delta x = \frac{b-a}{n}$, $x_k = a + k\Delta x$, k an integer and n an even integer. Moreover, the larger the value for n, the better the approximation.

22.
$$\int_0^1 [f(x)]^2\{2[f(x)]^2 - 3\}\,dx$$

$$= \int_0^1 \{2[f(x)]^4 - 3[f(x)]^2\}\,dx$$

$$= 2\int_0^1 [f(x)]^4\,dx - 3\int_0^1 [f(x)]^2\,dx$$

$$= 2(\tfrac{1}{5}) - 3(\tfrac{1}{3})$$

$$= -\tfrac{3}{5}$$

23. $F'(x) = \dfrac{d}{dx}\displaystyle\int_3^x t^5\,\sqrt{\cos(2t+1)}\,dt$

$$= x^5\sqrt{\cos(2x+1)}$$

24. $\displaystyle\int \dfrac{dx}{1+4x^2} = \tfrac{1}{2}\int \dfrac{du}{1+u^2} = \tfrac{1}{2}\tan^{-1}(2x) + C$

$\boxed{u = 2x,\ du = 2\,dx}$

25. $\displaystyle\int xe^{-x^2}\,dx = -\tfrac{1}{2}\int e^u\,du = -\tfrac{1}{2}e^{-x^2} + C$

$\boxed{u = -x^2;\ du = -2x\,dx}$

26. $\displaystyle\int_1^4 (x^{1/2} + x^{-3/2})\,dx = \dfrac{2x^{3/2}}{3} - \dfrac{2x^{-1/2}}{1}\Big|_1^4$

$$= \tfrac{16}{3} - 1 - \tfrac{2}{3} + 2$$

$$= \frac{17}{3}$$

27. $$\int_0^1 (2x - 6)(x^2 - 6x + 2)^2 \, dx$$

Let $u = x^2 - 6x + 2$; $du = (2x - 6) \, dx$,
If $x = 0$, $u = 2$, and if $x = 1$, then $u = -3$;

$$= \int_2^{-3} u^2 \, du$$

$$= \frac{u^3}{3} \Big|_2^{-3}$$

$$= -\frac{35}{3}$$

28. Let $u = 1 + \cos x$, $du = -\sin x \, dx$. For the new limits, when $x = 0$, $u = 2$, and when $x = \frac{\pi}{2}$, $u = 1$.

$$\int_0^{\pi/2} (1 + \cos x)^{-2} \sin x \, dx = -\int_2^1 u^{-2} \, du$$

$$= \frac{1}{u} \Big|_2^1$$

$$= 1 - \frac{1}{2}$$

$$= \frac{1}{2}$$

29. Let $u = 2x^2 + 2x + 5$, $du = (4x + 2) \, dx$. For new limits, when $x = -2$, $u = 9$, and when $x = 1$, $u = 9$.

$$\frac{1}{2} \int_9^9 u^{1/2} \, du = 0$$

30. $$A = \int_{-1}^3 (3x^2 + 2) \, dx$$

$$= \left[x^3 + 2x \right]\Big|_1^3$$

$$= 27 + 6 + 1 + 2$$

$$= 36$$

31. $$\int_1^2 e^{x^2} \, dx$$ cannot be written as a formula, so we must use approximate integration or technology to find

$$\int_1^2 e^{x^2} \, dx \approx 14.99$$

There is an interesting discussion of this integral in the May 1999 issue of *The American Mathematical Monthly* in an article entitled, "What Is a Closed-Form Number?" The function

$$\int e^{x^2} \, dx$$

is an example of a function that is *not* an elementary function (that we mentioned in Section 1.3).

32. Average value $$= \frac{1}{\frac{\pi}{2} - 0} \int_0^{\pi/2} \cos 2x \, dx$$

$$= \frac{2}{\pi} \int_0^{\pi/2} \frac{1}{2} \cos 2x \, (2 \, dx)$$

$$= \frac{1}{\pi} (\sin 2x) \Big|_0^{\pi/2}$$

$$= \tfrac{1}{\pi}(0)$$

$$= 0$$

33. The half life for ^{14}C is 5,730 years, so when using the decay formula

$$Q(t) = Q_0 e^{-kt}$$

to compute the percent of amount originally present, we find

$$e^{-(\ln 2/5730)(3,500,000)}$$

This is approximately 1.33×10^{-184} which exceeds the accuracy of most calculators and measuring devices. Other dating methods were used to date this artifact.

34.

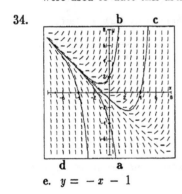

e. $y = -x - 1$

35. a. For the trapezoidal rule we need

$$\frac{(b-a)^3}{12n^2} M \le 0.0005 \text{ where } M \text{ is the}$$

maximum of $|f''(x)|$ on $[a, b]$.

$f(x) = \cos x, \quad f'(x) = -\sin x,$

$f''(x) = -\cos x, \text{ so on } [0, \tfrac{\pi}{2}],$

$$M = |-1| = 1.$$

$$\frac{(\tfrac{\pi}{2})^3}{12n^2}(1) \le 0.0005, \quad n^2 \ge 646, \quad n \ge 26.$$

b. For Simpson's rule we need

$$\frac{(b-a)^5}{180n^4} K \le 0.0005 \text{ where } K \text{ is the}$$

maximum of $|f^{(4)}(x)|$ on $[0, \tfrac{\pi}{2}]$.

$f'''(x) = \sin x, \quad f^{(4)}(x) = \cos x.$ So on

$[0, \tfrac{\pi}{2}], \quad M = 1.$

$$\frac{(\tfrac{\pi}{2})^5}{180n^4}(1) \le 0.0005, \quad n^4 \ge 106.3, \quad n \ge 4$$

Cumulative Review for Chapters 1-5, Pages 353-354

SURVIVAL HINT: *The first four **WHAT DOES THIS SAY?** problems in this cumulative review summarize the main ideas of calculus in the first five chapters of this book. If someone asks you, "What is calculus?" you would certainly want to include the responses to these four questions. Write them out in your own words and then look up each in the book.*

1. Formally, the limit statement $\lim\limits_{x \to c} f(x) = L$

means that for each $\epsilon > 0$, there corresponds a

number $\delta > 0$ with the property that

$$\left| f(x) - L \right| < \epsilon$$

whenever $0 < |x - c| < \delta$. The notation

whenever $0 < |x - c| < \delta$. The notation $\lim_{x \to c} f(x) = L$ is read "the limit of $f(x)$ as x approaches c is L" and means that the functional values $f(x)$ can be made arbitrarily close to L by choosing x sufficiently close to c (but not equal to c).

3. If f is defined on the closed interval $[a, b]$ we say f is integrable on $[a, b]$ if

$$I = \lim_{\|P\| \to 0} \sum_{k=1}^{n} f(\overset{*}{x}_k) \Delta x_k$$

exists. This limit is called the definite integral of f from a to b. The definite integral is denoted by

$$I = \int_a^b f(x)\, dx$$

5. $\lim_{x \to 2} \dfrac{3x^2 - 5x - 2}{3x^2 - 7x + 2} = \lim_{x \to 2} \dfrac{(3x + 1)(x - 2)}{(3x - 1)(x - 2)}$

$\qquad\qquad = \lim_{x \to 2} \dfrac{3x + 1}{3x - 1}$

$\qquad\qquad = \dfrac{7}{5}$

7. $\lim_{x \to +\infty} \left(\sqrt{x^2 + x} - x \right)$

$\quad = \lim_{x \to +\infty} \dfrac{\left(\sqrt{x^2 + x} - x\right)\left(\sqrt{x^2 + x} + x\right)}{\sqrt{x^2 + x} + x}$

$\quad = \lim_{x \to +\infty} \dfrac{x}{\sqrt{x^2 + x} + x}$

$\quad = \lim_{x \to +\infty} \dfrac{1}{\sqrt{1 + \frac{1}{x}} + 1}$

$\quad = \dfrac{1}{2}$

9. $\lim_{x \to 0} \dfrac{x \sin x}{x + \sin^2 x} = \lim_{x \to 0} \dfrac{x}{\frac{x}{\sin x} + \sin x}$

$\qquad\qquad = 0$

11. $\qquad L = \lim_{x \to +\infty} (1 + x)^{2/x}$

$\quad \ln L = \lim_{x \to +\infty} \dfrac{2\ln(1 + x)}{x}$

$\qquad\quad = \lim_{x \to +\infty} \dfrac{2}{1 + x} = 0$

Thus, $L = e^0 = 1$.

13. $\qquad L = \lim_{x \to 0^+} x^{\sin x}$

$\quad \ln L = \lim_{x \to 0^+} (\sin x) \ln x$

$\qquad\quad = \lim_{x \to 0^+} \dfrac{\ln x}{\csc x}$

$\qquad\quad = \lim_{x \to 0^+} \dfrac{1/x}{-\csc x \cot x}$

$\qquad\quad = \lim_{x \to 0^+} \dfrac{-\sin^2 x}{x \cos x}$

$\qquad\quad = \lim_{x \to 0^+} \left(\dfrac{-\sin x}{x} \right)\left(\dfrac{\sin x}{\cos x} \right)$

$\qquad\quad = -1 \cdot 0$

$\qquad\quad = 0$

Thus, $L = e^0 = 1$.

15. $y = (x^2 + 1)^3 (3x - 4)^2$

$y' = (x^2 + 1)^3 (2)(3x - 4)(3)$

$\qquad + 3(x^2 + 1)^2 (2x)(3x - 4)^2$

$\qquad = 6(x^2 + 1)^2 (3x - 4)[x^2 + 1 + 3x^2 - 4x]$

$\qquad = 6(x^2 + 1)^2 (3x - 4)(2x - 1)^2$

17. $y = \dfrac{x}{x + \cos x}$

$y' = \dfrac{x + \cos x - x(1 - \sin x)}{(x + \cos x)^2}$

$\qquad = \dfrac{\cos x + x \sin x}{(x + \cos x)^2}$

19. $y = \sec^2 3x$

$y' = 2 \sec 3x [\sec 3x \tan 3x (3)]$

$\qquad = 6 \sec^2 3x \tan 3x$

21. $y = \ln(5x^2 + 3x - 2)$

$y' = \dfrac{10x + 3}{5x^2 + 3x - 2}$

23. $\displaystyle\int_4^9 d\theta = (9 - 4)$

$\qquad = 5$

25. $\displaystyle\int_0^1 \dfrac{x\,dx}{\sqrt{9 + x^2}} = \dfrac{1}{2}\int_0^1 (9 + x^2)^{-1/2} (2x\,dx)$

$\qquad = 2(\tfrac{1}{2})(9 + x^2)^{1/2}\Big|_0^1$

$\qquad = \sqrt{10} - 3$

27. $\displaystyle\int \dfrac{e^x\,dx}{e^x + 2} = \ln\big| e^x + 2 \big| + C$

$\qquad\qquad = \ln(e^x + 2) + C$

29. $\displaystyle\int_0^4 \dfrac{dx}{\sqrt{1 + x^3}}$

$\Delta x = \dfrac{4 - 0}{6} = \dfrac{2}{3}$

$x_0 = 0.0000$	$f(x_0) = 1.0000$
$x_1 = 0.6667$	$f(x_1) = 0.8783$
$x_2 = 1.3333$	$f(x_2) = 0.5447$
$x_3 = 2.0000$	$f(x_3) = 0.3333$
$x_4 = 2.6667$	$f(x_4) = 0.2238$
$x_5 = 3.3333$	$f(x_5) = 0.1621$
$x_6 = 4.0000$	$f(x_6) = 0.1240$

$A \approx \tfrac{1}{3}[1(1) + 4(0.8783) + 2(0.5447)$

$\qquad + 4(0.3333) + 2(0.2238) + 4(0.1621)$

$\qquad + 1(0.124)](\tfrac{2}{3})$

$\qquad \approx \tfrac{1}{3}(8.1558)(\tfrac{2}{3})$

$\qquad \approx 1.812$

31. $y = \dfrac{4 + x^2}{4 - x^2} = \dfrac{8}{4 - x^2} - 1$

$\displaystyle\lim_{x \to \pm\infty} \dfrac{4 + x^2}{4 - x^2} = -1$ so $y = -1$ is a

horizontal asymptote;

$y' = \dfrac{(4 - x^2)2x - (4 + x^2)(-2x)}{(4 - x^2)^2}$

$\qquad = \dfrac{16x}{(4 - x^2)^2}$

$y' = 0$ when $x = 0$

$y'' = 16\left[\dfrac{(4 - x^2)^2 - x(2)(4 - x^2)(-2x)}{(4 - x^2)^4} \right]$

$$= \frac{16(3x^2 + 4)}{(4 - x^2)^3} \neq 0$$

$(0, 1)$ is a relative minimum;

vertical asymptotes when $4 - x^2 = 0$ or when

$x = \pm 2$

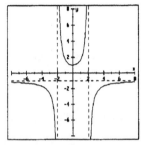

33.
$$\frac{dy}{dx} = x^2 y^2 \sqrt{4 - x^3}$$

$$\int y^{-2} \, dy = \int x^2 (4 - x^3)^{1/2} \, dx$$

$$-y^{-1} = -\frac{2}{9}[(4 - x^3)^{3/2} + C_1]$$

$$y = \frac{9}{2(4 - x^3)^{3/2} + C}$$

35.
$$\frac{dy}{dx} = 2(5 - y)$$

$$\int \frac{dy}{5 - y} = \int 2 \, dx$$

$$-\ln|5 - y| = 2x + C$$

Since $y = 3$ when $x = 0$, $C = -\ln 2$

$$-\ln|5 - y| = 2x - \ln 2$$

$$\ln \frac{2}{|5 - y|} = 2x$$

$$e^{2x} = \frac{2}{|5 - y|}$$

$$|5 - y| = 2e^{-2x}$$

$$y = 5 \pm 2e^{-2x}$$

37. a.
$$F(7) = \int_0^7 f(t) \, dt$$

$$= (\text{AREA OF QUARTER CIRCLE}$$
$$- \text{AREA OF TRIANGLE})$$

$$= \frac{1}{4}\pi(3)^2 - \frac{1}{2} \cdot 4 \cdot 4$$

$$= \frac{9}{4}\pi - 8$$

b. $F(x)$ has a relative maximum at $x = 3$ because $F'(x) = f(x)$ changes from positive to negative at $x = 3$.

c. $F(7) = \frac{9}{4}\pi - 8$

$F'(7) = f(7) = -4$

$y - (\frac{9}{4}\pi - 8) = -4(x - 7)$

$4x + y - 20 - \frac{9}{4}\pi = 0$

d. $F''(x) = f'(x)$ changes from increasing to decreasing at $x = 0$, and from decreasing to increasing at $x = 7$. Thus, the graph of F has points of inflection at $x = 0$ and $x = 7$.

39. Consider a triangle with legs 5,000 and y, where y is the altitude of the rocket. Use θ for the angle of elevation, opposite the y

leg.

$$y = 5,000 \tan \theta$$

$$\frac{dy}{dt} = 5,000 \sec^2\theta \, \frac{d\theta}{dt}$$

$$\frac{d\theta}{dt} = \frac{1}{5,000}\cos^2\theta \, \frac{dy}{dt}$$

Given $dy/dt = 850$; if $y = 4,000$,

$\cos^2\theta = 25/41$:

$$\frac{d\theta}{dt} = \frac{1}{5,000}\left(\frac{25}{41}\right)(850) = \frac{17}{164} \approx 0.1 \text{ rad/s}$$

41. a. $m = \dfrac{2 - 0}{0 - \frac{\pi}{2}} = -\dfrac{4}{\pi}$

$$y - 2 = -\tfrac{4}{\pi}(x - 0)$$

$$y = -\tfrac{4}{\pi}x + 2$$

b. $f'(x) = -2\sin x$

$$f'(\tfrac{\pi}{2}) = -2\sin\tfrac{\pi}{2} = -2$$

$$y - 0 = -2(x - \tfrac{\pi}{2})$$

$$y = -2x + \pi$$

c. $f'(x) = -2\sin x = -\dfrac{4}{\pi}$

$$\sin x = \tfrac{2}{\pi}$$

Thus, $x \approx 0.690$; mean value theorem

43. a. $f(x) = x^3 - 6x^2 + k$

$$f'(x) = 3x^2 - 12x;$$

$$f'(x) = 0 \text{ when } x = 0, 4$$

$$f''(x) = 6x - 12;$$

$f''(0) = -12,$

relative maximum at $x = 0$

$f(0) = k$, so relative maximum at $(0, k)$;

$f''(4) = 12$, relative minimum at $x = 4$

$$f(4) = k - 32,$$

relative minimum at $(4, k - 32)$

b. $f(x)$ has three distinct real roots when

$k > 0$ and $k - 32 < 0$, so $0 < k < 32$.

c. AVERAGE VALUE

$$= \frac{1}{2 - (-1)} \int_{-1}^{2} (x^3 - 6x^2 + k) \, dx = 2$$

$$\tfrac{1}{3}\left[3k - \tfrac{57}{4}\right] = 2, \text{ so } k = \tfrac{27}{4}$$

45.

$$\frac{dy}{y^2} = \sin 3x \, dx$$

$$\int y^{-2} dy = \tfrac{1}{3}\int \sin 3x \, (3 \, dx)$$

$$-\tfrac{1}{y} = -\tfrac{1}{3}\cos 3x + C_1$$

$$y = \frac{3}{\cos 3x + C}$$

CHAPTER 6

Additional Applications of the Integral

SURVIVAL HINT: *In this chapter we move, for the first time, from two dimensions to three dimensions. We refer to two dimensions as $\mathbf{R}^2$ and three dimensions as $\mathbb{R}^3$. Although we will formally introduce three dimensional coordinates in Chapter 10, you should begin to nurture your skills at visualizations in three dimensions. Today, many software packages and calculators are available to help this visualization. What we draw on our papers is often a two-dimensional cross-section or some boundary conditions.*

6.1 Area Between Two Curves, Pages 361-362

SURVIVAL HINT: *When finding area between curves, you should be neat, organized, and follow these steps:*
- *Draw a careful sketch, find the coordinates of the points of intersection.*
- *Decide whether horizontal or vertical strips will be most efficient. Draw the strip.*
- *Be careful about which is the leading curve, and express it properly; y as a function of x, or x as a function of y.*
- *Sum the strips between the appropriate values; x values for vertical strips, and y values for horizontal strips.*

1. $-x^2 + 6x - 5 = \frac{3}{2}x - \frac{3}{2}$

$2x^2 - 9x + 7 = 0$

$(2x - 7)(x - 1) = 0$

$x = \frac{7}{2}, 1$

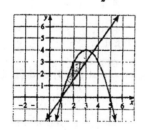

$$\int_{1}^{7/2} [(-x^2 + 6x - 5) - (\tfrac{3}{2}x - \tfrac{3}{2})]\, dx$$

$$= \int_{1}^{7/2} (-x^2 + \tfrac{9}{2}x - \tfrac{7}{2})\, dx$$

$$= (-\tfrac{1}{3}x^3 + \tfrac{9}{4}x^2 - \tfrac{7}{2}x)\Big|_{1}^{7/2}$$

$$= \frac{125}{48}$$

3. $\sin 2x = 0$

$x = 0, \frac{\pi}{2}, \pi$

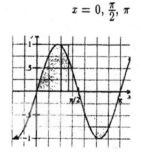

$$\int_{0}^{\pi/2} \sin 2x\, dx + \int_{\pi/2}^{\pi} (-\sin 2x)\, dx$$

$$= \left(-\tfrac{1}{2}\cos 2x \right)\Big|_0^{\pi/2} + \tfrac{1}{2}\cos 2x \Big|_{\pi/2}^{\pi}$$

$$= 1 + 1$$

$$= 2$$

5. $\quad y^2 - 5y = 0$

$$y(y - 5) = 0$$

$$y = 0,\ 5$$

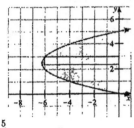

$$\int_0^5 [0 - (y^2 - 5y)]\ dy$$

$$= \int_0^5 (5y - y^2)\,dy$$

$$= \left[\frac{5y^2}{2} - \frac{y^3}{3} \right]\Bigg|_0^5$$

$$= \frac{125}{6}$$

7. $\quad y = x^2$ and $y = x$, so

$$x^2 = x$$

$$x^2 - x = 0$$

$$x(x - 1) = 0$$

$$x = 0,\ 1$$

The curves $y = x^2$ and $y = x$ intersect at

$(0,\ 0)$ and $(1,\ 1)$.

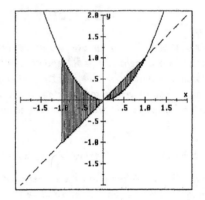

$$A = \int_{-1}^0 (x^2 - x)\ dx + \int_0^1 (x - x^2)\ dx$$

$$= \left(\tfrac{1}{3}x^3 - \tfrac{1}{2}x^2 \right)\Big|_{-1}^0 + \left(\tfrac{1}{2}x^2 - \tfrac{1}{3}x^3 \right)\Big|_0^1$$

$$= 1$$

9. $\quad y = x^2$ and $y = x^3$, so

$$x^2 = x^3$$

$$x^3 - x^2 = 0$$

$$x^2(x - 1) = 0$$

$$x = 0,\ 1$$

The curves intersect at $(0,\ 0)$ and $(1,\ 1)$.

$$A = \int\limits_{-1}^{1} (1 - x^2)\, dx + \int\limits_{1}^{2} (x^2 - 1)\, dx$$

$$= \left(-\tfrac{1}{3}x^3 + x\right)\Big|_{-1}^{1} + \left(\tfrac{1}{3}x^3 - x\right)\Big|_{1}^{2}$$

$$= \tfrac{8}{3}$$

13. $y = x^4 - 3x^2$ and $y = 6x^2$, so

$$x^4 - 3x^2 = 6x^2$$

$$x^4 - 9x^2 = 0$$

$$x^2(x^2 - 9) = 0$$

$$x^2(x - 3)(x + 3) = 0$$

$$x = 0, 3, -3$$

The curves intersect at $(0, 0)$ and $(-3, 54)$

and $(3, 54)$.

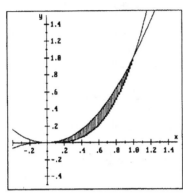

$$A = \int\limits_{0}^{1} (x^2 - x^3)\, dx$$

$$= \left(\tfrac{1}{3}x^3 - \tfrac{1}{4}x^4\right)\Big|_{0}^{1}$$

$$= \tfrac{1}{12}$$

11. $y = x^2 - 1$ and $y = 0$, so

$$x^2 - 1 = 0$$

$$(x - 1)(x + 1) = 0$$

$$x = 1, -1$$

The curves $y = x^2 - 1$ and $y = 0$ intersect at $(-1, 0)$ and $(1, 0)$.

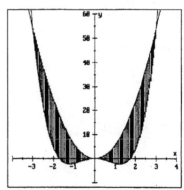

$$A = 2\int\limits_{0}^{3} (9x^2 - x^4)\, dx \qquad \textit{By symmetry}$$

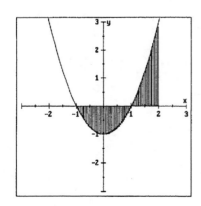

$$= 2(\tfrac{9}{3}x^3 - \tfrac{1}{5}x^5)\Big|_0^3$$

$$= \tfrac{324}{5}$$

15. $x = 2 - y^2$ and $x = y$, so

$$2 - y^2 = y$$

$$y^2 + y - 2 = 0$$

$$(y - 1)(y + 2) = 0$$

$$y = 1, \, -2$$

The curves intersect at $(1, 1)$ and $(-2, -2)$.

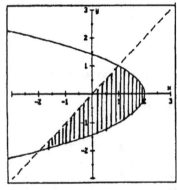

$$A = \int_{-2}^{1} (2 - y^2 - y)\, dy$$

$$= (2y - \tfrac{1}{3}y^3 - \tfrac{1}{2}y^2)\Big|_{-2}^{1}$$

$$= \tfrac{9}{2}$$

17. $y = 2x^3 + x^2 - x - 1$ and

$y = x^3 + 2x^2 + 5x - 1$, so

$$2x^3 + x^2 - x - 1 = x^3 + 2x^2 + 5x - 1$$

$$x^3 - x^2 - 6x = 0$$

$$x(x - 3)(x + 2) = 0$$

$$x = 0, \, -2, \, 3$$

The curves intersect at $(0, \, -1)$

$(-2, \, -11)$ and $(3, 59)$.

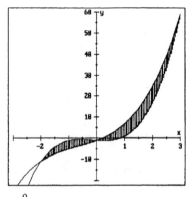

$$A = \int_{-2}^{0} (x^3 - x^2 - 6x)\, dx$$

$$+ \int_{0}^{3} (-x^3 + x^2 + 6x)\, dx$$

$$= (\tfrac{1}{4}x^4 - \tfrac{1}{3}x^3 - 3x^2)\Big|_{-2}^{0}$$

$$+ (-\tfrac{1}{4}x^4 + \tfrac{1}{3}x^3 + 3x^2)\Big|_{0}^{3}$$

$$= \tfrac{253}{12}$$

19. $y = \sin x;\ y = \sin 2x;$ on $[0, \pi]$ solve

$$\sin x = \sin 2x$$

$$\sin 2x\ -\ \sin x = 0$$

$$2\sin x \cos x\ -\ \sin x = 0$$

$$\sin x(2 \cos x\ -\ 1) = 0$$

$$\sin x = 0 \qquad 2 \cos x\ -\ 1 = 0$$

$$x = 0, \pi \qquad \cos x = \tfrac{1}{2}$$

$$x = \frac{\pi}{3}$$

The curves $y = \sin x$ and $y = \sin 2x$ intersect
when $x = 0,\ \pi/3,$ and $\pi.$

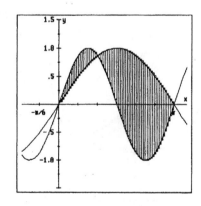

$$A = \int_0^{\pi/3} (\sin 2x - \sin x)\ dx + \int_{\pi/3}^{\pi} (\sin x - \sin 2x)\ dx$$

$$= \left(-\tfrac{1}{2}\cos 2x + \cos x\right)\Big|_0^{\pi/3} + \left(-\cos x + \tfrac{1}{2}\cos 2x\right)\Big|_{\pi/3}^{\pi}$$

$$= \frac{5}{2}$$

21.   The curves $y = |4x\ -\ 1|$ and $y = x^2\ -\ 5$ do
not intersect on $[0, 4]$, but the absolute value
function causes the equation of the line to
change at $x = 1/4.$

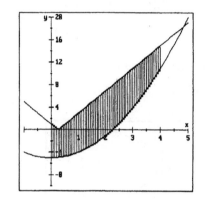

$$A = \int_0^{1/4} (-4x + 1\ -\ x^2 + 5)\ dx$$

$$+ \int_{1/4}^{4} (4x\ -\ 1\ -\ x^2 + 5)\ dx$$

$$= \left(-2x^2 + x\ -\ \tfrac{1}{3}x^3 + 5x\right)\Big|_0^{1/4}$$

$$+ \left(2x^2\ -\ x\ -\ \tfrac{1}{3}x^3 + 5x\right)\Big|_{1/4}^{4}$$

$$= \frac{323}{12}$$

23. $x = 0$ and $x = y^3 - 3y^2 - 4y + 12$, so

$$y^3 - 3y^2 - 4y + 12 = 0$$

$$y^2(y - 3) - 4(y - 3) = 0$$

$$(y^2 - 4)(y - 3) = 0$$

$$(y - 2)(y + 2)(y - 3) = 0$$

$$y = 2, -2, 3$$

The curves intersect at $(0, 2)$, $(0, -2)$, and $(0, 3)$.

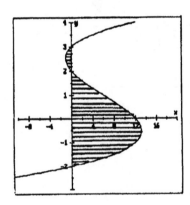

$$A = \int_{-2}^{2} (y^3 - 3y^2 - 4y + 12)\,dy$$

$$+ \int_{2}^{3} (-y^3 + 3y^2 + 4y - 12)\,dy$$

$$= (\tfrac{1}{4}y^4 - y^3 - 2y^2 + 12y)\Big|_{-2}^{2}$$

$$+ (-\tfrac{1}{4}y^4 + y^3 + 2y^2 - 12y)\Big|_{2}^{3}$$

$$= \frac{131}{4}$$

25. $y = \dfrac{1}{\sqrt{1 - x^2}}$ and $y = \dfrac{2}{x + 1}$, so

$$\frac{1}{\sqrt{1 - x^2}} = \frac{2}{x + 1}$$

$$(x + 1)^2 = 4(1 - x^2)$$

$$x^2 + 2x + 1 = 4 - 4x^2$$

$$5x^2 + 2x - 3 = 0$$

$$(5x - 3)(x + 1) = 0$$

$$x = \tfrac{3}{5}, -1$$

In the domain, the curves intersect at $x = 0.6$

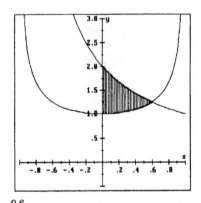

$$\int_{0}^{0.6} \left[\frac{2}{x + 1} - \frac{1}{\sqrt{1 - x^2}} \right] dx$$

$$= \left[2\ln(x + 1) - \sin^{-1}x \right]\Big|_{0}^{0.6}$$

$$= 2\ln 1.6 - \sin^{-1} 0.6$$

27.

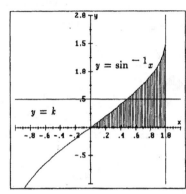

We want

$$\int_0^k (1 - \sin y)\, dy = \frac{1}{2} \int_0^{\pi/2} (1 - \sin y)\, dy$$

$$[y + \cos y]\Big|_0^k = \frac{1}{2} [y + \cos y]\Big|_0^{\pi/2}$$

$$(k + \cos k) - 1 = \frac{1}{2}\Big[\Big(\frac{\pi}{2} + 0\Big) - (0 + 1)\Big]$$

$$k + \cos k = \frac{\pi}{4} + \frac{1}{2}$$

$$k \approx 0.34 \quad \textit{By calculator}$$

29.

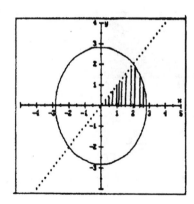

This is one-eighth of a circle with radius $\sqrt{8}$.

$$A = \tfrac{1}{8} \pi (\sqrt{8})^2 = \pi$$

31. a. A horizontal strip has area

$$dA = 2x\, dy = 2\sqrt{b^2 - y^2}\, dy$$

These strips lie between $y = -b$ and $y = b$

$$V = 2L \int_{-b}^{b} \sqrt{b^2 - y^2}\, dy$$

b. When the tank is filled to a height of $y = h$

$$V = 2L \int_{-b}^{h} \sqrt{b^2 - y^2}\, dy$$

c. The V values are approximately 72.5299, 196.539, 344.337, 502.655, 660.972, 808.77, 932.78, 1,005.31

6.2 Volume, Pages 372-375

1.

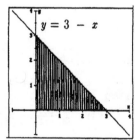

$$V = \int_0^3 (3 - x)^2\, dx$$

$$= -\tfrac{1}{3}(3 - x)^3 \Big|_0^3$$

$$= 9 \text{ cubic units}$$

3.

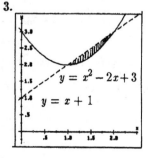

The curves intersect when:

$$x^2 - 2x + 3 = x + 1$$

$$x^2 - 3x + 2 = 0$$

$$(x - 2)(x - 1) = 0$$

$$x = 2, 1$$

$$V = \int_{1}^{2} [(x + 1) - (x^2 - 2x + 3)]^2 \, dx$$

$$= \int_{1}^{2} (x^4 - 6x^3 + 13x^2 - 12x + 4) \, dx$$

$$= (\tfrac{1}{5}x^5 - \tfrac{3}{2}x^4 + \tfrac{13}{3}x^3 - 6x^2 + 4x)\Big|_{1}^{2}$$

$$= \tfrac{1}{30} \text{ cubic units}$$

In Problems 5 and 7, we note that an equilateral triangle of side a has area $\frac{1}{4}\sqrt{3}\,a^2$.

5.

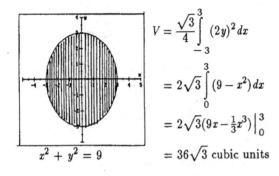

$x^2 + y^2 = 9$

$$V = \frac{\sqrt{3}}{4}\int_{-3}^{3} (2y)^2 \, dx$$

$$= 2\sqrt{3}\int_{0}^{3} (9 - x^2) \, dx$$

$$= 2\sqrt{3}(9x - \tfrac{1}{3}x^3)\Big|_{0}^{3}$$

$$= 36\sqrt{3} \text{ cubic units}$$

7. $y = \sqrt{\cos x}$

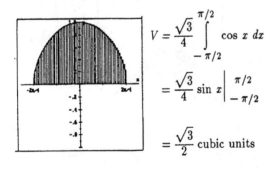

$$V = \frac{\sqrt{3}}{4}\int_{-\pi/2}^{\pi/2} \cos x \, dx$$

$$= \frac{\sqrt{3}}{4} \sin x \Big|_{-\pi/2}^{\pi/2}$$

$$= \frac{\sqrt{3}}{2} \text{ cubic units}$$

In Problems 9 and 11, we note that a semicircle of diameter a has area $\frac{1}{2}\pi\left(\frac{a}{2}\right)^2 = \frac{1}{8}\pi a^2$.

9.

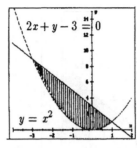

$2x + y - 3 = 0$

$y = x^2$

The curves intersect when

$$x^2 = -2x + 3$$
$$x^2 + 2x - 3 = 0$$
$$(x + 3)(x - 1) = 0$$
$$x = -3, 1$$

$$V = \frac{\pi}{8}\int_{-3}^{1} (3 - 2x - x^2)^2 \, dx$$

$$= \frac{\pi}{8}\int_{-3}^{1} (x^4 + 4x^2 + 9 + 4x^3 - 6x^2 - 12x) \, dx$$

$$= \frac{\pi}{8}\int_{-3}^{1} (x^4 + 4x^3 - 2x^2 - 12x + 9) \, dx$$

$$= \frac{\pi}{8}\left(\tfrac{1}{5}x^5 + x^4 - \tfrac{2}{3}x^3 - 6x^2 + 9x\right)\Big|_{-3}^{1}$$

$$= \frac{64\pi}{15} \text{ cubic units}$$

11. $y = \tan x$

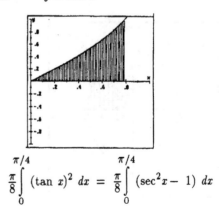

$$\frac{\pi}{8}\int_{0}^{\pi/4} (\tan x)^2 \, dx = \frac{\pi}{8}\int_{0}^{\pi/4} (\sec^2 x - 1) \, dx$$

$$= \frac{\pi}{8}(\tan x - x)\Big|_0^{\pi/4}$$

$$= \frac{\pi}{8}(1 - \frac{\pi}{4}) \text{ cubic units}$$

SURVIVAL HINT: *Success in finding volumes is highly dependent upon good visualization in $\mathbb{R}^3$ (called spatial perception). Practice drawing a good sketch for each problem. Draw the slice in $\mathbb{R}^2$ and write the appropriate **general formula**: πr^2 for disks, $\pi(R^2 - r^2)$ for washers, and $2\pi rh$ for shells. **Then** substitute the functional values for r, R, and h, and sum over the correct limits. The given volumes are all in cubic units.*

13. $y = \sqrt{x}$

Use disks;

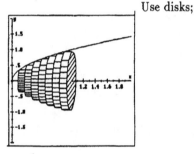

$$V = \pi \int_0^1 (x^{1/2})^2 \, dx = \frac{1}{2}\pi x^2\Big|_0^1$$

$$= \frac{\pi}{2} \text{ cubic units}$$

15. $y = 2x;\ y = x$

Use washers;

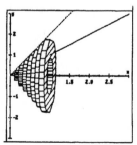

$$V = \pi \int_0^1 [(2x)^2 - x^2] \, dx$$

$$= \pi \text{ cubic units}$$

17. $y = x^2 + x^3$

Use disks;

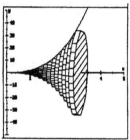

$$V = \pi \int_0^\pi (x^2 + x^3)^2 \, dx$$

$$= \pi \int_0^\pi (x^4 + 2x^5 + x^6) \, dx$$

$$= \frac{\pi^8}{7} + \frac{\pi^7}{3} + \frac{\pi^6}{5}$$

$$\approx 2{,}555 \text{ cubic units}$$

19. $y = \sin x;\ y = \cos x$

Use washers;

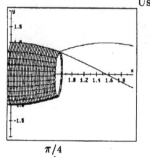

$$V = \pi \int_0^{\pi/4} (\cos^2 x - \sin^2 x) \, dx$$

$$= \pi \int_{0}^{\pi/4} \cos 2x \; dx = \pi \left[\tfrac{1}{2} \sin 2x \right] \Big|_{0}^{\pi/4}$$

$$= \tfrac{\pi}{2} \text{ cubic units}$$

$$V = 2\pi \int_{0}^{1} x(1 - x^2) \; dx$$

$$= \tfrac{\pi}{2} \text{ cubic units}$$

21. $y = 2x$

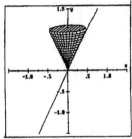

Use shells;

If $y = 1$, then $x = \tfrac{1}{2}$;

$$V = 2\pi \int_{0}^{1/2} x(1 - 2x) \; dx$$

$$= 2\pi \left(\frac{x^2}{2} - \frac{2}{3}x^3 \right) \Big|_{0}^{1/2}$$

$$= \tfrac{\pi}{12} \text{ cubic units}$$

23. $y = 1 - x^2$

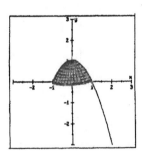

Use shells;

25. $y = e^{-x^2}$

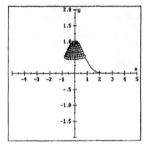

Use shells;

$$V = 2\pi \int_{0}^{\sqrt{\ln 2}} x\left(e^{-x^2} - \tfrac{1}{2} \right) dx$$

$$= \left[-\pi e^{-x^2} - \tfrac{\pi}{2}x^2 \right] \Big|_{0}^{\sqrt{\ln 2}}$$

$$= \tfrac{\pi}{2}(1 - \ln 2) \text{ cubic units}$$

In Problems 27-45, it is understood that the given volumes are all in cubic units.

27. $y = 4 - x, \; 0 \le x \le 4$

a. disks: $\pi \int_{0}^{4} (4 - x)^2 \; dx$

b. shells: $2\pi \int_{0}^{4} x(4 - x) \; dx$

c. washers: $\pi \displaystyle\int_0^4 [(y+1)^2 - 1^2]\, dx$

$$= \pi \int_0^4 [24 - 10x + x^2]\, dx$$

d. shells: $2\pi \displaystyle\int_0^4 (x+2)(4-x)\, dx$

29. $y = \sqrt{4 - x^2};\ 0 \le x \le 2$

a. shells: $2\pi \displaystyle\int_0^2 y\sqrt{4 - y^2}\, dy$

b. disks: $\pi \displaystyle\int_0^2 (4 - y^2)\, dy$

c. shells: $2\pi \displaystyle\int_0^2 (y+1)\sqrt{4 - y^2}\, dy$

d. washers: $\pi \displaystyle\int_0^2 [(x+2)^2 - 2^2]\, dy$

$$= \pi \int_0^2 [4 - y^2 + 4\sqrt{4 - y^2}]\, dy$$

31. $y = e^{-x};\ 0 \le x \le 1$

a. disks: $\pi \displaystyle\int_0^1 (e^{-x})^2\, dx$

b. shells: $2\pi \displaystyle\int_0^1 xe^{-x}\, dx$

c. washers: $\pi \displaystyle\int_0^1 \left[(e^{-x}+1)^2 - 1^2\right]\, dx$

$$= \pi \int_0^1 (e^{-2x} + 2e^{-x})\, dx$$

d. shells: $2\pi \displaystyle\int_0^1 (x+2)e^{-x}\, dx$

33. $y = \sin^{-1} x,\ 0 \le x \le 1$

a. shells: $2\pi \displaystyle\int_0^{\pi/2} y(1 - \sin y)\, dy$

b. washers: $\pi \displaystyle\int_0^{\pi/2} [1^2 - \sin^2 y]\, dy$

c. shells: $2\pi \displaystyle\int_0^{\pi/2} (y+1)(1 - \sin y)\, dy$

d. washers: $\pi \displaystyle\int_0^{\pi/2} [(1+2)^2 - (x+2)^2]\, dy$

$$= \pi \int_0^{\pi/2} [5 - \sin^2 y - 4\sin y]\, dy$$

35. $x = y^2;\ y = x^2$

The curves intersect where

$$x = y^2 = (x^2)^2 = x^4$$

$$x^4 - x = 0$$

$$x(x^3 - 1) = 0$$

$$x = 0, 1$$

a. By washers: $V = \pi \displaystyle\int_0^1 [(\sqrt{x})^2 - (x^2)^2]\, dx$

b. By washers: $V = \pi \displaystyle\int_0^1 [(\sqrt{y})^2 - (y^2)^2]\, dy$

37. $y = 1, x = 2, y = x^2 + 1$

a. By washers: $V = \pi \displaystyle\int_0^2 [(x^2 + 1)^2 - 1^2]\, dx$

b. By washers: $V = \pi \displaystyle\int_1^5 \left[(2)^2 - (\sqrt{y - 1})^2\right] dy$

39. $y = 0.1x^2$; $y = \ln x$; the curves intersect when

$$0.1x^2 = \ln x$$

Use computer software or a graphing

calculator to find the curves intersect at

(1.138, 0.130) and (3.566, 1.271).

a. By washers:

$$V = \pi \int_{1.138}^{3.566} \left[(\ln x)^2 - (0.1x^2)^2\right] dx$$

b. By washers:

$$V = \pi \int_{0.130}^{1.271} \left[(\sqrt{10y})^2 - (e^y)^2\right] dy$$

41. $y = x^2$ and $y = x^3$. The curves intersect when

$$x^2 = x^3$$

$$x^3 - x^2 = 0$$

$$x^2(x - 1) = 0$$

$$x = 0, 1$$

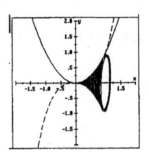

a. $V = \pi \displaystyle\int_0^1 [(x^2)^2 - (x^3)^2]\, dx$

$$= \pi \int_0^1 (x^4 - x^6)\, dx$$

$$= \pi\left(\tfrac{1}{5}x^5 - \tfrac{1}{7}x^7\right)\Big|_0^1$$

$$= \tfrac{2\pi}{35}$$

b. $V = 2\pi \displaystyle\int_0^1 y(y^{1/3} - y^{1/2})\, dy$

$$= 2\pi \int_0^1 (y^{4/3} - y^{3/2})\, dy$$

$$= 2\pi\left[\frac{3}{7}y^{7/3} - \frac{2}{5}y^{5/2}\right]\Big|_0^1$$

$$= \frac{2\pi}{35}$$

43. $y = x$, $y = 2x$, and $y = 1$. The curves intersect at when

$$x = 2x \qquad \text{or} \qquad 2x = 1$$

$$x = 0 \qquad\qquad x = \frac{1}{2}$$

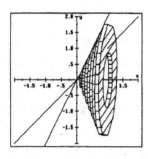

a. $V = \pi\int\limits_0^{1/2}[(2x)^2 - x^2]\,dx + \pi\int\limits_{1/2}^1 (1^2 - x^2)\,dx$

$$= \pi\int\limits_0^{1/2} 3x^2\,dx + \pi\int\limits_{1/2}^1 (1 - x^2)\,dx$$

$$= \pi x^3\Big|_0^{1/2} + \pi\left(x - \frac{x^3}{3}\right)\Big|_{1/2}^1$$

$$= \pi\left(\frac{1}{8} + 1 - \frac{1}{3} - \frac{1}{2} + \frac{1}{24}\right)$$

$$= \frac{\pi}{3}$$

b. $V = 2\pi\int\limits_0^1 y\left(y - \frac{y}{2}\right)\,dy$

$$= 2\pi\int\limits_0^1 \frac{y^2}{2}\,dy$$

$$= \pi\frac{y^3}{3}\Big|_0^1$$

$$= \frac{\pi}{3}$$

45. A sphere centered at the origin with radius r has the equation

$$x^2 + y^2 = r^2$$

$$y^2 = r^2 - x^2$$

Thus, the upper semicircle is

$$y = \sqrt{r^2 - x^2}$$

$$\int\limits_{-r}^r \pi\left(\sqrt{r^2 - x^2}\right)^2\,dx = 2\pi(r^2 x - \frac{1}{3}x^3)\Big|_0^r$$

$$= \frac{4}{3}\pi r^3$$

47. The cross section is an equilateral triangle of side $2y$ and area $\frac{1}{4}\sqrt{3}(2y)^2 = \sqrt{3}y^2$.

$$V = \sqrt{3}\int\limits_{-3}^3 (9 - x^2)\,dx$$

$$= 2\sqrt{3}(9x - \frac{1}{3}x^3)\Big|_0^3$$

$$= 36\sqrt{3} \text{ cubic units}$$

49. The cross section is a semicircle with radius y and area $\frac{1}{2}\pi y^2$.

$$V = \frac{1}{2}\pi \int_{-3}^{3} (9 - x^2)\, dx$$

$$= \pi \int_{0}^{3} (9 - x^2)\, dx$$

$$= \pi(9x - \tfrac{1}{3}x^3)\Big|_{0}^{3}$$

$$= 18\pi \text{ cubic units}$$

51. Suppose the triangular base has one leg $\overline{AC}$ on the y-axis and a vertex B on the positive x-axis. $M_{AB} = -\tan\frac{\pi}{6} = -1/\sqrt{3}$ and the equation of the line $\overline{AB}$ is

$$y - 2 = -\frac{1}{\sqrt{3}}x$$

The cross section is a square with side $2y$, and area $4y^2 = 4(2 - \frac{1}{\sqrt{3}}x)^2$.

$$V = 4\int_{0}^{2\sqrt{3}} (2 - \frac{1}{\sqrt{3}}x)^2\, dx$$

$$= -4\sqrt{3}\int_{0}^{2\sqrt{3}} (2 - \frac{1}{\sqrt{3}}x)^2(-\frac{1}{\sqrt{3}}\, dx)$$

$$= -4\sqrt{3}(\tfrac{1}{3})(2 - \frac{1}{\sqrt{3}}x)^3\Big|_{0}^{2\sqrt{3}}$$

$$= \frac{32\sqrt{3}}{3}$$

53.

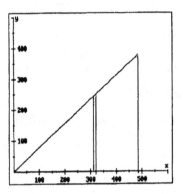

Put the vertex of the pyramid at the origin. Its side will be along the line $y = (375/480)\,x$. Each square slice perpendicular to the x-axis will have height of y and width $2y$. Twice the sum of all the rectangles from $x = 0$ to $x = 480$ will give the volume.

$$V = 2\int_{0}^{480} 2y^2\, dx$$

$$= 4\int_{0}^{480} \left(\frac{375}{480}x\right)^2\, dx$$

$$= 4\left(\frac{375}{480}\right)^2 \frac{x^3}{3}\Big|_{0}^{480}$$

$$= 4(375)^2 \frac{480}{3}$$

$$= 90{,}000{,}000 \text{ ft}^3$$

55. $y = x^{-1/2}$

a. Use disks;

$$V = \pi \int_1^4 (x^{-1/2})^2 \, dx$$

$$= \pi \ln|x| \Big|_1^4$$

$$= \pi \ln 4$$

b. Use shells;

$$V = 2\pi \int_1^4 x^{1/2} \, dx = 2\pi \left(\tfrac{2}{3}\right) x^{3/2} \Big|_1^4$$

$$= \frac{28\pi}{3}$$

c. Use washers;

$$dV = \pi[(y+2)^2 - 2^2] dx$$

$$= \pi[(x^{-1/2}+2)^2 - 4] dx$$

$$V = \pi \int_1^4 [x^{-1} + 4x^{-1/2} + 4 - 4] \, dx$$

$$= \pi(\ln|x| + 8\sqrt{x}) \Big|_1^4$$

$$= \pi(8 + \ln 4)$$

57. Use shells and double the portion (by symmetry) above the x-axis.

$$V = 2 \int_1^3 2\pi xy \, dx$$

$$= 2 \int_1^3 2\pi x \sqrt{4 - \frac{4x^2}{9}} \, dx$$

$$= \frac{8\pi}{3} \int_1^3 x\sqrt{9 - x^2} \, dx$$

$$= -\frac{8\pi}{9}(9 - x^2)^{3/2} \Big|_1^3$$

$$= \frac{128\sqrt{2}\pi}{9}$$

$$\approx 63.1877 \text{ ft}^3$$

59.
$$V \approx \tfrac{1}{3}[1(1.12)+4(1.09)+2(1.05)+4(1.03)$$
$$+2(0.99)+4(1.01)+2(0.98)+4(0.99)$$
$$+2(0.96)+4(0.93)+1(0.91)](2.0)$$
$$= \tfrac{2}{3}(30.19)$$
$$\approx 20.13$$

The volume is approximately 20.13 m^3.

61. The cross section is a semicircle with radius y, area $\tfrac{1}{2}\pi y^2$ and volume $dV = \tfrac{1}{2}\pi y^2 \, dx$;

$$V = \tfrac{1}{2}\pi \int_{-r}^r (r^2 - x^2) \, dx$$

$$= \frac{\pi}{2}(r^2 x - \tfrac{1}{3}x^3) \Big|_{-r}^r$$

$$= \frac{\pi}{6}(6r^3 - 2r^3) = \tfrac{2}{3}\pi r^3$$

Thus, the volume of the entire sphere is

$$V = 2(\tfrac{2\pi}{3} r^3) = \tfrac{4}{3}\pi r^3.$$

63. Position the equilateral triangular base to reflect symmetry as in the figure below.

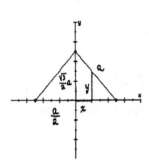

Using similar triangles, we see that

$$\frac{y}{\sqrt{3}a/2} = \frac{a/2 - x}{a/2}$$

$$y = \frac{\sqrt{3}a}{2}\left(\frac{a}{2} - x\right)\left(\frac{2}{a}\right)$$

$$= \sqrt{3}\left(\frac{a}{2} - x\right)$$

$$= \frac{\sqrt{3}}{2}a - \sqrt{3}x$$

Thus, the cross-sectional triangle at x has area

$$A(x) = \frac{\sqrt{3}}{4}y^2$$

$$= \frac{\sqrt{3}}{4}\left(\frac{\sqrt{3}}{2}a - \sqrt{3}x\right)^2$$

and the total volume of the solid figure is given by:

$$V = 2\int_0^{a/2} \frac{\sqrt{3}}{4}\left(\frac{\sqrt{3}}{2} - \sqrt{3}a\right)^2 dx$$

Let $u = \frac{\sqrt{3}}{2}a - \sqrt{3}x$; $du = -\sqrt{3}\,dx$

$$= \frac{\sqrt{3}}{2}\int_{x=0}^{x=a/2} u^2 \frac{du}{-\sqrt{3}}$$

$$= -\frac{1}{2}\left(\frac{u^3}{3}\right)\Bigg|_{x=0}^{x=a/2}$$

$$= -\frac{1}{6}\left(\frac{\sqrt{3}}{2}a - \sqrt{3}x\right)\Bigg|_0^{a/2}$$

$$= -\frac{1}{6}\left[\left(\frac{\sqrt{3}}{2}a - \frac{\sqrt{3}}{2}a\right)^3 - \left(\frac{\sqrt{3}}{2}a - 0\right)^3\right]$$

$$= -\frac{1}{6}\left(-\frac{\sqrt{3}}{2}a\right)^3$$

$$= \frac{\sqrt{3}\,a^3}{16}$$

This is not the same as the volume of a tetrahedron of side a, which is (see Problem 62) $\frac{\sqrt{2}\,a^3}{12}$. Thus, the figure cannot be a tetrahedron, and the conjecture is false.

65. Use disks with horizontal strips for $h \leq y \leq R$.

$$dV = \pi x^2\,dy = \pi(R^2 - y^2)dy$$

$$V = \pi\int_h^R (R^2 - y^2)\,dy$$

$$= \pi(R^2 y - \tfrac{1}{3}y^3)\Big|_h^R$$

$$= \tfrac{1}{3}\pi(2R^3 - 3R^2h + h^3)$$

6.3 Polar Forms and Area, Pages 384–385

SURVIVAL HINT: *You should remember the conversion formulas on page 376. Some calculators automatically do these conversions, but if you* *remember the formulas it is often easier to use them directly than to use the calculator conversions.*

1. *Step 1.* Find the simultaneous solution of the given system of equations.

    Step 2. Determine whether the pole lies on the two graphs.

    Step 3. Graph the curves to look for other points of intersection.

3. **a.** lemniscate   **b.** circle

    c. rose (3 petals)   **d.** none (spiral)

    e. cardioid   **f.** line

    g. lemniscate   **h.** limaçon

5. **a.** rose (4 petals)   **b.** circle

    c. limaçon   **d.** cardioid

    e. line   **f.** line

    g. rose (5 petals)   **h.** line

SURVIVAL HINT: *For a more complete discussion of the polar coordinate system see Chapter 5 of the Student Mathematics Handbook. Also see the summary of polar-form curves in Table 6.2 in the text.*

7. $\theta = -\frac{\pi}{2}$ is a line (the y-axis), and the limitation on r defines a line segment.

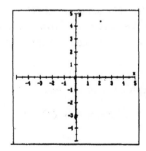

9. $r = 2\theta$ is a spiral, and the limitation on θ means the spiral rotates in a counterclockwise direction.

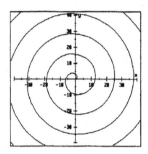

SURVIVAL HINT: *Study Page 379 until you can identify a circle, limaçon, cardioid, rose curve, and lemniscate from its equation. Once the general shape is identified, a decent sketch can usually be drawn by plotting the four intercepts (if they exist). Using the polar mode on your calculator, you can sketch most of these using a calculator.*

11. $r = 5\sin 3\theta$ is a rose curve with 3 leaves (since 3 is an odd number) and the first leaf is rotated $\frac{\pi}{2}$ divided by 3 (or $\frac{\pi}{6}$); the length of each leaf is 5 units.

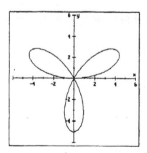

13. $r = 3\cos 3(\theta - \frac{\pi}{3})$ is a rose curve with 3 leaves (since 3 is an odd number) and the first

leaf is rotated $\frac{\pi}{3}$; the length of each leave is 3 units.

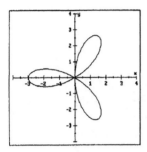

15. $r = \sin(2\theta + \frac{\pi}{3}) = \sin 2(\theta + \frac{\pi}{6})$ is a rose curve with 4 leaves (2 is an even number) which is first rotated $\frac{\pi}{2}$ divided by 2 (or $\frac{\pi}{4}$); from this position there is a further rotation of $-\frac{\pi}{6}$; the length of each leave is 1 unit.

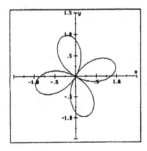

17. $r = 2 + \cos\theta$ is a limaçon with $a = 1$ and $b = 2$.

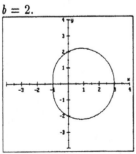

19. $r\cos\theta = 2$ or $x = 2$ (since $x = r\cos\theta$); we see this is a vertical line.

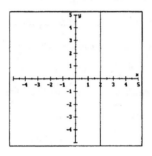

21. $r = -2\sin\theta$; this is a circle which, first, has been rotated $\frac{\pi}{2}$, and then the negative values reflect the circle below the x-axis.

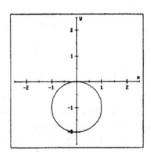

SURVIVAL HINT: *When solving trigonometric equations to find a point of intersection, rather than calculator approximations, use those values in Table 1.2 (page 10). The sides of triangles involving exact values are usual sides of a 45°-45°-90° triangle or a 30°-60°-90° triangle. The sides are in the ratio of 1, 1, $\sqrt{2}$ or 1, 2, $\sqrt{3}$, respectively. Proper placement of the triangle on a coordinate system often leads to an exact value.*

23. First, we graph each of the equations;

$r = 4\sin\theta$ is a 1 leaved rose (rotated $\frac{\pi}{2}$;

circular leaf), and the length of the leaf is 4 units;

$r = 2$ is a circle centered at the origin with radius 2.

The pole is not a point of intersection.

$$r_1 = r_2$$
$$4 \sin \theta = 2$$
$$\sin \theta = \tfrac{1}{2}$$
$$\theta = \tfrac{\pi}{6}, \tfrac{5\pi}{6}$$
$$P_1(2, \tfrac{\pi}{6}) \text{ and } P_2(2, \tfrac{5\pi}{6})$$

25. First, we graph each of the equations;

$r = 2(1 + \sin \theta)$ is a cardioid (rotated $-\tfrac{\pi}{2}$)

and

$r = 2(1 - \sin \theta)$ is a cardioid (rotated $\tfrac{\pi}{2}$).

The pole, P_1, is a point of intersection.

$$r_1 = r_2$$
$$1 + \sin \theta = 1 - \sin \theta$$
$$\sin \theta = 0$$
$$\theta = 0, \pi$$
$$P_2(2, 0), \ P_3(2, \pi)$$

27. First, we graph each of the equations;

$r^2 = \sin 2\theta$ is a lemniscate (rotated $\tfrac{\pi}{4}$)

and

$r = \sqrt{2} \sin \theta$ is a one leaved rose (circular leaf) that has been rotated $\tfrac{\pi}{2}$.

The pole, P_1, is a point of intersection.

$$r_1{}^2 = r_2{}^2$$
$$\sin 2\theta = 2 \sin^2\theta$$
$$2 \sin \theta \cos \theta = 2 \sin^2\theta$$
$$\sin \theta = 0 \quad \cos \theta = \sin \theta$$
$$\theta = 0, \pi \quad \theta = \tfrac{\pi}{4}, \tfrac{5\pi}{4}$$
$$P_2 = (1, \tfrac{\pi}{4})$$

29. First, graph each of the equations;

$r = \dfrac{4}{1 - \cos \theta}$ can be done using a graphing utility; it is a parabola with vertex at $(-2, 0)$

and

$r = 2\cos \theta$ is a one leaved rose (circular leaf) with the length of the leaf 2.

The pole is not is a point of intersection.

$$r_1 = r_2$$
$$2 \cos \theta = \frac{4}{1 - \cos \theta}$$
$$2\cos \theta - 2\cos^2\theta = 4$$
$$2\cos^2\theta - 2 \cos \theta + 4 = 0$$

There are no intersection points.

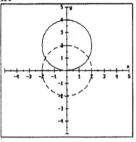

31. $A = \dfrac{1}{2} \displaystyle\int_0^{\pi/6} \cos^2\theta \ d\theta$

$= \tfrac{1}{2}\left(\tfrac{1}{2}\theta + \tfrac{1}{4}\sin 2\theta\right)\Big|_0^{\pi/6}$ *Cosine square formula*

$= \left[\tfrac{1}{4}\cdot\tfrac{\pi}{6} + \tfrac{1}{8}\sin\tfrac{\pi}{3}\right] - \left[\tfrac{1}{4}(0) + \tfrac{1}{8}\sin 0\right]$

$$= \frac{\pi}{24} + \frac{1}{8}\left(\frac{\sqrt{3}}{2}\right)$$

$$= \frac{\pi}{24} + \frac{\sqrt{3}}{16}$$

$$\approx 0.2392$$

33. $A = \frac{1}{2}\displaystyle\int_{\pi/6}^{\pi/2} \sin\theta\ d\theta$

$$= -\frac{1}{2}\cos\theta\ \Big|_{\pi/6}^{\pi/2}$$

$$= -\frac{1}{2}\cos\frac{\pi}{2} + \frac{1}{2}\cos\frac{\pi}{6}$$

$$= \frac{1}{2}\left(\frac{\sqrt{3}}{2}\right)$$

$$= \frac{\sqrt{3}}{4}$$

$$\approx 0.4330$$

35. $A = \frac{1}{2}\displaystyle\int_{0}^{\pi/4} (\sin\theta + \cos\theta)\ d\theta$

$$= \frac{1}{2}\int_{0}^{\pi/4} (\sin^2\theta + 2\sin\theta\cos\theta + \cos^2\theta)\ d\theta$$

$$= \frac{1}{2}\int_{0}^{\pi/4} (1 + 2\sin\theta\cos\theta)\ d\theta$$

$$= \frac{1}{2}\int_{0}^{\pi/4} (1 + \sin 2\theta)\ d\theta$$

> Let $u = 2\theta;\ du = 2\ d\theta$

$$= \frac{1}{2}(\theta - \frac{1}{2}\cos 2\theta)\ \Big|_{0}^{\pi/4}$$

$$= \frac{1}{2}\left[\left(\frac{\pi}{4} - \frac{1}{2}\cos\frac{\pi}{2}\right) - \left(0 - \frac{1}{2}\cos 0\right)\right]$$

$$= \frac{\pi}{8} + \frac{1}{4}$$

$$\approx 0.6427$$

37. $A = \frac{1}{2}\displaystyle\int_{0}^{2\pi} \left(\frac{\theta^2}{\pi}\right)^2\ d\theta$

$$= \frac{1}{2}\int_{0}^{2\pi} \frac{\theta^4}{\pi^2}\ d\theta$$

$$= \frac{1}{2\pi^2} \frac{\theta^5}{5}\ \Big|_{0}^{2\pi}$$

$$= \frac{16\pi^3}{5}$$

$$\approx 99.2201$$

39. Use a graphing utility:

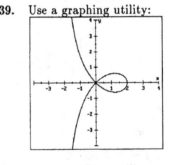

41. Use a graphing utility:

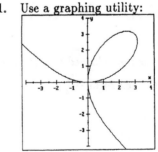

43. We will find the area from $\theta = 0$ to $\theta = \frac{\pi}{3}$ and multiply by 3.

$$A = (3)\frac{1}{2}\int_0^{\pi/3} a^2 \sin^2 3\theta \; d\theta$$

$$= \frac{3}{2}a^2 \int_0^{\pi/3} \left(\frac{1 - \cos 6\theta}{2}\right) d\theta$$

$$= \frac{3a^2}{4}\left(\theta - \frac{1}{6}\sin 6\theta\right)\Big|_0^{\pi/3}$$

$$= \frac{\pi a^2}{4}$$

45. The curves intersect when $\cos \theta = 0$; $\theta = \pi/2$, $3\pi/2$. Using symmetry,

$$A = 2\left(\frac{a^2}{2}\right)\int_0^{\pi/2} [1^2 - (1 - \cos \theta)^2] \; d\theta$$

$$= a^2 \int_0^{\pi/2} (1 - 1 + 2\cos\theta - \cos^2\theta) \; d\theta$$

$$= a^2 \int_0^{\pi/2} (2\cos\theta - \cos^2\theta) \; d\theta$$

$$= a^2\left[2\sin\theta - \left(\frac{1}{2}\theta + \frac{1}{4}\sin 2\theta\right)\right]\Big|_0^{\pi/2}$$

$$= a^2[(2\sin\tfrac{\pi}{2} - \tfrac{1}{2}\left(\tfrac{\pi}{2}\right) - \tfrac{1}{4}\sin \pi)$$
$$- (2\sin 0 - \tfrac{1}{2}(0) - \tfrac{1}{4}\sin 0)]$$

$$= a^2\left[2 - \frac{\pi}{4}\right]$$

47. Solving simultaneously:

$$6\cos\theta = 2 + 2\cos\theta$$

$$4\cos\theta = 2$$

$$\cos\theta = \frac{1}{2}$$

$$\theta = \frac{\pi}{3}, \frac{5\pi}{3}$$

Also note that the pole, $(0, 0)$, is a solution.

$$A = (2)\frac{1}{2}\int_0^{\pi/3} [36\cos^2\theta - 4(1 + \cos\theta)^2] \; d\theta$$

$$= 4\int_0^{\pi/3} [9\cos^2\theta - 1 - 2\cos\theta - \cos^2\theta] \; d\theta$$

$$= 4\int_0^{\pi/3} [8\cos^2\theta - 2\cos\theta - 1] \; d\theta$$

$$= 4\left[8\left(\frac{1}{2}\theta + \frac{1}{4}\sin 2\theta\right) - 2\sin\theta - \theta\right]\Big|_0^{\pi/3}$$

$$= 4\left[4\left(\frac{\pi}{3}\right) + 2\sin\frac{2\pi}{3} - 2\sin\frac{\pi}{3} - \frac{\pi}{3} - 0\right]$$

$$= 4\left[\frac{4\pi}{3} + 2\left(\frac{\sqrt{3}}{2}\right) - 2\left(\frac{\sqrt{3}}{2}\right) - 2\left(\frac{\sqrt{3}}{2}\right) - \frac{\pi}{3}\right]$$

$$= 4\pi$$

49. The inner loop is traced out for $\pi/6 \le \theta \le 5\pi/6$ and the outer loop for $5\pi/6 \le \theta \le 13\pi/6$. Thus, the area between the loops is

$$A = \int_{5\pi/6}^{13\pi/6} \frac{1}{2}(2 - 4\sin\theta)^2 \; d\theta$$

$$- \int_{\pi/6}^{5\pi/6} \frac{1}{2}(2 - 4\sin\theta)^2 \; d\theta$$

$$= \left[6\theta + 8\cos\theta - 2\sin 2\theta\right]\Big|_{5\pi/6}^{13\pi/6}$$

$$- \left[6\theta + 8\cos\theta - 2\sin 2\theta\right]\Big|_{\pi/6}^{5\pi/6}$$

$$= (8\pi + 6\sqrt{3}) - (4\pi - 6\sqrt{3})$$

$$= 4\pi + 12\sqrt{3} \approx 33.3510$$

51. The line and the lemniscate intersect where:

$$\frac{1}{(\cos\theta)^2} = 2\sin 2\theta$$

$$\theta \approx 0.2874, \, 0.7854$$

Thus, the area is

$$A = \int_{0.2874}^{0.7854} \frac{1}{2}\left[2\sin 2\theta - \frac{1}{\cos^2\theta}\right] d\theta$$

$$= \left[\frac{-\cos 2\theta}{2} - \frac{\tan\theta}{2}\right]\Big|_{0.2874}^{0.7854}$$

$$\approx 0.0674$$

53. At each point P on $r = 3 + 3\sin\theta$, the x-coordinate is

$$x = r\cos\theta = (3 + 3\sin\theta)\cos\theta$$

The maximum value of x occurs when $x'(\theta) = 0$ for $-\pi/2 \le \theta \le \pi/2$. Since

$$x'(\theta) = -3\sin\theta + 3(\cos^2\theta - \sin^2\theta)$$

$x'(\theta) = 0$ when $\theta = \frac{\pi}{6}$ so the maximum value of x is $9\sqrt{3}/4 \approx 3.8971$.

55. Use a graphing utility to graph this curve.

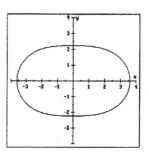

57. Because $r = f(\theta)$ and $x = r\cos\theta$, $y = r\sin\theta$, we have $x = f(\theta)\cos\theta$ and $y = f(\theta)\sin\theta$. Using the chain rule, we find that

$$\frac{dy}{d\theta} = \left(\frac{dy}{dx}\right)\left(\frac{dx}{d\theta}\right) \text{ or, equivalently,}$$

$$\frac{dy}{dx} = \frac{\dfrac{dy}{d\theta}}{\dfrac{dx}{d\theta}}$$

Because $x = f(\theta)\cos\theta$ and $y = f(\theta)\sin\theta$, it follows that

$$\frac{dx}{d\theta} = f(\theta)\frac{d}{d\theta}(\cos\theta) + \cos\theta\,\frac{df}{d\theta}$$

$$= -f(\theta)\sin\theta + f'(\theta)\cos\theta$$

$$\frac{dy}{d\theta} = f(\theta)\frac{d}{d\theta}(\sin\theta) + \sin\theta\,\frac{df}{d\theta}$$

$$= f(\theta)\cos\theta + f'(\theta)\sin\theta$$

and the slope of the tangent line is given by:

$$m = \frac{dy}{dx}$$

$$= \frac{\frac{dy}{d\theta}}{\frac{dx}{d\theta}}$$

$$= \frac{f(\theta)\cos\theta + f'(\theta)\sin\theta}{-f(\theta)\sin\theta + f'(\theta)\cos\theta}$$

59. **a.** $\tan\alpha = \frac{a\cos\theta}{-a\sin\theta}$

$$= -\cot\theta$$

b. $\frac{dr}{d\theta} = 2\sin\theta$; $\tan\alpha = \frac{1 - \cos\theta}{\sin\theta}$

c. $\frac{dr}{d\theta} = 6e^{3\theta}$; $\tan\alpha = \frac{2e^{3\theta}}{6e^{3\theta}} = \frac{1}{3}$

6.4 Arc Length and Surface Area, Pages 393-395

SURVIVAL HINT: *Formulas do not need to be "memorized" if you understand the concepts. The length of the arc is the sum (integral) of a collection of oblique line segments — each of which can be considered as the hypotenuse of a right triangle with base 1 and height equal to the slope of the line, namely $f'(x)$. This hypotenuse has length found by the Pythagorean theorem, namely $\sqrt{1 + [f'(x)]^2}$.*

1. $f(x) = 3x + 2$; $f'(x) = 3$

$$\sqrt{1 + [f'(x)]^2} = \sqrt{1 + 3^2} = \sqrt{10}$$

$$s = \int_{-1}^{2} \sqrt{10}\, dx$$

$$= 3\sqrt{10}$$

3. $f(x) = 1 - 2x$; $f'(x) = -2$

$$\sqrt{1 + [f'(x)]^2}\, dx = \sqrt{1 + (-2)^2} = \sqrt{5}$$

$$s = \int_{1}^{3} \sqrt{5}\, dx$$

$$= 2\sqrt{5}$$

5. $f(x) = \frac{2}{3}x^{3/2} + 1$; $f'(x) = x^{1/2}$

$$\sqrt{1 + [f'(x)]^2} = \sqrt{1 + x}$$

$$s = \int_{0}^{4} \sqrt{1 + x}\, dx$$

$$= \frac{2}{3}(1 + x)^{3/2}\Big|_{0}^{4}$$

$$= \frac{2}{3}\big[5^{3/2} - 1^{3/2}\big]$$

$$= \frac{10\sqrt{5}}{3} - \frac{2}{3}$$

7. $f(x) = \frac{1}{12}x^5 + \frac{1}{5}x^{-3}$; $f'(x) = \frac{5}{12}x^4 - \frac{3}{5}x^{-4}$

$$\sqrt{1 + [f'(x)]^2} = \sqrt{1 + (\frac{5}{12}x^4 - \frac{3}{5}x^{-4})^2}$$

$$= \sqrt{(\frac{5}{12}x^4)^2 + \frac{1}{2} + (\frac{3}{5}x^{-4})^2}$$

$$= \sqrt{(\frac{5}{12}x^4 + \frac{3}{5}x^{-4})^2}$$

$$s = \int_1^2 \left(\tfrac{5}{12}x^4 + \tfrac{3}{5}x^{-4}\right)\, dx$$

$$= \left(\tfrac{1}{12}x^5 - \tfrac{1}{5}x^{-3}\right)\Big|_1^2$$

$$= \tfrac{331}{120}$$

$$= \tfrac{2}{3}\left[4^{3/2} - 1^{3/2}\right]$$

$$= \tfrac{2}{3}[8 - 1]$$

$$= \tfrac{14}{3}$$

9. $f(x) = \tfrac{1}{4}x^4 + \tfrac{1}{8}x^{-2}; \; f'(x) = x^3 - \tfrac{1}{4}x^{-3}$

$$\sqrt{1 + [f'(x)]^2} = \sqrt{1 + (x^3 - \tfrac{1}{4}x^{-3})^2}$$

$$= \sqrt{1 + x^6 - \tfrac{1}{2} + \tfrac{1}{16}x^{-6}}$$

$$= \sqrt{(x^3 + \tfrac{1}{4}x^{-3})^2}$$

$$s = \int_1^2 (x^3 + \tfrac{1}{4}x^{-3})\, dx$$

$$= \left(\tfrac{1}{4}x^4 - \tfrac{1}{8}x^{-2}\right)\Big|_1^2$$

$$= \tfrac{123}{32}$$

11. Since $9x^2 = 4y^3$, we see $y = \sqrt[3]{\dfrac{9x^2}{4}}$.

Since this does not have a very nice derivative, solve for $x = g(y)$ and use a dy integral.

$$x = \sqrt{\tfrac{4y^3}{9}} = \tfrac{2y^{3/2}}{3}, \; g'(y) = y^{1/2}$$

$$\sqrt{1 + g'(y)^2} = \sqrt{1 + y}$$

$$s = \int_0^3 \sqrt{1 + y}\, dy$$

$$= \tfrac{2}{3}(1 + y)^{3/2}\Big|_0^3$$

SURVIVAL HINT: *Understand the **concept** of the surface area formula as the sum of the areas of small surfaces, each of which is the surface of the frustum of a cone:* $2\pi r\ell$, *where r is the distance from the axis of revolution to the bounding curve, and ℓ is the slant height of the frustum, namely* $\sqrt{1 + [f'(x)]^2}$.

13. $f(x) = 2x + 1; \; f'(x) = 2$

$$S = 2\pi \int_0^2 (2x + 1)\sqrt{1 + (2)^2}\, dx$$

$$= 2\pi\sqrt{5} \int_0^2 (2x + 1)\, dx$$

$$= 2\pi\sqrt{5}(x^2 + x)\Big|_0^2$$

$$= 12\pi\sqrt{5}$$

15. $y = \tfrac{1}{3}x^3 + \tfrac{1}{4}x^{-1}; \; \dfrac{dy}{dx} = x^2 - \tfrac{1}{4}x^{-2}$

$$dy = (x^2 - \tfrac{1}{4}x^{-2})dx$$

$$ds = \sqrt{1 + (x^2 - \tfrac{1}{4}x^{-2})^2}\, dx$$

$$= \sqrt{1 + (x^2)^2 - \tfrac{1}{2} + (\tfrac{1}{4}x^{-2})^2}\, dx$$

$$= \sqrt{(x^2)^2 + \tfrac{1}{2} + (\tfrac{1}{4}x^{-2})^2} \; dx$$

$$= \sqrt{(x^2 + \tfrac{1}{4}x^{-2})^2} \; dx$$

$$= (x^2 + \tfrac{1}{4}x^{-2}) \; dx$$

$$dS = 2\pi y \; ds$$

$$= 2\pi(\tfrac{1}{3}x^3 + \tfrac{1}{4}x^{-1})(x^2 + \tfrac{1}{4}x^{-2}) \, dx$$

$$= 2\pi(\tfrac{1}{3}x^5 + \tfrac{1}{12}x + \tfrac{1}{4}x + \tfrac{1}{16}x^{-3}) \; dx$$

$$S = 2\pi \int_1^2 (\tfrac{1}{3}x^5 + \tfrac{1}{3}x + \tfrac{1}{16}x^{-3}) \; dx$$

$$= 2\pi(\tfrac{1}{18}x^6 + \tfrac{1}{6}x^2 - \tfrac{1}{32}x^{-2})\Big|_1^2$$

$$= \frac{515\pi}{64}$$

$$\approx 25.28$$

17. The circle $r = \cos\theta$ is traced out for $0 \le \theta < \pi$.

$$s = \int_0^\pi \sqrt{(\cos\theta)^2 + (-\sin\theta)^2} \; d\theta$$

$$= \int_0^\pi d\theta = \pi$$

Alternatively, write

$$r = \cos\theta$$

$$r^2 = r\cos\theta$$

$$x^2 + y^2 = x$$

$$x^2 - x + y^2 = 0$$

$$(x - \tfrac{1}{2})^2 + y^2 = \tfrac{1}{4}$$

This is a circle with radius $r = \tfrac{1}{2}$, so the circumference is $2\pi r = 2\pi(\tfrac{1}{2}) = \pi$

19. $r = e^{3\theta}$;

$$s = \int_0^{\pi/2} \sqrt{(e^{3\theta})^2 + (3e^{3\theta})^2} \; d\theta$$

$$= \int_0^{\pi/2} \sqrt{10}\; e^{3\theta} \; d\theta$$

$$= \frac{\sqrt{10}}{3}\Big[e^{3\pi/2} - 1\Big]$$

21. $r = \theta^2$;

$$s = \int_0^1 \sqrt{(\theta^2)^2 + (2\theta)^2} \; d\theta$$

$$= \int_0^1 \theta\sqrt{\theta^2 + 4} \; d\theta$$

$$\boxed{\begin{array}{l} u = \theta^2 + 4; \; du = 2\theta \, d\theta \\ \text{If } \theta = 0, \text{ then } u = 4; \\ \text{if } \theta = 1, \text{ then } u = 5. \end{array}}$$

$$= \frac{1}{2}\int_4^5 u^{-1/2} \; du$$

$$= \frac{1}{2}\left(\frac{2}{3}\right)u^{3/2}\Big|_4^5$$

$$= \frac{1}{3}\left(5^{3/2} - 4^{3/2}\right)$$

$$= \frac{5}{3}\sqrt{5} - \frac{8}{3}$$

23. $r = 5$;

$$S = \int_0^{\pi/3} 2\pi(5)\sin\theta\sqrt{(5)^2 + [(5)']^2}\; d\theta$$

$$= \int_0^{\pi/3} 2\pi(5)^2\sin\theta\; d\theta$$

$$= 50\pi(-\cos\theta)\Big|_0^{\pi/3}$$

$$= 25\pi$$

25. $r = \csc\theta$;

$$S = \int_{\pi/4}^{\pi/3} 2\pi\csc\theta\sin\theta\sqrt{(\csc\theta)^2 + [(\csc\theta)']^2}\; d\theta$$

$$= \int_{\pi/4}^{\pi/3} 2\pi\sqrt{\csc^2\theta + (-\csc\theta\cot\theta)^2}\; d\theta$$

$$= \int_{\pi/4}^{\pi/3} 2\pi\csc\theta\sqrt{1 + \cot^2\theta}\; d\theta$$

$$= \int_{\pi/4}^{\pi/3} 2\pi\csc^2\theta\; d\theta$$

$$= 2\pi(-\cot\theta)\Big|_{\pi/4}^{\pi/3}$$

$$= \frac{2\pi}{3}(3 - \sqrt{3})$$

27. We show the result of numerical integration using one software program.

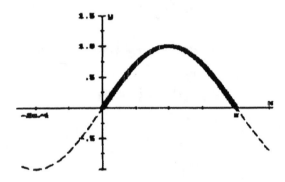

# OF TERMS	ARC LENGTH OVER [0, π]
2	3.724192
4	3.790091
8	3.812529
16	3.818275
32	3.819717
64	3.820077
128	3.820168
256	3.82019
512	3.820196
THIS IS THE FINAL ESTIMATE	

It appears the arc length is approximately 3.82.

29. $y = f(x) = \frac{1}{3}x^3 + (4x)^{-1}$;

$$f'(x) = \frac{1}{3}(3x^2) + \frac{1}{4}(-x^{-2})$$

$$= x^2 - \frac{1}{4x^2}$$

$$1 + [f'(x)]^2 = 1 + x^4 - \frac{1}{2} + \frac{1}{16x^4}$$

$$= (x^2 + \frac{1}{4x^2})^2$$

a. $S = 2\pi \int_1^3 [\frac{1}{3}x^3 + (4x)^{-1}](x^2 + \frac{1}{4x^2}) \, dx$

$$= 2\pi \int_1^3 \left(\frac{x^3}{3} + \frac{1}{4x}\right)\left(\frac{x^2}{1} + \frac{1}{4x^2}\right) dx$$

$$= 2\pi \int_1^3 \left[\frac{x^5}{3} + \frac{x}{12} + \frac{x}{4} + \frac{1}{16x^3}\right] dx$$

$$= 2\pi \left[\frac{x^6}{18} + \frac{x^2}{6} - \frac{1}{32x^2}\right]\Big|_1^3$$

$$= \frac{1{,}505\pi}{18}$$

$$\approx 262.67$$

b. $S = 2\pi \int_1^3 x\sqrt{1 + [f'(x)]^2} \, dx$

$$= 2\pi \int_1^3 x(x^2 + \frac{1}{4x^2}) dx$$

$$= 2\pi \left[\frac{1}{4}x^4 + \frac{1}{4}\ln x\right]\Big|_1^3$$

$$= 2\pi[20 + \frac{1}{4}\ln 3]$$

Numerical approximation is 127.4.

31. $y = f(x) = \frac{1}{3}(12 - x)$

$$dy = -\frac{1}{3} \, dx;$$

$$ds = \sqrt{1 + (-\frac{1}{3})^2} \, dx$$

$$= \frac{1}{3}\sqrt{10} \, dx$$

$$dS = 2\pi x \, ds$$

$$S = \frac{2}{3}\sqrt{10}\pi \int_0^3 x \, dx$$

$$= \frac{2}{3}\sqrt{10}\pi\left(\frac{x^2}{2}\right)\Big|_0^3$$

$$= 3\pi\sqrt{10}$$

$$\approx 29.8038$$

33. $y = f(x) = \frac{1}{3}\sqrt{x}(3 - x);$

$$dy = (\frac{1}{2}x^{-1/2} - \frac{1}{2}x^{1/2}) \, dx$$

$$ds = \sqrt{1 + (\frac{1}{2}x^{-1/2} - \frac{1}{2}x^{1/2})^2} \, dx$$

$$= (\frac{1}{2}x^{-1/2} + \frac{1}{2}x^{1/2}) \, dx$$

$$S = 2\pi \int_1^3 x(\frac{1}{2}x^{-1/2} + \frac{1}{2}x^{1/2}) \, dx$$

$$= \pi \int_1^3 (x^{1/2} + x^{3/2}) \, dx$$

$$= \pi(\frac{2}{3}x^{3/2} + \frac{2}{5}x^{5/2})\Big|_1^3$$

$$= \pi\left(\frac{28\sqrt{3}}{5} - \frac{16}{15}\right)$$

$$\approx 27.12$$

35. $x^{2/3} + y^{2/3} = 1$

$$y^{2/3} = 1 - x^{2/3}$$

$$y = (1 - x^{2/3})^{3/2}$$

Then $y = f(x) = (1 - x^{2/3})^{3/2}$;

$$f'(x) = \tfrac{3}{2}(1 - x^{2/3})^{1/2}(-\tfrac{2}{3}x^{-1/3})$$

$$= -x^{-1/3}(1 - x^{2/3})^{1/2}$$

$$ds = \sqrt{1 + x^{-2/3}(1 - x^{2/3})}\ dx$$

$$= x^{-1/3}\ dx$$

$$s = 4\int_{0}^{1} x^{-1/3}\ dx$$

$$= 6x^{2/3}\Big|_{0}^{1}$$

$$= 6$$

37. $y = \tan x,\ y' = \sec^2 x,$

$$ds = \sqrt{1 + \sec^4 x}\ dx.$$

$$S = 2\pi \int_{0}^{1} \tan x\ \sqrt{1 + \sec^4 x}\ dx$$

$$\approx 8.6322 \qquad \textit{By calculator}$$

39. $S = \displaystyle\int_{0}^{3} 2\pi x \sqrt{1 + [f'(x)]^2}\ dx$

x_k	$f'(x_k)$	$x_k\sqrt{1 + [f'(x)]^2}$
0.0	3.7	0.0000
0.3	3.9	1.2079
0.6	4.1	2.5321
0.9	4.1	3.7982
1.2	4.2	5.1809
1.5	4.4	6.7683
1.8	4.6	8.4733
2.1	4.9	10.5021
2.4	5.2	12.7087
2.7	5.5	15.0935
3.0	6.0	18.2484

$$S \approx \tfrac{2\pi}{3}[0 + 4(1.2079) + 2(2.5321) + \cdots$$
$$+ 4(15.0935) + 18.2484](0.3)$$

$$\approx 141.6974$$

41. Let the arc of the graph be subdivided into small elements, each approximated by Δx. Each of these is the hypotenuse of a right triangle with legs Δx and Δy. Then,

$$\Delta s = \sqrt{(\Delta x)^2 + (\Delta y)^2} = \sqrt{1 + \left(\frac{\Delta y}{\Delta x}\right)^2}\ \Delta x$$

and

$$\Delta S = 2\pi x_k^{*}\sqrt{1 + \left(\frac{\Delta y_k}{\Delta x_k}\right)^2}\ \Delta x_k$$

Thus,

$$S = \lim_{n \to +\infty} \sum_{k=1}^{n} 2\pi x_k^{*}\sqrt{1 + \left(\frac{\Delta y_k}{\Delta x_k}\right)^2}\ \Delta x_k$$

$$= 2\pi \lim_{n \to +\infty} \sum_{k=1}^{n} x_k^{*}\sqrt{1 + \left(\frac{\Delta y_k}{\Delta x_k}\right)^2}\ \Delta x_k$$

$$= 2\pi \int_{a}^{b} x\sqrt{1 + \left(\frac{dy}{dx}\right)^2}\ dx$$

$$= 2\pi \int_a^b x\sqrt{1+[f'(x)]^2}\,dx$$

43. **a.** Let P_{k-1} and P_k be the points with coordinates (x_{k-1}, y_{k-1}) and (x_k, y_k), respectively. When the line segment $P_{k-1}P_k$ is revolved about the x-axis, it generates the frustum of a cone with radii $r_1 = y_{k-1}$ and $r_2 = y_k$ and slant height $\ell = \ell_k$. In the text, we showed that such a frustum has surface area

$$\pi(r_1 + r_2)\ell = \pi(y_{k-1} + y_k)\ell_k$$

b. Assume $r_1 \leq r_2$. Then

$$r_1 = \tfrac{1}{2}r_1 + \tfrac{1}{2}r_1 \leq \tfrac{1}{2}r_1 + \tfrac{1}{2}r_2$$

and

$$r_2 = \tfrac{1}{2}r_2 + \tfrac{1}{2}r_2 \geq \tfrac{1}{2}r_1 + \tfrac{1}{2}r_2$$

so that

$$r_1 \leq \tfrac{1}{2}(r_1 + r_2) \leq r_2$$

A similar argument applies if $r_2 \geq r_1$. Since $r_1 = y_{k-1}$ and $r_2 = y_k$, we see that $\tfrac{1}{2}(y_{k-1} + y_k)$ is a number between y_{k-1} and y_k, and by applying the intermediate value theorem,

$$\tfrac{1}{2}(y_{k-1} + y_k) = f(c_k)$$

for some number c_k between x_{k-1} and x_k. Thus,

$$y_{k-1} + y_k = 2f(c_k)$$

c. The slant height ℓ_k of the line segment $P_{k-1}P_k$ is

$$\ell_k = \sqrt{(x_k - x_{k-1})^2 + (y_k - y_{k-1})^2}$$

$$= \sqrt{(\Delta x_k)^2 + [f(x_k) - f(x_{k-1})]^2}$$

MVT, we have

$$\frac{f(x_k) - f(x_{k-1})}{x_k - x_{k-1}} = f'(x_k^*)$$

for some number x_k^* between x_{k-1} and x_k. Thus,

$$f(x_k) - f(x_{k-1}) = f'(x_k^*)(x_k - x_{k-1})$$

$$= f'(x_k^*)\Delta x_k$$

and

$$\ell_k = \sqrt{(\Delta x_k)^2 + [f(x_k) - f(x_{k-1})]^2}$$

$$= \sqrt{1 + [f'(x_k^*)]^2}\,\Delta x_k$$

It follows that the surface area generated by revolving the line segment $P_{k-1}P_k$ about the x-axis is

$$\Delta S_k = \pi(y_{k-1} + y_k)\ell_k$$

$$= \pi[2f(c_k)]\sqrt{1 + [f'(x_k^*)]^2}\,\Delta x_k$$

where c_k is the number formed in part **b**. By adding all such areas associated with the partitioin P, we obtain an approximation for the total surface area of revolution S and by taking the limit as $\|P\| \to 0$, we are let the define S by

$$S = \lim_{\|P\| \to 0} \sum_{k=1}^n 2\pi f(c_k)\sqrt{1+[f'(x_k^*)]^2}\,\Delta x_k$$

d. $x_{k-1} < c_k \leq x_k \to x_k = x$ so $c_k \to x$ and $n \to \infty$ as $\|P\| \to 0$,

$$S = \lim_{n \to +\infty} \sum_{k=1}^{n} 2\pi f(c_k) \sqrt{1 + [f'(x)]^2} \, dx$$

$$= 2\pi \int_a^b f(x) \sqrt{1 + [f'(x)]^2} \, dx$$

45. $y = Cx^{2n} + Dx^{2(1-n)}$

$$y' = 2nCx^{2n-1} + 2(1-n)Dx^{1-2n}$$

$$1 + (y')^2 = 1 + [2nCx^{2n-1} + 2(1-n)Dx^{1-2n}]^2$$

$$= 1 + 4n^2C^2x^{2(2n-1)} - 8n(n-1)CD$$

$$+ 4(1-n)^2D^2x^{2(1-2n)}$$

$$= 4n^2C^2x^{2(2n-1)} + \frac{1}{2}$$

$$+ 4(1-n)^2D^2x^{2(1-2n)}$$

$$\textit{Since } 8n(n-1)CD = \frac{1}{2}$$

$$= [2nCx^{2n-1} + 2(n-1)Dx^{1-2n}]^2$$

Thus, the arc length is

$$L = \int_a^b [2nCx^{2n-1} + 2(n-1)Dx^{1-2n}] \, dx$$

$$= [Cx^{2n} - Dx^{2(1-n)}] \Big|_a^b$$

$$= C[b^{2n} - a^{2n}] - D[b^{2(1-n)} -$$

$$a^{2(1-n)}]$$

47. We are given

$$L(x) = \int_0^x \sqrt{1 + [f'(x)]^2} \, dt = \ln(\sec x + \tan x)$$

for all x in $[0, 1]$. Differentiating both sides of this equation, we obtain

$$\sqrt{1 + [f'(x)]^2} = \frac{\sec x \tan x + \sec^2 x}{\sec x + \tan x} = \sec x$$

$$1 + [f'(x)]^2 = \sec^2 x$$

$$[f'(x)]^2 = \sec^2 x - 1 = \tan^2 x$$

$$f'(x) = \pm \tan x$$

$$f(x) = \mp \ln|\cos x| + C$$

Since $f(0) = 0$ (the curve $y = f(x)$ passes through the origin), $C = 0$ and

$$f(x) = \pm \ln|\cos x|$$

6.5 Physical Applications: Work, Liquid Force, and Centroids, Pages 405–407

1. The work W done on a object moving a distance d in a straight line against a constant force F is $W = Fd$. If the force varies with x for $a \le x \le b$, the work is given by the integral

$$W = \int_a^b F(x) \, dx$$

3. Density ρ is a ratio of mass per unit length, or per unit area, or per unit volume. If the density varies in terms of a variable, find the element of mass and then sum up (integrate). Let f and g be continuous and satisfy $f(x) \ge g(x)$ on the interval $[a, b]$, and consider a thin plate (lamina) of uniform density ρ that covers the region R between the graphs of $y = f(x)$ and $y = g(x)$ on the interval $[a, b]$. Then the centroid of R is the point $(\overline{x}, \overline{y})$ such that

$$\overline{x} = \frac{M_y}{m} = \frac{\rho \int_a^b x[f(x) - g(x)] \, dx}{\rho \int_a^b [f(x) - g(x)] \, dx} \quad \text{and}$$

$$\bar{y} = \frac{M_x}{m} = \frac{\frac{1}{2}\rho \int_a^b \{[f(x)]^2 - [g(x)]^2\}\, dx}{\rho \int_a^b [f(x) - g(x)]\, dx}$$

5. $W = (850)(15)$

   $\quad = 12{,}750$ ft-lb

7.   The difference in F is 65 lb, the distance 100 ft for each. The additional work for the full bucket: $W = (65)(100) = 6{,}500$ ft-lb

9. $F(x) = kx;\ F(0.75) = 5;\ 0.75k = 5$ or $k = \frac{20}{3}$

   $$W = \frac{20}{3} \int_0^1 x\, dx$$

   $$= \frac{20}{3} \cdot \frac{1}{2} x^2 \Big|_0^1$$

   $$= \frac{10}{3}\ \text{ft-lb}$$

11.   The cable weighs 20/50 lb/ft, so

   $F = \frac{20}{50}\frac{\text{lbs}}{\text{ft}} = \frac{2}{5}$ lbs/ft. Let $(50 - x)$ be the length of cable hanging over the cliff. The work of the chain is

   $$W = \frac{2}{5} \int_0^{50} (50 - x)\, dx$$

   $$= 500$$

   The work done by the ball is

   $\quad W = (30)(50)$

   $\quad\quad = 1{,}500$

   Thus, the total work is

   $\quad 500 + 1{,}500 = 2{,}000$ ft-lb

13. $$W = \int_0^\pi \sin x\, dx - \int_\pi^{2\pi} \sin x\, dx$$

   $$= 2 \int_0^\pi \sin x\, dx$$

   $$= -2 \cos x \Big|_0^\pi$$

   $$= 4\ \text{ergs}$$

15.
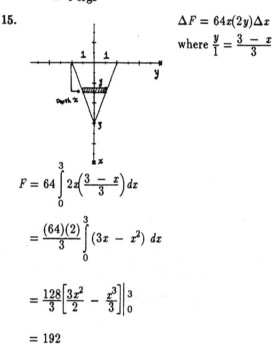

$\Delta F = 64x(2y)\Delta x$

where $\dfrac{y}{1} = \dfrac{3 - x}{3}$

$$F = 64 \int_0^3 2x\left(\frac{3 - x}{3}\right) dx$$

$$= \frac{(64)(2)}{3} \int_0^3 (3x - x^2)\, dx$$

$$= \frac{128}{3}\left[\frac{3x^2}{2} - \frac{x^3}{3}\right]\Big|_0^3$$

$$= 192$$

The force is 192 lb.

17.
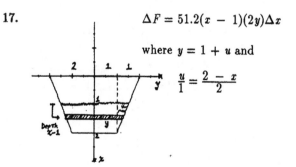

$\Delta F = 51.2(x - 1)(2y)\Delta x$

where $y = 1 + u$ and

$\dfrac{u}{1} = \dfrac{2 - x}{2}$

$$F = 51.2 \int_1^2 2(x - 1)\left[1 + \left(\frac{2 - x}{2}\right)\right] dx$$

$$= 51.2 \int_1^2 (x - 1)(4 - x) \, dx$$

$$= 51.2 \int_1^2 (-x^2 + 5x - 4) \, dx$$

$$= 51.2\left(-\frac{x^3}{3} + \frac{5x^2}{2} - 4x\right)\Big|_1^2$$

$$= 51.2\left(\frac{7}{6}\right)$$

$$\approx 59.7$$

The force is 59.7 lb.

21.

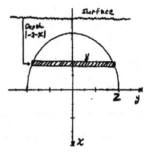

$$\Delta F = 51.2|-3 - x|(2y)\Delta x$$

where $y = \sqrt{4 - x^2}$

$$F = 51.2 \int_{-2}^0 2(x + 3)\sqrt{4 - x^2} \, dx$$

19.

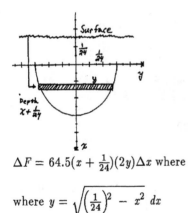

$$\Delta F = 64.5(x + \tfrac{1}{24})(2y)\Delta x \text{ where}$$

where $y = \sqrt{\left(\frac{1}{24}\right)^2 - x^2} \, dx$

$$F = 64.5 \int_0^{1/24} 2\left(x + \frac{1}{24}\right)\sqrt{\left(\frac{1}{24}\right)^2 - x^2} \, dx$$

SURVIVAL HINT: *To find $\overline{x}$ you use the moment about the y-axis, and to find $\overline{y}$ you use M_x. You will not confuse the formulas if you remember that the moment about the y-axis can be found by placing all the mass at $\overline{x}$: $M_y = m\overline{x}$, and the moment about the x-axis is found by placing all the mass at $\overline{y}$: $M_x = m\overline{y}$.*

23. The curves intersect where

$$x^2 - 9 = 0$$

$$x = -3, 3$$

$$m = 2 \int_0^3 [0 - (x^2 - 9)] \, dx$$

$$= -2(\tfrac{1}{3}x^3 - 9x)\Big|_0^3$$

$$= 36$$

$$M_y = 0 \text{ (by symmetry)}$$

$$M_x = 2 \int_0^3 \frac{1}{2}[0^2 - (x^2 - 9)^2] \, dx$$

$$= \int_0^3 (18x^2 - x^4 - 81) \, dx$$

$$= (6x^3 - \tfrac{1}{5}x^5 - 81x)\Big|_0^3$$

$$= -\frac{648}{5}$$

$$(\overline{x}, \overline{y}) = \left(0, \frac{-\frac{648}{5}}{36}\right) = \left(0, -\frac{18}{5}\right)$$

25. $\quad m = \int_1^2 x^{-1} dx = \ln x \Big|_1^2$

$$= \ln 2$$

$$M_y = \int_1^2 (x)x^{-1} dx$$

$$= 2 - 1$$

$$= 1$$

$$M_x = \frac{1}{2} \int_1^2 (x^{-1})^2 dx$$

$$= \frac{1}{2}(-x^{-1})\Big|_1^2$$

$$= \frac{1}{4}$$

$$(\overline{x}, \overline{y}) = \left(\frac{1}{\ln 2}, \frac{\frac{1}{4}}{\ln 2}\right) = \left(\frac{1}{\ln 2}, \frac{1}{4 \ln 2}\right)$$

27. $\quad m = \int_1^4 x^{-1/2} \, dx$

$$= 2\sqrt{x}\Big|_1^4$$

$$= 2$$

$$M_y = \int_1^4 x(x^{-1/2}) \, dx$$

$$= \tfrac{2}{3}x^{3/2}\Big|_1^4$$

$$= \frac{14}{3}$$

$$M_x = \int_1^4 \frac{1}{2}(x^{-1/2})^2 \, dx$$

$$= \tfrac{1}{2} \ln x \Big|_1^4$$

$$= \ln 2$$

$$(\overline{x}, \overline{y}) \approx \left(\frac{\frac{14}{3}}{2}, \frac{\ln 2}{2}\right) = \left(\tfrac{7}{3}, \tfrac{1}{2} \ln 2\right)$$

29. $\quad A = \tfrac{1}{2}(5)(5) = \frac{25}{2}$

The centroid of a triangle is located where the three medians meet, at the point $(\overline{x}, \overline{y})$ located $\frac{2}{3}$ of the distance from each vertex to the midpoint of the opposite side. The midpoint of the side with vertices $(-3, 0)$, and $(2, 0)$ is $(-\frac{1}{2}, 0)$, so

$$(\overline{x}, \overline{y}) = (0, 5) + \tfrac{2}{3}(-\tfrac{1}{2} - 0, 0 - 5)$$

$$= (-\tfrac{1}{3}, \tfrac{5}{3})$$

the distance from $(\overline{x}, \overline{y})$ to the line $y = -1$ is $s_1 = \frac{5}{3} + 1 = \frac{8}{3}$, so the volume of the solid formed by revolving the triangle about

$y = -1$ is

$$V = 2\pi s_1 A$$

$$= 2\pi(\tfrac{8}{3})(\tfrac{25}{2})$$

$$= \tfrac{200\pi}{3}$$

Using calculus is considerably more difficult

because there are two different curves for the

upper bound.

$$m = \int_{-3}^{0} \tfrac{5}{3}(x+3)\,dx + \int_{0}^{2} -\tfrac{5}{2}(x-2)\,dx$$

$$= \tfrac{5}{3}\left(\tfrac{x^2}{2} + 3x\right)\Big|_{-3}^{0} - \tfrac{5}{2}\left(\tfrac{x^2}{2} - 2x\right)\Big|_{0}^{2}$$

$$= \tfrac{25}{2}$$

$$M_x = \tfrac{1}{2}\int_{-3}^{0}\left[\tfrac{5}{3}(x+3)\right]^2 dx + \tfrac{1}{2}\int_{0}^{2}\left[-\tfrac{5}{2}(x-2)\right]^2 dx$$

$$= \tfrac{1}{2}(\tfrac{5}{3})^2\,\tfrac{(x+3)^3}{3}\Big|_{-3}^{0} + \tfrac{1}{2}(-\tfrac{5}{2})^2\tfrac{(x-2)^3}{3}\Big|_{0}^{2}$$

$$= \tfrac{25}{2\cdot 3^3}(3)^3 - 0 + 0 - \tfrac{1}{2}(\tfrac{25}{4})\left[\tfrac{(-2)^3}{3}\right]$$

$$= \tfrac{25}{2} + \tfrac{25}{3}$$

$$= \tfrac{125}{6}$$

$$\bar{y} = \tfrac{M_x}{M} = \tfrac{\frac{125}{6}}{\frac{25}{2}} = \tfrac{5}{3}$$

$$V = 2\pi\left(\tfrac{5}{3} + 1\right)\left(\tfrac{25}{2}\right) = \tfrac{200\pi}{3} \quad \text{(as before)}$$

31. $A = \tfrac{1}{2}\pi(2)^2 = 2\pi$

$$\bar{x} = \tfrac{1}{2\pi}\int_{-2}^{2} \tfrac{1}{2}\left[\sqrt{4-y^2}\right]^2 dy$$

$$= \tfrac{1}{4\pi}\int_{-2}^{2} (4 - y^2)\,dy$$

$$= \tfrac{1}{4\pi}\left(4y - \tfrac{y^3}{3}\right)\Big|_{-2}^{2}$$

$$= \tfrac{1}{4\pi}\left[(8 - \tfrac{8}{3}) - (-8 + \tfrac{8}{3})\right]$$

$$= \tfrac{1}{4\pi}\left[16 - \tfrac{16}{3}\right]$$

$$= \tfrac{8}{3\pi}$$

$$V = (2\pi)\left[2\pi\left(\tfrac{8}{3\pi} + 2\right)\right] = \tfrac{8\pi}{3}(4 + 3\pi)$$

33. a.

When the tank is full, its volume is $\tfrac{1}{3}\pi(3)^2(6) = 18\pi$ Let h be the height of water and r the corresponding radius when the tank is half full. Then,

$$\tfrac{h}{6} = \tfrac{r}{3} \text{ and } \tfrac{1}{3}\pi r^2 h = 9\pi,$$

so $h = \sqrt[3]{108} \approx 4.7622$

To compute the work W required to

pump all the water over the top of the tank, we proceed as in Example 3. Since the radius x of the disk of water satisfies

$$\frac{x}{3} = \frac{y}{6}$$

$$x = \frac{1}{2}y$$

and the distance moved by the disk is $6 - y$, we have

$$\Delta W = \underbrace{62.4\pi x^2 \Delta y}_{weight} \underbrace{(6-y)}_{distance}$$

$$= 62.4\pi(\tfrac{1}{2}y)^2(6-y)\Delta y$$

and

$$W = 62.4\pi \int_0^{4.7622} \left(\tfrac{1}{2}y\right)^2(6-y)\,dy$$

$$\approx 4,284 \qquad By\ calculator$$

The amount of work is approximately 4,284 ft-lb.

b. Think of all the weight concentrated at the center of gravity. Thus,

$$W = \int_0^{4.7622} 62.4\pi\left(\tfrac{1}{2}y^2\right)y\,dy = 12,603 \text{ ft-lb}$$

$$By\ calculator$$

35.
$$W = 62.4\pi \int_4^{10} y(100 - y^2)\,dy$$

$$= 62.4\pi(50y^2 - \tfrac{1}{4}y^4)\Big|_4^{10}$$

$$\approx 345,800 \text{ ft-lb.}$$

37.
$$W = 40 \int_0^2 \pi(3)^2(12 - y)\,dy$$

$$= 360\pi\left(12y - \tfrac{1}{2}y^2\right)\Big|_0^2$$

$$= 7,920\pi$$

$$\approx 24,881 \text{ ft-lb}$$

39. The horizontal force is 9 ft below the surface, so the force on this face is

$$F_1 = 62.4(1)^2(9) = 561.6 \text{ lb}$$

The force on each of the four vertical faces is

$$F_2 = 62.4 \int_9^{10} x(1)\,dx = 592.8 \text{ lb}$$

Thus, the total force on the five exposed sides is

$$F = F_1 + 4F_2 = 2,932.8 \text{ lb}$$

41. On the earth's surface, $F = 800$ lb and $x = 4,000$ mi. Thus,

$$800 = -\frac{1}{4,000^2}k$$

or

$$k = -800(4,000)^2$$

$$W = -800(4,000)^2 \int_{4,000}^{4,200} x^{-2}\, dx$$

$$\approx 152,381 \text{ mi-lb}$$

43. $F = \dfrac{k}{x^2}$, so $12 = \dfrac{k}{5^2}$ and $k = 300$

a. $W = \displaystyle\int_{10}^{8} 300x^{-2}\, dx$

$= -7.5 \text{ ergs}$

b. $W = \displaystyle\lim_{s \to +\infty} \int_{s}^{8} 300x^{-2}\, dx$

$= \displaystyle\lim_{s \to +\infty} [-300x^{-1}]\Big|_{s}^{8}$

$= \displaystyle\lim_{s \to +\infty}\left[\dfrac{-300}{8} + \dfrac{300}{s}\right]$

$\approx -37.5 \text{ ergs}$

45. The force on the bottom of the flat shallow end is

$$F_1 = 62.4(3)[(4)(12)] = 8,985.6 \text{ lb}$$

To find the force on the inclined plane part of the bottom, put the x-axis along one edge of the bottom as shown:

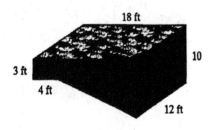

Note that the length of the incline is

$$\sqrt{14^2 + 7^2} = 7\sqrt{5}$$

If $h(x)$ is the depth of the water above point x on the bottom edge, then by similar triangles,

$$\frac{h(x) - 3}{7} = \frac{x}{7\sqrt{5}}$$

or

$$h(x) = \frac{1}{\sqrt{5}}x + 3$$

Thus, the fluid force on the inclined plane part of the bottom is

$$F_2 = 62.4 \int_{0}^{7\sqrt{5}} \left(\frac{1}{\sqrt{5}}x + 3\right)(12)\, dx$$

$$\approx 76,183.7 \qquad \textit{By calculator}$$

$$F = F_1 + F_2$$

$$\approx 8,985.6 + 76,183.7$$

$$= 85,169.3$$

The total force on the bottom is about 85,169 lb.

47. $y = \sqrt{x}$ or $x = y^2$;

$$I_x = \int_{0}^{2} y^2[(4 - y^2)\, dy]$$

$$= \int_{0}^{2} (4y^2 - y^4)\, dy$$

$$= \left(\frac{4y^3}{3} - \frac{y^5}{5}\right)\Bigg|_{0}^{2}$$

$$= \frac{64}{15}$$

$$I_y = \int_0^4 x^2 [(\sqrt{x})\,dx]$$

$$= \frac{x^{7/2}}{\frac{7}{2}} \Big|_0^4$$

$$= \frac{256}{7}$$

49. $y = 4 - x$;

$$I_y = \int_0^4 x^2 [y\,dx]$$

$$= \int_0^4 x^2 (4 - x)\,dx$$

$$= \int_0^4 (4x^2 - x^3)\,dx$$

$$= (\tfrac{4}{3}x^3 - \tfrac{1}{4}x^4)\Big|_0^4$$

$$= \frac{64}{3}$$

Since $A = \frac{1}{2}(4)(4) = 8$ and $I_y = \rho^2 A$ we see

$$\rho = \frac{2\sqrt{2}}{\sqrt{3}}$$

$$= \tfrac{2}{3}\sqrt{6}$$

51. $\quad m = A = \int_0^r x\,dx = \tfrac{1}{2}r^2$

$$\overline{y} = \frac{\frac{1}{2}\int_0^r x^2\,dx}{m}$$

$$= \frac{\frac{1}{2}\frac{r^3}{3}}{\frac{1}{2}r^2}$$

$$= \frac{r}{3}$$

Distance traveled by centroid is $2\pi\overline{y} = 2\pi\left(\frac{r}{3}\right)$

Pappus: $V = \left(2\pi\,\frac{r}{3}\right)\left(\frac{1}{2}\,r^2\right) = \frac{\pi r^3}{3}$

53. The centroid is at a distance of $s = L/2$ from the axis of rotation. $A = L^2$.

By Pappus' theorem,

$$V = 2\pi(s + \tfrac{1}{2}L)L^2$$

$$= \pi(2sL^2 + L^3)$$

55. Assume the region is under a curve $y = f(x) > 0$ over $[0, a_2]$. Then,

$$\overline{x} = \frac{\displaystyle\int_0^{a_2} xy\,dx}{\displaystyle\int_0^{a_2} y\,dx} = \frac{\displaystyle\int_0^{a_1} xy\,dx + \int_{a_1}^{a_2} xy\,dx}{\displaystyle\int_0^{a_1} y\,dx + \int_{a_1}^{a_2} y\,dx}$$

$$= \frac{1}{A_1 + A_2}\left[A_1 \frac{\displaystyle\int_0^{a_1} xy\,dx}{A_1} + A_2 \frac{\displaystyle\int_{a_1}^{a_2} xy\,dx}{A_2}\right]$$

$$= \frac{1}{A_1 + A_2} \left[A_1 \frac{\int_0^{a_1} xy\, dx}{\int_0^{a_1} y\, dx} + A_2 \frac{\int_{a_1}^{a_2} xy\, dx}{\int_{a_1}^{a_2} y\, dx} \right]$$

$$= \frac{A_1 \overline{x}_1 + A_2 \overline{x}_2}{A_1 + A_2}$$

Similarly, $\overline{y} = \dfrac{A_1 \overline{y}_1 + A_2 \overline{y}_2}{A_1 + A_2}$. Any region can be subdivided into subregions whose sum and/or differences are under a curve $y = f(x)$ over some interval $[0, a_2]$.

6.6 Applications to Business, Economics, and Life Sciences, Pages 414-417

1. For a quantity $Q(x)$, the net change, or *cumulative change* from $x = a$ to $x = b$ is given by the definite integral

    $$\int_a^b Q'(x)\, dx$$

3. If q_0 units of a commodity are sold at a price of p_0 dollars per unit, and if $p = D(q)$ is the consumer's demand function for the commodity, then

$$\left[\begin{array}{c} \text{CONSUMER'S} \\ \text{SURPLUS} \end{array} \right] =$$

$$\left[\begin{array}{c} \text{TOTAL AMOUNT CONSUM-} \\ \text{ERS ARE WILLING TO} \\ \text{SPEND FOR } q_0 \text{ UNITS} \end{array} \right] - \left[\begin{array}{c} \text{ACTUAL CONSUMER} \\ \text{EXPENDITURE FOR} \\ q_0 \text{ UNITS} \end{array} \right]$$

$$= \int_0^{q_0} D(q)\, dq - p_0 q_0$$

If q_0 units of a commodity are sold at a price of p_0 dollars per unit, and if $p = S(q)$ is the producer's supply function for the commodity, then

$$\left[\begin{array}{c} \text{PRODUCER'S} \\ \text{SURPLUS} \end{array} \right] =$$

$$\left[\begin{array}{c} \text{ACTUAL CONSUMER} \\ \text{EXPENDITURE} \\ \text{FOR } q_0 \text{ UNITS} \end{array} \right] - \left[\begin{array}{c} \text{TOTAL AMT PRODUCERS} \\ \text{RECEIVE WHEN} \\ q_0 \text{ UNITS ARE SUPPLIED} \end{array} \right]$$

$$= p_0 q_0 - \int_0^{q_0} S(q)\, dq$$

5. **a.** $q_0 = 1$, $p_0 = 2.5 - (1.5)(1) = 1$;

$$\text{C.S.} = \int_0^{q_0} D(q)\, dq - p_0 q_0$$

$$= \int_0^1 (2.5 - 1.5q)\, dq - 1$$

$$= (2.5q - 0.75q^2)\Big|_0^1 - 1$$

$$= 2.5 - 0.75 - 1 = 0.75$$

b. $q_0 = 0$, $p_0 = 2.5 - 0 = 2.5$;

$$\text{C.S.} = \int_0^0 (2.5 - 1)\, dq - 0 = 0$$

7. a. $q_0 = 5$, $p_0 = 150 - (6)(5) = 120$;

$$\text{C.S.} = \int_0^{q_0} D(q)\, dq - p_0 q_0$$

$$= \int_0^5 (150 - 6q)\, dq - (120)(5)$$

$$= (150q - 3q^2)\Big|_0^5 - 600$$

$$= 750 - 75 - 600$$

$$= 75$$

b. $q_0 = 12$, $p_0 = 150 - 6(12) = 78$;

$$\text{C.S.} = \int_0^{12} (150 - 6q)\, dq - (78)(12)$$

$$= (150q - 3q^2)\Big|_0^{12} - 936$$

$$= 1{,}800 - 432 - 936$$

$$= 432$$

9. a. The consumers' demand function is

$$D(q) = 150 - 2q - 3q^2$$

dollars per unit. For the market price of 6 units,

$$p_0 = 150 - 12 - 108$$

$$= 30$$

$$\text{C.S.} = \int_0^6 (150 - 2q - 3q^2)\, dq - 30(6)$$

$$= (150q - q^2 - q^3)\Big|_0^6 - 180$$

$$= 150(6) - 6^2 - 6^3 - 180$$

$$= 468$$

Thus, the consumer's surplus is $468.00.

b. The consumer's surplus in part **a** is the area of the region under the demand curve from $q = 0$ to $q = 6$ from which the actual spending is subtracted.

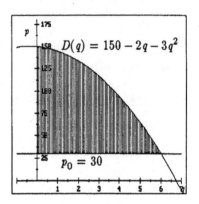

11. a. The consumer's demand function is $D(q) = 75e^{-0.04q}$ dollars per unit. The market price for 3 units is

$$D(3) = \$75e^{-0.04(3)} \approx \$66.52$$

$$\text{C.S.} = 75\int_0^3 e^{-0.04q}\, dq - 3(66.52)$$

$$= -\frac{75}{0.04}e^{-0.04q}\Big|_0^3 - 199.56$$

$$\approx 12.46$$

Thus, the consumer's surplus is $12.46.

b. The consumer's surplus in part **a** is the area of the region under the demand curve from $q = 0$ to $q = 3$.

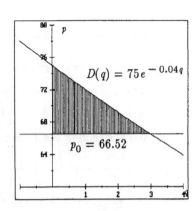

$$\text{P.S.} = 7(28.8) - \int_0^7 (17 + 11e^{0.01q}) \, dq$$

$$= 201.6 - (17q + 1,100e^{0.01q}2)\Big|_0^7$$

$$= 201.6 - 119 - 1,100e^{0.07} + 1,100$$

$$\approx 2.84$$

The producer's surplus for $q_0 = 7$ is $2.84.

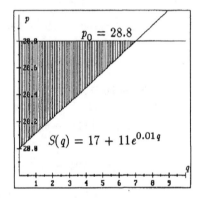

13. $S(q) = 0.5q + 15$, $p_0 = S(5) = 17.5$.

$$\text{P.S.} = 5(17.5) - \int_0^5 (0.5q + 15) \, dq$$

$$= 87.5 - \left(\frac{0.5q^2}{2} + 15q\right)\Big|_0^5$$

$$= 87.5 - 6.25 - 75$$

$$= 6.25$$

The producer's surplus for $q_0 = 5$ is $6.25.

17. Equilibrium occurs when $D(q) = S(q)$.

$$25 - q^2 = 5q^2 + 1$$

$$6q^2 = 24$$

$$q = 2 \quad \text{(disregard negative)}$$

$$p = D(2) = 21$$

Find the consumer's surplus:

$$\int_0^2 (25 - q^2) \, dq - 2(21) = \left[25q - \frac{q^3}{3}\right]\Big|_0^2 - 42$$

$$= 50 - \frac{8}{3} - 42$$

$$\approx 5.33$$

The consumer's surplus is $5.33.

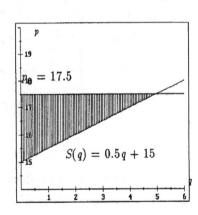

15. $S(q) = 17 + 11e^{0.01q}$, $p_0 = S(7) \approx 28.8$

19. Equilibrium occurs when $D(q) = S(q)$.

$$27 - q^2 = \tfrac{1}{4}q^2 + \tfrac{1}{2}q + 5$$

$$5q^2 + 2q - 88 = 0$$

$$(q - 4)(5q + 22) = 0$$

$$q = 4 \quad \text{(disregard negative)}$$

$$p = D(4) = 11$$

Find the consumer's surplus:

$$\int_0^4 (27 - q^2)\, dq - 4(11)$$

$$= \left[27q - \frac{q^3}{3} \right]\Big|_0^4 - 44$$

$$= 108 - \frac{64}{3} - 44$$

$$\approx 42.67$$

The consumer's surplus is \$42.67.

21. a. $FV = \int_0^3 \$3,000\, e^{0.08(3 - t)}\, dt$

$$= \frac{\$3,000\, e^{0.08(3 - t)}}{0.08}\Big|_0^3$$

$$\approx \$10,171.84$$

b. $PV = \int_0^3 \$3,000\, e^{-0.08t}\, dt$

$$= \frac{\$3,000\, e^{-0.08t}}{-0.08}\Big|_0^3$$

$$\approx \$8,001.46$$

23. a. The machine will be used until

$$R'(x) = C'(x):$$

$$6,025 - 10x^2 = 4,000 + 15x^2$$

$$25x^2 = 2,025$$

$$x^2 = 81$$

$$x = \pm 9$$

The machine will be sold in 9 years. (The negative value is not in the domain.)

b. The difference $R'(x) - C'(x)$ represents the rate of change of the net earnings generated by the machine at time x. Using integration to "add up" the net earnings over the period of profitability ($0 \le x \le 9$), we obtain

$$\int_0^9 [R'(x) - C'(x)]\, dx$$

$$= \int_0^9 [(6,025 - 10x^2) - (4,000 + 15x^2)]\, dx$$

$$= \int_0^9 (2,025 - 25x^2)\, dx$$

$$= \left(2,025x - \frac{25}{3}x^3 \right)\Big|_0^9$$

$$= 12,150$$

In geometric terms, the net earnings is represented by the area of the region between the curves $y = R'(x)$ and $y = C'(x)$ from $x = 0$ to $x = 9$. We show the result below using one common software program, called *Converge*, but there are many available products that you might want to use from time to time. Notice that this software approximates the result by computing Riemann sums; the final result for 64 rectangles is 12,291.64. More rectangles should give better approximations.

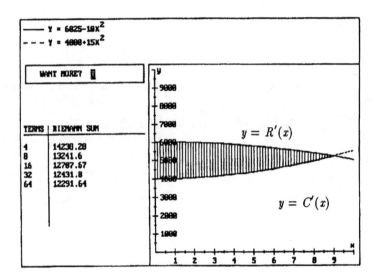

25. a. Integrating, we find

$$P_1(x) = 100x + \tfrac{1}{3}x^3$$

and

$$P_2(x) = 190x + x^2$$

Since $P_1(0) = P_2(0) = 0$. Solving $P_1'(x) = P_2'(x)$, we obtain $x = 18$. The second plan is more profitable for the first 18 years.

b.

$$\text{Extra profit} = \int_0^{18} [(190x + x^2) - (100x + \tfrac{1}{3}x^3)]\, dx$$

$$= \int_0^{18} (90x + x^2 - \tfrac{1}{3}x^3)\, dx$$

$$= \left[45x^2 + \frac{x^3}{3} - \frac{x^4}{12}\right]\Bigg|_0^{18}$$

$$= 7{,}776$$

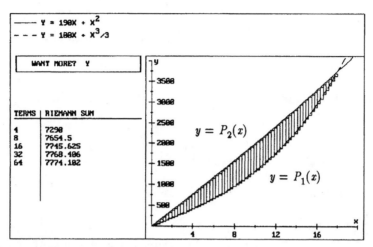

In geometric terms, the net excess profit generated by the second plan is the area of the region between the curves $y = P_2(x)$ and $y = P_1(x)$ from $x = 0$ to $x = 18$.

27.　a.　$P(x) = xp(x) - C(x)$

$$= (124 - 2x)x - (2x^3 - 59x^2 + 4x + 76)$$

$$= -2x^3 + 57x^2 + 120x - 76$$

$$P'(x) = -6x^2 + 114x + 120$$

$$= -6(x - 20)(x + 1)$$

$P'(x) = 0$ when $x = 20$ (disregard negative root);

$P''(x) = -12x + 114 < 0$ when $x = 20$, so $x = 20$ is a maximum.

b.　$\text{Consumer's surplus} = \int_0^{20} (124 - 2x)\,dx - 20(84)$

$$= (124x - x^2)\Big|_0^{20} - 1{,}680$$

$$= 400$$

29. a. The cost function is

$$C(q) = \int \left(\tfrac{3}{4}q^2 + 5\right) \, dq$$

$$= \tfrac{1}{4}q^3 + 5q + K$$

$K = 0$ since there is no overhead.

The revenue function is

$$R(q) = qp(q) = \tfrac{1}{4}q(10 - q)^2$$

The marginal revenue function is

$$R'(q) = \tfrac{1}{4}(100 - 40q + 3q^2)$$

$$= \tfrac{1}{4}(10 - q)(10 - 3q)$$

b. The profit function is

$$P(q) = R(q) - C(q)$$

$$= \tfrac{1}{4}(100q - 20q^2 + q^3) - \tfrac{1}{4}q^3 - 5q$$

$$= 20q - 5q^2$$

$$P'(q) = 20 - 10q = 0 \text{ when } q = 2$$

c. The consumer's surplus is

$$\tfrac{1}{4}\int_0^2 (100 - 20q + q^2) \, dq - 2(16)$$

$$= \tfrac{1}{4}(100q - 10q^2 + \tfrac{1}{3}q^3)\Big|_0^2 - 32$$

$$= \tfrac{26}{3} \approx 8.67$$

The consumer's surplus is $8.67.

31. $R(x) = q(20 - 4q^2) - (q^2 + 6q)$

$$= -4q^3 - q^2 + 14q$$

$$R'(x) = -12q^2 - 2q + 14 = -2(q - 1)(6q + 7)$$

$$R'(x) = 0 \text{ when } q = 1 \text{ (disregard}$$

negative)

The consumer's surplus is

$$\int_0^1 (20 - 4q^2) \, dq - (1)(16) = \left(20q - \tfrac{4}{3}q^3\right)\Big|_0^1 - 16$$

$$= 20 - \tfrac{4}{3} - 16$$

$$= \tfrac{8}{3}$$

33. At 9:00 A.M., $t = 0$; at 10:00 A.M., $t = 1$, and at noon, $t = 3$; we have

$$Q'(t) = -4(t + 2)^3 + 54(t + 2)^2$$

$$Q(t) = \int \left[-4(t + 2)^3 + 54(t + 2)^2\right] \, dt$$

$$= -(t + 2)^4 + 18(t + 2)^3 + Q_0$$

$$Q(3) = -5^4 + 18(5)^3 + Q_0;$$

$$Q(1) = -3^4 + 18(3)^3 + Q_0; \text{ so}$$

$$Q(3) - Q(1) = 5^3(18 - 5) - 3^3(18 - 3)$$

$$= 1,220$$

The number of people entering the fair during the prescribed time is 1,220 people.

35. Let $P(t)$ denote the price of the oil after t months. Then, $0 \le t \le 24$ (in months) and

$$P'(t) = 400(18 + 0.03t)$$

$$P(t) = 400 \int_0^{24} (18 + 0.03t) \, dt$$

$$= 400\left(18t + 0.015t^2\right)\Big|_0^{24}$$

$$= 172{,}800 + 6(24)^2$$

$$= 176{,}256$$

The total future revenue is \$176,256.

37. Let $D(t)$ denote the demand for the product. Since the current demand is 5,000 and the demand increases exponentially,

$$D(t) = 5{,}000\,e^{0.02t}$$

units/yr. Let $R(t)$ denote the total revenue t years from now. Then, the rate of change of revenue is

$$\frac{dR}{dt} = 400(5{,}000\,e^{0.02t})$$

The increase in revenue over the next two years is

$$R(2) \;-\; R(0) = \int_0^2 \$2{,}000{,}000\,e^{0.02t}\; dt$$

$$\approx \$4{,}081{,}077$$

39. a. The first plan generates profit at the rate of

$$P_1{}'(t) = 100 + t^2$$

dollars/yr and the second generates profit at the rate of

$$P_2{}'(t) = 220 + 2t$$

dollars/yr. The second plan will be more

profitable until

$$P_1{}'(t) = P_2{}'(t)$$

that is, until

$$100 + t^2 = 220 + 2t$$

$$(t - 12)(t + 10) = 0$$

$$t = 12,\; -10$$

The negative value is not in the domain, so the solution is that the second plan will be more profitable for 12 years.

b. For $0 \le t \le 12$, the rate at which the profit generated by the second plan exceeds that of the first plan is

$$P_2{}'(t) \;-\; P_1{}'(t)$$

Hence, the net excess profit,

$$E(t) = P_2(t) \;-\; P_1(t)$$

generated by the second plan over the twelve-year period is given by

$$E(12) \;-\; E(0) = \int_0^{12} [P_2{}'(t) - P_1{}'(t)]\; dt$$

$$= \int_0^{12} [220 + 2t - (100 + t^2)]\; dt$$

$$= \left(220t + t^2 - 100t - \frac{t^3}{3}\right)\Bigg|_0^{12}$$

$$= 220(12) + 12^2 - 100(12) - \frac{12^3}{3}$$

$$= 1{,}008.00$$

This means that the excess profit is

$100{,}800.

c. In geometric terms, the net excess profit generated by the second plan is the area of the region between the curves $y = P_2{'}(t)$ and $y = P_1{'}(t)$ from $t = 0$ to $t = 12$.

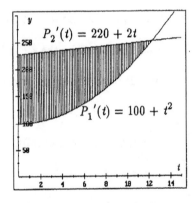

41. a. The first plan generates profit at the rate

$$P_1{'}(t) = 60 e^{0.12t}$$

thousand dollars/yr and the second generates profit at the rate of

$$P_2{'}(t) = 160 e^{0.08t}$$

thousand dollars/yr. The second plan will be more profitable until

$$P_1{'}(t) = P_2{'}(t)$$

that is, until

$$60 e^{0.12t} = 160 e^{0.08t}$$

$$e^{0.04t} = \frac{8}{3}$$

$$0.04t = \ln \frac{8}{3}$$

$$t \approx 24.52$$

The solution is that the second plan will be more profitable for about 24.5 years.

b. For $0 \leq t \leq 24.52$, the rate at which the profit generated by the second plan exceeds that of the first plan is

$$P_2{'}(t) - P_1{'}(t)$$

Hence, the net excess profit,

$$E(t) = P_2(t) - P_1(t)$$

generated by the second plan over the time period is given by

$$E(24.52) - E(0) = \int_0^{24.52} [P_2{'}(t) - P_1{'}(t)] \, dt$$

$$= \int_0^{24.52} [160 e^{0.08t} - 60 e^{0.12t}) \, dt$$

$$\approx 3{,}240.74 \qquad \textit{By calculator}$$

This means that the excess profit is about $3.2 million.

c. In geometric terms, the net excess profit generated by the second plan is the area of the region between the curves $y = P_2'(t)$ and $y = P_1'(t)$ from $t = 0$ to $t = 24.52$.

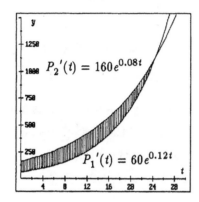

43. a. The machine generates revenue at the rate of $R'(t) = 6{,}025 - 8t^2$ dollars/yr, and costs accumulate at the rate of

$$C'(t) = 4{,}681 + 13t^2$$

dollars/yr. The use of the machine will be profitable as long as the rate at which revenue is generated is greater than the rate at which costs accumulate; that is, until

$$R'(t) = C'(t)$$

$$6{,}025 - 8t^2 = 4{,}681 + 13t^2$$

$$21t^2 = 1{,}344$$

$$t^2 = 64$$

$$t = \pm 8$$

Since the negative value is not in the domain, we see that the machine will be profitable for 8 years.

b. The difference $R'(t) - C'(t)$ represents the rate of change of the net earnings generated by the machine. Hence, the net earnings over the next eight years is given by the definite integral

$$\int_0^8 [R'(t) - C'(t)]\ dt$$

$$= \int_0^8 [(6{,}025 - 8t^2) - (4{,}681 + 13t^2)]\ dt$$

$$= \int_0^8 (1{,}344 - 21t^2)\ dt$$

$$= (1{,}344t - 7t^3)\Big|_0^8$$

$$= 7{,}168$$

Thus, the net earnings are \$7,168.

c. In geometric terms, the net earnings in part b is the area of the region between the curves $y = R'(t)$ and $y = C'(t)$ from $t = 0$ to $t = 8$.

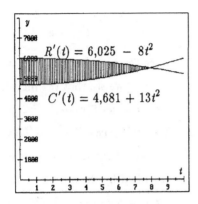

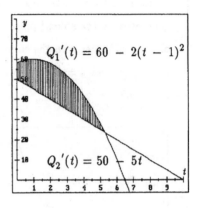

45. **a.** $Q_1'(t) = 60 - 2(t - 1)^2$

$Q_2'(t) = 50 - 5t$

The excess production is given by

$$\int_0^4 [60 - 2(t - 1)^2 - 50 + 5t] \, dt$$

$$= \int_0^4 (-2t^2 + 9t + 8) \, dt$$

$$= \left(-\frac{2}{3}t^3 + \frac{9}{2}t^2 + 8t \right)\Big|_0^4$$

$$= -\frac{128}{3} + 72 + 32$$

$$= \frac{184}{3}$$

b. The excess production is the difference between the areas under the production curves.

47. The density at a point r miles from the center is $D(r) = 25,000e^{-0.05r^2}$ people/mi^2. The area of the ring of thickness dr at r miles from the center is $2\pi r \, dr$, so

$$(25,000e^{-0.05r^2})(2\pi r \, dr) = 50,000\pi r e^{-0.05r^2} \, dr$$

people live in that ring. The total number, N, of people living between one and two miles from the center is

$$N = \int_1^2 50,000\pi r e^{-0.05r^2} \, dr$$

Let

$$u = -0.05r^2$$

$$du = -0.1r \, dr$$

If $r = 1$, $u = -0.05$ and if $r = 2$, $u = -0.2$:

$$N = \int_{-0.05}^{-0.2} 50,000\pi r e^u \frac{du}{-0.1r}$$

$$\approx 208,128 \qquad \textit{By calculator}$$

There would be approximately 208,128 people living in the ring.

49. The rate of flow through a ring of area $2\pi r\, dr$ is

$$2\pi r(8 - 800r^2)\, dr$$

The rate though the artery is

$$16\pi \int_0^{0.1} (r - 100r^3)\, dr = 16\pi\left(\frac{r^2}{2} - \frac{100r^4}{4}\right)\Big|_0^{0.1}$$

$$= 0.04\pi$$

This is about 0.126 cm^3/s.

51. Let $P(x)$ denote the population of the town x months from now. Then,

$$\frac{dP}{dx} = 5 + 3x^{2/3}$$

and the number by which the population will increase during the next eight months is

$$P(8) - P(0) = \int_0^8 (5 + 3x^{2/3})\, dx$$

$$= \left(5x + \frac{9}{5}x^{5/3}\right)\Big|_0^8$$

$$= 40 + \frac{288}{5}$$

$$\approx 97.6$$

The population will increase by about 98 people.

53. Let $D(t)$ denote the demand for oil after t years. Then,

$$D'(t) = D_0 e^{0.1t} \text{ and } D_0 = 30$$

and

$$D(t) = 30\int_0^{10} e^{0.1t}\, dt = 300(e - 1)$$

This is approximately 515.48 billion barrels.

55. Of the Spy's 6 friends, $6F(10)$ will be left after 10 years. To approximate the number of other spies left, divide the 10-year time interval $0 \le t \le 10$ into n equal subintervals of length $\Delta t = (10 - 0)/n$ and let t_j^* denote the beginning of the jth time period. Since new spies are accepted into the service each year, the number accepted during the jth subinterval is $30\Delta t$. When $t = 10$, approximately $10 - t_j^*$ years will have elapsed since these $30\Delta t$ spies began spying, so $(30\Delta t)F(10 - t_j^*)$ of them will still be alive, and the total number N still alive after 10 years is estimated by the sum

$$N \approx 6F(10) + \sum_{j=1}^n 30F(10 - t_j^*)\Delta t$$

As $n \to \infty$ the approximation improves and approaches the true value of N. Since this is a Riemann sum, we have

$$P = 6F(10) + \lim_{n\to\infty} \sum_{j=1}^n 30F(10 - t_j^*)\Delta t$$

$$= 6F(10) + \int_0^{10} 30F(10 - t)\, dt$$

$$= 6e^{-10/20} + \int_0^{10} 30e^{-(10-t)/20}\, dt$$

$$= 6e^{-1/2} + \left[600\,e^{(-10+t)/20}\right]\Big|_0^{10}$$

$$\approx 239.7$$

Thus, there are about 240 spies still alive.

57. Since $f(t) = e^{-t/10}$ is the fraction of members after t months, and since there are 8,000 charter members, the number of charter members still active at the end of 10 months is

$$8,000 f(10) = 8,000 e^{-1}$$

Now divide the interval $0 \le t \le 10$ into n equal subintervals of length Δt years and let t_{j-1} denote the beginning of the jth subinterval. During the jth subinterval, $200\Delta t$ members join, and at the end of 10 months, $(10 - t_{j-1})$ months later, the number of these retaining memberships is

$$200 f(10 - t_{j-1}) = 200 e^{-(10 - t_{j-1})/10}\Delta t$$

Hence, the number of new residents still active 10 months from now is approximately

$$\lim_{n \to \infty} \sum_{j=1}^{n} 200 e^{-(10 - t_{j-1})/10}\Delta t$$

$$= 200 \int_0^{10} e^{-(10 - t)/10}\, dt$$

Thus, the total number, N, of active members 10 months from now is

$$N = 8,000 e^{-1} + 200 \int_0^{10} e^{-(10 - t)/10}\, dt$$

$$\approx 2,943 + 200 e^{-1}\int_0^{10} e^{t/10}\, dt$$

$$\approx 4,207$$

There should be approximately 4,207 active members.

59. a. $f(t_k) e^{-rt_k - 1}$

  b. $\displaystyle\sum_{k=1}^{n} f(t_k) e^{-rt_k - 1}\Delta_n t$ where $\Delta_n t = \frac{N}{n}$

  c. $\mathrm{PV} = \displaystyle\lim_{n \to \infty} \sum_{k=1}^{n} f(t_k) e^{-rt_k - 1}\Delta_n t$

  $$= \int_0^{N} f(t) e^{-rt}\, dt$$

CHAPTER 6 REVIEW

Proficiency Examination, Pages 417–418

SURVIVAL HINT: *To help you review the concepts of this chapter, handwrite the answers to each of these questions onto your own paper.*

1. If f and g are continuous and satisfy $f(x) \ge g(x)$ on the closed interval $[a, b]$, then the area between the two curves $y = f(x)$ and $y = g(x)$ is given by

$$A = \int_a^b [f(x) - g(x)]\, dx$$

2. Take cross sections perpendicular to a convenient line — say the x-axis. If $A(x)$ is the area of the cross section at x and the base region extends from $x = a$ to $x = b$, the volume is given by

$$V = \int_a^b A(x) \ dx$$

3. Use disks or washers when the approximating strip is perpendicular to the axis of revolution. Use shells when the strip is parallel to the axis.

4. $x = r \cos \theta$, $y = r \sin \theta$

$r = \sqrt{x^2 + y^2};\ \bar{\theta} = \tan^{-1}\left|\frac{y}{x}\right|$

5. Answers vary; see Table 6.2.

6. **a.** rose curve (one circular leaf)

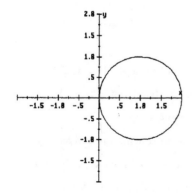

b. cardioid

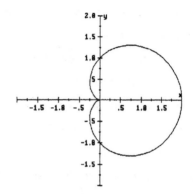

c. spiral

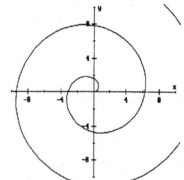

d. lemniscate

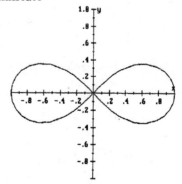

7. *Step 1.* Find the simultaneous solution of the given system of equations.

   Step 2. Determine whether the pole lies on the two graphs.

   Step 3. Graph the curves to look for other points of intersection. Some may require the alternate representation for (r, θ); namely, $(-r, \theta + \pi)$.

8. Let $r = f(\theta)$ define a polar curve, where f is continuous and $f(\theta) \geq 0$ on the closed interval $0 \leq \alpha \leq \theta \leq \beta \leq 2\pi$. Then the region bounded by the curve $r = f(\theta)$ and the rays $\theta = \alpha$ and $\theta = \beta$ has area

$$A = \frac{1}{2}\int_\alpha^\beta r^2 \, d\theta = \frac{1}{2}\int_\alpha^\beta [f(\theta)]^2 \, d\theta$$

9. Let f be a function whose derivative f' is continuous on the interval $[a, b]$. Then the arc length, s, of the graph of $y = f(x)$ between $x = a$ to $x = b$ is given by the integral

$$s = \int_a^b \sqrt{1 + [f'(x)]^2} \, dx$$

Similarly, for the graph of $x = g(y)$, where g' is continuous on the interval $[c, d]$, the arc length from $y = c$ to $y = d$ is

$$s = \int_c^d \sqrt{1 + [g'(y)]^2} \, dy$$

10. Suppose f' is continuous on the interval $[a, b]$. Then the surface generated by revolving about the x-axis the arc of the curve $y = f(x)$ on $[a, b]$ has surface area

$$S = 2\pi \int_a^b f(x)\sqrt{1 + [f'(x)]^2} \, dx$$

11. The work done by the variable force $F(x)$ as an object moves along the x-axis from $x = a$ to $x = b$ is given by

$$W = \int_a^b F(x) \, dx$$

12. Suppose a flat surface (a plate) is submerged vertically in a fluid of weight density ρ (lb/ft^3) and that the submerged portion of the plate extends from $x = a$ to $x = b$ on a vertical axis. Then the total force, F, exerted by the fluid is given by

$$F = \int_a^b \rho \, h(x) L(x) \, dx$$

where $h(x)$ is the depth at x and $L(x)$ is the corresponding length of a typical horizontal approximating strip.

13. Let f and g be continuous and satisfy $f(x) \geq g(x)$ on the interval $[a, b]$, and consider a thin plate (lamina) of uniform density ρ that covers the region R between the graphs of $y = f(x)$ and $y = g(x)$ on the interval

[a, b]. Then

The mass of R is: $m = \rho \int_a^b [f(x) - g(x)]\, dx$

The centroid of R is the point $(\overline{x}, \overline{y})$ such that

$$\overline{x} = \frac{M_y}{m} = \frac{\int_a^b x[f(x) - g(x)]\, dx}{\int_a^b [f(x) - g(x)]\, dx} \quad \text{and}$$

$$\overline{y} = \frac{M_x}{m} = \frac{\frac{1}{2}\int_a^b \{[f(x)]^2 - [g(x)]^2\}\, dx}{\int_a^b [f(x) - g(x)]\, dx}$$

14. The solid generated by revolving a region R about a line outside its boundary (but in the same plane) has volume $V = As$, where A is the area of R and s is the distance traveled by the centroid of R.

15. Let $f(t)$ be the amount of money deposited at time t over the time period $[0, T]$ in an account that earns interest at the annual rate r compounded continuously. Then the *future value* of the income flow over the time period is given by

$$FV = \int_0^T f(t)\, e^{r(T - t)}\, dt$$

and the *present value* of the same income flow over the time period is

$$PV = \int_0^T f(t)\, e^{-rt}\, dt$$

16. $f'(x)$ is the rate of change of a quantity $f(x)$ and you are required to find the net change $f(b) - f(a)$ in $f(x)$ as x varies from $x = a$ to $x = b$. But since $f(x)$ is an antiderivative of $f'(x)$, the fundamental theorem of calculus tells us that the net change is given by the definite integral

$$f(b) - f(a) = \int_a^b f'(x)\, dx$$

17. If q_0 units of a commodity are sold at a price of p_0 dollars per unit, and if $p = D(q)$ is the consumers' demand function and $p = S(q)$ is the supply function for the commodity, then

$$\begin{bmatrix} \text{CONSUMER'S} \\ \text{SURPLUS} \end{bmatrix} =$$

$$\begin{bmatrix} \text{TOTAL AMOUNT CONSUM-} \\ \text{ERS ARE WILLING TO} \\ \text{SPEND FOR } q_0 \text{ UNITS} \end{bmatrix} - \begin{bmatrix} \text{ACTUAL CONSUMER} \\ \text{EXPENDITURE FOR} \\ q_0 \text{ UNITS} \end{bmatrix}$$

$$= \int_0^{q_0} D(q)\, dq - p_0 q_0$$

$$\begin{bmatrix} \text{PRODUCER'S} \\ \text{SURPLUS} \end{bmatrix} =$$

$$\begin{bmatrix} \text{ACTUAL CONSUMER} \\ \text{EXPENDITURE} \\ \text{FOR } q_0 \text{ UNITS} \end{bmatrix} - \begin{bmatrix} \text{TOTAL AMOUNT} \\ \text{PRODUCERS RECEIVE AS} \\ q_0 \text{ UNITS ARE SUPPLIED} \end{bmatrix}$$

$$= p_0 q_0 - \int_0^{q_0} S(q) \, dq$$

18. A survival function gives the fraction of individuals of a population that can be expected to remain in the group and a renewal function gives the rate at which new members can be expected to join the group.

19.
$$\text{RATE OF FLOW} = \int_0^R 2\pi k (R^2 r - r^3) \, dr$$
$$= \frac{\pi k R^4}{2} \text{ in cm}^3/\text{s}$$

20. The definite integrals could represent the following:

A. Disks revolved about the y-axis.

B. Disks revolved about the x-axis.

C. Slices taken perpendicular to the x-axis.

D. Slices taken perpendicular to the y-axis.

E. Mass of a lamina with density π.

F. Washers taken along the x-axis.

G. Washers taken along the y-axis.

a. All but E are formulas for volumes of solids.

b. A, B, F, G

c. F, G

d. C, D

e. B, F

f. A, G

21. We graph the given curve and shade the area.

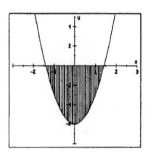

By symmetry
$$A = 2 \int_0^{\sqrt{2}} [0 - (3x^2 - 6)] \, dx$$
$$= 2(6x - x^3) \Big|_0^{\sqrt{2}}$$
$$= 8\sqrt{2}$$

22. $x = y^3 = (x^2)^3 = x^6$, so
$$x^6 = x$$
$$x^6 - x = 0$$
$$x(x^5 - 1) = 0$$
$$x = 0, 1$$
$$A = \int_0^1 (x^{1/3} - x^2) \, dx$$
$$= \left(\tfrac{3}{4} x^{4/3} - \tfrac{1}{3} x^3 \right) \Big|_0^1$$
$$= \tfrac{5}{12}$$

23. $V = 2\int_0^2 2\sqrt{4 - x^2}\ 4\sqrt{4 - x^2}\ dx$

$= 16\int_0^2 (4 - x^2)\ dx$

$= 16\left(4x - \dfrac{x^3}{3}\right)\Big|_0^2$

$= \dfrac{256}{3}$

24. a.

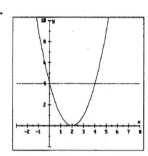

$A = 2\int_0^2 [4 - (x - 2)^2]\ dx$

$= 2\int_0^2 (4x - x^2)\ dx$

$= 2\left(2x^2 - \dfrac{x^3}{3}\right)\Big|_0^2$

$= \dfrac{32}{3}$

b. $V = 2\pi\int_0^2 [R^2 - r^2]\ dx$

$= 2\pi\int_0^2 [4^2 - (x - 2)^4]\ dx$

$= 2\pi\int_0^2 (-x^4 + 8x^3 - 24x^2 + 32x)\ dx$

$= 2\pi\left(-\dfrac{x^5}{5} + 2x^4 - 8x^3 + 16x^2\right)\Big|_0^2$

$= \dfrac{256\pi}{5}$

c. $V = \int_0^4 2\pi rh\ dx$

$= 2\pi\int_0^4 x[4 - (x - 2)^2]\ dx$

$= 2\pi\int_0^4 x(4x - x^2)\ dx$

$= 2\pi\int_0^4 (4x^2 - x^3)\ dx$

$= 2\pi\left(\dfrac{4x^3}{3} - \dfrac{x^4}{4}\right)\Big|_0^4$

$= \dfrac{128\pi}{3}$

25. Solving simultaneously, or by symmetry, we see that the intersection is at $(\sqrt{2}\,a, \frac{\pi}{4})$.

$A = \dfrac{1}{2}\int_0^{\pi/4} (2a\sin\theta)^2\ d\theta + \dfrac{1}{2}\int_{\pi/4}^{\pi/2} (2a\cos\theta)^2\ d\theta$

$= 2a^2\int_0^{\pi/4} \left(\dfrac{1 - \cos 2\theta}{2}\right) d\theta + 2a^2\int_{\pi/4}^{\pi/2} \left(\dfrac{1 + \cos 2\theta}{2}\right) d\theta$

$= 2a^2\left[\dfrac{\theta}{2} - \dfrac{\sin 2\theta}{4}\right]\Big|_0^{\pi/4} + 2a^2\left[\dfrac{\theta}{2} + \dfrac{\sin 2\theta}{4}\right]\Big|_{\pi/4}^{\pi/2}$

$$= 2a^2\left[\left(\frac{\pi}{8} - \frac{1}{4}\right) + \left(\frac{\pi}{4} - \frac{\pi}{8} - \frac{1}{4}\right)\right]$$

$$= 2a^2\left(\frac{\pi}{4} - \frac{1}{2}\right)$$

$$= a^2\left(\frac{\pi}{2} - 1\right)$$

$$\approx 0.5708a^2$$

26. $\dfrac{dy}{dt} = -\dfrac{3}{2}x^{1/2}$; $ds = \sqrt{1 + \dfrac{9}{4}x}\, dx$

$$s = \int_0^1 \sqrt{1 + \frac{9}{4}x}\, dx$$

$$= \frac{8}{27}\left(1 + \frac{9}{4}x\right)^{3/2}\Big|_0^1$$

$$= \frac{1}{27}(13^{3/2} - 8)$$

$$\approx 1.4397$$

27. $s = 2\pi \displaystyle\int_0^1 \sqrt{x}\,\sqrt{1 + \left(\dfrac{1}{2\sqrt{x}}\right)^2}\, dx$

$$= 2\pi \int_0^1 \sqrt{x}\,\sqrt{\frac{4x + 1}{4x}}\, dx$$

$$= \pi \int_0^1 \sqrt{4x + 1}\, dx$$

$$= \frac{\pi}{4} \int_0^1 (4x + 1)^{1/2}(4\, dx)$$

$$= \frac{\pi}{4} \frac{(4x + 1)^{3/2}}{\frac{3}{2}}\Big|_0^1$$

$$= \frac{\pi}{6}(5\sqrt{5} - 1)$$

$$\approx 5.33$$

28. $m = \displaystyle\int_0^1 [(x - x^3) - (x^2 - x)]\, dx$

$$= \int_0^1 (2x - x^2 - x^3)\, dx$$

$$= \left[x^2 - \frac{x^3}{3} - \frac{x^4}{4}\right]\Big|_0^1$$

$$= \frac{5}{12}$$

$$M_y = \int_0^1 x(2x - x^2 - x^3)\, dx$$

$$= \int_0^1 (2x^2 - x^3 - x^4)\, dx$$

$$= \left(\frac{2x^3}{3} - \frac{x^4}{4} - \frac{x^5}{5}\right)\Big|_0^1$$

$$= \frac{13}{60}$$

$$M_x = \frac{1}{2} \int_0^1 [(x - x^3)^2 - (x^2 - x)^2]\, dx$$

$$= \frac{1}{2} \int_0^1 (x^6 - 3x^4 + 2x^3)\, dx$$

$$= \frac{1}{2}\left(\frac{x^7}{7} - \frac{3x^5}{5} + \frac{x^4}{2}\right)\Big|_0^1$$

$$= \frac{3}{140}$$

$$\bar{x} = \frac{M_y}{M} = \frac{\frac{13}{60}}{\frac{5}{12}} = \frac{13}{25}$$

$$\bar{y} = \frac{M_x}{M} = \frac{\frac{3}{140}}{\frac{5}{12}} = \frac{9}{175}$$

The centroid is $\left(\frac{13}{25}, \frac{9}{175}\right)$.

29.

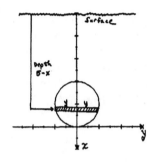

$$\Delta F = 64(5 - x)(2y)\Delta x$$

where $y = \sqrt{1 - x^2} + 1$

$$F = 128 \int_0^2 (5 - x)(\sqrt{1 - x^2} + 1) \, dx$$

30. This is the future value (FV) of money transferred into an account.

$$FV = \int_0^5 1{,}200\, e^{0.08(5 - t)} \, dt$$

$$= 1{,}200\, e^{0.4} \int_0^5 e^{-0.08t} \, dt$$

$$= 1{,}200\, e^{0.4} \frac{e^{-0.08t}}{-0.08} \Big|_0^5$$

$$\approx 7{,}377.37$$

The amount in the account will be \$7,377.37.

CHAPTER 7

Methods of Integration

SURVIVAL HINT: *In this chapter you will learn additional techniques for integration. The use of technology is both a blessing and a curse. It is a blessing because it relieves us of a great deal of drudgery of evaluating definite integrals and finding indefinite integrals. However, technology often fails, or provides a form which is unduly complicated or even worse, provides the same expression that you input! As you work your way through this chapter, keep in mind that you are learning worthwhile techniques, and in the process you can use technology to check your work, but you should not depend on the technology to do your work for you.*

7.1 Review of Substitution and Integration by Table, Pages 433-435

SURVIVAL HINT: *You will make fewer errors with substitution if you take the time to write down $u = \cdots$, $du = \cdots$ each time, and then carefully make the substitutions. In this manual, we show this information enclosed in boxes. For definite integrals, do not forget to change the limits when you make a u-substitution.*

1.
$$\int \frac{2x + 5}{\sqrt{x^2 + 5x}}\, dx = \int u^{-1/2}\, du$$

$$\boxed{\text{Let } u = x^2 + 5x;\ du = (2x + 5)\, dx}$$

$$= \frac{u^{1/2}}{\frac{1}{2}} + C$$

$$= 2(x^2 + 5x)^{1/2} + C$$

SURVIVAL HINT: *When making a u-substitution (as shown here with Problem 1, we often carry out the actual substitution steps mentally, which means that many will work Problem 1 as follows:*

$$\int \frac{2x + 5}{\sqrt{x^2 + 5x}}\, dx = \int (x^2 + 5x)^{-1/2}[(2x + 5)\, dx]$$

$$= 2(x^2 + 5x)^{1/2} + C$$

3.
$$\int \frac{dx}{x\ln x} = \int \frac{1}{\ln x} \frac{dx}{x}$$

$$\boxed{\text{Let } u = \ln x;\ du = \frac{1}{x}\, dx}$$

$$= \ln|u| + C = \ln|\ln x| + C$$

SURVIVAL HINT: *Were you able to carry out this substitution mentally? Here is what we did:*

$$\int \frac{1}{\ln x} \frac{dx}{x} = \int \frac{1}{u}\, du = \ln|u| + C$$

5.
$$\int \frac{x\, dx}{4 + x^4} = \frac{1}{4} \int \frac{x\, dx}{1 + (\frac{1}{2}x^2)^2}$$

$$\boxed{\text{Let } u = \frac{1}{2}x^2;\ du = x\, dx}$$

$$= \frac{1}{4} \int \frac{du}{1 + u^2}$$

$$= \frac{1}{4} \tan^{-1} \frac{1}{2} x^2 + C$$

$$= \frac{1}{4} \tan^{-1} \frac{x^2}{2} + C$$

7. $\int (1 + \cot x)^4 \csc^2 x \, dx = \int - u^4 \, du$

Let $u = 1 + \cot x \, dx = - \csc^2 x \, dx$

$$= - \frac{1}{5}(1 + \cot x)^5 + C$$

9. $\int \frac{x^3 - x}{(x^4 - 2x^2 + 3)^2} \, dx$

Let $u = x^4 - 2x^2 + 3; \, du = (4x^3 - 4x) \, dx$

$$= \frac{1}{4} \int (x^4 - 2x^2 + 3)^{-2}[(4x^3 - 4x) \, dx]$$

$$= \frac{1}{4} \int u^{-2} \, du$$

$$= - \frac{1}{4}(x^4 - 2x^2 + 3)^{-1} + C$$

11. $\int \frac{2x + 1}{x^2 + x + 1} \, dx$

Let $u = x^2 + x + 1; \, du = (2x + 1) \, dx$

$$= \int (x^2 + x + 1)^{-1}[(2x + 1) \, dx]$$

$$= \ln(x^2 + x + 1) + C$$

SURVIVAL HINT: Note that $x^2 + x + 1 > 0$ so you don't need the absolute values.

13. $\int \frac{dx}{x^2 \sqrt{x^2 - a^2}} = \frac{\sqrt{x^2 - a^2}}{a^2 x} + C$

Formula 201

15. $\int x \ln x \, dx = \frac{1}{2}x^2(\ln x - \frac{1}{2}) + C$

Formula 502

17. $\int xe^{ax} \, dx = a^{-1}e^{ax}(x - a^{-1}) + C$

Formula 484

19. $\int \frac{x^2 \, dx}{\sqrt{x^2 + 1}}$ Formula 174

$$= \frac{1}{2}x\sqrt{x^2 + 1} - \frac{1}{2}\ln\left| x + \sqrt{x^2 + 1} \right| + C$$

21. $\int \frac{x \, dx}{\sqrt{4x^2 + 1}} = \frac{1}{4}\sqrt{4x^2 + 1} + C$

Formula 173; $u = 2x$

23. $\int e^{-4x}\sin 5x \, dx$ Formula 492

$$= (16 + 25)^{-1}e^{-4x}(-4\sin 5x - 5\cos 5x) + C$$

$$= \frac{-4 \sin 5x - 5 \cos 5x}{41 e^{4x}} + C$$

25. $\int (1 + bx)^{-1} \, dx = b^{-1}\ln|1 + bx| + C$

Formula 34

27. $\int x(1 + x)^3 \, dx = \frac{(x + 1)^5}{5} - \frac{(x + 1)^4}{4} + C$

Formula 31

$$= \frac{1}{20}(x + 1)^4(4x - 1) + C$$

29. $\int xe^{4x}dx = \frac{1}{4}e^{4x}(x - \frac{1}{4}) + C$

Formula 484

31. $\int \frac{dx}{1 + e^{2x}} = x - \frac{1}{2}\ln\left|1 + e^{2x}\right| + C$

$$\boxed{\text{Formula 489}}$$

33. $\int \frac{x^3 \, dx}{\sqrt{4x^4 + 1}} = \frac{1}{16}\int u^{-1/2} \, du$

$$\boxed{u = 4x^4 + 1; \ du = 16x^3 \, dx}$$

$$= \frac{1}{8}(4x^4 + 1)^{1/2} + C$$

35. $\int \sec^3\left(\frac{x}{2}\right) dx \quad \boxed{\text{Formula 427}}$

$= \sec \frac{x}{2} \tan \frac{x}{2} + \ln\left|\sec \frac{x}{2} + \tan \frac{x}{2}\right| + C$

37. $\int \frac{dx}{9x^2 + 6x + 1} = \int \frac{dx}{(3x + 1)^2}$

$$\boxed{\text{Formula 41}}$$

$$= -\frac{1}{3(3x + 1)} + C$$

39. $\int \frac{\sin^2 x}{\cos x} \, dx = \int \frac{1 - \cos^2 x}{\cos x} \, dx$

$$= \int \sec x \, dx - \int \cos x \, dx$$

$$= \ln|\sec x + \tan x| - \sin x + C$$

You will obtain a different (but equivalent) form if you use $\boxed{\text{Formula 382}}$.

$$\int \frac{\sin^2 x}{\cos x} \, dx = -\sin x + \ln\left|\tan\left(\frac{x}{2} + \frac{\pi}{4}\right)\right| + C$$

SURVIVAL HINT: *It is common to integrate either a $\sin^2 x$ or $\cos^2 x$ and Problems 40 and 41 give these forms. You should remember or otherwise mark these problems for future use. They are found as Formulas*

317 and 348 in the table of integrals.

41. $\int \cos^2 x \, dx = \int \frac{1 + \cos 2x}{2} \, dx$

$$= \frac{1}{2}\int dx + \frac{1}{2}\int \cos 2x \, dx$$

$$\boxed{u = 2x; \ du = 2 \, dx}$$

$$= \frac{1}{2}x + \frac{1}{2}\frac{\sin 2x}{2} + C$$

$$= \frac{1}{2}x + \frac{1}{4}\sin 2x + C$$

SURVIVAL HINT: *The procedure shown in Problems 42 and 43 is common when working with powers of sines and cosines. It anticipates a method which is presented formally in Section 7.3.*

43. $\int \sin^3 x \cos^4 x \, dx = \int \sin^2 x \cos^4 x \, (\sin x \, dx)$

$$= \int (1 - \cos^2 x)\cos^4 x \, (\sin x \, dx)$$

$$= \int \cos^4 x \, (\sin x \, dx) - \int \cos^6 x \, (\sin x \, dx)$$

$$\boxed{u = \cos x, \ du = -\sin x \, dx}$$

$$= -\int u^4 \, du + \int u^6 \, du$$

$$= -\frac{u^5}{5} + \frac{u^7}{7} + C$$

$$= -\frac{1}{5}\cos^5 x + \frac{1}{7}\cos^7 x + C$$

45. If m is odd, let $u = \cos x$. If n is odd, let $u = \sin x$. If both m and n are even, use the

identities shown in Problems 40 and 41 until one exponent is odd.

47. $\displaystyle \int \frac{dx}{x^{1/2} + x^{1/4}} = \int \frac{4u^3\,du}{u^2 + u}$

$$\boxed{\text{Let } u = x^{1/4};\ du = \tfrac{1}{4}x^{-3/4}dx}$$

$$= 4 \int \frac{u^2\,du}{u + 1} \qquad \boxed{\text{Formula 36}}$$

$$= 4\left[\frac{(u+1)^2}{2} - \frac{2(u+1)}{1} + \ln|u+1|\right] + C_1$$

$$= 4\left[\tfrac{1}{2}u^2 + u + \tfrac{1}{2} - 2u - 2 + \ln|u+1|\right] + C_1$$

$$= 4\left[\tfrac{1}{2}u^2 - u + \ln|u+1|\right] + C$$

$$= 4\left[\frac{x^{1/2}}{2} - x^{1/4} + \ln(x^{1/4} + 1)\right] + C$$

49. $\displaystyle \int \frac{18 \tan^2 t \sec^2 t}{(2 + \tan^3 t)^2}\,dt$

$$\boxed{\text{Let } u = 2 + \tan^3 t;\ du = 3\tan^2 t \sec^2 t\,dt}$$

$$= \int \frac{6\,du}{u^2} = -6u^{-1} + C$$

$$= -6(2 + \tan^3 t)^{-1} + C$$

51. $\displaystyle \int \frac{e^{-x} - e^x}{e^{2x} + e^{-2x} + 2}\,dx = \int \frac{-(e^x - e^{-x})}{(e^x + e^{-x})^2}\,dx$

$$\boxed{\text{Let } u = e^x + e^{-x};\ du = \left(e^x - e^{-x}\right)dx}$$

$$= \int \frac{(-1)\,du}{u^2} = \frac{1}{u} + C = \frac{1}{e^x + e^{-x}} + C$$

53. $\displaystyle V = \pi \int_0^9 \left(\frac{x^{3/2}}{\sqrt{x^2 + 9}}\right)^2 dx$

$$= \pi \int_0^9 \frac{x^3}{x^2 + 9}\,dx$$

$$= \pi\left[\int_0^9 x\,dx - \int_0^9 \frac{9x}{x^2 + 9}\,dx\right]$$

$$\boxed{\text{Let } u = x^2 + 9,\ du = 2x\,dx}$$
$$\boxed{\text{If } x = 0, \text{ then } u = 9 \text{ and if } x = 9, \text{ then } u = 90}$$

$$= \pi \frac{x^2}{2}\bigg|_0^9 - \pi \int_9^{90} \frac{9x\left(\frac{du}{2x}\right)}{u}$$

$$= \tfrac{81}{2}\pi - \tfrac{9}{2}\pi \ln u \bigg|_9^{90}$$

$$= \tfrac{81}{2}\pi - \tfrac{9}{2}\pi (\ln 90 - \ln 9)$$

$$= \tfrac{81}{2}\pi - \tfrac{9}{2}\pi \ln 10$$

$$\approx 94.68$$

55. $\displaystyle V = 2\pi \int_1^4 x\left[\frac{1}{\sqrt{x}}(1 + \sqrt{x})^{1/3}\right] dx$

$$= 2\pi \int_1^4 \sqrt{x}(1 + \sqrt{x})^{1/3}\,dx$$

$$\boxed{u = 1 + \sqrt{x};\ du = \tfrac{1}{2}x^{-1/2}dx\ ;\ x = (u-1)^2}$$
$$\boxed{\text{If } x = 1, \text{ then } u = 2 \text{ and if } x = 4, \text{ then } u = 3}$$

$$= 2\pi \int_2^3 \sqrt{x}\ u^{1/3}(2\sqrt{x}\ du)$$

$$= 4\pi \int_2^3 (u - 1)^2 \, u^{1/3} \, du$$

$$= 4\pi \int_2^3 (u^{7/3} - 2u^{4/3} + u^{1/3}) \, du$$

$$= 4\pi \left(\frac{3}{10} u^{10/3} - \frac{6}{7} u^{7/3} + \frac{3}{4} u^{4/3} \right) \Big|_2^3$$

$$= \frac{\pi}{35} (369 \sqrt[3]{3} - 66 \sqrt[3]{2})$$

Or, by calculator $V \approx 40.3053$

57. $xy' = \sqrt{(\ln x)^2 - x^2}$

$$y' = \frac{1}{x} \sqrt{(\ln x)^2 - x^2}$$

$$= \sqrt{\frac{(\ln x)^2 - x^2}{x^2}}$$

$$s = \int_{1/4}^{1/2} \sqrt{1 + \frac{(\ln x)^2}{x^2} - 1}$$

$$\boxed{u = \ln x; \ du = \frac{1}{x} \, dx}$$

$$= -\int_{1/4}^{1/2} \frac{\ln x}{x} \, dx$$

$$= -\frac{1}{2} \ln^2 x \Big|_{1/4}^{1/2}$$

$$= \frac{3}{2} \ln^2 2$$

$$\approx 0.7207$$

59. $ds = \sqrt{1 + 4x^2} \, dx; \ dS = 2\pi y \, ds$

$$S = 2\pi \int_0^1 x^2 \sqrt{1 + 4x^2} \, dx$$

$$\boxed{\text{Formula 170; } u = 2x}$$

$$= 2\pi \int_0^2 \frac{u^2}{4} \sqrt{1 + u^2} \left(\frac{du}{2} \right)$$

$$= \frac{\pi}{4} \left(\frac{u(u^2 + 1)^{3/2}}{4} - \frac{u(u^2 + 1)^{1/2}}{8} \right.$$

$$\left. - \frac{1}{8} \ln\left[u + (u^2 + 1^2)^{1/2} \right] \right) \Big|_0^2$$

$$= \frac{9\sqrt{5}\pi}{16} - \frac{\pi \ln(\sqrt{5} + 2)}{32}$$

Or, by calculator $S \approx 3.8097$

61. $\int \csc x \, dx = -\int \frac{-\csc x (\csc x + \cot x)}{\csc x + \cot x} \, dx$

$$\boxed{\text{Let } u = \csc x + \cot x; \ du = (-\csc^2 x - \csc x \cot x) \, dx}$$

$$= \int \frac{du}{u}$$

$$= -\ln|\csc x + \cot x| + C$$

63. Let $u = \pi - x$, $du = -dx$. Then $\sin x = \sin u$

$$\int_0^\pi x f(\sin x) \, dx = \int_\pi^0 (\pi - u) f(\sin u)(-du)$$

$$= \int_0^\pi (\pi - u) \, f(\sin u) \, du$$

$$= \int_0^\pi \pi f(\sin u)\, du - \int_0^\pi u f(\sin u)\, du$$

$$= \tfrac{1}{2}x^2 \ln x - \tfrac{1}{4}x^2 + C$$

The value of the integral is independent of the variable used so

$$= \pi \int_0^\pi f(\sin x)\, dx - \int_0^\pi x f(\sin x)\, dx$$

Add the second integral to both sides and divide by 2 to obtain the desired result.

5. $\displaystyle \int \sin^{-1} x\, dx$

$$\boxed{u = \sin^{-1} x;\ dv = dx}$$
$$\boxed{du = \frac{dx}{\sqrt{1-x^2}};\ v = x}$$

$$= x \sin^{-1} x - \int \frac{x\, dx}{\sqrt{1-x^2}} \quad \textit{By parts}$$

$$\boxed{t = 1 - x^2;\ dt = -2x\, dx}$$

$$= x \sin^{-1} x + \sqrt{1-x^2} + C$$

7.2 Integration by Parts, Pages 439–441

SURVIVAL HINT: *Along with substitution, integration by parts is the most common technique of integration. When using parts, a necessary criterion is the integrability of dv; a desirable, but not necessary criterion is that the derivative of u be simple.*

Remember: $\int u\, dv = uv - \int v\, du$

1. $\displaystyle \int x e^{-2x}\, dx$ $\boxed{u = x;\ dv = e^{-2x}\, dx}$
$$\boxed{du = dx;\ v = -\tfrac{1}{2}e^{-2x}}$$

$$= -\tfrac{1}{2}x e^{-2x} + \tfrac{1}{2}\int e^{-2x}\, dx \quad \textit{By parts}$$

$$= -\tfrac{1}{2}x e^{-2x} - \tfrac{1}{4}e^{-2x} + C$$

3. $\displaystyle \int x \ln x\, dx$ $\boxed{u = \ln x;\ dv = x\, dx}$
$$\boxed{du = \tfrac{1}{x}\, dx;\ v = \tfrac{1}{2}x^2}$$

$$= \tfrac{1}{2}x^2 \ln x - \tfrac{1}{2}\int x\, dx \quad \textit{By parts}$$

SURVIVAL HINT: *Problem 7 illustrates the common variation of integration by parts illustrated by Example 4. You will use this technique when integration by parts gives a form which is a multiple of the original integral.*

7. $\displaystyle I = \int e^{-3x} \cos 4x\, dx$

$$\boxed{u = e^{-3x};\ dv = \cos 4x\, dx}$$
$$\boxed{du = -3e^{-3x}\, dx;\ v = \tfrac{1}{4}\sin 4x}$$

$$= \tfrac{1}{4}e^{-3x} \sin 4x + \tfrac{3}{4}\int e^{-3x} \sin 4x\, dx$$

$$\boxed{u = e^{-3x},\ dv = \sin 4x\, dx}$$
$$\boxed{du = -3e^{-3x}\, dx;\ v = -\tfrac{1}{4}\cos 4x}$$

$$= \tfrac{1}{4}e^{-3x} \sin 4x - \tfrac{3}{16}(e^{-3x} \cos 4x)$$

$$\qquad - \tfrac{9}{16}\int e^{-3x} \cos 4x\, dx$$

$$\tfrac{25}{16}I = \tfrac{1}{4}e^{-3x} \sin 4x - \tfrac{3}{16}e^{-3x} \cos 4x + C$$

$$I = \tfrac{4}{25} e^{-3x} \sin 4x - \tfrac{3}{25} e^{-3x} \cos 4x + C$$

SURVIVAL HINT: *You could have used* $u = \cos 4x$ *and* $dy = e^{-3x} dx$ *the first time, and then* $u = \sin 4x$ *and* $dv = e^{-3x} dx$ *the second time. But if you had used* $u = e^{-3x}$ *and* $dv = \cos 4x \, dx$ *the first time and* $u = \sin 4x$ *and* $dv = e^{-3x} dx$ *the second time, you would have obtained* $0 = 0$.

9. $\displaystyle \int x^2 \ln x \, dx$ $\boxed{u = \ln x; \; dv = x^2 \, dx}$

$\boxed{du = \tfrac{dx}{x}; \; v = \tfrac{1}{3} x^3}$

$= \tfrac{1}{3} x^3 \ln x - \tfrac{1}{3} \displaystyle \int x^2 \, dx$ *By parts*

$= \tfrac{1}{3} x^3 \ln x - \tfrac{1}{9} x^3 + C$

11. $\displaystyle \int \sin(\ln x) \, dx$

$\boxed{u = \sin(\ln x), \; dv = dx}$
$\boxed{du = \tfrac{1}{x} \cos(\ln x) \, dx; \; v = x}$

$= x \sin(\ln x) - \displaystyle \int \cos(\ln x) \, dx$ *By parts*

$\boxed{u = \cos(\ln x), \; dv = dx}$
$\boxed{du = -\tfrac{1}{x} \sin(\ln x); \; v = x}$

$= x \sin(\ln x) - x \cos(\ln x) - \displaystyle \int \sin(\ln x) \, dx$

By parts again

$2 \displaystyle \int \sin(\ln x) \, dx = x \big[\sin(\ln x) - \cos(\ln x) \big] + C_1$

Add $\int sin(\ln x) \, dx$ to both sides.

$\displaystyle \int \sin(\ln x) \, dx = \tfrac{x}{2} \big[\sin(\ln x) - \cos(\ln x) \big] + C$

Divide both sides by 2.

13. $\displaystyle \int \ln(x^2 + 1) \, dx$ $\boxed{u = \ln(x^2 + 1); \; dv = dx}$

$\boxed{du = \dfrac{2x \, dx}{x^2 + 1}; \; v = x}$

$= x \ln(x^2 + 1) - \displaystyle \int \frac{2x^2}{x^2 + 1} \, dx$ *By parts*

$= x \ln(x^2 + 1) - 2 \displaystyle \int \left(1 - \frac{1}{x^2 + 1} \right) dx$

By long division

$= x \ln(x^2 + 1) - 2x + 2 \tan^{-1} x + C$

15. $\displaystyle \int \frac{x e^{-x}}{(x - 1)^2} \, dx$ $\boxed{u = x e^{-x}; \; dv = dx/(x - 1)^2}$

$\boxed{\begin{array}{l} du = (1 - x) e^{-x} dx; \\ \qquad v = -1/(x - 1) \end{array}}$

$= \dfrac{-x e^{-x}}{x - 1} - \displaystyle \int \frac{-1}{x - 1} (1 - x) e^{-x} \, dx$

By parts

$= \dfrac{-x e^{-x}}{x - 1} + e^{-x} + C$

17. $\displaystyle \int_1^4 \sqrt{x} \ln x \, dx$ $\boxed{u = \ln x; \; dv = \sqrt{x} \, dx}$

$\boxed{du = \dfrac{dx}{x}; \; v = \tfrac{2}{3} x^{3/2}}$

$= \tfrac{2}{3} x^{3/2} \ln x \Big|_1^4 - \tfrac{2}{3} \displaystyle \int_1^4 x^{1/2} \, dx$ *By parts*

$= \tfrac{16}{3} \ln 4 - \tfrac{4}{9} x^{3/2} \Big|_1^4$

$= \tfrac{32}{3} \ln 2 - \tfrac{28}{9}$

By calculator, the value is 4.28246.

19. $\displaystyle\int_1^e (\ln x)^2 \, dx$ $\boxed{u = \ln^2 x; \; dv = dx}$

$\boxed{du = 2(\ln x)\left(\frac{1}{x}\right) dx; \; v = x}$

$\displaystyle = x(\ln x)^2 \Big|_1^e - 2\int_1^e \ln x \, dx$ *By parts*

$\boxed{u = \ln x; \; dv = dx}$
$\boxed{du = \frac{dx}{x}; \; v = x}$

$\displaystyle = e(\ln e)^2 - 2\left[x \ln x \Big|_1^e - \int_1^e dx \right]$

By parts again

$\displaystyle = e - 2\left[e \ln e - 1 \ln 1 - x \Big|_1^e \right]$

$\displaystyle = e - 2$

By calculator, the value is 0.71828.

21. $\displaystyle I = \int_0^\pi e^{2x} \cos 2x \, dx$ $\boxed{u = e^{2x}; \; dv = \cos 2x \, dx}$

$\displaystyle = \frac{1}{2} e^{2x} \sin 2x \Big|_0^\pi - \int_0^\pi e^{2x} \sin 2x \, dx$

$\boxed{u = e^{2x}; \; dv = \sin 2x \, dx}$

$\displaystyle = \frac{1}{2} e^{2x} \sin 2x \Big|_0^\pi$

$\displaystyle \quad - \left[-\frac{1}{2} e^{2x} \cos 2x \Big|_0^\pi + \int_0^\pi e^{2x} \cos 2x \, dx \right]$

$\displaystyle = \frac{1}{2} e^{2\pi}(0) - \frac{1}{2}(1)(0) + \frac{1}{2} e^{2\pi}(1) - \frac{1}{2}(1)(1) - I$

$\displaystyle 2I = \frac{1}{2} e^{2\pi} - \frac{1}{2}$

$\displaystyle I = \frac{1}{4}(e^{2\pi} - 1)$ Calculator value: 133.62291

23. Integration by parts is the application of the

formula

$$\int u \, dv = uv - \int v \, du$$

The u factor is a part of the integrand that is differentiated and dv is the part that is integrated. Generally, pick dv as complicated as possible yet still integrable, so that the integral on the right is easier to integrate than the original integral.

25. $\displaystyle \int [\sin 2x \ln(\cos x)] \, dx = \int 2 \sin x \cos x \ln(\cos x) \, dx$

$\boxed{t = \cos x; \; dt = -\sin x \, dx}$

$\displaystyle = -2 \int t \ln t \, dt$ $\boxed{u = \ln t; \; dv = t \, dt}$

$\displaystyle = -2 \left[\frac{1}{2} t^2 \ln t - \frac{1}{2} \int t \, dt \right]$

$\displaystyle = -t^2 \ln t + \frac{1}{2} t^2 + C$

$\displaystyle = \frac{1}{2} \cos^2 x - (\cos^2 x) \ln(\cos x) + C$

27. $\displaystyle \int [\sin x \ln(2 + \cos x)] \, dx = -\int \ln t \, dt$

$\boxed{t = 2 + \cos x; \; dt = -\sin x \, dx}$
$\boxed{u = \ln t; \; dv = dt}$

$\displaystyle = -t \ln t + t + C$

$\displaystyle = -(2 + \cos x) \ln(2 + \cos x) + (2 + \cos x) + C$

29. $\displaystyle I = \int \cos^2 x \, dx$ $\boxed{u = \cos x; \; dv = \cos x \, dx}$

$\displaystyle = \sin x \cos x + \int \sin^2 x \, dx$

$\displaystyle = \sin x \cos x + \int (1 - \cos^2 x) \, dx$

$$= \sin x \cos x + x - \int \cos^2 x \, dx$$

$$2I = \sin x \cos x + x + C_1$$

$$I = \tfrac{1}{2}(x + \sin x \cos x) + C$$

$$= \tfrac{x}{2} + \tfrac{1}{4}\sin 2x + C$$

31. $\displaystyle\int x \cos^2 x \, dx \qquad \boxed{u = x; \; dv = \cos^2 x \, dx}$

$$\boxed{du = dx; \; v = \tfrac{x}{2} + \tfrac{\sin 2x}{4} \text{ from Problem 29}}$$

$$= x\left(\tfrac{x}{2} + \tfrac{\sin 2x}{4}\right) - \int \left(\tfrac{x}{2} + \tfrac{\sin 2x}{4}\right) dx$$

$$= \tfrac{1}{2}x^2 + \tfrac{1}{4}x \sin 2x - \tfrac{1}{4}x^2 + \tfrac{1}{8}\cos 2x + C$$

$$= \tfrac{1}{4}x^2 + \tfrac{1}{4}x \sin 2x + \tfrac{1}{8}\cos 2x + C$$

33. $\displaystyle\int x^n \ln x \, dx \qquad \boxed{u = \ln x; \; dv = x^n \, dx}$

$$= \frac{x^{n+1}}{n+1}\ln x - \frac{1}{n+1}\int x^n \, dx$$

$$= \frac{x^{n+1}}{n+1}\ln x - \frac{x^{n+1}}{(n+1)^2} + C$$

35. Let $Q(t)$ be the amount of money raised in t weeks.

$$\frac{dQ}{dt} = 100te^{-0.5t}$$

$$Q(t) = \int_0^3 100te^{-0.5t}dt \qquad \boxed{u = t; \; dv = e^{-0.5t}\, dt}$$

$$= -200te^{-0.5t})\Big|_0^3 - 400\int_0^3 e^{-0.5t}\left(-\tfrac{1}{2}\right) dt$$

$$= \left[-200te^{-0.5t} - 400e^{-0.5t}\right]\Big|_0^3$$

$$= -200e^{-0.5t}(t+2)\Big|_0^3$$

$$= -1{,}000e^{-1.5} + 400$$

$$\approx 177$$

37. $\displaystyle V = 2\pi \int_0^2 xe^{-x}dx \qquad \boxed{u = x; \; dv = e^{-x}dx}$

$$= 2\pi\left(-xe^{-x}\Big|_0^2 + \int_0^2 e^{-x}dx\right)$$

$$= 2\pi(-xe^{-x} - e^{-x})\Big|_0^2$$

$$= 2\pi(1 - 3e^{-2})$$

$$\approx 3.73218$$

39. a. $\displaystyle V = \pi \int_1^e (\ln x)^2 dx \qquad \boxed{u = (\ln x)^2; \; dv = dx}$

$$= \pi x(\ln x)^2\Big|_1^e - 2\pi\int_1^e x \ln x \left(\tfrac{dx}{x}\right)$$

$$= \left[\pi x(\ln x)^2 - 2\pi(x \ln x - x)\right]\Big|_1^e$$

$$= \pi(e - 2e + 2e - 2)$$

$$= \pi(e - 2)$$

$$\approx 2.2565$$

b. $\displaystyle V = 2\pi \int_1^e x \ln x \, dx = 2\pi \left[\tfrac{x^2}{4}(2 \ln x - 1)\right]\Big|_1^e$

See Problem 3

$$= 2\pi\left[\tfrac{e^2}{4} + \tfrac{1}{4}\right]$$

$$= \tfrac{\pi}{2}(e^2 + 1)$$

$$\approx 13.1775$$

41. The curves intersect when $x = 0$. Assume the density is 1.

$$A = \int_0^1 (e^x - e^{-x})\,dx$$

$$= (e^x + e^{-x})\big|_0^1$$

$$\approx 1.0862$$

$$M_x = \tfrac{1}{2}\int_0^1 (e^x - e^{-x})(e^x + e^{-x})\,dx$$

$$= \tfrac{1}{2}\int_0^1 (e^{2x} - e^{-2x})\,dx = \tfrac{1}{4}(e^{2x} + e^{-2x})\big|_0^1$$

$$\approx 1.3811$$

$$M_y = \int_0^1 x(e^x - e^{-x})\,dx$$

$$\boxed{u = x;\ dv = (e^x - e^{-x})\,dx}$$

$$= x(e^x + e^{-x})\big|_0^1 - \int_0^1 (e^x + e^{-x})\,dx$$

$$= [x(e^x + e^{-x}) - (e^x - e^{-x})]\big|_0^1$$

$$= \tfrac{2}{e}$$

$$\approx 0.7358$$

$$(\overline{x}, \overline{y}) \approx \left(\frac{0.7358}{1.0862}, \frac{1.3811}{1.0862}\right)$$

$$\approx (0.68,\ 1.27)$$

43. $\dfrac{dy}{dx} = \sqrt{xy}\ \ln x$

$$\int y^{-1/2}\,dy = \int x^{1/2}\ \ln x\,dx$$

$$\boxed{u = \ln x,\ dv = x^{1/2}\,dx}$$

$$2y^{1/2} = \tfrac{2}{3}x^{3/2}\ln x - \tfrac{2}{3}\int x^{1/2}\,dx$$

$$2y^{1/2} = \tfrac{2}{3}x^{3/2}\ln x - \tfrac{4}{9}x^{3/2} + C$$

$$2y^{1/2} = \tfrac{2}{3}x^{3/2}(\ln x - \tfrac{2}{3}) + C$$

45. $\dfrac{dy}{dx} = \dfrac{-xy}{\sec x}$

$$-\int y^{-1}\,dy = \int x \cos x\,dx$$

$$\boxed{\text{By parts or by Formula 312}}$$

$$-\ln y = \cos x + x \sin x + C$$

At $(0, 1)$, $0 = 1 + C$, so $C = -1$.

$$-\ln y = \cos x + x \sin x - 1$$

$$y = \exp(1 - \cos x - x \sin x)$$

47. $v(t) = -r\ln \dfrac{w - kt}{w} - gt$

$$s(t) = \int v(t)\,dt$$

$$= \frac{r(w - kt)}{k}\ln\left(\frac{w - kt}{w}\right) - \tfrac{1}{2}gt^2 + rt$$

Since $s(0) = 0$ (the rocket starts at ground level), we find (by substituting the give information), $s(120)$

$$= \frac{8,000[30,000 - 200(120)]}{200}\ln\left(\frac{30,000 - 200(120)}{30,000}\right)$$
$$- \frac{1}{2}(32)(120)^2 + 8,000(120)$$

$\approx 343,335$ ft $\quad$ 5,280 ft = 1 mi

≈ 65 miles

49. Average displacement is:

$$\frac{1}{\frac{\pi}{5}}\int_0^{\pi/5} 2.3e^{-0.25t}\cos 5t \, dt$$

$$= \frac{5}{\pi}(2.3)\left[\frac{e^{-0.25t}(-0.25\cos 5t + 5\sin 5t)}{(-0.25)^2 + (5)^2}\right]\Bigg|_0^{\pi/5}$$

≈ 0.07

51. $\quad$ Average Work $= \dfrac{1}{9V_1}\displaystyle\int_{V_1}^{10V_1} nRT\ln\frac{V}{V_1}\, dV$

$$= \frac{nRT}{9V_1}\int_{V_1}^{10V_1}(\ln V - \ln V_1)dV.$$

$$\boxed{\text{By parts or by Formula 499}}$$

$$= \frac{nRT}{9V_1}(V\ln V - V - V\ln V_1)\Bigg|_{V_1}^{10V_1}$$

$$= \frac{nRT}{9}(10\ln 10 - 9)$$

$$\approx 1.558nRT$$

53. $\displaystyle\int x^n e^x \, dx \quad \boxed{u = x^n;\ dv = e^x\, dx}$

$$= x^n e^x - n\int x^{n-1}e^x \, dx$$

55. $\quad I = \displaystyle\int_0^{\pi/2} \sin^n x \, dx$, n is even;

$$\boxed{u = (\sin x)^{n-1};\ dv = \sin x \, dx}$$

$$= -\sin^{n-1}x\cos x\Big|_0^{\pi/2}$$

$$+ \int_0^{\pi/2}(n-1)(\sin x)^{n-2}\cos^2 x\, dx$$

$$= 0 + (n-1)\int_0^{\pi/2}[(\sin x)^{n-2}(1 - \sin^2 x)]\, dx$$

$$= (n-1)\int_0^{\pi/2}(\sin x)^{n-2}\, dx - (n-1)I$$

$$nI = (n-1)\int_0^{\pi/2}(\sin x)^{n-2}\, dx$$

$$I = \frac{n-1}{n}\int_0^{\pi/2}(\sin x)^{n-2}\, dx$$

This recursive formula will be used repeatedly with values of n that decrease by 2.

$$I = \frac{n-1}{n}\int_0^{\pi/2}\sin^{n-2}x\, dx$$

$$= \frac{n-1}{n}\left[-\frac{\sin^{n-3}x\cos x}{n-2}\right]\Bigg|_0^{\pi/2}$$

$$+ \frac{n-1}{n}\frac{n-3}{n-2}\int_0^{\pi/2}\sin^{n-4}x\, dx$$

$$= \frac{(n-1)(n-3)}{n(n-2)}\int_0^{\pi/2}\sin^{n-4}x\, dx$$

$$= \cdots = \frac{(n-1)(n-3)\cdots(3)}{n(n-2)\cdots(4)} \int_0^{\pi/2} \sin^2 x \, dx$$

$$= \frac{(n-1)(n-3)\cdots(3)(1)}{n(n-2)\cdots(4)(2)} \int_0^{\pi/2} (1+\cos 2x) \, dx$$

$$= \frac{1\cdot 3\cdot 5\cdots(n-3)(n-1)}{2\cdot 4\cdot 6\cdots(n-2)n} \cdot \frac{\pi}{2}$$

For $\int_0^{\pi/2} \cos^n x \, dx$, proceed as shown above.

7.3 Trigonometric Methods, Page 447

1. Convert $\sin^2 x = \frac{1}{2}(1 - \cos 2x)$ and $\cos^2 x = \frac{1}{2}(1 + \cos 2x)$ and substitute into the given integral.

3. Substitute $u = a \tan \theta$ and then integrate the resulting integral in trigonometric form. It may help to draw a reference triangle.

5. $\displaystyle \int \cos^3 x \, dx = \int \cos^2 x \, (\cos x \, dx)$

$$= \int (1 - u^2)(du) \quad \boxed{u = \sin x}$$

$$= u - \frac{u^3}{3} + C$$

$$= \sin x - \frac{1}{3}\sin^3 x + C$$

7. $\displaystyle \int \sin^2 x \cos^3 dx = \int \sin^2 x \cos^2 x \, (\cos x \, dx)$

$$= \int (\sin^2 x - \sin^4 x)(\cos x \, dx)$$

$$= \int (u^2 - u^4) \, du \quad \boxed{u = \sin x}$$

$$= \frac{1}{3}u^3 - \frac{1}{5}u^5 + C$$

$$= \frac{1}{3}\sin^3 x - \frac{1}{5}\sin^5 x + C$$

SURVIVAL HINT: *You might wish to compare Problem 7 with Problem 43, Section 7.1. Notice that Problem 7 is nothing more than a variation of substitution, in which you "peel off" one of the factors of the odd power.*

9. $\displaystyle \int \sqrt{\cos t} \, \sin t \, dt = \int u^{1/2}(-du) \quad \boxed{u = \cos t}$

$$= -\frac{2}{3}u^{3/2} + C$$

$$= -\frac{2}{3}(\cos t)^{3/2} + C$$

11. $\displaystyle \int e^{\cos x}\sin x \, dx = \int e^u \, (-du) \quad \boxed{u = \cos x}$

$$= -e^{\cos x} + C$$

13. $\displaystyle \int \sin^2 x \cos^2 x \, dx = \int \left[\frac{1-\cos 2x}{2}\right]\left[\frac{1+\cos 2x}{2}\right] dx$

$$= \frac{1}{4}\int (1 - \cos^2 2x) \, dx$$

$$= \frac{1}{4}x - \frac{1}{4}\int \left[\frac{1+\cos 4x}{2}\right] dx$$

$$= \frac{1}{4}x - \frac{1}{8}x - \frac{1}{8}\left[\frac{\sin 4x}{4}\right] + C$$

$$= \frac{1}{8}x - \frac{1}{32}\sin 4x + C$$

15. $\displaystyle \int \tan 2\theta \, d\theta = -\frac{1}{2}\ln|\cos 2\theta| + C \quad \boxed{u = 2\theta}$

17. $\displaystyle \int \tan^3 x \sec^4 x \, dx = \int \tan^3 x \sec^2 x(\sec^2 x \, dx)$

$$= \int \tan^3 x(\tan^2 x + 1)(\sec^2 x \, dx)$$

$$\boxed{u = \tan x}$$

$$= \int (u^5 + u^3) \, du$$

$$= \tfrac{1}{6} u^6 + \tfrac{1}{4} u^4 + C$$

$$= \tfrac{1}{6} \tan^6 x + \tfrac{1}{4} \tan^4 x + C$$

19. $\displaystyle \int (\tan^2 x + \sec^2 x) \, dx = \int (2\sec^2 x - 1) \, dx$

$$= 2\tan x - x + C$$

21. $\displaystyle \int \tan^2 u \sec u \, du = \int (\sec^2 u - 1)\sec u \, du$

$$= \int (\sec^3 u - \sec u) \, du \quad \boxed{\text{Formula 428}}$$

$$= \frac{\sec u \tan u}{2} + \tfrac{1}{2} \int \sec u \, du - \int \sec u \, du$$

$$= \tfrac{1}{2} \sec u \tan u - \tfrac{1}{2} \ln|\sec u + \tan u| + C$$

23. $\displaystyle \int \sqrt[3]{\tan x} \, \sec^2 x \, dx = \int u^{1/3} du \quad \boxed{u = \tan x}$

$$= \tfrac{3}{4} (\tan x)^{4/3} + C$$

25. $\displaystyle \int x \sin x^2 \cos x^2 \, dx = \tfrac{1}{2} \int \sin u \, (\cos u \, du)$

$$\boxed{u = x^2}$$

$$= \tfrac{1}{4} \sin^2 x^2 + C$$

27. $\displaystyle \int \tan^4 t \sec t \, dt = \int (\sec^2 t - 1)^2 \sec t \, dt$

$$= \int (\sec^5 t - 2\sec^3 t + \sec t) \, dt$$

$$= \frac{\sec^3 t \tan t}{4} + \tfrac{3}{4} \int \sec^3 t \, dt$$

$$\boxed{\text{Formula 428}}$$

$$- 2 \int \sec^3 t \, dt + \int \sec t \, dt$$

$$= \tfrac{1}{4} \sec^3 t \tan t - \tfrac{5}{4} \int \sec^3 t \, dt$$

Combine similar terms

$$+ \ln|\sec t + \tan t|$$

$$= \tfrac{1}{4} \sec^3 t \tan t - \tfrac{5}{4} \left(\frac{\sec t \tan t}{2} \right.$$

$$\left. + \tfrac{1}{2} \ln|\sec t + \tan t| \right)$$

$$\boxed{\text{Formula 427}}$$

$$+ \ln|\sec t + \tan t| + C$$

$$= \tfrac{1}{4} \sec^3 t \tan t - \tfrac{5}{8} \sec t \tan t$$

$$+ \tfrac{3}{8} \ln|\sec t + \tan t| + C$$

29. $\displaystyle \int \csc^3 x \cot x \, dx = - \int \csc^2 x \, (-\csc x \cot x \, dx)$

$$\boxed{u = \csc x; \; du = -\csc x \cot x \, dx}$$

$$= -\tfrac{1}{3} \csc^3 x + C$$

31. $\displaystyle \int \csc^2 x \cos x \, dx = = \int \frac{\cos x \, dx}{\sin^2 x}$

$$\boxed{u = \sin x}$$

$$= -(\sin x)^{-1} + C$$

$$= -\csc x + C$$

33. $\displaystyle \int \sqrt{4 - t^2} \, dt = \int \sqrt{4 - 4\sin^2 \theta} \, (2\cos \theta \, d\theta)$

$$\boxed{t = 2\sin \theta; \; dt = 2\cos \theta \, d\theta}$$

$$= 4 \int \cos^2\theta \, d\theta$$

$$= 4\left[\frac{\theta}{2} + \frac{\sin 2\theta}{4}\right] + C$$

Cosine squared formula

$$= 2\sin^{-1}\frac{t}{2} + \frac{1}{2}t\sqrt{4 - t^2} + C$$

35. $\int \frac{x+1}{\sqrt{4+x^2}} \, dx = \int \frac{x \, dx}{\sqrt{4+x^2}} + \int \frac{dx}{\sqrt{4+x^2}}$

$$= \frac{1}{2}\int \frac{2x \, dx}{\sqrt{4+x^2}} + \int \frac{dx}{\sqrt{4+x^2}}$$

$$= \sqrt{4+x^2} + \ln\left|x + \sqrt{4+x^2}\right| + C$$

Formula 172

This can also be done by trigonometric

substitution: let $x = 2\tan\theta$. Then

$dx = 2\sec^2\theta \, d\theta$, and

$$\sqrt{4+x^2} = \sqrt{4 + 4\tan^2\theta} = 2\sec\theta;$$

$$\int \frac{x+1}{\sqrt{4+x^2}} \, dx = \int \frac{2\tan\theta + 1}{2\sec\theta}(2\sec^2\theta \, d\theta)$$

$$= 2\int \tan\theta \sec\theta \, d\theta + \int \sec\theta \, d\theta$$

$$= 2\sec\theta + \ln|\sec\theta + \tan\theta| + C_1$$

$$= \sqrt{4+x^2} + \ln\left|\frac{1}{2}\sqrt{4+x^2} + \frac{x}{2}\right| + C_1$$

$$= \sqrt{4+x^2} + \ln\left|\sqrt{4+x^2} + x\right| + C$$

37. $\int \frac{dx}{\sqrt{x^2-7}} = \ln\left|x + \sqrt{x^2-7}\right| + C$

Formula 196

This can also be done by trigonometric

substitution: let $x = \sqrt{7}\sec\theta$. Then $dx = \sqrt{7}$

$\sec\theta\tan\theta \, d\theta$, and

$$\sqrt{x^2-7} = \sqrt{7\sec^2\theta - 7} = \sqrt{7}\tan\theta;$$

$$\int \frac{dx}{\sqrt{x^2-7}} = \int \frac{\sqrt{7}\sec\theta\tan\theta \, d\theta}{\sqrt{7}\tan\theta}$$

$$= \int \sec\theta \, d\theta$$

$$= \ln|\sec\theta + \tan\theta| + C_1$$

$$= \ln\left|\frac{x}{\sqrt{7}} + \frac{\sqrt{x^2-7}}{\sqrt{7}}\right| + C_1$$

$$= \ln\left|x + \sqrt{x^2-7}\right| - \frac{1}{2}\ln 7 + C_1$$

$$= \ln\left|x + \sqrt{x^2-7}\right| + C$$

39. $\int \frac{dx}{\sqrt{5-x^2}} = \sin^{-1}\frac{x}{\sqrt{5}} + C$

41. $\int \frac{dx}{x^2\sqrt{4-x^2}}$ $x = 2\sin\theta$

$$= \int \frac{2\cos\theta \, d\theta}{(4\sin^2\theta)\sqrt{4 - 4\sin^2\theta}}$$

$$= \frac{1}{4}\int \csc^2\theta \, d\theta$$

$$= -\frac{1}{4}\cot\theta + C$$

$$= -\frac{\sqrt{4-x^2}}{4x} + C$$

43. $\int \frac{\sqrt{x^2-4}}{x} \, dx$

$x = 2\sec\theta; \ dx = 2\sec\theta\tan\theta \, d\theta$

$$= \int \frac{\sqrt{4\sec^2\theta - 4}}{2\sec\theta}(2\sec\theta\tan\theta \ d\theta)$$

$$= \int 2\tan^2\theta \ d\theta$$

$$= 2\int (\sec^2\theta - 1) \ d\theta$$

$$= 2\tan\theta - 2\theta + C$$

$$= \sqrt{x^2 - 4} - 2\sec^{-1}\frac{x}{2} + C$$

45. $\displaystyle\int \frac{dx}{9 - (x+1)^2}$

$\boxed{\text{Let } x + 1 = 3\sin\theta; \ dx = 3\cos\theta \ d\theta \text{ or Formula 93}}$

$$= \int \frac{3\cos\theta \ d\theta}{9 - 9\sin^2\theta}$$

$$= \frac{1}{3}\int \frac{d\theta}{\cos\theta}$$

$$= \frac{1}{3}\int \sec\theta \ d\theta$$

$$= \frac{1}{3}\ln|\sec\theta + \tan\theta| + C$$

$$= \frac{1}{3}\ln\left|\frac{3}{\sqrt{9 - (x+1)^2}} + \frac{x+1}{\sqrt{9 - (x+1)^2}}\right| + C$$

$$= \frac{1}{3}\ln\left|\frac{x+4}{\sqrt{9 - (x+1)^2}}\right| + C$$

47. $\displaystyle\int \frac{dx}{\sqrt{x^2 - 2x + 6}} = \int \frac{dx}{\sqrt{(x-1)^2 + 5}}$

$$= \int \frac{\sqrt{5}\sec^2\theta \ d\theta}{\sqrt{5\tan^2\theta + 5}} \quad \boxed{\text{Let } x - 1 = \sqrt{5}\tan\theta}$$

$$= \int \sec\theta \ d\theta = \ln|\sec\theta + \tan\theta| + C_1$$

$$= \ln\left|\frac{\sqrt{x^2 - 2x + 6}}{\sqrt{5}} + \frac{x-1}{\sqrt{5}}\right| + C_1$$

This can be written

$$\ln\left|\sqrt{x^2 - 2x + 6} + x - 1\right| + C$$

49. $\displaystyle\int \frac{\sin^3 u}{\cos^5 u} \ du = \int \tan^3 u \sec^2 u \ du$

$\boxed{w = \tan u}$

$$= \frac{1}{4}\tan^4 u + C$$

51. $\displaystyle A = \frac{1}{\pi - 0}\int_0^\pi \sin^2 x \ dx$

$$= \frac{1}{\pi}\left(\frac{1}{2}x - \frac{1}{4}\sin 2x\right)\Big|_0^\pi$$

$$= \frac{1}{\pi}\left(\frac{\pi}{2} + 0 - 0 - 0\right)$$

$$= \frac{1}{2}$$

53. $\displaystyle V = 2\pi \int_0^\pi x\sin^2 x \ dx$

$\boxed{u = x, \ du = dx; \ dv = \sin^2 x \ dx, \ v = \frac{1}{2}x - \frac{1}{4}\sin 2x}$

$$= 2\pi\left[x\left(\frac{1}{2}x - \frac{1}{4}\sin 2x\right)\Big|_0^\pi - \int_0^\pi \left(\frac{1}{2}x - \frac{1}{4}\sin 2x\right) dx\right]$$

$$= 2\pi \left[\left(\tfrac{1}{2}x^2 - \tfrac{1}{4}x \sin 2x - \tfrac{1}{4}x^2 \right) \Big|_0^\pi + \tfrac{1}{4} \int_0^\pi \sin 2x \, dx \right]$$

$$= 2\pi \left[\left(\tfrac{1}{4}x^2 - \tfrac{1}{4}x \sin 2x \right) + \tfrac{1}{4} \left(-\tfrac{1}{2} \cos 2x \right) \right] \Big|_0^\pi$$

$$= 2\pi \left[\tfrac{1}{4}x^2 - \tfrac{1}{4}x \sin 2x - \tfrac{1}{8} \cos 2x \right] \Big|_0^\pi$$

$$= 2\pi \left(\tfrac{1}{4}\pi^2 - \tfrac{1}{8} + \tfrac{1}{8} \right)$$

$$= \frac{\pi^3}{2}$$

You can verify this result by calculator,

$$\frac{\pi^3}{2} \approx 15.5031.$$

55. Since $\cos^2 \tfrac{x}{2} = \dfrac{1 + \cos x}{2}$ *Half-angle identity*

we see $1 + \cos x = 2\cos^2 \tfrac{x}{2}$. Thus,

$$\int \sqrt{1 + \cos x} \, dx = \int \sqrt{2 \cos^2 \tfrac{x}{2}} \, dx$$

$$= \int \sqrt{2} \left| \cos \tfrac{x}{2} \right| \, dx$$

$$= \pm 2\sqrt{2} \sin \tfrac{x}{2} + C$$

57. Since the product-to-sum identity

$$2 \cos \tfrac{x}{2} \sin 2x = \sin\left(\tfrac{x}{2} + 2x \right) - \sin\left(\tfrac{x}{2} - 2x \right)$$

we see

$$\cos \tfrac{x}{2} \sin 2x = \tfrac{1}{2} \left[\sin\left(\tfrac{5}{2}x \right) - \sin\left(-\tfrac{3}{2}x \right) \right]$$

$$= \tfrac{1}{2} \left[\sin\left(\tfrac{3}{2}x \right) + \sin\left(\tfrac{5}{2}x \right) \right]$$

$$\int \cos \tfrac{x}{2} \sin 2x \, dx = \int \tfrac{1}{2} [\sin(\tfrac{3}{2}x) + \sin(\tfrac{5}{2}x)] \, dx$$

$$= -\tfrac{1}{3} \cos(\tfrac{3}{2}x) - \tfrac{1}{5} \cos(\tfrac{5}{2}x) + C$$

59. $\displaystyle \int \sin^2 3x \cos 4x \, dx = \int \tfrac{1}{2}(1 - \cos 6x) \cos 4x \, dx$

Double-angle identity

$$= \tfrac{1}{2} \int \cos 4x \, dx - \tfrac{1}{2} \int \cos 6x \cos 4x \, dx$$

$$= \tfrac{1}{2} \int \cos 4x \, dx - \tfrac{1}{2} \int \tfrac{1}{2}[\cos(6x - 4x) + \cos(6x + 4x)] \, dx$$

Product-to-sum identity

$$= \tfrac{1}{2} \int \cos 4x \, dx - \tfrac{1}{4} \int (\cos 2x + \cos 10x) \, dx$$

$$= \tfrac{1}{8} \sin 4x - \tfrac{1}{8} \sin 2x - \tfrac{1}{40} \sin 10x + C$$

61. $\displaystyle \int \frac{x \, dx}{9 - x^2 - \sqrt{9 - x^2}} = \int \frac{-u \, du}{u^2 - u}$

$$\boxed{\text{Let } u = \sqrt{9 - x^2};\ u^2 = 9 - x^2}$$
$$\boxed{2u \, du = -2x \, dx}$$

$$= -\int \frac{du}{u - 1}$$

$$= -\ln \left| \sqrt{9 - x^2} - 1 \right| + C$$

7.4 Method of Partial Fractions, Pages 455–457

SURVIVAL HINT: *Many calculators will do a partial fraction decomposition. Simply enter the fraction and use the EXPAND function.*

1. $\dfrac{1}{x(x - 3)} = \dfrac{A_1}{x} + \dfrac{A_2}{x - 3}$

Multiply both sides by $x(x-3)$:

$$1 = A_1(x-3) + A_2 x$$

If $x = 0$, then $A_1 = -\frac{1}{3}$; if $x = 3$, then $A_2 = \frac{1}{3}$.

Thus,

$$\frac{1}{x(x-3)} = \frac{-1}{3x} + \frac{1}{3(x-3)}$$

3. $\dfrac{3x^2 + 2x - 1}{x(x+1)} = 3 - \dfrac{x+1}{x(x+1)}$ *Long division*

$$= 3 - \frac{1}{x} \qquad x \neq -1$$

5. $\dfrac{4}{2x^2 + x} = \dfrac{4}{x(2x+1)}$

$$= \frac{A_1}{x} + \frac{A_2}{2x+1}$$

Multiply both sides by $x(2x+1)$

$$4 = A_1(2x+1) + A_2 x$$

If $x = 0$, then $A_1 = 4$; if $x = -\frac{1}{2}$, then $A_2 = -8$.

$$\frac{4}{2x^2 + x} = \frac{4}{x} + \frac{-8}{2x+1}$$

7. $F = \dfrac{4x^3 + 4x^2 + x - 1}{x^2(x+1)^2}$

$$= \frac{A_1}{x} + \frac{A_2}{x^2} + \frac{A_3}{x+1} + \frac{A_4}{(x+1)^2}$$

Multiply both sides by $x^2(x+1)^2$:

$$4x^3 + 4x^2 + x - 1 = A_1 x(x^2 + 2x + 1)$$
$$+ A_2(x^2 + 2x + 1) + A_3(x^3 + x^2) + A_4 x^2$$

If $x = 0$, then $A_2 = -1$; if $x = -1$, then $A_4 = -2$; finally equate the coefficient of x^3 and those of x to find $4 = A_1 + A_3$ and

$1 = A_1 + 2A_2$, so $A_1 = 3$ and $A_3 = 1$. Thus,

$$F = \frac{3}{x} + \frac{-1}{x^2} + \frac{1}{x+1} + \frac{-2}{(x+1)^2}$$

9. $F = \dfrac{x^3 + 3x^2 + 3x - 4}{x^2(x+3)^2}$

$$= \frac{A_1}{x^2} + \frac{A_2}{x} + \frac{A_3}{(x+3)^2} + \frac{A_4}{x+3}$$

Multiply both sides by $x^2(x+3)^2$:

$$x^3 + 3x^2 + 3x - 4 = A_1(x+3)^2$$
$$+ A_2 x(x+3)^2 + A_3 x^2 + A_4 x^2(x+3)$$
$$= (A_2 + A_4)x^3 + (A_1 + 6A_2 + A_3 + 3A_4)x^2$$
$$+ (6A_1 + 9A_2)x + 9A_1$$

If $x = 0$, then $A_1 = -\frac{4}{9}$

if $x = -3$, then $A_3 = -\frac{13}{9}$

coefficients of x^3: $A_2 + A_4 = 1$

coefficients of x^2: $A_1 + 6A_2 + A_3 + 3A_4 = 3$

coefficients of x: $6A_1 + 9A_2 = 3$

Solving this system of equations, using previously found values, we find:

$A_2 = \frac{17}{27}$ and $A_4 = \frac{10}{27}$.

$$F = \frac{-\frac{4}{9}}{x^2} + \frac{\frac{17}{27}}{x} + \frac{-\frac{13}{9}}{(x+3)^2} + \frac{\frac{10}{27}}{x+3}$$

$$= \frac{-4}{9x^2} + \frac{17}{27x} - \frac{13}{9(x+3)^2} + \frac{10}{27(x+3)}$$

11. $\dfrac{1}{1 - x^4} = \dfrac{1}{(1-x)(1+x)(1+x^2)}$

$$= \frac{A_1}{1-x} + \frac{A_2}{1+x} + \frac{A_3 x + B_1}{1+x^2}$$

Multiply both sides by $(1-x)(1+x)(1+x^2)$:

$$1 = A_1(1+x)(1+x^2) + A_2(1-x)(1+x^2)$$
$$+ A_3 x(1-x^2) + B_1(1-x^2)$$
$$= A_1(x^3 + x^2 + x + 1) + A_2(-x^3 + x^2 - x + 1)$$
$$+ A_3(x - x^3) + B_1 - B_1 x^2$$

After solving, we find

$$\frac{1}{1-x^4} = \frac{1}{4(1-x)} + \frac{1}{4(1+x)} + \frac{1}{2(1+x^2)}$$

13. $\quad \frac{x^2 + x - 1}{x(x+1)(2x-1)} = \frac{A_1}{x} + \frac{A_2}{x+1} + \frac{A_3}{2x-1}$

Multiply both sides by $x(x+1)(2x-1)$:

$$A_1(x+1)(2x-1) + A_2 x(2x-1) + A_3 x(x+1)$$
$$= x^2 + x - 1$$

If $x = 0$: $A_1(1)(-1) = -1$, so $A_1 = 1$
If $x = -1$: $A_2(-1)(-3) = -1$, so $A_2 = -\frac{1}{3}$
If $x = \frac{1}{2}$: $A_3(\frac{1}{2})(\frac{3}{2}) = -\frac{1}{4}$, so $A_3 = -\frac{1}{3}$

Thus,

$$\frac{x^2 + x - 1}{2x^3 + x^2 - x} = \frac{1}{x} + \frac{-\frac{1}{3}}{x+1} + \frac{-\frac{1}{3}}{2x-1}$$
$$= \frac{1}{x} - \frac{1}{3(x+1)} - \frac{1}{3(2x-1)}$$

15. $\quad \frac{1}{x(x-3)} = \frac{A_1}{x} + \frac{A_2}{x-3}$

Multiply both sides by $x(x-3)$:

$$1 = A_1(x-3) + A_2 x$$

If $x = 3$, then $A_2 = \frac{1}{3}$, and so $A_1 = -\frac{1}{3}$.

$$\int \frac{dx}{x(x-3)} = -\frac{1}{3}\int \frac{dx}{x} + \frac{1}{3}\int \frac{dx}{x-3}$$
$$= -\frac{1}{3}\ln|x| + \frac{1}{3}\ln|x-3| + C$$
$$= \frac{1}{3}\ln\left|\frac{x-3}{x}\right| + C$$

17. $\frac{3x^2 + 2x - 1}{x(x+1)} = 3 + \frac{-(x+1)}{x(x+1)}$ *By long division*
$$= 3 - \frac{1}{x}$$

$$\int \frac{3x^2 + 2x - 1}{x(x+1)} dx = 3\int dx - \int \frac{1}{x} dx$$
$$= 3x - \ln|x| + C$$

19. $\quad \frac{4}{2x^2 + x} = \frac{A_1}{x} + \frac{A_2}{2x+1}$

Multiply both sides by $x(2x+1)$:

$$4 = A_1(2x+1) + A_2 x$$

If $x = -\frac{1}{2}$, then $A_2 = -8$ so $A_1 = 4$.

$$\int \frac{4\,dx}{2x^2 + x} = \int \frac{4}{x} dx - \int \frac{8\,dx}{2x+1}$$
$$= 4\ln|x| - 4\ln|2x+1| + C$$

21. $\frac{2x^3 + 9x - 1}{x^2(x^2-1)} = \frac{A_1}{x} + \frac{A_2}{x^2} + \frac{A_3}{x+1} + \frac{A_4}{x-1}$

Multiply both sides by $x^2(x^2-1)$:

$$2x^3 + 9x - 1 = A_1 x(x^2-1) + A_2(x^2-1)$$
$$+ A_3 x^2(x-1) + A_4 x^2(x+1)$$

If $x = 0$, then $-1 = -A_2$ so $A_2 = 1$.

If $x = 1$, then $2 + 9 - 1 = 2A_4$, so $A_4 = 5$

Substituting $A_2 = 1$ and $A_4 = 5$, we obtain

$2x^3 + 9x - 1 = A_1 x(x^2 - 1) + x^2 - 1$

$\qquad + A_3 x^2(x - 1) + 5x^3 + 5x^2$

$-3x^3 - 6x^2 + 9x$

$\qquad = A_1 x(x^2 - 1) + A_3 x^2(x - 1)$

If $x = -1$, then

$3 - 6 - 9 = A_3(-1 - 1)$, so $A_3 = 6$

Finally, by substitution we find $A_1 = -9$:

$\dfrac{2x^3 + 9x - 1}{x^2(x^2 - 1)} = \dfrac{-9}{x} + \dfrac{1}{x^2} + \dfrac{6}{x + 1} + \dfrac{5}{x - 1}$

$\displaystyle\int \dfrac{2x^3 + 9x - 1}{x^2(x^2 - 1)}\, dx$

$= -9 \displaystyle\int \dfrac{dx}{x} + \int x^{-2} dx + 6 \int \dfrac{dx}{x + 1} + 5 \int \dfrac{dx}{x - 1}$

$= -9 \ln|x| - x^{-1} + 6\ln|x + 1| + 5\ln|x - 1| + C$

23. $\dfrac{x^2 + 1}{x^2 + x - 2} = 1 + \dfrac{-x + 3}{x^2 + x - 2}$

$\qquad = 1 + \dfrac{-x + 3}{(x - 1)(x + 2)}$

$\qquad = 1 + \dfrac{A_1}{x - 1} + \dfrac{A_2}{x + 2}$

Partial fraction decomposition gives

$A_1 = \frac{2}{3}, A_2 = -\frac{5}{3}$.

$\displaystyle\int \dfrac{x^2 + 1}{x^2 + x - 2}\, dx$

$= \displaystyle\int dx - \frac{5}{3} \int \dfrac{dx}{x + 2} + \frac{2}{3} \int \dfrac{dx}{x - 1}$

$= x - \frac{5}{3} \ln|x + 2| + \frac{2}{3} \ln|x - 1| + C$

25. Use the results of Problem 11.

$\displaystyle\int \dfrac{x^4 + 1}{x^4 - 1}\, dx = \int \left[1 + \dfrac{2}{x^4 - 1}\right] dx$

$= \displaystyle\int dx + 2\left[\frac{1}{4}\int \dfrac{dx}{x - 1} - \frac{1}{4}\int \dfrac{dx}{x + 1} - \frac{1}{2}\int \dfrac{dx}{x^2 + 1}\right]$

$= x + \frac{1}{2} \ln|x - 1| - \frac{1}{2} \ln|x + 1| - \tan^{-1}x + C$

27. $\dfrac{x}{(x + 1)^2} = \dfrac{A_1}{x + 1} + \dfrac{A_2}{(x + 1)^2}$

Partial fraction decomposition gives $A_1 = 1$

and $A_2 = -1$.

$\displaystyle\int \dfrac{x\, dx}{(x + 1)^2} = \int \dfrac{dx}{x + 1} - \int \dfrac{dx}{(x + 1)^2}$

$\qquad = \ln|x + 1| + (x + 1)^{-1} + C$

29. $\dfrac{1}{x(x + 1)(x - 2)} = \dfrac{A_1}{x} + \dfrac{A_2}{x + 1} + \dfrac{A_3}{x - 2}$

Partial fraction decomposition gives

$A_1 = -\frac{1}{2}, A_2 = \frac{1}{3}, A_3 = \frac{1}{6}$.

$\displaystyle\int \dfrac{dx}{x(x + 1)(x - 2)}$

$= -\frac{1}{2}\displaystyle\int \dfrac{dx}{x} + \frac{1}{3}\int \dfrac{dx}{x + 1} + \frac{1}{6}\int \dfrac{dx}{x - 2}$

$$= -\tfrac{1}{2}\ln|x| + \tfrac{1}{3}\ln|x+1| + \tfrac{1}{6}\ln|x-2| + C$$

31. $\dfrac{x}{(x+1)(x+2)^2} = \dfrac{A_1}{x+1} + \dfrac{A_2}{(x+2)^2} + \dfrac{A_3}{x+2}$

Partial fraction decomposition gives
$A_1 = -1$, $A_2 = 2$, $A_3 = 1$.

$$\int \frac{x\,dx}{(x+1)(x+2)^2}$$

$$= -\int \frac{dx}{x+1} + 2\int \frac{dx}{(x+2)^2} + \int \frac{dx}{x+2}$$

$$= -\ln|x+1| - \frac{2}{x+2} + \ln|x+2| + C$$

$$= \ln\left|\frac{x+2}{x+1}\right| - \frac{2}{x+2} + C$$

33. $\dfrac{5x+7}{x^2+2x-3} = \dfrac{5x+7}{(x-1)(x+3)} = \dfrac{A_1}{x-1} + \dfrac{A_2}{x+3}$

Partial fraction decomposition gives
$A_1 = 3$, $A_2 = 2$.

$$\int \frac{5x+7}{x^2+2x-3}\,dx = 3\int \frac{dx}{x-1} + 2\int \frac{dx}{x+3}$$

$$= 3\ln|x-1| + 2\ln|x+3| + C$$

35. $\displaystyle\int \frac{3x^2 - 2x + 4}{x^3 - x^2 + 4x - 4}\,dx$ $\boxed{u = x^3 - x^2 + 4x - 4}$

$$\boxed{du = (3x^2 - 2x + 4)\,dx}$$

$$= \int \frac{du}{u}$$

$$= \ln|x^3 - x^2 + 4x - 4| + C$$

37. Answers vary; expand the integrands as a sum of partial fractions, then integrate each term separately.

39. $\displaystyle\int \frac{e^x\,dx}{2e^{2x} - 5e^x - 3}$ $\boxed{u = e^x;\ du = e^x\,dx}$

$$= \int \frac{du}{2u^2 - 5u - 3}$$

$$= \int \frac{du}{(2u+1)(u-3)}$$

Now, $\dfrac{1}{(2u+1)(u-3)} = \dfrac{A_1}{2u+1} + \dfrac{A_2}{u-3}$

Partial fraction decomposition gives
$A_1 = -\tfrac{2}{7}$, $A_2 = \tfrac{1}{7}$.

$$\int \frac{du}{(2u+1)(u-3)} = -\frac{1}{7}\int \frac{2\,du}{2u+1} + \frac{1}{7}\int \frac{du}{u-3}$$

$$= \frac{1}{7}\ln|u-3| - \frac{1}{7}\ln|2u+1| + C$$

$$= \frac{1}{7}\ln|e^x - 3| - \frac{1}{7}\ln(2e^x + 1) + C$$

41. $\displaystyle\int \frac{\sin x\,dx}{(1+\cos x)^2} = -\int (1+\cos x)^{-2}(-\sin x\,dx)$

$$\boxed{u = 1 + \cos x;\ du = -\sin x\,dx}$$

$$= -\int u^{-2}\,du$$

$$= \frac{1}{1+\cos x} + C$$

43. $\displaystyle\int \frac{\sec^2 x\,dx}{\tan x + 4}$ $\boxed{u = \tan x + 4;\ du = \sec^2 x\,dx}$

$$= \int \frac{du}{u}$$

$$= \ln|u| + C$$

$$= \ln|\tan x + 4| + C$$

45. $\displaystyle\int \frac{dx}{x^{2/3} - x^{1/2}} = \int \frac{6u^5 \, du}{u^4 - u^3}$

$$\boxed{x = u^6; \; dx = 6u^5 \, du}$$

$$= \int \frac{6u^2 \, du}{u - 1}$$

$$= 6\int \left[u + 1 + \frac{1}{u - 1} \right] du$$

$$= 6\left[\frac{u^2}{2} + u + \ln|u - 1| \right] + C$$

$$= 3x^{1/3} + 6x^{1/6} + 6\ln\left| x^{1/6} - 1 \right| + C$$

SURVIVAL HINT: *Check with your instructor concerning Weierstrass substitutions. Find out if you need to memorize them or if you will have access to them on exams.*

47. $\displaystyle\int \frac{dx}{3\cos x + 4\sin x}$ $\boxed{\text{Weierstrass substitution}}$

$$= \int \frac{\dfrac{2 \, du}{1 + u^2}}{3\left(\dfrac{1 - u^2}{1 + u^2}\right) + 4\left(\dfrac{2u}{1 + u^2}\right)}$$

$$= \int \frac{-2 \, du}{3u^2 - 8u - 3}$$

Now, $\displaystyle \frac{-2}{3u^2 - 8u - 3} = \frac{-2}{(u - 3)(3u + 1)}$

$$= \frac{A_1}{u - 3} + \frac{A_2}{3u + 1}$$

Partial fraction decomposition gives

$A_1 = -\frac{1}{5}, \; A_2 = \frac{3}{5}.$

$$\int \frac{-2 \, du}{3u^2 - 8u - 3} = -\frac{1}{5}\int \frac{du}{u - 3} + \frac{1}{5}\int \frac{3 \, du}{3u + 1}$$

$$= -\frac{1}{5}\ln|u - 3| + \frac{1}{5}\ln|3u + 1| + C$$

$$= -\frac{1}{5}\ln\left| \tan \frac{x}{2} - 3 \right| + \frac{1}{5}\ln\left| 3\tan \frac{x}{2} + 1 \right| + C$$

49. $\displaystyle\int \frac{\sin x - \cos x}{\sin x + \cos x} \, dx = -\int \frac{dt}{t} = -\ln|t| + C$

$$\boxed{t = \sin x + \cos x; \; dt = (\cos x - \sin x) \, dx}$$

$$= -\ln|\sin x + \cos x| + C$$

51. $\displaystyle\int \frac{dx}{\sec x - \tan x} = \int \frac{dx}{\dfrac{1}{\cos x} - \dfrac{\sin x}{\cos x}}$

$$= \int \frac{dx}{\dfrac{1 - \sin x}{\cos x}}$$

$$= \int \frac{\cos x \, dx}{1 - \sin x}$$

$$\boxed{u = 1 - \sin x; \; du = -\cos x \, dx}$$

$$= -\ln(1 - \sin x) + C$$

53. $\displaystyle\int \frac{dx}{4\sin x - 3\cos x - 5}$ $\boxed{\text{Weierstrass substitution}}$

$$= \int \frac{\dfrac{2 \, du}{1 + u^2}}{4\left(\dfrac{2u}{1 + u^2}\right) - 3\left(\dfrac{1 - u^2}{1 + u^2}\right) - 5}$$

$$= \int \frac{2 \, du}{8u - 3(1 - u^2) - 5(1 + u^2)}$$

$$= \int \frac{2\,du}{-2u^2 + 8u - 8}$$

$$= -\int \frac{du}{u^2 - 4u + 4}$$

$$= -\int \frac{du}{(u-2)^2}$$

$$= \frac{1}{u-2} + C$$

$$= \frac{1}{\tan\frac{x}{2} - 2} + C$$

55. $\displaystyle\int \frac{dx}{x(3 - \ln x)(1 - \ln x)} = \int \frac{dt}{(3 - t)(1 - t)}$

$$\boxed{t = \ln x;\ dt = x^{-1}dx}$$

Now, $\displaystyle\frac{1}{(3 - t)(1 - t)} = \frac{A_1}{t - 3} + \frac{A_2}{t - 1}$

Partial fraction decomposition gives $A_1 = \frac{1}{2}$,

$A_2 = -\frac{1}{2}$.

$$\int \frac{dt}{(3 - t)(1 - t)} = \frac{1}{2}\int \frac{dt}{t - 3} - \frac{1}{2}\int \frac{dt}{t - 1}$$

$$= \frac{1}{2}\ln|t - 3| - \frac{1}{2}\ln|t - 1| + C$$

$$= \frac{1}{2}\ln|\ln x - 3| - \frac{1}{2}\ln|\ln x - 1| + C$$

57. $\displaystyle\frac{1}{x^2 - 5x + 6} = \frac{A_1}{x - 3} + \frac{A_2}{x - 2}$

Partial fraction decomposition gives
$A_1 = 1,\ A_2 = -1$

$$A = \int_{4/3}^{7/4} \frac{dx}{x^2 - 5x + 6}$$

$$= \int_{4/3}^{7/4} \frac{dx}{(x - 3)(x - 2)}.$$

$$= \int_{4/3}^{7/4} \frac{1}{x - 3}\,dx - \int_{4/3}^{7/4} \frac{dx}{x - 2}$$

$$= \left[\ln|x - 3| - \ln|x - 3|\right]\Big|_{4/3}^{7/4}$$

$$= \ln 2$$

$$\approx 0.6931$$

SURVIVAL HINT: *You need to develop an ability to know when technology is appropriate and when it is not. The following problem presents definite integrals which have fairly complicated antiderivatives, so you should use technology to give you approximate values.*

59. We use technology to evaluate these definite integrals.

a. $\displaystyle V = \pi \int_0^3 \left(\frac{1}{\sqrt{x^2 + 4x + 3}}\right)^2 dx$

$$\approx 1.0888$$

b. $\displaystyle V = 2\pi \int_0^3 \frac{x\,dx}{\sqrt{x^2 + 4x + 3}}$

$$\approx 7.6402$$

61. Substitution:

$$\int \frac{x\,dx}{x^2 - 9} \qquad \boxed{u = x^2 - 9}$$

$$= \frac{1}{2}\int \frac{du}{u} = \frac{1}{2}\ln|x^2 - 9| + C$$

Partial fractions:

$$\int \frac{x\,dx}{x^2 - 9} = \int \frac{dx}{2(x-3)} + \int \frac{dx}{2(x+3)}$$

$$= \frac{1}{2}\ln|x - 3| + \frac{1}{2}\ln|x + 3| + C$$

$$= \frac{1}{2}\ln|x^2 - 9| + C$$

Trigonometric substitution:

$$\int \frac{x\,dx}{x^2 - 9} \qquad \boxed{x = 3\sin\theta}$$

$$= \int \frac{3\sin\theta\,(3\cos\theta\,d\theta)}{9\sin^2\theta - 9}$$

$$= -\int \frac{\sin\theta}{\cos\theta}\,d\theta$$

$$= -\int \tan\theta\,d\theta$$

$$= \ln|\cos\theta| + C$$

$$= \ln\sqrt{9 - x^2} + C$$

$$= \frac{1}{2}\ln|x^2 - 9| + C$$

63. a. $\displaystyle \int k\,dt = \int \frac{dx}{(a-x)(b-x)}; \; x(0) = 0$

$$= \frac{1}{a - b}\int \left[\frac{1}{x - a} - \frac{1}{x - b}\right] dx$$

$$kt = \frac{1}{a - b}\Big[\ln|x - a| - \ln|x - b|\Big] + C_1$$

$$(a - b)kt = \ln\left|\frac{x - a}{x - b}\right| + C_1$$

$x(0) = 0$, so $0 = \ln\left|\frac{a}{b}\right| + C_1$, and

$$(a - b)kt = \ln\left|\frac{x - a}{x - b}\right| - \ln\left|\frac{a}{b}\right| = \ln\left|\frac{b(x - a)}{a(x - b)}\right|$$

Solving for $x(t)$, we obtain

$$x(t) = \frac{ab[\exp(b - a)kt - 1]}{[b\exp(b - a)kt] - a}$$

If $b < a$, then $e^{(b - a)kt} \to 0$ as $t \to +\infty$ and

$$x(t) \to \frac{ab[0 - 1]}{0 - a} = b$$

If $a < b$, then $e^{(b - a)kt} \to \infty$ as $t \to +\infty$ and

$$x(t) \to \frac{ab}{b} = a$$

b. If $a = b$,

$$\int k\,dt = \int \frac{dx}{(a - x)^2}$$

$$kt = \frac{1}{a - x} + C$$

and since $x(0) = 0$, $0 = \frac{1}{a} + C$; $C = -\frac{1}{a}$, so

$$kt = \frac{1}{a - x} - \frac{1}{a}$$

$$x(t) = \frac{a^2 kt}{1 + akt}$$

As $t \to +\infty$, $x(t) \to a$.

65. $\dfrac{1}{x(ax + b)} = \dfrac{A_1}{x} + \dfrac{A_2}{ax + b}$

Partial fraction decomposition gives
$A_1 = 1/b$, $A_2 = -a/b$.

$$\int \frac{dx}{x(ax + b)} = \frac{1}{b}\int \left[\frac{1}{x} - \frac{a}{ax + b}\right] dx$$

$$= \frac{1}{b}\Big[\ln|x| - \ln|ax + b|\Big] + C$$

$$= \frac{1}{b}\ln\left|\frac{x}{ax + b}\right| + C$$

67. $\displaystyle\int \sec x \, dx = \int \frac{dx}{\cos x}$

$\boxed{\text{Weierstrass substitution}}$

$$= \int \frac{\dfrac{2\,du}{1 + u^2}}{\dfrac{1 - u^2}{1 + u^2}}\, du = \int \frac{2\,du}{1 - u^2}$$

$$= \int \left(\frac{1}{1 + u} + \frac{1}{1 - u}\right) du$$

$$= \ln|1 + u| - \ln|1 - u| + C$$

$$= \ln\left|\frac{1 + u}{1 - u}\right| + C = \ln\left|\frac{1 + \tan\frac{x}{2}}{1 - \tan\frac{x}{2}}\right| + C$$

$$= \ln\left|\frac{\cos\frac{x}{2} + \sin\frac{x}{2}}{\cos\frac{x}{2} - \sin\frac{x}{2}}\right| + C$$

$$= \ln\left|\frac{\left(\cos\frac{x}{2} + \sin\frac{x}{2}\right)^2}{\cos^2\frac{x}{2} - \sin^2\frac{x}{2}}\right| + C$$

$$= \ln\left|\frac{1 + 2\sin\frac{x}{2}\cos\frac{x}{2}}{\cos x}\right| + C$$

$$= \ln\left|\frac{1 + \sin x}{\cos x}\right| + C$$

$$= \ln|\sec x + \tan x| + C$$

7.5 Summary of Integration Techniques, Pages 460–461

SURVIVAL HINT: *It would take an unreasonable amount of time to do all the problems in this problem set. However, it would be an excellent use of time for you, or a small study group, to look at each problem and decide how to proceed. Identify one or more techniques that look promising.*

1. $\displaystyle\int \frac{2x - 1}{(x - x^2)^3}\, dx$ $\boxed{u = x - x^2;\ du = (1 - 2x)\,dx}$

$$= -\int u^{-3}\, du = \frac{1}{2}u^{-2} + C$$

$$= \frac{1}{2(x - x^2)^2} + C$$

$$= \frac{1}{2x^2(x - 1)^2} + C$$

3. $\displaystyle\int (x \sec 2x^2)\, dx$ $\boxed{u = 2x^2;\ du = 4x\,dx}$

$$= \frac{1}{4}\int \sec u \, du = \frac{1}{4}\ln|\sec u + \tan u| + C$$

$$= \tfrac{1}{4}\ln\left|\sec 2x^2 + \tan 2x^2\right| + C$$

$$= \tfrac{1}{2}\tan^{-1}e^{2t} + C$$

5. $\displaystyle\int (e^x \cot e^x)\, dx$ $\boxed{u = e^x;\ du = e^x\, dx}$

$$= \int \cot u\, du$$

$$= \ln|\sin e^x| + C$$

7. $\displaystyle\int \frac{\tan(\ln x)\, dx}{x}$ $\boxed{u = \ln x;\ du = x^{-1}\, dx}$

$$= \int \tan u\, du$$

$$= -\ln|\cos u| + C$$

$$= -\ln\left|\cos(\ln x)\right| + C$$

9. $\displaystyle\int \frac{(3 + 2\sin t)}{\cos t}\, dt$

$$= \int 3\sec t\, dt + \int 2\tan t\, dt$$

$$= 3\ln|\sec t + \tan t| + 2\ln|\sec t| + C$$

11. $\displaystyle\int \frac{e^{2t}\, dt}{1 + e^{4t}}$ $\boxed{u = e^{2t};\ du = 2e^{2t}\, dt}$

$$= \tfrac{1}{2}\int \frac{du}{1 + u^2}$$

$$= \tfrac{1}{2}\tan^{-1}u + C$$

13. $\displaystyle\int \frac{x^2 + x + 1}{x^2 + 9}\, dx$ *Long division*

$$= \int\left[1 + \frac{x - 8}{x^2 + 9}\right] dx \quad \textit{Use partial fractions}$$

$$= \int dx + \int \frac{x\, dx}{x^2 + 9} - 8\int \frac{dx}{x^2 + 3^2}$$

$$\boxed{u = x^2 + 9;\ du = 2x\, dx}$$

$$= x + \tfrac{1}{2}\ln(x^2 + 9) - \tfrac{8}{3}\tan^{-1}\tfrac{x}{3} + C$$

15. $\displaystyle\int \frac{1 + e^x}{1 - e^x}\, dx$ *Long division*

$$= \int\left[-1 + \frac{2}{1 - e^x}\right] dx$$

$$= \int\left[-1 + \frac{2e^{-x}}{e^{-x} - 1}\right] dx \quad \boxed{\begin{array}{l}\text{Let } u = e^{-x} - 1 \\ du = -e^{-x}\, dx\end{array}}$$

$$= -x + 2\int \frac{-du}{u}$$

$$= -x - 2\ln|e^{-x} - 1| + C$$

17. $\displaystyle\int \frac{2t^2\, dt}{\sqrt{1 - t^6}}$ $\boxed{u = t^3;\ du = 3t^2\, dt}$

$$= \tfrac{2}{3}\int \frac{du}{\sqrt{1 - u^2}}$$

$$= \tfrac{2}{3}\sin^{-1}u + C$$

$$= \tfrac{2}{3} \sin^{-1} t^3 + C$$

19. $\displaystyle \int \frac{dx}{1 + e^{2x}}$

$$= \int \frac{e^{-2x}\, dx}{e^{-2x} + 1} \quad \boxed{u = e^{-2x} + 1; \; du = -2e^{-2x}\, dx}$$

$$= -\tfrac{1}{2} \int \frac{du}{u}$$

$$= -\tfrac{1}{2} \ln|u| + C$$

$$= -\tfrac{1}{2} \ln\left| e^{-2x} + 1 \right| + C$$

$$= -\tfrac{1}{2} \ln\left| \frac{1 + e^{2x}}{e^{2x}} \right| + C$$

$$= -\tfrac{1}{2}\left[\ln(1 + e^{2x}) - \ln e^{2x} \right] + C$$

$$= \tfrac{1}{2}(2x) - \tfrac{1}{2}\ln(1 + e^{2x}) + C$$

$$= x - \tfrac{1}{2} \ln(e^{2x} + 1) + C$$

21. $\displaystyle \int \frac{dx}{x^2 + 2x + 2}$

$$= \int \frac{dx}{(x + 1)^2 + 1}$$

$$= \tan^{-1}(x + 1) + C$$

23. $\displaystyle \int \frac{dx}{x^2 + x + 1} = \int \frac{dx}{(x + \tfrac{1}{2})^2 + \tfrac{3}{4}}$

$$\boxed{u = x + \tfrac{1}{2}; \; du = dx}$$

$$= \int \frac{du}{u^2 + \tfrac{3}{4}}$$

$$= \frac{2}{\sqrt{3}} \tan^{-1} \frac{2u}{\sqrt{3}} + C$$

$$= \frac{2}{\sqrt{3}} \tan^{-1} \frac{2}{\sqrt{3}}(x + \tfrac{1}{2}) + C$$

$$= \frac{2}{\sqrt{3}} \tan^{-1} \frac{1}{\sqrt{3}}(2x + 1) + C$$

25. $\displaystyle \int \tan^{-1} x \, dx = x\tan^{-1} x - \tfrac{1}{2} \ln(x^2 + 1) + C$

$$\boxed{\text{Formula 457}}$$

27. $\displaystyle \int e^{-x} \cos x \, dx \quad \boxed{\text{Parts or Formula 493}}$

$$= \frac{e^{-x}(-\cos x + \sin x)}{2} + C$$

$$= \tfrac{1}{2} e^{-x}(\sin x - \cos x) + C$$

29. $\displaystyle \int \cos^{-1}(-x) \, dx \quad \boxed{\text{Parts or Formula 445}}$

$$= -\left[(-x) \cos^{-1}(-x) - \sqrt{1 - x^2} \right] + C$$

$$= x \cos^{-1}(-x) + \sqrt{1 - x^2} + C$$

$$\text{or } x \sin^{-1} x + \tfrac{\pi x}{2} + \sqrt{1 - x^2} + C$$

31. $\displaystyle\int \sin^3 x\ dx$

$$= \int \sin^2 x \sin x\ dx$$

$$= -\int (1 - \cos^2 x)(-\sin x)\ dx \quad \boxed{u = \cos x}$$

$$= -\cos x + \tfrac{1}{3}\cos^3 x + C$$

33. $\displaystyle\int \sin^3 x \cos^2 x\ dx$

$$= -\int (1 - \cos^2 x)\cos^2 x(-\sin x\ dx)$$

$$= \int (\cos^4 x - \cos^2 x)(-\sin x\ dx) \quad \boxed{u = \cos x}$$

$$= \tfrac{1}{5}\cos^5 x - \tfrac{1}{3}\cos^3 x + C$$

35. $\displaystyle\int \sin^2 x \cos^4 x\ dx$

$$= \tfrac{1}{8}\int (1 - \cos 2x)(1 + \cos 2x)^2\ dx$$

$$= \tfrac{1}{8}\int [(1 - \cos 2x)(1 + \cos 2x)](1 + \cos 2x)\ dx$$

$$= \tfrac{1}{8}\int (1 - \cos^2 2x)(1 + \cos 2x)\ dx$$

$$= \tfrac{1}{8}\int \sin^2 2x(1 + \cos 2x)\ dx$$

$$= \tfrac{1}{8}\int \sin^2 2x\ dx + \tfrac{1}{8}\int \sin^2 2x\ (\cos 2x)\ dx$$

Sine squared formula

$$\boxed{u = \sin 2x;\ du = 2\cos 2x\ dx}$$

$$= \tfrac{1}{16}\Big[x - \tfrac{1}{4}\sin 4x\Big] + \tfrac{1}{16}\cdot\tfrac{1}{3}\sin^3 2x + C$$

$$= \tfrac{1}{16}x - \tfrac{1}{64}\sin 4x + \tfrac{1}{48}\sin^3 2x + C$$

37. $\displaystyle\int \sin^5 x \cos^4 x\ dx$

$$= -\int (1 - \cos^2 x)^2 \cos^4 x(-\sin x\ dx)$$

$$= -\int (\cos^8 x - 2\cos^6 x + \cos^4 x)(-\sin x\ dx)$$

$$\boxed{u = \cos x;\ du = -\sin x\ dx}$$

$$= -\tfrac{1}{9}\cos^9 x + \tfrac{2}{7}\cos^7 x - \tfrac{1}{5}\cos^5 x + C$$

39. $\displaystyle\int \tan^5 x \sec^4 x\ dx = \int \tan^5 x(\tan^2 x + 1)\sec^2 x\ dx$

$$= \int \tan^7 x \sec^2 x\ dx + \int \tan^5 x \sec^2 x\ dx$$

$$\boxed{u = \tan x;\ du = \sec^2 x\ dx}$$

$$= \tfrac{1}{8}\tan^8 x + \tfrac{1}{6}\tan^6 x + C$$

41. $\int \frac{\sqrt{1-x^2}}{x} dx$ $\boxed{\text{Formula 235 or } x = \sin\theta}$

$$= \sqrt{1-x^2} - \ln\left|\frac{1+\sqrt{1-x^2}}{x}\right| + C$$

43. $\int \frac{2x+3}{\sqrt{2x^2-1}} dx = \int \frac{2x \, dx}{\sqrt{2x^2-1}} + \int \frac{3 \, dx}{\sqrt{2x^2-1}}$

$$= \frac{1}{2}\int \frac{4x \, dx}{\sqrt{2x^2-1}} + \frac{3}{\sqrt{2}}\int \frac{dx}{\sqrt{x^2-\frac{1}{2}}}$$

$$\boxed{u = 2x^2 - 1; \ du = 4x \, dx} \quad \boxed{\text{Formula 196}}$$

$$= \frac{1}{2}\int u^{-1/2} \, du + \frac{3}{\sqrt{2}} \ln\left|x + \sqrt{x^2 - \frac{1}{2}}\right| + C_1$$

$$= \frac{1}{2}\cdot\frac{2}{1} u^{1/2} + \frac{3}{\sqrt{2}} \ln\left|\sqrt{2}\,x + \sqrt{2x^2 - 1}\right| + C$$

$$= \sqrt{2x^2 - 1} + \frac{3}{\sqrt{2}} \ln\left|\sqrt{2}\,x + \sqrt{2x^2 - 1}\right| + C$$

45. $\int \frac{dx}{x\sqrt{x^2+1}}$ $\boxed{\text{Formula 176 or } x = \tan\theta}$

$$= -\ln\left|\frac{1+\sqrt{x^2+1}}{x}\right| + C$$

$$= \ln|x| - \ln\left|1 + \sqrt{x^2+1}\right| + C$$

47. $\int \frac{(2x+1)\,dx}{\sqrt{4x-x^2-2}}$ *Partial fractions*

$$= -\int \frac{(-2x+4)\,dx}{\sqrt{4x-x^2-2}} + 5\int \frac{dx}{\sqrt{2-(x-2)^2}}$$

$$\boxed{u = 4x - x^2 - 2; \ du = (-2x+4)\,dx}$$

$$\boxed{t = x - 2; \ dt = dx}$$

$$= -\int u^{-1/2} \, du + 5\int \frac{dt}{\sqrt{2-t^2}}$$

$$= -2u^{1/2} + 5 \sin^{-1}\frac{t}{\sqrt{2}} + C$$

$$= -2\sqrt{4x - x^2 - 2} + 5 \sin^{-1}\frac{x-2}{\sqrt{2}} + C$$

49. $\int \frac{\cos x \, dx}{\sqrt{1+\sin^2 x}}$ $\boxed{u = \sin x; \ du = \cos x \, dx}$

$$= \int \frac{du}{\sqrt{1+u^2}}$$ $\boxed{\text{Formula 172 or } u = \tan\theta}$

$$= \ln\left|u + \sqrt{1+u^2}\right| + C$$

$$= \ln\left(\sin x + \sqrt{1+\sin^2 x}\right) + C$$

51. $\int \sin^5 x \, dx = -\int (\sin^4 x)(-\sin x \, dx)$

$$= -\int (1 - \cos^2 x)^2 (-\sin x \, dx)$$

$$= -\int (\cos^4 x - 2\cos^2 x + 1)(-\sin x \, dx)$$

$$\boxed{u = \cos x; \ du = -\sin x \, dx}$$

$$= -\frac{\cos^5 x}{5} + \frac{2\cos^3 x}{3} - \cos x + C$$

53. $\int \tan^4 x \, dx = \int (\sec^2 x - 1)\tan^2 x \, dx$

$$= \int \tan^2 x \sec^2 x \, dx - \int (\sec^2 x - 1) \, dx$$

$$\boxed{u = \tan x; \ du = \sec^2 x \, dx}$$

$$= \tfrac{1}{3}\tan^3 x - \tan x + x + C$$

$$= \tfrac{1}{9}\Big[-\tfrac{1}{3}\csc^3\theta + \csc\theta \Big]\Big|_{x=1}^{x=2}$$

55. $\displaystyle\int_0^2 \sqrt{4 - x^2}\, dx$ $\boxed{\text{Formula 231 or } x = 2\sin\theta}$

$$= \tfrac{1}{9}\Big(\frac{\sqrt{3 + x^2}}{x} \Big)\Big[-\tfrac{1}{3}\frac{3 + x^2}{x^2} + 1 \Big]\Big|_1^2$$

$$= \frac{x\sqrt{4 - x^2}}{2} + \frac{4}{2}\sin^{-1}\frac{x}{2}\Big|_0^2$$

$$= \tfrac{1}{9}\Big[\frac{\sqrt{7}}{2}\Big(-\tfrac{1}{3}\cdot\tfrac{7}{4} + 1 \Big) - 2\Big(-\tfrac{1}{3}\cdot 4 + 1 \Big) \Big]$$

$$= \pi$$

$$= \tfrac{1}{27}\big(\tfrac{5}{8}\sqrt{7} + 2\big) \approx 0.1353$$

57. $\displaystyle\int_0^{\ln 2} e^t\sqrt{1 + e^{2t}}\, dt = \int_1^2 \sqrt{1 + u^2}\, du$ $\boxed{\text{Let } u = e^t}$

61. $\displaystyle\int_{-2}^{2\sqrt{3}} x^3\sqrt{x^2 + 4}\, dx$ $\boxed{\text{Formula 171 or } x = 2\tan\theta}$

$$= \frac{u\sqrt{1 + u^2}}{2} + \tfrac{1}{2}\ln\big(u + \sqrt{1 + u^2}\big)\Big|_1^2$$

$$= \Big[\frac{(x^2 + 4)^{5/2}}{5} - \frac{4(x^2 + 4)^{3/2}}{3} \Big]\Big|_{-2}^{2\sqrt{3}}$$

$$\boxed{\text{Formula 168 or } u = \tan\theta}$$

$$= \tfrac{1}{2}\ln(\sqrt{5} + 2) + \tfrac{1}{2}\ln(\sqrt{2} - 1) + \sqrt{5} - \frac{\sqrt{2}}{2}$$

$$= \frac{1{,}792 - 64\sqrt{2}}{15}$$

$$\approx 1.8101$$

$$\approx 113.4327$$

59. $\displaystyle\int_1^2 \frac{dx}{x^4\sqrt{x^2 + 3}} = \int_{x=1}^{x=2} \frac{\sqrt{3}\,\sec^2\theta\, d\theta}{9\tan^4\theta\,\sqrt{3}\,\sec\theta}$

63. $\displaystyle\int \frac{e^x\, dx}{\sqrt{1 + e^{2x}}} = \int \frac{du}{\sqrt{1 + u^2}}$

$$\boxed{u = e^x;\ du = e^x dx}$$

$$\boxed{x = \sqrt{3}\tan\theta,\ dx = \sqrt{3}\sec^2\theta\, d\theta}$$

$$= \ln(\sqrt{u^2 + 1} + u) + C$$

$$= \tfrac{1}{9}\int_{x=1}^{x=2} \frac{(1 - \sin^2\theta)(\cos\theta\, d\theta)}{\sin^4\theta}$$

$$= \ln\big(\sqrt{1 + e^{2x}} + e^x\big) + C$$

65. $\displaystyle\int \frac{x^2 + 4x + 3}{x^3 + x^2 + x}\, dx = \int \frac{x^2 + 4x + 3}{x(x^2 + x + 1)}\, dx$

$$\frac{x^2 + 4x + 3}{x(x^2 + x + 1)} = \frac{A_1}{x} + \frac{A_2 x + B_1}{x^2 + x + 1}$$

Partial fraction decomposition gives

$A_1 = 3, A_2 = -2, B_1 = 1.$

$$\int \frac{x^2 + 4x + 3}{x(x^2 + x + 1)} \, dx = 3 \int x^{-1} dx + \int \frac{-2x + 1}{x^2 + x + 1} \, dx$$

Since $-2x + 1 = -2x - 1 + 2 = -(2x+1) + 2$

$$= 3 \int x^{-1} dx - \int \frac{2x + 1}{x^2 + x + 1} \, dx + 2 \int \frac{dx}{x^2 + x + 1}$$

$$= 3 \int x^{-1} dx - \int \frac{2x + 1}{x^2 + x + 1} \, dx + 2 \int \frac{dx}{(x + \frac{1}{2})^2 + \frac{3}{4}}$$

$$\boxed{u = x^2 + x + 1; \ du = (2x + 1) \, dx}$$

$$= 3 \ln|x| - \ln|x^2 + x + 1|$$

$$+ 2\left(\frac{2}{\sqrt{3}}\right)\left(\tan^{-1} \frac{2(x + \frac{1}{2})}{\sqrt{3}}\right) + C$$

$$= 3 \ln|x| - \ln|x^2 + x + 1|$$

$$+ \frac{4}{\sqrt{3}} \tan^{-1} \frac{\sqrt{3}}{3}(2x + 1) + C$$

67. $\displaystyle\int \frac{5x^2 + 18x + 34}{(x - 7)(x + 2)^2} \, dx$

$$\frac{5x^2 + 18x + 34}{(x - 7)(x + 2)^2} = \frac{A_1}{x - 7} + \frac{A_2}{x + 2} + \frac{A_3}{(x + 2)^2}$$

Partial fraction decomposition gives

$A_1 = 5, A_2 = 0, A_3 = -2$

$$\int \frac{5x^2 + 18x + 34}{(x - 7)(x + 2)^2} \, dx = 5 \int \frac{dx}{x - 7} - 2 \int \frac{dx}{(x + 2)^2}$$

$$= 5 \ln|x - 7| + 2(x + 2)^{-1} + C$$

69. $\displaystyle\int \frac{3x + 5}{x^2 + 2x + 1} \, dx$ *Note:* $3x + 5 = 3x + 3 + 2.$

$$= 3 \int \frac{x + 1}{(x + 1)^2} \, dx + 2 \int \frac{dx}{(x + 1)^2}$$

$$= 3 \ln|x + 1| - 2(x + 1)^{-1} + C$$

71. $\displaystyle\int \frac{x \, dx}{(x + 1)(x + 2)(x + 3)}$

$$\frac{x}{(x + 1)(x + 2)(x + 3)} = \frac{A_1}{x + 1} + \frac{A_2}{x + 2} + \frac{A_3}{x + 3}$$

Partial fraction decomposition gives

$A_1 = -\frac{1}{2}, A_2 = 2; A_3 = -\frac{3}{2}.$

$$\int \frac{x \, dx}{(x + 1)(x + 2)(x + 3)}$$

$$= -\frac{1}{2} \int \frac{dx}{x + 1} + 2 \int \frac{dx}{x + 2} - \frac{3}{2} \int \frac{dx}{x + 3}$$

$$= -\frac{1}{2} \ln|x + 1| + 2 \ln|x + 2| - \frac{3}{2} \ln|x + 3| + C$$

73. $\displaystyle \text{AV} = \frac{1}{1 - 0} \int_0^1 x \sin^3 x^2 \, dx$

$$= \tfrac{1}{2} \int_0^1 (1 - \cos^2 x^2) \sin x^2 (2x \ dx)$$

$$= \tfrac{1}{2} \int_0^1 \sin x^2 (2x \ dx) - \tfrac{1}{2} \int_0^1 \cos^2 x^2 \sin x^2 (2x \ dx)$$

$$\boxed{u = x^2; \ du = 2x \ dx} \ \boxed{w = \cos x^2; \ dw = -\sin x^2 (2x \ dx)}$$

$$= \tfrac{1}{2} \left[-\cos x^2 + \frac{\cos^3 x^2}{3} \right] \Big|_0^1$$

$$= \tfrac{1}{6} \cos^3 1 - \tfrac{1}{2} \cos 1 + \tfrac{1}{2} - \tfrac{1}{6}$$

$$\approx 0.09$$

75. $I = \displaystyle\int \cot^m x \ \csc^n x \ dx$

   a. m is odd; separate out

      $\csc x \cot x \ dx$ and then use the identity

      $\cot^2 x = \csc^2 x - 1$ and the substitution

      $u = \csc x$.

   b. n is even; separate out $\csc^2 x \ dx$ and use

      the identity $\cot^2 x + 1 = \csc^2 x$ and the

      substitution $u = \cot x$.

   c. m is even and n is odd; use the identity

      $\cot^2 x + 1 = \csc^2 x$ to convert the

      integrand into a power of $\csc x$. Then use

      reduction formula 438.

77. Since $y' = 2x$, the arc length is given by

$$s = \int_{-1}^1 \sqrt{1 + (2x)^2} \ dx$$

$$\boxed{\text{Let } 2x = \tan \theta, \ 2 \ dx = \sec^2 \theta \ d\theta}$$

$$\int \sqrt{1 + (2x)^2} \ dx = \int \sqrt{1 + \tan^2 \theta} \ (\tfrac{1}{2} \sec^2 \theta \ d\theta)$$

$$= \tfrac{1}{2} \int \sec^3 \theta \ d\theta$$

$$= \tfrac{1}{2} \left[\frac{\sec \theta \tan \theta}{2} + \tfrac{1}{2} \ln |\sec \theta + \tan \theta| \right] + C$$

Formula 427

$$= \tfrac{1}{4} \sqrt{1 + 4x^2} (2x) + \tfrac{1}{4} \ln \left| \sqrt{1 + 4x^2} + 2x \right| + C$$

Thus, the arc length is

$$s = \int_{-1}^1 \sqrt{1 + (2x)^2} \ dx$$

$$= \left[\tfrac{1}{4}(2x)\sqrt{1 + 4x^2} + \tfrac{1}{4} \ln \left| \sqrt{1 + 4x^2} + 2x \right| \right] \Big|_{-1}^1$$

$$\approx 2.9579$$

79. $A = \displaystyle\int_0^4 \frac{2x \ dx}{\sqrt{x^2 + 9}}$ $\boxed{u = x^2 + 9; \ du = 2x \ dx}$

$$= 2\sqrt{x^2 + 9} \ \Big|_0^4$$

$$= 4$$

81.

$$V = \pi \int_0^3 \left(x \sqrt{9 - x^2} \right)^2 \, dx$$

$$= \pi \int_0^3 (9x^2 - x^4) \, dx$$

$$= \pi \left[3x^3 - \frac{x^5}{5} \right] \Big|_0^3$$

$$= \frac{162\pi}{5}$$

$$\approx 101.7876$$

83. With centroid it is assumed $\rho = 1$;

$$m = \int_0^1 x^2 e^{-x} dx$$

$$= \left[-x^2 e^{-x} - 2xe^{-x} - 2e^{-x} \right] \Big|_0^1$$

$$= 2 - 5e^{-1}$$

$$\approx 0.1606$$

$$M_y = \int_0^1 x(x^2 e^{-x}) \, dx$$

$$= 6 - 16e^{-1}$$

$$\approx 0.1139$$

$$M_x = \frac{1}{2} \int_0^1 (x^2 e^{-x})^2 \, dx$$

$$= \frac{3}{8} - \frac{21e^{-2}}{8}$$

$$\approx 0.0197$$

Since $\overline{x} = \frac{1}{m} M_y$ and $\overline{y} = \frac{1}{m} M_x$, we have

$$(\overline{x}, \overline{y}) \approx (0.71, \, 0.12)$$

85. $\displaystyle\int_a^b xf''(x) \, dx$ $\boxed{u = x; \; dv = f''(x) \, dx}$

$$= xf'(x) \Big|_a^b - \int_a^b f'(x) \, dx$$

$$= bf'(b) - af'(a) - f(x) \Big|_a^b$$

$$= bf'(b) - af'(a) - f(b) + f(a)$$

87. $\displaystyle I = \int \sin^n Ax \, dx$

$$\boxed{u = \sin^{n-1} Ax; \; dv = \sin Ax \, dx}$$

$$= -\frac{1}{A} \sin^{n-1} Ax \cos Ax$$

$$+ (n-1) \int \sin^{n-2} Ax \cos^2 Ax \, dx$$

$$= -\frac{1}{A} \sin^{n-1} Ax \cos Ax$$

$$+ (n-1) \int \sin^{n-2} Ax (1 - \sin^2 Ax) \, dx$$

$$= -\frac{1}{A}\sin^{n-1}Ax\cos Ax$$

$$+ (n-1)\int \sin^{n-2}Ax\, dx$$

$$- (n-1)\int \sin^{n}Ax\, dx$$

$$= -\frac{1}{A}\sin^{n-1}Ax\cos Ax$$

$$+ (n-1)\int \sin^{n-2}Ax\, dx - (n-1)I$$

$$I + (n-1)I = -\frac{1}{A}\sin^{n-1}Ax\cos Ax$$

$$+ (n-1)\int \sin^{n-2}Ax\, dx$$

$$I = -\frac{1}{nA}\sin^{n-1}Ax\cos Ax$$

$$+ \frac{n-1}{n}\int \sin^{n-2}Ax\, dx + C$$

For example, if $n = 4$ and $A = 4$, we have

$$\int \sin^{4}4x\, dx = -\frac{1}{16}\sin^{3}4x\cos 4x + \frac{3}{4}\int \sin^{2}4x\, dx$$

$$= -\frac{1}{16}\sin^{3}4x\cos 4x + \frac{3}{4}\left(\frac{x}{2} - \frac{\sin 8x}{16}\right) + C$$

$$= -\frac{1}{16}\sin^{3}4x\cos 4x + \frac{3}{8}x - \frac{3}{64}\sin 8x + C$$

7.6 First-Order Differential Equations, Pages 469–472

1. $P(x) = \frac{3}{x};\ Q(x) = x;$

$$I(x) = e^{\int \frac{3}{x}\, dx} = e^{3\ln x} = e^{\ln x^{3}} = x^{3}$$

$$y = \frac{1}{x^{3}}\left(\int x(x^{3})\, dx + C\right) = \frac{1}{x^{3}}\left(\frac{x^{5}}{5} + C\right)$$

$$= \frac{1}{5}x^{2} + Cx^{-3}$$

3. Divide both sides by x^{4}: $\dfrac{dy}{dx} + \dfrac{2}{x}y = 5x^{-4}$

$$P(x) = \frac{2}{x};\ Q(x) = 5x^{-4};$$

$$I(x) = e^{\int \frac{2}{x}\, dx} = e^{2\ln x} = e^{\ln x^{2}} = x^{2}$$

$$y = \frac{1}{x^{2}}\left(\int x^{2}(5x^{-4})\, dx + C\right)$$

$$= \frac{1}{x^{2}}\left(-5x^{-1} + C\right)$$

$$= -5x^{-3} + Cx^{-2}$$

5. Divide both sides by x: $\dfrac{dy}{dx} + \dfrac{2}{x}y = e^{x^{3}}$

$$P(x) = \frac{2}{x};\ Q(x) = e^{x^{3}};$$

$$I(x) = e^{\int \frac{2}{x}\, dx} = e^{2\ln x} = x^{2}$$

$$y = \frac{1}{x^{2}}\left[\int x^{2}(e^{x^{3}})\, dx + C\right]$$

$$= \frac{1}{x^2}\left[\frac{1}{3}e^{x^3} + C\right]$$

$$= \frac{1}{3}x^{-2}e^{x^3} + Cx^{-2}$$

7. $P(x) = \frac{1}{x};\ Q(x) = \tan^{-1}x;$

$$I(x) = e^{\int \frac{1}{x}\,dx} = e^{\ln x} = x$$

$$y = \frac{1}{x}\left[\int x\tan^{-1}x\,dx + C\right]$$

$$\boxed{\text{Formula 458}}$$

$$= \left(\frac{1}{2}x + \frac{1}{2x}\right)\tan^{-1}x - \frac{1}{2} + \frac{C}{x}$$

9. $P(x) = \tan x;\ Q(x) = \sin x;$

$$I(x) = e^{\int \tan x\,dx} = e^{-\ln|\cos x|} = \frac{1}{\cos x}$$

$$y = \cos x\left[\int \sin x\,\frac{1}{\cos x}\,dx + C\right]$$

$$= \cos x\left[-\ln|\cos x| + C\right]$$

$$= -\cos x\,\ln|\cos x| + C\cos x$$

11. $P(x) = \frac{x}{1+x};\ Q(x) = x(1+x);$

$$I(x) = e^{\int \frac{x}{1+x}\,dx} = e^{x - \ln(x+1)} = \frac{e^x}{x+1}$$

$$y = \frac{x+1}{e^x}\left[\int x(1+x)\cdot\frac{e^x}{1+x}\,dx + C\right]$$

$$\boxed{\text{Formula 484}}$$

$$= \frac{x+1}{e^x}\left[e^x(x-1) + C\right]$$

$$= x^2 - 1 + C\left(\frac{x+1}{e^x}\right)$$

Since $y = -1$ when $x = 0$, $C = 0$; thus,

$$y = x^2 - 1 \text{ for } x > -1.$$

13. Divide both sides by x: $\frac{dy}{dx} - \frac{2}{x}y = 2x^2$

$$P(x) = -\frac{2}{x};\ Q(x) = 2x^2;$$

$$I(x) = e^{\int -\frac{2}{x}\,dx} = e^{-2\ln x} = x^{-2}$$

$$y = \frac{1}{x^{-2}}\left(\int (2x^2)(x^{-2})\,dx + C\right)$$

$$= x^2(2x + C)$$

$$= 2x^3 + Cx^2$$

Since $y = 0$ when $x = 3$, $0 = 2(27) + C(9)$ or $C = -6$; thus, $y = 2x^3 - 6x^2$

15. $$y^2 = 4kx$$

$$2yy' = 4k$$

$$y' = \frac{2k}{y} = \frac{2\frac{y^2}{4x}}{y} = \frac{y}{2x}$$

The slope of the orthogonal trajectory is the negative reciprocal:

$$\frac{dY}{dX} = -\frac{2X}{Y}$$

$$\int Y \, dY = \int -2X \, dX$$

$$\tfrac{1}{2}Y^2 = -X^2 + K$$

$$2X^2 + Y^2 = C$$

17.
$$x^2 + y^2 = r^2$$

$$2x + 2yy' = 0$$

$$y' = -\frac{x}{y}$$

The slope of the orthogonal trajectory is the negative reciprocal:

$$\frac{dY}{dX} = \frac{Y}{X}$$

$$\int Y^{-1} \, dY = \int X^{-1} \, dX$$

$$\ln Y = \ln X + K$$

$$Y = CX$$

19. Assume the growth rate for the period 1999 to 2010 remains at 6.42%.

$$Q(t) = 9{,}382 e^{0.0642t}$$

$$Q(10) = 9{,}382 e^{10(0.0642)}$$

$$\approx 17{,}828$$

The GDP in the year 2010 will be about $17,828 billion.

21. Assume that the divorce rate remains constant between 1990 and 2005.

$$Q(t) = 1{,}175 e^{0.047t}$$

$$Q(15) = 1{,}175 e^{15(0.047)}$$

$$\approx 2{,}378.02$$

The predicted number of divorces in 2005 is about 2,378,000.

23. a. Let $Q(t)$ be the amount of salt in the solution at time t (minutes). Then

$$\frac{dQ}{dt} = \underbrace{(1)(2)}_{\text{inflow}} - \underbrace{\frac{Q}{30}(2)}_{\text{outflow}}$$

$$\frac{dQ}{dt} + \frac{Q}{15} = 2$$

The integrating factor is $e^{\int 1/15 \, dt} = e^{t/15}$

$$Q(t) = e^{-t/15}\left[\int 2 e^{t/15} dt + C\right]$$

$$= e^{-t/15}\left[2\,\frac{e^{t/15}}{\frac{1}{15}} + C\right]$$

$$= 30 + C e^{-t/15}$$

$$Q(0) = 10 = 30 + C e^0 \text{ or } C = -20$$

$$Q(t) = 30 - 20 e^{-t/15}$$

b. When $Q = 15$, we have

$$15 = 30 - 20 e^{-t/15}$$

$$-\frac{t}{15} = \ln \frac{15}{20}$$

$$t = -15 \ln \frac{3}{4} \approx 4.31523 \text{ min}$$

This is about 4 minutes, 19 seconds.

25.
$$\frac{db}{dt} = \alpha - \beta b$$

$$\frac{db}{dt} + \beta b = \alpha$$

The integrating factor is $e^{\int \beta \, dt} = e^{\beta t}$

$$b = e^{-\beta t}\left[\int \alpha e^{\beta t} + C\right]$$

$$= e^{-\beta t}\left[\frac{\alpha e^{\beta t}}{\beta} + C\right]$$

$$= \frac{\alpha}{\beta} + Ce^{-\beta t}$$

$b(0) = 0$ implies $0 = \frac{\alpha}{\beta} + C$, so $C = -\frac{\alpha}{\beta}$

$$b(t) = \frac{\alpha}{\beta}(1 - e^{-\beta t})$$

In the "long run" (as $t \to +\infty$), the concentration will be

$$\lim_{t \to +\infty} \frac{\alpha}{\beta}(1 - e^{-\beta t}) = \frac{\alpha}{\beta}$$

The "half-way point" is reached when

$$b(t) = \frac{1}{2} \cdot \frac{\alpha}{\beta} = \frac{\alpha}{\beta}(1 - e^{-\beta t})$$

$$e^{-\beta t} = \frac{1}{2}$$

$$-\beta t = \ln \frac{1}{2}$$

$$t = \frac{\ln \frac{1}{2}}{-\beta}$$

$$= \frac{\ln 2}{\beta}$$

27. With the uninhibited growth model where t is the time after 1990 in years and $Q(t)$ is the number of Hispanics in year t, and $Q_0 = 16.1$.

$$Q(10) = 32.8 = 16.1e^{10k}$$

$$e^{10k} = \frac{32.8}{16.1}$$

$$k = \frac{1}{10} \ln\left(\frac{32.8}{16.1}\right)$$

$$\approx 0.0711609243$$

Thus,

$$Q(15) = 16.1e^{15k}$$

$$\approx 46.816$$

We would expect the Hispanic population to be about 46.8 million in 2005.

29.
$$\frac{dM}{dt} = r\left(\frac{k}{r} - M\right)$$

$$\frac{dM}{\frac{k}{r} - M} = r\, dt$$

$$\int \frac{dM}{\frac{k}{r} - M} = \int r\, dt$$

$$-\ln\left|\frac{k}{r} - M\right| = rt + K$$

$$\frac{k}{r} - M = e^{-rt}e^{-K} = Ce^{-rt}$$

$$M(t) = \frac{k}{r} - Ce^{-rt}$$

$$M(0) = 0 = \frac{k}{r} - C, \text{ so } C = \frac{k}{r}$$

Thus, $M(t) = \frac{k}{r}(1 - e^{-rt})$

31.
$$V = 5h$$

$$\frac{dV}{dt} = 5\frac{dh}{dt}$$

$$-4.8(0.07)\sqrt{h} = 5\frac{dh}{dt} \qquad \textit{by Torricelli's law}$$

$$-0.0672\, dt = \frac{1}{\sqrt{h}}\, dh$$

$$\int -0.0672\, dt = \int \frac{1}{\sqrt{h}}\, dh$$

$$2h^{1/2} + C = -0.0672\, t$$

When $t = 0$, $h = 4$, so $4 + C = 0$ or $C = -4$.

$$2h^{1/2} - 4 = -0.0672t$$

When $h = 0$, $t \approx 59.5$.

It will take about an hour to drain.

33. a. We have $y^{(4)} = -k$ with $y(0) = y(L) = 0$

$$y^{(4)} = -k$$

$$y''' = -kx + C_1$$

$$y'' = -\frac{k}{2}x^2 + C_1 x + C_2$$

$$y' = -\frac{k}{6}x^3 + \frac{C_1}{2}x^2 + C_2 x + C_3$$

$$y = -\frac{k}{24}x^4 + \frac{C_1}{6}x^3 + \frac{C_2}{2}x^2 + C_3 x + C_4$$

Since $y(0) = 0$, and $y''(0) = y''(L) = 0$, we have $C_4 = 0$ and $C_2 = 0$, and the conditions $y(L) = 0$ and $y''(L) = 0$ tell us that

$$-\frac{kL^4}{24} + \frac{C_1 L^3}{6} + C_3 L = 0$$

$$-\frac{kL^2}{2} + C_1 L = 0$$

$$C_1 = \frac{kL}{2}$$

Thus, $C_3 = -\frac{kL^3}{24}$, so

$$y = -\frac{k}{24}(x^4 - 2Lx^3 + L^3 x)$$

b. Maximum deflection occurs where $y' = 0$. Solve

$$y' = -\frac{k}{24}[4x^3 - 6Lx^2 + L^3] = 0$$

to obtain $x = L/2$; reject $(1 \pm \sqrt{3})L/2$

because these points lie outside the interval $[0, L]$. This is the maximum deflection since y is minimized at $x = L/2$; note $y''(L/2) = kL^2/8 > 0$. The maximum deflection is

$$y_m = y(L/2) = -\frac{k}{24}\left[\frac{L^4}{16} - 2L\left(\frac{L^3}{8}\right) + L^3\left(\frac{L}{2}\right)\right]$$

$$\approx -0.0130kL^4$$

c. For a cantilevered beam, we have

$$y = -\frac{k}{24}x^4 + \frac{C_1}{6}x^3 + \frac{C_2}{2}x^2 + C_3x + C_4$$

with boundary conditions $y(0) = y(L) = 0$ and $y''(0) = y'(L) = 0$. The conditions $y(0) = 0$ and $y''(0) = 0$ imply $C_2 = C_4 = 0$, and the other two conditions yield $C_1 = \frac{3}{8}kL$ and

$C_3 = -\frac{1}{48}kL^3$. Thus,

$$y = -\frac{k}{48}(2x^4 - 3Lx^3 + L^3x)$$

To find where maximum deflection occurs, we find

$$y' = -\frac{k}{48}(8x^3 - 9Lx^2 + L^3)$$

$$y' = 0 \text{ when } x = L, \left(\frac{1 \pm \sqrt{33}}{16}\right)L$$

Checking, we find that the maximum deflection occurs at

$$x = \left(\frac{1 + \sqrt{33}}{16}\right)L$$

$$\approx 0.4215L$$

and the maximum deflection is

$$y_m = y(0.4215L)$$

$$\approx -0.0054kL^4$$

The maximum deflection in the cantilevered case is less than that in part b.

35. Let $S_1(t)$ and $S_2(t)$ be the amounts of salt in the first and second tanks, respectively, at time t.

a.
$$\frac{dS_1}{dt} = (2)(1) - \frac{S_1}{100}(1)$$

$$\frac{dS_1}{dt} + \frac{S_1}{100} = 2$$

$$I(t) = e^{\int (1/100) \, dt} = e^{t/100}$$

$$S_1(t) = e^{-t/100}\left[\int 2e^{t/100} \, dt + C\right]$$

$$= 200 + Ce^{-t/100}$$

Since $S_1(0) = 0$ (only pure water at $t = 0$), we obtain

$$S_1(t) = 0 = 200 + C \text{ so that } C = -200$$

and

$$S_1(t) = 200(1 - e^{-t/100})$$

b.
$$\frac{dS_2}{dt} = \frac{S_1}{100}(1) - \frac{S_2}{100}(1)$$

$$\frac{dS_2}{dt} + \frac{S_2}{100} = \frac{1}{100}\left[200(1 - e^{-t/100})\right]$$

from part a. The integrating factor is

$$I(t) = e^{\int (1/100)\, dt}$$

$$= e^{t/100}$$

$$S_2(t) = e^{-t/100}\left[\int e^{t/100}(2 - 2e^{-t/100})\, dt + C_1\right]$$

$$= 200 - 2te^{-t/100} + C_2 e^{-t/100}$$

Since $S_2(0) = 0$, we obtain $0 = 200 + C_2$

or $C_2 = -200$, and

$$S_2(t) = 200 - 2te^{-t/100} - 200e^{-t/100}$$

c. The excess is

$$S(t) = S_1 - S_2$$

$$= 2te^{-t/100}$$

which is maximized when

$$S'(t) = (-0.02t + 2)e^{-t/100} = 0$$

or $t = 100$ minutes. The maximum excess is

$$S(100) \approx 73.58 \text{ lbs}$$

37. $\frac{dP}{dt} = kP(B - \ln P)$

Let $P = e^u$, $\frac{dP}{dt} = e^u \frac{du}{dt}$

$$e^u \frac{du}{dt} = ke^u(B - u)$$

$$\int \frac{du}{B - u} = \int k\, dt$$

$$-\ln|B - u| = kt + C_1$$

$$B - u = Ce^{-kt}$$

$$u = B - Ce^{-kt}$$

Thus,

$$P(t) = e^{B - Ce^{-kt}} = e^B e^{-Ce^{-kt}}$$

Applying the initial conditions,

$$P_0 = P(0) = e^{B-C} \text{ or } B - C = \ln P_0$$

$$P_\infty = \lim_{t \to +\infty} e^{B - Ce^{-kt}} = e^B$$

Thus, $B = \ln P_\infty$ and

$$C = \ln P_\infty - \ln P_0 = \ln \frac{P_\infty}{P_0}$$

so

$$P(t) = P_\infty \exp\left[-\left(\ln \frac{P_\infty}{P_0}\right)e^{-kt}\right]$$

39. $\frac{dy}{dx} = \frac{1 + y}{xy + e^y(1 + y)}$

$$\frac{dx}{dy} = \frac{xy + e^y(1 + y)}{1 + y} = \frac{y}{1 + y}x + e^y$$

$$\frac{dx}{dy} - \frac{y}{1 + y}x = e^y$$

This is a first order linear differential equation in x. The integrating factor is

$$I(t) = e^{\int -y/(1+y)\ dy} = (y+1)e^{-y}$$

so

$$x = \frac{1}{(y+1)e^{-y}}\left[\int (y+1)e^{-y}e^{y}\ dy + C\right]$$

$$= \frac{e^{y}}{y+1}\left[\tfrac{1}{2}(y+1)^2 + C\right]$$

$$= \tfrac{1}{2}(y+1)e^{y} + \frac{Ce^{y}}{y+1}$$

41. The governing differential equation is

$$5\frac{dI}{dt} + 10I = E(t)$$

$$\frac{dI}{dt} + 2I = \tfrac{1}{5}E(t)$$

The integrating factor is $e^{\int 2\ dt} = e^{2t}$. Then,

$$I(t) = e^{-2t}\left[\int \tfrac{1}{5}e^{2t}E(t)\ dt + C\right]$$

a. If $E = 15$, then

$$I(t) = e^{-2t}\left[\int \tfrac{15}{5}e^{2t}\ dt + C\right]$$

$$= e^{-2t}\left[\tfrac{15}{10}e^{2t} + C\right]$$

$$= \tfrac{3}{2} + Ce^{-2t}$$

Since $I(0) = 0$, $C = -\tfrac{3}{2}$, so

$$I(t) = \tfrac{3}{2}(1 - e^{-2t})$$

b. If $E = 5e^{-2t}\sin t$, then

$$I(t) = e^{-2t}\left[\int \tfrac{1}{5}e^{2t}(5e^{-2t}\sin t)\ dt + C\right]$$

$$= e^{-2t}[-\cos t + C]$$

Since $I(0) = 0$, $C = 1$, so

$$I(t) = e^{-2t}(1 - \cos t)$$

43. Let $Q(t)$ be the amount of pollutant in the lake at time t. Then

$$\frac{dQ}{dt} = (0.0006)(350) - \frac{Q}{6,000}(350)$$

where units are in millions of ft^3. Thus,

$$\frac{dQ}{dt} + 0.0583Q = 0.21.$$ The integrating factor is

$$I(t) = e^{\int 0.0583\ dt} = e^{0.0583t}$$

so

$$Q(t) = e^{-0.0583t}\left[\int e^{0.0583t}(0.21)\ dt + C\right]$$

$$= 3.602 + Ce^{-0.0583t}$$

Since the lake initially contains $Q(0) \approx 13.2$ million cubic feet of pollutant, we find $C \approx 9.598$, so

$$Q(t) = 3.602 + 9.598e^{-0.0583t}$$

The lake will contain 0.15% pollutant when

$$0.0015(6,000) = 3.602 + 9.598e^{-0.0583t}$$

Solving this equation, we obtain $t \approx 9.872$ days.

45.
$$\frac{dS}{dt} = -pS + \left(\frac{S}{N}\right)\frac{dN}{dt} + \frac{pS^2}{mN}$$

Let $y = \frac{S}{N}$, so $S = yN$ and the differential equation becomes

$$\frac{dS}{dt} = -pS + y\frac{dN}{dt} + \frac{pS}{m}\left(\frac{S}{N}\right)$$

$$\frac{d(yN)}{dt} = -pS + y\frac{dN}{dt} + \frac{pS}{m}\left(\frac{S}{N}\right)$$

$$y\frac{dN}{dt} + N\frac{dy}{dt} = -pS + y\frac{dN}{dt} + \frac{pS}{m}\left(\frac{S}{N}\right)$$

$$N\frac{dy}{dt} = -pS + \frac{pS}{m}y$$

$$\frac{dy}{dt} = -py + \frac{p}{m}y\left(\frac{S}{N}\right)$$

$$\frac{dy}{dt} + py = \frac{p}{m}y^2$$

This is a Bernoulli equation (see Problem 44). Let $u = y^{1-2} = y^{-1}$. Then

$$y = \frac{1}{u} \text{ and } \frac{dy}{dt} = \frac{-1}{u^2}\frac{du}{dt}$$

and the differential equation is

$$-\frac{1}{u^2}\frac{du}{dt} + p\left(\frac{1}{u}\right) = \frac{p}{m}\left(\frac{1}{u}\right)^2$$

$$\frac{du}{dt} - pu = \frac{-p}{m}$$

The integrating factor is

$$I(t) = e^{\int -p\, dt} = e^{-pt}$$

so

$$u = e^{pt}\left[\int\left(\frac{-p}{m}\right)e^{-pt}\, dt + C\right]$$

$$= \frac{1}{m} + Ce^{pt}$$

and since $u = \frac{1}{y} = \frac{N}{S}$, we have

$$\frac{N}{S} = \frac{1}{m} + Ce^{pt}$$

$$N = \left(\frac{1}{m} + Ce^{pt}\right)S$$

47. a.
$$m\frac{dv}{dt} = -mg - kv$$

$$\frac{dv}{dt} + \frac{k}{m}v = -g$$

The integrating factor is

$$I(t) = e^{\int k/m\, dt} = e^{kt/m}$$

so

$$v(t) = e^{-kt/m}\left[\int e^{kt/m}(-g)\, dt + C\right]$$

$$= -\frac{mg}{k} + Ce^{-kt/m}$$

Since $v(0) = v_0$, we have

$$v_0 = \frac{-mg}{k} + C, \text{ so } C = v_0 + \frac{mg}{k}$$

and

$$v(t) = \frac{-mg}{k} + \left(\frac{mg}{k} + v_0\right)e^{-kt/m}$$

Integrating and using $s(0) = 0$ (the object begins at ground level), we obtain

$$s(t) = \int v(t)\, dt$$

$$= \frac{-mg}{k}t + \left(\frac{mg}{k} + v_0\right)\left(\frac{-m}{k}\right)e^{-kt/m}$$

$$+ \left(\frac{mg}{k} + v_0\right)\left(\frac{m}{k}\right)$$

$$= \frac{-mg}{k}t + \frac{m}{k}\left(\frac{mg}{k} + v_0\right)\left(1 - e^{-kt/m}\right)$$

b. The object reaches its maximum height when $v(t) = 0$.

$$\frac{-mg}{k} + \left(\frac{mg}{k} + v_0\right)e^{-kt/m} = 0$$

$$e^{-kt/m} = \frac{mg}{mg + v_0 k}$$

$$t_{max} = \frac{m}{k}\ln\left(1 + \frac{kv_0}{mg}\right)$$

The maximum height is

$$s_{max} = s(t_{max})$$

$$= \frac{-m^2 g}{k^2}\ln\left(1 + \frac{kv_0}{mg}\right)$$

$$+ \left(\frac{mg}{k} + v_0\right)\left(\frac{-m}{k}\right)\left(\frac{mg}{mg + v_0 k}\right)$$

$$+ \left(\frac{mg}{k} + v_0\right)\left(\frac{m}{k}\right)$$

$$= \frac{mv_0}{k} - \frac{m^2 g}{k^2}\ln\left(1 + \frac{kv_0}{mg}\right)$$

c. With $k = 0.75$, $v_0 = 150$, and $m = \frac{20}{32} = 0.625$ slugs, we find that the maximum height is

$$s_{max} = \frac{0.625(150)}{0.75}$$

$$- \frac{(0.625)^2(32)}{(0.75)^2}\ln\left(1 + \frac{0.75(150)}{0.625(32)}\right)$$

$$\approx 82.98 \text{ ft}$$

The object hits the ground when $s(t) = 0$:

$$0 = \left[\frac{-0.625(32)}{0.75}\right]t$$

$$+ \left(\frac{0.625}{0.75}\right)\left[\frac{0.625(32)}{0.75} + 150\right]\left(1 - e^{-0.75t/0.625}\right)$$

We solve to find $t \approx 5.51$ seconds.

d. With $k = 0$, we have

$$v = -gt + v_0$$

$$s = -\frac{gt^2}{2} + v_0 t$$

The object hits the ground when $s(t) = 0$:

$$\frac{-32t^2}{2} + 150t = 0$$

$$t \approx 9.38 \text{ seconds}$$

7.7 Improper Integrals, Pages 479-481

1. An improper integral is a definite integral whose interval of integration is unbounded or whose integrand is unbounded on the interval of integration.

3.
$$\int_{1}^{+\infty} \frac{dx}{x^3} = \lim_{N \to +\infty} \int_{1}^{N} \frac{dx}{x^3}$$

$$= \lim_{N \to +\infty} -\frac{1}{2x^2}\Big|_{1}^{N}$$

$$= \lim_{N \to +\infty} \left(-\frac{1}{2N^2} + \frac{1}{2}\right)$$

$$= \frac{1}{2}$$

Converges

SURVIVAL HINT: *Pay attention to the exponent on the variable in Problems 5 and 7 and note that even though they are very similar, the conclusions are very different.*

5.
$$\int_{1}^{+\infty} \frac{dx}{x^{0.99}} = \lim_{N \to +\infty} \int_{1}^{N} \frac{dx}{x^{0.99}}$$

$$= 100 \lim_{N \to +\infty} x^{0.01}\Big|_{1}^{N}$$

$$= 100 \lim_{N \to +\infty} (N^{0.01} - 1)$$

$$= +\infty$$

Diverges

7.
$$\int_{1}^{+\infty} \frac{dx}{x^{1.1}} = \lim_{N \to +\infty} \int_{1}^{N} \frac{dx}{x^{1.1}}$$

$$= \lim_{N \to +\infty} \left[-\frac{1}{0.1x^{0.1}}\right]\Big|_{1}^{N}$$

$$= \lim_{N \to +\infty} \left[-\frac{10}{N^{0.1}} + 10\right]$$

$$= 10$$

Converges

9.
$$\int_{3}^{+\infty} \frac{dx}{2x - 1} = \frac{1}{2} \lim_{N \to +\infty} \int_{3}^{N} \frac{2\,dx}{2x - 1}$$

$$= \frac{1}{2} \lim_{N \to +\infty} \ln|2x - 1|\Big|_{3}^{N}$$

$$= +\infty$$

Diverges

11.
$$\int_{3}^{+\infty} \frac{dx}{(2x - 1)^2} = \lim_{N \to +\infty} \int_{3}^{N} \frac{dx}{(2x - 1)^2}$$

$$= -\frac{1}{2} \lim_{N \to +\infty} (2x - 1)^{-1}\Big|_{3}^{N}$$

$$= \lim_{N \to +\infty} \left[-\frac{1}{4N - 2} + \frac{1}{10}\right]$$

$$= \frac{1}{10}$$

Converges

13. $\displaystyle\int_0^{+\infty} 5e^{-2x}\, dx = 5 \lim_{N\to+\infty} \int_0^N e^{-2x}\, dx$

$\displaystyle = -\frac{5}{2} \lim_{N\to+\infty} e^{-2x}\Big|_0^N$

$\displaystyle = \frac{5}{2}$

Converges

15. $\displaystyle\int_1^{+\infty} \frac{x^2\, dx}{(x^3+2)^2} = \frac{1}{3} \lim_{N\to+\infty} \int_1^N \frac{3x^2\, dx}{(x^3+2)^2}$

$\displaystyle = \lim_{N\to+\infty} \frac{1}{3}\left(-\frac{1}{x^3+2}\right)\Big|_1^N$

$\displaystyle = \lim_{N\to+\infty}\left(-\frac{1}{3N^3+6}+\frac{1}{9}\right)$

$\displaystyle = \frac{1}{9}$

Converges

17. $\displaystyle\int_1^{+\infty} \frac{x^2\, dx}{\sqrt{x^3+2}} = \frac{1}{3} \lim_{N\to+\infty} \int_1^N (x^3+2)^{-1/2}(3x^2\, dx)$

$\displaystyle = \frac{2}{3} \lim_{N\to+\infty} \sqrt{x^3+2}\Big|_1^N$

$\displaystyle = +\infty$

Diverges

19. $\displaystyle\int_1^{+\infty} \frac{e^{-\sqrt{x}}}{\sqrt{x}}\, dx = \lim_{N\to+\infty} \int_1^N \frac{e^{-\sqrt{x}}}{\sqrt{x}}\, dx$

$\displaystyle = -2 \lim_{N\to+\infty} e^{-\sqrt{x}}\Big|_1^N$

$\displaystyle = -2 \lim_{N\to+\infty}\left(\frac{1}{e^{\sqrt{N}}} - \frac{1}{e}\right)$

$\displaystyle = \frac{2}{e}$

Converges

21. $\displaystyle\int_0^{+\infty} 5xe^{10-x}\, dx \quad \boxed{u = x,\; dv = e^{-x}\, dx}$

$\displaystyle = \lim_{N\to+\infty}\left[-5(x+1)e^{10-x}\Big|_0^N\right] = 5e^{10}$

Note: $\displaystyle\lim_{x\to+\infty} \frac{x}{e^x} = \lim_{x\to+\infty} \frac{1}{e^x} = 0$

by l'Hôpital's rule; converges

23. $\displaystyle\int_2^{+\infty} \frac{dx}{x\ln x} = \lim_{N\to+\infty} \int_2^N (\ln x)^{-1}\frac{dx}{x}$

$\displaystyle = \lim_{N\to+\infty} \ln(\ln x)\Big|_2^N$

$\displaystyle = \lim_{N\to+\infty}[\ln(\ln N) - \ln(\ln 2)]$

$\displaystyle = +\infty$

Diverges

25. $\displaystyle\int_{0}^{+\infty} x^2 e^{-x}\, dx$ $\boxed{\text{Parts or Formula 485}}$

$$= \lim_{N\to+\infty} \frac{e^{-x}}{-1}\left(x^2 - \frac{2x}{-1} + \frac{2}{1}\right)\Big|_{0}^{N}$$

$$= \lim_{N\to+\infty}\left[2 - e^{-N}(N^2 + 2N + 2)\right]$$

$$= 2$$

Converges

27. $\displaystyle\int_{-\infty}^{0} \frac{2x\, dx}{x^2 + 1}$ $= \displaystyle\lim_{t\to-\infty} \int_{t}^{0} \frac{2x\, dx}{x^2 + 1}$

$$= \lim_{t\to-\infty} \ln(x^2 + 1)\Big|_{t}^{0}$$

$$= \lim_{t\to-\infty} -\ln(t^2 + 1)$$

$$= -\infty$$

   Diverges

29. $\displaystyle\int_{-\infty}^{0} \frac{dx}{\sqrt{2 - x}}$ $= \displaystyle\lim_{t\to-\infty} \int_{t}^{0} (2 - x)^{-1/2}\, dx$

$$= \lim_{t\to-\infty} -2\sqrt{2 - x}\,\Big|_{t}^{0}$$

$$= +\infty$$

Diverges

31. $\displaystyle\int_{-\infty}^{\infty} x e^{-|x|}\, dx$

$$= \int_{-\infty}^{0} x e^{-|x|}\, dx + \int_{0}^{\infty} x e^{-|x|}\, dx$$

$$= \lim_{t\to-\infty} \int_{t}^{0} x e^{x}\, dx + \lim_{N\to\infty} \int_{0}^{N} x e^{-x}\, dx$$

$$= \lim_{t\to-\infty}\left[e^{x}(x - 1)\right]\Big|_{t}^{0} + 1 \quad \textit{See Problem 20.}$$

$$= -1 + 1 = 0$$

Converges

33. $\displaystyle\int_{0}^{1} \frac{dx}{x^{1/5}} = \lim_{t\to0^+} \int_{t}^{1} x^{-1/5}\, dx$

$$= \frac{5}{4} \lim_{t\to0^+} x^{4/5}\Big|_{t}^{1}$$

$$= \frac{5}{4}$$

Converges

35. $\displaystyle\int_{0}^{1} \frac{dx}{(1 - x)^{1/2}} = \lim_{N\to1^-} \int_{0}^{N} (1 - x)^{-1/2}\, dx$

$$= -2 \lim_{N\to1^-} \sqrt{1 - x}\,\Big|_{0}^{N}$$

$$= \lim_{N\to1^-} -2(\sqrt{1 - N} - 1)$$

$$= 2$$

Converges

37. $\displaystyle\int_{-1}^{1} \frac{e^x}{\sqrt[3]{1-e^x}}\,dx = \int_{-1}^{0} \frac{e^x\,dx}{\sqrt[3]{1-e^x}} + \int_{0}^{1} \frac{e^x\,dx}{\sqrt[3]{1-e^x}}$

$\displaystyle = \lim_{N \to 0^-} \int_{-1}^{N} \frac{e^x dx}{(1-e^x)^{1/3}} + \lim_{N \to 0^+} \int_{N}^{1} \frac{e^x dx}{(1-e^x)^{1/3}}$

$\displaystyle = \lim_{N \to 0^-} \frac{3}{2}(-1)(1-e^x)^{2/3}\Big|_{-1}^{N}$

$\displaystyle \qquad + \lim_{N \to 0^+} \frac{3}{2}(-1)(1-e^x)^{2/3}\Big|_{N}^{1}$

$\displaystyle = \frac{3}{2}\Big[(1-e^{-1})^{2/3} - (1-e)^{2/3}\Big]$

Converges

39. $\displaystyle\int_{0}^{1} \ln x\,dx = \lim_{t \to 0^+} \int_{t}^{1} \ln x\,dx$

$\boxed{\text{parts or Formula 499}}$

$\displaystyle = \lim_{t \to 0^+} (x \ln x - x)\Big|_{t}^{1}$

$\displaystyle = -1 - \lim_{t \to 0^+} (t \ln t - t)$

By l'Hôpital's rule

$= -1$

Converges

41. $\displaystyle\int_{e}^{+\infty} \frac{dx}{x(\ln x)^2} = \lim_{N \to +\infty} \left[-\frac{1}{\ln x}\right]\Big|_{e}^{N}$

$\displaystyle = \lim_{N \to \infty} \left(-\frac{1}{\ln N} + 1\right)$

$= 1$

Converges

43. $\displaystyle\int_{0}^{1} e^{-(1/2)\ln x}\,dx = \lim_{t \to 0^+} \int_{t}^{1} x^{-1/2}\,dx$

$\displaystyle = \lim_{t \to 0^+} 2\sqrt{x}\,\Big|_{t}^{1}$

$= 2$

Converges

45. The integrand becomes discontinuous where the denominator $1 - \tan x$ is 0; that is, at $x = \pi/4$.

$\displaystyle\int_{0}^{\pi/3} \frac{\sec^2 x}{1 - \tan x}\,dx$

$\displaystyle = \lim_{t \to (\pi/4)^-} \int_{0}^{t} \frac{\sec^2 x\,dx}{1 - \tan x} + \lim_{t \to (\pi/4)^+} \int_{t}^{\pi/3} \frac{\sec^2 x\,dx}{1 - \tan x}$

provided both limits exist. However,

$\displaystyle \lim_{t \to (\pi/4)^-} \int_{0}^{t} \frac{\sec^2 x\,dx}{1 - \tan x}$

$\boxed{u = 1 - \tan x;\ du = -\sec^2 x\,dx}$

$$= \lim_{t \to (\pi/4)} - (-1)\ln|1 - \tan x| \Big|_0^t$$

$$= +\infty$$

Since the limit does not exist, the improper integral diverges.

47. $A = \displaystyle\int_6^{+\infty} \frac{2\,dx}{(x-4)^3}$

$$= 2 \lim_{N \to +\infty} (-\tfrac{1}{2})(x-4)^{-2} \Big|_6^N$$

$$= \lim_{N \to +\infty} [-(N-4)^{-2} + (6-4)^{-2}]$$

$$= \tfrac{1}{4}$$

49. $A = \displaystyle\lim_{T \to +\infty} \int_0^T 200 e^{-0.002t}\,dt$

$$= -100,000 \lim_{T \to +\infty} e^{-0.002t} \Big|_0^T$$

$$= 100,000$$

There will be 100,000 millirads.

51.

$$I = \int_{-\infty}^{+\infty} f(x)\,dx$$

$$= \lim_{N \to -\infty} \int_N^{-1} e^{x+1}\,dx + \int_{-1}^1 dx + \lim_{N \to +\infty} \int_1^N x^{-2}\,dx$$

$$= 1 - \lim_{N \to -\infty} e^{N+1} + 2 + \lim_{N \to +\infty} \frac{-1}{N} + 1$$

$$= 4$$

53. $I = \displaystyle\int_0^1 \frac{dx}{x^p}$

$$= \lim_{t \to 0^+} \int_t^1 x^{-p}\,dx$$

$$= \lim_{t \to 0^+} \frac{x^{-p+1}}{1-p} \Big|_t^1$$

$$= (1-p)^{-1} \text{ if } p < 1, \text{ then } I \text{ converges and}$$

if $p > 1$, I is infinite so it diverges.

If $p = 1$,

$$I = \int_t^1 \frac{dx}{x}$$

$$= \lim_{t \to 0^+} \ln x \Big|_t^1$$

$$= -\infty$$

Thus, I converges if $p < 1$, and diverges if $p \geq 1$.

55. The integration is correct, but has overlooked the fact that the function is undefined (has a vertical asymptote) at $x = 0$. It should be written as:

$$\lim_{t \to 0^-} \int_{-1}^{t} \frac{dx}{x^2} + \lim_{t \to 0^+} \int_{t}^{1} \frac{dx}{x^2}$$

We note that

$$\lim_{t \to 0^+} \int_{t}^{1} \frac{dx}{x^2} = \lim_{t \to 0^+} -\frac{1}{x}\Big|_{t}^{1}$$

$$= \infty$$

Since at least one of these integrals diverges, the original integral diverges.

57. $I = \int_{0}^{2} f(x)\, dx$

$$= 4 \lim_{t \to 0^+} x^{1/4}\Big|_{t}^{1} + 4 \lim_{t \to 1^+} (x - 1)^{1/4}\Big|_{t}^{2}$$

$$= 8$$

59. $A = \int_{0}^{\infty} [\frac{1}{2}(e^{-2\theta})^2 - \frac{1}{2}(e^{-5\theta})^2]\, d\theta$

$$= \lim_{M \to +\infty} \int_{0}^{M} \frac{1}{2}(e^{-4\theta} - e^{-10\theta})\, d\theta$$

$$= \lim_{M \to +\infty} \left[\frac{1}{2}\left(-\frac{1}{4}\right)e^{-4\theta} - \frac{1}{2}\left(-\frac{1}{10}\right)e^{-10\theta}\right]\Big|_{0}^{M}$$

$$= \lim_{M \to +\infty} \left[\left(-\frac{1}{8}e^{-4M} + \frac{1}{20}e^{-10M}\right) - \left(-\frac{1}{8} + \frac{1}{20}\right)\right]$$

$$= 0 - \left(-\frac{1}{8} + \frac{1}{20}\right)$$

$$= \frac{3}{40}$$

61. $\mathcal{L}\{af + bg\} = \int_{0}^{+\infty} e^{-st}(af + bg)\, dt$

$$= a\int_{0}^{+\infty} e^{-st}f\, dt + b\int_{0}^{+\infty} e^{-st}g\, dt$$

$$= a\mathcal{L}\{f\} + b\mathcal{L}\{g\}$$

63. $F(s) = \mathcal{L}\{f(t)\}$

$$= \int_{0}^{+\infty} e^{-st}f(t)\, dt$$

$$\mathcal{L}\{f(t)e^{at}\} = \int_{0}^{+\infty} e^{-st}e^{at}f(t)\, dt$$

$$= \int_{0}^{+\infty} e^{-(s-a)t}f(t)\, dt$$

Let $s_1 = s - a$; then

$$\mathcal{L}\{f(t)e^{at}\} = \int_{0}^{+\infty} e^{-s_1 t}f(t)\, dt = F(s_1)$$

$$= F(s - a)$$

65. a. $F'(s) = \frac{d}{ds} F(s)$

$$= \frac{d}{ds} \int_0^{+\infty} e^{-st} f(t) \, dt$$

$$= \int_0^{+\infty} \frac{d}{ds}[e^{-st} f(t)] \, dt$$

$$= \int_0^{+\infty} [-te^{-st}] f(t) \, dt$$

$$= -\int_0^{+\infty} [te^{-st}] f(t) \, dt$$

$$= -\mathcal{L}\{tf(t)\}$$

b. Since $\mathcal{L}\{\cos 2t\} = \dfrac{s}{s^2 + 4}$, we have

$$\mathcal{L}\{t \cos 2t\} = -\frac{d}{ds}\left(\frac{s}{s^2 + 4}\right)$$

$$= -\left[\frac{(s^2 + 4) - s(2s)}{(s^2 + 4)^2}\right]$$

$$= \frac{s^2 - 4}{(s^2 + 4)^2}$$

67. a. $\mathcal{L}(f') = \displaystyle\int_0^\infty e^{-st} f'(t) \, dt$

$$= \lim_{k \to +\infty} \int_0^k e^{-st} f'(t) \, dt$$

$$\boxed{\begin{array}{l} u = e^{-st}, \; du = -se^{-st} \, dt \\ dv = f'(t) \, dt, \; v = f(t) \end{array}}$$

$$= \lim_{k \to +\infty}\left[e^{-st} f(t)\right]\Big|_0^k + \lim_{k \to +\infty} \int_0^k se^{-st} f(t) \, dt$$

$$= \lim_{k \to +\infty}[e^{-sk} f(k) - f(0)] + s \lim_{k \to +\infty} \int_0^k e^{-st} f(t) \, dt$$

$$= -f(0) + s\mathcal{L}(f) \quad \textit{since} \lim_{k \to +\infty} e^{-sk} f(k) = 0$$

b. Taking Laplace transforms on both sides
of the differential equation
$y' + 3y = \sin 2t$, we obtain

$$\mathcal{L}\{y'\} + 3\mathcal{L}\{y\} = \mathcal{L}\{\sin 2t\}$$

$$s\mathcal{L}\{y\} - 2 + 3\mathcal{L}\{y\} = \frac{2}{s^2 + 4} \quad \textit{by part a}$$

$$\mathcal{L}\{y\} = \frac{\dfrac{2}{s^2 + 4} + 2}{s + 3}$$

$$= \frac{2s^2 + 10}{(s^2 + 4)(s + 3)}$$

$$= \frac{-\dfrac{2}{13}s}{s^2 + 4} + \frac{\dfrac{6}{13}}{s^2 + 4} + \frac{\dfrac{28}{13}}{s + 3} \quad \textit{Partial fractions}$$

c. Finally, by taking inverse Laplace
transforms on both sides of this equation,

$$y = -\frac{2}{13}\mathcal{L}^{-1}\left(\frac{s}{s^2 + 4}\right)$$

$$+ \frac{6}{13}\mathcal{L}^{-1}\left(\frac{1}{s^2 + 4}\right) + \frac{28}{13}\mathcal{L}^{-1}\left(\frac{1}{s + 3}\right)$$

$$= -\frac{2}{13}\cos 2t + \frac{3}{13}\sin 2t + \frac{28}{13} e^{-3t}$$

7.8 Hyperbolic and Inverse Hyperbolic Functions, Pages 486–487

SURVIVAL HINT: *The hyperbolic functions are good news and bad news. The good news is that the identities, derivatives, and integrals are almost the same as the trigonometric functions. The bad news is the almost. Try to learn these function definitions by concentrating on which one are different by a sign.*

SURVIVAL HINT: *In Problems 1-12, make sure you find the appropriate keys on your calculator to duplicate the given answers.*

1. $\sinh 2 \approx 3.6269$

3. $\tanh(-1) \approx -0.7616$

5. $\coth 1.2 \approx 1.1995$

7. $\cosh^{-1} 1.5 \approx 0.9624$

9. $\cosh(\ln 3) \approx 1.6667$

11. $\text{sech } 1 \approx 2.2924$

13. $y = \sinh 3x$

$y' = (\cosh 3x)(3)$

$= 3\cosh 3x$

15. $y = \cosh(2x^2 + 3x)$

$y' = (4x + 3)\sinh(2x^2 + 3x)$

17. $y = \sinh x^{-1}$

$y' = (\cosh x^{-1})(-x^{-2})$

$= -x^{-2}\cosh x^{-1}$

19. $y = \sinh^{-1}(x^3)$

$y' = \dfrac{3x^2}{\sqrt{1 + x^6}}$

21. $y = \sinh^{-1}(\tan x)$

$y' = \dfrac{\sec^2 x}{\pm\sqrt{1 + \tan^2 x}}$

$= \dfrac{\sec^2 x}{\sec x}$

$= \sec x$

23. $y = \tanh^{-1}(\sin x)$

$y' = \dfrac{\cos x}{1 - \sin^2 x}$

$= \dfrac{\cos x}{\cos^2 x}$

$= \sec x$

25. $y = \dfrac{\sinh^{-1} x}{x}$

$y' = \dfrac{x\left(\dfrac{1}{\sqrt{1 + x^2}}\right) - \sinh^{-1} x}{x^2}$

$= \dfrac{x - \sqrt{1 + x^2}\,\sinh^{-1} x}{x^2\sqrt{1 + x^2}}$

27. $y = x\cosh^{-1} x - \sqrt{x^2 - 1}$

$y' = \cosh^{-1} x + \dfrac{x}{\sqrt{x^2 - 1}} - \frac{1}{2}(x^2 - 1)^{-1/2}(2x)$

$= \cosh^{-1} x$

29. $e^x\sinh^{-1}x + e^{-x}\cosh^{-1}y = 1$

$\cosh^{-1}y = e^x(1 - e^x\sinh^{-1}x)$

$\quad y = \cosh\left[e^x(1 - e^x\sinh^{-1}x)\right]$

$\quad = \cosh(e^x - e^{2x}\sinh^{-1}x)$

$y' = \sinh(e^x - e^{2x}\sinh^{-1}x)[e^x - (2e^{2x}\sinh^{-1}x$

$\qquad + e^{2x}\frac{1}{\sqrt{1+x^2}})]$

$\quad = \Big(e^x - 2e^{2x}\sinh^{-1}x$

$\qquad -\frac{e^x}{\sqrt{1+x^2}}\Big)\sinh(e^x - e^{2x}\sinh^{-1}x)$

31. $\int\frac{\sinh\frac{1}{x}\,dx}{x^2} = -\int\sinh(x^{-1})(-x^{-2}\,dx)$

$\qquad = -\cosh x^{-1} + C$

33. $\int\coth x\,dx = \int\frac{\cosh x\,dx}{\sinh x}$

$\qquad = \ln|\sinh x| + C$

35. $\int\frac{dt}{\sqrt{9t^2-16}} = \frac{1}{3}\int\frac{\left(\frac{3\,dt}{4}\right)}{\sqrt{\left(\frac{3t}{4}\right)^2-1}}$

$\qquad = \frac{1}{3}\cosh^{-1}\frac{3t}{4} + C$

37. $\int\frac{\cos x\,dx}{\sqrt{1+\sin^2x}} = \sinh^{-1}(\sin x) + C$

39. $\int\frac{x^2\,dx}{1-x^6} = \frac{1}{3}\int\frac{3x^2\,dx}{1-(x^3)^2}$

$\qquad = \frac{1}{3}\tanh^{-1}x^3 + C$

41. $\int_2^3\frac{dx}{1-x^2} = \coth^{-1}x\Big|_2^3$

$\qquad = \coth^{-1}3 - \coth^{-1}2$

$\qquad \approx -0.2027$

43. $\int_1^2\frac{e^x\,dx}{\sqrt{e^{2x}-1}} = \cosh^{-1}e^x\Big|_1^2$

$\qquad = \cosh^{-1}e^2 - \cosh^{-1}e$

$\qquad \approx 1.0311$

45. $\int_0^1 x\,\text{sech}^2x^2\,dx = \frac{1}{2}\int_0^1\text{sech}^2x^2\,(2x\,dx)$

$\qquad = \frac{1}{2}\tanh x^2\Big|_0^1$

$\qquad = \frac{1}{2}\tanh 1$

$\qquad \approx 0.3808$

47. By l'Hôpital's rule,

$\lim_{x\to+\infty}\frac{\sinh^{-1}x}{\cosh^{-1}x} = \lim_{x\to+\infty}\frac{\ln(x+\sqrt{x^2+1})}{\ln(x+\sqrt{x^2-1})}$

$$= \lim_{x \to +\infty} \frac{\frac{1}{\sqrt{x^2+1}}}{\frac{1}{\sqrt{x^2-1}}}$$

$$= \lim_{x \to +\infty} \sqrt{\frac{1-\frac{1}{x^2}}{1+\frac{1}{x^2}}} = 1$$

49. $y' = \text{sech}^2 x \geq 0$

The curve is rising

for all x.

$y'' = -2\,\text{sech}^2 x \tanh x$

$y'' = 0$ when $x = 0$

The curve is concave

up on $(-\infty, 0)$, and concave down on

$(0, +\infty)$.

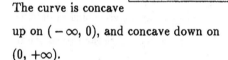

51. $(\cosh x + \sinh x)^n = [\frac{1}{2}(e^x + e^{-x} + e^x - e^{-x})]^n$

$$= (e^x)^n$$

$$= e^{nx}$$

so by taking nth roots, $\cosh x + \sinh x = e^x$

Replacing x by nx, we obtain

$\cosh nx + \sinh nx = e^{nx} = (\cosh x + \sinh x)^n$

53. **a.** $y' = ac \sinh cx + bc \cosh cx$

$y'' = c^2(a \cosh cx + b \sinh cx) = c^2 y$

Thus, $y'' - c^2 y = 0$

b. Let $c = 2$; then

$$y = a \cosh 2x + b \sinh 2x$$

is a solution of

$$y'' - 4y = 0$$

Since $y'(0) = 2$, $2 = 2b$ or $b = 1$. Since

$y = 1$ when $x = 0$, $a = 1$ and

$$y = \cosh 2x + \sinh 2x$$

55.

$$s = \int_{-a}^{a} \sqrt{1 + (y')^2}\, dx$$

$$= 2 \int_{0}^{a} \sqrt{1 + \sinh^2 \frac{x}{a}}\, dx$$

$$= 2a \int_{0}^{a} \cosh \frac{x}{a} \left(\frac{1}{a}\, dx\right)$$

$$= 2a \sinh \frac{x}{a}\Big|_{0}^{a}$$

$$= 2a(\sinh 1 - \sinh 0)$$

$$= (e - e^{-1})a$$

$$\approx 2.3504a$$

57. $y = \cosh x$

$$ds = \sqrt{1 + \sinh^2 x}\, dx = \cosh x\, dx$$

$$S = 2 \int_{0}^{1} 2\pi \cosh^2 x\, dx$$

$$= 2\pi \int_0^1 (\cosh 2x + 1)\, dx$$

$$= 2\pi(\tfrac{1}{2} \sinh 2x + x)\Big|_0^1$$

$$= \pi \sinh 2 + 2\pi$$

$$= \tfrac{\pi}{2}(e^2 - e^{-2} + 4)$$

$$\approx 17.68$$

59. $y = \cosh u$

$$= \tfrac{1}{2}(e^u + e^{-u})$$

$$\frac{dy}{dx} = \tfrac{1}{2}(e^u - e^{-u})\frac{du}{dx}$$

$$= \sinh u \frac{du}{dx}$$

$y = \tanh u$

$$= \frac{\sinh u}{\cosh u}$$

$$\frac{dy}{dx} = \frac{\cosh u(\cosh u \frac{du}{dx}) - \sinh u(\sinh u \frac{du}{dx})}{\cosh^2 u}$$

$$= \frac{\cosh^2 u - \sinh^2 u}{\cosh^2 u}\frac{du}{dx}$$

$$= \frac{1}{\cosh^2 u}\frac{du}{dx}$$

$$= \operatorname{sech}^2 u \frac{du}{dx}$$

$y = \operatorname{sech} u = (\cosh u)^{-1}$

$$\frac{dy}{dx} = -(\cosh u)^{-2} \sinh u \frac{du}{dx}$$

$$= -\frac{\sinh u}{\cosh u}\frac{1}{\cosh u}\frac{du}{dx}$$

$$= -\tanh u \operatorname{sech} u \frac{du}{dx}$$

61.   a. $y = \cosh^{-1}x$, so $x = \cosh y = \dfrac{e^y + e^{-y}}{2}$

$$x = \frac{e^y + e^{-y}}{2}$$

$$2x = e^y + e^{-y}$$

$$e^{2y} - 2xe^y + 1 = 0$$

$$e^y = \frac{2x \pm \sqrt{4x^2 - 4}}{2}$$

Reject negative choice since it corresponds to the second branch of solution $y = \cosh^{-1}x$. Thus,

$$y = \ln(x + \sqrt{x^2 - 1})$$

b. $x = \tanh y$

$$x = \frac{e^y - e^{-y}}{e^y + e^{-y}}$$

$$xe^y + xe^{-y} = e^y - e^{-y}$$

$$e^y(x - 1) = -e^{-y}(x + 1)$$

$$e^{2y} = \frac{1 + x}{1 - x}$$

$$2y = \ln\frac{1 + x}{1 - x}$$

$$y = \tfrac{1}{2}\ln\frac{1 + x}{1 - x}$$

Thus $\tan^{-1}x = \tfrac{1}{2}\ln\dfrac{1 + x}{1 - x}$

CHAPTER 7 REVIEW

Proficiency Examination, Pages 487–488

SURVIVAL HINT: *To help you review the concepts of this chapter, handwrite the answers to each of these questions onto your own paper.*

1. Let u replace a more complicated expression in the variable of integration, say x. Obtain all forms of x in terms of u. Substitute, integrate, return the form of answers from u back to x. In the case of a definite integral transform the limits of x into limits for u and evaluate using the fundamental theorem.

2. **a.** $\int u \, dv = uv - \int v \, du$

   b. A reduction formula for integration expresses an integral involving a power of a particular function in terms of an integral involving a lower power of the same function. By using the formula repeatedly, the given integral can be reduced to one that is more manageable.

3. **a.** A trigonometric substitution may be handy when the integrand contains one of the following forms:

   $$\sqrt{x^2 + a^2} \qquad \sqrt{x^2 - a^2} \qquad \sqrt{a^2 - x^2}$$

   b. The Weierstrass substitutions are:

$$u = \tan \frac{x}{2}, \qquad \sin x = \frac{2u}{1 + u^2};$$

$$\cos x = \frac{1 - u^2}{1 + u^2}, \text{ and } dx = \frac{2 \, du}{1 + u^2}.$$

4. The method of partial fractions may be handy when integrating a rational function.

5. See Pages 457–458:

   Step 1: simplify

   Step 2: use basic formulas (check table)

   Step 3: substitute

   Step 4: classify; parts, trig powers, Weierstrass substitution, trig substitutions, or partial fractions

   Step 5: try again

6. Equations that can be expressed in the form

   $$\frac{dy}{dx} + P(x)y = Q(x)$$

   are called first-order linear differential equations. The general solution is given by

   $$y = \frac{1}{I(x)}\left[\int Q(x)I(x) \, dx + C \right]$$

   where $I(x) = e^{\int P(x)\,dx}$.

7. An improper integral is one in which a limit of integration is infinite and/or at least one value in the interval of integration leads to an undefined integrand.

8. $\sinh x = \frac{1}{2}(e^x - e^{-x})$; $\cosh x = \frac{1}{2}(e^x + e^{-x})$;

    $\tanh x = \dfrac{e^x - e^{-x}}{e^x + e^{-x}}$

9. Let u be a differentiable function of x. Then:

    $\dfrac{d}{dx}(\sinh u) = \cosh u \dfrac{du}{dx}$

    $\dfrac{d}{dx}(\cosh u) = \sinh u \dfrac{du}{dx}$

    $\dfrac{d}{dx}(\tanh u) = \operatorname{sech}^2 u \dfrac{du}{dx}$

    $\dfrac{d}{dx}(\coth u) = -\operatorname{csch}^2 u \dfrac{du}{dx}$

    $\dfrac{d}{dx}(\operatorname{sech} u) = -\operatorname{sech} u \tanh u \dfrac{du}{dx}$

    $\dfrac{d}{dx}(\operatorname{csch} u) = -\operatorname{csch} u \coth u \dfrac{du}{dx}$

    $\displaystyle\int \sinh u \, du = \cosh u + C$

    $\displaystyle\int \cosh u \, du = \sinh u + C$

    $\displaystyle\int \operatorname{sech}^2 u \, du = \tanh u + C$

    $\displaystyle\int \operatorname{csch}^2 u \, du = -\coth u + C$

    $\displaystyle\int \operatorname{sech} u \tanh u \, du = -\operatorname{sech} u + C$

    $\displaystyle\int \operatorname{csch} u \coth u \, du = -\operatorname{csch} u + C$

10. $\sinh^{-1} x = \ln(x + \sqrt{x^2 + 1})$, all x

$\operatorname{csch}^{-1} x = \ln\left(\dfrac{1}{x} + \dfrac{\sqrt{1 + x^2}}{|x|}\right)$, $x \neq 0$

$\cosh^{-1} x = \ln(x + \sqrt{x^2 - 1})$, $x \geq 1$

$\operatorname{sech}^{-1} x = \ln\left(\dfrac{1 + \sqrt{1 - x^2}}{x}\right)$, $0 < x \leq 1$

$\tanh^{-1} x = \frac{1}{2}\ln\dfrac{1 + x}{1 - x}$, $|x| < 1$

$\coth^{-1} x = \frac{1}{2}\ln\dfrac{x + 1}{x - 1}$, $|x| > 1$

11. $\dfrac{d}{dx}(\sinh^{-1} u) = \dfrac{1}{\sqrt{1 + u^2}}\dfrac{du}{dx}$

$\dfrac{d}{dx}(\cosh^{-1} u) = \dfrac{1}{\sqrt{u^2 - 1}}\dfrac{du}{dx}$

$\dfrac{d}{dx}(\tanh^{-1} u) = \dfrac{1}{1 - u^2}\dfrac{du}{dx}$

$\dfrac{d}{dx}(\operatorname{csch}^{-1} u) = \dfrac{-1}{|u|\sqrt{1 + u^2}}\dfrac{du}{dx}$

$\dfrac{d}{dx}(\operatorname{sech}^{-1} u) = \dfrac{-1}{u\sqrt{1 - u^2}}\dfrac{du}{dx}$

$\dfrac{d}{dx}(\coth^{-1} u) = \dfrac{1}{1 - u^2}\dfrac{du}{dx}$

$\displaystyle\int \dfrac{du}{\sqrt{1 + u^2}} = \sinh^{-1} u + C$

$$\int \frac{du}{\sqrt{u^2 - 1}} = \cosh^{-1} u + C$$

$$\int \frac{du}{1 - u^2} = \tanh^{-1} u + C$$

$$\int \frac{du}{u\sqrt{1 + u^2}} = -\operatorname{csch}^{-1}|u| + C$$

$$\int \frac{du}{u\sqrt{1 - u^2}} = -\operatorname{sech}^{-1}|u| + C$$

$$\int \frac{du}{1 - u^2} = \coth^{-1} u + C$$

12. a. $\tanh^{-1}(0.5) = \frac{1}{2} \ln \frac{1 + 0.5}{1 - 0.5}$

$$\approx 0.5493$$

b. $\sinh(\ln 3) = \dfrac{e^{\ln 3} - e^{-\ln 3}}{2}$

$$= \frac{4}{3}$$

c. $\coth^{-1} 2 = \frac{1}{2} \ln \frac{2 + 1}{2 - 1}$

$$= \frac{1}{2} \ln 3$$

$$\approx 0.5493$$

13. $\displaystyle\int \frac{2x + 3}{\sqrt{x^2 + 1}} \, dx = \int \frac{2x \, dx}{\sqrt{x^2 + 1}} + \int \frac{3 \, dx}{\sqrt{x^2 + 1}}$

$$\boxed{\text{Formula 172}}$$

$$= 2\sqrt{x^2 + 1} + 3 \sinh^{-1} x + C$$

14. $\displaystyle\int x \sin 2x \, dx$ $\boxed{u = x, \; dv = \sin 2x \, dx}$

$$= -\frac{x}{2} \cos 2x - \int -\frac{1}{2} \cos 2x \, dx$$

$$= -\frac{x}{2} \cos 2x + \frac{1}{4} \sin 2x + C$$

15. $\displaystyle\int \sinh(1 - 2x) \, dx = -\frac{1}{2} \int \sinh(1 - 2x) \, (-2 \, dx)$

$$= -\frac{1}{2} \cosh(1 - 2x) + C$$

16. $\displaystyle\int \frac{dx}{\sqrt{4 - x^2}} = \sin^{-1} \frac{x}{2} + C$

17. $\dfrac{x^2}{(x^2 + 1)(x - 1)} = \dfrac{A_1 x + B_1}{x^2 + 1} + \dfrac{A_2}{x - 1}$

Partial fraction decomposition gives

$A_1 = \frac{1}{2}, \; A_2 = \frac{1}{2}, \; B_1 = \frac{1}{2}.$

$$\int \frac{x^2 \, dx}{(x^2 + 1)(x - 1)} = \frac{1}{2} \int \frac{x + 1}{x^2 + 1} \, dx + \frac{1}{2} \int \frac{dx}{x - 1}$$

$$= \frac{1}{4} \int \frac{2x \, dx}{x^2 + 1} + \frac{1}{2} \int \frac{dx}{x^2 + 1} + \frac{1}{2} \int \frac{dx}{x - 1}$$

$$= \frac{1}{4} \ln(x^2 + 1) + \frac{1}{2} \tan^{-1} x + \frac{1}{2} \ln|x - 1| + C$$

18. $\displaystyle\int \frac{x^3 \, dx}{x^2 - 1} = \int \left(x + \frac{x}{x^2 - 1} \right) dx$

$$= \int x \, dx + \frac{1}{2} \int \frac{2x \, dx}{x^2 - 1}$$

$$= \frac{x^2}{2} + \frac{1}{2} \ln|x^2 - 1| + C$$

19. $\displaystyle\int_1^2 x \ln x^3 \, dx = 3 \int_1^2 x \ln x \, dx$

$$\boxed{\text{Parts or Formula 502}}$$

$$= 3\left(\frac{x^2}{2} \ln x - \frac{x^2}{4}\right)\Big|_1^2$$

$$= 6 \ln 2 - \frac{9}{4}$$

$$\approx 1.9089$$

20. $\displaystyle\frac{1}{(x-1)^2(x+2)} = \frac{A_1}{(x-1)^2} + \frac{A_2}{x-1} + \frac{A_3}{x+2}$

Partial fraction decomposition gives

$A_1 = \frac{1}{3}$, $A_2 = -\frac{1}{9}$, $A_3 = \frac{1}{9}$

$$\int_2^3 \frac{dx}{(x-1)^2(x+2)}$$

$$= \frac{1}{9}\int_2^3 \left(\frac{3}{(x-1)^2} - \frac{1}{x-1} + \frac{1}{x+2}\right) dx$$

$$= \frac{1}{9}\left[-\frac{3}{x-1} - \ln(x-1) + \ln(x+2)\right]\Big|_2^3$$

$$= \frac{1}{9}\left(\ln \frac{5}{8} + \frac{3}{2}\right)$$

$$\approx 0.1144$$

21. $\displaystyle\int_3^4 \frac{dx}{2x - x^2} = \int_3^4 \left(\frac{\frac{1}{2}}{x} + \frac{\frac{1}{2}}{2-x}\right) dx$

$$= \frac{1}{2} \ln x - \frac{1}{2} \ln|2 - x|\Big|_3^4$$

$$= \frac{1}{2} \ln \frac{2}{3} = -0.2027$$

22. $\displaystyle\int_0^{\pi/4} (\sec^2 x)(\sec x \tan x \, dx) = \frac{\sec^3 x}{3}\Big|_0^{\pi/4}$

$$= \frac{2\sqrt{2} - 1}{3}$$

$$\approx 0.6095$$

23. $\displaystyle\lim_{N\to+\infty} \int_0^N x e^{-2x} \, dx$ $\boxed{\text{parts or Formula 484}}$

$$= \lim_{N\to+\infty}\left[-\frac{x}{2} e^{-2x} - \frac{1}{4} e^{-2x}\right]\Big|_0^N$$

$$= \lim_{N\to+\infty}\left[-\frac{1}{4} e^{-2N}(2N + 1) + \frac{1}{4}\right]$$

$$= \frac{1}{4}$$

24. $\displaystyle\int_0^{\pi/4} \frac{\sec^2 x}{\sqrt{\tan x}} \, dx$ $\boxed{u = \tan x}$

$$= \int_0^1 \frac{du}{\sqrt{u}}$$

$$= \lim_{t\to 0^+} 2\sqrt{u}\Big|_t^1$$

$$= 2$$

25. $\dfrac{2x + 3}{x^2(x - 2)} = \dfrac{A_1}{x^2} + \dfrac{A_2}{x} + \dfrac{A_3}{x - 2}$

Multiply both sides by $x^2(x - 2)$:

$2x + 3 = A_1(x - 2) + A_2 x(x - 2) + A_3 x^2$

If $x = 2$, then $7 = 4A_3$, so $A_3 = \frac{7}{4}$.

If $x = 0$, then $3 = -2A_1$, so $A_1 = -\frac{3}{2}$.

By substitution, $A_2 = -\frac{7}{4}$.

$\displaystyle\int_0^1 \dfrac{2x + 3}{x^2(x - 2)}\, dx$

$= \displaystyle\lim_{t \to 0^+} \int_t^1 \left(\dfrac{-\frac{3}{2}}{x^2} + \dfrac{-\frac{7}{4}}{x} + \dfrac{\frac{7}{4}}{x - 2} \right) dx$

$= \displaystyle\lim_{t \to 0^+} \left[\dfrac{3}{2x} + \dfrac{7}{4} \ln\left| \dfrac{x - 2}{x} \right| \, \Big|_t^1 \right]$

$= \displaystyle\lim_{t \to 0^+} \left[\dfrac{3}{2} + \dfrac{7}{4} \ln 1 - \dfrac{3}{2t} - \dfrac{7}{4} \ln\left| \dfrac{t - 2}{t} \right| \right]$

$= -\infty$; diverges

26. $\displaystyle\int_0^{+\infty} e^{-x} \sin x \, dx = \lim_{t \to +\infty} \int_0^t e^{-x} \sin x \, dx$

$\boxed{\text{Formula 492}}$

$= \displaystyle\lim_{t \to +\infty} \left[\dfrac{e^{-x}(-\sin x - \cos x)}{2} \, \Big|_0^t \right]$

$= \displaystyle\lim_{t \to +\infty} \left[\dfrac{e^{-t}(-\sin t - \cos t)}{2} + \dfrac{1}{2} \right]$

$= \dfrac{1}{2}$

27. $y' = \dfrac{1}{2\sqrt{\tanh^{-1} 2x}} \left[\dfrac{2}{1 - 4x^2} \right]$

$= \dfrac{1}{\sqrt{\tanh^{-1} 2x}} \left[\dfrac{1}{1 - 4x^2} \right]$

28. $2\pi \displaystyle\int_0^2 x \left(\dfrac{1}{\sqrt{9 - x^2}} \right) dx = -\pi \int_0^2 \dfrac{-2x \, dx}{\sqrt{9 - x^2}}$

$= -2\pi \left(\sqrt{9 - x^2} \right) \Big|_0^2$

$= -2\pi \left(\sqrt{5} - 3 \right)$

≈ 4.7999

29. $\dfrac{dy}{dx} + \left(\dfrac{x}{x + 1} \right) y = e^{-x}$

$P(x) = \dfrac{x}{x + 1}; \ Q(x) = e^{-x}$

$I(x) = e^{\int x/(x+1) \, dx} = \dfrac{e^x}{x + 1}$

$y = \dfrac{x + 1}{e^x} \left[\displaystyle\int \dfrac{e^x}{x + 1}(e^{-x}) \, dx + C \right]$

$= \left(\dfrac{x + 1}{e^x} \right) [\ln|x + 1| + C]$

Since $y = 1$ when $x = 0$, $C = 1$; thus,

$y = \dfrac{x + 1}{e^x}(\ln|x + 1| + 1)$

30. Let $S(t)$ be the amount of salt in the solution at time t.

$\dfrac{dS}{dt} = (1.3)(5) - \left(\dfrac{S}{200 + 2t} \right)(3)$

so

$$\frac{dS}{dt} + \frac{3S}{200 + 2t} = 6.5$$

$$I(t) = e^{\int 3/(200+2t)\ dt}$$

$$= (t + 100)^{3/2}$$

$$S(t) = (t + 100)^{-3/2}\left[\int 6.5(t + 100)^{3/2}\ dt + C\right]$$

$$= (t + 100)^{-3/2}\left[2.6(t + 100)^{5/2} + C\right]$$

$$= 2.6(t + 100) + C(t + 100)^{-3/2}$$

Since $S(0) = 200(2) = 400$, we find

$$C = -140,000 \text{ so}$$

$$S(t) = 2.6(t + 100) - 140,000(t + 100)^{-3/2}$$

After one hour, $S(60) \approx 485.2$ lb

CHAPTER 8

Infinite Series

8.1 Sequences and Their Limits, Pages 503-505

1. The limit of a sequence is that unique number that the elements of a sequence approaches as more and more numbers in the sequence are considered.

3. $\{1 + (-1)^n\}$;

   If $n = 1$, then $a_1 = 1 + (-1)^1 = 0$;

   if $n = 2$, then $a_2 = 1 + (-1)^2 = 2$;

   if $n = 3$, then $a_3 = 1 + (-1)^3 = 0$;

   if $n = 4$, then $a_4 = 1 + (-1)^4 = 2$;

   if $n = 5$, then $a_5 = 1 + (-1)^5 = 0$.

   The first five terms of the sequence are

   $0, 2, 0, 2, 0$.

5. $\left\{\dfrac{\cos 2n\pi}{n}\right\}$;

   If $n = 1$, then $a_1 = \dfrac{\cos 2\pi}{1} = 1$;

   if $n = 2$, then $a_2 = \dfrac{\cos 4\pi}{2} = \dfrac{1}{2}$;

   if $n = 3$, t4en $a_3 = \dfrac{\cos 6\pi}{3} = \dfrac{1}{3}$;

   if $n = 4$, then $a_4 = \dfrac{\cos 8\pi}{4} = \dfrac{1}{4}$;

   if $n = 5$, then $a_5 = \dfrac{\cos 4\pi}{5} = \dfrac{1}{5}$.

The first five terms of the sequence are

$1, \frac{1}{2}, \frac{1}{3}, \frac{1}{4}, \frac{1}{5}$.

7. $\left\{\dfrac{3n + 1}{n + 2}\right\}$;

   If $n = 1$, the $a_1 = \dfrac{3(1) + 1}{1 + 2} = \dfrac{4}{3}$;

   if $n = 2$, the $a_2 = \dfrac{3(2) + 1}{2 + 2} = \dfrac{7}{4}$;

   if $n = 3$, the $a_3 = \dfrac{3(3) + 1}{3 + 2} = \dfrac{10}{5} = 2$;

   if $n = 4$, the $a_4 = \dfrac{3(4) + 1}{4 + 2} = \dfrac{13}{6}$;

   if $n = 5$, the $a_5 = \dfrac{3(5) + 1}{5 + 2} = \dfrac{16}{7}$;

The first five terms of the sequence are

$\frac{4}{3}, \frac{7}{4}, 2, \frac{13}{6}, \frac{16}{7}$.

9. $a_n = \sqrt{a_{n-1}}$, $n \geq 2$

   $a_1 = 256$ (given)

   $a_2 = \sqrt{256} = 16$

   $a_3 = \sqrt{16} = 4$

   $a_4 = \sqrt{4} = 2$

   $a_5 = \sqrt{2}$

The first five terms of the sequence are

$256, 16, 4, 2, \sqrt{2}$.

11. $a_n = (a_{n-1})^2 + a_{n-1} + 1, \ n \geq 2$

$a_1 = 1$ (given)

$a_2 = 1^2 + 1 + 1 = 3$

$a_3 = 3^2 + 3 + 1 = 13$

$a_4 = 13^2 + 13 + 1 = 183$

$a_5 = 183^2 + 183 + 1 = 33{,}673$

The first five terms of the sequence are

1, 3, 13, 183, 33673.

SURVIVAL HINT: *Many limits (such as those in Problems 12-19) can be evaluated by multiplying by 1, written as*

$$\frac{\frac{1}{n^k}}{\frac{1}{n^k}}$$

where k is the largest power of n in the expression. For example, in Problem 17 (shown below) we multiply by 1 written as

$$\frac{\frac{1}{n^2}}{\frac{1}{n^2}}$$

13. $\displaystyle\lim_{n\to\infty} \frac{5n}{n+7} = \lim_{n\to\infty} \frac{5}{1 + \frac{7}{n}}$

$\qquad = 5$

15. $\displaystyle\lim_{n\to\infty} \frac{4 - 7n}{8 + n} = \lim_{n\to\infty} \frac{\frac{4}{n} - 7}{\frac{8}{n} + 1}$

$\qquad = -7$

17. $\displaystyle\lim_{n\to\infty} \frac{100n + 7{,}000}{n^2 - n - 1} = \lim_{n\to\infty} \frac{\frac{100}{n} + \frac{7{,}000}{n^2}}{1 - \frac{1}{n} - \frac{1}{n^2}}$

$\qquad = 0$

19. $\displaystyle\lim_{n\to\infty} \frac{n^3 - 6n^2 + 85}{2n^3 - 5n + 170} = \lim_{n\to\infty} \frac{\frac{n^3}{n^3} - \frac{6n^2}{n^3} + \frac{85}{n^3}}{\frac{2n^3}{n^3} - \frac{5n}{n^3} + \frac{170}{n^3}}$

$\qquad = \displaystyle\lim_{n\to\infty} \frac{1 - \frac{6}{n} + \frac{85}{n^3}}{2 - \frac{5}{n^2} + \frac{170}{n^3}}$

$\qquad = \dfrac{1 - 0 + 0}{2 - 0 + 0}$

$\qquad = \dfrac{1}{2}$

SURVIVAL HINT: *l'Hôpital's rule can be used as an alternate method for Problem 19 (here we apply it three times):*

$\displaystyle\lim_{n\to\infty} \frac{n^3 - 6n^2 + 85}{2n^3 - 5n + 170} = \lim_{n\to\infty} \frac{3n^2 - 12n}{6n^2 - 5}$

$\qquad = \displaystyle\lim_{n\to\infty} \frac{6n - 12}{12n}$

$\qquad = \displaystyle\lim_{n\to\infty} \frac{6}{12}$

$\qquad = \dfrac{1}{2}$

Never use l'Hôpital's rule without first verifying that the limit has the form 0/0 or ∞/∞.

21. $\displaystyle\lim_{n\to\infty} \frac{8n - 500\sqrt{n}}{2n + 800\sqrt{n}} = \lim_{n\to\infty} \frac{8 - \dfrac{500}{\sqrt{n}}}{2 + \dfrac{800}{\sqrt{n}}}$

$$= \frac{8}{2}$$

$$= 4$$

23. l'Hôpital's rule:

$$\lim_{n\to\infty} \frac{\ln n}{n^2} = \lim_{n\to\infty} \frac{\dfrac{1}{n}}{2n}$$

$$= \lim_{n\to\infty} \frac{1}{2n^2}$$

$$= 0$$

25. l'Hôpital's rule is used: Let $L = \displaystyle\lim_{n\to\infty} n^{3/n}$.

$$\ln L = \lim_{n\to\infty} \frac{3 \ln n}{n}$$

$$= \lim_{n\to\infty} \frac{3\left(\dfrac{1}{n}\right)}{1}$$

$$= 0; \quad L = e^0 = 1$$

27. l'Hôpital's rule is used:

Let $L = \displaystyle\lim_{n\to\infty} (n + 4)^{1/n}$

$$\ln L = \lim_{n\to\infty} \frac{\ln(n + 4)}{n}$$

$$= \lim_{n\to\infty} \frac{\dfrac{1}{n + 4}}{1}$$

$$= 0$$

$$L = e^0 = 1$$

29. l'Hôpital's rule is used:

Let $L = \displaystyle\lim_{n\to\infty} (\ln n)^{1/n}$

$$\ln L = \lim_{n\to\infty} \frac{\ln(\ln n)}{n}$$

$$= \lim_{n\to\infty} \frac{\dfrac{1}{\ln n}\left(\dfrac{1}{n}\right)}{1}$$

$$= 0$$

$$L = e^0 = 1$$

31. $\displaystyle\lim_{n\to\infty} (\sqrt{n^2 + n} - n)$

$$= \lim_{n\to\infty} \frac{(\sqrt{n^2 + n} - n)(\sqrt{n^2 + n} + n)}{\sqrt{n^2 + n} + n}$$

$$= \lim_{n\to\infty} \frac{n^2 + n - n^2}{\sqrt{n^2 + n} + n}$$

$$= \lim_{n\to\infty} \frac{n}{\sqrt{n^2 + n} + n}$$

$$= \lim_{n\to\infty} \frac{1}{\sqrt{1 + \dfrac{1}{n}} + 1}$$

$$= \frac{1}{2}$$

33. l'Hôpital's rule is used: Let $L = \displaystyle\lim_{n\to\infty} \sqrt[n]{n}$

$$\ln L = \lim_{n\to\infty} \frac{\ln n}{n}$$

$$= \lim_{n\to\infty} \frac{\dfrac{1}{n}}{1}$$

$$= 0$$

$$L = e^0 = 1$$

35. l'Hôpital's rule is used:

Let $L = \lim\limits_{n \to \infty} (an + b)^{1/n}$

$\ln L = \lim\limits_{n \to \infty} \dfrac{\ln(an + b)}{n}$

$= \lim\limits_{n \to \infty} \dfrac{\frac{a}{an + b}}{1}$

$= 0$

$L = e^0 = 1$

37. The elements of $\{a_n\} = \left\{ \ln\left(\frac{n+1}{n}\right) \right\}$

lie on the curve $f(x) = \ln(x + 1) - \ln x$.

$f'(x) = (x + 1)^{-1} - x^{-1} < 0$. Thus, $f(x)$

and $\{a_n\}$ are both decreasing. $M = 0$ is a lower

bound of the sequence (the elements are positive

since $\ln(n + 1) > \ln n$), so $\{a_n\}$ converges.

39. The elements of $\{a_n\} = \left\{ \frac{4n + 5}{n} \right\}$

lie on the curve $f(x) = 4 + 5x^{-1}$.

$f'(x) = -5x^{-2} < 0$. Thus, $f(x)$ and $\{a_n\}$ are

both decreasing. $M = 4$ is a lower bound of the

sequence, so $\{a_n\}$ converges.

41. The elements of $\{a_n\} = \left\{ \sqrt[n]{n} \right\}$ lie on the curve

$f(x) = \sqrt[x]{x} = \exp[\ln x^{1/x}]$

$= \exp[x^{-1}\ln x].$

$f'(x) = \exp[\frac{\ln x}{x}](x^{-2} - x^{-2}\ln x) < 0$ (when

$e < x$). Note: if you use technology for the

derivative you may obtain the following

equivalent form: $f'(x) = x^{(1 - 2x)/x}(1 - \ln x)$

Thus, $f(x)$ and $\{a_n\}$ are both decreasing. $M = 0$

is a lower bound of the sequence since $n^{1/n} > 0$

for all n, so $\{a_n\}$ converges.

43. The elements $\{a_n\} = \{\cos n\pi\}$ alternate between

-1 and 1, so the sequence diverges by

oscillation.

45. The sequence $\{a_n\} = \{\sqrt{n}\}$ diverges because

$\lim\limits_{n \to \infty} a_n = \infty.$

47. $a_1 = \frac{1}{2}, \ a_2 = \frac{1}{4}, \ a_3 = \frac{1}{8}, \ a_4 = \frac{1}{16}, \cdots, \ a_n = \left(\frac{1}{2}\right)^n$

$a_4 = 6.25\%, \ a_n = 100\left(\frac{1}{2}\right)^n\%$

49. $\left| \dfrac{n}{n + 1} - 1 \right| < 0.01$

$\left| \dfrac{n - n - 1}{n + 1} \right| < 0.01$

$n + 1 > 100$

$n > 99$

Choose $N = 100$.

51. $\left| \dfrac{n^2 + 1}{n^3} - 0 \right| < 0.001$

$\dfrac{n^2 + 1}{n^3} < 0.001$

$n^3 - 1{,}000n^2 - 1{,}000 > 0$

Since $n^3 - 1,000n^2 - 1,000 < 0$ when

$n = 1,000$, and > 0 when $n = 1,001$, we

choose $N = 1,001$.

53. **a.** Let A be the least upper bound of the

nondecreasing sequence $\{a_n\}$. Since $\epsilon > 0$,

it follows that $A - \epsilon < A$ and

$$A - \epsilon < a_N < A$$

for some integer N. If $n > N$, then

$a_n \geq a_N$ since the sequence is

nondecreasing, and

$$A - \epsilon < a_N \leq a_n \leq A$$

b. The inequality $A - \epsilon < a_n \leq A$ from

part **a** can be written as

$$A - \epsilon < a_n \leq A + \epsilon$$

since $\epsilon > 0$. Thus,

$$\left| a_n - A \right| < \epsilon$$

for all $n > N$ and $\lim_{n \to \infty} a_n = A$.

8.2 Introduction to Infinite Series: Geometric Series, Pages 511-514

SURVIVAL HINT: *The last section dealt with sequences (range values of a function whose domain is the set of natural numbers). and this section deals with series (the sum of the terms of a sequence). A sequence may converge, while its related series diverges. An example of this situation is given as a warning on page 511 of the text.*

1. A sequence is a succession of elements. A series is a sum of the elements of the sequence.

3. $S = \sum_{k=0}^{\infty} \left(\frac{4}{5}\right)^k$

$$= \frac{1}{1 - \frac{4}{5}}$$

$$= 5$$

5. $S = \sum_{k=0}^{\infty} \frac{2}{3^k}$

$$= \frac{2}{1 - \frac{1}{3}}$$

$$= 3$$

7. $S = \sum_{k=1}^{\infty} \left(\frac{3}{2}\right)^k$; this is a geometric series with

$r = \frac{3}{2} > 1$, so it diverges.

9. $S = \sum_{k=0}^{\infty} \frac{(-2)^k}{5^{2k+1}} = \frac{1}{5} - \frac{2}{5^3} + \frac{2^2}{5^5} - \frac{2^3}{5^7} + \cdots$

$$= \frac{1}{5}\left[1 - \frac{2}{5^2} + \left(\frac{2}{5^2}\right)^2 - \left(\frac{2}{5^2}\right)^3 + \cdots\right]$$

$$= \frac{1}{5}\left[\frac{1}{1 - \left(-\frac{2}{5^2}\right)}\right]$$

$$= \frac{5}{27}$$

11. $S = \sum_{k=1}^{\infty} e^{-0.2k}$

$$= e^{-0.2}(1 - e^{-0.2})^{-1}$$

$$\approx 4.51665$$

13. $S = \sum_{k=2}^{\infty} \frac{(-2)^{k-1}}{3^{k+1}}$

$$= -\frac{2}{27}\left(\frac{1}{1+\frac{2}{3}}\right)$$

$$= -\frac{2}{45}$$

15. $S = \frac{1}{2} - \frac{1}{2^2} + \frac{1}{2^3} - \frac{1}{2^4} + \cdots$

$$= \frac{1}{2}\left(\frac{1}{1+\frac{1}{2}}\right)$$

$$= \frac{1}{3}$$

17. $S = \frac{1}{4} + \left(\frac{1}{4}\right)^4 + \left(\frac{1}{4}\right)^7 + \left(\frac{1}{4}\right)^{10} + \cdots$

$$= \frac{1}{4}\left(\frac{1}{1-\frac{1}{64}}\right)$$

$$= \frac{16}{63}$$

19. $S = 2 + \sqrt{2} + 1 + \frac{1}{\sqrt{2}} + \cdots$

$$= 2\left[1 + \frac{1}{\sqrt{2}} + \left(\frac{1}{\sqrt{2}}\right)^2 + \left(\frac{1}{\sqrt{2}}\right)^3 + \cdots\right]$$

$$= 2\left(\frac{1}{1 - 1/\sqrt{2}}\right)$$

$$= 2(2 + \sqrt{2})$$

21. $S = (1 + \sqrt{2}) + 1 + (-1 + \sqrt{2}) + \cdots$

$$= (1 + \sqrt{2})\left[1 + \left(\frac{1}{1+\sqrt{2}}\right) + \left(\frac{1}{1+\sqrt{2}}\right)^2 + \cdots\right]$$

$$= (1 + \sqrt{2})\left[\frac{1}{1 - 1/(1 + \sqrt{2})}\right]$$

$$= \frac{(1 + \sqrt{2})^2}{\sqrt{2}} = \frac{1}{2}(4 + 3\sqrt{2})$$

23. $S_n = \sum_{k=1}^{n}\left[\frac{1}{k^{0.1}} - \frac{1}{(k+1)^{0.1}}\right]$

$$= \left(1 - \frac{1}{2^{0.1}}\right) + \left(\frac{1}{2^{0.1}} - \frac{1}{3^{0.1}}\right) + \cdots$$

$$+ \left[\frac{1}{n^{0.1}} - \frac{1}{(n+1)^{0.1}}\right]$$

$$= 1 - \frac{1}{(n+1)^{0.1}}$$

$$S = \lim_{n\to\infty} S_n$$

$$= \sum_{k=1}^{\infty}\left[\frac{1}{k^{0.1}} - \frac{1}{(k+1)^{0.1}}\right]$$

$$= \lim_{n\to\infty}\left[1 - \frac{1}{(n+1)^{0.1}}\right] = 1$$

The series converges to 1.

25. Note the starting value of k, so we let

$$S_n = \sum_{k=0}^{n} \frac{1}{(k+1)(k+2)}$$

$$= \sum_{k=0}^{n}\left[\frac{1}{k+1} - \frac{1}{k+2}\right] \quad \textit{Partial fractions}$$

$$S_n = (1 - \tfrac{1}{2}) + (\tfrac{1}{2} - \tfrac{1}{3}) + \cdots$$

$$+ [(n+1)^{-1} - (n+2)^{-1}]$$

$$= 1 - (n + 2)^{-1}$$

$$S = \lim_{n \to \infty} S_n = 1$$

The series converges to 1.

27. $S_n = \sum_{k=1}^{n} \ln\left(1 + \frac{1}{k}\right)$

$$= \sum_{k=1}^{n} [\ln(k + 1) - \ln k]$$
$$= (\ln 2 - \ln 1) + (\ln 3 - \ln 2)$$
$$\quad + (\ln 4 - \ln 3) + \cdots$$
$$\quad + [\ln(n + 1) - \ln n]$$
$$= \ln(n + 1) - \ln 1 = \ln(n + 1)$$

$$S = \lim_{n \to \infty} \ln(n + 1); \text{ series diverges}$$

29. $S_n = \sum_{k=1}^{n} \frac{2k + 1}{k^2(k + 1)^2}$

$$= \sum_{k=1}^{n} \left[\frac{1}{k^2} - \frac{1}{(k + 1)^2}\right] \text{ Partial fractions}$$

$$S_n = \left(1 - \tfrac{1}{4}\right) + \left(\tfrac{1}{4} - \tfrac{1}{9}\right) + \cdots$$
$$\quad + [n^{-2} - (n + 1)^{-2}]$$
$$= 1 - (n + 1)^{-2}$$

$$S = \lim_{n \to \infty} S_n = 1; \text{ The series converges to 1.}$$

31. $0.0101010\cdots = \frac{1}{100} + \frac{1}{10,000} + \frac{1}{1,000,000} + \cdots,$

is a geometric series with $r = \frac{1}{100}$.

$$S = \frac{\frac{1}{100}}{1 - \frac{1}{100}}$$

$$= \frac{\frac{1}{100}}{\frac{99}{100}}$$

$$= \frac{1}{99}$$

SURVIVAL HINT: *The procedure shown in Problem 33 illustrates a procedure for changing any repeating decimal into fractional form.*

33. $1.405405\cdots$

$$= 1 + 0.405[1 + \frac{1}{1,000} + (\frac{1}{1,000})^2 + \cdots]$$

is a geometric series with $r = \frac{1}{1,000}$.

$$S = 1 + \frac{\frac{405}{1,000}}{1 - \frac{1}{1,000}}$$

$$= 1 + \frac{405}{999}$$

$$= \frac{52}{37}$$

35. **a.** Adding on the right:

$$\frac{Ak}{2^k} - \frac{B(k + 1)}{2^{k+1}} = \frac{2Ak - B(k + 1)}{2^{k+1}}$$
$$= \frac{k - 1}{2^{k+1}}$$

So $2Ak - Bk = k$ and $-B = -1$; thus,
$A = B = 1.$

b. $S_n = \sum_{k=1}^{n} \frac{k}{2^k} - \sum_{k=1}^{n} \frac{k + 1}{2^{k+1}}$

$$= \left[\frac{1}{2} - \frac{2}{2^2}\right] + \left[\frac{2}{2^2} - \frac{3}{2^3}\right]$$

$$+ \left[\frac{3}{2^3} - \frac{4}{2^4}\right] + \cdots + \left[\frac{n}{2^n} - \frac{n+1}{2^{n+1}}\right]$$

$$= \frac{1}{2} - \frac{n+1}{2^{n+1}}$$

$$S = \lim_{n \to \infty} S_n$$

$$= \lim_{n \to \infty}\left(\frac{1}{2}\right) - \lim_{n \to \infty}\left(\frac{n+1}{2^{n+1}}\right)$$

$$= \frac{1}{2} - 0$$

$$= \frac{1}{2}$$

SURVIVAL HINT: *The solution to Problem 37 uses properties of logarithms. You might wish to review these properties in Theorem 2.9 on Page 83 of the text.*

37. $\dfrac{\ln\left[\dfrac{n^{n+1}}{(n+1)^n}\right]}{n(n+1)} = \dfrac{1}{n(n+1)}[(n+1)\ln n - n\ln(n+1)]$

$$S_N = \sum_{k=1}^{N}\left[\frac{\ln k}{k} - \frac{\ln(k+1)}{k+1}\right]$$

$$= \left(\frac{\ln 1}{1} - \frac{\ln 2}{2}\right) + \left(\frac{\ln 2}{2} - \frac{\ln 3}{3}\right) + \cdots$$

$$+ \left(\frac{\ln N}{N} - \frac{\ln(N+1)}{N+1}\right)$$

$$= \frac{-\ln(N+1)}{N+1}$$

$$S = \lim_{N \to \infty}\left[\frac{-\ln(N+1)}{N+1}\right] = 0$$

39. $2\displaystyle\sum_{k=0}^{\infty} a_k + \sum_{k=0}^{\infty}\frac{1}{2^k} = 2(3.57) + \frac{1}{1 - \frac{1}{2}}$

$$= 9.14$$

41. $\displaystyle\sum_{k=0}^{\infty}(2^{-k} + 3^{-k})^2$

$$= \sum_{k=0}^{\infty}[4^{-k} + 2(6^{-k}) + 9^{-k}]$$

$$= \frac{1}{1 - \frac{1}{4}} + 2\left(\frac{1}{1 - \frac{1}{6}}\right) + \frac{1}{1 - \frac{1}{9}}$$

$$= \frac{583}{120}$$

$$\approx 4.858$$

43. $S_n = \displaystyle\sum_{k=0}^{n}\frac{1}{(a+k)(a+k+1)}$

$$= \sum_{k=0}^{n}\left(\frac{1}{a+k} - \frac{1}{a+k+1}\right)$$

$$= \left(\frac{1}{a+0} - \frac{1}{a+1}\right) + \left(\frac{1}{a+1} - \frac{1}{a+2}\right)$$

$$+ \cdots + \left(\frac{1}{a+n} - \frac{1}{a+n+1}\right)$$

$$= \frac{1}{a} - \frac{1}{a+n+1}$$

$$S = \lim_{n \to \infty} S_n = \frac{1}{a}$$

45. $\displaystyle\sum_{k=0}^{\infty}(a_k - b_k)^2 = \sum_{k=0}^{\infty}(a_k^2 - 2a_k b_k + b_k^2)$

$$= \sum_{k=0}^{\infty} a_k^2 - 2 \sum_{k=0}^{\infty} a_k b_k + \sum_{k=0}^{\infty} b_k^2$$

$$= 4 - 2(3) + 4$$

$$= 2$$

47.
$$N = 500 + \tfrac{2}{3}(500) + \left(\tfrac{2}{3}\right)^2 (500) + \cdots$$
$$+ \left(\tfrac{2}{3}\right)^k (500) + \cdots$$
$$= 500 \left(\frac{1}{1 - \tfrac{2}{3}} \right)$$
$$= 1,500$$

The flywheel will make 1,500 revolutions.

49.
$$D = h + 2(0.75)(h) + 2(0.75)^2(h) + \cdots$$
$$= h + 0.75(2h)[1 + 0.75 + 0.75^2 + \cdots]$$
$$= h + 0.75(2h)\left(\frac{1}{1 - 0.75} \right)$$
$$= h + 0.75(2h)(4)$$
$$= 7h$$

If $7h = 21$, then $h = 3$ ft.

51.
$$D = (10,000)(0.2) + (10,000)(1 - 0.2) +$$
$$+ (10,000)(0.2)(1 - 0.2)^2 + \cdots$$
$$+ (10,000)(0.2)(1 - 0.2)^3 + \cdots$$
$$= 10,000(0.2)(1 + 0.8 + 0.8^2 + \cdots)$$
$$= 10,000(0.2)\left(\frac{1}{1 - 0.8} \right) = 10,000$$

The total depreciation is $10,000.

53. Just before the second injection there will be $20e^{-1/2}$ units; before the third,

$20e^{-2/2} + 20e^{-1/2}$, and so on. Just before the nth injection there will be

$$S = 20e^{-1/2} + 20e^{-1} + \cdots$$
$$= 20e^{-1/2}(1 + e^{-1/2} + \cdots)$$
$$= 20e^{-1/2}\left(\frac{1}{1 - e^{-1/2}} \right)$$
$$\approx 30.8$$

The patient will have about 30.8 units of the drug.

55. Let a_n be the number of trustees on December 31 of the nth year.

$$N = 6 + 6e^{-0.2} + 6e^{-0.2(2)} + 6e^{-0.2(3)} + \cdots$$
$$= \frac{6}{1 - e^{-0.2}}$$
$$\approx 33$$

In the long run there will be 33 members on the board.

SURVIVAL HINT: *Problem 57 is one of those little problems that are part of the history of mathematics with which every literate mathematics student should have some contact. Even if your instructor does not assign this problem, you might want to at least read the problem and its solution.*

57. Let T be the time it takes to run the first half of the course. The total time is

$$T + \tfrac{1}{2}T + \tfrac{1}{4}T + \tfrac{1}{8}T + \cdots = T(1 + \tfrac{1}{2} + \tfrac{1}{4} + \cdots)$$
$$= T\left(\frac{1}{1 - \tfrac{1}{2}} \right)$$

$$= 2T$$

59. a. Each train travels 2.5 miles before the collision, and since they travel at 10 mi/h, it takes

$$T = \frac{2.5 \text{ mi}}{10 \text{ mi/hr}}$$

$$= 0.25 \text{ hr}$$

before they collide. Since the fly travels at 16 mi/hr, it travels a total distance for

$$s = (16 \text{ mi/hr})(0.25 \text{ hr})$$

$$= 4 \text{ mi}$$

before collision.

b. Let t_1 be the time it takes the fly to fly from the first train to the second. Then

DISTANCE BY FLY + DISTANCE BY TRAIN II $= 5$

$$16t_1 + 10t_1 = 5$$

so $t_1 = \frac{5}{26}$ and the distance traveled by the fly is $d_1 = 16(\frac{5}{26}) = \frac{80}{26}$. If t_2 is the time it takes for the fly to return to Train I, then

$$10(t_1 + t_2) + 16t_2 = 5 - 10t_1$$

Solving, we obtain $t_2 = \frac{30}{26^2}$ and

$$d_2 = 16\left(\frac{30}{26^2}\right)$$

$$= \frac{6}{26}\left(\frac{80}{26}\right)$$

Similarly, the time needed for the fly to fly from the present position of Train I to Train II is $t_3 = \frac{180}{26^3}$, and the distance traveled by the fly on this leg of its journey is $d_3 = 16t_3 = \left(\frac{6}{26}\right)^2\left(\frac{80}{26}\right)$. In general, the distance traveled by the fly on the k leg is $d_k = \left(\frac{6}{26}\right)^{k-1}\left(\frac{80}{26}\right)$, and the total distance traveled is

$$s = \sum_{k=1}^{\infty} d_k = \sum_{k=1}^{\infty} \left(\frac{6}{26}\right)^{k-1}\left(\frac{80}{26}\right)$$

$$= \left(\frac{80}{26}\right)\left[\frac{1}{1 - \frac{6}{26}}\right]$$

$$= 4 \text{ miles}$$

61. $A = \sum_{k=1}^{\infty} a_k, \ A_n = \sum_{k=1}^{n} a_k$

$B = \sum_{k=1}^{\infty} b_k, \ B_n = \sum_{k=1}^{n} b_k$

Given $\epsilon > 0$

$$\left|A - A_n\right| < \frac{\epsilon}{2} \text{ for } n > N_1$$

$$\left|B - B_n\right| < \frac{\epsilon}{2} \text{ for } n > N_2$$

$$\left|(A + B) - (A_n + B_n)\right|$$

$$= \left|(A - A_n) + (B - B_n)\right|$$

$$\leq \left|A + A_n\right| + \left|B - B_n\right| < \epsilon$$

which is true if $n > \max(N_1, N_2)$.

Thus, $A + B = \displaystyle\sum_{k=1}^{\infty} (a_k + b_k)$

63. **a.** False; counterexample: let $\Sigma a_k = \Sigma(\frac{4}{3})^k$

which diverges, and $\Sigma b_k = \Sigma\left[\left(\frac{1}{2}\right)^k - \left(\frac{4}{3}\right)^k\right]$

which also diverges; then

$\Sigma(a_k + b_k) = \Sigma(\frac{1}{2})^k$ which converges.

b. True; use Theorem 8.6 (linearity).

65. **a.** $\displaystyle\sum_{k=1}^{n} (a_k - a_{k+2}) = (a_1 - a_3) + (a_2 - a_4)$

$+ (a_3 - a_5) + \cdots + (a_{n-2} - a_n)$

$+ (a_{n-1} - a_{n+1}) + (a_n - a_{n+2})$

$= a_1 - a_3 + a_2 - a_4 + a_3 - a_5 + \cdots$

$+ a_{n-2} - a_n + a_{n-1} - a_{n+1}$

$+ a_n - a_{n+2}$

$= (a_1 + a_2) + 0 + \cdots + 0 - (a_{n+1} + a_{n+2})$

b. $\displaystyle\sum_{k=1}^{\infty} (a_k - a_{k+2}) = \lim_{n\to\infty} \sum_{k=1}^{n} (a_k - a_{k+2})$

$= a_1 + a_2 - \lim_{n\to\infty} a_{n+1} - \lim_{n\to\infty} a_{n+2}$

$= a_1 + a_2 - A - A$

$= a_1 + a_2 - 2A$

c. Let $a_k = k^{1/k}$, and $L = \lim_{k\to\infty} k^{1/k}$

Then

$\ln L = \lim_{k\to\infty} \dfrac{\ln k}{k}$

$= \lim_{k\to\infty} \dfrac{1/k}{1}$

$= 0$

so

$A = \lim_{k\to\infty} a_k$

$= \lim_{k\to\infty} k^{1/k}$

$= e^0$

$= 1$

and by part b.

$\displaystyle\sum_{k=1}^{\infty} (a_k - a_{k+2}) = a_1 + a_2 - 2A$

$= 1 + 2^{1/2} - 2(1)$

$= \sqrt{2} - 1$

d. $\displaystyle\sum_{k=2}^{\infty} \dfrac{1}{k^2 - 1} = \sum_{k=1}^{\infty} \dfrac{1}{k(k+2)}$

$= \dfrac{1}{2} \lim_{k\to\infty} \sum_{k=1}^{n}\left[\dfrac{1}{k} - \dfrac{1}{k+2}\right]$

Let $a_k = \frac{1}{k}$, so $\lim_{k\to\infty} a_k = 0$ and

$\displaystyle\sum_{k=2}^{\infty} \dfrac{1}{k^2 - 1} = \dfrac{1}{2}\left[1 + \dfrac{1}{2} - 2(0)\right]$

$= \dfrac{3}{4}$

67. $A = 1$ and $B = -3$

$S_n = \displaystyle\sum_{k=1}^{n}\left[\dfrac{3^k}{4^k - 3^k} - \dfrac{3^{k+1}}{4^{k+1} - 3^{k+1}}\right]$

$$= \left(\frac{3}{4-3} - \frac{3^2}{4^2 - 3^2} \right)$$

$$+ \left(\frac{3^2}{4^2 - 3^2} - \frac{3^3}{4^3 - 3^3} \right)$$

$$+ \cdots + \left(\frac{3^n}{4^n - 3^n} - \frac{3^{n+1}}{4^{n+1} - 3^{n+1}} \right)$$

To find the sum, we find

$$\lim_{n \to \infty} \left[\frac{3}{4-3} - \frac{3^{n+1}}{4^{n+1} - 3^{n+1}} \right]$$

$$= 3 - \lim_{n \to \infty} \frac{\left(\frac{3}{4} \right)^{n+1}}{1 - \left(\frac{3}{4} \right)^{n+1}} = 3$$

Note: This is interesting because CAS has trouble.

8.3 The Integral Test: *p*-series, Pages 520-521

SURVIVAL HINT: *The convergence of the sequence* a_n *to 0 is a necessary, but not sufficient, condition for the convergence of a series. If the sequence does not converge to 0, then the series **must diverge**. However, if the sequence does converge to 0 the series may either converge or diverge. Developing tests to decide the question of convergence or divergence is one of the main objectives of this chapter.*

*Study the **p-series test** and remember that* $\sum_{k=1}^{\infty} \frac{1}{k^p}$ *converges if p > 1 and diverges if p ≤ 1. We show the solutions to Problems 3 and 5, but you should look at Problems 4 and 6 and check out that the answers are found by identifying a p-series.*

1. The *p*-series is one of the form $\sum_{k=1}^{\infty} \frac{1}{k^p}$

(with *p* a constant).

3. $\sum_{k=1}^{\infty} \frac{1}{k^3}$; $p = 3$; converges

5. $\sum_{k=1}^{\infty} \frac{1}{\sqrt[3]{k}}$; since $\sqrt[3]{k} = k^{1/3}$ we see $p = \frac{1}{3}$;

diverges

7. $S = \sum_{k=1}^{\infty} \frac{1}{(2 + 3k)^2}$

$f > 0$, continuous on $[1, \infty)$, and decreasing.

$$I = \int_1^{\infty} \frac{dx}{(2 + 3x)^2} = \lim_{b \to \infty} \int_1^b (2 + 3x)^{-2} dx$$

$$= -\frac{1}{3} \lim_{b \to \infty} (2 + 3x)^{-1} \Big|_1^b = \frac{1}{15}$$

I converges, so *S* converges.

9. $S = \sum_{k=2}^{\infty} \frac{\ln k}{k}$

$f > 0$, continuous on $[2, \infty)$, and decreasing for $k \geq 3$ (but the term with $k = 2$ cannot change the divergence or convergence).

$$I = \int_2^{\infty} \frac{\ln x}{x} dx = \lim_{b \to \infty} \int_2^b \ln x \frac{dx}{x}$$

$$= \frac{1}{2} \lim_{b \to \infty} (\ln x)^2 \Big|_2^b = \infty$$

I diverges, so *S* diverges.

11. $S = \sum_{k=1}^{\infty} \frac{(\tan^{-1} k)^2}{1 + k^2}$

$f > 0$, continuous on $[1, \infty)$, and decreasing.

$$I = \int_1^\infty \frac{(\tan^{-1}x)^2}{1 + x^2}\, dx$$

$$= \frac{1}{3} \lim_{b \to \infty} (\tan^{-1}x)^3 \Big|_1^b = \frac{\pi^3}{24} - \frac{\pi^3}{192}$$

I converges, so S converges.

13. $S = \sum_{k=1}^\infty \frac{k}{k+1}$ diverges by the divergence

test since $\lim_{k \to \infty} \frac{k}{k+1} = 1 \neq 0$

15. $S = \sum_{k=2}^\infty \frac{k}{\sqrt{k^2 - 1}}$; diverges by the

divergence test since $\lim_{k \to \infty} \frac{k}{\sqrt{k^2 - 1}} = 1 \neq 0$

17. $S = \sum_{k=0}^\infty \cos(k\pi)$ diverges by the divergence

test since $\lim_{k \to \infty} \cos(k\pi)$ does not exist

(divergence by oscillation)

19. $S = \sum_{k=1}^\infty \frac{\ln k}{k^2}$; apply the integral test:

$$I = \lim_{n \to \infty} \int_1^n \frac{\ln x}{x^2}\, dx \qquad \boxed{\text{Formula 505}}$$

$$= \lim_{n \to \infty} \left[-\frac{1}{x}(\ln x + 1) \right]\Big|_1^n$$

$$= \lim_{n \to \infty} \left(-\frac{\ln n}{n} - \frac{1}{n} + 1 \right)$$

$$= 1$$

The series converges.

21. $S = \sum_{k=1}^\infty \left(2 + \frac{3}{k} \right)^k$; $\lim_{k \to \infty} \left(2 + \frac{3}{k} \right)^k = \infty$

because $\lim_{k \to \infty} 2^k = \infty$, so the series diverges.

23. $S = \sum_{k=1}^\infty \frac{1}{k^4}$; this is a p-series where

$p = 4 > 1$, so S converges.

25. $S = \sum_{k=1}^\infty k^{-3/4}$; this is a p-series where

$p = \frac{3}{4} < 1$, so S diverges.

27. $S = \sum_{k=1}^\infty \frac{k}{k^2 + 1}$

$$I = \lim_{n \to \infty} \int_1^n \frac{x\, dx}{x^2 + 1} = \lim_{n \to \infty} \frac{1}{2} \ln(x^2 + 1) \Big|_1^n$$

$$= \lim_{n \to \infty} \frac{1}{2} \left[\ln(n^2 + 1) - \ln 2 \right]$$

$$= \infty$$

The series diverges.

29. $S = \sum_{k=1}^\infty k e^{-k^2}$

$$I = \lim_{n \to \infty} \int_1^n x e^{-x^2}\, dx$$

$$= \lim_{n \to \infty} -\tfrac{1}{2} e^{-x^2} \Big|_1^n$$

$$= \frac{1}{2e}$$

The series converges.

31. $S = \sum_{k=2}^{\infty} \frac{\ln \sqrt{k}}{\sqrt{k}}$ diverges by the integral test:

$$\int_2^{\infty} \frac{\ln \sqrt{x}}{\sqrt{x}} \, dx = \lim_{n \to \infty} \left[\left(2\sqrt{x} \ln \sqrt{x} \right) \Big|_2^n - \int_2^n \frac{dx}{\sqrt{x}} \right]$$

$$\boxed{u = \ln \sqrt{x}, \ dv = dx/\sqrt{x}}$$

$$= \lim_{n \to \infty} \left[\sqrt{x} (\ln x - 2) \right] \Big|_2^n$$

$$= \infty$$

33. $S = \sum_{k=1}^{\infty} \frac{k}{e^k}$

$$I = \lim_{n \to \infty} \int_1^n \frac{x}{e^x} \, dx \quad \boxed{u = x, \ dv = e^{-x} \, dx}$$

$$= \lim_{n \to \infty} \left[-\frac{x}{e^x} \Big|_1^n + \int_1^n e^{-x} dx \right]$$

$$= \lim_{n \to \infty} -\frac{1}{e^x} (x + 1) \Big|_1^n$$

$$= \lim_{n \to \infty} \left(-\frac{n}{e^n} - \frac{1}{e^n} + \frac{2}{e} \right)$$

$$= \frac{2}{e}$$

The series converges.

35. $S = \sum_{k=1}^{\infty} \frac{\tan^{-1} k}{1 + k^2}$

$$I = \lim_{n \to \infty} \int_1^n \frac{\tan^{-1} x \, dx}{1 + x^2}$$

$$= \lim_{n \to \infty} \tfrac{1}{2} (\tan^{-1} x)^2 \Big|_1^n$$

$$= \frac{\pi^2}{8} - \frac{\pi^2}{32}$$

The series converges.

37. $S = \sum_{k=1}^{\infty} \tan^{-1} k$ diverges by the divergence

test: $\lim_{n \to \infty} \tan^{-1} k = \frac{\pi}{2} \neq 0$

39. $S = \sum_{k=1}^{\infty} \frac{k - \sin k}{3k + 2 \sin k}$ diverges by the

divergence test: $\lim_{k \to \infty} \frac{k - \sin k}{3k + 2 \sin k}$

$$= \lim_{k \to \infty} \frac{1 - \frac{\sin k}{k}}{3 + \frac{2 \sin k}{k}} = \frac{1}{3} \neq 0$$

41. $S = \sum_{k=1}^{\infty} k \sin \frac{1}{k}$ diverges by the divergence

test: $\lim_{k \to \infty} k \sin \frac{1}{k}$; let $u = \frac{1}{k}$ so that $u \to 0$

as $k \to \infty$; thus, $\lim_{u \to 0} \frac{\sin u}{u} = 1 \neq 0$.

43. $S = \sum_{k=1}^{\infty} \frac{k - 1}{k + 1}$

Diverges because $\lim_{k \to \infty} \frac{k - 1}{k + 1} = 1 \neq 0$

The necessary condition for convergence is not

satisfied.

45. $S = \sum\limits_{k=1}^{\infty} \dfrac{k^2 + 1}{k^3} = \sum\limits_{k=1}^{\infty} \left(\dfrac{1}{k} + \dfrac{1}{k^3} \right)$

The first term is the divergent harmonic series, and the second terms is a convergent p-series, so S diverges (by Theorem 8.7).

47. $S = \sum\limits_{k=1}^{\infty} \dfrac{1}{k^2}$

This is a p-series with $p = 2 > 1$, so S converges.

SURVIVAL HINT: *Compare and contrast Problems 49 and 50. The issue here is not one of calculator accuracy, but of recognizing when $p > 1$. Note in Problem 49, $2.7321 > \sqrt{3} + 1$.*

49. $S = \sum\limits_{k=1}^{\infty} \dfrac{k^{\sqrt{3}} + 1}{k^{2.7321}}$

$= \sum\limits_{k=1}^{\infty} \left[\dfrac{1}{k^{2.7321 - \sqrt{3}}} + \dfrac{1}{k^{2.7321}} \right]$

Both of these are p-series with $p > 1$, so the series converges.

51. $S = \sum\limits_{k=1}^{\infty} \left[\dfrac{1}{k} + \dfrac{k+1}{k+2} \right]$

S diverges because $\lim\limits_{k\to\infty} \left[\dfrac{1}{k} + \dfrac{k+1}{k+2} \right] = 1 \neq 0$

53. $S = \sum\limits_{k=1}^{\infty} \left[\dfrac{1}{2^k} + \dfrac{2k+3}{3k+4} \right]$

S diverges because

$\lim\limits_{k\to\infty} \left(\dfrac{1}{2^k} + \dfrac{2k+3}{3k+4} \right) = \dfrac{2}{3} \neq 0$

55. $S = \sum\limits_{k=2}^{\infty} \dfrac{k}{(k^2 - 1)^p}$

$I = \lim\limits_{n\to\infty} \dfrac{1}{2} \int\limits_{2}^{n} (x^2 - 1)^{-p} (2x\ dx)$

$= \lim\limits_{n\to\infty} \dfrac{1}{2} \left. \dfrac{(x^2 - 1)^{-p+1}}{-p+1} \right|_{2}^{n}$

$= \dfrac{1}{2(1 - p)} \lim\limits_{n\to\infty} \left[(n^2 - 1)^{1-p} - 3^{1-p} \right]$

The improper integral converges if $p > 1$, so the series also converges for $p > 1$, and diverges if $p \leq 1$.

57. $S = \sum\limits_{k=3}^{\infty} \dfrac{1}{k^p \ln k}$

If $p \leq 1$, the series diverges since

$\int\limits_{3}^{\infty} \dfrac{dx}{x^p \ln x} \geq \int\limits_{3}^{\infty} \dfrac{dx}{x \ln x} = \infty$

If $p > 1$, the series converges since

$\int\limits_{1}^{\infty} \dfrac{dx}{x^p \ln x} \leq \int\limits_{1}^{\infty} \dfrac{dx}{x^p} = \dfrac{1}{p - 1}$

Thus, the series converges if $p > 1$ and diverges if $p \leq 1$.

59. $S = \sum_{k=3}^{\infty} \frac{1}{k \ln k[\ln(\ln k)]^p}$

$$I = \lim_{n \to \infty} \int_3^n [\ln (\ln x)]^{-p} \frac{dx}{x \ln x}$$

$$= \lim_{n \to \infty} \frac{[\ln (\ln x)]^{-p+1}}{1 - p} \Big|_3^n$$

$$= \frac{1}{1 - p} \lim_{n \to \infty} \left\{ [\ln(\ln n)]^{1-p} - [\ln(\ln 3)]^{1-p} \right\}$$

I converges if $p > 1$ and diverges if $p \le 1$, so the series converges if $p > 1$ and diverges if $p \le 1$.

61. For $a_k = \frac{1}{k^{1.1}}$ and $b_k = \frac{1}{k^{0.1}}$, then

$$\lim_{k \to \infty} a_k = \lim_{k \to \infty} b_k = 0 \text{ and } \Sigma a_k b_k = \Sigma k^{-p}$$

where $p = 1.2 > 1$, so it converges. However, $\Sigma(1/k^{1.1})$ is a convergent *p*-series while $\Sigma(1/k^{0.1})$ diverges.

63. From the proof of the integral test, we have

$$\int_1^{n+1} f(x) \, dx < S_n < a_1 + \int_1^n f(x) \, dx$$

where $S_n = \sum_{k=1}^{n} a_k$ is the *n*th partial sum of $\sum_{k=1}^{\infty} a_k$. Since $\sum_{k=1}^{\infty} a_k$ converges, we know that $\lim_{n \to \infty} S_n$ exists (and is finite). Thus, by the squeeze theorem

$$\lim_{n \to \infty} \int_1^{n+1} f(x) \, dx \le \lim_{n \to \infty} S_n \le a_1 + \lim_{n \to \infty} \int_1^n f(x) \, dx$$

So we have

$$\int_1^{\infty} f(x) \, dx \le \sum_{k=1}^{\infty} a_k \le a_1 + \int_1^{\infty} f(x) \, dx$$

as required.

8.4 Comparison Tests, Pages 526-527

SURVIVAL HINT: *Pay attention to Problems 1 and 2. These are both series you need to remember.*

1. The geometric series converges when $|r| < 1$.

3. $\sum_{k=1}^{\infty} \cos^k(\frac{\pi}{6})$ is a convergent geometric series with $r = \cos \frac{\pi}{6} < 1$.

5. $\sum_{k=0}^{\infty} 1.5^k$ is a divergent geometric series with $r = 1.5 > 1$.

7. $\sum_{k=1}^{\infty} \frac{1}{k}$ is a divergent *p*-series since $p = 1 \le 1$.

9. $\sum_{k=1}^{\infty} \frac{1}{k^{3/2}}$ is a convergent *p*-series since $p = 1.5 > 1$.

11. $\sum_{k=0}^{\infty} 1^k$ is a divergent geometric series with $r = 1 \ge 1$.

SURVIVAL HINT: *In order to use the comparison tests, you need to know some convergent and divergent series with which to compare. It is a good idea to keep a list of series for that purpose. In addition to the p, geometric, and harmonic series, add other general types as you determine their convergence or divergence.*

13. $\sum\limits_{k=1}^{\infty} \frac{1}{k^2 + k}$ is dominated by the convergent

p-series $\sum\limits_{k=1}^{\infty} \frac{1}{k^2}$, so the series converges.

15. $\sum\limits_{k=1}^{\infty} \frac{1}{\sqrt{k}} = \sum\limits_{k=1}^{\infty} \frac{1}{k^{1/2}}$ is the divergent p-series

with $p = 1/2$.

17. $\sum\limits_{k=1}^{\infty} \frac{1}{\sqrt{2k+3}}$ behaves like the divergent p-

series $\sum\limits_{k=1}^{\infty} \frac{1}{k^{1/2}}$.

19. $\sum\limits_{k=1}^{\infty} \frac{1}{\sqrt{k^3+2}}$ behaves like the convergent p-

series $\sum\limits_{k=1}^{\infty} \frac{1}{k^{3/2}}$.

21. $\sum\limits_{k=1}^{\infty} \frac{2k^2}{k^4 - 4}$ behaves like the convergent p-

series $\sum\limits_{k=1}^{\infty} \frac{1}{k^2}$.

23. $\sum\limits_{k=1}^{\infty} \frac{(k+2)(k+3)}{k^{7/2}}$ behaves like the

convergent p-series $\sum\limits_{k=1}^{\infty} \frac{1}{k^{3/2}}$.

25. $\sum\limits_{k=1}^{\infty} \frac{2k+3}{k^2 + 3k + 2}$ behaves like the divergent

p-series $\sum\limits_{k=1}^{\infty} \frac{2}{k}$.

27. $\sum\limits_{k=1}^{\infty} \frac{k}{(k+2)2^k}$ behaves like the convergent

geometric series $\sum\limits_{k=1}^{\infty} \frac{1}{2^k}$.

29. $\sum\limits_{k=1}^{\infty} \frac{1}{k(k+2)}$ behaves like the convergent

p-series $\sum\limits_{k=1}^{\infty} \frac{1}{k^2}$.

31. $\sum\limits_{k=1}^{\infty} \frac{1}{\sqrt{k}\, 2^k}$ is dominated by $\sum\limits_{k=1}^{\infty} \frac{1}{2^{k+1}}$ since

$\frac{1}{\sqrt{k}2^k} < \frac{1}{2^{k+1}}$ for $k > 4$. Since $\sum\limits_{k=1}^{\infty} \frac{1}{2^{k+1}}$

converges (geometric series with $r = \frac{1}{2} < 1$), so

the series also converges.

33. $\sum\limits_{k=1}^{\infty} \frac{|\sin(k!)|}{k^2}$ is dominated by the convergent

p-series $\sum\limits_{k=1}^{\infty} \frac{1}{k^2}$.

35. $\sum\limits_{k=1}^{\infty} \frac{2k^3 + k + 1}{k^3 + k^2 + 1}$ diverges since

$\lim\limits_{k \to \infty} \frac{2k^3 + k + 1}{k^3 + k^2 + 1} = 2 \neq 0$; the necessary

condition for convergence is not satisfied.

37. $\sum\limits_{k=1}^{\infty} \frac{k}{4k^3 - 5}$ behaves like the convergent

p-series $\sum\limits_{k=1}^{\infty} \frac{1}{k^2}$.

39. $\sum\limits_{k=1}^{\infty} \frac{k^2 + 1}{(k^2 + 2)k^2}$ behaves like the convergent

p-series $\sum\limits_{k=1}^{\infty} \dfrac{1}{k^2}.$

41. $\sum\limits_{k=1}^{\infty} \dfrac{6k^2 + 2k + 1}{k^{1.1}(4k^2 + k + 4)}$ behaves like the

convergent p-series $\sum\limits_{k=1}^{\infty} \dfrac{1}{k^{1.1}}.$

43. $\sum\limits_{k=1}^{\infty} \dfrac{\sqrt[6]{k}}{\sqrt[4]{k^3+2}\ \sqrt[8]{k}} = \sum\limits_{k=1}^{\infty} \dfrac{k^{4/24}}{(k^3+2)^{6/24} k^{3/24}}$

$$= \sum\limits_{k=1}^{\infty} \dfrac{k^{1/24}}{(k^3+2)^{6/24}}$$

behaves like the divergent p-series

$\sum\limits_{k=1}^{\infty} \dfrac{1}{k^{17/24}}.$

45. $\sum\limits_{k=1}^{\infty} \dfrac{1}{k^3 + 4}$ behaves like the convergent

p-series $\sum\limits_{k=1}^{\infty} \dfrac{1}{k^3}.$

47. $\sum\limits_{k=1}^{\infty} \dfrac{\ln(k+1)}{(k+1)^3}$ behaves like the convergent

series $\Sigma(\ln k)/k^3$ (q-log series with $q = 3 > 1$),

so the series converges.

49. $\sum\limits_{k=2}^{\infty} \dfrac{1}{(k+3)(\ln k)^{1.1}}$ behaves like

$\sum\limits_{k=1}^{\infty} \dfrac{1}{k(\ln k)^{1.1}}$ which converges by the integral

test since

$$\int\limits_{2}^{\infty} \dfrac{dx}{x(\ln x)^{1.1}} \approx 10.37$$

Thus, the series converges.

51. $\sum\limits_{k=1}^{\infty} k^{(1-k)/k} = \sum\limits_{k=1}^{\infty} \dfrac{1}{k^{1-1/k}}$

Compare with a divergent harmonic series
($p = 1 - k^{-1} \leq 1$). Thus, the series
diverges.

53. Compare $\sum\limits_{k=1}^{\infty} \dfrac{k^2}{(k+3)!}$ with the convergent

series $\sum\limits_{k=1}^{\infty} \dfrac{1}{(k+1)!}$ (see Example 3). Since

$$\lim_{k\to\infty} \dfrac{\dfrac{k^2}{(k+3)!}}{\dfrac{1}{(k+1)!}} = \lim_{k\to\infty} \dfrac{k^2(k+1)!}{(k+3)(k+2)(k+1)!}$$

$$= 1$$

we see the series also converges by the limit
comparison test.

55. $\sum\limits_{k=2}^{\infty} \dfrac{1}{(\ln k)^{\ln k}}$

$$= \sum\limits_{k=2}^{N} \dfrac{1}{(\ln k)^{\ln k}} + \sum\limits_{k=N+1}^{\infty} \dfrac{1}{(\ln k)^{\ln k}}$$

for some number N. Since $\ln k > e^2$ for
$k > e^{e^2} \approx 1620$, we have

$$\sum\limits_{k=2}^{\infty} \dfrac{1}{(\ln k)^{\ln k}} < \sum\limits_{k=2}^{1620} \dfrac{1}{(\ln k)^{\ln k}} + \sum\limits_{k=1621}^{\infty} \dfrac{1}{(e^2)^{\ln k}}$$

The first sum on the right is a finite number
a and the second is the convergent series

$$\sum\limits_{k=1621}^{\infty} \dfrac{1}{(e^2)^{\ln k}} = \sum\limits_{k=1621}^{\infty} \dfrac{1}{k^2}$$

so the given series converges by the direct
comparison test.

57. Let $\lim_{n \to \infty} b_n^* = L$ where $0 < L < \infty$. Then,

$$\lim_{n \to \infty} \frac{a_n b_n}{a_n} = \lim_{n \to \infty} b_n = L$$

implies $\Sigma a_n b_n$ and Σa_n either both converge or both diverge by the limit comparison test.

59. If Σa_k converges, then $\lim_{k \to \infty} a_k = 0$,

$$\lim_{k \to \infty} a_k^{-1} \neq 0, \text{ so } \Sigma a_k^{-1} \text{ diverges}$$

61. Since $\lim_{k \to \infty} \frac{a_k}{b_k} = 0$, it follows that there is an integer N so that

$$\left| \frac{a_k}{b_k} - 0 \right| < 1 \text{ if } k > N$$

(definition of limit with $\epsilon = 1$). Thus, $a_k < b_k$ for $k > N$ and

$$\sum_{k=N}^{\infty} a_k < \sum_{k=N}^{\infty} b_k$$

Thus, Σa_k converges by direct comparison with the convergent series Σb_k.

8.5 The Ratio Test and the Root Test, Pages 532-533

1. Given the series Σa_k with $a_k > 0$, suppose that

$$\lim_{k \to \infty} \frac{a_{k+1}}{a_k} = L$$

The ratio test states the following:

If $L < 1$, then Σa_k converges.

If $L > 1$ or if L is infinite, then Σa_k diverges.

If $L = 1$, the test is inconclusive.

3. $a_k = \frac{1}{k!}$; use the ratio test.

$$\lim_{k \to \infty} \frac{\frac{1}{(k+1)!}}{\frac{1}{k!}} = \lim_{k \to \infty} \frac{k!}{(k+1)!}$$

$$= \lim_{k \to \infty} \frac{1}{k+1}$$

$$= 0 < 1$$

The series converges.

5. $a_k = \frac{k!}{2^{3k}}$; use the ratio test.

$$\lim_{k \to \infty} \frac{(k+1)! 2^{3k}}{2^{3k+3} k!} = \frac{1}{8} \lim_{k \to \infty} (k+1)$$

$$= \infty > 1$$

The series diverges.

7. $a_k = \frac{k}{2^k}$; use the ratio test.

$$\lim_{k \to \infty} \frac{(k+1) 2^k}{2^{k+1} k} = \frac{1}{2} < 1$$

The series converges.

9. $a_k = \frac{k^{100}}{e^k}$; use the ratio test.

$$\lim_{k \to \infty} \frac{(k+1)^{100} e^k}{e^{k+1} k^{100}} = e^{-1} < 1$$

The series converges.

11. $a_k = k\left(\frac{4}{3}\right)^k$; diverges since

$$\lim_{k \to \infty} k \frac{4^k}{3^k} = \infty \neq 0$$

Following the directions, we use the ratio test:

$$\lim_{k \to \infty} \frac{(k+1)4^{k+1}3^k}{3^{k+1}k4^k} = \frac{4}{3} > 1$$

The series diverges.

13. $a_k = \left(\frac{2}{k}\right)^k$; use the root test.

$$\lim_{k \to \infty} \sqrt[k]{\left(\frac{2}{k}\right)^k} = \lim_{k \to \infty} \frac{2}{k} = 0 < 1$$

The series converges.

15. $\displaystyle\sum_{k=1}^{\infty} \frac{k^5}{10^k}$; use the ratio test.

$$\lim_{k \to \infty} \frac{10^k(k+1)^5}{k^5 10^{k+1}} = \frac{1}{10} \lim_{k \to \infty} \frac{(k+1)^5}{k^5}$$

$$= \frac{1}{10} < 1$$

The series converges.

17. $a_k = \left(\frac{k}{3k+1}\right)^k$; use the root test.

$$\lim_{k \to \infty} \sqrt[k]{\left(\frac{k}{3k+1}\right)^k} = \lim_{k \to \infty} \frac{k}{3k+1}$$

$$= \frac{1}{3} < 1$$

The series converges.

19. $a_k = \dfrac{k!}{(k+2)^4}$; use the ratio test.

$$\lim_{k \to \infty} \frac{(k+2)^4(k+1)!}{k!(k+3)^4} = \lim_{k \to \infty} \frac{(k+2)^4(k+1)}{(k+3)^4}$$

$$= \infty > 1$$

The series diverges.

21. $a_k = \dfrac{(k!)^2}{(2k)!}$; use the ratio test.

$$\lim_{k \to \infty} \frac{(2k)![(k+1)!]^2}{(k!)^2(2k+2)!} = \lim_{k \to \infty} \frac{(k+1)^2}{(2k+2)(2k+1)}$$

$$= \frac{1}{4} < 1; \text{ the series converges.}$$

23. $a_k = k^2 2^{-k}$; use the ratio test.

$$\lim_{k \to \infty} \frac{2^k(k+1)^2}{k^2 2^{k+1}} = \frac{1}{2} \lim_{k \to \infty} \frac{(k+1)^2}{k^2}$$

$$= \frac{1}{2} < 1$$

The series converges.

25. $a_k = \left(\frac{k-2}{k}\right)^{k^2}$; use the root test.

$$\lim_{k \to \infty} \left[\left(\frac{k-2}{k}\right)^{k^2}\right]^{1/k} = \lim_{k \to \infty} \left(\frac{k-2}{k}\right)^k$$

$$= \lim_{k \to \infty} \left(1 - \frac{2}{k}\right)^k$$

$$= e^{-2} \quad \textit{l'Hôpital's rule}$$

Since $e^{-2} < 1$, the series converges.

27. The ratio test is inconclusive; the series

$$\sum_{k=1}^{\infty} \frac{1,000}{k} \text{ diverges by direct comparison with}$$

the divergent p-series $\displaystyle\sum_{k=1}^{\infty} \frac{1}{k}$.

29. $\displaystyle\sum_{k=1}^{\infty} \frac{5k+2}{k2^k}$; use the ratio test.

$$\lim_{k \to \infty} \frac{(5k+7)(k)2^k}{(k+1)2^{k+1}(5k+2)}$$

$$= \tfrac{1}{2} \lim_{k \to \infty} \frac{(5k+7)(k)}{(k+1)(5k+2)}$$

$$= \tfrac{1}{2} < 1$$

The given series converges.

31. $\displaystyle\sum_{k=1}^{\infty} \frac{\sqrt{k!}}{2^k}$; use the ratio test

$$\lim_{k \to \infty} \frac{\sqrt{(k+1)!}\, 2^k}{2^{k+1}\sqrt{k!}} = \tfrac{1}{2} \lim_{k \to \infty} \sqrt{k+1}$$

$$= \infty > 1$$

The given series diverges.

33. $\displaystyle\sum_{k=1}^{\infty} \frac{2^k k!}{k^k}$; use the ratio test.

$$\lim_{k \to \infty} \frac{2^{k+1}(k+1)! k^k}{(k+1)^{k+1} 2^k k!} = 2 \lim_{k \to \infty} \frac{k^k}{(k+1)^k}$$

$$= 2 \lim_{k \to \infty} \frac{1}{(1+\tfrac{1}{k})^k}$$

$$= \tfrac{2}{e} < 1$$

The given series converges.

35. $\displaystyle\sum_{k=1}^{\infty} \frac{\sqrt{k+1}}{k^{k+0.5}}$; use the ratio test.

$$\lim_{k \to \infty} \frac{\sqrt{k+2}\, k^{k+0.5}}{(k+1)^{k+1.5}\sqrt{k+1}}$$

$$= \lim_{k \to \infty} \frac{1}{\left(1+\tfrac{1}{k}\right)^k} \frac{k^{0.5}}{(k+1)^{1.5}}\sqrt{\frac{k+2}{k+1}}$$

$$= (\tfrac{1}{e})(0)(1)$$

$$= 0 < 1$$

The given series converges.

37. Compare $\displaystyle\sum_{k=1}^{\infty} \frac{k!}{(k+1)!} = \sum_{k=1}^{\infty} \frac{1}{k+1}$ directly compare with the harmonic series to see that it diverges. (The integral test also works.)

39. $\displaystyle\sum_{k=1}^{\infty} \left(1+\tfrac{1}{k}\right)^{-k^2}$; use the root test.

$$\lim_{k \to \infty} \sqrt[k]{\left(1+\tfrac{1}{k}\right)^{-k^2}} = \lim_{k \to \infty} \left(1+\tfrac{1}{k}\right)^{-k}$$

$$= \lim_{k \to \infty} \frac{1}{(1+\tfrac{1}{k})^k}$$

$$= \tfrac{1}{e} < 1$$

The given series converges.

41. Compare $\displaystyle\sum_{k=1}^{\infty} \left|\frac{\cos k}{2^k}\right|$ directly with the convergent geometric series $\displaystyle\sum_{k=1}^{\infty} \frac{1}{2^k}$:

$$\sum_{k=1}^{\infty} \left|\frac{\cos k}{2^k}\right| \le \sum_{k=1}^{\infty} \frac{1}{2^k} = 1$$

The given series converges.

43. $\displaystyle\sum_{k=2}^{\infty} \left(\frac{\ln k}{k}\right)^k$; use the root test.

$$\lim_{k \to \infty} \sqrt[k]{\left(\frac{\ln k}{k}\right)^k} = \lim_{k \to \infty} \frac{\ln k}{k}$$

$$= \lim_{k \to \infty} \frac{\tfrac{1}{k}}{1}$$

$$= 0 < 1$$

The given series converges.

45. $\sum_{k=1}^{\infty} k^2 x^k$; use the ratio test.

$$\lim_{k \to \infty} \frac{(k+1)^2 x^{k+1}}{k^2 x^k} = x$$

By the ratio test, the given series converges when $x < 1$ and diverges when $x > 1$. Since the ratio test fails when the ratio equals 1, investigate

$$\sum_{k=1}^{\infty} k^2$$

separately to see that it diverges at $x = 1$. Thus, the given series converges for $0 \le x < 1$.

47. $\sum_{k=1}^{\infty} \frac{(x+0.5)^k}{k\sqrt{k}}$; use the ratio test.

$$\lim_{k \to \infty} \frac{\dfrac{(x+0.5)^{k+1}}{(k+1)^{3/2}}}{\dfrac{(x+0.5)^k}{k^{3/2}}} = \lim_{k \to \infty} \frac{(x+0.5)k^{3/2}}{(k+1)^{3/2}}$$

$$= x + 0.5$$

By the ratio test, the given series converges when $x + 0.5 < 1$ or $x < 0.5$ and diverges when $x > 0.5$. Since the ratio test fails when the ratio equals 1, investigate

$$\sum_{k=1}^{\infty} k^{-3/2}$$

separately to see that it converges if $x = 0.5$. Thus, the given series converges for $0 \le x \le 0.5$.

49. $\sum_{k=1}^{\infty} \frac{x^k}{k!}$; use the ratio test.

$$\lim_{k \to \infty} \frac{x^{k+1} k!}{(k+1)! x^k} = \lim_{k \to \infty} \frac{x}{k+1}$$

$$= 0 < 1$$

for all x. Thus, the given series converges for all $x \ge 0$.

51. $\sum_{k=1}^{\infty} (ax)^k$; use the ratio test.

$$\lim_{k \to \infty} \frac{(ax)^{k+1}}{(ax)^k} = ax$$

By the ratio test, the given series converges when $ax < 1$ or when $x < a^{-1}$ and diverges when $x > a^{-1}$. Since the ratio test fails when the ratio is 1, investigate

$$\sum_{k=1}^{\infty} 1^k$$

separately to see that this series diverges at $x = a^{-1}$. Thus, the given series converges for $0 \le x < a^{-1}$.

53. $\sum_{k=1}^{\infty} k^p e^{-k}$; use the root test.

$$\lim_{k \to \infty} \frac{k^{p/k}}{e} = e^{-1} < 1 \text{ for all } p$$

Thus, the given series converges for all p, and the integral test shows that

$$\int_1^{\infty} x^p e^{-x} \, dx$$

also converges for all p.

55. a. Since $L < 1$ and $\lim\limits_{k \to \infty} \frac{a_{k+1}}{a_k} = L$

the series converges and the necessary

condition for convergence is satisfied; that is,

$$\lim_{k \to \infty} a_k = 0$$

b. Since $\sum\limits_{k=1}^{\infty} \frac{x^k}{k!}$; use the ratio test.

$$\lim_{k \to \infty} \frac{x^{k+1} k!}{(k+1)! x^k} = \lim_{k \to \infty} \frac{x}{k+1}$$

$$= 0$$

This series converges for all x so that

$$\lim_{k \to \infty} \frac{x^k}{k!} = 0$$

57. a. Since $\lim\limits_{n \to \infty} \sqrt[n]{a_n} = R < 1$, all but a finite

number of terms in the sequence $\{ \sqrt[n]{a_n} \}$

must be "close" to R. Hence, if

$R < x < 1$, there is an N such that

$0 \leq \sqrt[n]{a_n} \leq x$ for all $n > N$, and

$$\sum_{n=N}^{\infty} a_n = \sum_{n=N}^{\infty} \left(\sqrt[n]{a_n} \right)^n \leq \sum_{n=N}^{\infty} x^n$$

Since $|x| < 1$, the geometric series on the

right converges, and by comparison, so

does $\sum\limits_{n=N}^{\infty} a_n$. This, in turn, implies the

convergence of $\sum\limits_{n=1}^{\infty} a_n$.

b. If $R > 1$, then all but a finite number of

terms in the sequence $\{ \sqrt[n]{a_n} \}$ satisfy

$\sqrt[n]{a_n} > 1$. Hence, $a_n > 1$ for infinitely

many values of n, so $\{ a_n \}$ cannot tend to

0. It follows from the divergence test

that $\sum\limits_{n=1}^{\infty} a_n$ must diverge.

c. $\sum\limits_{n=1}^{\infty} \frac{1}{n}$ diverges and $\sum\limits_{n=1}^{\infty} \frac{1}{n^2}$ converges,

yet $\lim\limits_{n \to \infty} \sqrt[n]{a_n} = 1$ for both.

8.6 Alternating Series; Absolute and Conditional Convergence, Pages 542-544

SURVIVAL HINT: *When using the alternating series test, it does not matter whether the first term is positive or negative. However, when writing a particular series in sigma notation, take care to use $(-1)^k$ or $(-1)^{k+1}$, whichever gives the first term the correct sign.*

1. Consider the alternating series $A = \Sigma (-1)^k a_k$ (with positive a_k). If Σa_k converges, then A converges absolutely. If Σa_k diverges, then A may converge conditionally.

SURVIVAL HINT: *Look at Table 8.1 and make sure you understand what is being said. Do not move quickly past these pages.*

3. The steps are outlined in Table 8.1, pp. 540-541.

5. $\displaystyle\sum_{k=1}^{\infty} \frac{(-1)^{k+1}k^2}{k^3+1}$ does not converge absolutely

since $\displaystyle\sum_{k=1}^{\infty} \frac{k^2}{k^3+1}$ behaves like the divergent

p-series $\Sigma(1/k)$. To apply the alternating

series test, first note that $\left\{\dfrac{k^2}{k^3+1}\right\}$ is a

decreasing sequence since

$$\frac{d}{dx}\left(\frac{x^2}{x^3+1}\right) = \frac{x(2-x^3)}{(x^3+1)^2} < 0 \text{ for } x > 2$$

Since $\displaystyle\lim_{k\to\infty} \frac{k^2}{k^3+1} = 0$ (l'Hôpital's rule), it

follows that the given series converges

conditionally.

7. $\displaystyle\sum_{k=1}^{\infty} \frac{(-1)^{k+1}k^2}{k^2+1}$ diverges because

$$\lim_{k\to\infty} \frac{k^2}{k^2+1} = 1 \neq 0$$

9. $\displaystyle\sum_{k=1}^{\infty} \frac{(-1)^{k+1}k}{2^k}$;

Apply the ratio test to the series of absolute
values:

$$\lim_{k\to\infty} \frac{\dfrac{k+1}{2^{k+1}}}{\dfrac{k}{2^k}} = \lim_{k\to\infty} \frac{1}{2}\frac{k+1}{k}$$

$$= \frac{1}{2} < 1$$

The given series is absolutely convergent.

11. $\displaystyle\sum_{k=1}^{\infty} \frac{(-1)^k}{\sqrt{k}}$ does not converge absolutely since

$\displaystyle\sum_{k=1}^{\infty} \frac{1}{\sqrt{k}}$ is a divergent p-series $(p = \frac{1}{2} < 1)$.

To apply the alternating series test, note that

$\left\{\dfrac{1}{\sqrt{k}}\right\}$ is a decreasing sequence since

$$\frac{d}{dx}\left(\frac{1}{\sqrt{x}}\right) = -\frac{1}{2}x^{-3/2} < 0$$

Since $\displaystyle\lim_{k\to\infty} \frac{1}{\sqrt{k}} = 0$, it follows that the given

series converges conditionally.

13. $\displaystyle\sum_{k=1}^{\infty} \frac{(-1)^{k+1}k!}{k^k}$

Apply the ratio test to the series of absolute
values:

$$\lim_{k\to\infty} \frac{(k+1)!k^k}{(k+1)^{k+1}k!} = \lim_{k\to\infty} \frac{1}{\left(1+\frac{1}{k}\right)^k}$$

$$= \frac{1}{e} < 1$$

The given series is absolutely convergent.

15. $\displaystyle\sum_{k=1}^{\infty} \frac{(-1)^k(2k)!}{k^k}$ diverges because

$\displaystyle\lim_{k\to\infty} \frac{(2k)!}{k^k} \neq 0$. To show this, note that the

sequence $\{a_n\}$ with $a_n = \dfrac{(2k)!}{k^k}$ is increasing

since

$$\frac{a_{k+1}}{a_k} = \frac{(2k+2)!k^k}{(k+1)^{k+1}(2k)!}$$

$$= \frac{(2k+2)(2k+1)k^k}{(k+1)^{k+1}}$$

$$= \frac{2(2k+1)}{\left(1+\frac{1}{k}\right)^k} > 1$$

Clearly an increasing sequence of positive terms cannot possibly tend toward 0.

17. $\sum_{k=1}^{\infty} \frac{(-3)^k(k+1)}{k!}$

Apply the ratio test to the series of absolute values:

$$\lim_{k\to\infty} \frac{3^{k+1}(k+2)k!}{3^k(k+1)(k+1)!} = \lim_{k\to\infty} \frac{3(k+2)}{(k+1)^2}$$

$$= 0$$

$$< 1$$

The given series is absolutely convergent.

19. $\sum_{k=2}^{\infty} \frac{(-1)^{k+1}}{\ln k}$ does not converge absolutely

since $\Sigma(1/\ln k)$ diverges by comparison with the divergent p-series $\Sigma(1/k)$, since $k > \ln k$ for all k. To test for conditional convergence, first note that $\left\{\frac{1}{\ln k}\right\}$ is decreasing to zero since

$$\frac{d}{dx}\left(\frac{1}{\ln x}\right) = \frac{-1}{x\ln^2 x} < 0 \quad \text{and} \quad \lim_{k\to\infty} \frac{1}{\ln k} = 0$$

It follows that the given series does converge conditionally.

21. $\sum_{k=2}^{\infty} \frac{(-1)^{k+1}}{(\ln k)^4}$ does not converge absolutely.

To see this, use the integral test:

$$\int_2^{\infty} \frac{dx}{(\ln x)^4} = \int_2^{\infty} \frac{e^u}{u^4}\,du \quad \boxed{u = \ln x}$$

$$= \infty \text{ since } e^u > u^4 \text{ for } u > 9$$

To apply the alternating series test, note that $\left\{\frac{1}{(\ln x)^4}\right\}$ is decreasing since

$$\frac{d}{dx}\left[\frac{1}{(\ln x)^4}\right] = \frac{-4}{x(\ln x)^5} < 0$$

Since $\lim_{k\to\infty} \frac{1}{(\ln k)^4} = 0$, it follows that the series converges conditionally.

23. $\sum_{k=2}^{\infty} \frac{(-1)^{k+1}}{k\ln k}$ does not converge absolutely

since $\sum_{k=2}^{\infty} \frac{1}{k\ln k}$ diverges by the integral test:

$$\int_2^{\infty} \frac{dx}{x\ln x} = \lim_{N\to\infty} [\ln(\ln x)]\Big|_2^N = \infty$$

To apply the alternating series test, note that $\left\{\frac{1}{k\ln k}\right\}$ is decreasing because

$$\frac{d}{dx}\left(\frac{1}{x\ln x}\right) = \frac{-1(1+\ln x)}{x^2(\ln x)^2} < 0 \text{ for } x > e$$

and $\lim_{k\to\infty} \frac{1}{k\ln k} = 0$; thus, the given series converges conditionally.

25. $\sum_{k=2}^{\infty} \frac{(-1)^{k+1}k}{\ln k}$

$$\lim_{k \to \infty} \frac{k}{\ln k} = \lim_{k \to \infty} \frac{1}{\frac{1}{k}} \qquad \textit{l'Hôpital's rule}$$

$$= \infty \neq 0$$

The given series is divergent.

27. $\sum_{k=1}^{\infty} \frac{(-1)^{k+1} k}{(k+2)^2}$ does not converge absolutely

since $\sum_{k=1}^{\infty} \frac{k}{(k+2)^2}$ behaves like the divergent

p-series $\Sigma(1/k)$. The sequence $\left\{ \frac{k}{(k+2)^2} \right\}$ is

decreasing because

$$\frac{d}{dx} \frac{x}{(x+2)^2} = \frac{(x+2)^2 - 2x(x+2)}{(x+2)^4}$$

$$= \frac{-(x-2)}{(x+2)^3} < 0 \text{ for } x > 2$$

and $\lim_{k \to \infty} \frac{k}{(k+2)^2} = 0$; thus, the given series

converges conditionally.

29. $\sum_{k=2}^{\infty} (-1)^{k+1} \frac{\ln(\ln k)}{k \ln k}$ does not converge

absolutely since $\sum_{k=2}^{\infty} \frac{\ln(\ln k)}{k \ln k}$ diverges by the

integral test:

$$\int_{2}^{\infty} \frac{\ln(\ln x)}{x \ln x} \, dx = \lim_{N \to \infty} \left[\frac{1}{2} (\ln (\ln x)^2) \right] \Big|_{2}^{N}$$

$$= \infty$$

To apply the alternating series test, note that

$$\frac{d}{dx} \left(\frac{\ln[(\ln x)]}{x \ln x} \right)$$

$$= \frac{(x \ln x)(x \ln x)^{-1} - \ln(\ln x)(1 + \ln x)}{(x \ln x)^2}$$

$$< 0 \text{ for } x > 5 \qquad \text{and}$$

$$\lim_{k \to \infty} \frac{\ln(\ln k)}{k \ln k} = \lim_{k \to \infty} \frac{\frac{1}{k \ln k}}{\ln k + 1}$$

$$= 0 \qquad \textit{l'Hôpital's rule}$$

Thus, the given series converges conditionally.

31. $\sum_{k=1}^{\infty} (-1)^{k+1} \frac{k^5 \, 5^{k+2}}{2^{3k}}$; use the ratio test.

$$\lim_{k \to \infty} \frac{2^{3k}(k+1)^5 5^{k+3}}{k^5 5^{k+2} 2^{3k+3}} = \frac{5}{8} \lim_{k \to \infty} \frac{(k+1)^5}{k^5}$$

$$= \frac{5}{8} < 1$$

This series is absolutely convergent.

33. $S = \sum_{k=1}^{\infty} \frac{(-1)^{k+1}}{k!}$

a. $S_4 = 1 - \frac{1}{2} + \frac{1}{6} - \frac{1}{24} = \frac{5}{8}$

$$\left| S - S_4 \right| < a_5 = \frac{1}{120} \approx 0.00833$$

b. $\qquad \frac{1}{n!} < 0.0005$

$$n! > 2,000$$

$$n > 6$$

Choose $n = 7$; $S_7 = \frac{177}{280} \approx 0.632$

35. $S = \sum_{k=1}^{\infty} \dfrac{(-1)^{k+1}}{3^{k+1}}$

a. $S_4 = \dfrac{1}{9} - \dfrac{1}{27} + \dfrac{1}{81} - \dfrac{1}{243}$

$= \dfrac{20}{243} \approx 0.0823$

$|S - S_4| < a_5 = \dfrac{1}{729} \approx 0.001372$

b. $\dfrac{1}{3^{n+1}} < 0.0005$

$3^{n+1} > 2{,}000$

$n + 1 > \log_3 2{,}000$

$n > 5.918$

Choose $n = 6$; $S_6 \approx 0.083$

37. $S = \sum_{k=1}^{\infty} \left(\dfrac{-1}{5} \right)^k$

a. $S_4 = -\dfrac{1}{5} + \dfrac{1}{25} - \dfrac{1}{125} + \dfrac{1}{625}$

$= -\dfrac{104}{625} \approx -0.1664$

$|S - S_4| < a_5 = \dfrac{1}{3{,}125} = 0.00032$

b. $\dfrac{1}{5^n} < 0.0005$

$5^n > 2{,}000$

$n > 4.72$

Choose $n = 5$; $S_5 = -\dfrac{521}{3{,}125} \approx -0.167$

39. $\sum_{k=1}^{\infty} \dfrac{(2x)^k}{\sqrt{k}}$; use generalized ratio test.

$\lim_{k \to \infty} \left| \dfrac{(2x)^{k+1}\sqrt{k}}{\sqrt{k+1}(2x)^k} \right| = |2x|$

Converges if $|x| < \dfrac{1}{2}$ and diverges if $|x| > \dfrac{1}{2}$.

For $x = \dfrac{1}{2}$, $\sum_{k=1}^{\infty} \dfrac{1}{k^{1/2}}$ diverges.

For $x = -\dfrac{1}{2}$, $\sum_{k=1}^{\infty} \dfrac{(-1)^{k+1}}{\sqrt{k}}$ converges.

The interval of convergence is $\left[-\dfrac{1}{2}, \dfrac{1}{2} \right)$.

41. $\sum_{k=1}^{\infty} \dfrac{(k+2)x^k}{k^2(k+3)}$; use the generalized ratio test.

$\lim_{k \to \infty} \left| \dfrac{k^2(k+3)(k+3)x^{k+1}}{(k+1)^2(k+2)(k+4)x^k} \right| = |x|$

Converges if $|x| < 1$ and diverges if $|x| > 1$.

If $x = 1$, $\sum_{k=1}^{\infty} \dfrac{k+2}{k^3 + 3k}$ converges.

If $x = -1$, $\sum_{k=1}^{\infty} \dfrac{(-1)^k(k+2)}{k^2(k+3)}$ converges.

The interval of convergence is $[-1, 1]$.

43. $\sum_{k=1}^{\infty} (-1)^k k^p x^k$ for $p > 0$; use generalized ratio test.

$\lim_{k \to \infty} \left| \dfrac{(k+1)^p x^{k+1}}{k^p x^k} \right| = |x|$

Converges if $|x| < 1$ and diverges if $|x| > 1$.

For $x = \pm 1$, $\sum_{k=1}^{\infty} k^p$ diverges because the necessary condition for convergence is not satisfied. The interval of convergence is $(-1, 1)$ for all $p > 0$.

45. $E_{\max} = \dfrac{1}{6^2} \approx 0.0278$

47. $E_{\max} = \dfrac{7}{2^7} \approx 0.0547$

49. $\displaystyle\sum_{k=1}^{\infty} \left| \dfrac{\sin \sqrt[k]{2}}{k^2} \right| \le \sum_{k=1}^{\infty} \dfrac{1}{k^2}$, a convergent p-series.

Thus, the given series converges absolutely.

SURVIVAL HINT: *Problem 51 is another of those problems that introduces a new number, known as Euler's constant. Not of earthshaking importance, but these ideas help to broaden your mathematical background.*

51. Consider the alternating series

$$S = \sum_{n=1}^{\infty} (-1)^{n+1} a_n, \text{ where } a_1 = 1,$$

$$a_2 = \int_1^2 \frac{dx}{x}, \; a_3 = \frac{1}{2}, \; a_4 = \int_2^3 \frac{dx}{x}, \cdots$$

The $(2n-1)$st partial sum of the series is

$$S_{2n-1} = 1 - \int_1^2 \frac{dx}{x} + \frac{1}{2} - \int_2^3 \frac{dx}{x} + \cdots$$

$$+ \frac{1}{n-1} - \int_{n-1}^{n} \frac{dx}{x} + \frac{1}{n}$$

$$= 1 + \frac{1}{2} + \cdots + \frac{1}{n} - \int_1^n \frac{dx}{x}$$

$$= 1 + \frac{1}{2} + \cdots + \frac{1}{n} - \ln n$$

Since $\dfrac{1}{k+1} \le \displaystyle\int_k^{k+1} \frac{dx}{x} \le \dfrac{1}{k}$, the sequence $\{a_n\}$ is

decreasing and clearly $\lim\limits_{n \to \infty} a_n = 0$. Thus, S converges and we have

$$S = \lim_{n \to \infty} S_{2n-1}$$

$$= \lim_{n \to \infty} \left(1 + \frac{1}{2} + \cdots + \frac{1}{n} - \ln n \right)$$

The limit S is called the *Euler constant* and has the approximate value $S \approx 0.577216\cdots$.

53. **a.** Suppose n is a multiple of 3, say $n = 3m$. Then the partial sum S_{3m} has $2m$ positive terms and m negative terms; we can write

$$S_{3m} = \sum_{k=1}^{2m} \frac{1}{2k-1} - \sum_{k=1}^{m} \frac{1}{2k}$$

For example, with $m = 2$, we have $n = 3m = 6$ and

$$S_6 = 1 + \frac{1}{3} - \frac{1}{2} + \frac{1}{5} + \frac{1}{7} - \frac{1}{4}$$

$$= \left(1 + \frac{1}{3} + \frac{1}{5} + \frac{1}{7} \right) - \left(\frac{1}{2} + \frac{1}{4} \right)$$

We find that

$$S_{3m} = \sum_{k=1}^{2m} \frac{1}{2k-1} - \sum_{k=1}^{m} \frac{1}{2k}$$

$$= \left(\sum_{k=1}^{4m} \frac{1}{k} - \sum_{k=1}^{2m} \frac{1}{2k} \right) - \frac{1}{2} \sum_{k=1}^{m} \frac{1}{k}$$

$$= \sum_{k=1}^{4m} \frac{1}{k} - \frac{1}{2} \sum_{k=1}^{2m} \frac{1}{k} - \frac{1}{2} \sum_{k=1}^{m} \frac{1}{k}$$

$$= H_{4m} - \frac{1}{2} H_{2m} - \frac{1}{2} H_m$$

b. $\lim\limits_{m \to \infty} S_{3m} = \lim\limits_{m \to \infty} (H_{4m} - \ln 4m)$

$$-\tfrac{1}{2}\lim_{m\to\infty}(H_{2m}-\ln 2m)$$

$$-\tfrac{1}{2}\lim_{m\to\infty}(H_m - \ln m)$$

$$+\ln 4m - \tfrac{1}{2}\ln 2m - \tfrac{1}{2}\ln m$$

$$=\gamma - \tfrac{1}{2}\gamma - \tfrac{1}{2}\gamma + \ln 4m - \tfrac{1}{2}\ln 2m - \tfrac{1}{2}\ln m$$

$$=\ln 4 + \ln m - \tfrac{1}{2}\ln 2 - \tfrac{1}{2}\ln m - \tfrac{1}{2}\ln m$$

$$=2\ln 2 - \tfrac{1}{2}\ln 2 = \tfrac{3}{2}\ln 2$$

So the rearranged series converges with sum $\tfrac{3}{2}\ln 2$.

55. Conditionally convergent series should not be rearranged.

57. False; for a counterexample, let $a_k = k^{-4}$;

$S = \displaystyle\sum_{k=1}^{\infty} k^{-4}$ is a convergent p-series and so is $S = \displaystyle\sum_{k=1}^{\infty} k^{-2}$.

59. $\displaystyle\sum_{k=1}^{\infty} a_k$; by the ratio test, $\displaystyle\lim_{k\to\infty}\frac{a_{k+1}}{a_k} = L < 1$

since the series converges, by hypothesis. Then,

$$T = \sum_{k=1}^{\infty} a_k{}^2 \text{ converges}$$

because $\displaystyle\lim_{k\to\infty}\frac{a_{k+1}{}^2}{a_k{}^2} = L^2 < 1.$

61. Σk^{-1} diverges and Σk^{-2} converges.

63. Consider

$$(2-1) + (1 - \tfrac{1}{2}) + (\tfrac{2}{3} - \tfrac{1}{3}) + (\tfrac{1}{2} - \tfrac{1}{4}) + \cdots$$

The even terms $\{2, 1, \tfrac{2}{3}, \tfrac{1}{2}, \tfrac{2}{5}, \cdots\}$ decrease to 0, as do the odd terms $\{1, \tfrac{1}{2}, \tfrac{1}{3}, \tfrac{1}{4}, \tfrac{1}{5}, \cdots\}$.

However, the overall sequence

$$2, 1, 1, \tfrac{1}{2}, \tfrac{1}{3}, \tfrac{1}{2}, \tfrac{1}{4}, \tfrac{2}{5}, \tfrac{1}{5}, \cdots$$

is not decreasing. The series diverges because the sequence of even partial sums

$$S_2 = 1, \; S_4 = 1 + \tfrac{1}{2}, \; S_6 = 1 + \tfrac{1}{2} + \tfrac{1}{3} + \cdots$$

is the sequence of partial sums for the harmonic series, which diverges.

8.7 Power Series, Pages 552-553

SURVIVAL HINT: *Power series and their associated intervals of convergence are the basis for the development of Taylor series in the next section. Taylor series are a powerful tool for the computation of values for various transcendental functions. To gain a better perspective about our interest in power series and their convergence it might be a good idea to read the next section at this point. This process increases your understanding and allows you to ask perceptive questions.*

1.
$$L = \lim_{k\to\infty}\left|\frac{\dfrac{(k+1)x^{k+1}}{k+2}}{\dfrac{kx^k}{k+1}}\right|$$

$$= \lim_{k\to\infty}\left|\frac{(k+1)^2}{(k+2)k}\right||x|$$

$$= |x|$$

The interval of absolute convergence is $|x| < 1$.

Endpoints:

$x = 1$: $\displaystyle\sum_{k=1}^{\infty} \frac{k}{k+1}$ diverges by the divergence test

$x = -1$: $\displaystyle\lim_{k \to \infty} \sum_{k=1}^{\infty} \frac{(-1)^k k}{k+1}$ does not exist, so

it diverges. The convergence set is $(-1, 1)$.

SURVIVAL HINT: *The interval of convergence may be open, closed, or half-open. Always remember to test both endpoints.*

3. $\displaystyle L = \lim_{k \to \infty} \left| \frac{\dfrac{(k+1)(k+2)x^{k+1}}{k+3}}{\dfrac{k(k+1)x^k}{k+2}} \right|$

$\displaystyle = \lim_{k \to \infty} \left| \frac{(k+1)(k+2)^2}{(k+3)(k+1)(k)} \right| |x|$

$$= |x|$$

The interval of absolute convergence is $|x| < 1$.

Endpoints:

$x = 1$: $\displaystyle\sum_{k=1}^{\infty} \frac{k(k+1)}{k+2}$ diverges by the

divergence test: $\displaystyle\lim_{k \to \infty} \frac{k(k+1)}{k+2} \neq 0$

$x = -1$: $\displaystyle\sum_{k=1}^{\infty} \frac{k(k+1)}{k+2}(-1)^k$ diverges

since $\displaystyle\lim_{k \to \infty} \frac{k(k+1)(-1)^k}{k+2}$ does not exist.

The convergence set is $(-1, 1)$.

5. $\displaystyle L = \lim_{k \to \infty} \left| \frac{(k+1)^2 3^{k+1}(x-3)^{k+1}}{k^2 3^k (x-3)^k} \right|$

$\displaystyle = \lim_{k \to \infty} \left| \frac{(k+1)^2 3^{k+1}}{k^2 3^k} \right| |x-3|$

$$= 3|x-3|$$

The interval of convergence is

$$3|x-3| < 1$$

$$|x-3| < \tfrac{1}{3}$$

$$-\tfrac{1}{3} < x - 3 < \tfrac{1}{3}$$

$$\tfrac{8}{3} < x < \tfrac{10}{3}$$

Endpoints:

$x = \tfrac{8}{3}$: $\displaystyle\sum_{k=1}^{\infty} (-1)^k k^2$ diverges by the

divergence test.

$x = \tfrac{10}{3}$: $\displaystyle\sum_{k=1}^{\infty} k^2$ diverges by the divergence

test. The convergence set is $\left(\tfrac{8}{3}, \tfrac{10}{3}\right)$.

7. $\displaystyle L = \lim_{k \to \infty} \left| \frac{\dfrac{3^{k+1}(x+3)^{k+1}}{4^{k+1}}}{\dfrac{3^k(x+3)^k}{4^k}} \right|$

$\displaystyle = \lim_{k \to \infty} \left| \frac{3^{k+1} 4^k}{4^{k+1} 3^k} \right| |x+3|$

$$= \tfrac{3}{4}|x+3|$$

The interval of absolute convergence is

$|x + 3| < \frac{4}{3}$, or $-\frac{13}{3} < x < -\frac{5}{3}$.

Endpoints:

$x = -\frac{13}{3}$: $\displaystyle\sum_{k=1}^{\infty} \frac{3^k(\frac{4}{3})^k(-1)^k}{4^k} = \sum_{k=1}^{\infty}(-1)^k$ which

diverges by the divergence test.

$x = -\frac{5}{3}$: $\displaystyle\sum_{k=1}^{\infty} 1$ also diverges by the divergence

test.

The convergence set is $(-\frac{13}{3}, -\frac{5}{3})$.

9. $L = \displaystyle\lim_{k\to\infty} \left| \frac{\frac{(k+1)!(x-1)^{k+1}}{5^{k+1}}}{\frac{k!(x-1)^k}{5^k}} \right|$

$= \displaystyle\lim_{k\to\infty} \left| \frac{(k+1)!5^k}{5^{k+1}k!} \right| |x-1|$

$= \displaystyle\lim_{k\to\infty} \left| \frac{k+1}{5} \right| |x-1|$

This limit does not exist unless $x = 1$. The convergence set is the single point $x = 1$.

11. $L = \displaystyle\lim_{k\to\infty} \left| \frac{\frac{(k+1)^2(x-1)^{k+1}}{2^{k+1}}}{\frac{k^2(x-1)^k}{2^k}} \right|$

$= \displaystyle\lim_{k\to\infty} \left| \frac{(k+1)^2 2^k}{2^{k+1}k^2} \right| |x-1|$

$= \displaystyle\lim_{k\to\infty} \left| \frac{(k+1)^2}{2k^2} \right| |x-1|$

The interval of absolute convergence is $|x - 1| < 2$, or $-1 < x < 3$.

Endpoints:

$x = -1$: $\displaystyle\sum_{k=1}^{\infty}(-1)^k k^2$, which diverges by the divergence test.

$x = 3$: $\displaystyle\sum_{1}^{\infty} k^2$, which diverges by the divergence test.

The convergence set is $(-1, 3)$.

13. $L = \left| \frac{\frac{(k+1)(3x-4)^{k+1}}{(k+2)^2}}{\frac{k(3x-4)^k}{(k+1)^2}} \right|$

$= \displaystyle\lim_{k\to\infty} \left| \frac{(k+1)^3}{(k+2)^2(k)} \right| |3x - 4|$

$= |3x - 4|$

The interval of absolute convergence is $|3x - 4| < 1$, or $1 < x < \frac{5}{3}$.

Endpoints:

$x = 1$: $\displaystyle\sum_{k=1}^{\infty} \frac{(-1)^k k}{(k+1)^2}$ converges by the

alternating series test:

$\displaystyle\lim_{k\to\infty} \frac{k}{(k+1)^2} = 0$ and

$f(x) = \frac{x}{(x+1)^2}$

$$f'(x) = \frac{-(x-1)}{(x+1)^3} < 0 \text{ (for } x > 1)$$

$x = \frac{5}{3}$: $\displaystyle\sum_{k=1}^{\infty} \frac{k}{(k+1)^2}$ diverges by the comparison

test; behaves like $\displaystyle\sum_{k=1}^{\infty} \frac{1}{k}$ (p-series; $p = 1$).

The convergence set is $[1, \frac{5}{3})$.

15. $L = \left| \dfrac{\dfrac{(k+1)x^{k+1}}{7^{k+1}}}{\dfrac{kx^k}{7^k}} \right|$

$\quad = \displaystyle\lim_{k\to\infty} \left| \frac{(k+1)7^k}{7^{k+1}k} \right| |x|$

$\quad = \displaystyle\lim_{k\to\infty} \left| \frac{k+1}{7k} \right| |x|$

$\quad = \frac{1}{7}|x|$

The interval of absolute convergence is
$-7 < x < 7$.

Endpoints:

$x = -7$: $\displaystyle\sum_{k=1}^{\infty} k(-1)^k$ diverges by the

divergence test.

$x = 7$: $\displaystyle\sum_{k=1}^{\infty} k$, which diverges by the divergence

test.

The convergence set is $(-7, 7)$.

17. $L = \displaystyle\lim_{k\to\infty} \left| \dfrac{\dfrac{[(k+1)!]^2 x^{k+1}}{(k+1)^{k+1}}}{\dfrac{(k!)^2 x^k}{k^k}} \right|$

$\quad = \displaystyle\lim_{k\to\infty} \left| \frac{[(k+1)!]^2 k^k}{(k+1)^{k+1}(k!)^2} \right| |x|$

$\quad = \displaystyle\lim_{k\to\infty} \left[\frac{k+1}{\left(1 + \frac{1}{k}\right)^k} \right] |x|$

$\quad = \infty$

The convergence set is the point $x = 0$.

19. $L = \left| \dfrac{\dfrac{(-1)^{k+1}x^{k+1}}{(k+1)[\ln(k+1)^2]}}{\dfrac{(-1)^k x^k}{k(\ln k)^2}} \right|$

$\quad = \displaystyle\lim_{k\to\infty} \left| \frac{k(\ln k)^2}{(k+1)[\ln(k+1)]^2} \right| |x|$

$\quad = |x|$

The interval of absolute convergence is $|x| < 1$.

Endpoints:

$x = -1$: $\displaystyle\sum_{k=2}^{\infty} \frac{(-1)^k(-1)^k}{k(\ln k)^2} = \sum_{k=2}^{\infty} \frac{1}{k(\ln k)^2}$

converges by the integral test

where $f(x) = \dfrac{1}{x(\ln x)^2}$ is positive,

continuous and decreasing on

$(2, \infty)$:

$$\int_2^{\infty} \frac{dx}{x(\ln x)^2} = \lim_{N\to\infty} \frac{-1}{\ln x}\bigg|_2^N$$

$$= \frac{1}{\ln 2}$$

$x = 1$: $\displaystyle\sum_{k=2}^{\infty} \frac{(-1)^k}{k(\ln k)^2}$ converges absolutely by the

integral test.

The convergence set is $[-1, 1]$.

21. $L = \displaystyle\lim_{k \to \infty} \left| \frac{\dfrac{(2x)^{2(k+1)}}{(k+1)!}}{\dfrac{(2x)^{2k}}{k!}} \right|$

$= \displaystyle\lim_{k \to \infty} \left| \frac{2^{2k+2} k!}{(k+1)! 2^{2k}} \right| |x|^2$

$= \displaystyle\lim_{k \to \infty} \left| \frac{2^2}{k+1} \right| |x|$

$= 0$

The power series converges for all x. The convergence set is $(-\infty, \infty)$.

23. $L = \displaystyle\lim_{k \to \infty} \left| \frac{\dfrac{(k+1)!(3x)^{3(k+1)}}{2^{k+1}}}{\dfrac{k!(3x)^{3k}}{2^k}} \right|$

$= \displaystyle\lim_{k \to \infty} \left| \frac{(k+1)! 3^{3k+3} 2^k}{2^{k+1} k! 3^{3k}} \right| |x|^3$

$= \displaystyle\lim_{k \to \infty} \left| \frac{(k+1)}{2} \right| |3x|^3$

$= \infty$

The powers series converges only for $x = 0$.

25. $L = \displaystyle\lim_{k \to \infty} \left| \frac{2^{k+1}(2x-1)^{2(k+1)} k!}{(k+1)! 2^k (2x-1)^{2k}} \right|$

$= \displaystyle\lim_{k \to \infty} \frac{2}{k+1} |2x - 1|^2$

$= 0 < 1$

so the series converges for all x.

the convergence set is $(-\infty, \infty)$.

27. $L = \displaystyle\lim_{k \to \infty} \left| \frac{\dfrac{x^{k+1}}{(k+1)^{3/2}}}{\dfrac{x^k}{k^{3/2}}} \right|$

$= \displaystyle\lim_{k \to \infty} \left| \frac{k\sqrt{k}}{(k+1)\sqrt{k+1}} \right| |x|$

$= |x|$

The interval of absolute convergence is

$-1 < x < 1$.

Endpoints:

$x = -1$: $\displaystyle\sum_{k=1}^{\infty} \frac{(-1)^k}{k\sqrt{k}}$ converges absolutely

(p-series with $p = \frac{3}{2} > 1$).

$x = 1$: $\displaystyle\sum_{k=1}^{\infty} \frac{1^k}{k\sqrt{k}}$ converges absolutely

(p-series with $p = \frac{3}{2} > 1$).

The convergence set is $[-1, 1]$.

29. $\displaystyle\lim_{k \to \infty} \left| \frac{(k+1)^2(x+1)^{2k+3}}{k^2(x+1)^{2k+1}} \right| = |x+1|^2 < 1$,

so $R = 1$.

31. $\lim\limits_{k \to \infty} \left| \dfrac{(k+1)! k^k x^{k+1}}{k! (k+1)^{k+1} x^k} \right| = \lim\limits_{k \to \infty} \left| \dfrac{(k+1) k^k}{(k+1)^{k+1}} \right| |x|$

$\qquad\qquad = \lim\limits_{k \to \infty} \left| \dfrac{1}{\left(1 + \frac{1}{k}\right)^k} \right| |x|$

$\qquad\qquad = \dfrac{1}{e} |x| < 1$

when $|x| < e$. Thus, $R = e$.

33. $\lim\limits_{k \to \infty} \left| \dfrac{(k+1)(ax)^{k+1}}{k(ax)^k} \right| = |ax| < 1$

when $|x| < \dfrac{1}{|a|}$. Thus, $R = \dfrac{1}{|a|}$.

35. $f(x) = 1 + \dfrac{x}{2} + \dfrac{x^2}{4} + \dfrac{x^3}{8} + \dfrac{x^4}{16} + \cdots$

$f'(x) = (1)\dfrac{1}{2} + (2)\dfrac{x}{4} + (3)\dfrac{x^2}{8} + (4)\dfrac{x^3}{16} + \cdots$

$\qquad = \sum\limits_{k=1}^{\infty} \dfrac{k x^{k-1}}{2^k}$

37. $f(x) = 2 + 3x + 4x^2 + 5x^3 + \cdots$

$f'(x) = 3 + 8x + 15x^2 + \cdots$

$\qquad = \sum\limits_{k=1}^{\infty} k(k+2) x^{k-1}$

39. $f(x) = 1 + \dfrac{x}{2} + \dfrac{x^2}{4} + \dfrac{x^3}{8} + \dfrac{x^4}{16} + \cdots$

$F(x) = \int\limits_0^u f(u) \, du$

$\qquad = x + \dfrac{x^2}{2(2)} + \dfrac{x^3}{3(2)^2} + \dfrac{x^4}{4(2)^3} + \cdots$

$\qquad = \sum\limits_{k=1}^{\infty} \dfrac{x^k}{k(2)^{k-1}}$

Alternatively, we can write $f(x) = \sum\limits_{k=0}^{\infty} \left(\dfrac{x}{2}\right)^k$

$F(x) = \sum\limits_{k=0}^{\infty} \int\limits_0^x \dfrac{u^k}{2^k} \, du$

$\qquad = \sum\limits_{k=0}^{\infty} \dfrac{x^{k+1}}{(k+1)2^k}$

41. $f(x) = \sum\limits_{k=0}^{\infty} (k+2)x^k;$

$F(x) = \sum\limits_{k=0}^{\infty} \int\limits_0^x (k+2)u^k \, du$

$\qquad = \sum\limits_{k=0}^{\infty} \dfrac{(k+2)x^{k+1}}{(k+1)}$

43. $S = \sum\limits_{k=1}^{\infty} \left| \dfrac{\sin(k!x)}{k^2} \right| \le \sum\limits_{k=1}^{\infty} \dfrac{1}{k^2}$

S converges absolutely. Let

$T = S' = \sum\limits_{k=1}^{\infty} \dfrac{k! \cos(k!x)}{k^2}$

$\lim\limits_{k \to \infty} \dfrac{k! \cos(k!x)}{k^2} \ne 0$ (the limit does not exist),

so T is always divergent. Theorem 8.23 does not

apply to S since S is not a power series.

45. Let $S = \sum\limits_{k=1}^{\infty} a_k x^{kp}$; by the ratio test

$\lim\limits_{k \to \infty} \left| \dfrac{a_{k+1} x^{kp+p}}{a_k x^{kp}} \right| = |x|^p \lim\limits_{k \to \infty} \left| \dfrac{a_{k+1}}{a_k} \right|$

$\qquad = |x|^p \left(\dfrac{1}{R}\right) < 1$ if $|x|^p < R$ or $|x| < R^{1/p}$.

Thus, the radius of convergence $R^{1/p}$.

47. Let $S = \sum\limits_{k=1}^{\infty} \dfrac{k}{x^k}$; let $t = x^{-1}$ so $S = \sum\limits_{k=1}^{\infty} kt^k$

which converges (ratio test) if $|t| < 1$ and diverges if $|t| \geq 1$. This means S converges if $|x| > 1$.

49. Since $\dfrac{1}{1-x} = \sum\limits_{k=0}^{\infty} x^k$ for $|x| < 1$, we can

differentiate term by term to obtain

$$\frac{1}{(1-x)^2} = \sum_{k=1}^{\infty} kx^{k-1} \text{ for } |x| < 1$$

And again to obtain

$$\frac{2}{(1-x)^3} = \sum_{k=2}^{\infty} k(k-1)x^{k-2} \text{ for } |x| < 1$$

Thus, a power series for $f(x) = \dfrac{1}{(1-x)^3}$ is

$$\sum_{k=2}^{\infty} \tfrac{1}{2} k(k-1)x^{k-2}$$

$$= 1 + 3x + 6x^2 + 10x^3 + \cdots$$

This is absolutely convergent for $|x| < 1$.

Endpoints:

$x = -1$: $\sum\limits_{k=2}^{\infty} \tfrac{1}{2}k(k-1)(-1)^{k-2}$ diverges

$x = 1$: $\sum\limits_{k=2}^{\infty} \tfrac{1}{2}k(k-1)(-1)^{k-2}$ also diverges.

The interval of convergence is $(-1, 1)$.

51. $L = \lim\limits_{k \to \infty} \left| \dfrac{\dfrac{1 \cdot 2 \cdot 3 \cdot \cdots k(k+1)(-x)^{2(k+1)-1}}{1 \cdot 3 \cdot 5 \cdot \cdots (2k-1)(2k+1)}}{\dfrac{1 \cdot 2 \cdot 3 \cdot \cdots k(-x)^{2k-1}}{1 \cdot 3 \cdot 5 \cdot \cdots (2k-1)}} \right|$

$$= \lim_{k \to \infty} \left| \frac{k+1}{2k+1} \right| |x|^2$$

$$= \tfrac{1}{2}x^2$$

The power series converges absolutely for $\tfrac{1}{2}x^2 < 1$; that is, for $-\sqrt{2} < x < \sqrt{2}$. The radius of convergence is $R = \sqrt{2}$.

53. $f(x) = \sum\limits_{k=0}^{\infty} \dfrac{x^{2k}}{(2k)!} = 1 + \dfrac{x^2}{2!} + \dfrac{x^4}{4!} + \cdots$

$$f'(x) = x + \frac{x^3}{3!} + \frac{x^5}{5!} + \cdots$$

$$f''(x) = 1 + \frac{x^2}{2!} + \frac{x^4}{4!} + \cdots$$

$$= f(x)$$

8.8 Taylor and Maclaurin Series, Pages 564-566

SURVIVAL HINT: *The power and utility of the Taylor series is well illustrated by the list of functions on Page 563.*

1. Suppose there is an open interval I containing c throughout which the function f and all its derivatives exist. Then the power series

$$f(c) + \frac{f'(c)}{1!}(x - c) + \frac{f''(c)}{2!}(x - c)^2$$

$$+ \frac{f'''(c)}{3!}(x - c)^3 + \cdots$$

is called the Taylor series of f at c. The special case where $c = 0$ is called the Maclaurin series of f:

$$f(0) + \frac{f'(0)}{1!}x + \frac{f''(0)}{2!}x^2 + \frac{f'''(0)}{3!}x^3 + \cdots$$

3. $e^{2x} = 1 + 2x + \dfrac{1}{2!}(2x)^2 + \cdots + \dfrac{1}{k!}(2x)^k + \cdots$

$$= \sum_{k=0}^{\infty} \frac{(2x)^k}{k!}$$

5. $e^{x^2} = 1 + x^2 + \frac{1}{2!}(x^2)^2 + \cdots + \frac{1}{k!}(x^2)^k + \cdots$

$$= \sum_{k=0}^{\infty} \frac{x^{2k}}{k!}$$

7. $\sin x^2 = x^2 - \frac{1}{3!}(x^2)^3 + \frac{1}{5!}(x^2)^5 - \cdots$

$$= \sum_{k=0}^{\infty} \frac{(-1)^k x^{4k+2}}{(2k+1)!}$$

9. $\sin ax = ax - \frac{1}{3!}(ax)^3 + \frac{1}{5!}(ax)^5 - \cdots$

$$= \sum_{k=0}^{\infty} \frac{(-1)^k(ax)^{2k+1}}{(2k+1)!}$$

11. $\cos 2x^2 = 1 - \frac{1}{2!}(2x^2)^2 + \frac{1}{4!}(2x^2)^4 - \cdots$

$$= \sum_{k=0}^{\infty} \frac{(-1)^k(2x^2)^{2k}}{(2k)!}$$

13. $x^2\cos x = x^2 - \frac{1}{2!}(x)^4 + \frac{1}{4!}(x)^6 - \cdots$

$$= \sum_{k=0}^{\infty} \frac{(-1)^k(x)^{2k+2}}{(2k)!}$$

15. $x^2 + 2x + 1$ is its own Maclaurin series.

17. $xe^x = x + x^2 + \frac{1}{2!}(x)^3 + \cdots$

$$= \sum_{k=0}^{\infty} \frac{(x)^{k+1}}{k!}$$

19. $e^x + \sin x = \left[1 + x + \frac{x^2}{2!} + \frac{x^3}{3!} + \cdots\right]$

$$+ \left[x - \frac{x^3}{3!} + \frac{x^5}{5!} - \cdots\right]$$

$$= 1 + 2x + \frac{x^2}{2!} + \frac{x^4}{4!} + \frac{2x^5}{5!} + \frac{x^6}{6!} + \frac{x^8}{8!} + \frac{2x^9}{9!} + \cdots$$

21. $\frac{1}{1+4x} = \frac{1}{1-(-4x)}$

$$= 1 + (-4x) + (-4x)^2 + \cdots$$

$$= \sum_{k=0}^{\infty} (-4)^k x^k$$

23. $\frac{1}{a+x} = \frac{1}{a\left(1 + \frac{x}{a}\right)}$

$$= \frac{1}{a\left(1 - \left[-\frac{x}{a}\right]\right)}$$

$$= \frac{1}{a} \sum_{k=0}^{\infty} \left(-\frac{x}{a}\right)^k$$

$$= \sum_{k=0}^{\infty} \frac{(-1)^k x^k}{a^{k+1}}$$

25. $\ln(3 + x) = \ln 3\left(1 + \frac{x}{3}\right)$

$$= \ln 3 + \ln\left(1 + \frac{x}{3}\right)$$

$$= \ln 3 + \frac{1}{3}x - \frac{1}{2(3^2)}x^2 + \frac{1}{3(3^3)}x^3 - \frac{1}{4(3^4)}x^4 + \cdots$$

$$= \ln 3 + \sum_{k=0}^{\infty} \frac{(-1)^k x^{k+1}}{(k+1)3^{k+1}}$$

27. $\tan^{-1}(2x) = \sum_{k=0}^{\infty} \frac{(-1)^k(2x)^{2k+1}}{2k+1}$

$$= \sum_{k=0}^{\infty} \frac{(-1)^k 2^{2k+1} x^{2k+1}}{2k+1}$$

29. $e^{-x^2} = 1 - x^2 + \frac{1}{2!}(x^2)^2 - \cdots$

$$= \sum_{k=0}^{\infty} \frac{(-1)^k x^{2k}}{k!}$$

31. $e^x \approx e + e(x-1) + \frac{1}{2!}e(x-1)^2 + \frac{1}{3!}e(x-1)^3$

33. $\cos x \approx \cos \frac{\pi}{3} - (x - \frac{\pi}{3})\sin \frac{\pi}{3}$

$$- \frac{(x - \frac{\pi}{3})^2}{2!}\cos \frac{\pi}{3} + \frac{(x - \frac{\pi}{3})^3}{3!}\sin \frac{\pi}{3}$$

35. $f(x) = \tan x;\ f(0) = 0$

$f'(x) = \sec^2 x;\ f'(0) = 1$

$f''(x) = 2\sec^2 x \tan x;\ f''(0) = 0$

$f'''(x) = 2\sec^2 x(\sec^2 x) + 4\sec x(\sec x \tan x)\tan x$

$\qquad = 2\sec^4 x + 4\sec^2 x \tan^2 x$

$f'''(0) = 2$

$\tan x \approx 0 + 1 \cdot x + \frac{0x^2}{2!} + \frac{2x^3}{3!}$

Note: you could use the *Student Mathematics*

Handbook, Formula 35, page 211.

37. $f(x) = x^3 - 2x^2 + x - 5$

$\qquad = [(x-2) + 2]^3 - 2[(x-2) + 2]^2$

$\qquad\qquad + [(x-2) + 2] - 5$

$\qquad = (x-2)^3 + 4(x-2)^2 + 5(x-2) - 3$

39. $f(x) = (2-x)^{-1};\ f(5) = -\frac{1}{3}$

$f'(x) = (2-x)^{-2};\ f'(5) = \frac{1}{9}$

$f''(x) = 2(2-x)^{-3};\ f''(5) = -\frac{2}{27}$

$f'''(x) = 6(2-x)^{-4};\ f'''(5) = \frac{6}{81}$

$f(x) \approx -\frac{1}{3} + \frac{1}{9}(x-5) - \frac{1}{27}(x-5)^2 + \frac{1}{81}(x-5)^3$

41. $f(x) = \dfrac{3}{2x - 1}$

$\qquad = \dfrac{3}{2(x-2) + 3}$

$\qquad = \dfrac{1}{1 + \frac{2}{3}(x-2)}$

$\qquad = 1 + [-\frac{2}{3}(x - 2)] + [-\frac{2}{3}(x - 2)]^2$

$\qquad\qquad + [-\frac{2}{3}(x - 2)]^3 + \cdots$

$\qquad \approx 1 - \frac{2}{3}(x-2) + \frac{4}{9}(x-2)^2 - \frac{8}{27}(x-2)^3$

43. $f(x) = (1 + x)^{1/2}$

$\qquad = 1 + \frac{1}{2}x + \dfrac{\frac{1}{2}(\frac{1}{2} - 1)x^2}{2!} + \dfrac{\frac{1}{2}(\frac{1}{2} - 1)(\frac{1}{2} - 2)x^3}{3!}$

$\qquad = 1 + \frac{1}{2}x - \frac{1}{8}x^2 + \frac{1}{16}x^3 - \frac{5}{128}x^4 + \cdots$

Since the exponent, $\frac{1}{2}$, is greater than 0 and

not an integer, the interval of convergence is

$[-1, 1]$.

45. $f(x) = (1 + x)^{2/3}$

$\qquad = 1 + \frac{2}{3}x + \dfrac{\frac{2}{3}(-\frac{1}{3})}{2!}x^2 + \dfrac{\frac{2}{3}(-\frac{1}{3})(-\frac{4}{3})}{3!}x^3 + \cdots$

$\qquad = 1 + \frac{2}{3}x - \frac{1}{9}x^2 + \frac{4}{81}x^3 + \cdots$

Since the exponent, $\frac{2}{3}$, is greater than 0 and

not an integer, the interval of convergence is

$[-1, 1]$.

47. $f(x) = x(1 - x^2)^{-1/2}$

$$= x\left[1 - (-\tfrac{1}{2})x^2 + \frac{(-\tfrac{1}{2})(-\tfrac{3}{2})x^4}{2!}\right.$$

$$\left. - \frac{(-\tfrac{1}{2})(-\tfrac{3}{2})(-\tfrac{5}{2})x^6}{3!} + \cdots\right]$$

$$= x + \tfrac{1}{2}x^3 + \tfrac{3}{8}x^5 + \tfrac{5}{16}x^7 + \cdots$$

Since the exponent, $-\tfrac{1}{2}$, is between -1 and

1, the interval of convergence is

$$-1 < -x^2 \le 1$$

$$-1 \le x^2 < 1$$

$x^2 < 1$ implies $-1 < x < 1$, or $(-1, 1)$.

49. $\quad \ln(1 + x) = \int\limits_0^x \frac{1}{1 + t}\, dt$

$$= \int\limits_0^x [1 - t + t^2 - t^3 + \cdots + (-1)^{k-1}t^{k-1} + \cdots]\, dt$$

$$= [t - \tfrac{1}{2}t^2 + \tfrac{1}{3}t^3 - \cdots + \tfrac{1}{k}(-1)^{k-1}t^k + \cdots]\Big|_0^x$$

$$= \sum_{k=1}^{\infty} \frac{(-1)^{k-1}x^k}{k}$$

The interval of convergence $(-1, 1)$.

51. $\quad \sinh x = \tfrac{1}{2}(e^x - e^{-x})$

$$= \tfrac{1}{2}\Big([1 + x + \tfrac{1}{2!}x^2 + \tfrac{1}{3!}x^3 + \cdots]$$

$$- [1 - x + \tfrac{1}{2!}x^2 - \tfrac{1}{3!}x^3 + \cdots]\Big)$$

$$= x + \tfrac{1}{3!}x^3 + \tfrac{1}{5!}x^5 + \cdots$$

$$= \sum_{k=0}^{\infty} \frac{x^{2k+1}}{(2k+1)!}$$

53. $\quad R_n(\tfrac{1}{3}) \le \frac{e^{z_n}}{(n+1)!}\left(\tfrac{1}{3}\right)^{n+1}$ for $0 < z_n < \tfrac{1}{3} < 1$

$$\frac{e^{z_n}}{(n+1)!3^{n+1}} < \frac{3}{3^{n+1}(n+1)!} < 0.0005$$

Thus, $3^{n+1}(n+1)! > 6,000$ or choose $n = 4$.

Thus,

$$e^{1/3} \approx \frac{\left(\tfrac{1}{3}\right)^0}{0!} + \frac{\left(\tfrac{1}{3}\right)^1}{1!} + \frac{\left(\tfrac{1}{3}\right)^2}{2!} + \frac{\left(\tfrac{1}{3}\right)^3}{3!}$$

$$\approx 1.40$$

55. $f(x) = \dfrac{1}{x^2 - 3x + 2}$

$$= \frac{1}{(x - 2)(x - 1)}$$

$$= \frac{1}{1 - x} - \frac{\tfrac{1}{2}}{1 - \tfrac{x}{2}}$$

$$= [1 + x + x^2 + \cdots]$$

$$- \tfrac{1}{2}[1 + \tfrac{1}{2}x + (\tfrac{1}{2}x)^2 + \cdots]$$

$$= \sum_{k=0}^{\infty}\left[1 - \frac{1}{2^{k+1}}\right]x^k$$

57. $f(x) = \sin x \cos x = \tfrac{1}{2}\sin 2x$

$$= \tfrac{1}{2}\sum_{k=0}^{\infty}(-1)^k \frac{(2x)^{2k+1}}{(2k+1)!}$$

$$= \sum_{k=0}^{\infty} \frac{(-1)^k 2^{2k}x^{2k+1}}{(2k+1)!}$$

59. $f(x) = \left(\cos \frac{3x}{2}\right)\left(\cos \frac{x}{2}\right)$

$\qquad = \frac{1}{2}(\cos 2x + \cos x)$

$\qquad = \frac{1}{2}\left[\sum_{k=0}^{\infty}(-1)^k \frac{(2x)^{2k}}{(2k)!} + \sum_{k=0}^{\infty}(-1)^k \frac{x^{2k}}{(2k)!}\right]$

$\qquad = \frac{1}{2}\sum_{k=0}^{\infty}(-1)^k\left[\frac{1+2^{2k}}{(2k)!}\right]x^{2k}$

61. $f(x) = \ln\left[\frac{1+2x}{1-3x+2x^2}\right]$

$\qquad = \ln(1+2x) - \ln(1-2x) - \ln(1-x)$

$\qquad = \sum_{k=0}^{\infty}(-1)^k\left[\frac{2^{k+1}-(-2)^{k+1}-(-1)^{k+1}}{k+1}\right]x^{k+1}$

$\qquad = \sum_{k=0}^{\infty}[(-1)^k 2^{k+1} + 2^{k+1} + 1]\frac{x^{k+1}}{k+1}$

63. $g(x) = e^x - 1$

$\qquad = x + \frac{x^2}{2!} + \frac{x^3}{3!} + \cdots$

$\qquad \lim_{x \to 0}\frac{e^x - 1}{x} = \lim_{x \to 0}\left[1 + \frac{x}{2!} + \frac{x^2}{3!} + \cdots\right]$

$\qquad\qquad\qquad = 1$

65. Let $f(x) = \cos x$. Then $f^{(n)}(x) =$ is either

$\pm \sin x$ or $\pm \cos x$, so $\left|f^{(n+1)}(x)\right| \leq 1$ for all x and

$\left|R_n(x)\right| = \left|\frac{f^{(n+1)}(c)}{(n+1)!}x^{n+1}\right| \leq \frac{|x|^{n+1}}{(n+1)!}$

for $0 < c < x$

67. Let $n = f'(x)$, $n = f''(x)$, $n = f'''(x)$,

$n = f^{(4)}(x)$ in turn, to obtain

$f(x) - f(0) = f'(x)x - \frac{f''(x)}{2!}x^2 + \frac{f'''(x)}{3!}x^3$

$\qquad\qquad - \frac{f^{(4)}(x)}{4!}x^4 + \cdots$

$f'(x) - f'(0) = f''(x)x - \frac{f'''(x)}{2!}x^2$

$\qquad\qquad + \frac{f^{(4)}(x)}{3!}x^3 + \cdots$

$f''(x) - f''(0) = f'''(x)x - \frac{f^{(4)}(x)}{2!}x^2 + \cdots$

$f'''(x) - f'''(0) = f^{(4)}(x)x - \cdots$

Finally, eliminate $f'(x)$, $f''(x)$, $f'''(x)$ successively to obtain Taylor's series where $c = 0$.

69. We know $\frac{1}{1-x} = 1 + x + x^2 + x^3 + \cdots$, so

if we replace x by $-x^7$, we obtain:

$\frac{1}{1-(-x^7)} = 1 - x^7 + x^{14} - x^{21} + \cdots$

We integrate, term-by-term:

$\int_0^{1/2}\frac{dx}{1+x^7} = \int_0^{1/2}\left[1 - x^7 + x^{14} - x^{21} + \cdots\right]dx$

$\qquad = \left[x - \frac{x^8}{8} + \frac{x^{15}}{15} - \frac{x^{22}}{22} + \cdots\right]\Big|_0^{1/2}$

$\qquad = \frac{1}{2} - \frac{1}{8 \cdot 2^8} + \frac{1}{15 \cdot 2^{15}} - \frac{1}{22 \cdot 2^{22}} + \cdots$

This series converges for $\left| -x^7 \right| < 1$ or $|x| < 1$.

Since we want six decimal place accuracy, the error cannot exceed 0.0000005. If we stop with $n = 3$, the error is smaller than the term with $n = 4$:

$$\frac{1}{22 \cdot 2^{22}} \approx 1.08372 \times 10^{-8}$$

which is well within our desired accuracy.

Thus,

$$\int_0^{1/2} \frac{dx}{1 + x^7} \approx \frac{1}{2} - \frac{1}{8 \cdot 2^8} + \frac{1}{15 \cdot 2^{15}}$$

$$\approx 0.499514$$

71. a. $J_0(x) = \sum_{k=0}^{\infty} \frac{(-1)^k x^{2k}}{(k!)^2 2^{2k}}$; ratio test:

$$\lim_{k \to \infty} \left| \frac{(k!)^2 2^{2k} x^{2k+2}}{[(k+1)!]^2 2^{2k+2} x^{2k}} \right| = \lim_{k \to \infty} \frac{x^2}{2^2 (k+1)^2}$$

$$= 0 < 1$$

Thus, $J_0(x)$ converges absolutely for all x.

$$J_1(x) = \sum_{k=0}^{\infty} \frac{(-1)^k x^{2k+1}}{k!(k+1)! 2^{2k+1}}; \text{ ratio test:}$$

$$\lim_{k \to \infty} \left| \frac{k!(k+1)! 2^{2k+1} x^{2k+3}}{(k+1)!(k+2)! 2^{2k+3} x^{2k+1}} \right|$$

$$= \lim_{k \to \infty} \frac{x^2}{4(k+1)(k+2)} = 0 < 1$$

and J_1 also converges absolutely for all x.

b. $J_0'(x) = \sum_{k=1}^{\infty} \frac{(-1)^k (2k) x^{2k-1}}{(k!)^2 2^{2k}}$

$$= \frac{-x}{1!0!2} + \frac{x^3}{2!1!2^3} + \frac{-x^5}{3!2!2^5} + \cdots$$

$$= \sum_{k=0}^{\infty} \frac{(-1)^{k+1} x^{2k+1}}{k!(k+1)! 2^{2k+1}} = -J_1(x)$$

73. $f'(x) = 1 + x + x^2 + \cdots + x^{k-1} + \cdots$

$$= \frac{1}{1 - x};$$

$$f(x) = \int_0^x f'(t)\, dt$$

$$= \int_0^x (1 + t + t^2 + \cdots + t^{k-1} + \cdots)\, dt$$

$$= -\ln|1 - x| \text{ for } x \text{ in } (-1, 1)$$

75. The solution is found on p. 197 of the 1985 issue (Vol. 1) of *Pi Mu Epsilon Journal*.

CHAPTER 8 REVIEW

Proficiency Examination, Pages 566–567

SURVIVAL HINT: *To help you review the concepts of this chapter,* **handwrite the answers to each of these** *questions onto your own paper.*

1. A sequence is a succession of numbers that are listed according to a given prescription or rule.

2. If the terms of the sequence $\{a_n\}$ approach the number L as n increases without bound, we say that the sequence $\{a_n\}$ *converges to the limit* L and write $L = \lim\limits_{n \to \infty} a_n$.

3. A sequence converges if the limit of the nth element is finite (and unique). If not, it diverges.

4. a. The elements of a bounded sequence lie within a finite range.

   b. A sequence is monotonic if it is nondecreasing or nonincreasing.

   c. A sequence is strictly monotonic if it is increasing or decreasing.

5. A monotonic sequence $\{a_n\}$ converges if it is bounded and diverges otherwise.

6. An infinite series is a sum of infinitely many terms.

7. A sequence $\{a_n\}$ converges if $\lim\limits_{n \to \infty} a_n = L$ is finite (and unique). A series $\sum\limits_{k=0}^{\infty} a_k$ converges if the sequence $\{S_n\}$ of partial sums $S_n = \sum\limits_{k=1}^{n} a_k$ converges.

8. The middle terms of a telescoping series vanish (by addition and subtraction of the same numbers). Specifically, the series $S = \Sigma a_k$ telescopes if $a_k = b_k - b_{k-1}$.

9. The harmonic series is a p-series with $p = 1$. It diverges, but the alternating harmonic series converges.

10. The ratio of consecutive terms of a geometric series is a constant, r. That is, a geometric series is one of the form Σar^k. It diverges if $|r| \ge 1$ and converges to $S = a/(1-r)$ if $|r| < 1$.

11. $S = \sum\limits_{k=0}^{\infty} a_k$ diverges if $\lim\limits_{k \to \infty} a_k \neq 0$.

12. If $a_k = f(k)$ for $k = 1, 2, \ldots$, where f is a positive, continuous, and decreasing function of x for $x \ge 1$, then
$$\sum_{k=1}^{\infty} a_k \quad \text{and} \quad \int_{1}^{\infty} f(x)\, dx$$
either both converge or both diverge.

13. $\sum\limits_{k=1}^{\infty} \frac{1}{k^p}$ is a convergent p-series if $p > 1$. It diverges when $p \le 1$.

14. Let $0 \le a_k \le c_k$ for all k. If $\sum\limits_{k=1}^{\infty} c_k$ converges, then $\sum\limits_{k=1}^{\infty} a_k$ also converges.

   Let $0 \le d_k \le a_k$ for all k. If $\sum\limits_{k=1}^{\infty} d_k$ diverges, then $\sum\limits_{k=1}^{\infty} a_k$ also diverges.

15. Suppose $a_k > 0$ and $b_k > 0$ for all sufficiently large k and that
$$\lim_{k \to \infty} \frac{a_k}{b_k} = L$$
where L is finite and positive ($0 < L < \infty$). Then Σa_k and Σb_k either both converge or both diverge.

16. Suppose $a_k > 0$ and $b_k > 0$ for all sufficiently large k. Then,

   If $\lim\limits_{k \to \infty} \frac{a_k}{b_k} = 0$ and Σb_k converges, the series Σa_k converges.

If $\displaystyle\lim_{k \to \infty} \frac{a_k}{b_k} = \infty$ and Σb_k diverges,

the series Σa_k diverges.

17. Given the series Σa_k with $a_k > 0$, suppose

that $\displaystyle\lim_{k \to \infty} \frac{a_{k+1}}{a_k} = L$. The ratio test states

the following:

If $L < 1$, then Σa_k converges.

If $L > 1$ or if L is infinite, then Σa_k diverges.

If $L = 1$, the test is inconclusive.

18. Given the series Σa_k with $a_k \geq 0$, suppose

that $\displaystyle\lim_{k \to \infty} \sqrt[k]{a_k} = L$. The root test states the
following:

If $L < 1$, then Σa_k converges.

If $L > 1$ or if L is infinite, then Σa_k diverges.

If $L = 1$, the root test is inconclusive.

19. If $a_k > 0$, then an alternating series

$$\sum_{k=1}^{\infty} (-1)^k a_k \qquad \text{or} \qquad \sum_{k=1}^{\infty} (-1)^{k+1} a_k$$

converges if both of the following two

conditions are satisfied:

1. $\displaystyle\lim_{k \to \infty} a_k = 0$

2. $\{a_k\}$ is a decreasing sequence; that is,

$\quad a_{k+1} < a_k$ for all k.

20. Suppose an alternating series

$$\sum_{k=1}^{\infty} (-1)^k a_k \qquad \text{or} \qquad \sum_{k=1}^{\infty} (-1)^{k+1} a_k$$

satisfies the conditions of the alternating

series test; namely, $\displaystyle\lim_{k \to \infty} a_k = 0$ and $\{a_k\}$ is

a decreasing sequence $(a_{k+1} < a_k)$. If the

series has sum S, then $\left| S - S_n \right| < a_{n+1}$,

where S_n is the nth partial sum of the series.

21. A series of real numbers Σa_k must converge if

the related absolute value series $\Sigma |a_k|$

converges.

22. The series Σa_k is absolutely convergent if the

related series $\Sigma |a_k|$ converges. The series Σa_k

is conditionally convergent if it converges but

$\Sigma |a_k|$ diverges.

23. For the series Σa_k, suppose $a_k \neq 0$ for $k \geq 1$

and that

$$\lim_{k \to \infty} \left| \frac{a_{k+1}}{a_k} \right| = L$$

where L is a real number or ∞. Then:

If $L < 1$, the series Σa_k converges absolutely

and hence converges.

If $L > 1$ or if L is infinite, the series Σa_k

diverges.

If $L = 1$, the test is inconclusive.

24. An infinite series of the form

$$\sum_{k=0}^{\infty} a_k (x - c)^k$$
$$= a_0 + a_1(x - c) + a_2(x - c)^2 + \dots$$

is called a power series in $(x - c)$.

25. Let $\Sigma a_k u^k$ be a power series, and consider

$$L = \lim_{k \to \infty} \left| \frac{a_{k+1}}{a_k} \right|$$

Then, If $L = \infty$, the power series converges

only at $u = 0$. If $L = 0$, the power series converges for all real u. If $0 < L < \infty$, let $R = 1/L$. Then the power series *converges absolutely* for $|u| < R$ (or $-R < u < R$) and *diverges* for $|u| > R$. This is called the *interval of absolute convergence*. Finally, check for convergence at the endpoints $u = -R$ and $u = R$. The resulting set (with convergent endpoints) is the *convergence set*.

26. The interval of convergence of a power series consists of those values of x for which the series converges. If the interval is $-R < x < R$, then R is the radius of convergence.

27. $P_n(x) = \sum_{k=1}^{n} a_k(x - c)^k$ is a Taylor polynomial if $a_k = \dfrac{f^{(k)}(c)}{k!}$.

28. If f and all its derivatives exist in an open interval I containing c, then for each x in I
$$f(x) = f(c) + \frac{f'(c)}{1!}(x - c) + \frac{f''(c)}{2!}(x - c)^2$$
$$+ \ldots + \frac{f^{(n)}(c)}{n!}(x - c)^n + R_n(x)$$
where the remainder function $R_n(x)$ is given by
$$R_n(x) = \frac{f^{(n+1)}(z_n)}{(n + 1)!}(x - c)^{n+1}$$
for some z_n that depends on x and lies between c and x.

29. The Taylor series is $T = \sum_{k=1}^{\infty} a_k(x - c)^k$ and is a Maclaurin series if $c = 0$.

30. The binomial function $(1 + x)^p$ is represented by its Maclaurin series
$$(1 + x)^p = 1 + px + \frac{p(p - 1)}{2!}x^2$$
$$+ \frac{p(p - 1)(p - 2)}{3!}x^3 + \ldots$$
$$+ \frac{p(p - 1)\cdots(p - k + 1)}{k!} x^k + \ldots$$
$-1 < x < 1$ if $p \leq -1$;
$-1 < x \leq 1$ if $-1 < p < 0$;
$-1 \leq x \leq 1$ if $p > 0$, p not an integer;
all x if p is a nonnegative integer.

31. This is the definition of e.

32. **a.** The sequence has an upper bound of 4, a lower bound of 0, and after $n = 2$ is monotonic decreasing:
$$\lim_{n \to \infty} \frac{\dfrac{e^{n+1}}{(n + 1)!}}{\dfrac{e^n}{n!}} = \lim_{n \to \infty} \frac{e}{n + 1}$$
$$= 0$$

Thus, the sequence converges by the BMCT.

b. $S = \sum_{k=1}^{\infty} \dfrac{e^k}{k!}$; use ratio test.
$$\lim_{k \to \infty} \frac{e^{k+1} k!}{(k + 1)! e^k} = \lim_{k \to \infty} \frac{e}{k + 1}$$
$$= 0$$
$$< 1$$

S converges.

c. Answers vary. In part **a**, we consider the limit of a sequence, and in part **b** we consider the convergence of a series.

33. $\sum\limits_{k=2}^{\infty} \dfrac{1}{k \ln k}$; Use the integral test.

Let $f(x) = \dfrac{1}{x \ln x}$; it is continuous, positive, and decreasing for $x > 1$.

$$\lim_{b\to\infty} \int_2^b (x \ln x)^{-1}\, dx = \lim_{b\to\infty} \ln|\ln x|\Big|_2^b$$

$$= \infty$$

The series diverges.

34. $\sum\limits_{k=1}^{\infty} \dfrac{\pi^k k!}{k^k}$; use the ratio test.

$$\lim_{k\to\infty} \frac{\pi^{k+1}(k+1)!\, k^k}{(k+1)^{k+1}\pi^k k!} = \pi \lim_{k\to\infty} \frac{1}{\left(1 + \frac{1}{k}\right)^k}$$

$$= \frac{\pi}{e}$$

$$> 1$$

The series diverges.

35. $\sum\limits_{k=2}^{\infty} \dfrac{1}{(\ln k)^{1/k}}$; use divergence test.

$$\lim_{k\to\infty} (\ln k)^{-1/k} = \lim_{k\to\infty} \exp[\ln(\ln k)^{-1/k}]$$

$$= \lim_{k\to\infty} e^{-\ln(\ln k)/k}$$

$$= 1$$

$$\neq 0$$

The series diverges.

36. $\sum\limits_{k=1}^{n} \dfrac{3k^2 - k + 1}{(1 - 2k)k}$; check the necessary condition:

$$\lim_{k\to\infty} \frac{3k^2 - k + 1}{(1 - 2k)k} = -\frac{3}{2}$$

$$\neq 0$$

The series diverges.

37. $\sum\limits_{k=1}^{\infty} \dfrac{(-1)^{k+1}}{k^2}$ converges absolutely when compared with the convergent p-series

$\Sigma(1/k^2)$ $(p = 2 > 1)$.

38. $\sum\limits_{k=0}^{\infty} (-1)^k (k+1) u^k$

$$L = \lim_{k\to\infty} \left| \frac{(-1)^{k+1}(k+2)u^{k+1}}{(-1)^k(k+1)u^k} \right|$$

$$= \lim_{k\to\infty} \frac{k+2}{k+1}|u|$$

$$= |u|$$

The series converges absolutely for

$-1 < u < < 1.$

Endpoints:

$u = -1$: $\sum\limits_{k=1}^{\infty} (-1)^{k+1} k(-1)^{k-1} = \sum\limits_{k=1}^{\infty} k$

diverges by the divergence test.

$u = 1$: $\sum\limits_{k=1}^{\infty} (-1)^{k+1} k$ diverges by the

divergence test. The

convergence set is $(-1, 1)$.

39. $\sin x = \sum_{k=0}^{\infty} \frac{(-1)^k x^{2k+1}}{(2k+1)!}$

$\sin 2x = \sum_{k=0}^{\infty} \frac{(-1)^k (2x)^{2k+1}}{(2k+1)!}$

40. $f(x) = \frac{1}{x-3}$ at $c = \frac{1}{2}$

$$= \frac{1}{x - \frac{1}{2} - \frac{5}{2}}$$

$$= \frac{-\frac{2}{5}}{1 - \frac{2}{5}\left(x - \frac{1}{2}\right)}$$

$$= -\frac{2}{5}\left[1 + \frac{2}{5}\left(x - \frac{1}{2}\right) + \cdots\right]$$

$$= -\frac{2}{5}\sum_{k=0}^{\infty}\left(\frac{2}{5}\right)^k\left(x - \frac{1}{2}\right)^k$$

$$= -\sum_{k=0}^{\infty}\left(\frac{2}{5}\right)^{k+1}\left(x - \frac{1}{2}\right)^k$$

Cumulative Review, Chapters 6-8, Page 571

1. a. The work done by the variable force $F(x)$ in moving an object along the x-axis from $x = a$ to $x = b$ is given by

$$W = \int_a^b F(x)\,dx$$

b. Suppose a flat surface (a plate) is submerged vertically in a fluid of weight density ρ (lb/ft^3) and that the submerged portion of the plate extends from $x = a$ to $x = b$ on a vertical axis. Then the total force F exerted by the fluid is given by

$$F = \int_a^b \rho\, h(x) L(x)\,dx$$

where $h(x)$ is the depth at x and $L(x)$ is the corresponding length of a typical horizontal approximating strip.

3. A *geometric series* is an infinite series in which the ratio of successive terms is constant. If this constant ratio is r, then the series has the form $(a \neq 0)$

$$\sum_{k=0}^{\infty} ar^k = a + ar + ar^2 + ar^3 + \cdots + ar^n + \cdots$$

This series diverges if $|r| \geq 1$ and converges if $|r| < 1$ with sum

$$\sum_{k=0}^{\infty} ar^k = \frac{a}{1-r}.$$

5. $\int \ln \sqrt{x}\,dx = \frac{1}{2}\int \ln x\,dx$

Let $u = \ln x$, $du = 1/x\,dx$; $dv = dx$, $v = x$

$$= \frac{1}{2}\left[x \ln x - \int x\left(\frac{1}{x}\right)dx\right]$$

$$= \frac{1}{2}x \ln x - \frac{1}{2}x + C$$

7. $\int \frac{\tan^{-1} x}{1 + x^2}\,dx = \int u\,du$

Let $u = \tan^{-1} x$; $du = (1 + x^2)^{-1}dx$

$$= \frac{1}{2}u^2 + C$$

$$= \frac{1}{2}(\tan^{-1} x)^2 + C$$

9. $\displaystyle\int \sqrt{4 - x^2}\, dx = \int \sqrt{4 - 4\sin^2\theta}\;(2\cos\theta\ d\theta)$

> Let $x = 2\sin\theta;\ dx = 2\cos\theta\ d\theta$

$$= \int 4\sqrt{1 - \sin^2\theta}\,\cos\theta\ d\theta$$

$$= \int 4\cos^2\theta\ d\theta$$

$$= 4\int \frac{1 + \cos 2\theta}{2}\ d\theta$$

$$= 2\Big[\theta + \frac{\sin 2\theta}{2}\Big] + C$$

$$= 2\theta + \sin 2\theta + C$$

$$= 2\sin^{-1}\!\Big(\tfrac{x}{2}\Big) + 2\Big(\tfrac{x}{2}\Big)\Big(\tfrac{1}{2}\sqrt{4 - x^2}\Big) + C$$

$$= 2\sin^{-1}\!\Big(\tfrac{x}{2}\Big) + \tfrac{1}{2}x\sqrt{4 - x^2} + C$$

11. a. $\displaystyle\frac{x^2 + 17x - 8}{(2x+3)(x-1)^2}$

$$= \frac{A}{x-1} + \frac{B}{(x-1)^2} + \frac{C}{2x+3}$$

$$A(x-1)(2x+3) + B(2x+3) + C(x-1)^2$$

$$= x^2 + 17x - 8$$

Thus, $\qquad 2A + C = 1$

$$A + 2B - 2C = 17$$

$$-3A + 3B + C = -8$$

Solving, we obtain $A = 3$, $B = 2$, and

$C = -5$

b. $\displaystyle\int \frac{x^2 + 17x - 8}{(2x+3)(x-1)^2}\ dx$

$$= \int\Big[\frac{3}{x-1} + \frac{2}{(x-1)^2} + \frac{-5}{2x+3}\Big]\ dx$$

$$= 3\ln|x - 1| - \frac{2}{x - 1} - \tfrac{5}{2}\ln|2x + 3| + C$$

c. $\displaystyle\int_2^\infty \frac{x^2 + 17x - 8}{(2x + 3)(x - 1)^2}\ dx$

$$= \lim_{N\to\infty}\Big[3\ln|x - 1| - \frac{2}{x - 1} - \tfrac{5}{2}\ln|2x + 3|\Big]\Big|_2^N$$

$$= \infty$$

The series diverges.

13. $\displaystyle\sum_{k=0}^{\infty} \frac{k^2 3^{k+1}}{4^k}$; ratio test

$$\lim_{k\to\infty} \frac{\dfrac{(k+1)^2 3^{k+2}}{4^{k+1}}}{\dfrac{k^2 3^{k+1}}{4^k}} = \lim_{k\to\infty} \frac{(k+1)^2 3^{k+2} 4^k}{k^2 3^{k+1} 4^{k+1}}$$

$$= \lim_{k\to\infty} \frac{(k+1)^2(3)}{k^2(4)}$$

$$= \frac{3}{4}$$

$$< 1$$

The series converges.

15. $\displaystyle\sum_{k=1}^{\infty} \frac{(-1)^{k+1} k}{k+1}$ diverges by the divergence

test: $\displaystyle\lim_{k\to\infty} \frac{(-1)^{k+1} k}{k + 1} \neq 0$

17. $\displaystyle\sum_{k=1}^{\infty} k e^{-k^2}$; integral test

$$\int_0^\infty x e^{-x^2} dx = \lim_{t\to\infty}\Big[-\tfrac{1}{2}e^{-t^2} + \tfrac{1}{2}\Big]$$

$$= \frac{1}{2}$$

The series converges.

19. $\displaystyle\sum_{k=1}^{\infty} 5\left(-\frac{2}{3}\right)^{2k-1}$

$$= 5\left[\frac{-2}{3} + \left(\frac{-2}{3}\right)^3 + \left(\frac{-2}{3}\right)^5 + \cdots\right]$$

$$= -\frac{10}{3}\left[1 + \left(\frac{2}{3}\right)^2 + \left(\frac{2}{3}\right)^4 + \cdots\right]$$

$$= \frac{-10}{3}\left[\frac{1}{1 - \frac{4}{9}}\right]$$

$$= -6$$

21. $\displaystyle\sum_{k=1}^{\infty} \frac{x^k}{k^3}$

$$\lim_{k \to \infty}\left|\frac{\dfrac{x^{k+1}}{(k+1)^3}}{\dfrac{x^k}{k^3}}\right| = \lim_{k \to \infty}\left|\frac{k^3}{(k+1)^3}\right| |x|$$

$$= |x|$$

$$< 1$$

Endpoints:

$x = 1$;

$\displaystyle\sum_{k=1}^{\infty} \frac{1}{k^3}$ converges (*p*-series with $p = 3 > 1$)

$x = -1$;

$\displaystyle\sum_{k=1}^{\infty} \frac{(-1)^k}{k^3}$ converges absolutely

Convergence set is $-1 \le x \le 1$.

23.

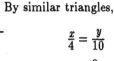

By similar triangles,

$$\frac{x}{4} = \frac{y}{10}$$

$$x = \frac{2}{5} y$$

$$\Delta F = \underbrace{62.4\,\pi x^2 \Delta y}_{\text{Force (Weight)}} \underbrace{(12 - y)}_{\text{Distance}}$$

$$F = \int_0^6 62.4\pi\left(\frac{2}{5}y\right)^2(12 - y)\ dy$$

$$\approx 16{,}937 \text{ ft-lb}$$

25. $y = \frac{3}{2} x^{2/3}$

$$y' = x^{-1/3}$$

$$s = \int_{1/8}^{8} \sqrt{1 + (x^{-1/3})^2}\ dx$$

$$= \int_{1/8}^{8} \frac{1}{x^{1/3}}\sqrt{x^{2/3} + 1}\ dx$$

Let $u = x^{2/3} + 1$; $du = \frac{2}{3}x^{-1/3}\ dx$

$$= \int_{5/4}^{5} \frac{3}{2}\sqrt{u}\ du$$

$$= u^{3/2}\Big|_{5/4}^{5}$$

$$= 5^{3/2} - \left(\frac{5}{4}\right)^{3/2}$$

$$= \frac{7}{8}(5)^{3/2}$$

$$\approx 9.7828$$

27. a. $y' - \frac{y}{x} = \frac{x}{1 + x^2}$

$u = e^{\int (-1/x)\, dx} = e^{-\ln x}$

$y = \frac{1}{1/x}\left[\int \frac{1}{x}\left(\frac{x}{1 + x^2}\right) dx + C\right]$

$= x\left[\tan^{-1}x + C\right]$

$= x\tan^{-1}x + Cx$

b. $y' + (\tan x)y = \tan x$

$u = e^{\int \tan x\, dx} = e^{\ln(\sec x)} = \sec x$

$y = \frac{1}{\sec x}\left[\int \sec x \tan x + C\right]$

$= \frac{1}{\sec x}[\sec x + c]$

$= 1 + C\cos x$

Now, $y(0) = 3 = 1 + C\cos 0$, so $C = 2$

The solution is $y = 1 + 2\cos x$.

29. a. $\sin u = u - \frac{u^3}{3!} + \frac{u^5}{5!} - \cdots$

$f(x) = \sin x^2$

$= x^2 - \frac{(x^2)^3}{3!} + \frac{(x^2)^5}{5!} - \cdots$

$= x^2 - \frac{x^6}{3!} + \frac{x^{10}}{5!} - \cdots$

$= \sum_{k=0}^{\infty} \frac{(-1)^k x^{4k+2}}{(2k + 1)!}$

b. $f(x) = \sqrt{e^x}$

$= e^{x/2}$

$= 1 + \left(\frac{x}{2}\right) + \frac{1}{2!}\left(\frac{x}{2}\right)^2 + \frac{1}{3!}\left(\frac{x}{2}\right)^3 + \cdots$

$= \sum_{k=0}^{\infty} \frac{1}{k!}\left(\frac{1}{2}\right)^k x^k$

c. $f(x) = \frac{3x - 2}{5 + 2x}$

$= \frac{3}{2} - \frac{\frac{19}{10}}{1 + \frac{2}{5}x} \qquad$ *by long division*

$= \frac{3}{2} - \frac{19}{10}\sum_{k=0}^{\infty}\left(-\frac{2}{5}x\right)^k$

CHAPTER 9

Vectors in the Plane and in Space

9.1 Vectors in $\mathbb{R}^2$, Pages 580-582

1. You can choose any point as the starting point. We choose $(0, 0)$. Count 3 units in the positive x-direction and then 4 units in the negative y-direction.

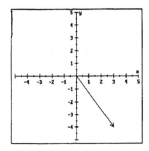

3. From $(0, 0)$ count $\frac{1}{2}$ unit in the negative x-direction and then $\frac{5}{2}$ units in the positive y-direction. A convenient scale is two squares for one unit.

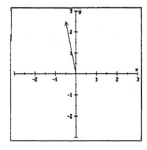

5. Connect the points P and Q.

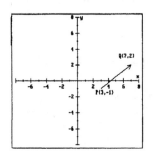

$$\mathbf{PQ} = (7 - 3)\mathbf{i} + (2 - (-1))\mathbf{j}$$

$$= 4\mathbf{i} + 3\mathbf{j} = \langle 4, 3 \rangle$$

$$\| \mathbf{PQ} \| = \sqrt{4^2 + 3^2}$$

$$= 5$$

7. Draw the vector from P to Q.

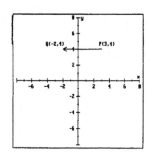

$$PQ = (-2 - 3)i + (4 - 4)j$$

$$= -5i = \langle -5, 0 \rangle$$

$$\|PQ\| = 5$$

9. $$PQ = (1 + 1)i + (-2 + 2)j$$

$$= 2i$$

$$\|PQ\| = \sqrt{(1 + 1)^2 + (-2 + 2)^2}$$

$$= 2$$

11. $$PQ = (0 + 4)i + (-1 + 3)j$$

$$= 4i + 2j$$

$$\|PQ\| = \sqrt{(0 + 4)^2 + (-1 + 3)^2}$$

$$= 2\sqrt{5}$$

13. Let $\mathbf{v} = \mathbf{i} + \mathbf{j}$

$$\|\mathbf{v}\| = \sqrt{1^2 + 1^2}$$

$$= \sqrt{2}$$

$$\mathbf{u} = \frac{\mathbf{v}}{\|\mathbf{v}\|}$$

$$= \frac{1}{\sqrt{2}}(\mathbf{i} + \mathbf{j})$$

15. Let $\mathbf{v} = 3\mathbf{i} - 4\mathbf{j}$

$$\|\mathbf{v}\| = \sqrt{3^2 + (-4)^2}$$

$$= 5$$

$$\mathbf{u} = \frac{\mathbf{v}}{\|\mathbf{v}\|}$$

$$= \tfrac{1}{5}(3\mathbf{i} - 4\mathbf{j})$$

$$= \tfrac{3}{5}\mathbf{i} - \tfrac{4}{5}\mathbf{j}$$

17. $su + tv = \langle -3s + t, 4s - t \rangle$ so that

$$-3s + t = 6 \text{ and } 4s - t = 0.$$

Add these equations:

$$s = 6$$

Substitute this value into either equation:

$$-3(6) + t = 6$$

$$t = 24$$

19. $$s\mathbf{v} + t\langle -2, 1 \rangle = \mathbf{u}$$

$$s\langle 1, -1 \rangle + t\langle -2, 1 \rangle = \langle -3, 4 \rangle$$

$$\langle s - 2t, -s + t \rangle = \langle -3, 4 \rangle$$

Thus, $s - 2t = -3$ and $-s + t = 4$

Add these equations:

$$-t = 1$$

$$t = -1$$

Substitute this value into either equation:

$$s - 2(-1) = -3$$

$$s = -5$$

21. $$2\mathbf{u} + 3\mathbf{v} - \mathbf{w} = (6 + 12 - 1)\mathbf{i} + (-8 - 9 - 1)\mathbf{j}$$

$$= 17\mathbf{i} - 18\mathbf{j}$$

23. $$\|\mathbf{v}\| = \sqrt{4^2 + (-3)^2}$$

$$= 5$$

$$\|\mathbf{u}\| = \sqrt{3^2 + (-4)^2}$$

$$= 5$$

$$\|\mathbf{v}\|\mathbf{u} + \|\mathbf{u}\|\mathbf{v} = 5(3\mathbf{i} - 4\mathbf{j}) + 5(4\mathbf{i} - 3\mathbf{j})$$

$$= 15\mathbf{i} - 20\mathbf{j} + 20\mathbf{i} - 15\mathbf{j}$$

$$= 35\mathbf{i} - 35\mathbf{j}$$

25. Since $(x - y - 1)\mathbf{i} + (2x + 3y - 12)\mathbf{j} = \mathbf{0}$,

$$x - y - 1 = 0$$

$$2x + 3y - 12 = 0$$

Multiply the first equation by 3

$$3x - 3y - 3 = 0$$

$$2x + 3y - 12 = 0$$

Add these equations:

$$5x - 15 = 0$$

$$x = 3$$

Substitute this value to find y:

$$2(3) + 3y - 12 = 0$$

$$3y = 6$$

$$y = 2$$

Thus, $(x, y) = (3, 2)$.

27. $(x^2 + y^2)\mathbf{i} + y\mathbf{j} = 20\mathbf{i} + (x + 2)\mathbf{j}$

Equate the components:

$$x^2 + y^2 = 20$$

$$y = x + 2$$

Substitute the second equation into the first:

$$x^2 + (x + 2)^2 = 20$$

$$x^2 + x^2 + 4x + 4 = 20$$

$$2x^2 + 4x + 4 = 20$$

$$x^2 + 2x + 2 = 10$$

$$x^2 + 2x - 8 = 0$$

$$(x - 2)(x + 4) = 0$$

$$x = 2, \ -4$$

We have $(2, 4)$ and $(-4, -2)$.

29. $\mathbf{u} = (\cos 30°)\mathbf{i} + (\sin 30°)\mathbf{j}$

$$= \frac{\sqrt{3}}{2}\mathbf{i} + \frac{1}{2}\mathbf{j}$$

31. Let $\mathbf{v} = -4\mathbf{i} + \mathbf{j}$ so

$$\|\mathbf{v}\| = \sqrt{(-4)^2 + 1^2}$$

$$= \sqrt{17}$$

Then

$$\mathbf{u} = \frac{\mathbf{v}}{\|\mathbf{v}\|}$$

$$= -\frac{1}{\sqrt{17}}(-4\mathbf{i} + \mathbf{j})$$

$$= \frac{4}{\sqrt{17}}\mathbf{i} - \frac{1}{\sqrt{17}}\mathbf{j}$$

33. $\mathbf{u} + \mathbf{v} = 5\mathbf{i} + \mathbf{j}$

$$\|\mathbf{u} + \mathbf{v}\| = \sqrt{5^2 + 1^2}$$

$$= \sqrt{26}$$

The desired unit vector is $\dfrac{5}{\sqrt{26}}\mathbf{i} + \dfrac{1}{\sqrt{26}}\mathbf{j}$.

35. If the vector is from (x_1, y_1) to (x_2, y_2), we are given that $(x_1, y_1) = (-2, 3)$. We need to find the terminal point, so that

$$(x_2 + 2)\mathbf{i} + (y_2 - 3)\mathbf{j} = 5\mathbf{i} + 7\mathbf{j}$$

so that (by equating the components)

$$x_2 + 2 = 5 \text{ or } x_2 = 3$$

and

$$y_2 - 3 = 7 \text{ or } y_2 = 10.$$

The terminal point is $(3, 10)$.

37. a. The midpoint of $\overline{PQ}$ is

$$M = \left(\frac{-3 + 9}{2}, \frac{-8 - 2}{2}\right) = (3, -5)$$

If the initial point of this vector is at $P(-3, -8)$, then its terminal point is at $(3, -5)$, so the vector is

$$[3 - (-3)]\mathbf{i} + [-5 - (-8)]\mathbf{j} = 6\mathbf{i} + 3\mathbf{j}$$

b. $\mathbf{PQ} = [9 - (-3)]\mathbf{i} + [2 - (-8)]\mathbf{j}$

$$= 12\mathbf{i} + 6\mathbf{j}$$

$$= \langle 12, 6 \rangle$$

$\frac{5}{6}\langle 12, 6 \rangle = \langle 10, 5 \rangle = 10\mathbf{i} + 5\mathbf{j}$. This vector has initial point at $P(-3, -8)$ and terminal point

$$(-3 + 10, -8 + 5) = (7, -3)$$

39. $\|\mathbf{v}\| = \sqrt{\cos^2\theta + \sin^2\theta} = 1$

41. Not necessarily equal. Equal magnitudes say nothing about their direction.

43. a. $\|\mathbf{u} - \mathbf{u_0}\| = \sqrt{(x - x_0)^2 + (y - y_0)^2} = 1$

This is the set of points on the circle with center (x_0, y_0) and radius 1.

b. $\|\mathbf{u} - \mathbf{u_0}\| \leq r$ is the set of points on or interior to the circle with center (x_0, y_0) and radius r.

45. $a\mathbf{u} + b(\mathbf{u} - \mathbf{v}) + c(\mathbf{u} + \mathbf{v}) = \mathbf{0}$

$$(a + b + c)\mathbf{u} = (b - c)\mathbf{v}$$

Since $\mathbf{u}$ and $\mathbf{v}$ are not parallel, we must have

$$a + b + c = 0$$

$$b - c = 0$$

Adding these equations, we obtain

$a + 2b = 0$. If we let $b = t$ (for some parameter t), then $a = -2t$, and $c = t$.

47. Let $\mathbf{F_3} = a\mathbf{i} + b\mathbf{j}$. We want

$$\mathbf{F_1} + \mathbf{F_2} + \mathbf{F_3} = \mathbf{0}$$

Thus, $3\mathbf{i} + 4\mathbf{j} + 3\mathbf{i} - 7\mathbf{j} + a\mathbf{i} + b\mathbf{j} = 0\mathbf{i} + 0\mathbf{j}$

$$3 + 3 + a = 0 \text{ or } a = -6$$

$4 - 7 + b = 0$ or $b = 3$

We have $\mathbf{F}_3 = -6\mathbf{i} + 3\mathbf{j}$.

49. The boat travels east to west, a distance of 2.1 mi in 0.5 hr, so its velocity relative to still water is $\mathbf{B} = 4.2(-\mathbf{i})$. The velocity of the river is $\mathbf{R} = -3.1\mathbf{j}$, so the actual velocity of the boat is

$$\mathbf{v} = \mathbf{B} + \mathbf{R} = -4.2\mathbf{i} - 3.1\mathbf{j}$$

The actual speed is

$$\|\mathbf{v}\| = \sqrt{(-4.2)^2 + (-3.1)^2} \approx 5.22 \text{ mi/h}$$

Let θ be the angle between the direction of the boat and still water. Then

$$\theta = \tan^{-1} \frac{3.1}{4.2} \approx 36.4°$$

so the direction is $90 - 36.4° = 53.6°$; that is N53.6°W.

51. Let P, Q, R, and S, be consecutive vertices in counterclockwise order of a parallelogram with diagonals $\overline{PR}$ and $\overline{QS}$. The point T is where the diagonals intersect. Then, for positive constants a and b,

$\mathbf{PT} = a\mathbf{PR}$ *Since these vectors have the same direction*

$\mathbf{QT} = b\mathbf{QS}$ *Since these vectors have the same direction*

$\mathbf{PT} + \mathbf{TQ} = \mathbf{PQ}$ *Definition of vector addition*

We see that,

$$a\mathbf{PR} - b\mathbf{QS} = \mathbf{PQ}$$

Note that $\mathbf{RT} = \mathbf{RP} - \mathbf{TP}$

$$= \mathbf{RP} + \mathbf{PT}$$

$$= \mathbf{RP} + a\mathbf{PR}$$

$$= (1 - a)\mathbf{RP}$$

and $\mathbf{TS} = \mathbf{QS} - \mathbf{QT}$

$$= \mathbf{QS} - b\mathbf{QS}$$

$$= (1 - b)\mathbf{QS}$$

$\mathbf{RS} = \mathbf{RT} + \mathbf{TS}$

$$= (1 - a)\mathbf{RP} + (1 - b)\mathbf{QS}$$

Since $\mathbf{PQ} = \mathbf{SR}$, then

$$a\mathbf{PR} - b\mathbf{QS} = (1 - a)\mathbf{PR} - (1 - b)\mathbf{QS}$$

it follows that $a = (1 - a)$ or $a = \frac{1}{2}$ and $b = (1 - b)$ or $b = \frac{1}{2}$, which means that the point T bisects the diagonals of the parallelogram.

53. a. $\mathbf{CN} = \mathbf{CA} + \mathbf{AN}$ $\mathbf{BM} = \mathbf{BA} + \mathbf{AM}$

$\quad = -\mathbf{AC} + \frac{1}{2}\mathbf{AB}$ $= -\mathbf{AB} + \frac{1}{2}\mathbf{AC}$

$\quad = \frac{1}{2}\mathbf{AB} - \mathbf{AC}$ $= \frac{1}{2}\mathbf{AC} - \mathbf{AB}$

b. $\mathbf{CP} = r\mathbf{CN} = r\left[\frac{1}{2}\mathbf{AB} - \mathbf{AC}\right]$

$$BP = sBM = s\left[\tfrac{1}{2}AC - AB\right]$$

$$CB = CP + PB = CP - BP$$

$$= \tfrac{r}{2}AB - rAC - \tfrac{s}{2}AC + sAB$$

$$= \left(\tfrac{r}{2} + s\right)AB + \left(-r - \tfrac{s}{2}\right)AC$$

Since $CB = CA + AB$ we see

$$\tfrac{r}{2} + s = 1 \quad \text{and} \quad -r - \tfrac{s}{2} = -1$$

Solving simultaneously, we find $r = \tfrac{2}{3}$,

$s = \tfrac{2}{3}$.

The same procedure applies to any other pair of medians.

c. Let $P(\overline{x}, \overline{y})$ be the centroid. The midpoint of $\overline{AB}$ is

$$M\left(\frac{x_1 + x_2}{2}, \frac{y_1 + y_2}{2}\right).$$

From part **b**, we have $CP = \tfrac{2}{3}CN$, so

$$\overline{x} - x_3 = \tfrac{2}{3}\left[\left(\frac{x_1 + x_2}{2}\right) - x_3\right]$$

$$\overline{x} = \tfrac{1}{3}(x_1 + x_2 + x_3)$$

and similarly,

$$\overline{y} = \tfrac{1}{3}(y_1 + y_2 + y_3)$$

55. $AB = AM + MN + NB$

$AD = AM + MN + ND$

$CB = CM + MN + NB$

$CD = CM + MN + ND$

Adding, we obtain

$$AB + AD + CB + CD$$

$$= 4MN + 2(AM + NB + ND + CM)$$

Since M and N are the midpoints of $\overline{AC}$ and $\overline{BD}$, then $AM = -CM$ and $NB = -ND$. Thus,

$$AB + AD + CB + CD = 4MN + 0$$

$$MN = \tfrac{1}{4}(AB + AD + CB + CD)$$

9.2 Coordinates and Vectors in $\mathbb{R}^3$, Page 587-588

1. $\mathbf{u} = \langle 4, -3, 1 \rangle$, $\mathbf{v} = \langle -2, 5, 3 \rangle$

   a. $\mathbf{u} + \mathbf{v} = \langle 4 + (-2), -3 + 5, 1 + 3 \rangle$
   $$= \langle 2, 2, 4 \rangle$$

   b. $\mathbf{u} - \mathbf{v} = \langle 4 - (-2), -3 - 5, 1 - 3 \rangle$
   $$= \langle 6, -8, -2 \rangle$$

   c. $\tfrac{5}{2}\mathbf{u} = \tfrac{5}{2}\langle 4, -3, 1 \rangle$
   $$= \langle 10, -\tfrac{15}{2}, \tfrac{5}{2} \rangle$$

   d. $2\mathbf{u} + 3\mathbf{v} = 2\langle 4, -3, 1 \rangle + 3\langle -2, 5, 3 \rangle$
   $$= \langle 2, 9, 11 \rangle$$

3. $\mathbf{u} = \langle 1, -2, 5 \rangle$, $\mathbf{v} = \langle 0, -1, 3 \rangle$

   a. $\mathbf{u} + \mathbf{v} = \langle 1 + 0, -2 + (-1), 5 + 3 \rangle$

$$= \langle 1, -3, 8 \rangle$$

b. $\quad \mathbf{u} - \mathbf{v} = \langle 1 - 0, -2 - (-1), 5 - 3 \rangle$

$$= \langle 1, -1, 2 \rangle$$

c. $\frac{5}{2}\mathbf{u} = \frac{5}{2}\langle 1, -2, 5 \rangle$

$$= \langle \frac{5}{2}, -5, \frac{25}{2} \rangle$$

d. $\quad 2\mathbf{u} + 3\mathbf{v} = 2\langle 1, -2, 5 \rangle + 3\langle 0, -1, 3 \rangle$

$$= \langle 2, -7, 19 \rangle$$

5. $\quad P(3, -4, 5), \ Q(1, 5, -3)$

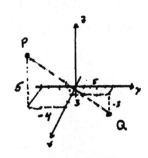

$$d = \sqrt{(1-3)^2 + (5+4)^2 + (-3-5)^2}$$

$$= \sqrt{149}$$

7. $\quad P(-3, -5, 8), \ Q(3, 6, -7)$

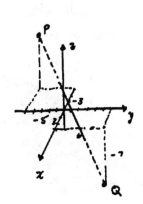

$$d = \sqrt{(3+3)^2 + (6+5)^2 + (-7-8)^2}$$

$$= \sqrt{6^2 + 11^2 + (-15)^2}$$

$$= \sqrt{382}$$

9. $\quad C(0, 0, 0), \ r = 1;$

$$(x - 0)^2 + (y - 0)^2 + (z - 0)^2 = 1$$

$$x^2 + y^2 + z^2 = 1$$

11. $\quad C(0, 4, -5), \ r = 3;$

$$(x - 0)^2 + (y - 4)^2 + (z + 5)^2 = 9$$

$$x^2 + (y - 4)^2 + (z + 5)^2 = 9$$

13. $\quad\quad x^2 + y^2 + z^2 - 2y + 2z - 2 = 0$

$$x^2 + (y^2 - 2y + 1^2) + (z^2 + 2z + 1^2) = 2 + 1 + 1$$

$$x^2 + (y - 1)^2 + (z + 1)^2 = 4$$

$$C(0, 1, -1); \ r = 2$$

15. $\quad x^2 + y^2 + z^2 - 6x + 2y - 2z + 10 = 0$

$$(x^2 - 6x + 3^2) + (y^2 + 2y + 1^2) + (z^2 - 2z + 1^2)$$

$$= -10 + 9 + 1 + 1$$

$$(x - 3)^2 + (y + 1)^2 + (z - 1)^2 = 1$$

$$C(3, -1, 1); \ r = 1$$

17. $\quad P(1, -1, 3), \ Q(-1, 1, 4);$

$$\mathbf{PQ} = (-1 - 1)\mathbf{i} + [1 - (-1)]\mathbf{j} + (4 - 3)\mathbf{k}$$

$$= -2\mathbf{i} + 2\mathbf{j} + \mathbf{k}$$

$$\|PQ\| = \sqrt{(-2)^2 + (2)^2 + (1)^2}$$

$$= 3$$

19. $P(1, 1, 1)$, $Q(-3, -3, -3)$;

$$PQ = (-3 - 1)i + (-3 - 1)j + (-3 - 1)k$$

$$= -4i - 4j - 4k;$$

$$\|PQ\| = 4\sqrt{1 + 1 + 1} = 4\sqrt{3}$$

21. $u = 2i - j + 3k = \langle 2, -1, 3 \rangle$,

$v = i + j - 5k = \langle 1, 1, -5 \rangle$, and

$w = 5i + 7k = \langle 5, 0, 7 \rangle$.

$u + v - 2w$

$$= \langle 2, -1, 3 \rangle + \langle 1, 1, -5 \rangle - 2\langle 5, 0, 7 \rangle$$

$$= \langle -7, 0, -16 \rangle$$

This is $-7i - 16k$.

23. $u = 2i - j + 3k = \langle 2, -1, 3 \rangle$,

$v = i + j - 5k = \langle 1, 1, -5 \rangle$, and

$w = 5i + 7k = \langle 5, 0, 7 \rangle$.

$4u + w = 4\langle 2, -1, 3 \rangle + \langle 5, 0, 7 \rangle$

$$= \langle 13, -4, 19 \rangle$$

This is $13i - 4j + 19k$.

25. $v = \langle 3, -2, 1 \rangle$

$$u = \frac{v}{\|v\|}$$

$$= \frac{\langle 3, -2, 1 \rangle}{\sqrt{9 + 4 + 1}}$$

$$= \frac{1}{\sqrt{14}}\langle 3, -2, 1 \rangle$$

27. $v = \langle -5, 3, 4 \rangle$

$$u = \frac{v}{\|v\|}$$

$$= \frac{\langle -5, 3, 4 \rangle}{\sqrt{25 + 9 + 16}}$$

$$= \frac{1}{5\sqrt{2}}\langle -5, 3, 4 \rangle$$

29. $x^2 + y^2 = 4$ is a cylinder with circular cross sections parallel to the xy-plane.

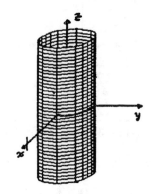

31. $y = 3z^2$ is a cylinder with parabolic cross sections parallel to the yz-plane.

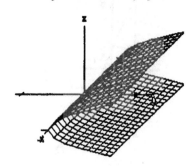

33. Let $A(1, 2, -3)$ and $B = (-2, 3, 3)$; the center

    of the circle is the midpoint of **AB**; namely

    $\left(\frac{1-2}{2}, \frac{2+3}{2}, \frac{-3+3}{2} \right) = (-\frac{1}{2}, \frac{5}{2}, 0)$

    The diameter is

    $\|\mathbf{AB}\| = \sqrt{(-3)^2 + 1^2 + 6^2}$

    $\qquad = \sqrt{46}$

    $r^2 = \frac{46}{4} = \frac{23}{2}$

    The equation of the sphere is

    $(x + \frac{1}{2})^2 + (y - \frac{5}{2})^2 + z^2 = \frac{23}{2}$

    or

    $(2x + 1)^2 + (2y - 5)^2 + 4z^2 = 46$

35. $\|\mathbf{i} - \mathbf{j} + \mathbf{k}\| = \sqrt{1 + 1 + 1}$

    $\qquad\qquad = \sqrt{3}$

37. $\|2(\mathbf{i} - \mathbf{j} + \mathbf{k}) - 3(2\mathbf{i} + \mathbf{j} - \mathbf{k})\|^2$

    $= \| -4\mathbf{i} - 5\mathbf{j} + 5\mathbf{k}\|^2$

    $= 16 + 25 + 25$

    $= 66$

39. $\mathbf{v} = \mathbf{i} - 2\mathbf{j} + 2\mathbf{k}; \mathbf{w} = 2\mathbf{i} + 4\mathbf{j} - \mathbf{k};$

    $\|\mathbf{v}\|\mathbf{w} = \sqrt{1 + 4 + 4}(2\mathbf{i} + 4\mathbf{j} - \mathbf{k})$

    $\qquad = 6\mathbf{i} + 12\mathbf{j} - 3\mathbf{k}$

41. $\mathbf{v} = \mathbf{i} - 2\mathbf{j} + 2\mathbf{k}; \mathbf{w} = 2\mathbf{i} + 4\mathbf{j} - \mathbf{k};$

    $\mathbf{v} - \mathbf{w} = (1 - 2)\mathbf{i} + (-2 - 4)\mathbf{j} + (2 + 1)\mathbf{k}$

$\qquad = -\mathbf{i} - 6\mathbf{j} + 3\mathbf{k}$

$\mathbf{v} + \mathbf{w} = (1 + 2)\mathbf{i} + (-2 + 4)\mathbf{j} + (2 - 1)\mathbf{k}$

$\qquad = 3\mathbf{i} + 2\mathbf{j} + \mathbf{k}$

$\|\mathbf{v} - \mathbf{w}\|(\mathbf{v} + \mathbf{w}) = \| -\mathbf{i} - 6\mathbf{j} + 3\mathbf{k}\|(3\mathbf{i} + 2\mathbf{j} + \mathbf{k})$

$\qquad\qquad = \sqrt{46}(3\mathbf{i} + 2\mathbf{j} + \mathbf{k})$

43. $\mathbf{v} = \langle -2, 6, -10 \rangle$

    $\quad = -2\langle 1, -3, 5 \rangle$

    $\quad \neq s\mathbf{u}$

    So **v** is not parallel to **u**.

45. $\mathbf{v} = \langle -1, \frac{3}{2}, -\frac{5}{2} \rangle$

    $\quad = -\frac{1}{2}\langle 2, -3, 5 \rangle$

    $\quad = -\frac{1}{2}\mathbf{u}$

    So **v** is parallel to **u**.

47. $A(1, 1, 1), B(3, 3, 2), C(3, -3, 5)$

    $\mathbf{AB} = (3 - 1)\mathbf{i} + (3 - 1)\mathbf{j} + (2 - 1)\mathbf{k}$

    $\qquad = 2\mathbf{i} - 2\mathbf{j} + \mathbf{k}$

    $\|\mathbf{AB}\|^2 = 4 + 4 + 1 = 9$

    $\mathbf{AC} = (3 - 1)\mathbf{i} + (-3 - 1)\mathbf{j} + (5 - 1)\mathbf{k}$

    $\qquad = 2\mathbf{i} - 4\mathbf{j} + 4\mathbf{k}$

    $\|\mathbf{AC}\|^2 = 4 + 16 + 16 = 36$

    $\mathbf{BC} = (3 - 3)\mathbf{i} + (-3 - 3)\mathbf{j} + (5 - 2)\mathbf{k}$

    $\qquad = -6\mathbf{j} + 3\mathbf{k}$

$\|\mathbf{BC}\|^2 = 0 + 36 + 9 = 45$

Since

$$\|\mathbf{AB}\|^2 + \|\mathbf{AC}\|^2 = 9 + 36$$

$$= 45$$

$$= \|\mathbf{BC}\|^2$$

we see that $\triangle ABC$ is right, but not isosceles.

49. $A(2, 4, 3)$, $B(-3, 2, -4)$, $C(-6, 8, -10)$

$\mathbf{AB} = (-3 - 2)\mathbf{i} + (2 - 4)\mathbf{j} + (-4 - 3)\mathbf{k}$

$\quad = -5\mathbf{i} - 2\mathbf{j} - 7\mathbf{k}$

$\|\mathbf{AB}\|^2 = 25 + 4 + 49$

$\quad = 78$

$\mathbf{AC} = (-6 - 2)\mathbf{i} + (8 - 4)\mathbf{j} + (-10 - 3)\mathbf{k}$

$\quad = -8\mathbf{i} + 4\mathbf{j} - 13\mathbf{k}$

$\|\mathbf{AC}\|^2 = 64 + 16 + 169 = 249$

$\mathbf{BC} = (-6 + 3)\mathbf{i} + (8 - 2)\mathbf{j} + (-10 + 4)\mathbf{k}$

$\quad = -3\mathbf{i} + 6\mathbf{j} - 6\mathbf{k}$

$\|\mathbf{BC}\|^2 = 9 + 36 + 36$

$\quad = 81$

$\triangle ABC$ is neither right nor isosceles

51. Let $A(2, 3, 2)$, $B(-1, 4, 0)$, and $C(-4, 5, -2)$.

Then

$\mathbf{AB} = \langle -3, 1, -2 \rangle$ and $\mathbf{AC} = \langle -6, 2, -4 \rangle$.

Since $\mathbf{AC}$ is a multiple of $\mathbf{AB}$, it follows that A, B, and C lie on the same line.

53. Let $\mathbf{F}_1$, $\mathbf{F}_2$, and $\mathbf{F}_3$ be the forces in the three legs, where for constants a, b, c, we have

$\mathbf{F}_1 = a\mathbf{PA} = a\langle 0, -2, -6 \rangle$

$\mathbf{F}_2 = b\mathbf{PB} = b\langle -\sqrt{3}, 1, -6 \rangle$

$\mathbf{F}_3 = c\mathbf{PB} = c\langle \sqrt{3}, 1, -6 \rangle$

The total force exerted by the seated official is

$\mathbf{F} = -150\mathbf{k}$, so we have

$\mathbf{F} = \mathbf{F}_1 + \mathbf{F}_2 + \mathbf{F}_3$

$-150\langle 0, 0, 1 \rangle = a\langle 0, -2, -6 \rangle + b\langle -\sqrt{3}, 1, -6 \rangle$
$\qquad\qquad + c\langle \sqrt{3}, 1, -6 \rangle$

Equating coefficients we find that

$$\begin{cases} -\sqrt{3}b + \sqrt{3}c = 0 \\ -2a + b + c = 0 \\ -6a - 6b - 6c = -150 \end{cases}$$

Solving, we obtain $a = b = c = \frac{25}{3}$, so

$\mathbf{F}_1 = \frac{25}{3}\langle 0, -2, -6 \rangle$

$\mathbf{F}_2 = \frac{25}{3}\langle -\sqrt{3}, 1, -6 \rangle$

$\mathbf{F}_3 = \frac{25}{3}\langle \sqrt{3}, 1, -6 \rangle$

55. The midpoint M of $\overline{BC}$ is

$M\left(\dfrac{-2 + 4}{2}, \dfrac{3 + 1}{2}, \dfrac{7 - 3}{2} \right) = (1, 2, 2)$

Let $P(x, y, z)$ so that

$AP = (x + 1)\mathbf{i} + (y - 3)\mathbf{j} + (z - 9)\mathbf{k}$

$AM = 2\mathbf{i} - \mathbf{j} - 7\mathbf{k}$

$AP = \frac{2}{3}AM$ so $3AP = 2AM$ or

$3(x + 1) = 4$ or $x = \frac{1}{3}$;

$3(y - 3) = -2$ or $y = \frac{7}{3}$

$3(z - 9) = -14$, so $z = \frac{13}{3}$

The desired point is $P(\frac{1}{3}, \frac{7}{3}, \frac{13}{3})$.

9.3 The Dot Product, Page 595-597

SURVIVAL HINT: *The motivation for the definition of the dot product comes from a desire to find the angle between two vectors. If we had done the theorem for the angle between two vectors (Theorem 9.3) before defining the dot product, cos θ would equal an expression whose numerator was what we have called the **dot product**.*

1. Answers vary;

    If $\mathbf{u} = \langle u_1, u_2, u_3 \rangle$ and $\mathbf{v} = \langle v_1, v_2, v_3 \rangle$, then

    $\mathbf{u} \cdot \mathbf{v} = u_1 v_1 + u_2 v_2 + u_3 v_3$. Work can be

    expressed as a dot product of force and

    displacement vectors.

3. $\mathbf{v} = \langle 3, -2, 4 \rangle$ and $\mathbf{w} = \langle 2, -1, -6 \rangle$;

    $\mathbf{v} \cdot \mathbf{w} = 3(2) + (-2)(-1) + 4(-6)$

    $\qquad = -16$

5. $\mathbf{v} = 2\mathbf{i} + 3\mathbf{j} - \mathbf{k}$; $\mathbf{w} = -3\mathbf{i} + 5\mathbf{j} + 4\mathbf{k}$;

    $\mathbf{v} \cdot \mathbf{w} = 2(-3) + 3(5) + (-1)(4)$

    $\qquad = 5$

7. $\mathbf{v} = \mathbf{i}$; $\mathbf{w} = \mathbf{k}$;

    $\mathbf{v} \cdot \mathbf{w} = 0$

    Orthogonal

9. $\mathbf{v} = 3\mathbf{i} - 2\mathbf{j}$; $\mathbf{w} = 6\mathbf{i} + 9\mathbf{j}$;

    $\mathbf{v} \cdot \mathbf{w} = 3(6) + (-2)(9)$

    $\qquad = 0$

    Orthogonal

11. $\mathbf{v} = 3\mathbf{i} - 2\mathbf{j} + \mathbf{k}$; $\mathbf{w} = \mathbf{i} + \mathbf{j} - \mathbf{k}$;

    $\mathbf{v} + \mathbf{w} = (3 + 1)\mathbf{i} + (-2 + 1)\mathbf{j} + (1 - 1)\mathbf{k}$

    $\qquad = 4\mathbf{i} - \mathbf{j}$

    $\mathbf{v} - \mathbf{w} = (3 - 1)\mathbf{i} + (-2 - 1)\mathbf{j} + (1 + 1)\mathbf{k}$

    $\qquad = 2\mathbf{i} - 3\mathbf{j} + 2\mathbf{k}$

    $(\mathbf{v} + \mathbf{w}) \cdot (\mathbf{v} - \mathbf{w})$

    $\qquad = 4(2) + (-1)(-3) + (0)(2)$

    $\qquad = 11$

13. $\mathbf{v} = 3\mathbf{i} - 2\mathbf{j} + \mathbf{k}$; $\mathbf{w} = \mathbf{i} + \mathbf{j} - \mathbf{k}$;

    $\| \mathbf{v} \| = \sqrt{3^2 + (-2)^2 + (1)^2}$

    $\qquad = \sqrt{14}$

    $\| \mathbf{w} \| = \sqrt{1^2 + 1^2 + (-1)^2}$

$$= \sqrt{3}$$

$$(\|\mathbf{v}\| \mathbf{w}) \cdot (\|\mathbf{w}\| \mathbf{v})$$

$$= [(\sqrt{14})(\mathbf{i} + \mathbf{j} - \mathbf{k})] \cdot [\sqrt{3}(3\mathbf{i} - 2\mathbf{j} + \mathbf{k})]$$

$$= \sqrt{14}\sqrt{3}(3 - 2 - 1)$$

$$= 0$$

15. $\mathbf{v} = \mathbf{i} + \mathbf{j} + \mathbf{k}; \mathbf{w} = \mathbf{i} - \mathbf{j} + \mathbf{k}$

$$\|\mathbf{v}\| = \sqrt{1^2 + 1^2 + 1^2}$$

$$= \sqrt{3}$$

$$\|\mathbf{w}\| = \sqrt{1^2 + (-1)^2 + 1^2}$$

$$= \sqrt{3}$$

$$\cos\theta = \frac{\mathbf{v} \cdot \mathbf{w}}{\|\mathbf{v}\|\|\mathbf{w}\|}$$

$$= \frac{1 - 1 + 1}{3}$$

$$= \frac{1}{3}$$

$$\theta = \cos^{-1}\frac{1}{3}$$

$$\approx 71°$$

17. $\mathbf{v} = 2\mathbf{j} + \mathbf{k}; \mathbf{w} = \mathbf{i} - 2\mathbf{k}$

$$\|\mathbf{v}\| = \sqrt{0^2 + 2^2 + 1^2}$$

$$= \sqrt{5}$$

$$\|\mathbf{w}\| = \sqrt{1^2 + 2^2}$$

$$= \sqrt{5}$$

$$\cos\theta = \frac{\mathbf{v} \cdot \mathbf{w}}{\|\mathbf{v}\|\|\mathbf{w}\|}$$

$$= \frac{-2}{\sqrt{5}\sqrt{5}}$$

$$= -\frac{2}{5}$$

$$\theta = \cos^{-1}\left(-\frac{2}{5}\right)$$

$$\approx 114°$$

SURVIVAL HINT: *Remember, the scalar projection is a number, whereas the vector projection is a **vector**.*

19. $\mathbf{v} = \mathbf{i} + \mathbf{j} + \mathbf{k}; \mathbf{w} = 2\mathbf{k}$

The scalar projection: $\frac{\mathbf{v} \cdot \mathbf{w}}{\|\mathbf{w}\|} = \frac{2}{2} = 1$

The vector projection:

$$\left(\frac{\mathbf{v} \cdot \mathbf{w}}{\mathbf{w} \cdot \mathbf{w}}\right)\mathbf{w} = \frac{2}{4}\mathbf{w}$$

$$= \frac{1}{2}(2)\mathbf{k}$$

$$= \mathbf{k}$$

21. $\mathbf{v} = \mathbf{i} + 2\mathbf{k}; \mathbf{w} = -3\mathbf{j}$

The scalar projection: $\frac{\mathbf{v} \cdot \mathbf{w}}{\|\mathbf{w}\|} = 0$

The vector projection: $\left(\frac{\mathbf{v} \cdot \mathbf{w}}{\mathbf{w} \cdot \mathbf{w}}\right)\mathbf{w} = \mathbf{0}$

23. For $\mathbf{u} = \langle u_1, u_2, u_3 \rangle$ to be orthogonal to $\mathbf{v} = \mathbf{i} + \mathbf{j} - \mathbf{k} = \langle 1, 1, -1 \rangle$
and
$\mathbf{w} = -\mathbf{i} + \mathbf{j} + \mathbf{k} = \langle -1, 1, 1 \rangle$
We must have

$$\mathbf{u} \cdot \mathbf{v} = u_1 + u_2 - u_3 = 0$$

$$\mathbf{u} \cdot \mathbf{w} = -u_1 + u_2 + u_3 = 0$$

Solving, we obtain $\mathbf{u} = \langle a, 0, a \rangle$ for an arbitrary a. Two unit vectors of this form are

$$\mathbf{u}_1 = \left\langle \frac{\sqrt{2}}{2}, 0, \frac{\sqrt{2}}{2} \right\rangle$$

and

$$\mathbf{u}_2 = \left\langle -\frac{\sqrt{2}}{2}, 0, -\frac{\sqrt{2}}{2} \right\rangle$$

25. $\mathbf{v} = 2\mathbf{i} + 3\mathbf{j} - 2\mathbf{k};$

$$\mathbf{u} = \frac{-\mathbf{v}}{\|\mathbf{v}\|}$$

$$= \frac{-(2\mathbf{i} + 3\mathbf{j} - 2\mathbf{k})}{\sqrt{2^2 + 3^2 + (-2)^2}}$$

$$= -\frac{1}{\sqrt{17}}(2\mathbf{i} + 3\mathbf{j} - 2\mathbf{k})$$

27. $x(\mathbf{i}+\mathbf{j}+\mathbf{k}) + y(\mathbf{i}-\mathbf{j}+2\mathbf{k}) + z(\mathbf{i}+\mathbf{k}) = 2\mathbf{i} + \mathbf{k}$

$$\begin{cases} x + y + z = 2 \\ x - y = 0 \\ x + 2y + z = 1 \end{cases}$$

Solving this system of equation simultaneously,

we see from the second equation that $y = x$,

so substituting into the other two equations,

we have

$$\begin{cases} 2x + z = 2 \\ 3x + z = 1 \end{cases}$$

Subtracting the first equation from the

second, we obtain

$$x = -1$$

so that $y = -1$, and consequently, $z = 4$.

29. Let $\mathbf{u} = 3\mathbf{i} - 2\mathbf{j} + \mathbf{k}$ and $\mathbf{v} = 2\mathbf{i} + a\mathbf{j} - 2a\mathbf{k}$

If these vectors are orthogonal, then $\mathbf{u} \cdot \mathbf{v} = 0$.

$\mathbf{u} \cdot \mathbf{v} = 6 - 2a - 2a = 0$ so $a = \frac{3}{2}$.

31. $\mathbf{v} = 2\mathbf{i} - 3\mathbf{j} - 5\mathbf{k};$

$$\|\mathbf{v}\| = \sqrt{2^2 + (-3)^2 + (-5)^2}$$

$$= \sqrt{38}$$

$v_1 = 2, \quad v_2 = -3, \quad v_3 = -5$

$\cos \alpha = \dfrac{2}{\sqrt{38}}, \ \alpha \approx 1.24 \text{ or } 71°;$

$\cos \beta = \dfrac{-3}{\sqrt{38}}, \ \beta \approx 2.08 \text{ or } 119°;$

$\cos \gamma = \dfrac{-5}{\sqrt{38}}, \ \gamma \approx 2.52 \text{ or } 144°$

33. $\mathbf{v} = 5\mathbf{i} - 4\mathbf{j} + 3\mathbf{k};$

$$\|\mathbf{v}\| = \sqrt{5^2 + (-4)^2 + (3)^2}$$

$$= 5\sqrt{2}$$

$v_1 = 5, \quad v_2 = -4, \quad v_3 = 3$

$\cos \alpha = \dfrac{5}{5\sqrt{2}} = \dfrac{1}{\sqrt{2}}, \ \alpha \approx 0.79 \text{ or } 45°;$

$\cos \beta = \dfrac{-4}{5\sqrt{2}}, \ \beta \approx 2.17 \text{ or } 124°;$

$\cos \gamma = \dfrac{3}{5\sqrt{2}}, \ \gamma \approx 1.13 \text{ or } 65°$

35. $\mathbf{v} = \mathbf{i} - \mathbf{j} + 2\mathbf{k}; \ \mathbf{w} = 2\mathbf{i} + \mathbf{j} - \mathbf{k}$

$$\cos \theta = \frac{\mathbf{v} \cdot \mathbf{w}}{\|\mathbf{v}\| \|\mathbf{w}\|}$$

$$= \frac{2 - 1 - 2}{6}$$

$$= -\frac{1}{6}$$

The vector projection of **v** onto **w** is

$$\mathbf{u} = \left(\frac{\mathbf{v} \cdot \mathbf{w}}{\mathbf{w} \cdot \mathbf{w}}\right)\mathbf{w}$$

$$= -\frac{1}{6}\mathbf{w}$$

$$= -\frac{1}{6}(2\mathbf{i} + \mathbf{j} - \mathbf{k})$$

37. $\mathbf{v} = 2\mathbf{i} - 3\mathbf{j} + 6\mathbf{k}$; $\mathbf{w} = 4\mathbf{i} + 3\mathbf{k}$

a. $\mathbf{v} \cdot \mathbf{w} = (2 \cdot 4) + (-3 \cdot 0) + (6 \cdot 3)$

$$= 8 + 18$$

$$= 26$$

b. $\|\mathbf{v}\| = \sqrt{2^2 + (-3)^2 + 6^2} = \sqrt{49} = 7$

$$\|\mathbf{w}\| = \sqrt{4^2 + 0^2 + 3^2} = \sqrt{25} = 5$$

$$\cos\theta = \frac{\mathbf{v} \cdot \mathbf{w}}{\|\mathbf{v}\|\,\|\mathbf{w}\|}$$

$$= \frac{26}{(7)(5)}$$

$$= \frac{26}{35}$$

c. $\mathbf{v} \cdot (\mathbf{v} - s\mathbf{w}) = 0$

$$s = \frac{\|\mathbf{v}\|^2}{\mathbf{v} \cdot \mathbf{w}} = \frac{49}{26}$$

d. $(\mathbf{v} + t\mathbf{w}) \cdot \mathbf{w} = \mathbf{v} \cdot \mathbf{w} + t\|\mathbf{w}\|^2 = 0$

$$t = -\frac{\mathbf{v} \cdot \mathbf{w}}{\mathbf{w} \cdot \mathbf{w}}$$

$$= \frac{-26}{25}$$

39. $\mathbf{F} = 2\mathbf{i} + 3\mathbf{j} + \mathbf{k}$;

$$\mathbf{PQ} = (3 - 1)\mathbf{i} + (1 - 0)\mathbf{j} + [2 - (-1)]\mathbf{k}$$

$$= 2\mathbf{i} + \mathbf{j} + 3\mathbf{k};$$

$$W = \mathbf{F} \cdot \mathbf{PQ}$$

$$= 4 + 3 + 3$$

$$= 10$$

41.   Consider a Cartesian coordinate system with the end of the log at the origin and units in feet. The position vector along Fred's rope is

$$\mathbf{L}_f = x_f\mathbf{i} - \mathbf{j} + 2\mathbf{k}$$

$$\|\mathbf{L}_f\|^2 = x_f^2 + 1 + 4$$

$$= 64$$

$$x_f = \sqrt{59}$$

The position vector along Sam's rope is

$$\mathbf{L}_S = x_s\mathbf{i} + \mathbf{j} + \mathbf{k};$$

$$\|\mathbf{L}_s\|^2 = x_s^2 + 1 + 1$$

$$= 64,$$

$$x_s = \sqrt{62}$$

Thus, the resultant force on the log is

$$F_x = 30\left(\frac{\mathbf{L}_f}{\|\mathbf{L}_f\|}\right) + 20\left(\frac{\mathbf{L}_s}{\|\mathbf{L}_s\|}\right)$$

$$= \frac{30}{8}(\sqrt{59}\,\mathbf{i} - \mathbf{j} + 2\mathbf{k}) + \frac{20}{8}(\sqrt{62}\,\mathbf{i} + \mathbf{j} + \mathbf{k})$$

$$\approx 48.49\mathbf{i} - 1.25\mathbf{j} + 10\mathbf{k}$$

The force has magnitude $\|\mathbf{F}\| \approx 49.53$ lbs

and points in the direction of the unit vector

$\langle 0.979, -0.025, 0.202 \rangle$.

43. **a.** The force vector is

$$\mathbf{F} = 50[(\cos \tfrac{\pi}{3})\mathbf{i} + (\sin \tfrac{\pi}{3})\mathbf{j}]$$

and the "drag" vector is $\mathbf{D} = 20\mathbf{i}$.

Work is

$$W = 50[(\cos \tfrac{\pi}{3})\mathbf{i} + (\sin \tfrac{\pi}{3})\mathbf{j}] \cdot 20\mathbf{i}$$

$$= 500 \text{ ft} \cdot \text{lb}$$

b. The force vector is

$$\mathbf{F} = 50[(\cos \tfrac{\pi}{4})\mathbf{i} + (\sin \tfrac{\pi}{4})\mathbf{j}]$$ and the

"drag" vector is $\mathbf{D} = 20\mathbf{i}$.

Work is

$$W = 50[(\cos \tfrac{\pi}{4})\mathbf{i} + (\sin \tfrac{\pi}{4})\mathbf{j}] \cdot 20\mathbf{i}$$

$$= 500\sqrt{2} \text{ ft} \cdot \text{lb}$$

45. **a.** $\displaystyle\int_{-b}^{b} x^2(x^3 - 5x)\, dx = \left(\frac{x^6}{6} - \frac{5x^4}{4} \right)\Bigg|_{-b}^{b}$

$$= 0$$

b. $\displaystyle\int_{-b}^{b} \sin kx \sin nx\, dx$

$$= \frac{1}{2}\int_{-\pi}^{\pi} [\cos(k-n)x]\, dx - \frac{1}{2}\int_{-\pi}^{\pi} [\cos(k+n)x]\, dx$$

$$= \frac{1}{2}\left[\frac{\sin(k-n)x}{k-n} - \frac{\sin(k+n)x}{k+n} \right]\Bigg|_{-\pi}^{\pi}$$

$$= 0$$

47. $\|\mathbf{u}_0 - \mathbf{u}\|^2 = (x-a)^2 + (y-b)^2 + (z-c)^2 < r^2$

The region inside a sphere with center

(a, b, c) and radius r.

49. The vector $\mathbf{B}$ bisects the angle between $\mathbf{u}$ and

$\mathbf{v}$. To prove this, let α be the angle between

$\mathbf{B}$ and $\mathbf{u}$, and let β be the angle between $\mathbf{B}$

and $\mathbf{v}$. Then

$$\mathbf{u} \cdot \mathbf{B} = \mathbf{u} \cdot (\|\mathbf{v}\|\mathbf{u} + \|\mathbf{u}\|\mathbf{v})$$

$$= \|\mathbf{v}\|(\mathbf{u} \cdot \mathbf{u}) + \|\mathbf{u}\|(\mathbf{u} \cdot \mathbf{v})$$

$$= \|\mathbf{v}\|\|\mathbf{u}\|^2 + \|\mathbf{u}\|(\mathbf{u} \cdot \mathbf{v})$$

$$\cos \alpha = \frac{\mathbf{u} \cdot \mathbf{B}}{\|\mathbf{u}\|\|\mathbf{B}\|}$$

$$= \frac{\|\mathbf{v}\|\|\mathbf{u}\|^2 + \|\mathbf{u}\|(\mathbf{u} \cdot \mathbf{v})}{\|\mathbf{u}\|\|\mathbf{B}\|}$$

$$= \frac{\|\mathbf{v}\|\|\mathbf{u}\| + (\mathbf{u} \cdot \mathbf{v})}{\|\mathbf{B}\|}$$

Similarly, $\mathbf{v} \cdot \mathbf{B} = \|\mathbf{u}\|\|\mathbf{v}\|^2 + \|\mathbf{v}\|(\mathbf{u} \cdot \mathbf{v})$

$$\cos \beta = \frac{\mathbf{v} \cdot \mathbf{B}}{\|\mathbf{v}\|\|\mathbf{B}\|}$$

$$= \frac{\|\mathbf{v}\|(\mathbf{v} \cdot \mathbf{u}) + \|\mathbf{u}\|\|\mathbf{v}\|^2}{\|\mathbf{v}\|\|\mathbf{B}\|}$$

$$= \frac{(\mathbf{v} \cdot \mathbf{u}) + \|\mathbf{u}\| \|\mathbf{v}\|}{\|\mathbf{B}\|}$$

Since $\cos \alpha = \cos \beta$, it follows that $\alpha = \beta$

and the vector **B** bisects the angle between **u**

and **v**.

SURVIVAL HINT: *The triangle inequality is a result you will run across again if you continue your mathematical studies.*

51.　a.　$(\mathbf{v} + \mathbf{w}) \cdot (\mathbf{v} + \mathbf{w})$

$$= \mathbf{v} \cdot \mathbf{v} + \mathbf{v} \cdot \mathbf{w} + \mathbf{w} \cdot \mathbf{v} + \mathbf{w} \cdot \mathbf{w}$$

$$= \|\mathbf{v}\|^2 + 2\mathbf{v} \cdot \mathbf{w} + \|\mathbf{w}\|^2$$

　　b.　$\|\mathbf{v} + \mathbf{w}\|^2 = (\mathbf{v} + \mathbf{w}) \cdot (\mathbf{v} + \mathbf{w})$

$$= \|\mathbf{v}\|^2 + \|\mathbf{w}\|^2 + 2\|\mathbf{v}\| \|\mathbf{w}\| \cos \theta$$

$$\leq \|\mathbf{v}\|^2 + \|\mathbf{w}\|^2 + 2\|\mathbf{v}\| \|\mathbf{w}\|$$

$$= (\|\mathbf{v}\| + \|\mathbf{w}\|)^2$$

Thus, $\|\mathbf{v} + \mathbf{w}\| \leq \|\mathbf{v}\| + \|\mathbf{w}\|$

53.　If $\mathbf{v} = v_1 \mathbf{i} + v_2 \mathbf{j} + v_3 \mathbf{k}$, then

$$\cos \alpha = \frac{v_1}{\|\mathbf{v}\|}, \cos \beta = \frac{v_2}{\|\mathbf{v}\|}, \cos \gamma = \frac{v_3}{\|\mathbf{v}\|}$$

so

$$\cos^2\alpha + \cos^2\beta + \cos^2\gamma = \frac{v_1^2}{\|\mathbf{v}\|^2} + \frac{v_2^2}{\|\mathbf{v}\|^2} + \frac{v_3^2}{\|\mathbf{v}\|^2}$$

$$= \frac{v_1^2 + v_2^2 + v_3^2}{\|\mathbf{v}\|^2}$$

$$= \frac{\|\mathbf{v}\|^2}{\|\mathbf{v}\|^2}$$

$$= 1$$

9.4　The Cross Product, Page 604-606

SURVIVAL HINT: *Remember that the dot product of two vectors is a **real number**, and the cross product of two vectors is **vector**. You will also want to remember that the cross product is a vector normal to the plane determined by the two give vectors, and is equal in magnitude to the area of the parallelogram determined by them.*

1.　$\mathbf{v} \times \mathbf{w} = \mathbf{i} \times \mathbf{j} = \mathbf{k}$

3.　$\mathbf{v} \times \mathbf{w} = \begin{vmatrix} \mathbf{i} & \mathbf{j} & \mathbf{k} \\ 3 & 0 & 2 \\ 2 & 1 & 0 \end{vmatrix}$

$$= -2\mathbf{i} + 4\mathbf{j} + 3\mathbf{k}$$

SURVIVAL HINT: *If you need to review determinants, see Section 2.8 of the Student Mathematics Handbook; however, the most common method for finding cross product is to use a calculator or computer. Individual vectors are input using a format similar to:*

$$\text{Cross}([3, 0, 2], [2, 1, 0])$$

Check your Owner's Manual.

5.　$\mathbf{v} \times \mathbf{w} = \begin{vmatrix} \mathbf{i} & \mathbf{j} & \mathbf{k} \\ 3 & -2 & 4 \\ 1 & 4 & -7 \end{vmatrix}$

$$= -2\mathbf{i} + 25\mathbf{j} + 14\mathbf{k}$$

7.　$\mathbf{v} \times \mathbf{w} = \begin{vmatrix} \mathbf{i} & \mathbf{j} & \mathbf{k} \\ 3 & -1 & 2 \\ 2 & 3 & -4 \end{vmatrix}$

$$= -2\mathbf{i} + 16\mathbf{j} + 11\mathbf{k}$$

9.　$\mathbf{v} \times \mathbf{w} = \begin{vmatrix} \mathbf{i} & \mathbf{j} & \mathbf{k} \\ 1 & -6 & 10 \\ -1 & 5 & -6 \end{vmatrix}$

$$= -14\mathbf{i} - 4\mathbf{j} - \mathbf{k}$$

11. $\mathbf{v} = \mathbf{i} + \mathbf{k}$; $\mathbf{w} = \mathbf{i} + \mathbf{j}$

$$\|\mathbf{v}\| = \sqrt{1^2 + 0^2 + 1^2} = \sqrt{2}$$

$$\|\mathbf{w}\| = \sqrt{1^2 + 1^2 + 0^2} = \sqrt{2}$$

$$\mathbf{v} \times \mathbf{w} = \begin{vmatrix} \mathbf{i} & \mathbf{j} & \mathbf{k} \\ 1 & 0 & 1 \\ 1 & 1 & 0 \end{vmatrix}$$

$$= -\mathbf{i} + \mathbf{j} + \mathbf{k}$$

$$\|\mathbf{v} \times \mathbf{w}\| = \sqrt{(-1)^2 + 1^2 + 1^2}$$

$$= \sqrt{3}$$

$$\sin \theta = \frac{\|\mathbf{v} \times \mathbf{w}\|}{\|\mathbf{v}\| \, \|\mathbf{w}\|}$$

$$= \frac{\sqrt{3}}{\sqrt{2} \, \sqrt{2}}$$

$$= \frac{\sqrt{3}}{2}$$

13. $\mathbf{v} = \mathbf{j} + \mathbf{k}$; $\mathbf{w} = \mathbf{i} + \mathbf{k}$

$$\|\mathbf{v}\| = \sqrt{0^2 + 1^2 + 1^2}$$

$$= \sqrt{2}$$

$$\|\mathbf{w}\| = \sqrt{1^2 + 0^2 + 1^2}$$

$$= \sqrt{2}$$

$$\mathbf{v} \times \mathbf{w} = \begin{vmatrix} \mathbf{i} & \mathbf{j} & \mathbf{k} \\ 0 & 1 & 1 \\ 1 & 0 & 1 \end{vmatrix}$$

$$= \mathbf{i} + \mathbf{j} - \mathbf{k}$$

$$\|\mathbf{v} \times \mathbf{w}\| = \sqrt{(1)^2 + 1^2 + (-1)^2}$$

$$= \sqrt{3}$$

$$\sin \theta = \frac{\|\mathbf{v} \times \mathbf{w}\|}{\|\mathbf{v}\| \, \|\mathbf{w}\|}$$

$$= \frac{\sqrt{3}}{\sqrt{2} \, \sqrt{2}}$$

$$= \frac{\sqrt{3}}{2}$$

15. $$\mathbf{v} \times \mathbf{w} = \begin{vmatrix} \mathbf{i} & \mathbf{j} & \mathbf{k} \\ 2 & 0 & 1 \\ 1 & -1 & -1 \end{vmatrix}$$

$$= \mathbf{i} + 3\mathbf{j} - 2\mathbf{k}$$

The unit normal is

$$\frac{\mathbf{i} + 3\mathbf{j} - 2\mathbf{k}}{\sqrt{1^2 + 3^2 + (-2)^2}}$$

$$= \frac{1}{\sqrt{14}}\mathbf{i} + \frac{3}{\sqrt{14}}\mathbf{j} - \frac{2}{\sqrt{14}}\mathbf{k}$$

17. $$\mathbf{v} \times \mathbf{w} = \begin{vmatrix} \mathbf{i} & \mathbf{j} & \mathbf{k} \\ 1 & 1 & 1 \\ 3 & 12 & -4 \end{vmatrix}$$

$$= -16\mathbf{i} + 7\mathbf{j} + 9\mathbf{k}$$

The unit normal is

$$\frac{-16\mathbf{i} + 7\mathbf{j} + 9\mathbf{k}}{\sqrt{(-16)^2 + 7^2 + 9^2}}$$

$$= \frac{-16}{\sqrt{386}}\mathbf{i} + \frac{7}{\sqrt{386}}\mathbf{j} + \frac{9}{\sqrt{386}}\mathbf{k}$$

19. $v \times w = \begin{vmatrix} i & j & k \\ 3 & 4 & 0 \\ 1 & 1 & -1 \end{vmatrix}$

$$= -4i + 3j - k$$

$$A = \| -4i + 3j - k \|$$

$$= \sqrt{16 + 9 + 1}$$

$$= \sqrt{26}$$

21. $v \times w = \begin{vmatrix} i & j & k \\ 4 & -1 & 1 \\ 2 & 3 & -1 \end{vmatrix}$

$$= -2i + 6j + 14k$$

$$= 2(-i + 3j + 7k)$$

$$A = 2 \| -i + 3j + 7k \|$$

$$= 2\sqrt{1 + 9 + 49}$$

$$= 2\sqrt{59}$$

23. $PQ = (1 - 0)i + (1 - 1)j + (0 - 1)k$

$$= i - k$$

$$PR = (1 - 0)i + (0 - 1)j + (1 - 1)k$$

$$= i - j$$

$$PQ \times PR = \begin{vmatrix} i & j & k \\ 1 & 0 & -1 \\ 1 & -1 & 0 \end{vmatrix}$$

$$= -i - j - k$$

$$A = \frac{1}{2} \| -i - j - k \|$$

$$= \frac{1}{2}\sqrt{1 + 1 + 1}$$

$$= \frac{1}{2}\sqrt{3}$$

25. $PQ = (2 - 1)i + (3 - 2)j + (1 - 3)k$

$$= i + j - 2k$$

$$PR = (3 - 1)i + (1 - 2)j + (2 - 3)k$$

$$= 2i - j - k$$

$$PQ \times PR = \begin{vmatrix} i & j & k \\ 1 & 1 & -2 \\ 2 & -1 & -1 \end{vmatrix}$$

$$= -3i - 3j - 3k$$

$$= 3(-i - j - k)$$

$$A = \frac{3}{2} \| -i - j - k \|$$

$$= \frac{3}{2}\sqrt{1 + 1 + 1}$$

$$= \frac{3}{2}\sqrt{3}$$

27. a. does not exist; vectors are required for cross product; $v \cdot w$ is a scalar, not a vector

b. scalar

29. a. scalar   b. vector

31. $V = \begin{vmatrix} 0 & 1 & 1 \\ 2 & 1 & 2 \\ 5 & 0 & 0 \end{vmatrix} = 5$

33. $V = \begin{vmatrix} 2 & 1 & -1 \\ 3 & 0 & 1 \\ 0 & 1 & 1 \end{vmatrix} = -8$

$$V = 8$$

35. The right hand rule allows finding the proper direction of the unit vector used in the cross product $u \times v$. Place your little finger of your right hand along u, turn your fingers toward v. Then the direction of your thumb is the direction of the unit vector.

37. $u = i, v = i + j + k, w = i + 2j + sk$;

n is normal to the plane of u and v if

$$n = u \times v = \begin{vmatrix} i & j & k \\ 1 & 0 & 0 \\ 1 & 1 & 1 \end{vmatrix} = -j + k$$

n must be orthogonal to every vector in the plane, so $n \cdot w = 0$, so $0 - 2 + s = 0$ or $s = 2$.

39. $n_1 = u \times v$

$$= \begin{vmatrix} i & j & k \\ 1 & 1 & 0 \\ 2 & -1 & 1 \end{vmatrix}$$

$$= i - j - 3k$$

$$n_1 \times w = \begin{vmatrix} i & j & k \\ 1 & -1 & -3 \\ 3 & 0 & 0 \end{vmatrix}$$

$$= -9j + 3k$$

Also, $n_2 = v \times w$

$$= \begin{vmatrix} i & j & k \\ 2 & -1 & 1 \\ 3 & 0 & 0 \end{vmatrix}$$

$$= 3j + 3k$$

$$u \times n_2 = \begin{vmatrix} i & j & k \\ 1 & 1 & 0 \\ 0 & 3 & 3 \end{vmatrix}$$

$$= 3i - 3j + 3k$$

Cross product is not an associative operation.

41. Since $v \times w$ is orthogonal to both v and w, we can have $v \times w = w$ only if $w \cdot w = 0$ because $0 = (v \times w) \cdot w = w \cdot w$. Thus, $w = 0$.

43. Let $F = u = -40k$ and
$$PQ = 2\left[(\cos \tfrac{\pi}{6})j + (\sin \tfrac{\pi}{6})k\right]$$

$$= \sqrt{3}j + k$$

$$T = PQ \times F$$

$$= \begin{vmatrix} i & j & k \\ 0 & \sqrt{3} & 1 \\ 0 & 0 & -40 \end{vmatrix}$$

$$= -40\sqrt{3}\,i$$

45. Let $v = 2i - j + k$

$u = PQ$
$$= (-1 - 1)i + (2 + 2)j + (3 - 3)k$$
$$= -2i + 4j;$$

$w = PR$
$$= (-1 - 1)i + (2 + 2)j + (-3 - 3)k$$
$$= 4j - 6k$$

n is normal to the plane of u and w if

$$\mathbf{n} = \mathbf{u} \times \mathbf{w}$$

$$= \begin{vmatrix} \mathbf{i} & \mathbf{j} & \mathbf{k} \\ -2 & 4 & 0 \\ 0 & 4 & -6 \end{vmatrix}$$

$$= -24\mathbf{i} - 12\mathbf{j} - 8\mathbf{k}$$

Now,

$$\cos\theta = \frac{\mathbf{n} \cdot \mathbf{v}}{\|\mathbf{n}\| \|\mathbf{v}\|}$$

$$= \frac{-48 + 12 - 8}{\sqrt{784}\sqrt{6}}$$

$$= \frac{-11}{7\sqrt{6}}$$

so that $\theta \approx 130°$. The acute angle is 50°, so the angle between the vector and the plane is

$$90° - 50° = 40°.$$

47. Let $P(x_1, y_1)$, $Q(x_2, y_2)$, and $R(x_3, y_3)$ be the vertices of the triangle. Also, let

$$\mathbf{v} = \mathbf{PR} = \langle x_3 - x_1, y_3 - y_1, 0 \rangle,$$

$$\mathbf{u} = \mathbf{PQ} = \langle x_2 - x_1, y_2 - y_1, 0 \rangle$$

$$\mathbf{u} \times \mathbf{v} = \begin{vmatrix} \mathbf{i} & \mathbf{j} & \mathbf{k} \\ x_2 - x_1 & y_2 - y_1 & 0 \\ x_3 - x_1 & y_3 - y_1 & 0 \end{vmatrix}$$

$$= (x_2 y_3 - x_2 y_1 - x_1 y_3 + x_1 y_1 - x_3 y_2$$

$$+ x_3 y_1 + x_1 y_2 - x_1 y_1)\mathbf{k}$$

$$A = \tfrac{1}{2}\|\mathbf{u} \times \mathbf{v}\|$$

$$= \tfrac{1}{2}|(x_2 y_3 - x_3 y_2) - (x_1 y_3 - x_3 y_1)$$

$$+ (x_1 y_2 - x_2 y_1)|$$

$$= \tfrac{1}{2}|D| \text{ where } D = \begin{vmatrix} x_1 & y_1 & 1 \\ x_2 & y_2 & 1 \\ x_3 & y_3 & 1 \end{vmatrix}$$

49. $(\mathbf{u} \times \mathbf{v}) \cdot \mathbf{w} = \begin{vmatrix} \mathbf{i} & \mathbf{j} & \mathbf{k} \\ u_1 & u_2 & u_3 \\ v_1 & v_2 & v_3 \end{vmatrix} \cdot (w_1\mathbf{i} + w_2\mathbf{j} + w_3\mathbf{k})$

$$= \left(\begin{vmatrix} u_2 & u_3 \\ v_2 & v_3 \end{vmatrix} \mathbf{i} - \begin{vmatrix} u_1 & u_3 \\ v_1 & v_3 \end{vmatrix} \mathbf{j} + \begin{vmatrix} u_1 & u_2 \\ v_1 & v_2 \end{vmatrix} \mathbf{k} \right)$$

$$\cdot (w_1\mathbf{i} + w_2\mathbf{j} + w_3\mathbf{k})$$

$$= \begin{vmatrix} u_2 & u_3 \\ v_2 & v_3 \end{vmatrix} w_1 - \begin{vmatrix} u_1 & u_3 \\ v_1 & v_3 \end{vmatrix} w_2 + \begin{vmatrix} u_1 & u_2 \\ v_1 & v_2 \end{vmatrix} w_3$$

$$= \begin{vmatrix} u_1 & u_2 & u_3 \\ v_1 & v_2 & v_3 \\ w_1 & w_2 & w_3 \end{vmatrix}$$

51. If $\mathbf{u} + \mathbf{v} + \mathbf{w} = \mathbf{0}$; then $(\mathbf{u} + \mathbf{v} + \mathbf{w}) \times \mathbf{v} = \mathbf{0}$

$\mathbf{u} \times \mathbf{v} + \mathbf{v} \times \mathbf{v} + \mathbf{w} \times \mathbf{v} = \mathbf{0}$. Thus, since

$\mathbf{v} \times \mathbf{v} = \mathbf{0}$, we have $\mathbf{u} \times \mathbf{v} = -(\mathbf{w} \times \mathbf{v}) = \mathbf{v} \times \mathbf{w}$.

Similarly, if we cross with $\mathbf{w}$, then

$\mathbf{u} \times \mathbf{w} + \mathbf{v} \times \mathbf{w} + \mathbf{w} \times \mathbf{w} = \mathbf{0}$ or

$\mathbf{v} \times \mathbf{w} = -(\mathbf{u} \times \mathbf{w}) = \mathbf{w} \times \mathbf{u}$ since $\mathbf{w} \times \mathbf{w} = \mathbf{0}$

Thus, $\mathbf{u} \times \mathbf{v} = \mathbf{v} \times \mathbf{w} = \mathbf{w} \times \mathbf{u}$.

53. Recall that $\cos\theta = \dfrac{\mathbf{v} \cdot \mathbf{w}}{\|\mathbf{v}\| \|\mathbf{w}\|}$, and

$\sin\theta = \dfrac{\|\mathbf{v} \times \mathbf{w}\|}{\|\mathbf{v}\| \|\mathbf{w}\|}$ so $\tan\theta = \dfrac{\|\mathbf{v} \times \mathbf{w}\|}{\mathbf{v} \cdot \mathbf{w}}$.

55. Let $\mathbf{a} = \langle a_1, a_2, a_3 \rangle$, $\mathbf{b} = \langle b_1, b_2, b_3 \rangle$,

$\mathbf{c} = \langle c_1, c_2, c_3 \rangle$. Then

$\mathbf{i} \times (\mathbf{b} \times \mathbf{c}) = (b_2 c_1 - b_1 c_2)\mathbf{j} + (b_3 c_1 - b_1 c_3)\mathbf{k}$

$\qquad = c_1 \mathbf{b} - b_1 \mathbf{c}$

Similarly,

$\qquad \mathbf{j} \times (\mathbf{b} \times \mathbf{c}) = c_2 \mathbf{b} - b_2 \mathbf{c},\ \mathbf{k} \times (\mathbf{b} \times \mathbf{c})$

$\qquad\qquad = c_3 \mathbf{b} - b_3 \mathbf{c}$

Thus,

$\mathbf{a} \times (\mathbf{b} \times \mathbf{c}) = [a_1 \mathbf{i} + a_2 \mathbf{j} + a_3 \mathbf{k}] \times (\mathbf{b} \times \mathbf{c})$

$\qquad = a_1[\mathbf{i} \times (\mathbf{b} \times \mathbf{c})] + a_2[\mathbf{j} \times (\mathbf{b} \times \mathbf{c})] + a_3[\mathbf{k} \times (\mathbf{b} \times \mathbf{c})]$

$\qquad = a_1[c_1 \mathbf{b} - b_1 \mathbf{c}] + a_2[c_2 \mathbf{b} - b_2 \mathbf{c}] + a_3[c_3 \mathbf{b} - b_3 \mathbf{c}]$

$\qquad = (a_1 c_1 + a_2 c_2 + a_3 c_3)\mathbf{b} - (a_1 b_1 + a_2 b_2 + a_3 b_3)\mathbf{c}$

$\qquad = (\mathbf{a} \cdot \mathbf{c})\mathbf{b} - (\mathbf{a} \cdot \mathbf{b})\mathbf{c}$

57. $\mathbf{u} \times (\mathbf{v} \times \mathbf{w}) + \mathbf{v} \times (\mathbf{w} \times \mathbf{u}) + \mathbf{w} \times (\mathbf{u} \times \mathbf{v})$

$\qquad = (\mathbf{u} \cdot \mathbf{w})\mathbf{v} - (\mathbf{u} \cdot \mathbf{v})\mathbf{w} + (\mathbf{v} \cdot \mathbf{u})\mathbf{w} - (\mathbf{v} \cdot \mathbf{w})\mathbf{u}$

$\qquad\qquad + (\mathbf{w} \cdot \mathbf{v})\mathbf{u} - (\mathbf{w} \cdot \mathbf{u})\mathbf{v}$

$\qquad = (\mathbf{v} \cdot \mathbf{u})\mathbf{w} - (\mathbf{u} \cdot \mathbf{v})\mathbf{w} - (\mathbf{v} \cdot \mathbf{w})\mathbf{u} + (\mathbf{w} \cdot \mathbf{v})\mathbf{u}$

$\qquad\qquad + (\mathbf{u} \cdot \mathbf{w})\mathbf{v} - (\mathbf{w} \cdot \mathbf{u})\mathbf{v}$

$\qquad = \mathbf{0}$

59. $\mathbf{u} \times (\mathbf{u} \times \mathbf{v}) = (\mathbf{v} \cdot \mathbf{u})\mathbf{u} - (\mathbf{u} \cdot \mathbf{u})\mathbf{v}$ and

$\mathbf{u} \times [\mathbf{u} \times (\mathbf{u} \times \mathbf{v})] = \mathbf{u} \times (\mathbf{v} \cdot \mathbf{u})\mathbf{u} - \mathbf{u} \times (\mathbf{u} \cdot \mathbf{u})\mathbf{v}$

$\qquad = \mathbf{0} - (\mathbf{u} \cdot \mathbf{u})(\mathbf{u} \times \mathbf{v})$

$\qquad = -\| \mathbf{u} \|^2 (\mathbf{u} \times \mathbf{v})$

Therefore,

$\mathbf{u} \times [\mathbf{u} \times (\mathbf{u} \times \mathbf{v})] \cdot \mathbf{w} = -\| \mathbf{u} \|^2 [(\mathbf{u} \times \mathbf{v}) \cdot \mathbf{w}]$

$\qquad = -\| \mathbf{u} \|^2 [\mathbf{u} \cdot (\mathbf{v} \times \mathbf{w})]$

9.5 Parametric Representation of Curves, Lines in $\mathbb{R}^3$, Pages 614–615

SURVIVAL HINT: *Parametric representation of a curve is an extremely important concept, one which is essential to the study of calculus. Parametric graphing is particularly easy when using a calculator.*

*In Problems 3–16, the domain of t is given, but beyond these problems the domain of the parameter is frequently not given. When you eliminate the parameter to obtain an explicit equation, it may not have the same domain as the original parametric equations. Always use the domain of the **original** equations; for instance see the solution to Problem 3 below and note that the domain for x is not the same as the domain for t.*

1. A parameter is an arbitrary constant or a variable in a mathematical expression, which distinguishes various specific cases.

3. $t = x - 1$, so

$y = (x - 1) - 1$

$\quad = x - 2;$

$1 \le x \le 3$

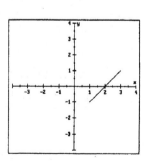

5. $t = \frac{x}{60}$

$y = \frac{80x}{60} - \frac{16x^2}{60^2}$

$= \frac{4}{3}x - \frac{1}{225}x^2$

$0 \le x \le 180$

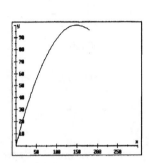

7. $t = x^{1/3}$

$x^{1/3} = y^{1/2}$

$y = x^{2/3}, \; x \ge 0$

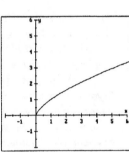

9. $\frac{x}{3} = \cos \theta;$

$\frac{y}{3} = \sin \theta$

$x^2 + y^2 = 9$

$-3 \le x \le 3$

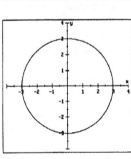

11. $\sin t = x - 1$

$\cos t = y + 2$

$(x-1)^2 + (y+2)^2 = 1$

$0 \le x \le 2$

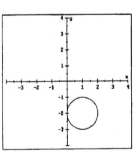

13. $\frac{x}{4} = \tan 2t$

$\frac{y}{3} = \sec 2t$

$\frac{x^2}{16} + 1 = \frac{y^2}{9}$

$\frac{y^2}{9} - \frac{x^2}{16} = 1$

$-\infty < x < \infty$

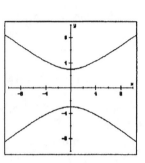

15. $t = x^{1/3}$

$y = 3 \ln x^{1/3}$

$= \ln x$

$x > 0$

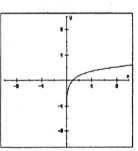

17. $x = 1 + 3t, \; y = -1 - 2t, \; z = -2 + 5t$

$\frac{x-1}{3} = \frac{y+1}{-2} = \frac{z+2}{5}$

19. $\mathbf{v} = (2 - 1)\mathbf{i} + (1 + 1)\mathbf{j} + (3 - 2)\mathbf{k}$

$= \mathbf{i} + 2\mathbf{j} + \mathbf{k}$

$\frac{x-1}{1} = \frac{y+1}{2} = \frac{z-2}{1}$

$x = 1 + t, \; y = -1 + 2t, \; z = 2 + t$

21. $\frac{x-1}{1} = \frac{y+3}{-3} = \frac{z-6}{-5}$

$x = 1 + t, \; y = -3 - 3t, \; z = 6 - 5t$

23. Note that $\langle 22, -6, 10 \rangle = 2 \langle 11, -3, 5 \rangle$ and

the line passes through $(0, 4, -3)$ so that

$\frac{x}{11} = \frac{y-4}{-3} = \frac{z+3}{5}$

$x = 11t, \; y = 4 - 3t, \; z = -3 + 5t$

25. $\dfrac{x+1}{3} = \dfrac{y-1}{1} = \dfrac{z-6}{-2}$

$x = -1 + 3t,\ y = 1 + t,\ z = 6 - 2t$

27. $x = 4 + 4t,\ y = -3 + 3t,\ z = -2 + t$

$x = 0;\ t = -1,$ for point $(0, -6, -3)$;

$y = 0;\ t = 1,$ for point $(8, 0, -1)$;

$z = 0;\ t = 2,$ for point $(12, 3, 0)$

29. $x = 6 - 2t,\ y = 1 + t,\ z = 3t$

$x = 0;\ t = 3,$ for point $(0, 4, 9)$;

$y = 0;\ t = -1,$ for point $(8, 0, -3)$;

$z = 0;\ t = 0,$ for point $(6, 1, 0)$

31. A vector parallel to the first line is

$\mathbf{v}_1 = 2\mathbf{i} - 3\mathbf{j} + 5\mathbf{k}$. A vector parallel to the

second line is $\mathbf{v}_2 = 4\mathbf{i} - 6\mathbf{j} + 10\mathbf{k}$

$= 2(2\mathbf{i} - 3\mathbf{j} + 5\mathbf{k})$; since these vectors have

the same direction numbers the lines are

either coincident or parallel; since

$P(4, 6, -2)$ is a point on the first line, but

not on the second line, and the first line is

parallel to the second.

33. A vector parallel to the first line is

is $\mathbf{v}_1 = 3\mathbf{i} - 4\mathbf{j} - 7\mathbf{k}$. A vector parallel to

the second line is $\mathbf{v}_2 = 3\mathbf{i} - 4\mathbf{j} - 7\mathbf{k}$. Since

$\mathbf{v}_2 = \mathbf{v}_1$, the lines are either coincident or

parallel; since $P(3, 1, -4)$ is a point on the

first line, but not the second, they are

parallel.

35. A vector parallel to the first line is

is $\mathbf{v}_1 = 2\mathbf{i} - \mathbf{j} + \mathbf{k}$. A vector parallel to the

second line is $\mathbf{v}_2 = 3\mathbf{i} - \mathbf{j} + \mathbf{k}$. Since these

lines do not have proportional direction

numbers, the lines are not parallel or

coincident. If they intersect, we must have:

for x: $3 + 2t_1 = -2 + 3t_2$ or $2t_1 - 3t_2 = -5$;

for y; $1 - t_1 = 3 - t_2$ or $t_1 - t_2 = -2$.

for z; $4 + t_1 = 2 + t_2$ or $t_1 - t_2 = -2$.

Solving this system simultaneously we find

$t_1 = -1,\ t_2 = 1$. Thus, $x = 3 - 2 = 1$,

$y = 1 - (-1) = 2$, and $z = 4 + (-1) = 3$.

The point of intersection is $(1, 2, 3)$.

37. A vector parallel to the given line is one with

direction $\pm [4, 2, 1]$. To find the unit vector,

we divide these vectors by their length.

$$\pm \frac{1}{\sqrt{21}}(4\mathbf{i} + 2\mathbf{j} + \mathbf{k})$$

39. The equation of the circle is

$$x^2 + y^2 = 3$$

$$\left(\frac{x}{\sqrt{3}}\right)^2 + \left(\frac{y}{\sqrt{3}}\right)^2 = 1$$

Use the identity $\cos^2 t + \sin^2 t = 1$

Let $\cos t = \dfrac{x}{\sqrt{3}}$ or $x = \sqrt{3}\cos t$

$\sin t = \dfrac{y}{\sqrt{3}}$ or $y = \sqrt{3}\sin t$

The orientation is counterclockwise because t

is measured in a positive direction for

$0 \le t \le 2\pi$.

41. The equation of the ellipse is

$$\left(\frac{x}{3}\right)^2 + \left(\frac{y}{2}\right)^2 = 1$$

Let $\cos t = \dfrac{x}{3}$ or $x = 3\cos t$

$\sin t = \dfrac{y}{2}$ or $y = 2\sin t$

The orientation is counterclockwise because t

is measured in a positive direction for

$0 \le t \le 2\pi$.

43. The equation of the hyperbola is

$$\left(\frac{x}{4}\right)^2 - \left(\frac{y}{3}\right)^2 = 1$$

Use the identity

$$1 + \tan^2 t = \sec^2 t \text{ or } \sec^2 t - \tan^2 t = 1$$

Let $\sec t = \dfrac{x}{4}$ or $x = \dfrac{4}{\cos t}$

$\tan t = \dfrac{y}{3}$ or $y = 3\tan t$ for $0 \le t \le 2\pi$.

45. Since $\cos 2u = 1 - 2\sin^2 u$, $y = 1 - 2x^2$ where

$0 \le x \le 1$; $-1 \le y \le 1$; this is a parabolic arc.

47. $\mathbf{PQ} = 2\mathbf{i} + \mathbf{j} - 2\mathbf{k}$; $\mathbf{v} \cdot \mathbf{PQ} = 6 - 4 - 2 = 0$,

so $\mathbf{v}$ and $\mathbf{PQ}$ are orthogonal.

49. The lines are parallel to each other since they

are both parallel to $\mathbf{v} = \langle 2, 4, 1 \rangle$, and they

will coincide if they also intersect. Solving

$$\begin{cases} 2t + a = 2 - 4s \\ 1 + 4t = b - 8s \\ -2 + t = -1 - 2s \end{cases}$$

We see that the lines intersect when $a = 0$,

$b = 5$. (Other choices of a and b will also

work.)

51. The given line is parallel to the vector

$\mathbf{v} = \langle -1, -2, 1 \rangle$ and contains the point

$Q(2, 1, 5)$. Then $\mathbf{QP} = \langle -3, 2, -4 \rangle$ is the

vector from Q to the given point $P(-1, 3, 1)$,

and the cross product $\mathbf{QP} \times \mathbf{v} = \langle -6, 7, 8 \rangle$ is

orthogonal to both the given line and the line

we seek. This means the cross product

$$\mathbf{v} \times (\mathbf{QP} \times \mathbf{v}) = \langle -23, 2, -19 \rangle$$

is parallel to the required line, which thus has

parametric equations

$$x = -1 - 23t, \ y = 3 + 2t, \ z = 1 - 19t$$

53. Suppose the helicopter is at point $P_1(a, b, c)$ when it is first sighted. Then we have

$$AP_1 = (a-7)\mathbf{i} + b\mathbf{j} + c\mathbf{k} = (-4\mathbf{i} + 2\mathbf{j} + 5\mathbf{k})s_1$$

$$BP_1 = a\mathbf{i} + (b-4)\mathbf{j} + c\mathbf{k} = (3\mathbf{i} - 2\mathbf{j} + 5\mathbf{k})s_2$$

for constants s_1 and s_2, so that from the first equation (AP_1),

$$a - 7 = -4s_1$$

$$b = 2s_1$$

and

$$c = 5s_1$$

and from the second equation (BP_1)

$$a = 3s_2$$

$$b - 4 = -2s_2$$

and

$$c = 5s_2$$

Solving simultaneously, we find $s_1 = s_2 = 1$ and $a = 3$, $b = 2$, $c = 5$, so the first point of sighting is $P_1(3, 2, 5)$. Similarly, we find the second point of sighting to be $P_2(13, 7, 5.005)$.

Blohardt's helicopter travels

$$\|P_1P_2\| = \sqrt{(13-3)^2 + (7-2)^2 + (5.005-5)^2}$$

$$\approx 11.180 \text{ thousand ft/min}$$

Suppose the Spy intercepts Blohardt at point $P(d, e, f)$. Then, since P lies on the line through $P_1(3, 2, 5)$ and $P_2(13, 7, 5.005)$, we have

$$d - 3 = (13 - 3)t$$

$$e - 2 = (7 - 2)t$$

and

$$f - 5 = (5.005 - 5)t$$

so that

$$d = 3 + 10t$$

$$e = 2 + 5t$$

$$f = 5 + 5.005t$$

Since Blohardt travels on the line through P_1 and P_2, we must have

$$d = 3 + 10s$$

$$e = 2 + 5s$$

$$f = 5 + 0.005s$$

for some s. We have seen that Blohardt travels at about 11.180 thousand ft/min and the Spy travels at

$$\frac{150 \text{ mi}}{\text{hr}} \frac{5,280 \text{ ft}}{\text{mi}} \frac{1 \text{ hr}}{60 \text{ min}} = 13.2 \text{ thousand ft/min}$$

Suppose Blohardt reaches the intercept point P at t minutes after noon. Then, the distance from P_1 to P_2 is

$$\sqrt{(d-3)^2 + (e-2)^2 + (f-5)^2} \approx 11.180t$$

$$\sqrt{(10s)^2 + (5s)^2 + (0.005)^2} \approx 11.180t$$

$$s = t$$

Since the Spy travels $t - 10$ minutes to reach the same point, using the distance from Q to P, we have

$$\sqrt{(d-0)^2 + (e-0)^2 + (f-1)^2} \approx 13.2(t-10)$$

and since

$$d = 3 + 10s$$

$$e = 2 + 5s$$

$$f = 4 + 0.005s$$

we have

$$\sqrt{(3 + 10s)^2 + (2 + 5s)^2 + (4 + 0.005s)^2}$$

$$= 13.2t$$

$$= 13.2(s - 10)$$

Solving (using a calculator)

$$\sqrt{(3 + 10s)^2 + (2 + 5s)^2 + (4 + 0.005s)^2}$$

$$= 13.2(s - 10)$$

for s, we obtain $s \approx 67.135$, so

$$t \approx 67.135$$

and the intercept point is approximately at $(674.35, 337.68, 5.34)$. Thus, the Spy should travel from $Q(0, 0, 1)$ in the direction of the vector

$$\mathbf{v} = 674.35\mathbf{i} + 337.68\mathbf{j} + 4.34\mathbf{k}$$

55. Consider a fixed circle of radius a with center at the origin O. Let E be the point with coordinates $(a, 0)$. A ray makes an angle θ with the positive x-axis and contains the center A of a moving circle of radius R which makes contact with the fixed circle at D. A ray from A to a point $P(x, y)$ − drawn to the right and below A for convenience − on this moving circle also makes an angle ϕ with respect to $\overline{OD}$. This point P was originally at E, before the second circle started moving. Arcs $\overline{DP}$ and $\overline{DE}$ are the same length. We have

$$a\theta = \phi R \text{ or } \phi = \frac{a\theta}{R}$$

Consider a right triangle with $\overline{AP}$ as

hypotenuse and third vertex above P and to the right of A (for convenience). Label the acute angle at A, α, and label the legs x_1 and y_1, respectively. Then,

$$\alpha = \phi - \theta$$
$$= \left(\frac{a - R}{R}\right)\theta$$

$$x_1 = R \cos \alpha$$
$$= R \cos \frac{(a - R)\theta}{R}$$

$$y_1 = R \sin \alpha$$
$$= R \sin \frac{(a - R)\theta}{R}$$

The distance x_2 from O to the projection of A on the x-axis is $x_2 = (a - R)\cos\theta$. The vertical distance y_1 from A to the x-axis is $y_2 = (a - R)\sin\theta$. Putting all this together, we have

$$x = x_2 + x_1$$
$$= (a - R)\cos\theta + R \cos \frac{(a - R)\theta}{R}$$

$$y = y_2 - y_1$$
$$= (a - R)\sin\theta - R \sin \frac{(a - R)\theta}{R}$$

9.6 Planes in $\mathbb{R}^3$, Pages 621-622

SURVIVAL HINT: *You should be able to represent a line in either parametric or symmetric form, and a plane is either normal or standard form.*

1. The plane containing $P_0(x_0,\ y_0,\ z_0)$ with normal vector $A\mathbf{i} + B\mathbf{j} + C\mathbf{k}$ has the equation
$$A(x - x_0) + B(y - y_0) + C(z - z_0) = 0$$
Conversely, if $Ax + By + Cz + D = 0$ is the equation of a plane, a vector normal to the plane is given by $\mathbf{N} = A\mathbf{i} + B\mathbf{j} + C\mathbf{k}$. Note that the coefficients of the variables are the direction numbers of the normal.

3. The distance from point P to the line L is given by the formula
$$d = \frac{\|\mathbf{v} \times \mathbf{QP}\|}{\|\mathbf{v}\|}$$
where $\mathbf{v}$ is a vector parallel to L and Q is any point on L.

5. $5(x - 2) - 3(y + 2) + 4(z + 3) = 0$
$$5x - 10 - 3y - 6 + 4z + 12 = 0$$
$$5x - 3y + 4z - 4 = 0$$

7. $-2(x + 1) + 4(y - 3) - 8 = 0$
$$-2x - 2 + 4y - 12 - 8z = 0$$
$$-2x + 4y - 8z - 14 = 0$$
$$x - 2y + 4z + 7 = 0$$

9. $P(0, -7, 1);\ \mathbf{N} = -\mathbf{i} + \mathbf{k} = \langle -1, 0, 1 \rangle$

$$-(x - 0) + 0(y + 7) + (z - 1) = 0$$
$$x - z + 1 = 0$$

11. $P(1, 1, -1);$

$$\mathbf{N} = -\mathbf{i} - 2\mathbf{j} + 3\mathbf{k} = \langle -1, -2, 3 \rangle$$
$$-(x - 1) - 2(y - 1) + 3(z + 1) = 0$$
$$x + 2y - 3z - 6 = 0$$

13. $P(0, 0, 0);\ \mathbf{N} = \mathbf{i}$

$$x = 0$$

15. The normal to the plane is $5\mathbf{i} - 3\mathbf{j} + 2\mathbf{k}$, and

to find a unit vector, we divide this vector by

its length: $\pm\dfrac{1}{\sqrt{38}}(5\mathbf{i} - 3\mathbf{j} + 2\mathbf{k})$

SURVIVAL HINT: *Problems 16-35 present several formulas involving distance in $\mathbb{R}^3$. Be sure to ask your instructor if you need to have any (or all) of these formulas committed to memory.*

17. $P(0, 0, 0);\ 2x - 3y + 5z = 10$

$$d = \frac{|-10|}{\sqrt{2^2 + 3^2 + 5^2}}$$
$$= \frac{10}{\sqrt{38}}$$
$$= \frac{10\sqrt{38}}{38}$$
$$= \frac{5\sqrt{38}}{19}$$

19. $P(2, 1, -2);\ 3x - 4y + z = -1$

$$d = \frac{|6 - 4 - 2 + 1|}{\sqrt{3^2 + 4^2 + 1^2}}$$
$$= \frac{1}{\sqrt{26}}$$
$$= \frac{\sqrt{26}}{26}$$

21. $P(a, 2a, 3a);\ 3x - 2y + z = -\dfrac{1}{a}$

$$d = \frac{\left|3a - 4a + 3a + \frac{1}{a}\right|}{\sqrt{9 + 4 + 1}}$$
$$= \frac{\left|2a + \frac{1}{a}\right|}{\sqrt{14}}$$
$$= \frac{2a^2 + 1}{|a|\sqrt{14}}$$

23. Name the points: $A(-1, 1, 1)$, $B(4, 3, 7)$, and $C(3, -1, 0)$. Then,

$$\mathbf{AB} = (4 + 1)\mathbf{i} + (3 - 1)\mathbf{j} + (7 - 1)\mathbf{k}$$
$$= 5\mathbf{i} + 2\mathbf{j} + 6\mathbf{k}$$
$$\mathbf{AC} = (3 + 1)\mathbf{i} + (-1 - 1)\mathbf{j} + (0 - 1)\mathbf{k}$$
$$= 4\mathbf{i} - 2\mathbf{j} - \mathbf{k}$$

A normal to the plane formed by $\mathbf{AB}$ and $\mathbf{AC}$ is

$$\mathbf{N} = \mathbf{AB} \times \mathbf{AC}$$
$$= \begin{vmatrix} \mathbf{i} & \mathbf{j} & \mathbf{k} \\ 5 & 2 & 6 \\ 4 & -2 & -1 \end{vmatrix}$$
$$= 10\mathbf{i} + 29\mathbf{j} - 18\mathbf{k}$$

The equation of the plane containing the

given points is $10x + 29y - 18z - 1 = 0$, so

$$d = \frac{|-10 + 58 - 18 - 1|}{\sqrt{10^2 + 29^2 + 18^2}}$$

$$= \frac{29}{\sqrt{1,265}}$$

25. The equation of the plane through $(1, 0, 1)$ with normal vector $2\mathbf{i} - \mathbf{j} + 2\mathbf{k}$ is

$$2x - y + 2z + D = 0$$

so

$$2(1) - (0) + 2(1) + D = 0$$
$$D = -4$$

Thus, the equation of the plane is

$$2x - y + 2z - 4 = 0$$

The distance from the point to this plane is

$$d = \frac{|-2 - 2 + 2 - 4|}{\sqrt{4 + 1 + 4}}$$

$$= \frac{6}{3}$$

$$= 2$$

27. $P(9, -3); 3x - 4y + 8 = 0$:

$$d = \frac{|3(9) - 4(-3) + 8|}{\sqrt{3^2 + (-4)^2}}$$

$$= \frac{47}{5}$$

29. $P(1, 0, -1); \frac{x - 2}{3} = \frac{y + 1}{1} = \frac{z - 1}{2}$

This line is parallel to the vector

$$\mathbf{v} = 3\mathbf{i} + \mathbf{j} + 2\mathbf{k}$$

It contains the point $Q(2, -1, 1)$. Thus,

$$\mathbf{PQ} = \mathbf{i} - \mathbf{j} + 2\mathbf{k}$$

and

$$\mathbf{PQ} \times \mathbf{v} = \begin{vmatrix} \mathbf{i} & \mathbf{j} & \mathbf{k} \\ 1 & -1 & 2 \\ 3 & 1 & 2 \end{vmatrix}$$

$$= -4\mathbf{i} + 4\mathbf{j} + 4\mathbf{k}$$

We now find the distance:

$$d = \frac{\|\mathbf{PQ} \times \mathbf{v}\|}{\|\mathbf{v}\|}$$

$$= \frac{\sqrt{(-4)^2 + 4^2 + 4^2}}{\sqrt{9 + 1 + 4}}$$

$$= \frac{4\sqrt{3}}{\sqrt{14}}$$

31. $P(1, -2, 2); \frac{x}{1} = \frac{2y}{1} = \frac{z}{-1}$

This line is parallel to $\mathbf{v} = \mathbf{i} + \frac{1}{2}\mathbf{j} - \mathbf{k}$ and contains $Q(0, 0, 0)$. Thus

$$\mathbf{PQ} = (0 - 1)\mathbf{i} + (0 + 2)\mathbf{j} + (0 - 2)\mathbf{k}$$

$$= -\mathbf{i} + 2\mathbf{j} - 2\mathbf{k}$$

and

$$\mathbf{PQ} \times \mathbf{v} = \begin{vmatrix} \mathbf{i} & \mathbf{j} & \mathbf{k} \\ -1 & 2 & -2 \\ 1 & \frac{1}{2} & -1 \end{vmatrix}$$

$$= -\mathbf{i} - 3\mathbf{j} - \frac{5}{2}\mathbf{k}$$

$$d = \frac{\| \mathbf{PQ} \times \mathbf{v} \|}{\| \mathbf{v} \|}$$

$$= \frac{\sqrt{(-1)^2 + (-3)^2 + (-\frac{5}{2})^2}}{\sqrt{1 + \frac{1}{4} + 1}}$$

$$= \frac{\sqrt{65}}{3}$$

33. The line $x = 2 - t$, $y = 5 + 2t$, $z = 3t$
is parallel to $\mathbf{v}_1 = -\mathbf{i} + 2\mathbf{j} + 3\mathbf{k}$ and contains
$P_1(2, 5, 0)$. The line $x = 2t$, $y = -1 - t$,
$z = 1 + 2t$ is parallel to $\mathbf{v}_2 = 2\mathbf{i} - \mathbf{j} + 2\mathbf{k}$
and contains $P_2(0, -1, 1)$. Then

$$\mathbf{P}_1\mathbf{P}_2 = (0 - 2)\mathbf{i} + (-1 - 5)\mathbf{j} + (1 - 0)\mathbf{k}$$

$$= -2\mathbf{i} - 6\mathbf{j} + \mathbf{k};$$

The normal, $\mathbf{N}$, to the plane determined by $\mathbf{v}_1$
and $\mathbf{v}_2$ is

$$\mathbf{N} = \mathbf{v}_1 \times \mathbf{v}_2$$

$$= \begin{vmatrix} \mathbf{i} & \mathbf{j} & \mathbf{k} \\ -1 & 2 & 3 \\ 2 & -1 & 2 \end{vmatrix}$$

$$= 7\mathbf{i} + 8\mathbf{j} - 3\mathbf{k}$$

$$d = \frac{| \mathbf{P}_1\mathbf{P}_2 \cdot \mathbf{N} |}{\| \mathbf{N} \|}$$

$$= \frac{| -2(7) + (-6)(8) + 1(-3) |}{\sqrt{7^2 + 8^2 + (-3)^2}}$$

$$= \frac{65}{\sqrt{122}}$$

35. The line $x = -1 + t$, $y = -2t$, $z = 3$
is parallel to $\mathbf{v}_1 = \mathbf{i} - 2\mathbf{j}$ and contains
$P_1(-1, 0, 3)$. The line through $(0, -1, 2)$
and $(1, -2, 3)$ is parallel to $\mathbf{v}_2 = \mathbf{i} - \mathbf{j} + \mathbf{k}$
and contains $P_2(0, -1, 2)$. Then

$$\mathbf{P}_1\mathbf{P}_2 = (0 + 1)\mathbf{i} + (-1 - 0)\mathbf{j} + (2 - 3)\mathbf{k}$$

$$= \mathbf{i} - \mathbf{j} - \mathbf{k}$$

The normal, $\mathbf{N}$, to the plane determined by $\mathbf{v}_1$
and $\mathbf{v}_2$ is

$$\mathbf{N} = \mathbf{v}_1 \times \mathbf{v}_2$$

$$= \begin{vmatrix} \mathbf{i} & \mathbf{j} & \mathbf{k} \\ 1 & -2 & 0 \\ 1 & -1 & 1 \end{vmatrix}$$

$$= -2\mathbf{i} - \mathbf{j} + \mathbf{k}$$

$$d = \frac{| \mathbf{P}_1\mathbf{P}_2 \cdot \mathbf{N} |}{\| \mathbf{N} \|}$$

$$= \frac{| -2 + 1 - 1 |}{\sqrt{4 + 1 + 1}}$$

$$= \frac{2}{\sqrt{6}}$$

37. **a.** A vector parallel to
$$\frac{x - 1}{3} = \frac{y}{-2} = \frac{z + 1}{1}$$
is

$$\mathbf{v} = 3\mathbf{i} - 2\mathbf{j} + \mathbf{k}$$

and the normal to $x + 2y + z = 1$ is

$N = i + 2j + k$. Then,

$$N \cdot v = 3 - 4 + 1$$

$$= 0$$

so the vectors are orthogonal which means that the line is parallel to the given plane.

b. The point $Q(1, 0, -1)$ is on the line, and the distance from the line to the plane is the same as the distance from Q to the plane, namely

$$d = \frac{|1 + 0 - 1 - 1|}{\sqrt{6}}$$

$$= \frac{1}{\sqrt{6}}$$

39. The distance from $P(x, y, z)$ to the plane

$2x - 5y + 3z = 7$ is

$$d = \frac{|2x - 5y + 3z - 7|}{\sqrt{4 + 25 + 9}}$$

Also,

$$\|P_0 P\|^2 = (x + 1)^2 + (y - 2)^2 + (z - 4)^2$$

Thus, $\|P_0 P\| = d$ when

$$(x + 1)^2 + (y - 2)^2 + (z - 4)^2$$

$$= \tfrac{1}{38}(2x - 5y + 3z - 7)^2$$

41. a. $N = ai + bj + ck$ is normal to the plane, and $v = Ai + Bj + Ck$ is parallel to the line. The cosine of the acute angle between the vectors is

$$\cos \theta = \frac{|v \cdot N|}{\|v\| \|N\|}$$

$$= \frac{aA + bB + cC}{\sqrt{a^2 + b^2 + c^2}\sqrt{A^2 + B^2 + C^2}}$$

b. The angle between $x + y + z = 10$ and

$$\frac{x - 1}{2} = \frac{y + 3}{3} = \frac{z - 2}{-1}$$

So we use the formula from part a:

$$\cos \theta = \frac{aA + bB + cC}{\sqrt{a^2 + b^2 + c^2}\sqrt{A^2 + B^2 + C^2}}$$

$$= \frac{(1)(2) + (1)(3) + (1)(-1)}{\sqrt{1^2 + 1^2 + 1^2}\sqrt{2^2 + 3^2 + (-1)^2}}$$

$$= \frac{4}{\sqrt{3}\sqrt{14}}$$

so $\quad \theta = \cos^{-1}\left(\frac{4}{\sqrt{3}\sqrt{14}}\right)$

$$\approx 52°$$

43. $2x + 3y = 0$; let $N_1 = 2i + 3j$

$3x - y + z = 1$; let $N_2 = 3i - j + k$

$$N_1 \times N_2 = \begin{vmatrix} i & j & k \\ 2 & 3 & 0 \\ 3 & -1 & 1 \end{vmatrix}$$

$$= 3i - 2j - 11k$$

45. $x + y + z = 3$; let $\mathbf{N_1} = \mathbf{i} + \mathbf{j} + \mathbf{k}$

    $2x + 3y - z = 4$; let $\mathbf{N_2} = 2\mathbf{i} + 3\mathbf{j} - \mathbf{k}$

    $$\mathbf{N_1} \times \mathbf{N_2} = \begin{vmatrix} \mathbf{i} & \mathbf{j} & \mathbf{k} \\ 1 & 1 & 1 \\ 2 & 3 & -1 \end{vmatrix}$$

    $$= -4\mathbf{i} + 3\mathbf{j} + \mathbf{k}$$

    $\|\mathbf{N_1} \times \mathbf{N_2}\|^2 = 16 + 9 + 1 = 26;$

    $\cos\alpha = \dfrac{-4}{\sqrt{26}}$ so $\alpha \approx 2.47$ or $142°;$

    $\cos\beta = \dfrac{3}{\sqrt{26}}$ so $\beta \approx 0.94$ or $54°;$

    $\cos\gamma = \dfrac{1}{\sqrt{26}}$ so $\gamma \approx 1.37$ or $79°$

47. A vector normal to $2x - 3y + z = 1$ is

    $\mathbf{N} = 2\mathbf{i} - 3\mathbf{j} + \mathbf{k}$, and this vector is parallel to

    the desired line. Thus,

    $$\frac{x-1}{2} = \frac{y+5}{-3} = \frac{z-3}{1}$$

49. A vector normal to $x + y + z = 3$ is

    $\mathbf{N_1} = \mathbf{i} + \mathbf{j} + \mathbf{k}$ and a vector normal to

    $x - y + z = 1$ is $\mathbf{N_2} = \mathbf{i} - \mathbf{j} + \mathbf{k}$. Then

    $$\mathbf{N_1} \times \mathbf{N_2} = \begin{vmatrix} \mathbf{i} & \mathbf{j} & \mathbf{k} \\ 1 & 1 & 1 \\ 1 & -1 & 1 \end{vmatrix}$$

    $$= 2\mathbf{i} - 2\mathbf{k}$$

    The unit vectors are $\mathbf{N} = \pm\dfrac{1}{\sqrt{2}}(\mathbf{i} - \mathbf{k})$.

51. Two parallel planes have the same normal

    vectors, so the desired equation is

    $$2x - y + 3z + D = 0$$

    The plane passes through $P(1, 2, -1)$, so

    $$2 - 2 - 3 + D = 0 \text{ or } D = 3$$

    Thus, $2x - y + 3z + 3 = 0$.

53. Parametric equations for $\dfrac{x-1}{2} = \dfrac{y+1}{-1} = \dfrac{z}{3}$

    are $x = 1 + 2t$, $y = -1 - t$, $z = 3t$. This

    line intersects the plane $3x + 2y - z = 5$

    when

    $$3(1 + 2t) + 2(-1 - t) - 3t = 5$$
    $$6t + 3 - 2t - 2 - 3t = 5$$
    $$t = 4$$

    Thus, the given point is $P(9, -5, 12)$.

55. $x + 2y - 3z = 4$; let $\mathbf{N_1} = \mathbf{i} + 2\mathbf{j} - 3\mathbf{k}$

    $x - 2y + z = 0$; let $\mathbf{N_2} = \mathbf{i} - 2\mathbf{j} + \mathbf{k}$

    Then

    $$\mathbf{N_1} \times \mathbf{N_2} = \begin{vmatrix} \mathbf{i} & \mathbf{j} & \mathbf{k} \\ 1 & 2 & -3 \\ 1 & -2 & 1 \end{vmatrix}$$

    $$= -4\mathbf{i} - 4\mathbf{j} - 4\mathbf{k}$$

    is parallel to the required line. Since the line

    passes through $P(2, 3, 1)$, its equation is

$$\frac{x-2}{1} = \frac{y-3}{1} = \frac{z-1}{1}$$

57. Since $N_1 = A_1\mathbf{i} + B_1\mathbf{j} + C_1\mathbf{k}$ is normal to the first plane and $N_2 = A_2\mathbf{i} + B_2\mathbf{j} + C_2\mathbf{k}$ is normal to the second, the angle between the planes is $\pi/2$ if and only if $N_1 \cdot N_2 = 0$; that is, $A_1A_2 + B_1B_2 + C_1C_2 = 0$.

59. $N_1 = \mathbf{v}_1 \times \mathbf{w}_1$ is normal to p_1 and

    $N_2 = \mathbf{v}_2 \times \mathbf{w}_2$ is normal to p_2. Then,

    $\mathbf{v} = N_1 \times N_2$ is parallel to both planes, or

    aligned with their line of intersection.

9.7 Quadric Surfaces, Page 626-627

SURVIVAL HINT: *When using the distance formula, it is immaterial as to which point you consider the first or the second, since*

$$(x_2 - x_1)^2 = (x_1 - x_2)^2$$

It is usually easiest to think in terms of the change Δx, form one point to the other:

$$d = \sqrt{(\Delta x)^2 + (\Delta y)^2 + (\Delta z)^2}$$

1. A quadric surface is the graph of an equation

   of the form

   $Ax^2 + By^2 + Cz^2 + Dxy + Exz + Fyz + Gx$

   $+ Hy + Iz + J = 0$

   where A, B, C, D, E, F, G, H, I, and J are

   constants.

SURVIVAL HINT: *In identifying quadrics, let each variable, one at a time, equal zero and identify the resulting second degree conic in the coordinate plane. Two or more conics of the same type give the quadric the "oid" name and remaining conic describes the type. For example, if two coordinate planes have parabolic traces, and the third an ellipse, the surface is called an elliptic paraboloid.*

3. circular cone; B

5. hyperboloid of one sheet; E

7. sphere; A

9. paraboloid, G

11. hyperboloid of two sheets; I

13. Ellipsoid

    xy-plane; ellipse

    yz-plane; ellipse

    xz-plane; ellipse

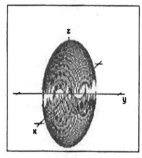

15. Hyperboloid of one sheet

    xy-plane; ellipse

    yz-plane; hyperbola

    xz-plane; hyperbola

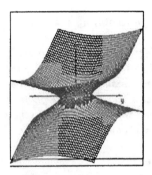

17. Elliptic paraboloid parallel to xy-plane;

$z > 0$; ellipse

yz-plane; parabola

xz-plane; parabola

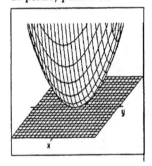

19. Elliptic cone

parallel to xy-plane; $z \neq 0$; ellipse

parallel to yz-plane; $x \neq 0$; hyperbola

parallel xz-plane; $z \neq 0$; hyperbola

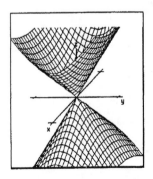

21. $y = x^2 + 9z^2$

elliptic paraboloid;

xy-plane; parabola

yz-plane; parabola

xz-plane; ellipse

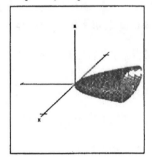

23. $y^2 = x^2 + 3z^2$

elliptic cone

xy-plane; hyperbola

yz-plane; hyperbola

xz-plane; ellipse

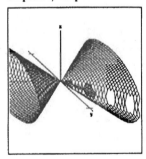

25. Ellipsoid centered at $(1, -2, 3)$ with

semi-axes 2, 1, and 3.

27. $$7x^2 + y^2 + 3z^2 - 9x + 4y + 7z = 1$$

$$7\left[x^2 - \tfrac{9}{7}x\right] + \left[y^2 + 4y\right] + 3\left[z^2 + \tfrac{7}{3}z\right] = 1$$

$$7\left(x - \tfrac{9}{14}\right)^2 + (y + 2)^2 + 3\left(z + \tfrac{7}{6}\right)^2 = \tfrac{503}{42}$$

Note: $\tfrac{503}{42} = 1 + 7\left(\tfrac{9}{14}\right)^2 + (2)^2 + 3\left(\tfrac{7}{6}\right)^2$

This is an ellipsoid centered at $\left(\tfrac{9}{14}, -2, -\tfrac{7}{6}\right)$

with semi-axes approximately 1.31, 3.46, and

2.00.

29. $x^2 - x - y + z^2 + z + 5 = 0$

$\left(x - \tfrac{1}{2}\right)^2 - y + \left(z + \tfrac{1}{2}\right)^2 = -5 + \tfrac{1}{4} + \tfrac{1}{4}$

$\qquad\qquad y - \tfrac{9}{2} = \left(x - \tfrac{1}{2}\right)^2 + \left(z + \tfrac{1}{2}\right)^2$

This is an elliptic paraboloid with vertex

$\left(\tfrac{1}{2}, \tfrac{9}{2}, -\tfrac{1}{2}\right)$ and the y-axis as the axis of

symmetry.

31. The ellipsoid $\dfrac{x^2}{9} + \dfrac{y^2}{4} + \dfrac{z^2}{5} = 1$ and the elliptic

cone $z^2 = \dfrac{x^2}{3} + \dfrac{y^2}{2}$ intersect where

$$\dfrac{x^2}{9} + \dfrac{y^2}{4} + \dfrac{1}{5}\left(\dfrac{x^2}{3} + \dfrac{y^2}{2}\right) = 1$$

$$\dfrac{8}{45}x^2 + \dfrac{7}{20}y^2 = 1$$

This is an ellipse in the xy-plane.

33. Each cross-section perpendicular to the z-axis

(for $|z| < c$) is an ellipse with equation

$$\dfrac{x^2}{a^2} + \dfrac{y^2}{b^2} = 1 - \dfrac{z^2}{c^2}$$

$$\dfrac{x^2}{\alpha^2} + \dfrac{y^2}{\beta^2} = 1$$

where $\alpha^2 = a^2\left(1 - \dfrac{z^2}{c^2}\right)$ and $\beta^2 = b^2\left(1 - \dfrac{z^2}{c^2}\right)$

The area of this cross section is

$$\pi\alpha\beta = \pi\left[a\sqrt{1 - \dfrac{z^2}{c^2}}\right]\left[b\sqrt{1 - \dfrac{z^2}{c^2}}\right]$$

$$= \pi ab\left(1 - \dfrac{z^2}{c^2}\right)$$

Integrating with respect to z, we find that the

volume of the ellipsoid is

$$V = \int_{-c}^{c} \pi ab\left(1 - \dfrac{z^2}{c^2}\right) dz$$

$$= \dfrac{\pi ab}{c^2}\left[c^2 z - \dfrac{z^3}{3}\right]\Big|_{-c}^{c}$$

$$= \dfrac{4}{3}\pi abc$$

35. The intersection of the surfaces is found by

solving the equations simultaneously:

$$x^2 + y^2 = 1 - y^2$$

$$x^2 + 2y^2 = 1$$

This is the equation of an ellipse.

37. **a.** The equatorial radius is $a = 6{,}378.2$ and the

polar radius is $b = 6{,}356.5$.

b. The volume is

$$V = \tfrac{4}{3}\pi a^2 b$$

$$= \tfrac{4}{3}\pi(6{,}378.2)^2(6{,}356.5) \text{ km}^3$$

$$\approx 1.08319 \times 10^{12} \text{ km}^3$$

39. If the point $P(x, y, z)$ is on the required surface, then

$$\sqrt{(x-1)^2 + y^2 + z^2} + \sqrt{(x+1)^2 + y^2 + z^2} = 3$$

$$\sqrt{(x-1)^2 + y^2 + z^2} = 3 - \sqrt{(x+1)^2 + y^2 + z^2}$$

$$(x-1)^2 + y^2 + z^2 = 9 - 6\sqrt{(x+1)^2 + y^2 + z^2} + (x+1)^2 + y^2 + z^2$$

$$6\sqrt{(x+1)^2 + y^2 + z^2} = 9 + 4x$$

$$36[(x+1)^2 + y^2 + z^2] = 81 + 72x + 16x^2$$

$$20x^2 + 36y^2 + 36z^2 = 45$$

This is an ellipsoid.

CHAPTER 9 REVIEW

Proficiency Examination, Pages 628-629

SURVIVAL HINT: *To help you review the concepts of this chapter, **handwrite** the answers to each of these questions onto your own paper.*

1. A vector is a directed line segment and a scalar is a real number.

2. If $v = \langle a_1, a_2, a_3 \rangle$, then $sv = \langle sa_1, sa_2, sa_3 \rangle$; Geometrically, sv is a $s\|v\|$ units long and points along v if $s > 0$ or in the opposite direction if $s < 0$.

3. If two vectors are arranged so that their initial points coincide, then the sum of the vectors is the diagonal of the parallelogram formed by the two vectors.

4. **a.** $u + v = v + u$

   b. $(u + v) + w = u + (v + w)$

   c. $u + 0 = u$

   d. $u + (-u) = 0$

   e. $\|v\| = (v \cdot v)^{1/2}$

   f. $v \cdot w = w \cdot v$

   g. $c(v \cdot w) = (cv) \cdot w = v \cdot (cw)$

   h. $u \cdot (v + w) = u \cdot v + u \cdot w$

   i. $v \times 0 = 0 \times v = 0$

   j. $v \times w = -(w \times v)$

   k. $u \times (v + w) = (u \times v) + (u \times w)$ or $(u + v) \times w = (u \times w) + (v \times w)$

5. If $u = sv$, then $u \times v = 0$.

6. $\|u\| = \|ai + bj + ck\| = \sqrt{a^2 + b^2 + c^2}$

7. $\|u + v\| \leq \|u\| + \|v\|$

8. i, j, k

9. $(x - a)^2 + (y - b)^2 + (z - c)^2 = r^2$

10. Draw a planar curve. Define a direction (generatrix). Move a line (directrix) along the curve parallel to the generatrix. The resulting surface is a cylinder.

11. The distance $\left| P_1 P_2 \right|$ between $P_1(x_1,\, y_1,\, z_1)$

    and $P_2(x_2,\, y_2,\, z_2)$ is

    $$\left| P_1 P_2 \right| = \sqrt{(x_2 - x_1)^2 + (y_2 - y_1)^2 + (z_2 - z_1)^2}$$

12. $\mathbf{u} = \dfrac{\mathbf{v}}{\|\mathbf{v}\|}$

13. $\|\mathbf{v} \times \mathbf{w}\|^2 = \|\mathbf{v}\|^2\|\mathbf{w}\|^2 - (\mathbf{v} \cdot \mathbf{w})^2$

14. The dot product of vectors

    $\mathbf{v} = a_1\mathbf{i} + a_2\mathbf{j} + a_3\mathbf{k}$ and $\mathbf{w} = b_1\mathbf{i} + b_2\mathbf{j} + b_3\mathbf{k}$

    is the scalar denoted by $\mathbf{v} \cdot \mathbf{w}$ and given by

    $$\mathbf{v} \cdot \mathbf{w} = a_1 b_1 + a_2 b_2 + a_3 b_3$$

    Alternately, $\mathbf{u} \cdot \mathbf{v} = \|\mathbf{u}\|\|\mathbf{v}\| \cos\theta$ where θ is

    the angle between the vectors $\mathbf{u}$ and $\mathbf{v}$.

15. If θ is the angle between the nonzero vectors $\mathbf{v}$

    and $\mathbf{w}$, then

    $$\cos\theta = \frac{\mathbf{v} \cdot \mathbf{w}}{\|\mathbf{v}\|\|\mathbf{w}\|}$$

16. $\mathbf{u}$ and $\mathbf{v}$ are orthogonal vectors if the lines

    determined by those vectors are

    perpendicular. Algebraically, $\mathbf{u} \cdot \mathbf{v} = 0$

17. The vector projection of $\mathbf{AB}$ onto $\mathbf{AC}$ is the

    vector from $\mathbf{A}$ to a point D on $\mathbf{AC}$ so $\overline{BD}$ is

    perpendicular to the line formed by $\mathbf{AC}$. The

    formula for the vector projection of $\mathbf{v}$ in the

    direction of $\mathbf{w}$ is the vector

$$\text{proj}_{\mathbf{w}}\mathbf{v} = \left(\frac{\mathbf{v} \cdot \mathbf{w}}{\mathbf{w} \cdot \mathbf{w}}\right)\mathbf{w}$$

18. The scalar projection of $\mathbf{AB}$ onto $\mathbf{AC}$ is the

    component of $\mathbf{v}$ in the direction of $\mathbf{w}$. It is

    the scalar

    $$\text{comp}_{\mathbf{w}}\mathbf{v} = \frac{\mathbf{v} \cdot \mathbf{w}}{\|\mathbf{w}\|}$$

19. An object moving with displacement $\mathbf{R}$ with a

    constant force $\mathbf{F}$ does work $W = \mathbf{F} \cdot \mathbf{R}$

20. If $\mathbf{v} = a_1\mathbf{i} + a_2\mathbf{j} + a_3\mathbf{k}$ and

    $\mathbf{w} = b_1\mathbf{i} + b_2\mathbf{j} + b_3\mathbf{k}$, the cross product,

    written $\mathbf{v} \times \mathbf{w}$, is the vector

    $$\mathbf{v} \times \mathbf{w} = (a_2 b_3 - a_3 b_2)\mathbf{i} + (a_3 b_1 - a_1 b_3)\mathbf{j} + (a_1 b_2 - a_2 b_1)\mathbf{k}$$

    These terms can be obtained by using a

    determinant

    $$\mathbf{v} \times \mathbf{w} = \begin{vmatrix} \mathbf{i} & \mathbf{j} & \mathbf{k} \\ a_1 & a_2 & a_3 \\ b_1 & b_2 & b_3 \end{vmatrix}$$

21. Place the little finger of your right hand along

    the positive x-axis and wrap your fingers

    around the positive y-axis. Then your thumb

    points in the direction of the positive z-axis.

22. **a.** If $\mathbf{v}$ and $\mathbf{w}$ are nonzero vectors in $\mathbb{R}^3$, that

    are not multiples of one another, then

$\mathbf{v} \times \mathbf{w}$ is orthogonal to both $\mathbf{v}$ and $\mathbf{w}$.

b. If $\mathbf{v}$ and $\mathbf{w}$ are nonzero vectors in $\mathbb{R}^3$ with

θ the angle between $\mathbf{v}$ and $\mathbf{w}$ ($0 \leq \theta \leq \pi$),

then $\| \mathbf{v} \times \mathbf{w} \| = \| \mathbf{v} \| \| \mathbf{w} \| \sin \theta$

23. If $\mathbf{u} = a_1\mathbf{i} + a_2\mathbf{j} + a_3\mathbf{k}$, $\mathbf{v} = b_1\mathbf{i} + b_2\mathbf{j} + b_3\mathbf{k}$,

and $\mathbf{w} = c_1\mathbf{i} + c_2\mathbf{j} + c_3\mathbf{k}$, then the triple

scalar product can be found by evaluating the

determinant

$$(\mathbf{u} \times \mathbf{v}) \cdot \mathbf{w} = \begin{vmatrix} a_1 & a_2 & a_3 \\ b_1 & b_2 & b_3 \\ c_1 & c_2 & c_3 \end{vmatrix}$$

24. The volume of a parallelepiped formed by $\mathbf{u}$,

$\mathbf{v}$, and $\mathbf{w}$ is the absolute value of the triple

scalar product, $|(\mathbf{u} \times \mathbf{v}) \cdot \mathbf{w}|$.

25. Suppose the line passes through the point

(x_0, y_0, z_0) and is parallel to

$\mathbf{v} = A\mathbf{i} + B\mathbf{j} + C\mathbf{k}$.

a. $x = x_0 + tA$, $y = y_0 + tB$, $z = z_0 + tC$

b. $\dfrac{x - x_0}{A} = \dfrac{y - y_0}{B} = \dfrac{z - z_0}{C}$

26. If $\mathbf{v} = a_1\mathbf{i} + a_2\mathbf{j} + a_3\mathbf{k}$ is a nonzero vector,

then the direction cosines of $\mathbf{v}$ are

$\cos \alpha = \dfrac{a_1}{\|\mathbf{v}\|}$; $\cos \beta = \dfrac{a_2}{\|\mathbf{v}\|}$; $\cos \gamma = \dfrac{a_3}{\|\mathbf{v}\|}$

27. The plane $Ax + By + Cz + D = 0$ has

normal vector $\mathbf{N} = A\mathbf{i} + B\mathbf{j} + C\mathbf{k}$

28. The plane containing the point $P_0(x_0, y_0, z_0)$

with normal vector $\mathbf{N} = A\mathbf{i} + B\mathbf{j} + C\mathbf{k}$ has

point-normal form

$A(x - x_0) + B(y - y_0) + C(z - z_0) = 0$

29. The standard form of a plane with normal

vector $\mathbf{N} = A\mathbf{i} + B\mathbf{j} + C\mathbf{k}$ is

$Ax + By + Cz + D = 0$

30. The distance from the point (x_0, y_0, z_0) to the

plane $Ax + By + Cz + D = 0$ is given by

$$d = \left| \frac{Ax_0 + By_0 + Cz_0 + D}{\sqrt{A^2 + B^2 + C^2}} \right|$$

31. The (shortest) distance from the point P to

the line L is given by the formula

$$d = \frac{\| \mathbf{v} \times \mathbf{QP} \|}{\| \mathbf{v} \|}$$

where $\mathbf{v}$ is a vector aligned with L and Q is

any point on L.

32. A quadric surface is the set of all (x, y, z) that

satisfy the equation

$Ax^2 + By^2 + Cz^2 + Dxy + Exz + Fyz + Gx$

$+ Hy + Iz + J = 0$ where A, B, C, D, E, F,

G, H, I, and J are constants.

33. $\mathbf{v} = 2\mathbf{i} - 3\mathbf{j} + \mathbf{k}$, $\mathbf{w} = 3\mathbf{i} - 2\mathbf{j}$

   a. $2\mathbf{v} + 3\mathbf{w} = [2(2) + 3(3)]\mathbf{i}$

   $+ [2(-3) + 3(-2)]\mathbf{j} + [2(1) + 3(0)]\mathbf{k}$

   $= 13\mathbf{i} - 12\mathbf{j} + 2\mathbf{k}$

   b. $\|\mathbf{v}\|^2 - \|\mathbf{w}\|^2 = [2^2 + (-3)^2 + 1^2]$

   $- [3^2 + (-2)^2] = 14 - 13 = 1$

   c. $\mathbf{v} \cdot \mathbf{w} = 2(3) + (-3)(-2) + 1(0) = 12$

   d. $\begin{vmatrix} \mathbf{i} & \mathbf{j} & \mathbf{k} \\ 2 & -3 & 1 \\ 3 & -2 & 0 \end{vmatrix} = 2\mathbf{i} + 3\mathbf{j} + 5\mathbf{k}$

   e. $\text{proj}_{\mathbf{u}}\mathbf{v} = \left(\dfrac{\mathbf{v} \cdot \mathbf{w}}{\mathbf{w} \cdot \mathbf{w}}\right)\mathbf{w}$

   $= \dfrac{12}{13}(3\mathbf{i} - 2\mathbf{j})$

   f. $\text{comp}_{\mathbf{v}}\mathbf{w} = \dfrac{\mathbf{w} \cdot \mathbf{v}}{\|\mathbf{v}\|}$

   $= \dfrac{12}{\sqrt{14}}$

   $= \dfrac{6\sqrt{14}}{7}$

34. $\mathbf{u} = 2\mathbf{i} - 3\mathbf{j} + \mathbf{k}$, $\mathbf{v} = \mathbf{i} + \mathbf{j} - 2\mathbf{k}$,

   $\mathbf{w} = 3\mathbf{i} + 5\mathbf{k}$

   a. $\mathbf{u} \times \mathbf{v} = \begin{vmatrix} \mathbf{i} & \mathbf{j} & \mathbf{k} \\ 2 & -3 & 1 \\ 1 & 1 & -2 \end{vmatrix}$

   $= 5\mathbf{i} + 5\mathbf{j} + 5\mathbf{k}$

$(\mathbf{u} \times \mathbf{v}) \cdot \mathbf{w} = (5\mathbf{i} + 5\mathbf{j} + 5\mathbf{k}) \cdot (3\mathbf{i} + 5\mathbf{k})$

$= 40$

b. Not possible to take the cross product of a scalar and a vector.

c. $\begin{vmatrix} \mathbf{i} & \mathbf{j} & \mathbf{k} \\ 2 & -3 & 1 \\ 1 & 1 & -2 \end{vmatrix} \times (3\mathbf{i} + 5\mathbf{k})$

$= (5\mathbf{i} + 5\mathbf{j} + 5\mathbf{k}) \times (3\mathbf{i} + 5\mathbf{k})$

$= \begin{vmatrix} \mathbf{i} & \mathbf{j} & \mathbf{k} \\ 5 & 5 & 5 \\ 3 & 0 & 5 \end{vmatrix}$

$= 25\mathbf{i} - 10\mathbf{j} - 15\mathbf{k}$

d. Not possible to take the dot product of a scalar and a vector.

35. $P(-1, 4, -3)$, $Q(0, -2, 1)$, so

   $\mathbf{PQ} = \langle 0 + 1, -2 - 4, 1 + 3 \rangle$

   $= \langle 1, -6, 4 \rangle$

   Using Q and the vector $\mathbf{PQ}$

   $\dfrac{x}{1} = \dfrac{y + 2}{-6} = \dfrac{z - 1}{4}$

36. Using point $P(1, 1, 3)$ and the direction numbers of $\mathbf{v} = 2\mathbf{i} + 3\mathbf{k}$, $[2, 0, 3]$:

   $2(x - 1) + 0(y - 1) + 3(z - 3) = 0$

   $2x + 3z - 11 = 0$

37. The normals to the plane

$2x + 3y + z = 2$ is $N_1 = 2i + 3j + k$

and to

$y - 3z = 5$ is $N_2 = j - 3k$

A vector parallel to the required line is

$$N_1 \times N_2 = \begin{vmatrix} i & j & k \\ 2 & 3 & 1 \\ 0 & 1 & -3 \end{vmatrix}$$

$$= \langle -10, 6, 2 \rangle$$

A point on the line of intersection of

$$\begin{cases} 2x + 3y + z = 2 \\ y - 3z = 5 \end{cases}$$

can be found by setting $z = 0$. Then

$$y = 5$$

so that

$$2x + 3(5) + 0 = 2$$

$$2x = -13$$

$$x = -\frac{13}{2}$$

Thus, using the point $P_0(-\frac{13}{2}, 5, 0)$ we find

the equation of the line in parametric form is:

$$x = -\frac{13}{2} - 10t, \ y = 5 + 6t, \ z = 2t$$

38. $P(0, 2, -1)$, $Q(1, -3, 5)$, and $R(3, 0, -2)$

We find

$$PQ = (1 - 0)i + (-3 - 2)j + (5 + 1)k$$

$$= i - 5j + 6k$$

$$PR = (3 - 0)i + (0 - 2)j + (-2 + 1)k$$

$$= 3i - 2j - k$$

We find the direction numbers of the line by

finding $PQ \times PR$.

$$PQ \times PR = \begin{vmatrix} i & j & k \\ 1 & -5 & 6 \\ 3 & -2 & -1 \end{vmatrix}$$

$$= \langle 17, 19, 13 \rangle$$

The equation of the plane is:

$$17x + 19(y - 2) + 13(z + 1) = 0$$

$$17x + 19y + 13z - 25 = 0$$

39. $\cos \alpha = \dfrac{-2}{\sqrt{(-2)^2 + 3^2 + 1^2}}$

$$= \dfrac{-2}{\sqrt{14}}$$

so $\alpha \approx 2.13$ or $122°$

$\cos \beta = \dfrac{3}{\sqrt{14}};$ so $\beta \approx 0.64$ or $37°$

$\cos \gamma = \dfrac{1}{\sqrt{14}};$ so $\gamma \approx 1.30$ or $74°$

40. a. The direction numbers are not scalar

multiples, so they are not parallel. From

the second equation, we see that $z = 3$,

which means that if the lines intersect

$t = \frac{3}{2}$ (from the first equation). In this

case, then $x = 0$ (from the first equation)

and $x = 7$ (from the second equation).

Thus, they are skew.

b. The direction numbers are not scalar multiples, so they are not parallel. To determine if they intersect we need to solve a system of equations.

1st line:

$$x = 7 + 5t_1, \ y = 6 + 4t_1, \ z = 8 + 5t_1$$

2nd line:

$$x = 8 + 6t_2, \ y = 6 + 4t_2, \ z = 9 + 6t_2$$

Solving these equations simultaneously, we find $t_1 = t_2 = -1$. Both equations contain the point $(2, 2, 3)$.

41. $\mathbf{u} = 2\mathbf{i} + \mathbf{j}$, $\mathbf{v} = \mathbf{i} - \mathbf{j} - \mathbf{k}$, and $\mathbf{w} = 3\mathbf{i} + \mathbf{k}$

a. $V = \left\| \begin{matrix} 2 & 1 & 0 \\ 1 & -1 & -1 \\ 3 & 0 & 1 \end{matrix} \right\|$

$\quad = 6$

(This is the absolute value of the determinant).

b. Volume of a tetrahedron is $\frac{1}{6}|(\mathbf{u} \times \mathbf{v}) \cdot \mathbf{w}|$

$$2 = \tfrac{1}{6} A^2 |(\mathbf{u} \times \mathbf{v}) \cdot \mathbf{w}|$$

$$12 = A^2 \left| \left| \begin{matrix} \mathbf{i} & \mathbf{j} & \mathbf{k} \\ 2 & 1 & 0 \\ 1 & -1 & -1 \end{matrix} \right| \cdot (3\mathbf{i} + \mathbf{k}) \right|$$

$$12 = A^2(6)$$

$$A^2 = 2$$

$$A = \sqrt{2}$$

42. Given $2x + 5y - z = 3$ with $P(-1, 1, 4)$

$$d = \left| \frac{Ax_0 + By_0 + Cz_0 + D}{\sqrt{A^2 + B^2 + C^2}} \right|$$

$$= \left| \frac{2(-1) + 5(1) + (-1)(4) - 3}{\sqrt{2^2 + 5^2 + (-1)^2}} \right|$$

$$= \frac{4}{\sqrt{30}}$$

$$= \frac{2\sqrt{30}}{15}$$

43. The first line contains $P_1(0, 0, -1)$ and is parallel to $\mathbf{v}_1 = \langle 1, 2, 3 \rangle$, and the second contains $P_2(1, 2, 0)$ and is parallel to $\mathbf{v}_2 = \langle -1, 1, 1 \rangle$. A normal to both lines is

$$\mathbf{N} = \mathbf{v}_1 \times \mathbf{v}_2$$

$$= \left| \begin{matrix} \mathbf{i} & \mathbf{j} & \mathbf{k} \\ 1 & 2 & 3 \\ -1 & 1 & 1 \end{matrix} \right|$$

$$= -\mathbf{i} - 4\mathbf{j} + 3\mathbf{k}$$

Then the distance between the lines is the

scalar projection of P_1P_2 on N:

$P_1P_2 = i + 2j + k$, so

$$d = \frac{|N \cdot P_1P_2|}{\|N\|}$$

$$= \frac{|\langle -1, -4, 3\rangle \cdot \langle 1, 2, 1\rangle|}{\sqrt{(-1)^2 + (-4)^2 + 3^2}}$$

$$= \frac{6}{\sqrt{26}}$$

44. The line contains the point $Q(2, 0, -1)$ and

is parallel to $v = 3i + 5j - k$. We have

$QP = 2i + 5j + k$ and

$$v \times QP = \begin{vmatrix} i & j & k \\ 3 & 5 & -1 \\ 2 & 5 & 1 \end{vmatrix}$$

$$= 10i - 5j + 5k$$

Thus, the distance from the point to the line

is

$$d = \frac{\|v \times QP\|}{\|v\|}$$

$$= \frac{\sqrt{100 + 25 + 25}}{\sqrt{9 + 25 + 1}}$$

$$= \frac{5\sqrt{6}}{\sqrt{35}}$$

45. The path of the airplane is represented by the

sum of the vector of the airplane,

$\langle 0, -200, 0\rangle$, and the vector of the wind,

$\langle 25\sqrt{2}, 25\sqrt{2}, 0\rangle$.

Path: $\langle 25\sqrt{2}, 25\sqrt{2} - 200, 0\rangle$. The

magnitude of this vector is the ground speed:

$$\sqrt{(25\sqrt{2})^2 + (25\sqrt{2} - 200)^2} \approx 168.4 \text{ mph}$$

46. The work performed on the sled is the

horizontal component of the force times the

displacement. In vector terms:

$$W = F \cdot PQ = \left\langle \frac{3\sqrt{3}}{2}, \frac{3}{2}\right\rangle \cdot \langle 50, 0\rangle$$

$$= 75\sqrt{3} \approx 130 \text{ ft} \cdot \text{lb}$$

47. a.

b. $\frac{y}{x} = \frac{2^{t+1}}{2^t} = 2$ or $y = 2x$, $x \geq 1$

48. a.

b. $\frac{x}{4} = t$, so $x = 4t$, $\frac{y}{-9} = t$, so $y = -9t$

49. **a.**

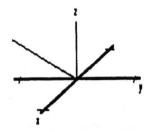

b. $x = 4t,\ y = -9t,\ z = t$

50. **a.** elliptic cone

b. hyperbolic paraboloid

c. hyperboloid of two sheets

d. hyperboloid of one sheet

e. plane

f. line

g. sphere

h. elliptic (circular) paraboloid

i. hyperboloid of two sheets

j. hyperboloid of two sheets

CHAPTER 10

Vector-Valued Functions

SURVIVAL HINT: *With this chapter, you begin a new calculus topic, one involving vectors. You might wish to review vectors in the plane and space as presented in Chapter 9; in particular, remember that cross product is not commutative. Generally speaking the operations on vector-valued functions are exactly what you would expect them to be. They are the same as the scalar functions applied to the component functions. The ideas presented in this section are essential to the ideas presented in Chapter 13.*

10.1 Introduction to Vector Functions, Pages 640-641

1. $t \neq 0$, because $1/t$ is not defined if $t = 0$.

3. $\sin t$ and $\cos t$ are defined for all t, but $\tan t$ is not defined for $\frac{\pi}{2}, \frac{3\pi}{2}$, and multiplies of those values, namely, $t \neq \frac{(2n + 1)\pi}{2}$, n an integer.

5. $\mathbf{F}$ is not defined whenever $\cos t$, $\sin t$, or $\tan t$ is equal to 0; $t = \cos^{-1} 0 = \frac{\pi}{2}$, $\sin^{-1} 0 = 0$, or $\tan^{-1} 0 = \frac{\pi}{2}$, and multiples of these values, namely, $t \neq \frac{n\pi}{2}$, n an integer.

7. $\ln t$ requires $t > 0$.

9. There is no **k**-component, so the graph is a cylinder parallel to the xy-plane where $x = 2t$, $y = t^2$. It can be described as a parabolic cylinder; in $\mathbf{R}^2$, the graph is a parabola in the xy-plane.

11. The is no **k**-component, so the graph is a cylinder parallel to the xy plane where $x = \sin t$, $y = -\cos t$; in $\mathbf{R}^2$, the graph is a circle in the xy-plane.

13. There is no **j**-component, and $z = -4$ (a constant), so the graph is a plane parallel to the xz plane where $x = t$; in $\mathbf{R}^2$, the graph is a line in the xz plane parallel to and four units below the x-axis.

15. This is a circular helix; in $\mathbf{R}^2$, the graph is a circle in the xy-plane.

17. The curve is in the intersection of the parabolic cylinder $y = (1 - x)^2$ with the plane $x + z = 1$; in $\mathbf{R}^2$, the graph is a parabola in the xy-plane.

19. The curve is the intersection of the cylinder $y = x^2 + 1$ and the plane $y = z + 1$; in $\mathbf{R}^2$, the graph is a parabola in the xy-plane.

21. $2\mathbf{F}(t) - 3\mathbf{G}(t)$
    $= [2(2t) - 3(1 - t)]\mathbf{i}$
    $\qquad + [2(-5) - 3(0)]\mathbf{j} + [2(t^2) - 3(\frac{1}{t})]\mathbf{k}$
    $= (7t - 3)\mathbf{i} - 10\mathbf{j} + (2t^2 - \frac{3}{t})\mathbf{k}$

23. $\mathbf{F}(t) \cdot \mathbf{G}(t)$

$= (2t\mathbf{i} - 5\mathbf{j} + t^2\mathbf{k}) \cdot [(1 - t)\mathbf{i} + (\frac{1}{t})\mathbf{k}]$

$= (2t - 2t^2) + t$

$= 3t - 2t^2$

25. $\mathbf{G}(t) \cdot \mathbf{H}(t) = [(1 - t)\mathbf{i} + (\frac{1}{t})\mathbf{k}] \cdot (\sin t\,\mathbf{i} + e^t\mathbf{j})$

$= (1 - t)\sin t$

27. $\mathbf{F}(t) \times \mathbf{H}(t) = \begin{vmatrix} \mathbf{i} & \mathbf{j} & \mathbf{k} \\ 2t & -5 & t^2 \\ \sin t & e^t & 0 \end{vmatrix}$

$= -t^2 e^t\mathbf{i} + (t^2\sin t)\mathbf{j} + (2te^t + 5\sin t)\mathbf{k}$

29. $2e^t\mathbf{F}(t) + t\mathbf{G}(t) + 10\mathbf{H}(t)$

$= [2e^t(2t) + t(1 - t) + 10\sin t]\mathbf{i}$

$\quad + [2e^t(-5) + 10e^t]\mathbf{j} + [2e^t t^2 + 1]\mathbf{k}$

$= (4te^t - t^2 + t + 10\sin t)\mathbf{i} + (2e^t t^2 + 1)\mathbf{k}$

31. $\mathbf{G}(t) \cdot [\mathbf{H}(t) \times \mathbf{F}(t)] = \begin{vmatrix} 1-t & 0 & \frac{1}{t} \\ \sin t & e^t & 0 \\ 2t & -5 & t^2 \end{vmatrix}$

$= (1 - t)(e^t t^2) - 0 + \frac{1}{t}(-5\sin t - 2te^t)$

$= t^2 e^t - t^3 e^t - \frac{5}{t}\sin t - 2e^t$

33. A vector-valued function $\mathbf{F}$ with domain D assigns a unique vector $\mathbf{F}(t)$ to each scalar t in the set D. The set of all vectors $\mathbf{v}$ of the form $\mathbf{v} = \mathbf{F}(t)$ for t in D is the range of $\mathbf{F}$. In this text we consider vector functions whose range

are vectors in $\mathbf{R}^2$ or $\mathbf{R}^3$; that is,

$\mathbf{F}(t) = f_1(t)\mathbf{i} + f_2(t)\mathbf{j}$ in $\mathbf{R}^2$ (plane), or

$\mathbf{F}(t) = f_1(t)\mathbf{i} + f_2(t)\mathbf{j} + f_3(t)\mathbf{k}$ in $\mathbf{R}^3$ (space)

where f_1, f_2, and f_3 are real-valued (scalar-valued) functions of the real number t defined on the domain set D. In this context f_1, f_2, and f_3 are called the components of $\mathbf{F}$.

35. $x = 2\sin t$, $y = 2\sin t$, $z = \sqrt{8}\cos t$.

$x^2 + y^2 + z^2 = 4\sin^2 t + 4\sin^2 t + 8\cos^2 t$

$= 8(\sin^2 t + \cos^2 t)$

$= 8$

This is a sphere with center at the origin and a radius of $2\sqrt{2}$.

37. Let t be the variable that is squared, so that $x(t) = t$, $y(t) = t^2$ and $z = 2$; we have

$\mathbf{F}(t) = t\mathbf{i} + t^2\mathbf{j} + 2\mathbf{k}$

39. $\mathbf{F}(t) = 2t\mathbf{i} + (1 - t)\mathbf{j} + (\sin t)\mathbf{k}$

41. Let t be the variable that is squared, so that $x(t) = t^2$, $y(t) = t$, and then

$z = \sqrt{9 - (t^2)^2 - (t)^2}$; we have

$\mathbf{F}(t) = t^2\mathbf{i} + t\mathbf{j} + \sqrt{9 - t^2 - t^4}\,\mathbf{k}$

43. a. $x^2 + y^2 = (\sin t)^2 + (\cos t)^2$

$= 1$

$\neq z$

No

b. $x^2 + y^2 = (e^{-t}\cos t)^2 + (e^{-t}\sin t)^2$

$= e^{-2t}(\cos^2 t + \sin^2 t)$

$= e^{-2t}$

$= z$

Yes

c. $x^2 + y^2 = \left(1 + \frac{1}{t}\right)^2 + \left(1 - \frac{1}{t}\right)^2$

$= 1 + \frac{2}{t} + \frac{1}{t^2} + 1 - \frac{2}{t} + \frac{1}{t^2}$

$= 2 + \frac{2}{t^2}$

$= z$

Yes

45. $\lim_{t \to 1}\left[2t\mathbf{i} - 3\mathbf{j} + e^t\mathbf{k}\right] = 2\mathbf{i} - 3\mathbf{j} + e\mathbf{k}$

47. $\lim_{t \to 0}\left[\frac{(\sin t)\mathbf{i} - t\mathbf{k}}{t^2 + t - 1}\right] = \frac{\lim_{t \to 0}[(\sin t)\mathbf{i} - t\mathbf{k}]}{\lim_{t \to 0}(t^2 + t - 1)}$

$= \frac{0}{-1}$

$= 0$

49. $\lim_{t \to 0}\left[\frac{te^t}{1 - e^t}\mathbf{i} + \frac{e^t - 1}{\cos t}\mathbf{j}\right]$

$= \left[\lim_{t \to 0}\frac{te^t + e^t}{-e^t}\right]\mathbf{i} + \left[\lim_{t \to 0}\frac{e^t - 1}{\cos t}\right]\mathbf{j}$

l'Hôpital's rule

$= \left[\lim_{t \to 0} -t - 1\right]\mathbf{i} + e^{-1}\mathbf{j}$

$= -\mathbf{i} + e^{-1}\mathbf{j}$

51. $\lim_{t \to 0^+}\left[\frac{\sin 3t}{\sin 2t}\mathbf{i} + \frac{\ln(\sin t)}{\ln(\tan t)}\mathbf{j} + \left(\frac{\ln t}{1/t}\right)\mathbf{k}\right]$

l'Hôpital's rule

$= \lim_{t \to 0^+}\left[\left(\frac{3\cos 3t}{2\cos 2t}\right)\mathbf{i} + \left(\frac{\cos t \tan t}{\sin t \sec^2 t}\right)\mathbf{j} + \left(\frac{1/t}{-1/t^2}\right)\mathbf{k}\right]$

$= \frac{3}{2}\mathbf{i} + \mathbf{j}$

53. Since t, 3, and $-(1 - t)$ are all continuous for all t, we say $\mathbf{F}$ is continuous for all t

55. Since $\frac{1}{t^2 + t}$ and $\frac{2}{t^2 + t}$ are both continuous for all t except 0 and -1, we say $\mathbf{G}$ is continuous for $t \neq 0$, $t \neq -1$.

57. Since te^t is continuous for all t, $\frac{e^t}{t}$ is continuous for all t except 0, and $3e^t$ is continuous for all t, we say $\mathbf{F}$ is continuous for all $t \neq 0$.

59. We show $\mathbf{R}(t) = \langle t, \frac{1 - t}{t}, \frac{1 - t^2}{t}\rangle$ lies on a plane by finding constants A, B, C, and D so that $Ax + By + Cz = D$. Thus,

$$At + B\left(\frac{1-t}{t}\right) + C\left(\frac{1-t^2}{t}\right) = D$$

$$(-B-D) + (A-C)t + (B+C)\left(\frac{1}{t}\right) = 0$$

Solving the system of equations

$$\begin{cases} -B - D = 0 \\ A - C = 0 \\ B + C = 0 \end{cases}$$

we obtain $A = C = -B = D$. Choosing

$A = 1$, $B = -1$, $C = 1$, and $D = 1$, we

conclude that $\mathbf{R}(t)$ lies on the plane

$x - y + z = 1$.

61. **a.** $\displaystyle\lim_{t \to 0} e^t \mathbf{F}(t) = \lim_{t \to 0} e^t(t\mathbf{i} + t^2\mathbf{j} + t^3\mathbf{k}) = \mathbf{0}$

$$\left[\lim_{t \to 0} e^t\right]\left[\lim_{t \to 0} \mathbf{F}(t)\right] = (1)(0\mathbf{i} + 0\mathbf{j} + 0\mathbf{k}) = \mathbf{0}$$

b. $\displaystyle\lim_{t \to 1} \mathbf{F}(t) \cdot \mathbf{G}(t)$

$$= \lim_{t \to 1}[(t\mathbf{i} + t^2\mathbf{j} + t^3\mathbf{k}) \cdot (\tfrac{1}{t}\mathbf{i} - e^t\mathbf{j})]$$

$$= \lim_{t \to 1}(1 - t^2 e^t)$$

$$= 1 - e$$

$$\left[\lim_{t \to 1} \mathbf{F}(t)\right] \cdot \left[\lim_{t \to 1} \mathbf{G}(t)\right]$$

$$= \left[\lim_{t \to 1}(t\mathbf{i} + t^2\mathbf{j} + t^3\mathbf{k})\right] \cdot \left[\lim_{t \to 1}(\tfrac{1}{t}\mathbf{i} - e^t\mathbf{j})\right]$$

$$= (\mathbf{i} + \mathbf{j} + \mathbf{k}) \cdot (\mathbf{i} - e\mathbf{j})$$

$$= 1 - e$$

c. $\displaystyle\lim_{t \to 1}[\mathbf{F}(t) \times \mathbf{G}(t)]$

$$= \lim_{t \to 1} \begin{vmatrix} \mathbf{i} & \mathbf{j} & \mathbf{k} \\ t & t^2 & t^3 \\ \frac{1}{t} & -e^t & 0 \end{vmatrix}$$

$$= \lim_{t \to 1}[t^3 e^t \mathbf{i} + t^2 \mathbf{j} + (-te^t - t)\mathbf{k}]$$

$$= e\mathbf{i} + \mathbf{j} + (-e - 1)\mathbf{k}$$

$$\left[\lim_{t \to 1} \mathbf{F}(t)\right] \times \left[\lim_{t \to 1} \mathbf{G}(t)\right]$$

$$= \begin{vmatrix} \mathbf{i} & \mathbf{j} & \mathbf{k} \\ 1 & 1 & 1 \\ 1 & -e & 0 \end{vmatrix}$$

$$= e\mathbf{i} + \mathbf{j} + (-e - 1)\mathbf{k}$$

63. $\displaystyle\lim_{t \to t_0}[\mathbf{F}(t) + \mathbf{G}(t)]$

$$= \lim_{t \to t_0}[f_1(t)\mathbf{i} + f_2(t)\mathbf{j} + f_3(t)\mathbf{k}$$

$$+ g_1(t)\mathbf{i} + g_2(t)\mathbf{j} + g_3(t)\mathbf{k}]$$

$$= \lim_{t \to t_0}[f_1(t)\mathbf{i} + f_2(t)\mathbf{j} + f_3(t)\mathbf{k}]$$

$$+ \lim_{t \to t_0}[g_1(t)\mathbf{i} + g_2(t)\mathbf{j} + g_3(t)\mathbf{k}]$$

$$= \lim_{t \to t_0} \mathbf{F}(t) + \lim_{t \to t_0} \mathbf{G}(t)$$

65. **a.** $3\mathbf{F}(t_0) + 5\mathbf{G}(t_0)$ exists and

$$\lim_{t \to t_0}[3\mathbf{F}(t) + 5\mathbf{G}(t)]$$

$$= 3 \lim_{t \to t_0} \mathbf{F}(t) + 5 \lim_{t \to t_0} \mathbf{G}(t)$$

$$= 3\mathbf{F}(t_0) + 5\mathbf{G}(t_0)$$

The function is continuous at t_0.

b. $\mathbf{F}(t_0) \cdot \mathbf{G}(t_0)$ exists and

$$\lim_{t \to t_0} [\mathbf{F}(t_0) \cdot \mathbf{G}(t_0)]$$

$$= \left[\lim_{t \to t_0} \mathbf{F}(t) \right] \cdot \left[\lim_{t \to t_0} \mathbf{G}(t_0) \right]$$

$$= \mathbf{F}(t_0) \cdot \mathbf{G}(t_0)$$

The function is continuous at t_0.

c. Let $h(t) = t$; for $t_0 = 0$;

$(h(t_0))^{-1} \mathbf{F}(t_0)$ is not continuous.

d. $\mathbf{F}(t_0) \times \mathbf{G}(t_0)$ exists and

$$\lim_{t \to t_0} [\mathbf{F}(t) \times \mathbf{G}(t)]$$

$$= \left[\lim_{t \to t_0} \mathbf{F}(t) \right] \times \left[\lim_{t \to t_0} \mathbf{G}(t) \right]$$

$$= \mathbf{F}(t_0) \times \mathbf{G}(t_0)$$

The function is continuous at t_0.

10.2 Differentiation and Integration of Vector Functions, Page 650-651

SURVIVAL HINT: *The rules for the derivatives of the product (scalar, dot, or cross) of vector functions are easy to remember because they all follow the same pattern as that of two real-valued functions: the product rule. Once again, we remind you to be careful with the cross product, as it is not commutative.*

1. $\mathbf{F}(t) = t\mathbf{i} + t^2\mathbf{j} + (t + t^3)\mathbf{k}$

$\mathbf{F}'(t) = \mathbf{i} + 2t\mathbf{j} + (1 + 3t^2)\mathbf{k}$

3. $\mathbf{F}(s) = (\ln s)[s\mathbf{i} + 5\mathbf{j} - e^s\mathbf{k}]$

We use the product rule:

$\mathbf{F}'(s) = \frac{1}{s}[s\mathbf{i} + 5\mathbf{j} - e^s\mathbf{k}] + (\ln s)[\mathbf{i} - e^s\mathbf{k}]$

$$= (1 + \ln s)\mathbf{i} + 5s^{-1}\mathbf{j} - e^s(\ln s + s^{-1})\mathbf{k}$$

5. $\mathbf{F}(t) = t^2\mathbf{i} + t^{-1}\mathbf{j} + e^{2t}\mathbf{k}$

$\mathbf{F}'(t) = 2t\mathbf{i} - t^{-2}\mathbf{j} + 2e^{2t}\mathbf{k}$

$\mathbf{F}''(t) = 2\mathbf{i} + 2t^{-3}\mathbf{j} + 4e^{2t}\mathbf{k}$

7. $\mathbf{F}(s) = (\sin s)\mathbf{i} + (\cos s)\mathbf{j} + s^2\mathbf{k}$

$\mathbf{F}'(s) = (\cos s)\mathbf{i} - (\sin s)\mathbf{j} + 2s\mathbf{k}$

$\mathbf{F}''(s) = (-\sin s)\mathbf{i} - (\cos s)\mathbf{j} + 2\mathbf{k}$

9. $f(x) = [x\mathbf{i} + (x + 1)\mathbf{j})] \cdot [(2x)\mathbf{i} - (3x^2)\mathbf{j}]$

$$= 2x^2 + (x + 1)(-3x^2)$$

$$= -3x^3 - x^2$$

$$f'(x) = -9x^2 - 2x$$

11. $g(x) = \| \langle \sin x, -2x, \cos x \rangle \|$

$$= \sqrt{(\sin x)^2 + (-2x)^2 + (\cos x)^2}$$

$$= \sqrt{1 + 4x^2}$$

$$g'(x) = \frac{4x}{\sqrt{1 + 4x^2}}$$

13. $\mathbf{R}(t) = t\mathbf{i} + t^2\mathbf{j} + 2t\mathbf{k}$

$$\mathbf{V}(t) = \mathbf{R}'(t)$$

$$= \mathbf{i} + 2t\mathbf{j} + 2\mathbf{k}$$

$$\mathbf{V}(1) = \mathbf{i} + 2\mathbf{j} + 2\mathbf{k}$$

$$\mathbf{A}(t) = \mathbf{V}'(t)$$

$$= \mathbf{R}''(t) = 2\mathbf{j}$$

$$\mathbf{A}(1) = 2\mathbf{j}$$

$$\text{Speed} = \|\mathbf{V}(1)\|$$

$$= \sqrt{1^2 + 2^2 + 2^2}$$

$$= 3$$

Direction of motion is that of the unit vector

$$\frac{\mathbf{V}}{\|\mathbf{V}\|} = \tfrac{1}{3}\mathbf{i} + \tfrac{2}{3}\mathbf{j} + \tfrac{2}{3}\mathbf{k}$$

15. $\mathbf{R}(t) = (\cos t)\mathbf{i} + (\sin t)\mathbf{j} + 3t\mathbf{k}$

$$\mathbf{V}(t) = \mathbf{R}'(t)$$

$$= -\sin t\mathbf{i} + \cos t\mathbf{j} + 3\mathbf{k}$$

$$\mathbf{V}(\tfrac{\pi}{4}) = -\frac{\sqrt{2}}{2}\mathbf{i} + \frac{\sqrt{2}}{2}\mathbf{j} + 3\mathbf{k}$$

$$\mathbf{A}(t) = -\cos t\mathbf{i} - \sin t\mathbf{j}$$

$$\mathbf{A}(\tfrac{\pi}{4}) = -\frac{\sqrt{2}}{2}\mathbf{i} - \frac{\sqrt{2}}{2}\mathbf{j}$$

$$\text{Speed} = \left\|\mathbf{V}(\tfrac{\pi}{4})\right\|$$

$$= \sqrt{\tfrac{2}{4} + \tfrac{2}{4} + 9} = \sqrt{10}$$

Direction of motion is that of the unit vector

$$\frac{\mathbf{V}}{\|\mathbf{V}\|} = -\frac{1}{2\sqrt{5}}\mathbf{i} + \frac{1}{2\sqrt{5}}\mathbf{j} + \frac{3}{\sqrt{10}}\mathbf{k}$$

17. $\mathbf{R}(t) = e^t\mathbf{i} + e^{-t}\mathbf{j} + e^{2t}\mathbf{k}$

$$\mathbf{V}(t) = e^t\mathbf{i} - e^{-t}\mathbf{j} + 2e^{2t}\mathbf{k}$$

$$\mathbf{V}(\ln 2) = 2\mathbf{i} - \tfrac{1}{2}\mathbf{j} + 8\mathbf{k}$$

$$\mathbf{A}(t) = e^t\mathbf{i} + e^{-t}\mathbf{j} + 4e^{2t}\mathbf{k}$$

$$\mathbf{A}(\ln 2) = 2\mathbf{i} + \tfrac{1}{2}\mathbf{j} + 16\mathbf{k}$$

$$\text{Speed} = \|\mathbf{V}(\ln 2)\|$$

$$= \sqrt{2^2 + (\tfrac{1}{2})^2 + 8^2}$$

$$= \frac{\sqrt{273}}{2}$$

Direction of motion is that of the unit vector

$$\frac{\mathbf{V}}{\|\mathbf{V}\|} = \frac{1}{\sqrt{273}}(4\mathbf{i} - \mathbf{j} + 16\mathbf{k})$$

19. $\mathbf{F}(t) = t^2\mathbf{i} + 2t\mathbf{j} + (t^3 + t^2)\mathbf{k}$

$$\mathbf{F}'(t) = 2t\mathbf{i} + 2\mathbf{j} + (3t^2 + 2t)\mathbf{k}$$

$$\mathbf{F}'(0) = 2\mathbf{j}$$

$$\mathbf{F}'(1) = 2\mathbf{i} + 2\mathbf{j} + 5\mathbf{k};$$

$$\mathbf{F}'(-1) = -2\mathbf{i} + 2\mathbf{j} + \mathbf{k}$$

21. $\mathbf{F}(t) = t^2\mathbf{i} + (\cos t)\mathbf{j} + (t^2\cos t)\mathbf{k}$

$$\mathbf{F}'(t) = 2t\mathbf{i} - \sin t\mathbf{j} + (-t^2\sin t + 2t\cos t)\mathbf{k}$$

$$\mathbf{F}'(0) = \mathbf{0};$$

$$\mathbf{F}'(\tfrac{\pi}{2}) = \pi\mathbf{i} - \mathbf{j} - \frac{\pi^2}{4}\mathbf{k}$$

23. $F(t) = \langle t^{-3}, t^{-2}, t^{-1} \rangle$

The point corresponding to $t_0 = -1$ is

$P_0 = \langle -1, 1, -1 \rangle$. The required line is

parallel to $F'(t) = \langle \frac{-3}{t^4}, \frac{-2}{t^3}, \frac{-1}{t^2} \rangle$ evaluated

at $t_0 = -1$; that is, parallel to

$F'(-1) = \langle -3, 2, -1 \rangle$. Thus, the line has

parametric equations

$x + 1 = -3t, \ y - 1 = 2t, \ z + 1 = -t$, or

$x = -1 - 3t, \ y = 1 + 2t, \ z = -1 - t$

25. $\int (t\mathbf{i} - e^{3t}\mathbf{j} + 3\mathbf{k}) \, dt = \frac{t^2}{2}\mathbf{i} - \frac{e^{3t}}{3}\mathbf{j} + 3t\mathbf{k} + \mathbf{C}$

27. $\int (\ln t \, \mathbf{i} - t\mathbf{j} + 3\mathbf{k}) \, dt$

$= (t \ln t - t)\mathbf{i} - \frac{1}{2}t^2\mathbf{j} + 3t\mathbf{k} + \mathbf{C}$

29. $\int [t \ln t \, \mathbf{i} - \sin(1 - t)\mathbf{j} + t\mathbf{k}] \, dt$

$= \frac{t^2}{2}(\ln t - \frac{1}{2})\mathbf{i} - \cos(1 - t)\mathbf{j} + \frac{t^2}{2}\mathbf{k} + \mathbf{C}$

31. $V(t) = t^2\mathbf{i} - e^{2t}\mathbf{j} + \sqrt{t}\mathbf{j}$

$R(t) = \int V(t) \, dt$

$= \int \langle t^2, -e^{2t}, \sqrt{t} \rangle \, dt$

$= \langle \frac{1}{3}t^3, -\frac{1}{2}e^{2t}, \frac{2}{3}t^{3/2} \rangle + \mathbf{C}$

$R(0) = \langle 0, -\frac{1}{2}, 0 \rangle + \langle c_1, c_2, c_3 \rangle$

$= \langle 1, 4, -1 \rangle$, so $\mathbf{C} = \langle 1, \frac{9}{2}, -1 \rangle$

$R(t) = (\frac{1}{3}t^3 + 1)\mathbf{i} + (-\frac{1}{2}e^{2t} + \frac{9}{2})\mathbf{j} + (\frac{2}{3}t^{3/2} - 1)\mathbf{k}$

33. $V(t) = 2\sqrt{t}\,\mathbf{i} + (\cos t)\mathbf{j}$

$R(t) = \int V(t) \, dt$

$= \int \langle 2t^{1/2}, \cos t, 0 \rangle \, dt$

$= \langle \frac{4}{3}t^{3/2}, \sin t, 0 \rangle + \mathbf{C}$

$R(0) = \langle 0, 0, 0 \rangle + \langle c_1, c_2, c_3 \rangle = \langle 1, 1, 0 \rangle$, so

$\mathbf{C} = \langle 1, 1, 0 \rangle$

$R(t) = (\frac{4}{3}t^{3/2} + 1)\mathbf{i} + (\sin t + 1)\mathbf{j}$

35. $A(t) = (\cos t)\mathbf{i} - (t \sin t)\mathbf{k}$

$V(t) = \int A(t) \, dt$

$= \int \langle \cos t, 0, -t\sin t \rangle \, dt$

$= \langle \sin t, 0, t\cos t - \sin t \rangle + \mathbf{C}_1$

$V(0) = \langle 0, 0, 0 \rangle + \mathbf{C}_1 = \langle 2, 0, 3 \rangle; \ \mathbf{C}_1 = \langle 2, 0, 3 \rangle$

$V(t) = (2 + \sin t)\mathbf{i} + (3 + t\cos t - \sin t)\mathbf{k}$

$R(t) = \int V(t) \, dt$

$= \int \langle \sin t + 2, 0, 3 + t\cos t - \sin t \rangle \, dt$

$$= \langle 2t - \cos t,\ 0,\ 3t + 2\cos t + t\sin t \rangle + \mathbf{C}_2$$

$$\mathbf{R}(0) = \langle -1,\ 0,\ 2 \rangle + \mathbf{C}_2 = \langle 1,\ -2,\ 1 \rangle$$

$$\mathbf{C}_2 = \langle 2,\ -2,\ -1 \rangle$$

$$\mathbf{R}(t) = \langle 2 + 2t - \cos t,\ -2,\ -1 + 3t + 2\cos t + t\sin t \rangle$$

37. $\mathbf{V}(t) = e^t\mathbf{i} + t^2\mathbf{j}$;

$$\mathbf{R}(t) = \int \mathbf{V}(t)\ dt$$

$$= \int \langle e^t,\ t^2 \rangle\ dt$$

$$= \langle e^t,\ \tfrac{t^3}{3} \rangle + \mathbf{C};$$

$$\mathbf{R}(0) = \langle 1,\ 0 \rangle + \mathbf{C} = \langle 1,\ -1 \rangle$$

Thus, $\mathbf{C} = \langle 0,\ -1 \rangle$. so

$$\mathbf{R}(t) = e^t\mathbf{i} + (\tfrac{1}{3}t^3 - 1)\mathbf{j}$$

39. $\mathbf{A}(t) = \langle 24t^2,\ 4 \rangle$

$$\mathbf{V}(t) = \int \mathbf{A}(t)\ dt$$

$$= \langle 8t^3,\ 4t \rangle + \mathbf{C}_1$$

$$\mathbf{V}(0) = \langle 0,\ 0 \rangle + \mathbf{C}_1 = \langle 0,\ 0 \rangle;\ \mathbf{C}_1 = \langle 0,\ 0 \rangle$$

$$\mathbf{R}(t) = \int \mathbf{V}(t)\ dt$$

$$= \langle 2t^4,\ 2t^2 \rangle + \mathbf{C}_2$$

$$\mathbf{R}(0) = \langle 0,\ 0 \rangle + \mathbf{C}_2 = \langle 1,\ 2 \rangle;\ \mathbf{C}_2 = \langle 1,\ 2 \rangle$$

$$\mathbf{R}(t) = \langle 2t^4 + 1,\ 2t^2 + 2 \rangle$$

41. Velocity is the time derivative of position $\mathbf{R}(t)$, and speed is the magnitude of the velocity. Thus, $\mathbf{V} = \mathbf{R}'(t)$ and speed $\|\mathbf{V}\|$.

43. $\mathbf{v} = 2\mathbf{i} - \mathbf{j} + 5\mathbf{k},\ \mathbf{w} = \mathbf{i} + 2\mathbf{j} - 3\mathbf{k}$;

$$t^4\mathbf{w} = t^4\mathbf{i} + 2t^4\mathbf{j} - 3t^4\mathbf{k}$$

Let

$$f(t) = \mathbf{v} \cdot t^4\mathbf{w}$$

$$= 2t^4 - 2t^4 - 15t^4$$

$$= -15t^4$$

$$\frac{d}{dt}f(t) = -60t^3$$

$$\frac{d^2}{dt^2}f(t) = -180t^2$$

45. $\mathbf{v} = 2\mathbf{i} - \mathbf{j} + 5\mathbf{k},\ \mathbf{w} = \mathbf{i} + 2\mathbf{j} - 3\mathbf{k}$;

$$\mathbf{v} \times \mathbf{w} = \begin{vmatrix} \mathbf{i} & \mathbf{j} & \mathbf{k} \\ 2 & -1 & 5 \\ 1 & 2 & -3 \end{vmatrix}$$

$$= -7\mathbf{i} + 11\mathbf{j} + 5\mathbf{k}$$

Let

$$f(t) = t\mathbf{v} \times t^2\mathbf{w}$$

$$= t^3\mathbf{v} \times \mathbf{w}$$

$$= -7t^3\mathbf{i} + 11t^3\mathbf{j} + 5t^3\mathbf{k}$$

$$\frac{d}{dt}f(t) = -21t^2\mathbf{i} + 33t^2\mathbf{j} + 15t^2\mathbf{k}$$

$$= 3t^2(-7\mathbf{i} + 11\mathbf{j} + 5\mathbf{k})$$

47. $F(t) = (3 + t^2)i + (-\cos 3t)j + t^{-1}k$

$G(t) = \sin(2 - t)i - e^{2t}k$

$(F \cdot G)'(t) = [(3 + t^2)\sin(2 - t) - t^{-1}e^{2t}]'$

$\qquad = -(3 + t^2)\cos(2 - t)$

$\qquad\quad + 2t\sin(2 - t) - 2t^{-1}e^{2t} + t^{-2}e^{2t}$

Also,

$F'(t) = 2ti + 3\sin 3tj - t^{-2}k;$

$G'(t) = -\cos(2 - t)i - 2e^{2t}k$

and

$(F' \cdot G)(t) + (F \cdot G')(t)$

$\quad = [2ti + 3\sin 3tj - t^{-2}k] \cdot [\sin(2 - t)i - e^{2t}k)]$

$\qquad + [(3 + t^2)i - \cos 3tj$

$\qquad\quad + t^{-1}k] \cdot [-\cos(2 - t)i - 2e^{2t}k]$

$\quad = 2t\sin(2 - t) + t^{-2}e^{2t}$

$\qquad - (3 + t^2)\cos(2 - t) - 2t^{-1}e^{2t}$

Therefore, $(F \cdot G)'(t) = (F' \cdot G)(t) + (F \cdot G')(t)$

49. $\displaystyle\int_0^{2a} [\cos t\, i + \sin t\, j + \sin t \cos t\, k]\, dt$

$\qquad = \left[\sin t\, i - \cos t\, j + (\tfrac{1}{2}\sin^2 t)k\right]\Big|_0^{2a}$

$\qquad = (\sin 2a)i + (1 - \cos 2a)j + (\tfrac{1}{2}\sin^2 2a)k$

Solve $(\sin 2a)i + (1 - \cos 2a)j + (\tfrac{1}{2}\sin^2 2a)k$

$\qquad = i + j + \tfrac{1}{2}k$

to find $a = \dfrac{\pi}{4}$

51. $F(t) = (\cos kt)i + (\sin kt)j$

$F'(t) = k[-\sin(kt)i + \cos(kt)j]$

$F''(t) = -k^2[\cos(kt)i + \sin(kt)j]$

$\qquad = -k^2 F(t)$

Thus, $F''(t)$ is parallel to $F(t)$ because $F''(t)$ is a scalar multiple of $F(t)$.

53. $F(t) = \langle e^t - t,\ t^{3/2},\ \cos t \rangle$

$F'(t) = \langle e^t - 1,\ \tfrac{3}{2}t^{1/2},\ -\sin t \rangle$

$F'(t) = 0$ when $t = 0$, so $F(t)$ is not smooth

on $[-1, 2]$, since 0 is in this interval.

55. Let $F(t) = u(t)i + v(t)j + w(t)k$

Then

$[h(t)F(t)]' = [h(t)u(t)]'i + [h(t)v(t)]'j + [h(t)w(t)]'k$

$\qquad = [h(t)u'(t) + h'(t)u(t)]i + [h(t)v'(t)$

$\qquad\qquad + h'(t)v(t)]j + [h(t)w'(t) + h(t)w'(t)]k$

$\qquad = [h(t)u'(t)i + h(t)v'(t)j + h(t)w'(t)k]$

$\qquad\qquad + [h'(t)u(t)i + h'(t)v(t)j + h'(t)w(t)k]$

$\qquad = h(t)F'(t) + h'(t)F(t)$

57. Let $\mathbf{F}(t) = u_1(t)\mathbf{i} + v_1(t)\mathbf{j} + w_1(t)\mathbf{k}$

and $\mathbf{G}(t) = u_2(t)\mathbf{i} + v_2(t)\mathbf{j} + w_2(t)\mathbf{k}$

$$[\mathbf{F} \times \mathbf{G}]'(t) = \frac{d}{dt}\begin{vmatrix} \mathbf{i} & \mathbf{j} & \mathbf{k} \\ u_1(t) & v_1(t) & w_1(t) \\ u_2(t) & v_2(t) & w_2(t) \end{vmatrix}$$

$$= [v_1(t)w_2(t) - v_2(t)w_1(t)]'\mathbf{i}$$

$$- [u_1(t)w_2(t) - u_2(t)w_1(t)]'\mathbf{j}$$

$$+ [u_1(t)v_2(t) - u_2(t)v_1(t)]'\mathbf{k}$$

$$= [v_1{}'(t)w_2(t) + v_1(t)w_2{}'(t)$$

$$- v_2(t)w_1{}'(t) - v_2{}'(t)w_1(t)]\mathbf{i}$$

$$- [u_1{}'(t)w_2(t) + u_1(t)w_2{}'(t)$$

$$- u_2(t)w_1{}'(t) - u_2{}'(t)w_1(t)]\mathbf{j}$$

$$+ [u_1{}'(t)v_2(t) + u_1(t)v_2{}'(t)$$

$$- u_2(t)v_1{}'(t) - u_2{}'(t)v_1(t)]\mathbf{k}$$

Also,

$$(\mathbf{F}' \times \mathbf{G})(t) + (\mathbf{F} \times \mathbf{G}')(t)$$

$$= \begin{vmatrix} \mathbf{i} & \mathbf{j} & \mathbf{k} \\ u_1{}'(t) & v_1{}'(t) & w_1{}'(t) \\ u_2(t) & v_2(t) & w_2(t) \end{vmatrix}$$

$$+ \begin{vmatrix} \mathbf{i} & \mathbf{j} & \mathbf{k} \\ u_1(t) & v_1(t) & w_1(t) \\ u_2{}'(t) & v_2{}'(t) & w_2{}'(t) \end{vmatrix}$$

$$= [v_1{}'(t)w_2(t) - v_2(t)w_1{}'(t)]\mathbf{i}$$

$$- [u_1{}'(t)w_2(t) - u_2(t)w_1{}'(t)]\mathbf{j}$$

$$+ [u_1{}'(t)v_2(t) - u_2(t)v_1{}'(t)]\mathbf{k}$$

$$+ [v_1(t)w_2{}'(t) - v_2{}'(t)w_1(t)]\mathbf{i}$$

$$- [u_1(t)w_2{}'(t) - u_2{}'(t)w_1(t)]\mathbf{j}$$

$$+ [u_1(t)v_2{}'(t) - u_2{}'(t)v_1(t)]\mathbf{k}$$

Thus,

$$[\mathbf{F} \times \mathbf{G}]'(t) = (\mathbf{F}' \times \mathbf{G})(t) + (\mathbf{F} \times \mathbf{G}')(t)$$

59. $$\frac{d}{dt}\Big(\|\mathbf{R}(t)\|\Big) = \frac{d}{dt}\Big[(\mathbf{R} \cdot \mathbf{R})^{1/2}\Big]$$

$$= \frac{1}{2}(\mathbf{R} \cdot \mathbf{R})^{-1/2}(2\mathbf{R} \cdot \mathbf{R}')$$

$$= \frac{1}{\|\mathbf{R}\|}\mathbf{R} \cdot \mathbf{R}'$$

61. Using the cab-bac formula,

$$\mathbf{F} \times (\mathbf{G} \times \mathbf{H}) = (\mathbf{H} \cdot \mathbf{F})\mathbf{G} - (\mathbf{G} \cdot \mathbf{F})\mathbf{H}$$

Thus,

$$[\mathbf{F} \times (\mathbf{G} \times \mathbf{H})]' = [(\mathbf{H} \cdot \mathbf{F})\mathbf{G}]' - [(\mathbf{G} \cdot \mathbf{F})\mathbf{H}]'$$

63. It is not true; here is a counterexample.

$\mathbf{F} = \langle t, 0, 0 \rangle$ and $\mathbf{G} = \langle -t, 0, 0 \rangle$ are both

smooth on $[1, 2]$, but $\mathbf{F} + \mathbf{G} = \langle 0, 0, 0 \rangle$ is

not.

10.3 Modeling Ballistics and Planetary Motion, Page 657-660

SURVIVAL HINT: *This section is basically the same material covered in Section 3.4, except that the path of the projectile has been broken into its horizontal and vertical components. This allows us to write a vector function, which gives the position, velocity, and acceleration at any time, t. A review of some of the ideas of Section 3.4 might be helpful.*

1. Use the formulas on page 654.

$$T_f = \frac{2}{g} v_0 \sin \alpha$$

$$= \frac{2}{32}(128)\sin 35°$$

$$\approx 4.6 \text{ sec}$$

$$R_f = \frac{v_0^2}{g}\sin 2\alpha$$

$$= \frac{128^2}{32}\sin 2(35°)$$

$$\approx 481 \text{ ft}$$

3. Use the formulas on page 654.

$$T_f = \frac{2}{g} v_0 \sin \alpha$$

$$= \frac{2}{9.8}(850)\sin 48.5°$$

$$\approx 129.9 \text{ sec}$$

$$R_f = \frac{v_0^2}{g}\sin 2\alpha$$

$$= \frac{850^2}{9.8}\sin 2(48.5°)$$

$$\approx 73,175 \text{ m}$$

5. Use the formulas on page 654.

$$T_f = \frac{2}{g} v_0 \sin \alpha$$

$$= \frac{2}{9.8}(23.3)\sin 23.74°$$

$$\approx 1.9 \text{ sec}$$

$$R_f = \frac{v_0^2}{g}\sin 2\alpha$$

$$= \frac{23.3^2}{9.8}\sin 2(23.74°)$$

$$\approx 41 \text{ m}$$

7. Use the formulas on page 654.

$$T_f = \frac{2}{g} v_0 \sin \alpha$$

$$= \frac{2}{32}(100)\sin 14.11°$$

$$\approx 1.5 \text{ sec}$$

$$R_f = \frac{v_0^2}{g}\sin 2\alpha$$

$$= \frac{100^2}{32}\sin 2(14.11°)$$

$$\approx 148 \text{ ft}$$

9. $\mathbf{R}(t) = 2t\mathbf{i} + t\mathbf{j}$

$$r = \| \mathbf{R}(t) \|$$

$$= \sqrt{(2t)^2 + t^2}$$

$$= t\sqrt{5}$$

$$\frac{dr}{dt} = \sqrt{5}$$

$$\frac{d^2 r}{dt^2} = 0$$

$$\theta = \tan^{-1}\left(\frac{y}{x}\right)$$

$$= \tan^{-1}\left(\frac{t}{2t}\right)$$

$$= \tan^{-1}\left(\frac{1}{2}\right)$$

$$\frac{d\theta}{dt} = \frac{d^2\theta}{dt^2} = 0$$

$$\mathbf{V} = \frac{dr}{dt}\mathbf{u}_r + r\frac{d\theta}{dt}\mathbf{u}_\theta$$

$$\mathbf{V}(t) = \sqrt{5}\,\mathbf{u}_r; \ \mathbf{A}(t) = \mathbf{0}$$

11. $r = \sin\theta; \ \theta = 2t$

$$r = \sin 2t$$

$\dfrac{dr}{dt} = 2 \cos 2t$

$\dfrac{d^2r}{dr^2} = -4 \sin 2t$

$\dfrac{d\theta}{dt} = 2$

$\dfrac{d^2\theta}{dt^2} = 0$

$\mathbf{V} = \dfrac{dr}{dt}\mathbf{u}_r + r\dfrac{d\theta}{dt}\mathbf{u}_\theta$

$\quad = (2 \cos 2t)\mathbf{u}_r + (2 \sin 2t)\mathbf{u}_\theta$

$\mathbf{A} = \left[\dfrac{d^2r}{dt^2} - r\left(\dfrac{d\theta}{dt}\right)^2\right]\mathbf{u}_r + \left[r\dfrac{d^2\theta}{dt^2} + 2\dfrac{dr}{dt}\dfrac{d\theta}{dt}\right]\mathbf{u}_\theta$

$\quad = [-4 \sin 2t - (\sin 2t)(4)]\mathbf{u}_r$

$\qquad + [(\sin 2t)(0) + 2(2)(2 \cos 2t)]\mathbf{u}_\theta$

$\quad = (-8 \sin 2t)\mathbf{u}_r + (8 \cos 2t)\mathbf{u}_\theta$

13. $r = 5(1 + \cos \theta); \; \theta = 2t + 1$

$r = 5 + 5[\cos(2t + 1)]$

$\dfrac{dr}{dt} = -10\sin(2t + 1)$

$\dfrac{d^2r}{dt^2} = -20\cos(2t + 1)$

$\dfrac{d\theta}{dt} = 2$

$\dfrac{d^2\theta}{dt^2} = 0$

$\mathbf{V} = \dfrac{dr}{dt}\mathbf{u}_r + r\dfrac{d\theta}{dt}\mathbf{u}_\theta$

$\quad = -10\sin(2t + 1)\mathbf{u}_r + 10[1 + \cos(2t + 1)]\mathbf{u}_\theta$

$\mathbf{A} = \left[\dfrac{d^2r}{dt^2} - r\left(\dfrac{d\theta}{dt}\right)^2\right]\mathbf{u}_r + \left[r\dfrac{d^2\theta}{dt^2} + 2\dfrac{dr}{dt}\dfrac{d\theta}{dt}\right]\mathbf{u}_\theta$

$\quad = \left\{-20\cos(2t + 1) - 20[1 + \cos(2t + 1)]\right\}\mathbf{u}_r$

$\qquad + 4\left[-10\sin(2t + 1)\right]\mathbf{u}_\theta$

$= [-40\cos(2t + 1) - 20]\mathbf{u}_r - 40\sin(2t + 1)\mathbf{u}_\theta$

15. $\alpha = 45°$, so we can use the formula for maximum range since this is on ground level:

$$R_m = \dfrac{v_0{}^2}{g}$$

$$2{,}000 = \dfrac{v_0{}^2}{9.8}$$

$$v_0{}^2 = 19{,}600$$

$$v_0 = 140 \text{ m/sec}$$

17. From the formula on page 654,

$$R_f = \dfrac{v_0{}^2}{g}\sin 2\alpha$$

$$\sin 2\alpha = \dfrac{R_f g}{v_0{}^2}$$

$$= \dfrac{600(32)}{(167.1)^2}$$

$$\approx 0.6876$$

Thus,

$$\alpha \approx 21.7°$$

19. The maximum height is reached when $y'(t) = 0$. We can use this equation to find the time at which this occurs, then use that time in the equation for $y(t)$ to find the height.

$y(t) = -16t^2 + (V_0 \sin \alpha)t + s_0$

$\quad = -16t^2 + (144)(\tfrac{1}{2})t + 4$

$y'(t) = -32t + 72$

$y'(t) = 0$ when $t = 2.25$ sec

$y(2.25) = -16(2.25)^2 + 72(2.25) + 4$

$\quad = 85 \text{ ft}$

The ball will land when $y = 0$. We can use $y(t) = 0$ to find the time of flight, and then use that time to find $x(t)$.

$$-16t^2 + 72t + 4 = 0$$

$$t \approx 4.5549 \text{ sec}$$

(reject negative solution)

$$x(t) = (v_0 \cos \alpha)t$$

$$= \frac{\sqrt{3}}{2}(144)t$$

$$= 72\sqrt{3}\ t$$

$$x(4.5549) \approx 568 \text{ ft}$$

To find the distance to the fence we will find t for $y(t) = 5$, then use that time in the equation $x(t)$.

$$-16t^2 + 72t + 4 = 5$$

$$t \approx 4.49 \text{ sec}$$

(reject negative solution)

$$x(4.49) = 72\sqrt{3}(4.49)$$

$$\approx 560 \text{ ft}$$

21. Let α be the nozzle angle. Then,

$$x = (50 \cos \alpha)t, \qquad y = -16t^2 + (50 \sin \alpha)t$$

For the range, $y = 0$ or $16t^2 = 50t\sin\alpha$.

$T_f = \frac{50 \sin \alpha}{16}$ and $t_h = \frac{25 \sin \alpha}{16}$, where t_h is the time at which the water reaches its highest point.

$y_{\max} = 5$, so

$$-16\left(\frac{25 \sin \alpha}{16}\right)^2 + 50 \sin \alpha\left(\frac{25 \sin \alpha}{16}\right) = 5$$

$$80 = 625 \sin^2\alpha$$

$$\alpha \approx 21°$$

The time of travel is $T_f = \frac{25 \sin 21°}{8} \approx 1.12$ seconds and $x \approx 50(\cos 21°)(1.12) \approx 52.3 \text{ ft}$

23. Evel's "range" was at least 4,700 ft. Using the formula $R = v_0{}^2\sin 2\alpha/g$, we find

$$\frac{v_0{}^2\sin 2(45°)}{32} = 4,700 \text{ or } v_0 \approx 387.8 \text{ ft/s}$$

(This is about 264 mi/h.)

25. $T_f = \frac{2v_0\sin \alpha}{g}$

$$= \frac{2(125)\sqrt{2}}{2(32)}$$

$$\approx 5.5 \text{ sec}$$

27. Let s and h be the horizontal and vertical distances, respectively, from the helicopter to the bunker. Then,

$$h = s \tan 20° \text{ and } 10,000 - h$$

$$= s \tan 15°$$

so that

$$s = \frac{10,000}{\tan 15° + \tan 20°}$$

$$\approx 15,824.8 \text{ ft}$$

and the height of the bunker above the level ground is

$h_1 = s \tan 15°$

$\approx 15{,}824.8 \tan 15°$

$\approx 4{,}240.2 \text{ ft}$

Since the canister is "dropped" (no initial vertical velocity), it will take t_1 seconds to reach the level of the bunker, where

$-16 t_1{}^2 + 10{,}000 = 4{,}240.2$

$t_1 \approx 18.97$

The time is about 19 seconds; the horizontal distance (in feet) traveled by the canister is

$s_1 = 200\, t_1$

$\approx 3{,}794$

Thus, the helicopter should travel

$s - s_1 \approx 15{,}824.8 - 3{,}794$

$\approx 12{,}030.8 \text{ ft}$

before the canister is released. At the rate of 200 ft/s, this takes

$t \approx \dfrac{12{,}030.8}{200}$

$\approx 60.1 \text{ sec}$

The Spy should wait one minute before releasing the canister.

29. $\mathbf{V} = v_0\cos\alpha\, \mathbf{i} + (-gt + v_0\sin\alpha)\mathbf{j}$

$\mathbf{R} = (v_0\cos\alpha)t\, \mathbf{i} + [-\dfrac{gt^2}{2} + (v_0\sin\alpha)t + s_0]\mathbf{j}$

$v_0 = 8,\ g = 32,\ s_0 = 120,\ \alpha = -\dfrac{\pi}{6}$

$-\tfrac{1}{2}gt^2 + (v_0\sin\alpha)t + s_0 = y(t)$

$-16t^2 - 4t + 120 = 0$

$4t^2 + t - 30 = 0$

$t = \dfrac{-1 \pm \sqrt{1 + 480}}{8}$

$t \approx 2.6165 \text{ sec}$ (disregard negative value);

$x = 4\sqrt{3}(2.6165)$

$\approx 18.13 \text{ ft}$

31. $\mathbf{R}(t) = (a\cos\omega t)\mathbf{i} + (a\sin\omega t)\mathbf{j};\ \omega = \dfrac{d\theta}{dt}$

$\mathbf{V} = \dfrac{d\mathbf{R}}{dt}$

$= (-a\omega\sin\omega t)\mathbf{i} + (a\omega\cos\omega t)\mathbf{j}$

$\mathbf{R}\cdot\mathbf{V} = -a^2\omega\sin\omega t\cos\omega t + a^2\omega\sin\omega t\cos\omega t$

$= 0$

33. $\mathbf{R}(t) = (a\cos\omega t)\mathbf{i} + (a\sin\omega t)\mathbf{j};\ \omega = \dfrac{d\theta}{dt}$

$\mathbf{V} = \dfrac{d\mathbf{R}}{dt}$

$= (-a\omega\sin\omega t)\mathbf{i} + (a\omega\cos\omega t)\mathbf{j};$

$\mathbf{A} = (-a\omega^2\cos\omega t)\mathbf{i} + (-a\omega^2\sin\omega t)\mathbf{j}$

$= -a\omega^2\mathbf{R}$

Since $\mathbf{R}$ points away from the center, $\mathbf{A} = -a\omega^2\mathbf{R}$ points toward the center.

35. The equations of motion for the shell are

$x = (100\cos 40°)t$

$\approx 76.6t$

$$y = -4.9t^2 + (100 \sin 40°)t$$

$$\approx -4.9t^2 + 64.28t$$

since $g \approx 9.8$, $s_0 = 0$, and $v_0 = 100$, with respect to a coordinate system with the mortar at the origin and the x-axis running horizontal. Since the slope is along a line with angle of declination 20°, an equation for this line is

$$y = -\tan 20°x \approx -0.364x$$

a. The shell strikes the slope when

$$-4.9t^2 + 64.28t = -0.364(76.6t)$$

$$t = 18.81 \text{ seconds}$$

so the point of impact has coordinates (x_0, y_0), where

$$x_0 \approx 76.6(18.81)$$

$$\approx 1,440.85 \text{ m}$$

$$y_0 = -0.364x_0$$

$$\approx -524.47 \text{ m}$$

Thus, the impact occurs a distance R down the slope, where

$$R = \sqrt{x_0{}^2 + y_0{}^2}$$

$$\approx 1,533.33 \text{ m}$$

b. The velocity vector is

$$\mathbf{V}(t) = \langle 76.6, \ -9.8t + 64.28 \rangle$$

so the velocity at impact is

$$\mathbf{V}(18.81) = \langle 76.6, \ -120.06 \rangle$$

with speed

$$s = \| \mathbf{V}(18.81) \|$$

$$= 142.41 \text{ m/s}$$

37. $\quad R = \dfrac{550^2}{32} \sin 2(22°)$

$$\approx 6,566.69$$

The desired range is about 6,516.69. This means

$$6,516.69 \approx \frac{550^2}{32} \sin 2\alpha$$

$$0.6893688595 \approx \sin 2\alpha$$

$$\alpha \approx 21.8°$$

39. We can find when the first shell is 50 ft above ground by solving the following equation.

$$50 = -16t^2 + (750 \sin 25°)t$$

$$t \approx 19.65 \text{ seconds} \quad \textit{By calculator}$$

At that time, its horizontal distance from the second gun is

$$s_1 = 20,000 - 750(\cos 25°)(19.65)$$

$$\approx 6,643.29 \text{ ft}$$

Since the second gun fires 2 seconds after the first, its shell will be in the air only 17.65 seconds before the collision. Thus,

$$-16(17.65)^2 + (v_0 \sin \alpha)(17.65) = 50$$

$$(v_0 \cos \alpha)(17.65) = 6,643.29$$

Solving, we obtain

$v_0 \sin \alpha = 285.23$ and $v_0 \cos \alpha = 376.39$

so

$$\alpha = \tan^{-1}\left(\frac{285.23}{376.39}\right)$$

$$\approx 37.2°$$

$$v_0 = \frac{285.23}{\sin 37.2°}$$

$$\approx 471.77 \text{ ft/s}$$

41. $\mathbf{A}(t) = \frac{d}{dt}\mathbf{V}$

$$= \frac{d}{dt}\left[\frac{dr}{dt}\mathbf{u}_r + r\frac{d\theta}{dt}\mathbf{u}_\theta\right]$$

$$= \frac{dr}{dt}\frac{d\mathbf{u}_r}{dt} + \mathbf{u}_r\frac{d^2r}{dt^2}$$

$$+ r\frac{d\theta}{dt}\frac{d\mathbf{u}_\theta}{dt} + r\frac{d^2\theta}{dt^2}\mathbf{u}_\theta + \frac{dr}{dt}\frac{d\theta}{dt}\mathbf{u}_\theta$$

$$= \frac{dr}{dt}\mathbf{u}_\theta\frac{d\theta}{dt} + \mathbf{u}_r\frac{d^2r}{dt^2} + r\left(\frac{d\theta}{dt}\right)^2(-1)\mathbf{u}_r$$

$$+ r\frac{d^2\theta}{dt^2}\mathbf{u}_\theta + \frac{dr}{dt}\frac{d\theta}{dt}\mathbf{u}_\theta$$

$$= \left[\frac{d^2r}{dt^2} - r\left(\frac{d\theta}{dt}\right)^2\right]\mathbf{u}_r$$

$$+ \left[2\frac{dr}{dt}\frac{d\theta}{dt} + r\frac{d^2\theta}{dt^2}\right]\mathbf{u}_\theta$$

43. The shell travels along a trajectory with parametric equations

$$x = (v_0\cos \alpha)t$$
$$y = -\frac{1}{2}gt^2 + (v_0\sin \alpha)t$$

The downward sloping hill has the equation

$$y = (-\tan \beta)x$$

so the shell strikes the hill when

$$-\frac{1}{2}g T_f^2 + (v_0\sin \alpha)T_f = (-\tan \beta)(v_0\cos \alpha)T_f$$

a. Thus, the time of flight is

$$T_f = \frac{v_0\sin \alpha + v_0\cos \alpha \tan \beta}{\frac{1}{2}g}$$

Impact occurs at (x_i, y_i), where

$$x_i = (v_0\cos \alpha)T_f$$

$$y_i = (-\tan \beta)x_i$$

$$= (-\tan \beta)(v_0\cos \alpha)T_f$$

The range is

$$R_f = \sqrt{x_i^2 + y_i^2}$$

$$= \sqrt{v_0^2\cos^2\alpha\, T_f^2 + \tan^2\beta\, v_0^2\cos^2\alpha\, T_f^2}$$

$$= v_0 \cos \alpha \sec \beta\, T_f$$

b. $R_f = v_0\cos \alpha \sec \beta\left[\dfrac{v_0\sin \alpha + v_0\cos \alpha \tan \beta}{\frac{1}{2}g}\right]$

$$\frac{dR_f}{d\alpha} = \frac{2}{g}v_0^2\sec \beta[\cos 2\alpha + 2\tan \beta \cos \alpha(-\sin \alpha)]$$

$$= 0 \text{ when } \quad \cos 2\alpha = \sin 2\alpha \tan \beta$$

$$\alpha = \frac{1}{2}\tan^{-1}(\cot \beta)$$

$$= \frac{1}{2}\left(\frac{\pi}{2} - \beta\right)$$

$$= \frac{\pi}{4} - \frac{\beta}{2}$$

10.4 Unit Tangent and Principal Unit Normal Vectors; Curvature, Page 671-673

1. $\mathbf{R}(t) = t^2\mathbf{i} + t^3\mathbf{j},\ t \neq 0$

$$R'(t) = 2t\mathbf{i} + 3t^2\mathbf{j}$$

$$\|R'(t)\| = t\sqrt{4 + 9t^2}$$

$$T(t) = \frac{R'(t)}{\|R'(t)\|}$$

$$= \frac{2}{\sqrt{4 + 9t^2}}\mathbf{i} + \frac{3t}{\sqrt{4 + 9t^2}}\mathbf{j}$$

$$T'(t) = -\frac{18t}{(4 + 9t^2)^{3/2}}\mathbf{i} + \frac{12}{(4 + 9t^2)^{3/2}}\mathbf{j}$$

$$\|T'(t)\| = \frac{6}{4 + 9t^2}$$

$$N(t) = \frac{T'(t)}{\|T'(t)\|}$$

$$= \frac{-3t}{\sqrt{4 + 9t^2}}\mathbf{i} + \frac{2}{\sqrt{4 + 9t^2}}\mathbf{j}$$

3. $R(t) = e^t\cos t\,\mathbf{i} + e^t \sin t\,\mathbf{j}$

$$R'(t) = e^t(-\sin t + \cos t)\mathbf{i} + e^t(\cos t + \sin t)\mathbf{j}$$

$$\|R'(t)\| = \sqrt{2}\,e^t$$

$$T(t) = \frac{\sqrt{2}}{2}[(\cos t - \sin t)\mathbf{i} + (\cos t + \sin t)\mathbf{j}]$$

$$T'(t) = \frac{\sqrt{2}}{2}[(-\sin t - \cos t)\mathbf{i} + (-\sin t + \cos t)]\mathbf{j}$$

$$\|T'(t)\| = 1$$

$$N(t) = \frac{T'(t)}{\|T'(t)\|}$$

$$= -\frac{\sqrt{2}}{2}[(\sin t + \cos t)\mathbf{i} + (\sin t - \cos t)\mathbf{j}]$$

SURVIVAL HINT: *The arc length formula is easily recalled if you think of it as the sum (definite integral) of lines in $\mathbb{R}^3$ (space diagonals);*

$$\sqrt{(\Delta x)^2 + (\Delta y)^2 + (\Delta z)^2}$$

where the delta change is found by using the derivative.

5. $R(t) = (\cos t)\mathbf{i} + (\sin t)\mathbf{j} + t\mathbf{k}$

$$R'(t) = (-\sin t)\mathbf{i} + (\cos t)\mathbf{j} + \mathbf{k}$$

$$\|R'(t)\| = \sqrt{2}$$

$$T(t) = \frac{1}{\sqrt{2}}(-\sin t\,\mathbf{i} + \cos t\,\mathbf{j} + \mathbf{k})$$

$$T'(t) = \frac{1}{\sqrt{2}}(-\cos t\,\mathbf{i} - \sin t\,\mathbf{j})$$

$$\|T'(t)\| = \tfrac{1}{2}\sqrt{2}$$

$$N(t) = \frac{T'(t)}{\|T'(t)\|}$$

$$= -(\cos t\,\mathbf{i} + \sin t\,\mathbf{j})$$

7. $R(t) = (\ln t)\mathbf{i} + t^2\mathbf{k}$

$$R'(t) = \tfrac{1}{t}\mathbf{i} + 2t\mathbf{k}$$

$$\|R'(t)\| = \tfrac{1}{t}\sqrt{1 + 4t^4}$$

$$T(t) = \frac{1}{\sqrt{1 + 4t^4}}(\mathbf{i} + 2t^2\mathbf{k})$$

$$T'(t) = \frac{-8t^3\mathbf{i} + 4t\mathbf{k}}{(1 + 4t^4)^{3/2}}$$

$$\|T'(t)\| = \sqrt{\frac{(-8t^3)^2 + (4t)^2}{(1 + 4t^4)^3}}$$

$$= \frac{4t\sqrt{4t^4 + 1}}{(1 + 4t^4)^{3/2}}$$

$$= \frac{4t}{1 + 4t^4}$$

$$N(t) = \frac{T'(t)}{\|T'(t)\|}$$

$$= \frac{1}{\sqrt{1 + 4t^4}}(-2t^2\mathbf{i} + \mathbf{k})$$

9. $R'(t) = 2\mathbf{i} + \mathbf{j}$

$$\frac{ds}{dt} = \|R'(t)\|$$

$$= \sqrt{2^2 + 1^2}$$

$$= \sqrt{5}$$

$$s = \int_0^4 \sqrt{5} \; dt$$

$$= 4\sqrt{5}$$

11. $R'(t) = 3\mathbf{i} - (3\sin t)\mathbf{j} + (3\cos t)\mathbf{k}$

$$\frac{ds}{dt} = \|R'(t)\|$$

$$= \sqrt{3^2 + 9\sin^2 t + 9\cos^2 t}$$

$$= 3\sqrt{2}$$

$$ds = 3\sqrt{2} \; dt$$

$$s = \int_0^{\pi/2} 3\sqrt{2} \; dt$$

$$= \frac{3\sqrt{2}\pi}{2}$$

13. $R(t) = (4\cos t)\mathbf{i} + (4\sin t)\mathbf{j} + 5t\mathbf{k}$

$\quad R'(t) = (-4\sin t)\mathbf{i} + (4\cos t)\mathbf{j} + 5\mathbf{k}$

$$\frac{ds}{dt} = \|R'(t)\|$$

$$= \sqrt{(-4\sin t)^2 + (4\cos t)^2 + 5^2}$$

$$= \sqrt{16(\sin^2 t + \cos^2 t) + 25}$$

$$= \sqrt{41}$$

$$ds = \sqrt{41} \; dt$$

$$s = \int_0^{\pi} \sqrt{41} \; dt$$

$$= \sqrt{41}\pi$$

SURVIVAL HINT: *There are many different formulas for curvature. The one you will use is dependent on the form of the given information (see page 671). You should ask your instructor if you are required to know (memorize) any of these formulas.*

15. $y = 4x - 2$

$$y' = 4$$

$$y'' = 0$$

$$\kappa = \frac{|f''(x)|}{(1 + [f'(x)]^2)^{3/2}}$$

$$\kappa = 0$$

This is a straight line, so the curvature is 0 everywhere.

17. $y = x + \frac{1}{x}$

$$y' = 1 - \frac{1}{x^2}$$

$$y'' = \frac{2}{x^3}$$

$$\kappa = \frac{|f''(x)|}{(1 + [f'(x)]^2)^{3/2}}$$

$$= \frac{\left|\frac{2}{x^3}\right|}{\left(1 + \left[1 - \frac{1}{x^2}\right]^2\right)^{3/2}}$$

$$\kappa(1) = \frac{2}{(1 + 0)^{3/2}}$$

$$= 2$$

19. $y = \ln x$

$$y' = \frac{1}{x}$$

$$y'' = \frac{-1}{x^2}$$

$$\kappa = \frac{|f''(x)|}{(1 + [f'(x)]^2)^{3/2}}$$

$$= \frac{\left|-\frac{1}{x^2}\right|}{\left(1 + \left[\frac{1}{x}\right]^2\right)^{3/2}}$$

$$\kappa(1) = \frac{1}{(1 + 1)^{3/2}}$$

$$= 2^{-3/2}$$

21. $R(t) = \langle e^{-t}, 1 - e^{-t}\rangle$

$$R'(t) = \langle -e^{-t}, e^{-t}\rangle$$

$$s = \int_0^t \|R'(u)\|\, du$$

$$= \int_0^t \sqrt{(e^{-u})^2 + (e^{-u})^2}\, du$$

$$= \sqrt{2}(1 - e^{-t})$$

Solve for e^{-t}: $e^{-t} = \frac{\sqrt{2} - s}{\sqrt{2}}$

Thus, $R(s) = \left\langle \frac{\sqrt{2} - s}{\sqrt{2}}, \frac{s}{\sqrt{2}}\right\rangle$

23. $R(t) = \langle 2 - 3t, 1 + t, -4t\rangle$

$$R'(t) = \langle -3, 1, -4\rangle$$

$$s = \int_0^t \|R'(u)\|\, du$$

$$= \int_0^t \sqrt{(-3)^2 + (1)^2 + (-4)^2}\, du$$

$$= \sqrt{26}\, t$$

Solve for t: $t = \frac{s}{\sqrt{26}}$

Thus, $R(s) = \left\langle 2 - \frac{3s}{\sqrt{26}}, 1 + \frac{s}{\sqrt{26}}, \frac{-4s}{\sqrt{26}}\right\rangle$

25. If u and v are constant vectors,

$R(t) = u + vt$, then $R'(t) = v$, and

$$R''(t) = 0$$

$$R' \times R'' = 0$$

so $\kappa = 0$ since $\kappa = \frac{\|R' \times R''\|}{\|R'\|^3}$.

27. $R = [\ln(\sin t)]i + [\ln(\cos t)]j$

$$\frac{dR}{dt} = \frac{\cos t}{\sin t}i - \frac{\sin t}{\cos t}j$$

$$= \cot t\, i - \tan t\, j$$

$$\left\| \frac{d\mathbf{R}}{dt} \right\| = \sqrt{\cot^2 t + \tan^2 t};$$

$$\mathbf{T}(t) = \frac{\mathbf{R}'(t)}{\|\mathbf{R}'(t)\|}$$

$$= \frac{\cot t\,\mathbf{i} - \tan t\,\mathbf{j}}{\sqrt{\cot^2 t + \tan^2 t}}$$

At $t = \frac{\pi}{3}$, the unit tangent vector is

$$\mathbf{T} = \frac{\frac{1}{\sqrt{3}}\mathbf{i} - \sqrt{3}\,\mathbf{j}}{\sqrt{\frac{10}{3}}}$$

$$= \frac{1}{\sqrt{10}}(\mathbf{i} - 3\mathbf{j})$$

29. $\mathbf{R}(t) = (\sin t)\mathbf{i} + (\cos t)\mathbf{j} + t\mathbf{k}$

$\mathbf{R}'(t) = (\cos t)\mathbf{i} + (-\sin t)\mathbf{j} + \mathbf{k}$

$\mathbf{R}''(t) = (-\sin t)\mathbf{i} - (\cos t)\mathbf{j}$

a. $\mathbf{T} = \dfrac{\mathbf{R}'(t)}{\|\mathbf{R}'(t)\|}$

$$= \frac{(\cos t)\mathbf{i} + (-\sin t)\mathbf{j} + \mathbf{k}}{\sqrt{\cos^2 t + \sin^2 t + 1}}$$

$$= \frac{\sqrt{2}}{2}\,[(\cos t)\mathbf{i} + (-\sin t)\mathbf{j} + \mathbf{k}]$$

$$\mathbf{T}(\pi) = \frac{\sqrt{2}}{2}\,(-\mathbf{i} + \mathbf{k})$$

b. $\mathbf{R}'(\pi) = (\cos \pi)\mathbf{i} + (-\sin \pi)\mathbf{j} + \mathbf{k}$

$$= -\mathbf{i} + \mathbf{k}$$

$$\mathbf{R}''(\pi) = \mathbf{j}$$

$$\kappa = \frac{\|\mathbf{R}' \times \mathbf{R}''\|}{\|\mathbf{R}'\|^3}$$

$$= \frac{\|(-\mathbf{i} + \mathbf{k}) \times \mathbf{j}\|}{(\sqrt{2})^3}$$

$$= \frac{\sqrt{2}}{2\sqrt{2}}$$

$$= \frac{1}{2}$$

c. $s = \displaystyle\int_0^\pi \|\mathbf{R}'\|\, dt$

$$= \int_0^\pi \sqrt{2}\, dt$$

$$= \sqrt{2}\,\pi$$

31. $9x^2 + 4y^2 = 36$

$$\frac{x^2}{4} + \frac{y^2}{9} = 1$$

Parameterize this curve by letting

$x = 2\cos t,\ y = 3\sin t.$

$x' = -2\sin t,\ y' = 3\cos t$

$x'' = -2\cos t,\ y'' = -3\sin t$

$$\kappa = \frac{|x'y'' - y'x''|}{[(x')^2 + (y')^2]^{3/2}}$$

$$= \frac{|(-2\sin t)(-3\sin t) - (3\cos t)(-2\cos t)|}{[(-2\sin t)^2 + (3\cos t)^2]^{3/2}}$$

$$= \frac{6}{(4\sin^2 t + 9\cos^2 t)^{3/2}}$$

$$= 6(4 + 5\cos^2 t)^{-3/2}$$

$\dfrac{d\kappa}{dt} = -\dfrac{6(3)}{2}(4+5\cos^2 t)^{-5/2}(-10\cos t \sin t)$

$\dfrac{d\kappa}{dt} = 0$ if $t = 0, \frac{\pi}{2}, \pi, \frac{3\pi}{2}, 2\pi$

At $t = 0$ and $t = \pi$, $\kappa = \dfrac{6}{(4+5)^{3/2}} = \dfrac{2}{9}$

(a minimum)

At $t = \frac{\pi}{2}$ and $t = \frac{3\pi}{2}$, $\kappa = \dfrac{6}{4^{3/2}} = \dfrac{3}{4}$

(a maximum)

The points at which a maximum occurs are $P(0, 3)$, and $Q(0, -3)$.

33. $y = x^6 - 3x^2$

$y' = 6x^5 - 6x$

Find where $y' = 0$:

$$6x^5 - 6x = 0$$
$$6x(x^4 - 1) = 0$$
$$6x(x^2 - 1)(x^2 + 1) = 0$$
$$6x(x - 1)(x + 1)(x^2 + 1) = 0$$
$$x = 0, 1, -1$$

Since $y'' = 30x^4 - 6$, we have

$\rho = \dfrac{1}{\kappa}$

$= \dfrac{[1 + (y')^2]^{3/2}}{|y''|}$

$= \dfrac{[1 + (6x^5 - 6x)^2]^{3/2}}{|30x^4 - 6|}$

$\rho(0) = \dfrac{1}{6}$

$\rho(1) = \rho(-1) = \dfrac{1}{24}$

35. $\mathbf{R}(t) = t\mathbf{i} + t^2\mathbf{j} + t^3\mathbf{k}$

$\mathbf{R}'(t) = \mathbf{i} + 2t\mathbf{j} + 3t^2\mathbf{k};$

$\mathbf{R}''(t) = 2\mathbf{j} + 6t\mathbf{k}$

$\mathbf{R}' \times \mathbf{R}'' = \begin{vmatrix} \mathbf{i} & \mathbf{j} & \mathbf{k} \\ 1 & 2t & 3t^2 \\ 0 & 2 & 6t \end{vmatrix}$

$= (12t^2 - 6t^2)\mathbf{i} - 6t\mathbf{j} + 2\mathbf{k}$

$\|\mathbf{R}' \times \mathbf{R}''\| = 2\sqrt{9t^4 + 9t^2 + 1};$

$\|\mathbf{R}'\| = \sqrt{1 + 4t^2 + 9t^4}$

$\kappa = \dfrac{\|\mathbf{R}' \times \mathbf{R}''\|}{\|\mathbf{R}'\|^3}$

$= \dfrac{2\sqrt{9t^4 + 9t^2 + 1}}{(1 + 4t^2 + 9t^4)^{3/2}}$

37. $y = x^2$

$y' = 2x$

$y'' = 2$

$\kappa = \dfrac{|y''|}{[1 + (y')^2]^{3/2}}$

$= \dfrac{2}{(1 + 4x^2)^{3/2}}$

39. $y = x^{-2}$

$y' = -2x^{-3}$

$y'' = 6x^{-4}$

$$\kappa = \frac{|y''|}{[1 + (y')^2]^{3/2}}$$

$$= \frac{6x^{-4}}{[1 + (-2x^{-3})^2]^{3/2}}$$

$$= \frac{6x^{-4}}{[1 + 4x^{-6}]^{3/2}}$$

$$= \frac{x^9(6x^{-4})}{(x^6 + 4)^{3/2}}$$

$$= \frac{6x^5}{(x^6 + 4)^{3/2}}$$

41. $r = r' = r'' = e^\theta$

$$\kappa = \frac{|r^2 + 2r'^2 - rr''|}{(r^2 + r'^2)^{3/2}}$$

$$= \frac{e^{2\theta} + 2e^{2\theta} - e^{2\theta}}{(e^{2\theta} + e^{2\theta})^{3/2}}$$

$$= \frac{2e^{2\theta}}{(2e^{2\theta})^{3/2}}$$

$$= (2e^{2\theta})^{-1/2}$$

$$= \frac{1}{\sqrt{2}e^\theta}$$

43. $R(t) = t\mathbf{u} + t^2\mathbf{v} + 2\left(\frac{2}{3}t\right)^{3/2}(\mathbf{u} \times \mathbf{v})$

$$R'(t) = \mathbf{u} + 2t\mathbf{v} + 2\left(\frac{3}{2}\right)\left(\frac{2}{3}t\right)^{1/2}\left(\frac{2}{3}\right)(\mathbf{u} \times \mathbf{v})$$

$$= \mathbf{u} + 2t\mathbf{v} + 2\left(\frac{2}{3}t\right)^{1/2}(\mathbf{u} \times \mathbf{v})$$

$$\|R'\|^2 = R' \cdot R'$$

$$= (\mathbf{u} \cdot \mathbf{u}) + 4t(\mathbf{u} \cdot \mathbf{v}) + 4\left(\frac{2}{3}t\right)^{1/2}(\mathbf{u} \times \mathbf{v} \cdot \mathbf{u})$$

$$+ 4t^2(\mathbf{v} \cdot \mathbf{v}) + 8t\left(\frac{2}{3}t\right)^{1/2}(\mathbf{u} \times \mathbf{v} \cdot \mathbf{v})$$

$$+ \frac{8}{3}t[(\mathbf{u} \times \mathbf{v}) \cdot (\mathbf{u} \times \mathbf{v})]$$

$$= \|\mathbf{u}\|^2 + 4t(\mathbf{u} \cdot \mathbf{v}) + 4t^2\|\mathbf{v}\|^2 + \frac{8}{3}t\|\mathbf{u} \times \mathbf{v}\|^2$$

since $\mathbf{u} \times \mathbf{v} \cdot \mathbf{u} = \mathbf{u} \times \mathbf{v} \cdot \mathbf{v} = 0$. Since $\mathbf{u}$ and $\mathbf{v}$ are unit vectors separated by 60°, this expression becomes

$$\|R'\|^2 = 1^2 + 4t(1)(1)\cos 60° + 4t^2(1)^2$$

$$+ \frac{8}{3}t[(1)(1)\sin 60°]^2$$

$$= 1 + 4t + 4t^2$$

The fly's speed at time t is $4t^2 + 4t + 1$. Also,

$$\frac{ds}{dt} = \|R'\|$$

$$= \sqrt{4t^2 + 4t + 1}$$

$$= 2t + 1$$

$$s = \int (2t + 1) \, dt$$

$$= t^2 + t + C$$

If $t = 0$, then $s = 0$, so $C = 0$.

$$s = t^2 + t$$

We now solve

$$t^2 + t = 20$$

$$t^2 + t - 20 = 0$$

$(t - 4)(t + 5) = 0$

$$t = 4, \ -5$$

We find $t = 4$ seconds for the pestus to fly 20 units.

45. See Table 10.2. Use the arc length parameter form of the formula for curvature if the unit vector **T** is easily differentiated. The vector/derivative form is used if curve is given as the graph of $\mathbf{R}(t)$. The parametric form is applicable when x and y can be written in terms of a parameter. The functional form is applied to curves for the form $y = f(x)$. The polar form is used when the equation of the curve is given in polar coordinates.

47. $x = f(\theta) \cos \theta, \ y = f(\theta) \sin \theta$

$x' = -f(\theta) \sin \theta + f'(\theta) \cos \theta$

$x'' = -f(\theta) \cos \theta - f'(\theta) \sin \theta - f'(\theta) \sin \theta$

$\qquad + f''(\theta) \cos \theta$

$y' = f(\theta) \cos \theta + f'(\theta) \sin \theta$

$y'' = -f(\theta) \sin \theta + f'(\theta) \cos \theta + f'(\theta) \cos \theta$

$\qquad + f''(\theta) \sin \theta$

Thus,

$x'y'' - y'x''$

$= [-f(\theta) \sin \theta + f'(\theta) \cos \theta][-f(\theta) \sin \theta$

$\qquad + 2f'(\theta) \cos \theta + f''(\theta) \sin \theta]$

$\quad - [f(\theta) \cos \theta + f'(\theta) \sin \theta][-f(\theta) \cos \theta$

$\qquad - 2f'(\theta) \sin \theta + f''(\theta) \cos \theta]$

$= f^2(\theta) - f(\theta)f''(\theta) + 2(f'(\theta))^2$

and

$(x')^2 + (y')^2 = [-f(\theta) \sin \theta + f'(\theta) \cos \theta]^2$

$\qquad + [f(\theta) \cos \theta + f'(\theta) \sin \theta]^2$

$\qquad = f^2(\theta) + (f'(\theta))^2$

so that

$$\kappa = \frac{|x'y'' - y'x''|}{[(x')^2 + (y')^2]^{3/2}}$$

$$= \frac{|r^2 + 2(r')^2 - rr''|}{[r^2 + (r')^2]^{3/2}}$$

where $r = f(\theta)$.

49. $\kappa = \dfrac{|r^2 + 2r'^2 - rr''|}{(r^2 + r'^2)^{3/2}}$

$\displaystyle \lim_{r \to 0} \kappa = \lim_{r \to 0} \frac{|r^2 + 2(r')^2 - rr''|}{[r^2 + (r')^2]^{3/2}}$

$\kappa = \dfrac{|2(r')^2|}{(r')^3}$

$\quad = \dfrac{2}{|r'|}$

51. **a.** $\dfrac{dx}{dt} = \cos\left(\dfrac{\pi t^2}{2}\right)$ and $\dfrac{dy}{dt} = \sin\left(\dfrac{\pi t^2}{2}\right)$

so

$$s = \int\limits_0^t \sqrt{\left(\dfrac{dx}{du}\right)^2 + \left(\dfrac{dy}{du}\right)^2}\, du$$

$$= \int\limits_0^t \sqrt{\cos^2\left(\dfrac{\pi u}{2}\right) + \sin^2\left(\dfrac{\pi u^2}{2}\right)}\, du$$

$$= \int\limits_0^t du = t$$

b. If $\mathbf{R}'(s) = \left\langle \dfrac{dx}{dt}, \dfrac{dy}{dt} \right\rangle$

$$= \left\langle \cos\left(\dfrac{\pi s^2}{2}\right), \sin\left(\dfrac{\pi s^2}{2}\right) \right\rangle$$

then

$$\mathbf{R}''(s) = \left\langle -\pi s \sin\left(\dfrac{\pi s^2}{2}\right), \pi s \cos\left(\dfrac{\pi s^2}{2}\right) \right\rangle$$

and $\kappa(s) = \left\| \mathbf{R}''(s) \right\| = \pi |s|$

c. $\lim\limits_{s \to +\infty} \kappa(s) = +\infty$, so the spiral keeps winding tighter and tighter ($\rho \to 0$).

53. $\dfrac{d\mathbf{T}}{ds} = \kappa\mathbf{N}$ follows from the definition of κ.

Since $\mathbf{N} \cdot \mathbf{T} = 0$ and $\mathbf{N} \cdot \mathbf{N} = 1$, we have

$$0 = \dfrac{d}{ds}(\mathbf{N} \cdot \mathbf{T})$$

$$= \mathbf{N} \cdot \dfrac{d\mathbf{T}}{ds} + \mathbf{T} \cdot \dfrac{d\mathbf{N}}{ds}$$

$$= \mathbf{N} \cdot (\kappa\mathbf{N}) + \mathbf{T} \cdot \dfrac{d\mathbf{N}}{ds}$$

$$\kappa = -\mathbf{T} \cdot \dfrac{d\mathbf{N}}{ds}$$

It follows that $\dfrac{d\mathbf{N}}{ds} = \kappa\mathbf{T}$ is orthogonal to $\mathbf{T}$ and $\mathbf{N}$:

$$\left(\dfrac{d\mathbf{N}}{ds} + \kappa\mathbf{T}\right) \cdot \mathbf{T} = \dfrac{d\mathbf{N}}{ds} \cdot \mathbf{T} + \kappa(\mathbf{T} \cdot \mathbf{T})$$

$$= -\kappa + \kappa(1)$$

$$= 0$$

$$\left(\dfrac{d\mathbf{N}}{ds} + \kappa\mathbf{T}\right) \cdot \mathbf{N} = \dfrac{d\mathbf{N}}{ds} \cdot \mathbf{N} + \kappa(\mathbf{T} \cdot \mathbf{N})$$

$$= 0 + 0$$

$$= 0$$

since $\mathbf{N} \cdot \mathbf{N} = 1$. Thus, the vector

$$\dfrac{d\mathbf{N}}{ds} + \kappa\mathbf{T}$$

is parallel to $\mathbf{B} = \mathbf{T} \times \mathbf{N}$. Define τ by

$$\dfrac{d\mathbf{N}}{ds} + \kappa\mathbf{T} = \tau\mathbf{B}$$

so

$$\dfrac{d\mathbf{N}}{ds} = -\kappa\mathbf{T} + \tau\mathbf{B}$$

Then

$$\dfrac{d\mathbf{B}}{ds} = \dfrac{d}{ds}(\mathbf{T} \times \mathbf{N})$$

$$= \dfrac{d\mathbf{T}}{ds} \times \mathbf{N} + \mathbf{T} \times \dfrac{d\mathbf{N}}{ds}$$

$$= (\kappa\mathbf{N}) \times \mathbf{N} + \mathbf{T} \times (-\kappa\mathbf{T} + \tau\mathbf{B})$$

$$= \kappa(\mathbf{N} \times \mathbf{N}) - \kappa(\mathbf{T} \times \mathbf{T}) + \tau(\mathbf{T} \times \mathbf{B})$$

$$= 0 - 0 + \tau(-\mathbf{N})$$

$$= -\tau\mathbf{N}$$

55. $y = \frac{1}{32}x^{5/2}$

$y' = \frac{5}{64}x^{3/2}$

$y'' = \frac{15}{128}x^{1/2}$

$\kappa = \dfrac{|y''|}{[1 + (y')^2]^{3/2}}$

$= \dfrac{\frac{15}{128}x^{1/2}}{\left[1 + \left(\frac{5}{64}x^{3/2}\right)^2\right]^{3/2}}$

$\kappa(4) \approx 0.1429$, by calculator and $\rho(4) \approx 7$, by calculator. The tangent vector at (x, y) on the curve is

$\mathbf{R}' = \langle 1, \frac{5}{64}x^{3/2}\rangle$

so $\mathbf{R}'(4) = \langle 1, \frac{5}{8}\rangle$ and a normal vector is $\langle -\frac{5}{8}, 1\rangle$. Thus, if $C(a, b)$ is the center of the osculating circle, we must have

$\langle a - 4, b - 1\rangle = s\langle -\frac{5}{8}, 1\rangle$

and

$\sqrt{(a - 4)^2 + (b - 1)^2} = 7$

Solving, we obtain $a \approx 0.29$, $b \approx 6.94$, and the osculating circle is

$(x - 0.29)^2 + (y - 6.94)^2 = 49$

When $y = 3$, we find that $x \approx 6.08$. The part of the ramp from $A(4, 1)$ to $B(6.08, 3)$ along the circle is an arc subtended by the angle θ between the vectors

$\mathbf{CA} = \langle 3.71, -5.94\rangle$ and $\mathbf{CB} = \langle 5.79, -3.94\rangle$

Since

$\cos\theta = \dfrac{\mathbf{CA}\cdot\mathbf{CB}}{\|\mathbf{CA}\|\|\mathbf{CB}\|} \approx 0.9151$

$\theta \approx 0.4153$

and the length of the arc is

$L_1 = 7\theta \approx 2.891$

The part of the ramp along the curve has length

$L_2 = \displaystyle\int_0^4 \sqrt{1 + \left(\frac{5}{64}x^{3/2}\right)^2}\, dx$

≈ 4.1857

so the total length of the ramp is

$L = L_1 + L_2 \approx 7.096$ units

57. $x = t - \sin t,\ y = 1 - \cos t$

$x' = 1 - \cos t,\ y' = \sin t$

$x'' = \sin t,\ y'' = \cos t$

$\kappa = \dfrac{|(1 - \cos t)(\cos t) - (\sin t)(\sin t)|}{\left[(1 - \cos t)^2 + (\sin t)^2\right]^{3/2}}$

$= \dfrac{|\cos t - \cos^2 t - \sin^2 t|}{\left[1 - 2\cos t + \cos^2 t + \sin^2 t\right]^{3/2}}$

$= \dfrac{|\cos t - 1|}{[2 - 2\cos t]^{3/2}}$

$$= \frac{1}{2^{3/2}} \frac{|\cos t - 1|}{|1 - \cos t|^{3/2}}$$

$$= \frac{1}{2^{3/2}} \frac{1}{\sqrt{1 - \cos t}}$$

$$\kappa_{\min} = \kappa(\pi)$$

$$= \frac{1}{2^{3/2} \sqrt{1 - (-1)}}$$

$$= \frac{1}{4}$$

$\kappa_{\max} = \kappa(0)$ is infinite, so there is no

maximum value.

10.5 Tangential and Normal Components of Acceleration, Page 678-679

SURVIVAL HINT: *Most of these problems involve finding several different derivatives, and the sums and products of several components. It is essential that your work be well organized and each step be properly labeled.*

1. $\mathbf{R}(t) = t\mathbf{i} + t^2\mathbf{j}$

    $\mathbf{V}(t) = \mathbf{R}'(t) = \mathbf{i} + 2t\mathbf{j}$

    $\mathbf{A}(t) = \mathbf{V}'(t) = 2\mathbf{j}$

    $\frac{ds}{dt} = \|\mathbf{V}(t)\| = \sqrt{1 + 4t^2}$

    $A_T = \frac{d^2s}{dt^2}$

$$= \frac{4t}{\sqrt{1 + 4t^2}};$$

$$A_N = \sqrt{\|\mathbf{A}\|^2 - A_T^2}$$

$$= \sqrt{4 - \frac{16t^2}{1 + 4t^2}}$$

$$= \frac{2}{\sqrt{1 + 4t^2}}$$

3. $\mathbf{R}(t) = (t \sin t)\mathbf{i} + (t \cos t)\mathbf{j}$

    $\mathbf{V}(t) = \mathbf{R}'(t) = (\sin t + t\cos t)\mathbf{i} + (\cos t - t\sin t)\mathbf{j},$

    $\mathbf{A}(t) = \mathbf{V}'(t) = (-t \sin t + 2 \cos t)\mathbf{i}$
    $$+ (-t \cos t - 2 \sin t)\mathbf{j}$$

    $\frac{ds}{dt} = \|\mathbf{V}(t)\| = \sqrt{1 + t^2}$

    $A_T = \frac{d^2s}{dt^2} = \frac{t}{\sqrt{1 + t^2}}$

    $A_N = \sqrt{\|\mathbf{A}\|^2 - A_T^2}$

    $$= \sqrt{t^2 + 4 - \frac{t^2}{1 + t^2}}$$

    $$= \frac{\sqrt{(t^2 + 4)(1 + t^2) - t^2}}{\sqrt{1 + t^2}}$$

    $$= \frac{t^2 + 2}{\sqrt{t^2 + 1}}$$

5. $\mathbf{R}(t) = t\mathbf{i} + t^2\mathbf{j} + t\mathbf{k}$

    $\mathbf{V}(t) = \mathbf{R}'(t) = \mathbf{i} + 2t\mathbf{j} + \mathbf{k}$

    $\mathbf{A}(t) = \mathbf{V}'(t) = 2\mathbf{j}$

$$\frac{ds}{dt} = \|\mathbf{V}(t)\| = \sqrt{2 + 4t^2}$$

$$A_T = \frac{d^2s}{dt^2} = \frac{4t}{\sqrt{2 + 4t^2}}$$

$$A_N = \sqrt{\|\mathbf{A}\|^2 - A_T{}^2}$$

$$= \sqrt{\frac{8 + 16t^2 - 16t^2}{2 + 4t^2}}$$

$$= \frac{2}{\sqrt{1 + 2t^2}}$$

7. $\mathbf{R}(t) = (\sin t)\mathbf{i} + (\cos t)\mathbf{j} + (\sin t)\mathbf{k}$

$\mathbf{V}(t) = \mathbf{R}'(t) = (\cos t)\mathbf{i} - (\sin t)\mathbf{j} + (\cos t)\mathbf{k}$

$\mathbf{A}(t) = \mathbf{V}'(t) = (-\sin t)\mathbf{i} - (\cos t)\mathbf{j} - (\sin t)\mathbf{k}$

$$\frac{ds}{dt} = \|\mathbf{V}(t)\| = \sqrt{1 + \cos^2 t}$$

$$A_T = \frac{d^2s}{dt^2} = -\frac{\sin t \cos t}{\sqrt{1 + \cos^2 t}}$$

$$A_N = \sqrt{\|\mathbf{A}\|^2 - A_T{}^2}$$

$$= \sqrt{1 + \sin^2 t - \frac{\sin^2 t \cos^2 t}{1 + \cos^2 t}}$$

$$= \sqrt{\frac{2}{1 + \cos^2 t}}$$

9. $\mathbf{V}_0 = \langle 1, -3 \rangle;\ \mathbf{A}_0 = \langle 2, 5 \rangle$

$$\|\mathbf{V}_0\| = \sqrt{1^2 + (-3)^2} = \sqrt{10}$$

$$\|\mathbf{A}_0\| = \sqrt{2^2 + 5^2} = \sqrt{29}$$

$$A_T = \frac{\langle 1, -3 \rangle \cdot \langle 2, 5 \rangle}{\sqrt{10}}$$

$$= \frac{-13}{\sqrt{10}}$$

$$A_N = \sqrt{\|\mathbf{A}_0\|^2 - A_T{}^2}$$

$$= \sqrt{29 - \frac{169}{10}}$$

$$= \frac{11}{\sqrt{10}}$$

11. $\mathbf{V}_0 = 2\mathbf{i} + 3\mathbf{j} - \mathbf{k};\ \mathbf{A}_0 = -\mathbf{i} - 5\mathbf{j} + 2\mathbf{k}$

$$\|\mathbf{V}_0\| = \sqrt{2^2 + 3^2 + (-1)^2} = \sqrt{14}$$

$$\|\mathbf{A}_0\| = \sqrt{(-1)^2 + (-5)^2 + 2^2} = \sqrt{30}$$

$$A_T = \frac{\langle 2, 3, -1 \rangle \cdot \langle -1, -5, 2 \rangle}{\sqrt{14}}$$

$$= \frac{-19}{\sqrt{14}}$$

$$A_N = \sqrt{\|\mathbf{A}_0\|^2 - A_T{}^2}$$

$$= \sqrt{30 - \left(\frac{19}{\sqrt{14}}\right)^2}$$

$$= \sqrt{\frac{59}{14}}$$

13. $\|\mathbf{V}\| = \sqrt{5t^2 + 3}$

$$A_T = \|\mathbf{V}\|' = \frac{5t}{\sqrt{5t^2 + 3}}$$

at $t = 1,\ A_T = \frac{5}{\sqrt{8}}$

15. $\|\mathbf{V}\| = \sqrt{\sin^2 t + \cos 2t}$

$A_T = \|\mathbf{V}\|' = \dfrac{-\sin 2t}{2\sqrt{\sin^2 t + \cos 2t}}$;

at $t = 0$, $A_T = 0$

17. $\mathbf{R}(t) = (4\sin 2t)\mathbf{i} - (3\cos 2t)\mathbf{j}$

$\mathbf{R}'(t) = (8\cos 2t)\mathbf{i} + (6\sin 2t)\mathbf{j}$

speed $= \dfrac{ds}{dt}$

$= \|\mathbf{R}'(t)\|$

$= \sqrt{64\cos^2 2t + 36\sin^2 2t}$

$= \sqrt{28\cos^2 2t + 36}$

$\dfrac{d^2 s}{dt^2} = \dfrac{-28\sin 4t}{\sqrt{28\cos^2 2t + 36}}$

$\dfrac{d^2 s}{dt^2} = 0$ when

$\dfrac{-28\sin 4t}{\sqrt{28\cos^2 2t + 36}} = 0$

$-28\sin 4t = 0$

$\sin 4t = 0$

$4t = 0, \pi, 2\pi, \cdots$

$t = 0, \dfrac{\pi}{4}, \dfrac{\pi}{2}, \cdots$

The maximum speed is 8 at $t = 0$, and the minimum speed is 6 at $t = \pi/4$.

19. $x = 1 + \cos 2t$, $y = \sin 2t$, so

$\mathbf{R}(t) = (1 + \cos 2t)\mathbf{i} + (\sin 2t)\mathbf{j}$

$\mathbf{V}(t) = \mathbf{R}'(t) = (-2\sin 2t)\mathbf{i} + (2\cos 2t)\mathbf{j}$

$\mathbf{A}(t) = \mathbf{V}'(t) = (-4\cos 2t)\mathbf{i} - (4\sin 2t)\mathbf{j}$

$\dfrac{ds}{dt} = \|\mathbf{V}(t)\|$

$= \sqrt{(-2\sin 2t)^2 + (2\cos 2t)^2}$

$= \sqrt{2^2(\sin^2 2t + \cos^2 2t)}$

$= 2$

$A_T = \dfrac{d^2 s}{dt^2} = 0$

$A_N = \sqrt{\|\mathbf{A}\|^2 - A_T{}^2}$

$= 4$

21. $y = 4x^2$

Parameterize the curve by $x = t$, $y = 4t^2$;

$\mathbf{R}(t) = t\mathbf{i} + 4t^2\mathbf{j}$

$\mathbf{V}(t) = \mathbf{i} + 8t\mathbf{j}$

$\mathbf{A}(t) = \mathbf{V}'(t) = 8\mathbf{j}$

$\|\mathbf{A}\| = 8$

$\dfrac{ds}{dt} = \|\mathbf{V}(t)\|$

$= \sqrt{1 + 64t^2}$

If $\dfrac{ds}{dt} = 20$

$$1 + 64t^2 = 400$$

$$64t^2 = 399$$

$$t = \frac{\sqrt{399}}{8}$$

(The negative value is not in the domain.)

$$A_T = \frac{d^2 s}{dt^2} = \frac{64\,t}{\sqrt{1 + 64t^2}}$$

$$A_N = \sqrt{\|\mathbf{A}\|^2 - A_T^2}$$

$$= \sqrt{64 - \frac{(64\,t)^2}{1 + 64\,t^2}}$$

$$= \frac{8}{\sqrt{1 + 64t^2}}$$

At $t = \frac{\sqrt{399}}{8}$,

$$A_T = \frac{8\sqrt{399}}{20}$$

$$\approx 7.98999$$

$$A_N = \frac{8}{20} = 0.4$$

23. Since the pail moves in a vertical plane with angular velocity ω rev/s, we have

$$\mathbf{R}(t) = \langle 3\cos 2\pi\omega t,\ 3\sin 2\pi\omega t\rangle$$

and

$$\mathbf{V}(t) = \langle -6\pi\omega\sin 2\pi\omega t,\ 6\pi\omega\cos 2\pi\omega t\rangle$$

$$\mathbf{A}(t) = \langle -12\pi^2\omega^2\cos 2\pi\omega t,\ -12\pi^2\omega^2\sin 2\pi\omega t\rangle$$

$$\frac{ds}{dt} = \|\mathbf{V}(t)\| = 6\pi\omega$$

$$\|\mathbf{A}(t)\| = 12\pi^2\omega^2$$

$$A_T = \|\mathbf{V}\|' = 0$$

$$A_N = \sqrt{\|\mathbf{A}\|^2 - A_T^2}$$

$$= 12\pi^2\omega^2$$

The force due to the motion of the pail with mass 2/32 slugs is

$$F_N = mA_N$$

$$= \frac{2}{32}(12\pi^2\omega^2)$$

$$\approx 7.4\ \omega^2\ \text{lbs}$$

The force on the bottom of the pail is greatest at the lowest point of the swing where it equals

$$7.4\omega^2 + 2$$

The force is $7.4\omega^2 - 2$ lbs at the top of the swing, and to keep the water from spilling, we must have

$$7.4\omega^2 - 2 = 0$$

$$\omega = 0.52\ \text{rev/s}\ \ (\text{or } 31.2\ \text{rev/min})$$

25. **a.** Since the car moves in a circle, its tangential acceleration is 0. The force required to keep the car moving in a circle has magnitude

$$m\kappa\left(\frac{ds}{dt}\right)^2 = \frac{m}{\rho}\left(\frac{ds}{dt}\right)^2$$

where $m = W/g$ is the car's mass and $\rho = 1/\kappa$. We want this force to be balanced by the frictional force keeping the car on the road, so

$$\frac{m}{\rho}\left(\frac{ds}{dt}\right)^2 = \mu W$$

$$= \mu(mg)$$

The m's cancel (so weight does not matter), and for $\mu = 0.47$, $\rho = 150$, and $g = 32$, the maximum safe speed is

$$\frac{ds}{dt} = \sqrt{\mu g \rho}$$

$$= \sqrt{(0.47)(32)(150)}$$

$$= 47 \text{ ft/s} \quad (\text{about } 32 \text{ mi/h})$$

b. With a roadway banked at angle θ, the horizontal force pulling the car toward the center of the circle has magnitude

$$\|\mathbf{F}_k\| = \|\mathbf{F}_N\|\sin\theta + \|\mathbf{F}_s\|\cos\theta$$

$$= \|\mathbf{F}_N\|\sin\theta + \mu\|\mathbf{F}_N\|\cos\theta$$

where $\mathbf{F}_N$ is the normal force and $\mathbf{F}_s$ is the friction force, which is directed downward along the bank. We also have

$$mg = \|\mathbf{F}_N\|\cos\theta - \|\mathbf{F}_s\|\sin\theta$$

$$= \|\mathbf{F}_N\|\cos\theta - \mu\|\mathbf{F}_N\|\sin\theta$$

so that

$$\|\mathbf{F}_N\| = \frac{mg}{(\cos\theta - \mu\sin\theta)}$$

and

$$\|\mathbf{F}_k\| = (\sin\theta + \mu\cos\theta)\|\mathbf{F}_N\|$$

$$= \frac{mg(\sin\theta + \mu\cos\theta)}{\cos\theta - \mu\sin\theta}$$

To find the optimal banking speed we set this expression for $\|\mathbf{F}_k\|$ equal to

$$\frac{m}{\rho}\left(\frac{ds}{dt}\right)^2 = \frac{mg(\sin\theta + \mu\cos\theta)}{\cos\theta - \mu\sin\theta}$$

$$\frac{ds}{dt} = \sqrt{\frac{\rho g(\sin\theta + \mu\cos\theta)}{\cos\theta - \mu\sin\theta}}$$

For $\mu = 0.47$, $\rho = 150$, $g = 32$, and $\theta = 17°$, we have

$$\frac{ds}{dt} = \sqrt{\frac{150(32)(\sin 17° + 0.47\cos 17°)}{\cos 17° - 0.47\sin 17°}}$$

$$= 65.94 \text{ ft/s} \quad (\text{about } 45 \text{ mi/h})$$

c. For the optimal safe speed to be 50 mi/h (73.33 ft/s), we want θ to satisfy

$$(73.33)^2 = \frac{150(32)(\sin\theta + 0.47\cos\theta)}{\cos\theta - 0.47\sin\theta}$$

$$\theta \approx 23.1°$$

27. Since the radius of curvature is $r = 15$,

$$A_N = (2\pi r\omega)^2\kappa = 4\pi^2 r^2\left(\frac{1}{r}\right)\omega^2 = 4\pi^2 r\omega^2$$

A person on the wheel "flies off when"

$$mA_N = W$$

$$\frac{W}{g}(4\pi^2 r\omega^2) = W$$

$$\omega^2 = \frac{g}{4\pi^2(15)}$$

$$= \frac{32}{60\pi^2}$$

$$\omega \approx 0.23 \text{ rev/s}$$

$$\approx 14 \text{ rev/min}$$

29. $\mathbf{R}(t) = t\mathbf{i} + 2t\mathbf{j} + t^2\mathbf{k}$

$$\mathbf{R}'(t) = \mathbf{i} + 2\mathbf{j} + 2t\mathbf{k}$$

$$\mathbf{V}'(t) = \mathbf{R}''(t) = 2\mathbf{k}$$

$$\|\mathbf{R}'(t)\| = \sqrt{1 + 4 + 4t^2}$$

$$= \sqrt{5 + 4t^2}$$

$$\mathbf{R}' \cdot \mathbf{R}'' = 4t$$

$$\mathbf{R}' \times \mathbf{R}'' = \begin{vmatrix} \mathbf{i} & \mathbf{j} & \mathbf{k} \\ 1 & 2 & 2t \\ 0 & 0 & 2 \end{vmatrix}$$

$$= 4\mathbf{i} - 2\mathbf{j}$$

$$\|\mathbf{R}' \times \mathbf{R}''\| = \sqrt{16 + 4}$$

$$= 2\sqrt{5}$$

$$A_T = \frac{\mathbf{R}' \cdot \mathbf{R}''}{\|\mathbf{R}'\|}$$

$$= \frac{4t}{\sqrt{5 + 4t^2}}$$

$$A_N = \frac{\|\mathbf{R}' \times \mathbf{R}''\|}{\|\mathbf{R}'\|}$$

$$= \frac{2\sqrt{5}}{\sqrt{5 + 4t^2}}$$

31. $\mathbf{R}(t) = e^t\cos t\,\mathbf{i} + e^t\sin t\,\mathbf{j} + e^t\,\mathbf{k}$

$$\mathbf{R}'(t) = e^t[(-\sin t + \cos t)\mathbf{i} + (\cos t + \sin t)\mathbf{j} + \mathbf{k}]$$

$$\mathbf{R}''(t) = e^t[-2\sin t\,\mathbf{i} + 2\cos t\,\mathbf{j} + \mathbf{k}]$$

$$\|\mathbf{R}'(t)\| = \sqrt{3e^{2t}} = \sqrt{3}e^t$$

$$\mathbf{R}' \cdot \mathbf{R}'' = 3e^{2t};$$

$$\mathbf{R}' \times \mathbf{R}'' = \langle \sin t - \cos t, -\sin t - \cos t, 2\rangle e^{2t}$$

$$A_T = \frac{\mathbf{R}' \cdot \mathbf{R}''}{\|\mathbf{R}'\|}$$

$$= \frac{3e^{2t}}{\sqrt{3}\,e^t}$$

$$= \sqrt{3}e^t$$

$$A_N = \frac{\|\mathbf{R}' \times \mathbf{R}''\|}{\|\mathbf{R}'\|}$$

$$= \frac{\sqrt{6}\,e^{2t}}{\sqrt{3}\,e^t}$$

$$= \sqrt{2}e^t$$

33. The curve $x = \dfrac{y^2}{120}$ has parametric form:

$$x = \frac{t^2}{120}, \; y = t$$

$$x' = \frac{t}{60}, \; y' = 1$$

$$x'' = \frac{1}{60}, \; y'' = 0$$

The curvature is

$$\kappa = \frac{|x'y'' - y'x''|}{[(x')^2 + (y')^2]^{3/2}}$$

$$= \frac{\left|\dfrac{t}{60}(0) - (1)\dfrac{1}{60}\right|}{\left[\left(\dfrac{t}{60}\right)^2 + 1\right]^{3/2}}$$

$$= \frac{60^2}{[t^2 + 60^2]^{3/2}}$$

At $t = 0$, $\kappa(0) = \frac{1}{60}$ and

$$A_N = \kappa\left(\frac{ds}{dt}\right)^2$$

$$= \frac{1}{60}\left(\frac{ds}{dt}\right)^2$$

Thus, the maximum safe speed $\frac{ds}{dt}$ satisfies

$$A_N = \frac{1}{60}\left(\frac{ds}{dt}\right)^2 = 30$$

$$\frac{ds}{dt} = 42.43 \text{ units/s}$$

35. Let $R(x) = x\mathbf{i} + f(x)\mathbf{j}$.

$$\mathbf{R}' = \mathbf{i} + f'(x)\mathbf{j}$$

$$\|\mathbf{R}'\| = \sqrt{1 + [f'(x)]^2}$$

$$\mathbf{R}'' = f''(x)\mathbf{j}$$

$$\mathbf{R}' \cdot \mathbf{R}'' = f'(x)f''(x)$$

$$\mathbf{R}' \times \mathbf{R}'' = \begin{vmatrix} \mathbf{i} & \mathbf{j} & \mathbf{k} \\ 1 & f'(x) & 0 \\ 0 & f''(x) & 0 \end{vmatrix}$$

$$= f''(x)\mathbf{k}$$

$$A_T = \frac{\mathbf{R}' \cdot \mathbf{R}''}{\|\mathbf{R}'\|}$$

$$= \frac{f'(x)f''(x)}{\sqrt{1 + [f'(x)]^2}}$$

$$A_N = \frac{\|\mathbf{R}' \times \mathbf{R}''\|}{\|\mathbf{R}'\|}$$

$$= \frac{|f''(x)|}{\sqrt{1 + [f'(x)]^2}}$$

37. $\mathbf{A} = -g\mathbf{j}$

$$\mathbf{V} = (v_0\cos\alpha)\mathbf{i} + (-gt + v_0\sin\alpha)\mathbf{j}$$

$$\|\mathbf{V}\| = \sqrt{v_0{}^2\cos^2\alpha + (-gt + v_0\sin\alpha)^2}$$

The maximum height occurs when the $\mathbf{j}$ component of velocity is 0:

$$-gt_m + v_0\sin\alpha = 0$$

$$t_m = \frac{v_0}{g}\sin\alpha$$

We find $\mathbf{V} \cdot \mathbf{A} = g^2 t - gv_0\sin\alpha$

and $\mathbf{V} \times \mathbf{A} = -(gv_0\cos\alpha)\mathbf{k}$

At time $t = t_m$, we have

$$\|\mathbf{V}(t_m)\| = |v_0\cos\alpha|; \ \mathbf{V} \cdot \mathbf{A} = 0;$$

$$\|\mathbf{V} \times \mathbf{A}\| = |gv_0\cos\alpha|, \text{ so that}$$

$$A_T = \frac{\mathbf{V} \cdot \mathbf{A}}{\|\mathbf{V}\|} = 0 \text{ and}$$

$$A_N = \frac{\|\mathbf{V} \times \mathbf{A}\|}{\|\mathbf{V}\|} = \frac{|gv_0\cos\alpha|}{|v_0\cos\alpha|} = g$$

39. a. Let $R_s = R + R_e$ be the distance of the satellite from the center of the earth. From Example 4,

$$v = \sqrt{\frac{GM}{R_s}}$$

$$T = \frac{2\pi R_s}{v}$$

$$= \frac{2\pi R_s^{3/2}}{\sqrt{GM}}$$

b. $T = (24)(3,600)$

$$= \frac{2\pi R_s^{3/2}}{\sqrt{398,600}}$$

$$R_s^{3/2} = \frac{(24)(3,600)\sqrt{398,600}}{2\pi}$$

$$\approx 8,681,655$$

$$R_s \approx 42,241 \text{ and}$$

$$R \approx 42,241 - 6,440$$

$$\approx 36,000 \text{ km}$$

c. The speed is

$$v = \sqrt{\frac{398,600}{35,801}}$$

$$\approx 3.3367 \text{ km/s}$$

This is about 200 km/h.

CHAPTER 10 REVIEW

Proficiency Examination, Pages 679–680

SURVIVAL HINT: *To help you review the concepts of this chapter,* handwrite *the answers to each of these questions onto your own paper.*

1. A vector-valued function (or, simply, a vector function) **F** with domain D assigns a unique vector $\mathbf{F}(t)$ to each scalar t in the set D. The set of all vectors **v** of the form $\mathbf{v} = \mathbf{F}(t)$ for t in D is the range of **F**.

2. $\mathbf{F}(t) = f_1(t)\mathbf{i} + f_2(t)\mathbf{j}$ in $\mathbb{R}^2$ (plane) or

   $\mathbf{F}(t) = f_1(t)\mathbf{i} + f_2(t)\mathbf{j} + f_3(t)\mathbf{k}$ in $\mathbb{R}^3$ (space)

   where f_1, f_2, and f_3 are real-valued (scalar-valued) functions of the real number t defined on the domain set D. In this context f_1, f_2, and f_3 are called the components of **F**.

3. The graph of $\mathbf{F}(t) = f_1(t)\mathbf{i} + f_2(t)\mathbf{j} + f_3(t)\mathbf{k}$ in $\mathbb{R}^3$ (space) is the graph of the parametric equations $x = f_1(t)$, $y = f_2(t)$, $z = f_3(t)$.

4. The limit of a vector function consists of the vector sum of the limits of its individual components.

5. $\mathbf{F}'(t) = f_1'(t)\mathbf{i} + f_2'(t)\mathbf{j} + f_3'(t)\mathbf{k}$

6. $\displaystyle\int \mathbf{F}(t)\,dt$

   $\displaystyle= \int f_1(t)\,dt\,\mathbf{i} + \int f_2(t)\,dt\,\mathbf{j} + \int f_3(t)\,dt\,\mathbf{k}$

7. The graph of the vector function defined by $\mathbf{F}(t)$ is said to be *smooth* on any interval of t where $\mathbf{F}'$ is continuous and $\mathbf{F}'(t) \neq \mathbf{0}$.

8. **a.** Linearity rule

   $(a\mathbf{F} + b\mathbf{G})'(t) = a\mathbf{F}'(t) + b\mathbf{G}'(t)$

   for constants a, b

b. Scalar multiple

$$(h\mathbf{F})'(t) = h'(t)\mathbf{F}(t) + h(t)\mathbf{F}'(t)$$

c. Dot product rule

$$(\mathbf{F} \cdot \mathbf{G})'(t) = (\mathbf{F}' \cdot \mathbf{G})(t) + (\mathbf{F} \cdot \mathbf{G}')(t)$$

d. Cross product rule

$$(\mathbf{F} \times \mathbf{G})'(t) = (\mathbf{F}' \times \mathbf{G})(t) + (\mathbf{F} \times \mathbf{G}')(t)$$

e. Chain rule

$$[\mathbf{F}(h(t))]' = h'(t)\mathbf{F}'(h(t))$$

9. If the nonzero vector function $\mathbf{F}(t)$ is differentiable and has constant length, then $\mathbf{F}(t)$ is orthogonal to the derivative vector $\mathbf{F}'(t)$.

10. $\mathbf{R}(t) = x(t)\mathbf{i} + y(t)\mathbf{j} + z(t)\mathbf{k}$ is the position;

$$\mathbf{V}(t) = \frac{d\mathbf{R}(t)}{dt} = \mathbf{R}'(t) \text{ is the velocity;}$$

$$\mathbf{A}(t) = \frac{d\mathbf{V}(t)}{dt} = \frac{d^2\mathbf{R}(t)}{dt^2} \text{ is the acceleration}$$

11. The speed is $\|\mathbf{V}(t)\|$.

12. Consider a projectile that travels in a vacuum in a coordinate plane, with the x-axis along level ground. If the projectile is fired from a height of s_0 with initial speed v_0 and angle of elevation α, then at time t $(t \geq 0)$ it will be at the point $(x(t), y(t))$, where

$$x(t) = (v_0\cos \alpha)t, \; y(t) = -\tfrac{1}{2}gt^2 + (v_0\sin \alpha)t + s_0$$

13. A projectile fired from ground level has time of flight T_f and range R given by the equations

$$T_f = \tfrac{2}{g}v_0\sin \alpha \text{ and } R = \frac{v_0^2}{g}\sin 2\alpha$$

The maximal range is $R_m = \frac{v_0^2}{g}$, and it occurs when $\alpha = \frac{\pi}{4}$.

14. **1.** The planets move about the sun in elliptical orbits, with the sun at one focus.

2. The radius vector joining a planet to the sun sweeps over equal areas in equal intervals of time.

3. The square of the time of one complete revolution of a planet about its orbit is proportional to the cube of the orbit's semimajor axis.

15. $\mathbf{u}_r = (\cos \theta)\mathbf{i} + (\sin \theta)\mathbf{j}$ and

$$\mathbf{u}_\theta = (-\sin \theta)\mathbf{i} + (\cos \theta)\mathbf{j}$$

16. If $\mathbf{R}(t)$ is a vector function that defines a smooth graph $(\mathbf{R}'(t) \neq \mathbf{0})$, then at each point a unit tangent is

$$\mathbf{T}(t) = \frac{\mathbf{R}'(t)}{\|\mathbf{R}'(t)\|}$$

and the principal unit normal vector is

$$\mathbf{N}(t) = \frac{\mathbf{T}'(t)}{\|\mathbf{T}'(t)\|}$$

17. Suppose the position of a moving object is $R(t)$, where $R'(t)$ is continuous on the interval $[a, b]$. Then the object has speed

$$\|V(t)\| = \|R'(t)\| = \frac{ds}{dt} \qquad \text{for } a \le t \le b$$

18. If C is a smooth curve defined by

$R(t) = x(t)i + y(t)j + z(t)k$ on an interval

$[a, b]$, then the arc length of C is given by

$$s = \int_a^b \|R'(t)\| \, dt$$

$$= \int_a^b \sqrt{[x'(t)]^2 + [y'(t)]^2 + [z'(t)]^2} \, dt$$

19. The curvature of a graph is an indication of how quickly the graph changes direction.

$$\kappa = \left\|\frac{dT}{ds}\right\|$$

where T is a unit vector.

20. If $V = R'$, $A = R''$, then $\kappa = \dfrac{\|V \times A\|}{\|V\|^3}$

21. $T = \dfrac{dR}{ds}$ and $N = \dfrac{1}{\kappa}\left(\dfrac{dT}{ds}\right)$

22. The radius of curvature is the reciprocal of curvature $\rho = 1/\kappa$.

23. The acceleration A of a moving object can be written as $A = A_T T + A_N N$, where

$A_T = \dfrac{d^2 s}{dt^2}$ is the tangential component; and

$A_N = \kappa\left(\dfrac{ds}{dt}\right)^2$ is the normal component.

24. This is a helix with radius of 3, climbing in a counter-clockwise direction.

$$s = \int_0^{2\pi} \|R'\| \, dt$$

$R = (3 \cos t)i + (3 \sin t)j + tk,$

$R' = (-3 \sin t)i + (3 \cos t)j + k,$

$\|R'\| = \sqrt{9 \sin^2 t + 9 \cos^2 t + 1} = \sqrt{10}$

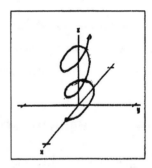

$$s = \int_0^{2\pi} \sqrt{10} \, dt = 2\pi\sqrt{10}$$

25. $F' = \dfrac{1}{(1 + t)^2}i + \dfrac{t \cos t - \sin t}{t^2}j + (-\sin t)k$

$F'' = -\dfrac{2}{(1 + t)^3}i + \dfrac{-t^2 \sin t - 2t\cos t + 2\sin t}{t^3}j$

$\qquad - (\cos t)k$

26. $\begin{vmatrix} i & j & k \\ 3t & 0 & 3 \\ 0 & \ln t & -t^2 \end{vmatrix} = (-3\ln t)i + (3t^3)j + (3t\ln t)k$

$$\int_1^2 [(-3\ln t)\mathbf{i} + (3t^3)\mathbf{j} + (3t\ln t)\mathbf{k}]\, dt$$

$$= \left[-3(t\ln t - t)\mathbf{i} + \frac{3t^4}{4}\mathbf{j} + 3\left(\frac{t^2}{2}\left[\ln t - \frac{1}{2}\right]\right)\mathbf{k} \right]\Bigg|_1^2$$

$$\boxed{\text{Formulas } 499\ 502}$$

$$= (6 - 6\ln 2)\mathbf{i} + 12\mathbf{j} + (6\ln 2 - 3)\mathbf{k}$$

$$- (3\mathbf{i} + \tfrac{3}{4}\mathbf{j} - \tfrac{3}{4}\mathbf{k})$$

$$= (3 - 6\ln 2)\mathbf{i} + \tfrac{45}{4}\mathbf{j} + (6\ln 2 - \tfrac{9}{4})\mathbf{k}$$

27. $\mathbf{F}'' = e^t\mathbf{i} - t^2\mathbf{j} + 3\mathbf{k},$

$\mathbf{F}' = e^t\mathbf{i} - \dfrac{t^3}{3}\mathbf{j} + 3t\mathbf{k} + \mathbf{C}_1,$ but $\mathbf{F}'(0) = 3\mathbf{k}$

so

$$\mathbf{F}' = (e^t - 1)\mathbf{i} - \frac{t^3}{3}\mathbf{j} + (3t + 3)\mathbf{k}$$

$\mathbf{F} = (e^t - t)\mathbf{i} - \dfrac{t^4}{12}\mathbf{j} + (\dfrac{3t^2}{2} + 3t)\mathbf{k} + \mathbf{C}_2,$ but

$$\mathbf{F}(0) = \mathbf{i} - 2\mathbf{j} \text{ so}$$

$$\mathbf{F} = (e^t - t)\mathbf{i} - \left(\frac{t^4}{12} + 2\right)\mathbf{j} + (\frac{3t^2}{2} + 3t)\mathbf{k}$$

28. $\mathbf{R} = t\mathbf{i} + 2t\mathbf{j} + te^t\mathbf{k}$

$$\mathbf{V} = \mathbf{R}' = \mathbf{i} + 2\mathbf{j} + e^t(t + 1)\mathbf{k}$$

$$\frac{ds}{dt} = \|\mathbf{V}\|$$

$$= \sqrt{1 + 4 + e^{2t}(t + 1)^2}$$

$$= \sqrt{5 + e^{2t}(t + 1)^2}$$

$$\mathbf{A} = \mathbf{V}'$$

$$= e^t(t + 2)\mathbf{k}$$

29. $\mathbf{R} = t^2\mathbf{i} + 3t\mathbf{j} - 3t\mathbf{k}$

$$\mathbf{T} = \frac{\mathbf{R}'}{\|\mathbf{R}'\|}$$

$$= \frac{2t\mathbf{i} + 3\mathbf{j} - 3\mathbf{k}}{\sqrt{4t^2 + 18}}$$

$$\mathbf{T}' = \frac{36\mathbf{i} - 12t\mathbf{j} + 12t\mathbf{k}}{(4t^2 + 18)^{3/2}}$$

$$\mathbf{N} = \frac{\mathbf{T}'}{\|\mathbf{T}'\|}$$

$$= \frac{3\mathbf{i} - t\mathbf{j} + t\mathbf{k}}{(2t^2 + 9)^{1/2}}$$

$$\mathbf{A} = \mathbf{R}''$$

$$= 2\mathbf{i}$$

$$\mathbf{R}' \times \mathbf{R}'' = \begin{vmatrix} \mathbf{i} & \mathbf{j} & \mathbf{k} \\ 2t & 3 & -3 \\ 2 & 0 & 0 \end{vmatrix}$$

$$= -6\mathbf{j} - 6\mathbf{k}$$

$$\kappa = \frac{\|\mathbf{R}' \times \mathbf{R}''\|}{\|\mathbf{R}'\|^3}$$

$$= \frac{\sqrt{36 + 36}}{(4t^2 + 18)^{3/2}}$$

$$= \frac{\sqrt{72}}{(4t^2 + 18)^{3/2}}$$

$$= \frac{3}{(2t^2 + 9)^{3/2}}$$

$$A_T = \frac{d^2s}{dt^2}$$

$$= \frac{4t}{(4t^2 + 18)^{1/2}}$$

$$= \frac{2\sqrt{2}t}{\sqrt{2t^2 + 9}}$$

$$A_N = \kappa\left(\frac{ds}{dt}\right)^2$$

$$= \frac{3}{(2t^2 + 9)^{3/2}}(4t^2 + 18)$$

$$= \frac{6}{\sqrt{2t^2 + 9}}$$

30. a. $\mathbf{R} = [(v_0\cos\alpha)t]\mathbf{i}$

$$+ [(v_0\sin\alpha)t - \tfrac{1}{2}gt^2 + s_0]\mathbf{j};$$

In our case:

$$\mathbf{R}(t) = (25\sqrt{3}\ t)\mathbf{i} + (25t - 16t^2)\mathbf{j}$$

$$\mathbf{R}'(t) = (25\sqrt{3})\mathbf{i} + (25 - 32t)\mathbf{j}$$

The maximum height is reached when the $\mathbf{j}$ component of $\mathbf{R}'$ is 0; that is, when

$$25 - 32t = 0$$

$$t = \frac{25}{32}$$

$$y_{\max} = 25\left(\tfrac{25}{32}\right) - 16\left(\tfrac{25}{32}\right)^2$$

$$\approx 9.77\ \text{ft}$$

b. $T_f = \frac{2}{g}\,v_0\sin\alpha$

$$= \tfrac{1}{16}(50)\left(\tfrac{1}{2}\right)$$

$$= \tfrac{25}{16}\ \text{sec}$$

$$\text{Range} = \frac{v_0^2}{g}\sin 2\alpha$$

$$= \frac{50^2}{32}\sin 60°$$

$$= \frac{625\sqrt{3}}{16}$$

$$\approx 67.7\ \text{ft}$$

Cumulative Review for Chapters 1-10, Pages 691-692

SURVIVAL HINT: *This is the third of four cumulative reviews included in the text. This one (Chapters 1-10) can be very valuable to refresh some of the skills and concepts that you may not have used for awhile. It is also a valuable tool in preparing for midterm or final examinations. If you do not have the time to actually do all of these problems, try looking at each one to see if you recall the concept involved and how to proceed with the solution. If you are confident about your ability to solve the problem, do not spend the time. If you feel a little uncertain about the problem, refer to the appropriate section, review the concepts, look in your old homework for a similar problem, and then see if you can work it. Be more concerned about understanding the concept than about obtaining a correct answer. Do not spend a lot of your time looking for algebraic or arithmetic errors, but rather focus on important ideas.*

1. Step 1. Simplify;

Step 2. Use basic formulas;

Step 3. Substitute;

Step 4. Classify.

Step 5. Try again. See Table 7.2, Pages 457-458.

3. A vector is a directed line segment. A vector function is a function whose range is a set of vectors for each point in its domain. Vector calculus involves differentiation and integration of vector functions.

5.
$$\lim_{x \to +\infty} \frac{(2 - \frac{3}{x} + \frac{1}{x^2})(\frac{1}{x} - 5)}{5 + \frac{4}{x^2} - \frac{9}{x^3}} = -\frac{10}{5}$$

$$= -2$$

7. Let $L = \lim_{x \to +\infty} \left(1 - \frac{2}{x}\right)^{3x}$

$$= \lim_{x \to +\infty} \left[\left(1 - \frac{2}{x}\right)^{-x/2}\right]^{-6}$$

$$\ln L = \lim_{x \to +\infty} \frac{3 \ln\left(1 - \frac{2}{x}\right)}{\frac{1}{x}}$$

$$= 3 \lim_{x \to +\infty} \frac{\left(\frac{1}{1 - \frac{2}{x}}\right)\left(\frac{2}{x^2}\right)}{\left(\frac{-1}{x^2}\right)} \quad \textit{l'Hôpital's rule}$$

$$= 3(-2) = -6$$

Thus, $L = e^{-6}$.

9. $\lim_{x \to 0} \frac{\cos x - 1}{e^x - x - 1} = \lim_{x \to 0} \frac{-\sin x}{e^x - 1}$

$$= \lim_{x \to 0} \frac{-\cos x}{e^x} \quad \textit{l'Hôpital's rule}$$

$$= -1$$

11. $y = \sin^3 x + 2 \tan x$

$$y' = 3 \sin^2 x \cos x + 2 \sec^2 x$$

13. $y = \sin^{-1} x + 2 \tan^{-1} \frac{1}{x}$

$$y' = \frac{1}{\sqrt{1 - x^2}} + \frac{2}{1 + \left(\frac{1}{x}\right)^2}\left(-\frac{1}{x^2}\right)$$

$$= \frac{1}{\sqrt{1 - x^2}} - \frac{2}{x^2 + 1}$$

15. $xy^3 + x^2 e^{-y} = 4$

$$x(3y^2 y') + y^3 + x^2(-e^{-y}y') + 2xe^{-y} = 0$$

$$(3xy^2 - x^2 e^{-y})y' = -y^3 - 2xe^{-y}$$

$$y' = \frac{-(y^3 + 2xe^{-y})}{3xy^2 - x^2 e^{-y}}$$

17. $\int \sin^2 x \cos^3 x \, dx = \int \sin^2 x (1 - \sin^2 x) \cos x \, dx$

$$= \int \sin^2 x (\cos x \, dx) - \int \sin^4 x (\cos x \, dx)$$

$$= \frac{1}{3} \sin^3 x - \frac{1}{5} \sin^5 x + C$$

19. $\int \frac{dx}{1 + \cos x} = \int \frac{1 - \cos x}{1 - \cos^2 x} \, dx$

$$= \int \frac{dx}{\sin^2 x} - \int \frac{\cos x \, dx}{\sin^2 x}$$

$$= \int \csc^2 x \, dx - \int \cot x \csc x \, dx$$

$$= -\cot x + \csc x + C$$

21. $\int \frac{dx}{\sqrt{2x - x^2}} = \int \frac{dx}{\sqrt{-(x^2 - 2x + 1) + 1}}$

$$= \int \frac{dx}{\sqrt{1 - (x - 1)^2}}$$

$$= \int \frac{du}{\sqrt{1 - u^2}}$$

$$= \sin^{-1} u + C$$

$$= \sin^{-1}(x - 1) + C$$

23. a. Yes; since $f'(x)$ can be negative and $f(0) = 0$, we know that f can decrease for $x \geq 0$.

b. Yes; since $f'(x)$ can be positive and $f(0) = 0$, we know that f can be negative at $x = -2$.

c. No, it need not have a critical point. For example, if $f(x) = x$, then the conditions of this problem are satisfied.

25. The curves intersect at $x = \frac{\pi}{4}$.

$$A = \int_0^{\pi/4} (\cos x - \sin x)\, dx + \int_{\pi/4}^1 (\sin x - \cos x)\, dx$$

$$= (\sin x + \cos x)\Big|_0^{\pi/4} - (\cos x + \sin x)\Big|_{\pi/4}^1$$

$$= 2\sqrt{2} - 1 - \cos 1 - \sin 1$$

$$\approx 0.4467$$

27. $V = \pi \int_1^2 [2(3x - 2)^{-1/2}]^2 \, dx$

$$\boxed{\text{Let } u = 3x - 2; \ du = 3\, dx}$$

$$= \pi \int_1^4 4u^{-1} \frac{du}{3}$$

$$= \frac{4\pi}{3} \ln|u|\Big|_1^4$$

$$= \frac{4\pi}{3} \ln 4 \approx 5.8069$$

29. a. Let $A(5, 1, 2)$, $B(3, 1, -2)$, $C(3, 2, 5)$

$$\mathbf{AB} = -2\mathbf{i} - 4\mathbf{k}$$

$$\mathbf{AC} = -2\mathbf{i} + \mathbf{j} + 3\mathbf{k}$$

$$\mathbf{N} = \begin{vmatrix} \mathbf{i} & \mathbf{j} & \mathbf{k} \\ -2 & 0 & -4 \\ -2 & 1 & 3 \end{vmatrix}$$

$$= 4\mathbf{i} + 14\mathbf{j} - 2\mathbf{k}$$

The desired plane is

$$4(x - 5) + 14(y - 1) - 2(z - 2) = 0$$

$$2x + 7y - z - 15 = 0$$

b. $P(-2, -1, 4)$; a vector normal to the plane is parallel to the line, so

$\mathbf{N} = 2\mathbf{i} + 5\mathbf{j} - 2\mathbf{k}$. The desired plane is

$$2(x + 2) + 5(y + 1) - 2(z - 4) = 0$$

$$2x + 5y - 2z + 17 = 0$$

31. $\displaystyle\sum_{k=1}^{\infty} \frac{1}{k \cdot 4^k}$ converges by the ratio test

$$\lim_{k \to \infty} \frac{k \, 4^k}{(k+1)4^{k+1}} = \frac{1}{4} < 1$$

$$= \frac{\pi}{2}$$

33. $\displaystyle\sum_{k=1}^{\infty} \frac{k!}{2^k \cdot k}$ diverges by the ratio test

Converges

$$\lim_{k \to \infty} \frac{(k+1)! \, 2^k \, k}{(k+1)2^{k+1}k!} = \lim_{k \to \infty} \frac{(k+1)k}{2(k+1)}$$

$$= \infty > 1$$

39. $\mathbf{F}(t) = 2t\mathbf{i} + e^{-3t}\mathbf{j} + t^4\mathbf{k}$

$\mathbf{F}'(t) = 2\mathbf{i} - 3e^{-3t}\mathbf{j} + 4t^3\mathbf{k}$

$\mathbf{F}''(t) = 9e^{-3t}\mathbf{j} + 12t^2\mathbf{k}$

35. The related absolute value series is $\displaystyle\sum \frac{1}{k^3}$ which converges (p-series, $p = 3 > 1$). Thus, the given series converges by the absolute convergence test.

41. $\mathbf{R}(t) = 2(\sin 2t)\mathbf{i} + (2 + 2\cos 2t)\mathbf{j} + 6t\mathbf{k}$

$\mathbf{R}'(t) = 4(\cos 2t)\mathbf{i} - (4\sin 2t)\mathbf{j} + 6\mathbf{k}$

$\|\mathbf{R}'\| = \sqrt{16 + 36} = 2\sqrt{13}$

$\mathbf{T} = \dfrac{1}{\sqrt{13}}[2(\cos 2t)\mathbf{i} - 2(\sin 2t)\mathbf{j} + 3\mathbf{k}]$

$\dfrac{d\mathbf{T}}{dt} = \dfrac{1}{\sqrt{13}}[-4(\sin 2t)\mathbf{i} - 4(\cos 2t)\mathbf{j}]$

$\mathbf{N} = -\sin 2t\,\mathbf{i} - \cos 2t\,\mathbf{j}$

37. **a.** $\displaystyle\int_{1}^{\infty} x^2 e^{-x}\, dx$ $\boxed{\text{Formula 485}}$

$$= \lim_{t \to +\infty} \int_{1}^{t} x^2 e^{-x}\, dx$$

$$= \lim_{t \to +\infty} \frac{e^{-x}}{-1}\left(x^2 + 2x + 2\right)\Big|_{1}^{t}$$

$$= \frac{1}{e}(1 + 2 + 2)$$

$$= \frac{5}{e}$$

43. **a.** $e^x = \displaystyle\sum_{k=0}^{\infty} \frac{x^k}{k!}$, so

$$e^{-x^2} = \sum_{k=0}^{\infty} \frac{(-x^2)^k}{k!}$$

$$= \sum_{k=0}^{\infty} \frac{(-1)^k x^{2k}}{k!}$$

and

Converges

b. $\displaystyle\int_{0}^{2} \frac{dx}{\sqrt{4 - x^2}} = \lim_{t \to 2^-} \int_{0}^{t} \frac{dx}{\sqrt{4 - x^2}}$

$$= \lim_{t \to 2^-} \sin^{-1}\frac{x}{2}\Big|_{0}^{t}$$

$$x^2 e^{-x^2} = x^2 \sum_{k=0}^{\infty} \frac{(-1)^k x^{2k}}{k!}$$

$$= \sum_{k=0}^{\infty} \frac{(-1)^k x^{2k+2}}{k!}$$

b. $\displaystyle\int_0^1 x^2 e^{-x^2}\,dx = \sum_{k=0}^{\infty} \frac{(-1)^k}{k!} \int_0^1 x^{2k+2}\,dx$

$$= \frac{x^3}{3} - \frac{x^5}{5(1!)} + \frac{x^7}{7(2!)} - \frac{x^9}{9(3!)} + \cdots \Big|_0^1$$

$$\approx \frac{1}{3} - \frac{1}{5} + \frac{1}{14} - \frac{1}{54} + \frac{1}{264} - \frac{1}{1,560} + \frac{1}{10,800}$$

$$\approx 0.189$$

45. $\dfrac{dy}{dx} + 2y = x^2$

The integrating factor is

$$I = e^{\int 2\,dx} = e^{2x}$$

$$y = \frac{1}{e^{2x}}\left[\int x^2 e^{2x}\,dx + C \right]$$

$$= \frac{1}{e^{2x}}\left[\frac{e^{2x}}{4}(2x^2 - 2x + 1) + C \right]$$

If $x = 0$, then $y = 2$, so

$$y = \frac{x^2}{2} - \frac{x}{2} + \frac{1}{4} + \frac{7}{4}e^{-2x}$$

47. $x = t^2 - 2t - 1,\ y = t^4 - 4t^2 + 2$

$\dfrac{dx}{dt} = 2t - 2;\ \dfrac{dy}{dt} = 4t^3 - 8t$

$\mathbf{R}'(t) = \langle 2t - 2,\ 4t^3 - 8t \rangle$

At $t = 1$, the point is $P(-2, -1)$, and the

tangent line is parallel to the vector

$\mathbf{R}'(1) = \langle 0,\ -4 \rangle$.

Thus, the tangent line is vertical with

equation $x = -2$.

49. The element of rope dx is lifted through a

distance $50 - x$, so $dF_r = 0.25\,dx$ and

$$W_r = \int_0^{50} 0.25(50 - x)\,dx$$

$$= -\frac{0.25}{2}(50 - x)^2 \Big|_0^{50}$$

$$= 312.5 \text{ ft-lb}$$

For the bucket, $W_b = 50(25) = 1,250$ ft-lb.

Thus, $W = W_b + W_r = 1,562.5$ ft-lb.

CHAPTER 11

Partial Differentiation

11.1 Functions of Several Variables, Pages 699-701

1. $f(x, y) = x^2 y + xy^2$

   a. $f(0, 0) = 0$

   b. $f(-1, 0) = 0$

   c. $f(0, -1) = 0$

   d. $f(1, 1) = 2$

   e. $\quad f(2, 4) = 2^2(4) + 2(4)^2$

            $= 48$

   f. $\quad f(t, t) = t^2(t) + t(t)^2$

            $= 2t^3$

   g. $\quad f(t, t^2) = t^2 t^2 + t(t^2)^2$

             $= t^4 + t^5$

   h. $f(1 - t, t) = (1 - t)^2 t + (1 - t)t^2$

               $= t - t^2$

3. A (real) function of two variables associates a real number with each point in the two-dimensional xy-plane; examples vary.

SURVIVAL HINT: *To find the domain of a function (in $\mathbb{R}^2$ or $\mathbb{R}^3$) look for values of the variable(s) that cause division by 0 or negative numbers under a square root (or even roots); these values are excluded. Other situations are possible, and would be found by looking for values of the variable that cause the expression to be undefined.*

	Domain	Range
5.	$x - y \geq 0$	$f \geq 0$
7.	$uv \geq 0$	$f \geq 0$
9.	$y - x > 0$	$\mathbb{R}$
11.	$\mathbb{R}^2$	$f \geq 0$
13.	$x^2 - y^2 > 0$	$f > 0$

15. We graph a few representatives (which may vary).

$$2x - 3y = 0$$
$$2x - 3y = 1$$
$$2x - 3y = 2$$
$$2x - 3y = 3$$

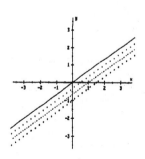

17. We graph a few representatives (which may vary).

$$x^3 - y = 0$$
$$x^3 - y = 1$$
$$x^3 - y = 2$$
$$x^3 - y = 3$$

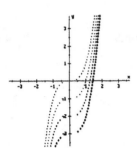

19. We graph a few representatives (which may vary).

$$x^2 + \frac{y^2}{4} = 1$$

$$x^2 + \frac{y^2}{4} = 2$$

$$x^2 + \frac{y^2}{4} = 3$$

$$x^2 + \frac{y^2}{4} = 4$$

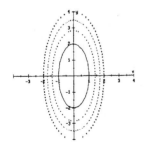

21. This is a cylinder $y^2 + z^2 = 1$, which has the x-axis as its axis.

23. This is a plane $x + y - z = 1$; its trace in the xy-plane is the line $x + y = 1$; its trace in the xz-plane is $x - z = 1$; and its trace in the yz-plane is $y - z = 1$.

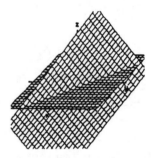

25. This is a sphere:

$$(x + 1)^2 + (y - 2)^2 + (z - 3)^2 = 4$$

its cross sections in the planes $x = -1$, $y = 2$, and $z = 3$ are circles.

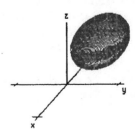

27. Ellipsoid; traces are ellipses in all three coordinate planes.

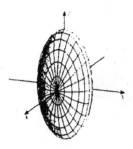

29. Hyperboloid (1 sheet); trace in the xy-plane is an ellipse; in the xz-and yz-planes the traces are hyperbolas.

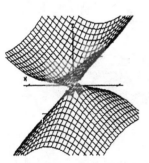

31 Elliptic paraboloid; traces in the xz- and yz-planes are parabolas; trace in the xy-plane is a point. In a plane parallel to the xy-plane the trace is an ellipse.

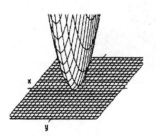

33. Elliptic cone; traces in the xz- and yz-planes are pairs of lines; trace in the xy-plane is the origin if $z = 0$, and is an ellipse if $z \neq 0$.

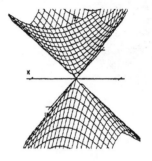

35. $f(x, y) = x^2 - y^2$

This is an equation of the form $z = \dfrac{x^2}{a^2} - \dfrac{y^2}{b^2}$ which is a hyperbolic paraboloid; the level curves are hyperbolas as shown, so the surface must be the one shown in D.

37. $f(x, y) = \dfrac{1}{x^2 + y^2}$

The level curves are circles, and the only surface shown that has circular level curves is the surface shown in B.

39. $f(x, y) = \dfrac{\cos xy}{x^2 + y^2}$

It is difficult to identify the level curves; but we do see that the center portion does not have any level curves shown. Also note (from the equation) that as we move away from the origin, there is a periodic effect. This looks most like the one shown in A.

41. $z = -4$; this is a plane parallel to the xy-plane.

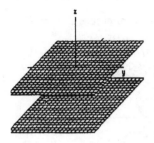

43. $z = y^2 + 1$; this is a cylinder which bends upward from the xy-plane.

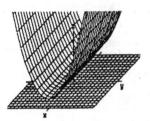

45. $z = 2x - 3y$; this is a plane which intersects the xy-plane along the line $2x - 3y = 0$.

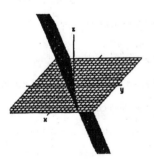

47. $z = 2x^2 + y^2$; this is an elliptic paraboloid with vertex at the origin which rises upward from the xy-plane.

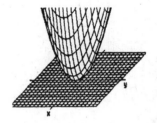

49. $z = \frac{x}{y}$; this is a surface that lies close to the xy-plane as it moves away from the origin, but "peels" upward close to the x-axis.

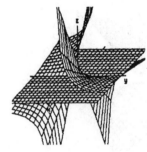

51. $z = x^2 + y^2 + 2$; this is a paraboloid with vertex at $(0, 0, 2)$ and rises upward from that point.

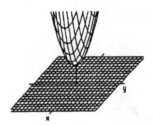

SURVIVAL HINT: *The trace $y = f(x)$ in $\mathbb{R}^2$ is a point on a line. The level curve for $z = f(x, y)$ in $\mathbb{R}^3$ is a curve or line in a plane, which is a slice of $\mathbb{R}^3$. If you can imagine a level surface for $w = f(x, y, z)$ in $\mathbb{R}^4$ you would see a surface in $\mathbb{R}^3$. Since we are trapped in a three-dimensional world we cannot visualize $\mathbb{R}^4$, but the mathematics in not restricted to three variables.*

53. $E(x, y) = \dfrac{7}{\sqrt{3 + x^2 + 2y^2}}$

Equipotential curves are ellipses.

$E = 1$: $x^2 + 2y^2 = 46$

$E = 2$: $x^2 + 2y^2 = \frac{37}{4}$

$E = 3$: $x^2 + 2y^2 = \frac{22}{9}$

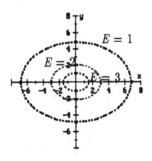

55. $\dfrac{1}{d_o} + \dfrac{1}{d_i} = \dfrac{1}{L}$

$$L = \frac{d_o d_i}{d_i + d_o} \qquad d_o > 0,\ d_i > 0$$

The level curves $L = K$ satisfy

$$\frac{d_o d_i}{d_i + d_o} = K \text{ or } d_o = \frac{K d_i}{d_i - K}$$

A typical level curve for $(d_o,\ d_i)$ is a hyperbola

with vertical asymptote $d_i = K$ and horizontal asymptote $d_o = K$.

57. $10xy = 1,000$ or $y = \frac{100}{x}$ (a hyperbola)

59. $z = xy$; we sketch a few representatives (answers vary).

$xy = 1$; $xy = 2$; $xy = 3$; $xy = 4$, and $xy = 5$

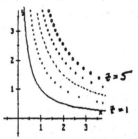

61. x machines sold at $60 - \frac{x}{5} + \frac{y}{20}$ thousand dollars apiece and y machines at $50 - \frac{x}{10} + \frac{y}{20}$ thousand dollars apiece yield revenue

$$R = \left(60 - \frac{x}{5} + \frac{y}{20}\right)x + \left(50 - \frac{x}{10} + \frac{y}{20}\right)y$$

11.2 Limits and Continuity, Pages 708-710

SURVIVAL HINT: *It might be a good idea to review the concepts and definitions of limit and continuity as presented in Chapter 2, since the concepts and definitions for functions of several variables are a direct extension from $\mathbb{R}^2$ to $\mathbb{R}^3$. If you covered the ϵ-δ definition of limit and the three-part definition of continuity, then the multivariable definitions are the obvious extensions.*

1. Pick a positive value ϵ. You are given the surface $z = f(x, y)$. Consider a fixed point (a, b) such that $f(a, b) = L$. Draw a circle around (a, b) with radius $\delta > 0$. If the vertical distance from $z = f(x, y)$ to $z = L$ is less than ϵ for every point in the circle, with the possible exception of the center (a, b), then

$$\lim_{(x,y)\to(a,b)} f(x, y) = L$$

If no such circle exists, then the limit does not exist. Formally,

$$\lim_{(x, y)\to(a, b)} f(x, y) = L$$

means that for any given $\epsilon > 0$ there exists a $\delta > 0$ so that

$$\bigl| f(x, y) - L \bigr| < \epsilon$$

whenever (x, y) is a point inside the punctured disk with radius δ.

3.
$$\lim_{(x, y)\to(-1, 0)} (xy^2 + x^3y + 5)$$
$$= (-1)0^2 + (-1)^3 0 + 5$$
$$= 5$$

5.
$$\lim_{(x,y)\to(1, 3)} \frac{x + y}{x - y} = \frac{1 + 3}{1 - 3}$$
$$= -2$$

7.
$$\lim_{(x,y)\to(1, 0)} e^{xy} = e^0$$
$$= 1$$

9.
$$\lim_{(x,y)\to(0, 1)} [e^{x^2+x}\ln(ey^2)] = e^0 \ln e$$
$$= 1$$

11. Note $x \neq y$.
$$\lim_{(x,y)\to(0, 0)} \frac{x^2 - 2xy + y^2}{x - y} = \lim_{(x,y)\to(0,0)} (x - y)$$
$$= 0$$

13. Note $y \neq 0$.
$$\lim_{(x,y)\to(0,0)} \frac{e^x \tan^{-1} y}{y} = \lim_{y\to0} \frac{\tan^{-1} y}{y}$$
$$= \lim_{y\to0} \frac{1}{1 + y^2}$$
$$= 1$$

15. Note $x \neq -y$.
$$\lim_{(x,y)\to(0,0)} \frac{\sin(x + y)}{x + y} = \lim_{t\to0} \frac{\sin t}{t}$$
$$= 1$$

17. Note $x^2 \neq y^2$.
$$\lim_{(x,y)\to(5,5)} \frac{x^4 - y^4}{x^2 - y^2}$$
$$= \lim_{(x,y)\to(5,5)} \frac{(x^2 + y^2)(x^2 - y^2)}{x^2 - y^2}$$
$$= \lim_{(x,y)\to(5,5)} (x^2 + y^2)$$
$$= 50$$

19. Note $x \neq 2y$.

$$\lim_{(x,y)\to(2,1)} \frac{x^2 - 4y^2}{x - 2y} = \lim_{(x,y)\to(2,1)}(x + 2y)$$

$$= 4$$

21.

$$\lim_{(x,y)\to(2, 1)}(xy^2 + x^3 y) = 2(1) + 8(1)$$

$$= 10$$

23. Note $x \neq y$. Along the line $x = 0$, we have:

$$\lim_{(x,y)\to(0,0)} \frac{x + y}{x - y} = \lim_{y\to0} \frac{y}{(-y)}$$

$$= \lim_{y\to0}(-1)$$

$$= -1$$

Along the line $y = 0$,

$$\lim_{(x,y)\to(0,0)} \frac{x + y}{x - y} = \lim_{x\to0} \frac{x}{x}$$

$$= \lim_{x\to0}(1)$$

$$= 1$$

The limit does not exist because these values are not the same.

25.

$$\lim_{(x,y)\to(0,0)} e^{xy} = e^0$$

$$= 1$$

27.

$$\lim_{(x,y)\to(0,0)}(\sin x - \cos y) = 0 - 1$$

$$= -1$$

29.

$$\lim_{(x,y)\to(0,0)} \frac{1 - \cos(x^2 + y^2)}{x^2 + y^2}$$

$$= \lim_{t\to0} \frac{1 - \cos t}{t}$$

$$= 0$$

31. $f(x, y) = \dfrac{x - y^2}{x^2 + y^2}$

Along the line $y = 0$, we have $\lim_{x\to0} \dfrac{x}{x^2}$, which is not defined, so the limit does not exist.

33. Along the line $y = mx$, we have

$$\lim_{(x,y)\to(0,0)} \frac{x^2 y^2}{x^4 + y^4} = \lim_{x\to0} \frac{m^2 x^4}{(1 + m^4)x^4}$$

$$= \frac{m^2}{1 + m^4}$$

which is not unique (the result varies with m). Thus, the limit does not exist.

35. Along the line $y = kx$,

$$\lim_{(x,y)\to(0,0)} \frac{xy^3}{x^2 + y^6} = \lim_{x\to0} \frac{k^3 x^4}{x^2 + k^6 x^6}$$

$$= \lim_{x\to0} \frac{k^3 x^2}{1 + k^6 x^4}$$

$$= 0$$

Along the path $x = y^3$,

$$\lim_{(x,y)\to(0,0)} \frac{xy^3}{x^2 + y^6} = \lim_{y\to0} \frac{y^6}{y^6 + y^6}$$

$$= \frac{1}{2}$$

Since these path limits are not the same,

$f(x, y)$ has no limiting value as

$(x, y) \to (0, 0)$. Thus, f is not continuous at

$(0, 0)$.

37. a.
$$\lim_{(x,y)\to(3,1)} \frac{x^2 + 2y^2}{x^2 + y^2} = \frac{9 + 2}{9 + 1}$$

$$= \frac{11}{10}$$

b. Along the line $x = 0$,

$$\lim_{y\to0} \frac{2y^2}{y^2} = 2$$

Along the line $y = 0$,

$$\lim_{x\to0} \frac{x^2}{x^2} = 1$$

Since these path limits are not the same,
the limit does not exist.

39.
$$\lim_{(x,y)\to(0,0)} \frac{3x^3 - 3y^3}{x^2 - y^2}$$

Along the line $y = 0$:

$$\lim_{(x,y)\to(0,0)} \frac{3x^3 - 3y^3}{x^2 - y^2} = \lim_{x\to0} \frac{3x^3}{x^2}$$

$$= 0$$

Since f is to be continuous at $(0, 0)$, we must

have $B = 0$.

41. Approach the origin along the line $x = t$,

$y = t$, $z = t$:

$$\lim_{(x,y,z)\to(0,0,0)} \frac{xyz}{x^2 + y^2 + z^2} = \lim_{t\to0} \frac{t^3}{3t^2}$$

$$= 0$$

43. In polar coordinates,

$$\lim_{(x,y)\to(0,0)} \frac{2x^2 - x^2y^2 + 2y^2}{x^2 + y^2}$$

$$= \lim_{r\to0} \frac{2r^2 - r^4\cos^2\theta\sin^2\theta}{r^2}$$

$$= 2$$

Thus, $A = 2$.

45. $\lim\limits_{r\to0} (1 + r^2)^{1/r^2} = L$, so

$$\ln L = \lim_{r\to0} \frac{\ln(1 + r^2)}{r^2}$$

$$= \lim_{r\to0} \frac{\frac{2r}{1 + r^2}}{2r} = 1 \quad \textit{l'Hôpital's rule}$$

Then $L = e^1 = e$.

47. $\lim\limits_{r\to0} (r\cos\theta)\ln r = \cos\theta \lim\limits_{r\to0} \frac{\ln r}{r^{-1}}$

$$= \cos\theta \lim_{r\to0} \frac{\frac{1}{r}}{-r^{-2}}$$

$$= 0 \quad \textit{l'Hôpital's rule}$$

49. False; let $f(x, y) = \dfrac{x^2y^2}{x^4 + y^4}$. Then $f(x, y)$ is

continuous for all $(x, y) \ne (0, 0)$ but

$\lim\limits_{(x,y)\to(0,0)} f(x, y)$ does not exist. *See Problem 33.*

51.
$$\left| 2x^2 + 3y^2 - 0 \right| \leq 3 \left| x^2 + y^2 \right|$$
$$< 3\sqrt{x^2 + y^2}$$

for $x^2 + y^2 < 1$, for a point (x, y) near $(0, 0)$. Thus, if $\sqrt{x^2 + y^2} < \delta$, we have

$$\left| (2x^2 + 3y^2) - 0 \right| < 3\sqrt{x^2 + y^2}$$
$$< 3\delta$$

and

$$\left| (2x^2 + 3y^2) - 0 \right| < \epsilon$$

if $\delta = \frac{\epsilon}{3}$.

53.
$$\left| \frac{x^2 - y^2}{x + y} - 0 \right| = |x - y| \text{ if } x \neq -y$$
$$\leq |x| + |y|$$
$$\leq \sqrt{2}\sqrt{x^2 + y^2}$$

Thus, if $\sqrt{x^2 + y^2} < \delta$, we have

$$|x - y| \leq \sqrt{2}\sqrt{x^2 + y^2} < \sqrt{2}\,\delta$$

and $\left| \dfrac{x^2 - y^2}{x + y} - 0 \right| < \epsilon$ if $\delta = \dfrac{\epsilon}{\sqrt{2}}$

55. Let $\epsilon = \dfrac{f(a, b)}{2}$; since f is continuous at (a, b), there exists a $\delta > 0$ so that

$$\left| f(x, y) - f(a, b) \right| < \epsilon$$

whenever

$$0 < \sqrt{(x - a)^2 + (y - b)^2} < \delta$$

Thus, if (x, y) is any point inside the punctured disk

$$0 < \sqrt{(x - a)^2 + (y - b)^2} < \delta$$

we must have

$$-\epsilon < f(x, y) - f(a, b) < \epsilon$$

which implies

$$f(a, b) - \epsilon < f(x, y)$$

and since $\epsilon = f(a, b)/2$, we have

$$f(x, y) > f(a, b) - \frac{f(a, b)}{2} = \frac{1}{2} f(a, b) > 0$$

57. Let $\epsilon > 0$ be given; there exist $\delta_1 > 0$ and $\delta_2 > 0$ so that

$$\left| f(x, y) - L \right| < \frac{\epsilon}{2} \text{ if}$$
$$0 < \sqrt{(x - x_0)^2 + (y - y_0)^2} < \delta_1$$

$$\left| g(x, y) - M \right| < \frac{\epsilon}{2} \text{ if}$$
$$0 < \sqrt{(x - x_0)^2 + (y - y_0)^2} < \delta_2$$

Let $\delta = \min(\delta_1, \delta_2)$; then if

$$0 < \sqrt{(x - x_0)^2 + (y - y_0)^2} < \delta$$

we have

$$\left| f(x, y) + g(x, y) - (L + M) \right|$$
$$\leq \left| f(x, y) - L \right| + \left| g(x, y) - M \right|$$
$$< \frac{\epsilon}{2} + \frac{\epsilon}{2}$$
$$= \epsilon$$

Thus,

$$\lim_{(x,\ y)\to(x_0,\ y_0)} [f(x,\ y) + g(x,\ y)] = L + M$$

11.3 Partial Derivatives, Pages 717-719

SURVIVAL HINT: *For a function of two variables, when you hold one variable fixed and find the derivative with respect to the other, you are reducing the surface in* $\mathbb{R}^3$ *to a curve in* $\mathbb{R}^2$. *The first and second derivatives give the slope and concavity as you move along the curve determined by the trace through a given point. The mixed partial indicates how the slope would change if you moved to another trace. The analogy of moving in various directions from a point on a hillside is useful.*

1. Given a function of two (or more) variables,

   hold all but one independent variable(s)

   constant and differentiate with respect to the

   remaining variable.

3. $f(x,\ y) = x^3 + x^2y + xy^2 + y^3$

   $f_x = 3x^2 + 2xy + y^2$

   $f_{xx} = 6x + 2y$

   $f_y = x^2 + 2xy + 3y^2$

   $f_{yx} = 2x + 2y$

5. $f(x,\ y) = \frac{x}{y}$

   $f_x = \frac{1}{y}$

   $f_{xx} = 0$

$f_y = \frac{-x}{y^2}$

$f_{yx} = \frac{-1}{y^2}$

7. $f(x,\ y) = \ln(2x + 3y)$

   $f_x = \frac{2}{2x + 3y}$

   $f_{xx} = \frac{-4}{(2x + 3y)^2}$

   $f_y = \frac{3}{2x + 3y}$

   $f_{yx} = \frac{-6}{(2x + 3y)^2}$

9. **SURVIVAL HINT:** *Contrast parts a and b: note that in part a, the function* sin x^2 *is multiplied by the function* cos y, *whereas in part b, the function is the sine of the product* x^2cos y.

   a. $f(x,\ y) = (\sin x^2)\cos y$

   Treat y as a constant (not a product rule

   because cos y is not considered a

   variable):

   $$f_x = (2x \cos x^2)(\cos y);$$

   Now, treat x as a constant:

   $$f_y = -(\sin x^2)(\sin y)$$

   b. $f(x,\ y) = \sin(x^2\cos y)$

   Treat y as a constant:

   $$f_x = \cos(x^2\cos y)(2x \cos y)$$

   Treat x as a constant:

$$f_y = \cos(x^2\cos y)(-x^2 \sin y)$$

11. $f(x, y) = \sqrt{3x^2 + y^4}$

$$= (3x^2 + y^4)^{1/2}$$

$$f_x = \tfrac{1}{2}(3x^2 + y^4)^{-1/2}(6x)$$

$$= \frac{3x}{(3x^2 + y^4)^{1/2}}$$

$$f_y = \tfrac{1}{2}(3x^2 + y^4)^{-1/2}(4y^3)$$

$$= \frac{2y^3}{(3x^2 + y^4)^{1/2}}$$

13. $f(x, y) = x^2 e^{x+y}\cos y$

$$f_x = (\cos y)(x^2 e^{x+y} + 2xe^{x+y})$$

$$= xe^{x+y}(x + 2)\cos y$$

$$f_y = -x^2 e^{x+y}\sin y + x^2 e^{x+y} \cos y$$

$$= x^2 e^{x+y}(\cos y - \sin y)$$

15. $f(x, y) = \sin^{-1}(xy)$

$$f_x = \frac{y}{\sqrt{1 - x^2 y^2}}$$

$$f_y = \frac{x}{\sqrt{1 - x^2 y^2}}$$

17. $f(x, y, z) = xy^2 + yz^3 + xyz$

$$f_x = y^2 + yz$$

$$f_y = 2xy + z^3 + xz$$

$$f_z = 3yz^2 + xy$$

19. $f(x, y, z) = \dfrac{x + y^2}{z}$

$$f_x = \tfrac{1}{z}$$

$$f_y = \frac{2y}{z}$$

$$f_z = -\frac{x + y^2}{z^2}$$

21. $f(x, y, z) = \ln(x + y^2 + z^3)$

$$f_x = \frac{1}{x + y^2 + z^3}$$

$$f_y = \frac{2y}{x + y^2 + z^3};$$

$$f_z = \frac{3z^2}{x + y^2 + z^3}$$

23. $\qquad \dfrac{x^2}{9} - \dfrac{y^2}{4} + \dfrac{z^2}{2} = 1$

$$\frac{\partial}{\partial x}\left(\frac{x^2}{9} - \frac{y^2}{4} + \frac{z^2}{2}\right) = \frac{\partial}{\partial x}(1)$$

$$\frac{2x}{9} - 0 + \frac{2z}{2}\,z_x = 0$$

$$\frac{2x}{9} + zz_x = 0$$

$$z_x = -\frac{2x}{9z}$$

$$\frac{\partial}{\partial y}\left(\frac{x^2}{9} - \frac{y^2}{4} + \frac{z^2}{2}\right) = \frac{\partial}{\partial y}(1)$$

$$-\frac{y}{2} + zz_y = 0$$

$$z_y = \frac{y}{2z}$$

25. $\qquad 3x^2 y + y^3 z - z^2 = 1$

$$\frac{\partial}{\partial x}(3x^2 y + y^3 z - z^2) = \frac{\partial}{\partial x}(1)$$

$$6xy + y^3 z_x - 2zz_x = 0$$

$$(y^3 - 2z)z_x = -6xy$$

$$z_x = -\frac{6xy}{y^3 - 2z}$$

$$\frac{\partial}{\partial y}(3x^2 y + y^3 z - z^2) = \frac{\partial}{\partial y}(1)$$

$$3x^2 + y^3 z_y + 3y^2 z - 2zz_y = 0$$

$$z_y = -\frac{3x^2 + 3y^2 z}{y^3 - 2z}$$

27. $\quad \sqrt{x} + y^2 + \sin xz = 2$

$$\frac{\partial}{\partial x}\left(\sqrt{x} + y^2 + \sin xz\right) = \frac{\partial}{\partial x}(2)$$

$$\frac{1}{2\sqrt{x}} + (\cos xz)(z + xz_x) = 0$$

$$(\cos xz)(z + xz_x) = -\frac{1}{2\sqrt{x}}$$

$$z + xz_x = -\frac{\frac{1}{2\sqrt{x}}}{\cos xz}$$

$$xz_x = -\frac{\frac{1}{2\sqrt{x}}}{\cos xz} - z$$

$$z_x = -\frac{\frac{1}{2\sqrt{x}} + z\cos xz}{x\cos xz}$$

$$= -\frac{1}{2x^{3/2}\cos xz} - \frac{z}{x}$$

Now, with respect to y:

$$2y + (\cos xz)(xz_y) = 0$$

$$z_y = -\frac{2y}{x\cos xz}$$

29. $\quad f(x, y) = xy^3 + x^3 y; \ P_0(1, -1, -2)$

    a. $f_x = y^3 + 3x^2 y; \ f_x(1, -1) = -4$

    b. $f_y = 3xy^2 + x^3; \ f_y(1, -1) = 4$

31. $\quad f(x, y) = x^2 \sin(x + y); \ P_0(\frac{\pi}{2}, \frac{\pi}{2}, 0)$

    a. $f_x = x^2 \cos(x + y) + 2x\sin(x + y);$

$$f_x(\tfrac{\pi}{2}, \tfrac{\pi}{2}) = -\frac{\pi^2}{4}$$

    b. $f_y = x^2 \cos(x + y); \ f_y(\tfrac{\pi}{2}, \tfrac{\pi}{2}) = -\frac{\pi^2}{4}$

33. $\quad f(x, y) = \int_x^y (t^2 + 2t + 1) \, dt$

$$= \int_0^y (t^2 + 2t + 1) \, dt - \int_0^x (t^2 + 2t + 1) \, dt$$

$$f_x = -(x^2 + 2x + 1)$$

$$f_y = y^2 + 2y + 1$$

35. $\quad f(x, y) = 3x^2 y - y^3$

$$f_x = 6xy$$

$$f_{xx} = 6y$$

$$f_y = 3x^2 - 3y^2$$

$$f_{yy} = -6y$$

Thus, $f_{xx} + f_{yy} = 0$.

37. $\quad f(x, y) = e^x \sin y$

$$f_x = e^x \sin y$$

$$f_{xx} = e^x \sin y$$

$$f_y = e^x \cos y$$

$$f_{yy} = - e^x \sin y$$

Thus, $f_{xx} + f_{yy} = 0$.

39. $f(x, y) = \cos xy^2$

$$f_x = - y^2 \sin xy^2$$

$$f_{xy} = - y^2(2xy) \cos xy^2 - 2y \sin xy^2$$

$$= - 2xy^3 \cos xy^2 - 2y \sin xy^2$$

$$f_y = - 2xy \sin xy^2;$$

$$f_{yx} = - 2y[(\sin xy^2) + x(\cos xy^2)(y^2)]$$

$$= - 2xy^3 \cos xy^2 - 2y \sin xy^2$$

They are the same.

41. $f(x, y, z) = x^2 + y^2 - 2xy \cos z$

$$f_x = 2x - 2y \cos z$$

$$f_{xz} = 2y \sin z$$

$$f_{xzy} = 2 \sin z$$

$$f_y = 2y - 2x \cos z$$

$$f_{yz} = 2x \sin z$$

$$f_{yzz} = 2x \cos z$$

$$f_{xzy} - f_{yzz} = 2(\sin z - x \cos z)$$

43. **a.** $\dfrac{\partial D_1}{\partial p_1} < 0$ and $\dfrac{\partial D_2}{\partial p_2} < 0$ since the demand

for a commodity decreases as its price

increases.

b. $\dfrac{\partial D_1}{\partial p_2} < 0$ and $\dfrac{\partial D_2}{\partial p_1} < 0$ since the demand

for Q_1 decreases as the price for the

complementary commodity Q_2 increases;

likewise the demand for Q_2 increases as

the price for Q_1 decreases.

c. cameras and film

45. $C(m, t, T) = \sigma(T - t)m^{-0.67}$

a. $C_m = \sigma(T - t)(-0.67m^{-1.67})$

$$= -0.67\sigma(T - t)m^{-1.67}$$

b. $C_T = \sigma m^{-0.67}$

c. $C_t = - \sigma m^{-0.67}$

47. **a.** $PV = kT$

$$P\frac{\partial V}{\partial T} = k$$

$$\frac{\partial V}{\partial T} = \frac{k}{P}$$

b. $PV = kT$

$$V\frac{\partial P}{\partial V} + P = 0$$

$$\frac{\partial P}{\partial V} = -\frac{P}{V}$$

c. $PV = kT$

$$V = k\frac{\partial T}{\partial P}; \frac{\partial T}{\partial P} = \frac{V}{k}$$

$$\frac{\partial P}{\partial V}\frac{\partial V}{\partial T}\frac{\partial T}{\partial P} = - \frac{P}{V}\frac{k}{P}\frac{V}{k} = - 1$$

49. $T(x, y) = x^3 + 2xy^2 + y$

a. Moving parallel to **j** means y is changing,

and x is constant.

$$\frac{\partial T}{\partial y} = 4xy + 1$$

$$\frac{\partial T}{\partial y}(2,\ 1) = 9$$

b. In this case x is changing and y is constant.

$$\frac{\partial T}{\partial x} = 3x^2 + 2y^2$$

$$\frac{\partial T}{\partial x}(2,\ 1) = 3(2)^2 + 2(1)^2$$

$$= 14$$

51. a. $u = e^{-x}\cos y,\ v = e^{x}\sin y$

$$\frac{\partial u}{\partial x} = -e^{-x}\cos y$$

$$\frac{\partial u}{\partial y} = -e^{-x}\sin y$$

$$\frac{\partial v}{\partial x} = -e^{-x}\sin y$$

$$\frac{\partial v}{\partial y} = e^{-x}\cos y$$

Thus, $\frac{\partial u}{\partial x} \neq \frac{\partial v}{\partial y}$ and $\frac{\partial u}{\partial y} \neq -\frac{\partial v}{\partial x}$

so the Cauchy-Riemann equations are not

satisfied.

b. $u = x^2 + y^2,\ v = 2xy$

$$\frac{\partial u}{\partial x} = 2x$$

$$\frac{\partial u}{\partial y} = 2y$$

$$\frac{\partial v}{\partial x} = 2y$$

$$\frac{\partial v}{\partial y} = 2x$$

Thus, $\frac{\partial u}{\partial x} = \frac{\partial v}{\partial y}$ but $\frac{\partial u}{\partial y} \neq -\frac{\partial v}{\partial x}$

so the Cauchy-Riemann equations are not

satisfied.

c. $u = \ln(x^2 + y^2),\ v = 2\tan^{-1}\left(\frac{y}{x}\right)$

$$\frac{\partial u}{\partial x} = \frac{2x}{x^2 + y^2}$$

$$\frac{\partial u}{\partial y} = \frac{2y}{x^2 + y^2}$$

$$\frac{\partial v}{\partial x} = \frac{-2y}{x^2 + y^2}$$

$$\frac{\partial v}{\partial y} = \frac{2x}{x^2 + y^2}$$

Thus, $\frac{\partial u}{\partial x} = \frac{\partial v}{\partial y}$ and $\frac{\partial u}{\partial y} = -\frac{\partial v}{\partial x}$

so the Cauchy-Riemann equations are

satisfied.

53. If $P(L,\ K) = L^{\alpha}K^{\beta}$, then $\frac{\partial P}{\partial L} = \alpha L^{\alpha-1}K^{\beta}$

and $\frac{\partial P}{\partial K} = \beta L^{\alpha}K^{\beta-1}$, so

$$L\frac{\partial P}{\partial L} + K\frac{\partial P}{\partial K} = L(\alpha L^{\alpha-1}K^{\beta}) + K(\beta L^{\alpha}K^{\beta-1})$$

$$= \alpha L^{\alpha}K^{\beta} + \beta L^{\alpha}K^{\beta}$$

$$= (\alpha + \beta)P$$

55. $P(x,\ t) = P_0 + P_1 e^{-kx}\sin(At - kx)$

$$\frac{\partial P}{\partial x} = -P_1 k e^{-kx}[\sin(At - kx) + \cos(At - kx)]$$

$$\frac{\partial^2 P}{\partial x^2} = 2k^2 P_1 e^{-kx}\cos(At - kx)$$

$$\frac{\partial P}{\partial t} = AP_1 e^{-kx}\cos(At - kx)$$

so $\frac{\partial P}{\partial t} = c^2 \frac{\partial^2 P}{\partial x^2}$, where $c^2 = \frac{A}{2k^2}$

57. Using the definition of the derivative, we find

$$f_x(0, 0) = \lim_{h \to 0} \frac{f(0 + h, 0) - f(0, 0)}{h}$$

$$= \lim_{h \to 0} \frac{h^2\sin(1/h^2) - 0}{h} = 0$$

since $\left|\sin\frac{1}{h}\right| \le 1$ for all $h \ne 0$.

$$f_y(0, 0) = \lim_{h \to 0} \frac{f(0, 0 + h) - f(0, 0)}{h}$$

$$= \lim_{h \to 0} \frac{h\sin(1/h^2)}{h}$$

$$= \lim_{h \to 0} \sin\frac{1}{h^2}$$

does not exist.

59. $A = \frac{1}{2}ab\sin\gamma$

a. $\frac{\partial A}{\partial a} = \frac{1}{2}b\sin\gamma$

$\frac{\partial A}{\partial b} = \frac{1}{2}a\sin\gamma$

$\frac{\partial A}{\partial \gamma} = \frac{1}{2}ab\cos\gamma$

b. $a = \frac{2A}{b\sin\gamma}$

$\frac{\partial a}{\partial \gamma} = -\frac{2A\cos\gamma}{b\sin^2\gamma}$

11.4 Tangent Planes, Approximations, and Differentiability, Pages 727-729

SURVIVAL HINT: *The equation of a tangent plane can be remembered as an extension of the point-slope formula for a line in* $\mathbb{R}^2$: $y - y_0 = m(x - x_0)$. *For the plane in* $\mathbb{R}^3$:

$$z - z_0 = m(x - x_0) + n(y - y_0)$$

where m and n are the slopes in the x and y directions, found with the partial derivatives.

1. $z = (x^2 + y^2)^{1/2}$; the equation of a tangent plane is given by the formula:

$$z - z_0 = f_x(x_0, y_0)(x - x_0) + f_y(x_0, y_0)(y - y_0)$$

For this problem, we let $z = f(x, y)$ so that

$$f_x = \frac{1}{2}(x^2 + y^2)^{-1/2}(2x)$$

$$= \frac{x}{\sqrt{x^2 + y^2}}$$

$$f_y = \frac{y}{\sqrt{x^2 + y^2}}$$

At $P_0(3, 1, \sqrt{10})$

$$f_x(3, 1) = \frac{3}{\sqrt{3^2 + 1^2}}$$

$$= \frac{3}{\sqrt{10}}$$

$$f_y(3, 1) = \frac{1}{\sqrt{3^2 + 1^2}}$$

$$= \frac{1}{\sqrt{10}}$$

The tangent plane is:

$$z - \sqrt{10} = \frac{3}{\sqrt{10}}(x - 3) + \frac{1}{\sqrt{10}}(y - 1)$$

$$3x + y - \sqrt{10}z = 0$$

3. $f(x, y) = x^2 + y^2 + \sin xy$

$f_x = 2x + y \cos xy$

$f_y = 2y + x \cos xy$

at P_0,

$$f_x(0, 2) = 2(0) + 2 \cos 0$$

$$= 2$$

$$f_y(0, 2) = 2(2) + 0 \cos 0$$

$$= 4$$

$$z - 4 = 2(x - 0) + 4(y - 2)$$

$$2x + 4y - z - 4 = 0$$

5. $z = \tan^{-1} \frac{y}{x}$

$$z_x = \frac{-y/x^2}{1 + (y/x)^2}$$

$$= \frac{-y}{x^2 + y^2}$$

$$z_y = \frac{1/x}{1 + (y/x)^2}$$

$$= \frac{x}{x^2 + y^2}$$

at $P_0(2, 2, \frac{\pi}{4})$

$$z_x = \frac{-2}{2^2 + 2^2}$$

$$= -\frac{1}{4}$$

$$z_y = \frac{2}{2^2 + 2^2}$$

$$= \frac{1}{4}$$

$$z - \frac{\pi}{4} = -\frac{1}{4}(x - 2) + \frac{1}{4}(y - 2)$$

$$x - y + 4z - \pi = 0$$

SURVIVAL HINT: *The total differential can be remembered as an* $\mathbb{R}^3$ *extension of the concepts of the differential in* $\mathbb{R}^2$; $dy = f'(x) \, dx$. *This says the change along the tangent line is the rate of change,* $f'(x)$, *multiplied by the change in the x-direction,* Δx, *or in the limiting case,* dx. *In* $\mathbb{R}^3$ *the change in z along the tangent plane, dz or df, is the rate of change in the x direction,* $\partial z/\partial x$, *multiplied by the change in the x-direction, dx, plus the same computation for change of z in the y-direction:*

$$df = \frac{\partial z}{\partial x} \, dx + \frac{\partial z}{\partial y} \, dy$$

7.   The total differential is given by the formula:

$$df = f_x \, dx + f_y \, dy$$

$$f(x, y) = 5x^2 y^3$$

$$df = 10xy^3 \, dx + 15x^2 y^2 \, dy$$

9. $f(x, y) = \sin xy$

$$df = y(\cos xy) \, dx + x(\cos xy) \, dy$$

11. $f(x, y) = \frac{y}{x}$

$$df = -\frac{y}{x^2} \, dx + \frac{1}{x} \, dy$$

13. $f(x, y) = ye^x$

$df = ye^x\, dx + e^x\, dy$

15. $f(x, y, z) = 3x^3 - 2y^4 + 5z$

$df = 9x^2 dx - 8y^3\, dy + 5\, dz$

17. $f(x, y, z) = z^2\sin(2x - 3y)$

$df = 2z^2\cos(2x - 3y)dx$

$\quad - 3z^2\cos(2x - 3y)dy$

$\quad + 2z\sin(2x - 3y)dz$

19. $f(x, y) = xy^3 + 3xy^2$

$f_x = y^3 + 3y^2$

$f_y = 3xy^2 + 6xy$

$f, f_x,$ and f_y are all continuous, so the function is differentiable.

21. $f(x, y) = e^{2x+y^2}$

$f_x = 2e^{2x+y^2}$

$f_y = 2ye^{2x+y^2}$

$f, f_x,$ and f_y are all continuous, so the function is differentiable.

23. $f(x, y) = 3x^4 + 2y^4$;

$x_0 = 1, y_0 = 2, \Delta x = 0.01, \Delta y = 0.03$

$f_x = 12x^3$

$f_y = 8y^3$

$f(1.01, 2.03)$

$\approx f(1, 2) + f_x(1, 2)\Delta x + f_y(1, 2)\Delta y$

$\approx 35 + 12(0.01) + 64(0.03)$

$= 37.04$

By calculator,

$f(1.01, 2.03) \approx 37.08544565$

25. $f(x, y) = \sin(x + y)$

$x_0 = \frac{\pi}{2}, y_0 = \frac{\pi}{2}, \Delta x = 0.01, \Delta y = -0.01$

$f_x = \cos(x + y)$

$f_y = \cos(x + y)$

$f(\frac{\pi}{2} + 0.01, \frac{\pi}{2} - 0.01)$

$\approx f(\frac{\pi}{2}, \frac{\pi}{2}) + f_x(\frac{\pi}{2}, \frac{\pi}{2})\Delta x + f_y(\frac{\pi}{2}, \frac{\pi}{2})\Delta y$

$\approx 0 + (-1)(0.01) + (-1)(-0.01)$

$= 0$

By calculator,

$f(\frac{\pi}{2} + 0.01, \frac{\pi}{2} - 0.01) = 0$

27. $f(x, y) = e^{xy}$

$x_0 = 1, y_0 = 1, \Delta x = 0.01, \Delta y = -0.02$

$f_x = ye^{xy}$

$f_y = xe^{xy}$

$f(1 + 0.01, 1 - 0.02)$

$\approx e + e(0.01) + e(-0.02)$

$$\approx 2.691$$

By calculator,

$$f(1.01, 0.98) \approx 2.6906963$$

29. $z = 5 - x^2 - y^2 + 4y$

$$z_x = -2x$$

$z_x = -2x$ is equal to 0 when $x = 0$.

$$z_y = -2y + 4$$

$z_y = -2y + 4$ is equal to 0 when $y = 2$.

The tangent plane is horizontal when

$z_x = z_y = 0$: Since

$$z(0, 2) = 5 - 0^2 - 2^2 + 4(2)$$

$$= 9$$

the equation of the horizontal plane is

$z = 9$.

31. a. Since x and y are very small, we can take

$\Delta x = x$ and $\Delta y = y$. Then

$$f(x, y) \approx f(0, 0) + x f_x(0, 0) + y f_y(0, 0)$$

b. $f(x, y) = \dfrac{1}{1 + x - y}; f(0, 0) = 1$

$$f_x = \frac{-1}{(1 + x - y)^2}; f_x(0, 0) = -1$$

$$f_y = \frac{1}{(1 + x - y)^2}; f_y(0, 0) = 1$$

Thus,

$$f(x, y) \approx f(0, 0) + x f_x(0, 0) + y f_y(0, 0)$$

$$\frac{1}{1 + x - y} \approx 1 + x(-1) + y(1)$$

$$= 1 - x + y$$

c. $f(x, y) = \dfrac{1}{(x + 1)^2 + (y + 1)^2}$

$$f_x = -\frac{2(x + 1)}{[(x + 1)^2 + (y + 1)^2]^2}$$

$$f_y = -\frac{2(y + 1)}{[(x + 1)^2 + (y + 1)^2]^2}$$

$f(0, 0) = \frac{1}{2}; f_x(0, 0) = -\frac{1}{2}; f_y(0, 0) = -\frac{1}{2}$

$$f(x, y) \approx f(0, 0) + x f_x(0, 0) + y f_y(0, 0)$$

$$= \tfrac{1}{2} - \tfrac{1}{2}x - \tfrac{1}{2}y$$

$$= \tfrac{1}{2}(1 - x - y)$$

33. Let x, y, and z be the length, width, and

height of the box, respectively. The total cost

is

$$C = 2xy + 1.50(xy + 2xz + 2yz)$$

$$= 3.5xy + 3xz + 3yz$$

$$C_x = 3.5y + 3z,$$

$$C_y = 3.5x + 3z,$$

$$C_z = 3x + 3y$$

$$\Delta C = (3.5y + 3z)\Delta x + (3.5x + 3z)\Delta y + (3x + 3y)\Delta z$$

Since $x = 2$, $y = 4$, $z = 3$ and $|\Delta x| \leq 0.02$,

$|\Delta y| \leq 0.02$, and $|\Delta z| \leq 0.02$, we have

$|\Delta C| \leq |3.5(4) + 3(3)|(0.02)$

$\qquad + |3.5(2) + 3(3)|(0.02) + |3(2) + 3(4)|(0.02)$

$\qquad = 1.14$

Thus, the maximum possible error will cost

$1.14.

35. $p(x) = 4{,}000 - 500x$, $q(y) = 3{,}000 - 450y$

a. $x = \dfrac{4{,}000 - p}{500}$

$\qquad y = \dfrac{3{,}000 - q}{450}$

$\qquad R = x(p) \cdot p + y(q) \cdot q$

$\qquad\quad = \left(\dfrac{4{,}000 - p}{500}\right)p + \left(\dfrac{3{,}000 - q}{450}\right)q$

b. $R_p = 8 - \dfrac{2p}{500}$; $R_q = \dfrac{20}{3} - \dfrac{2q}{450}$

$\qquad \Delta R \approx R_p \Delta p + R_q \Delta q$

$\qquad\quad = \left(8 - \dfrac{2p}{500}\right)\Delta p + \left(\dfrac{20}{3} - \dfrac{2q}{450}\right)\Delta q$

When $p = 500$, $q = 750$, $\Delta p = 20$, and

$\Delta q = 18$, we have

$\Delta R \approx \left[8 - \dfrac{2(500)}{500}\right](20) + \left[\dfrac{20}{3} - \dfrac{2(750)}{450}\right](18)$

$\qquad = 180$

That is, revenue is increased by approximately $180.

37. $dR = \dfrac{c\,dx}{r^4} - \dfrac{4cx}{r^5}\,dr$

$\qquad \dfrac{dR}{R} = \dfrac{cr^4\,dx}{cxr^4} - \dfrac{4cxr^4\,dr}{cxr^5}$

$\qquad\quad = \dfrac{dx}{x} - \dfrac{4\,dr}{r}$

$\qquad\quad \approx 0.03 - 4(-0.02)$

$\qquad\quad = 0.11$

Thus, R increases by approximately 11%.

39. $F(x, y) = \dfrac{1.786xy}{1.798x + y}$

$\qquad F_x = \dfrac{(1.798x + y)(1.786y) - (1.786xy)(1.798)}{(1.798x + y)^2}$

$\qquad F_y = \dfrac{(1.798x + y)(1.786x) - 1.786xy(1)}{(1.798x + y)^2}$

$\qquad x = 5$, $y = 4$, $dx = 0.1$, $dy = 0.04$

$\qquad F_x(5, 4) \approx 0.1693$

$\qquad F_y(5, 4) \approx 0.4758$

$\qquad \Delta F \approx F_x\,dx + F_y\,dy$

$\qquad\quad \approx (0.1693)(0.1) + (0.4758)(0.04)$

$\qquad\quad \approx 0.0360 \text{ cal}$

41. $f(x, y) = 10xy^{1/2}$

$x = 30$, $y = 36$, $dx = 1$

$$10xy^{1/2} = C$$

$$(\tfrac{1}{2})xy^{-1/2}dy + y^{1/2}dx = 0$$

$$dy = \frac{-\sqrt{y}\,dx\,(2)\sqrt{y}}{x}$$

$$= -\frac{2y\,dx}{x}$$

$$= -\frac{2(36)(1)}{30}$$

$$= -2.4$$

The manufacturer should decrease the level of unskilled labor by about 2.4 hours.

Alternatively, we can work with approximations

$$\Delta f \approx df = (10y^{1/2})\,dx + (5xy^{-1/2})dy$$

For f to stay constant when $x = 30$, $y = 36$, $dx = 1$, we want $\Delta f = 0$. Thus, dy must satisfy

$$10(36)^{1/2}(1) + 5(30)(36)^{-1/2}\,dy = 0$$

$$dy \approx \frac{-10(36)}{5(30)}$$

$$= -2.4$$

43. $V(H, R) = \pi R^2 H$

Given, $H = 12$, $R = 3$, $dH = -0.2$,

$dR = -0.3$; $V(12, 3) = 108\pi$

$$\Delta V \approx dV = \pi(R^2\,dH + 2RH\,dR)$$

$$= \pi[9(-0.2) + 2(3)(12)(-0.3)]$$

$$= -23.4\pi$$

$$\frac{\Delta V}{V} = \frac{-23.4\pi}{108\pi}$$

$$\approx -0.217$$

This is about 22% less volume.

45. $S = \dfrac{x}{x - y}$

$$S_x = \frac{-y}{(x - y)^2}$$

$$S_y = \frac{x}{(x - y)^2}$$

$$dS = \left[\frac{-y}{(x - y)^2}\right]dx + \left[\frac{x}{(x - y)^2}\right]dy$$

For $x = 1.2$, $y = 0.5$, $0 \leq dx \leq 0.01$,

$0 \leq dy \leq 0.01$, we have

$$\Delta S \approx ds$$

$$= \left[\frac{-0.5}{(1.2 - 0.5)^2}\right]dx + \left[\frac{1.2}{(1.2 - 0.5)^2}\right]dy$$

$$\leq (-1.02)(0.01) + (2.45)(0.01)$$

$$= 0.014$$

The maximum possible error in the measurement of S is 0.014.

47. Approach along the line $y = mx$;

$$\lim_{x \to 0} \frac{x(mx)}{x^2 + (mx)^2} = \lim_{x \to 0} \frac{m}{m^2 + 1} \neq 0$$

Thus, f is not continuous at $(0, 0)$, so it cannot be differentiable at $(0, 0)$ either.

49. $A = hb = ab \sin \frac{\pi}{6} = \frac{1}{2} ab$

$$\frac{da}{a} = 0.04, \frac{db}{b} = -0.03$$

$$dA = \frac{1}{2} b \; da + \frac{1}{2} a \; db$$

$$\frac{dA}{A} = \frac{b \; da}{ba} + \frac{a \; db}{ab}$$

$$= \frac{da}{a} + \frac{db}{b}$$

$$= 0.04 - 0.03$$

$$= 0.01$$

Thus, A increases by about 1%.

11.5 Chain Rules, Pages 735-737

1. Let $f(x, y)$ be a differentiable function of x and y, and let $x = x(t)$ and $y = y(t)$ be differentiable functions of t. Then $z = f(x, y)$ is a differentiable function of t, and

$$\frac{dz}{dt} = \frac{\partial z}{\partial x} \frac{dx}{dt} + \frac{\partial z}{\partial y} \frac{dy}{dt}$$

Recall the chain rule for a single variable:

$$\frac{dy}{dx} = \frac{dy}{du} \frac{du}{dx}$$

The corresponding rule for two variables is

essentially the same, the formula involves *both* variables. For two parameters, suppose $z = f(x, y)$ is differentiable at (x, y) and that the partial derivatives of $x = x(u, v)$ and $y = y(u, v)$ exist at (u, v). Then the composite function $z = f[x(u, v), y(u, v)]$ is differentiable at (u, v) with

$$\frac{\partial z}{\partial u} = \frac{\partial z}{\partial x} \frac{\partial x}{\partial u} + \frac{\partial z}{\partial y} \frac{\partial y}{\partial u} \text{ and } \frac{\partial z}{\partial v} = \frac{\partial z}{\partial x} \frac{\partial x}{\partial v} + \frac{\partial z}{\partial y} \frac{\partial y}{\partial v}$$

The chain rule allows us to compute dz/dt without determining $z(t)$ explicitly.

3. $z = f(x, y)$, $x = x(u, v, w)$, $y = y(u, v, w)$,

$$\frac{\partial z}{\partial u} = \frac{\partial f}{\partial x} \frac{\partial x}{\partial u} + \frac{\partial f}{\partial y} \frac{\partial y}{\partial u}$$

$$\frac{\partial z}{\partial v} = \frac{\partial f}{\partial x} \frac{\partial x}{\partial v} + \frac{\partial f}{\partial y} \frac{\partial y}{\partial v}$$

$$\frac{\partial z}{\partial w} = \frac{\partial f}{\partial x} \frac{\partial x}{\partial w} + \frac{\partial f}{\partial y} \frac{\partial y}{\partial w}$$

5. $f(x, y) = (4 + y^2)x$, $x = e^{2t}$ and $y = e^{3t}$

a. $f(t) = (4 + e^{6t})e^{2t}$

$$= 4e^{2t} + e^{8t}$$

$$f'(t) = 8e^{2t} + 8e^{8t}$$

$$= 8e^{2t}(1 + e^{6t})$$

b. $f'(t) = \frac{\partial z}{\partial x} \frac{dx}{dt} + \frac{\partial z}{\partial y} \frac{dy}{dt}$

$$= (4 + y^2)2e^{2t} + (2xy)3e^{3t}$$

In terms of t:

$$(4 + e^{6t})2e^{2t} + (2e^{2t}e^{3t})3e^{3t}$$

$$= 8e^{2t}(1 + e^{6t})$$

7. $f(x, y) = xy^2$, $x = \cos 3t$, $y = \tan 3t$

a. $f(t) = (\cos 3t)(\tan^2 3t)$

$$= \frac{\sin^2 3t}{\cos 3t}$$

$$f'(t) = \frac{\cos 3t(6\sin 3t)(\cos 3t) - \sin^2 3t(-3\sin 3t)}{\cos^2 3t}$$

$$= \frac{6\sin 3t(1 - \sin^2 3t) + 3\sin^3 3t}{\cos^2 3t}$$

$$= \frac{6\sin 3t - 3\sin^3 3t}{\cos^2 3t}$$

b. $f'(t) = \frac{\partial}{\partial x}(xy^2)\frac{d}{dt}(\cos 3t) + \frac{\partial}{\partial y}(xy^2)\frac{d}{dt}(\tan 3t)$

$$= y^2(-3\sin 3t) + 2xy(3\sec^2 t)$$

$$= -3\sin 3t\tan^2 3t + 6\cos 3t\tan 3t\sec^2 3t$$

$$= \frac{-3\sin^3 3t + 6\sin 3t}{\cos^2 3t}$$

SURVIVAL HINT: *Check to see if the given parameters restrict the domain of $f(x, y)$. In Problem 11, both x and y must be positive, so $f(x, y)$ has only first octant values.*

11. $F(x, y) = \ln xy$, where $x = e^{uv^2}$ and $y = e^{uv}$

a. $F(u, v) = \ln[e^{uv^2 + uv}]$

$$= uv^2 + uv$$

$$\frac{\partial F}{\partial u} = v^2 + v$$

$$\frac{\partial F}{\partial v} = 2uv + u$$

b. $\frac{\partial F}{\partial u} = \frac{\partial F}{\partial x}\frac{\partial x}{\partial u} + \frac{\partial F}{\partial y}\frac{\partial y}{\partial u}$

$$= \frac{1}{x}\left[e^{uv^2}(v^2)\right] + \frac{1}{y}\left[e^{uv}(v)\right]$$

$$= \frac{e^{uv^2}}{e^{uv^2}}v^2 + v$$

$$= v^2 + v$$

$$\frac{\partial F}{\partial v} = \frac{\partial F}{\partial x}\frac{\partial x}{\partial v} + \frac{\partial F}{\partial y}\frac{\partial y}{\partial v}$$

$$= \frac{1}{x}\left[e^{uv^2}(2uv)\right] + \frac{1}{y}\left[e^{uv}u\right]$$

$$= 2uv + u$$

13. $\frac{\partial w}{\partial s} = \frac{\partial w}{\partial x}\frac{\partial x}{\partial s} + \frac{\partial w}{\partial y}\frac{\partial y}{\partial s} + \frac{\partial w}{\partial z}\frac{\partial z}{\partial s}$

$$\frac{\partial w}{\partial t} = \frac{\partial w}{\partial x}\frac{\partial x}{\partial t} + \frac{\partial w}{\partial y}\frac{\partial y}{\partial t} + \frac{\partial w}{\partial z}\frac{\partial z}{\partial t}$$

15. $\frac{\partial w}{\partial s} = \frac{\partial w}{\partial x}\frac{\partial x}{\partial s} + \frac{\partial w}{\partial y}\frac{\partial y}{\partial s} + \frac{\partial w}{\partial z}\frac{\partial z}{\partial s}$

$$\frac{\partial w}{\partial t} = \frac{\partial w}{\partial x}\frac{\partial x}{\partial t} + \frac{\partial w}{\partial y}\frac{\partial y}{\partial t} + \frac{\partial w}{\partial z}\frac{\partial z}{\partial t}$$

$$\frac{\partial w}{\partial u} = \frac{\partial w}{\partial x}\frac{\partial x}{\partial u} + \frac{\partial w}{\partial y}\frac{\partial y}{\partial u} + \frac{\partial w}{\partial z}\frac{\partial z}{\partial u}$$

17. $w = \sin xyz$, where $x = 1 - 3t$, $y = e^{1-t}$, and $z = 4t$

$$\frac{dw}{dt} = \frac{\partial w}{\partial x}\frac{dx}{dt} + \frac{\partial w}{\partial y}\frac{dy}{dt} + \frac{\partial w}{\partial z}\frac{dz}{dt}$$

$$= [yz\cos(xyz)](-3) + [xz\cos(xyz)](-e^{1-t})$$

$$+ [xy\cos(xyz)](4)$$

$$= \cos(xyz)[-3yz - e^{1-t}xz + 4xy]$$

19. $w = e^{x^3+yz}$, where $x = \frac{2}{t}$, $y = \ln(2t - 3)$, and

$z = t^2$

$$\frac{dw}{dt} = \frac{\partial w}{\partial x}\frac{dx}{dt} + \frac{\partial w}{\partial y}\frac{dy}{dt} + \frac{\partial w}{\partial z}\frac{dz}{dt}$$

$$= e^{x^3+yz}\left[3x^2\left(\frac{-2}{t^2}\right) + \frac{2z}{2t-3} + y(2t)\right]$$

$$= (e^{x^3+yz})\left[\frac{-6x^2}{t^2} + 2ty + \frac{2z}{2t-3}\right]$$

21. $w = \dfrac{x+y}{2-z}$, where $x = 2rs$, $y = \sin rt$, and

$z = st^2$

$$\frac{\partial w}{\partial r} = \frac{\partial w}{\partial x}\frac{\partial x}{\partial r} + \frac{\partial w}{\partial y}\frac{\partial y}{\partial r} + \frac{\partial w}{\partial z}\frac{\partial z}{\partial r}$$

$$= \frac{2s}{2-z} + \frac{t\cos(rt)}{2-z} + 0$$

$$= \frac{2s + t\cos(rt)}{2-z}$$

$$\frac{\partial w}{\partial t} = \frac{\partial w}{\partial x}\frac{\partial x}{\partial t} + \frac{\partial w}{\partial y}\frac{\partial y}{\partial t} + \frac{\partial w}{\partial z}\frac{\partial z}{\partial t}$$

$$= 0 + \frac{r\cos(rt)}{2-z} + \frac{(x+y)(2st)}{(2-z)^2}$$

$$= \frac{(2-z)r\cos(rt) + 2st(x+y)}{(2-z)^2}$$

23. $F(x, y) = (x^2 - y)^{3/2} + x^2y - 2$

$$F_x = \tfrac{3}{2}(x^2 - y)^{1/2}(2x) + 2xy$$

$$F_y = \tfrac{3}{2}(x^2 - y)^{1/2}(-1) + x^2$$

$$\frac{dy}{dx} = \frac{-F_x}{F_y}$$

$$= \frac{-[3x(x^2-y)^{1/2} + 2xy]}{-\tfrac{3}{2}(x^2-y)^{1/2} + x^2}$$

25. $F(x, y) = x\cos y + y\tan^{-1}x - x$

$$F_x = \cos y + \frac{y}{1+x^2} - 1$$

$$F_y = -x\sin y + \tan^{-1}x$$

$$\frac{dy}{dx} = \frac{-F_x}{F_y}$$

$$= \frac{-\left[\cos y + \dfrac{y}{1+x^2} - 1\right]}{-x\sin y + \tan^{-1}x}$$

$$= \frac{(1 - \cos y)(1 + x^2) - y}{(1+x^2)(-x\sin y + \tan^{-1}x)}$$

27. $F(x, y) = \tan^{-1}\left(\frac{x}{y}\right) - \tan^{-1}\left(\frac{y}{x}\right)$

$$F_x = \frac{1}{1+\left(\frac{x}{y}\right)^2}\left(\frac{1}{y}\right) - \frac{1}{1+\left(\frac{y}{x}\right)^2}\left(-\frac{y}{x^2}\right)$$

$$= \frac{y}{x^2+y^2} + \frac{y}{x^2+y^2} = \frac{2y}{x^2+y^2}$$

$$F_y = \frac{1}{1 + \left(\frac{x}{y}\right)^2}\left(\frac{-x}{y^2}\right) - \frac{1}{1 + \left(\frac{y}{x}\right)^2}\left(\frac{1}{x}\right)$$

$$= \frac{-x}{x^2 + y^2} - \frac{x}{x^2 + y^2} = \frac{-2x}{x^2 + y^2}$$

$$\frac{dy}{dx} = \frac{-F_x}{F_y}$$

$$= \frac{-\left(\frac{2y}{x^2 + y^2}\right)}{\left(\frac{-2x}{x^2 + y^2}\right)}$$

$$= \frac{y}{x}$$

29. **a.** $z = \frac{2}{xy}$

$$z_x = -\frac{2}{x^2 y}$$

$$z_y = \frac{-2}{xy^2}$$

$$\frac{\partial^2 z}{\partial x \partial y} = \frac{\partial}{\partial x}\left(\frac{-2}{xy^2}\right)$$

$$= \frac{-2}{y^2}\left(\frac{-1}{x^2}\right)$$

$$= \frac{2}{x^2 y^2}$$

$$= \frac{z^2}{2}$$

b. $z_{xx} = \frac{4}{x^3 y}$

c. $z_{yy} = \frac{4}{xy^3}$

31. **a.** $x^{-1} + y^{-1} + z^{-1} = 3$

Take the derivative of both sides with respect to x:

$$-x^{-2} - z^{-2}z_x = 0$$

$$z_x = -\frac{z^2}{x^2}$$

Next, take the derivative of both sides with respect to y:

$$-y^{-2} - z^{-2}z_y = 0$$

$$z_y = \frac{-z^2}{y^2}$$

Now, find z_{yx}:

$$z_{yx} = \frac{\partial}{\partial x}\left(\frac{-z^2}{y^2}\right)$$

$$= \frac{-1}{y^2}(2zz_x)$$

$$= \frac{-2z}{y^2}\left(\frac{-z^2}{x^2}\right)$$

$$= \frac{2z^3}{x^2 y^2}$$

b. $z_{xx} = \frac{\partial^2 z}{\partial x^2}$

$$= \frac{\partial}{\partial x}\left(-\frac{z^2}{x^2}\right)$$

$$= -\frac{x^2(2z\frac{\partial z}{\partial x}) - z^2(2x)}{x^4}$$

$$= \frac{2z^2(x + z)}{x^4}$$

c. $\quad z_{yy} = \dfrac{\partial^2 z}{\partial y^2}$

$$= \frac{\partial}{\partial y}\left(-\frac{x^2}{y^2}\right)$$

$$= -\frac{y^2(2z\frac{\partial z}{\partial y}) - z^2(2y)}{y^4}$$

$$= \frac{2z^2(y + z)}{y^4}$$

33. a. $\quad z^2 + \sin x = \tan y$

$$z^2 = \tan y - \sin x$$

$$2zz_x = -\cos x$$

$$z_x = -\frac{\cos x}{2z}$$

Next, take the partial with respect to y:

$$2zz_y = \sec^2 y$$

$$z_y = \frac{\sec^2 y}{2z}$$

Finally, the mixed partial is:

$$z_{yx} = \frac{\partial}{\partial x}\left(\frac{\sec^2 y}{2z}\right)$$

$$= \frac{\sec^2 y}{2}\left(\frac{-1}{z^2}z_x\right)$$

$$= \frac{-\sec^2 y}{2z^2}\left(\frac{-\cos x}{2z}\right)$$

$$= \frac{\sec^2 y \cos x}{4z^3}$$

b. $\quad z_{xx} = \dfrac{\partial}{\partial x}\left(-\dfrac{\cos x}{2z}\right)$

$$= -\frac{1}{2}\frac{-z\sin x - z_x\cos x}{z^2}$$

$$= \frac{2z^2\sin x - \cos^2 x}{2z^3}$$

c. $\quad z_{yy} = \dfrac{\partial}{\partial x}\left(\dfrac{\sec^2 y}{2z}\right)$

$$= \frac{1}{2z^2}(2z\sec^2 y\tan y - \sec^2 y\, z_y)$$

$$= \frac{4z^2\sec^2 y\tan y - \sec^4 y}{4z^3}$$

35. $\quad \dfrac{\partial z}{\partial u} = \dfrac{\partial z}{\partial x}\dfrac{\partial x}{\partial u} + \dfrac{\partial z}{\partial y}\dfrac{\partial y}{\partial u}$

$$= \frac{\partial z}{\partial x}\cdot a + \frac{\partial z}{\partial y}\cdot 0$$

$$= a\frac{\partial z}{\partial x}$$

$$\frac{\partial z}{\partial v} = \frac{\partial z}{\partial x}\frac{\partial x}{\partial v} + \frac{\partial z}{\partial y}\frac{\partial y}{\partial v}$$

$$= \frac{\partial z}{\partial x}\cdot 0 + \frac{\partial z}{\partial y}\cdot b$$

$$= b\frac{\partial z}{\partial y}$$

$$\frac{\partial^2 z}{\partial u^2} = a\frac{\partial}{\partial u}z_x$$

$$= a[(z_x)_x x_u + (z_x)_y y_u]$$

$$= a^2 z_{xx}$$

$$\frac{\partial^2 z}{\partial v^2} = b\,\frac{\partial}{\partial v}\,z_y$$

$$= b[(z_y)_x x_v + (z_y)_y y_v]$$

$$= b^2 z_{yy}$$

37. $V = \ell w h$

$$\frac{dV}{dt} = (\ell w)\frac{dh}{dt} + (\ell h)\frac{dw}{dt} + (wh)\frac{d\ell}{dt}$$

When $t = 5$, $h = 20 - 3(5) = 5$;

$$\ell = 10 + 2(5) = 20$$

$$w = 8 + 2(5) = 18$$

$$V'(5) = (20)(18)(-3) + (20)(5)(2) + (18)(5)(2)$$

$$= -700 < 0 \text{ (decreasing)}$$

For the area, $S = 2(\ell h + \ell w + hw)$;

$$\frac{dS}{dt} = 2\left[(\ell + w)\frac{dh}{dt} + (\ell + h)\frac{dw}{dt} + (h + w)\frac{d\ell}{dt}\right]$$

$$S'(5) = 2[(20 + 18)(-3) + (20 + 5)(2)$$

$$+ (5 + 18)(2)]$$

$$= -36 < 0 \quad (\text{also decreasing})$$

39. a.

$$\frac{\partial C}{\partial a} = \frac{(b-a)(-te^{-at}) - (e^{-at} - e^{-bt})(-1)}{(b-a)^2}$$

$$= \frac{[1 - t(b-a)]e^{-at} - e^{-bt}}{(b-a)^2}$$

$$\frac{\partial C}{\partial b} = \frac{(b-a)(te^{-bt}) - (e^{-at} - e^{-bt})(1)}{(b-a)^2}$$

$$= \frac{[(b-a)t + 1]e^{-bt} - e^{-at}}{(b-a)^2}$$

$$\frac{\partial C}{\partial t} = \frac{1}{b-a}[-ae^{-at} + be^{-bt}]$$

b. Since $a = \dfrac{\ln b}{t}$, b constant, we have

$$e^{-at} = \frac{1}{b}$$

$$\frac{db}{dt} = 0$$

$$\frac{da}{dt} = \frac{-\ln b}{t^2},$$

and

$$(b-a)t = \left(b - \frac{\ln b}{t}\right)t$$

$$= bt - \ln b$$

Thus,

$$\frac{dC}{dt} = \frac{\partial C}{\partial a}\frac{da}{dt} + \frac{\partial C}{\partial b}\frac{db}{dt} + \frac{\partial C}{\partial t}\frac{dt}{dt}$$

First, we find

$$\frac{\partial C}{\partial a} = \frac{-te^{-at}}{b-a} + \frac{e^{-at} - e^{-bt}}{(b-a)^2}$$

$$= \frac{e^{-at} - t(b-a)e^{-at} - e^{-bt}}{(b-a)^2}$$

$$= \frac{[1 - (b-a)t]e^{-at} - e^{-bt}}{(b-a)^2}$$

Now we find:

$$\frac{dC}{dt} = \frac{\partial C}{\partial a}\frac{da}{dt} + \frac{\partial C}{\partial b}\frac{db}{dt} + \frac{\partial C}{\partial t}\frac{dt}{dt}$$

$$= \left[\frac{(1 - bt + \ln b)\left(\frac{1}{b}\right) - e^{-bt}}{\left(b - \frac{\ln b}{t}\right)^2}\right]\frac{(-\ln b)}{t^2}$$

$$+ \left(\frac{\partial C}{\partial b}\right)(0) + \frac{1}{b - a}\left[-ae^{-at} + be^{-bt}\right](1)$$

$$= \left[\frac{(1 - bt + \ln b)\left(\frac{1}{b}\right) - e^{-bt}}{(bt - \ln b)^2}\right](-\ln b)$$

$$+ \frac{t}{bt - \ln b}\left[\frac{-\ln b}{bt} + be^{-bt}\right]$$

41. $Q = 240 - 21\sqrt{x} + 4(0.2y + 12)^{3/2}$

$$Q_x = -21(\tfrac{1}{2})\frac{1}{\sqrt{x}}$$

$$= \frac{-10.5}{\sqrt{x}}$$

$$Q_y = 4(\tfrac{3}{2})(0.2y + 12)^{1/2}(0.2)$$

$$= 1.2(0.2y + 12)^{1/2}$$

Since $x = 120 + 6t$, $\frac{dx}{dt} = 6$.

Since $y = 80 + 10\sqrt{4t}$, $\frac{dy}{dt} = \frac{10}{\sqrt{t}}$.

$$\frac{dQ}{dt} = Q_x\frac{dx}{dt} + Q_y\frac{dy}{dt}$$

$$= \left(\frac{-10.5}{\sqrt{x}}\right)(6) + [1.2(0.2y + 12)^{1/2}]\left(\frac{10}{\sqrt{t}}\right)$$

When $t = 4$:

$$x = 120 + 6(4) = 144$$

and

$$y = 80 + 10\sqrt{4(4)} = 120$$

$$\frac{dQ}{dt}\bigg|_{t=4} = \frac{-10.5(6)}{\sqrt{144}} + \frac{1.2(10)}{\sqrt{4}}[0.2(120) + 12]^{1/2}$$

$$\approx 30.75$$

Thus, the monthly demand for bicycles will be increasing at the rate of about 31 bicycles per month (4 months from now).

43. $\frac{1}{R} = \frac{1}{R_1} + \frac{1}{R_2} + \frac{1}{R_3}$

$$-\frac{1}{R^2}\frac{\partial R}{\partial R_1} = -\frac{1}{R_1^2}$$

$$\frac{\partial R}{\partial R_1} = \left(\frac{R}{R_1}\right)^2$$

Similarly,

$$\frac{\partial R}{\partial R_2} = \left(\frac{R}{R_2}\right)^2$$

$$\frac{\partial R}{\partial R_3} = \left(\frac{R}{R_3}\right)^2$$

$$\frac{dR}{dt} = \left(\frac{R}{R_1}\right)^2\frac{dR_1}{dt} + \left(\frac{R}{R_2}\right)^2\frac{dR_2}{dt} + \left(\frac{R}{R_3}\right)^2\frac{dR_3}{dt}$$

When $R_1 = 100$, $R_2 = 200$, $R_3 = 300$

$$\frac{1}{R} = \frac{1}{100} + \frac{1}{200} + \frac{1}{300}$$

$$\frac{1}{R} = \frac{11}{600}$$

$$R \approx 54.545$$

With

$$\frac{dR_1}{dt} = -1.5$$

$$\frac{dR_2}{dt} = 2$$

$$\frac{dR_3}{dt} = -1.5$$

$$\frac{dR}{dt} \approx \left(\frac{54.545}{100}\right)^2(-1.5) + \left(\frac{54.545}{200}\right)^2(2)$$

$$+ \left(\frac{54.545}{300}\right)^2(-1.5)$$

$$\approx -0.3471$$

The joint resistance is decreasing at the

approximate rate of 0.3471 ohms/second.

45. Let $x = u - v$, $y = v - u$ so $z = f(x, y)$

$$\frac{\partial z}{\partial u} = \frac{\partial z}{\partial x}\frac{\partial x}{\partial u} + \frac{\partial z}{\partial y}\frac{\partial y}{\partial u}$$

$$= z_x(1) + z_y(-1)$$

$$\frac{\partial z}{\partial v} = \frac{\partial z}{\partial x}\frac{\partial x}{\partial v} + \frac{\partial z}{\partial y}\frac{\partial y}{\partial v}$$

$$= z_x(-1) + z_y(1)$$

$$z_u + z_v = (z_x - z_y) + (z_y - z_x)$$

$$= 0$$

47. Let $u = \frac{r - s}{s} = \frac{r}{s} - 1$, so $w = f(u)$

$$\frac{\partial w}{\partial r} = \frac{df}{du}\frac{\partial u}{\partial r}$$

$$= [f'(u)](s^{-1})$$

$$\frac{\partial w}{\partial s} = \frac{df}{du}\frac{\partial u}{\partial s}$$

$$= [f'(u)](-rs^{-2})$$

$$r\frac{\partial w}{\partial r} + s\frac{\partial w}{\partial s} = r\left[\frac{1}{s}f'(u)\right] + s\left[\frac{-r}{s^2}f'(u)\right]$$

$$= 0$$

49. $w = f(t)$, where $t = (x^2 + y^2 + z^2)^{1/2}$

$$\frac{\partial w}{\partial x} = \frac{df}{dt}\frac{\partial t}{\partial x}$$

$$= [f'(t)](\tfrac{1}{2})(x^2 + y^2 + z^2)^{-1/2}(2x)$$

$$\frac{\partial w}{\partial y} = \frac{df}{dt}\frac{\partial t}{\partial y}$$

$$= [f'(t)](\tfrac{1}{2})(x^2 + y^2 + z^2)^{-1/2}(2y)$$

$$\frac{\partial w}{\partial z} = \frac{df}{dt}\frac{\partial t}{\partial z}$$

$$= [f'(t)](\tfrac{1}{2})(x^2 + y^2 + z^2)^{-1/2}(2z)$$

$$\left(\frac{\partial w}{\partial x}\right)^2 + \left(\frac{\partial w}{\partial y}\right)^2 + \left(\frac{\partial w}{\partial z}\right)^2$$

$$= [f'(t)]^2(x^2 + y^2 + z^2)^{-1}(x^2 + y^2 + z^2)$$

$$= [f'(t)]^2$$

$$= \left(\frac{dw}{dt}\right)^2$$

51.

$$\frac{dz}{d\theta} = \frac{\partial f}{\partial x}\frac{dx}{d\theta} + \frac{\partial f}{\partial y}\frac{dy}{d\theta}$$

$$= (-\sin\theta)f_x + (\cos\theta)f_y$$

$$= -yf_x + xf_y$$

$$\frac{d^2z}{d\theta^2} = \frac{d}{d\theta}\left(\frac{dz}{d\theta}\right)$$

$$= \frac{d}{d\theta}[-yf_x + xf_y]$$

$$= \frac{\partial}{\partial x}[-yf_x + xf_y]\frac{dx}{d\theta} + \frac{\partial}{\partial y}[-yf_x + xf_y]\frac{dy}{d\theta}$$

$$= (-yf_{xx} + xf_{yx} + f_y)(-\sin\theta)$$

$$+ (-yf_{xy} - f_x + xf_{yy})(\cos\theta)$$

$$= y^2f_{xx} - yxf_{yx} - yf_y - xyf_{xy} - xf_x + x^2f_{yy}$$

$$= y^2f_{xx} + x^2f_{yy} - 2xyf_{xy} - xf_x - yf_y$$

53.

$$\frac{\partial z}{\partial r} = \frac{\partial z}{\partial x}\frac{\partial x}{\partial r} + \frac{\partial z}{\partial y}\frac{\partial y}{\partial r}$$

$$= f_x(e^r\cos\theta) + f_y(e^r\sin\theta)$$

$$= xf_x + yf_y$$

$$\frac{\partial^2 z}{\partial r^2} = \frac{\partial}{\partial r}\left(\frac{\partial z}{\partial r}\right)$$

$$= \frac{\partial}{\partial r}(xf_x + yf_y)$$

$$= \frac{\partial}{\partial x}(xf_x + yf_y)\frac{\partial x}{\partial r} + \frac{\partial}{\partial y}(xf_x + yf_y)\frac{\partial y}{\partial r}$$

$$= (xf_{xx} + yf_{yx} + f_x)(e^r\cos\theta)$$

$$+ (xf_{xy} + f_y + yf_{yy})(e^r\sin\theta)$$

$$= x^2f_{xx} + y^2f_{yy} + 2xyf_{xy} + xf_x + yf_y$$

$$\frac{\partial z}{\partial\theta} = \frac{\partial z}{\partial x}\frac{\partial x}{\partial\theta} + \frac{\partial z}{\partial y}\frac{\partial y}{\partial\theta}$$

$$= f_x(-e^r\sin\theta) + f_y(e^r\cos\theta)$$

$$= -yf_x + xf_y$$

$$\frac{\partial^2 z}{\partial\theta^2} = (-yf_{xx} + xf_{yx} + f_y)(-e^r\sin\theta)$$

$$+ (-yf_{xy} - f_x + xf_{yy})(e^r\cos\theta)$$

$$= y^2f_{xx} + x^2f_{yy} - 2xyf_{xy} - xf_x - yf_y$$

$$\frac{\partial^2 z}{\partial r^2} + \frac{\partial^2 z}{\partial\theta^2} = (x^2 + y^2)f_{xx} + (x^2 + y^2)f_{yy}$$

$$+ (2xy - 2xy)f_{xy} + (x - x)f_x$$

$$+ (y - y)f_y$$

$$= (x^2 + y^2)f_{xx} + (x^2 + y^2)f_{yy}$$

$$= e^{2r}(f_{xx} + f_{yy})$$

Thus,

$$\frac{\partial^2 z}{\partial x^2} + \frac{\partial^2 z}{\partial y^2} = e^{-2r}\left[\frac{\partial^2 z}{\partial r^2} + \frac{\partial^2 z}{\partial\theta^2}\right]$$

55. By the chain rule

$$\frac{\partial u}{\partial r} = \frac{\partial u}{\partial x}\cos\theta + \frac{\partial u}{\partial y}\sin\theta$$

and

$$\frac{\partial v}{\partial\theta} = -\frac{\partial v}{\partial x}(r\sin\theta) + \frac{\partial v}{\partial y}(r\cos\theta)$$

Substituting,

$$\frac{\partial u}{\partial x} = \frac{\partial v}{\partial y} \quad\text{and}\quad \frac{\partial u}{\partial y} = -\frac{\partial v}{\partial x}$$

into the first equation, we obtain

$$\frac{\partial u}{\partial r} = \frac{\partial v}{\partial y}\cos\theta - \frac{\partial v}{\partial x}\sin\theta$$

$$\frac{\partial u}{\partial r} = \frac{1}{r}\left[\frac{\partial v}{\partial y}(r\cos\theta) - \frac{\partial v}{\partial x}(r\sin\theta)\right]$$

Multiply by $\frac{r}{r} = 1$.

$$= \frac{1}{r}\frac{\partial v}{\partial\theta} \quad \text{Substitute from the second equation.}$$

The equation $\frac{\partial v}{\partial r} = -\frac{1}{r}\frac{\partial u}{\partial\theta}$ is obtained by

computing $\frac{\partial v}{\partial r}$ and $\frac{\partial u}{\partial\theta}$ and substituting

$\frac{\partial u}{\partial x} = \frac{\partial v}{\partial y}$ and $\frac{\partial u}{\partial y} = -\frac{\partial v}{\partial x}$.

57. $\qquad F(x, y, z) = C$

$$\frac{\partial F}{\partial x} + \frac{\partial F}{\partial y}\frac{\partial y}{\partial x} + \frac{\partial F}{\partial z}\frac{\partial z}{\partial x} = 0$$

$$\frac{\partial z}{\partial x} = -\frac{F_x}{F_z}$$

For the equation $F(x, y, z) = 4$, where

$F = x^2 + 2xyz + y^3 + e^z$, we have

$$F_x = 2x + 2yz$$

$$F_y = 2xz + 3y^2$$

$$F_z = 2xy + e^z$$

so

$$\frac{\partial z}{\partial x} = \frac{-F_x}{F_z} = \frac{-(2x + 2yz)}{2xy + e^z}$$

$$\frac{\partial z}{\partial y} = \frac{-F_y}{F_z} = \frac{-(2xz + 3y^2)}{2xy + e^z}$$

59. a. If $f(x, y) = x^2y + 2y^3$, then

$$f(tx, ty) = (tx)^2(ty) + 2(ty)^3$$

$$= t^3(x^2y + 2y^3)$$

$$= t^3 f(x, y)$$

Thus, the degree is $n = 3$.

b. Let $u = tx$ and $v = ty$, so

$f(u, v) = t^n f(x, y)$. Then,

$$\frac{\partial f}{\partial u}\frac{du}{dt} + \frac{\partial f}{\partial v}\frac{dv}{dt} = nt^{n-1}f(x, y)$$

$$x\frac{\partial f}{\partial u} + y\frac{\partial f}{\partial v} = nt^{n-1}f(x, y)$$

$$x\frac{\partial f}{\partial x} + y\frac{\partial f}{\partial y} = nf(x, y) \quad \text{Let } t = 1.$$

11.6 Directional Derivatives and the Gradient, Pages 747-749

1. $\quad f(x, y) = x^2 - 2xy$

$$f_x = 2x - 2y$$

$$f_y = -2x$$

$$\nabla f = f_x\mathbf{i} + f_y\mathbf{j}$$

$$= (2x - 2y)\mathbf{i} - 2x\mathbf{j}$$

3. $\quad f(x, y) = yx^{-1} + xy^{-1}$

$$f_x = -yx^{-2} + y^{-1}$$

$$f_y = x^{-1} - xy^{-2}$$

$$\nabla f = \left(-\frac{y}{x^2} + \frac{1}{y}\right)\mathbf{i} + \left(\frac{1}{x} - \frac{x}{y^2}\right)\mathbf{j}$$

5. $\quad f(x, y) = xe^{3-y}$

$$f_x = e^{3-y}$$

$$f_y = -xe^{3-y}$$

$$\nabla f = e^{3-y}\mathbf{i} - xe^{3-y}\mathbf{j}$$

$$= e^{3-y}(\mathbf{i} - x\mathbf{j})$$

7. $f(x, y) = \sin(x + 2y)$

$$f_x = \cos(x + 2y)$$

$$f_y = 2\cos(x + 2y)$$

$$\nabla f = \cos(x + 2y)\mathbf{i} + 2\cos(x + 2y)\mathbf{j}$$

$$= \cos(x + 2y)(\mathbf{i} + 2\mathbf{j})$$

9. $g(x, y, z) = xe^{y+3z}$

$$g_x = e^{y+3z}$$

$$g_y = xe^{y+3z}$$

$$g_z = 3xe^{y+3z}$$

$$\nabla g = e^{y+3z}\mathbf{i} + xe^{y+3z}\mathbf{j} + 3xe^{y+3z}\mathbf{k}$$

$$= e^{y+3z}(\mathbf{i} + x\mathbf{j} + 3x\mathbf{k})$$

11. $f(x, y) = x^2 + xy$

$$f_x = 2x + y$$

$$f_y = x$$

$$\mathbf{v} = \mathbf{i} + \mathbf{j}$$

$$\|\mathbf{v}\| = \sqrt{1^2 + 1^2} = \sqrt{2}$$

$$\mathbf{u} = \frac{\mathbf{v}}{\|\mathbf{v}\|}$$

$$= \frac{1}{\sqrt{2}}\mathbf{i} + \frac{1}{\sqrt{2}}\mathbf{j}$$

$$\nabla f = (2x + y)\mathbf{i} + x\mathbf{j}$$

At $(1, -2)$,

$$f_x(1, -2) = 2(1) + (-2) = 0$$

$$f_y(1, -2) = 1$$

$$\nabla f(1, -2) = 0\mathbf{i} + 1\mathbf{j} = \mathbf{j}$$

$$D_u f = \nabla f \cdot \mathbf{u}$$

$$= \frac{1}{\sqrt{2}}$$

$$= \frac{\sqrt{2}}{2}$$

13. $f(x, y) = \ln(x^2 + 3y)$

$$f_x = \frac{2x}{x^2 + 3y}$$

$$f_y = \frac{3}{x^2 + 3y}$$

$$\mathbf{v} = \mathbf{i} + \mathbf{j}$$

$$\|\mathbf{v}\| = \sqrt{1^2 + 1^2} = \sqrt{2}$$

$$\mathbf{u} = \frac{\mathbf{v}}{\|\mathbf{v}\|}$$

$$= \frac{1}{\sqrt{2}}\mathbf{i} + \frac{1}{\sqrt{2}}\mathbf{j}$$

$$= \frac{\sqrt{2}}{2}\mathbf{i} + \frac{\sqrt{2}}{2}\mathbf{j}$$

$$\nabla f = \frac{2x}{x^2 + 3y}\mathbf{i} + \frac{3}{x^2 + 3y}\mathbf{j}$$

At $(1, 1)$,

$$f_x(1, 1) = \frac{2(1)}{1^2 + 3(1)} = \frac{1}{2}$$

$$f_y(1, 1) = \frac{3}{4}$$

$$\nabla f(1, 1) = \frac{1}{2}\mathbf{i} + \frac{3}{4}\mathbf{j}$$

$$D_u f = \nabla f \cdot \mathbf{u}$$

$$= \frac{\sqrt{2}}{4} + \frac{3\sqrt{2}}{8}$$

$$= \frac{5\sqrt{2}}{8}$$

15. $f(x, y) = \sec(xy - y^3)$

$$f_x = \sec(xy - y^3)\tan(xy - y^3)y$$

$$f_y = \sec(xy - y^3)\tan(xy - y^3)(x - 3y^2)$$

$$\mathbf{v} = -\mathbf{i} - 3\mathbf{j}$$

$$\|\mathbf{v}\| = \sqrt{(-1)^2 + (-3)^2} = \sqrt{10}$$

$$\mathbf{u} = \frac{\mathbf{v}}{\|\mathbf{v}\|}$$

$$= -\frac{1}{\sqrt{10}}\mathbf{i} - \frac{3}{\sqrt{10}}\mathbf{j};$$

$$\nabla f = \sec(xy - y^3)\tan(xy - y^3)[y\mathbf{i}$$

$$+ (x - 3y^2)\mathbf{j}]$$

$$\nabla f(2, 0) = \mathbf{0}$$

$$D_u f = 0$$

17. $x^2 + y^2 + z^2 = 3$

$$f_x = 2x$$

$$f_y = 2y$$

$$f_z = 2z$$

$$\mathbf{N} = \nabla f$$

$$= 2x\mathbf{i} + 2y\mathbf{j} + 2z\mathbf{k}$$

$$= 2(x\mathbf{i} + y\mathbf{j} + z\mathbf{k})$$

$$\mathbf{N}(1, -1, 1) = 2(\mathbf{i} - \mathbf{j} + \mathbf{k})$$

$$\mathbf{N}_u = \pm \frac{\mathbf{N}}{2\sqrt{3}}$$

$$= \pm \frac{\sqrt{3}}{3}(\mathbf{i} - \mathbf{j} + \mathbf{k})$$

The tangent plane is:

$$(x - 1) - (y + 1) + (z - 1) = 0$$

$$x - y + z - 3 = 0$$

SURVIVAL HINT: *Using the hillside analogy: If you are standing on the hillside $z = f(x, y)$ at the point P, and walk in the compass direction indicated by the unit vector $\mathbf{u}$, the directional derivative will tell you the instantaneous rate of change in z.*

19. $\cos z = \sin(x + y)$

$$f(x, y, z) = \sin(x + y) - \cos z$$

$$f_x = \cos(x + y)$$

$$f_y = \cos(x + y)$$

$$f_z = \sin z$$

$$\mathbf{N} = \nabla f$$

$$= \cos(x + y)\mathbf{i} + \cos(x + y)\mathbf{j} + \sin z\,\mathbf{k}$$

$\mathbf{N}(\frac{\pi}{2}, \frac{\pi}{2}, \frac{\pi}{2}) = -\mathbf{i} - \mathbf{j} + \mathbf{k}$

$$\mathbf{N}_u = \pm \frac{\mathbf{N}}{\sqrt{3}}$$

$$= \pm \frac{\sqrt{3}}{3}(-\mathbf{i} - \mathbf{j} + \mathbf{k})$$

The tangent plane is:

$$-(x - \tfrac{\pi}{2}) - (y - \tfrac{\pi}{2}) + (z - \tfrac{\pi}{2}) = 0$$

$$x + y - z - \tfrac{\pi}{2} = 0$$

21. Write the function as

$$f(x, y, z) = \ln x - \ln(y - z)$$

$$f_x = \tfrac{1}{x}$$

$$f_y = \frac{-1}{y - z}$$

$$f_z = \frac{1}{y - z}$$

$$\mathbf{N} = \nabla f$$

$$= \tfrac{1}{x}\mathbf{i} - \frac{1}{y - z}\mathbf{j} + \frac{1}{y - z}\mathbf{k}$$

$\mathbf{N}(2, 5, 3) = \tfrac{1}{2}\mathbf{i} - \tfrac{1}{2}\mathbf{j} + \tfrac{1}{2}\mathbf{k}$

$$= \tfrac{1}{2}(\mathbf{i} - \mathbf{j} + \mathbf{k})$$

$$\mathbf{N}_u = \pm \frac{\mathbf{N}}{\tfrac{1}{2}\sqrt{3}}$$

$$= \pm \frac{\sqrt{3}}{3}(\mathbf{i} - \mathbf{j} + \mathbf{k})$$

The tangent plane is:

$$(x - 2) - (y - 5) + (z - 3) = 0$$

$$x - y + z = 0$$

23. $ze^{x+2y} = 3$

$$f_x = ze^{x+2y}$$

$$f_y = 2ze^{x+2y}$$

$$f_z = e^{x+2y}$$

$$\mathbf{N} = \nabla f$$

$$= e^{x+2y}(z\mathbf{i} + 2z\mathbf{j} + \mathbf{k})$$

$\mathbf{N}(2, -1, 3) = 3\mathbf{i} + 6\mathbf{j} + \mathbf{k}$

$$\mathbf{N}_u = \pm \frac{\mathbf{N}}{\sqrt{46}}$$

$$= \pm \frac{1}{\sqrt{46}}(3\mathbf{i} + 6\mathbf{j} + \mathbf{k})$$

The tangent plane is:

$$3x + 6y + z - 3 = 0$$

25. $f(x, y) = 3x + 2y - 1$

$$f_x = 3$$

$$f_y = 2$$

$$\nabla f = 3\mathbf{i} + 2\mathbf{j}$$

$$\nabla f(1, -1) = 3\mathbf{i} + 2\mathbf{j}$$

$$\|\nabla f\| = \sqrt{13}$$

27. $f(x, y) = x^3 + y^3$

$$f_x = 3x^2$$

$$f_y = 3y^2$$

$$\nabla f = 3x^2\mathbf{i} + 3y^2\mathbf{j}$$

$$\nabla f(3, -3) = 27(\mathbf{i} + \mathbf{j})$$

$$\|\nabla f\| = 27\sqrt{2}$$

29. $f(x, y, z) = ax^2 + by^2 + cx^2$

$$f_x = 2ax$$

$$f_y = 2by$$

$$f_z = 2cz$$

$$\nabla f = 2ax\mathbf{i} + 2by\mathbf{j} + 2cz\mathbf{k}$$

$$\nabla f(a, b, c) = 2(a^2\mathbf{i} + b^2\mathbf{j} + c^2\mathbf{k});$$

$$\|\nabla f\| = 2\sqrt{a^4 + b^4 + c^4}$$

31. $f(x, y) = \ln\sqrt{x^2 + y^2}$

$$f_x = \frac{2x}{2(x^2 + y^2)}$$

$$f_y = \frac{2y}{2(x^2 + y^2)}$$

$$\nabla f = \frac{1}{x^2 + y^2}(x\mathbf{i} + y\mathbf{j})$$

$$\nabla f(1, 2) = \tfrac{1}{5}(\mathbf{i} + 2\mathbf{j})$$

$$\|\nabla f\| = \frac{1}{\sqrt{5}} \text{ or } \frac{\sqrt{5}}{5}$$

33. $f(x, y, z) = (x + y)^2 + (y + z)^2 + (x + z)^2$

$$f_x = 2(x + y) + 2(x + z)$$

$$f_y = 2(x + y) + 2(y + z)$$

$$f_z = 2(y + z) + 2(x + z)$$

$$\nabla f = [2(x + y) + 2(x + z)]\mathbf{i}$$
$$+ [2(x + y) + 2(y + z)]\mathbf{j}$$
$$+ [2(x + z) + 2(y + z)]\mathbf{k}$$

$$\nabla f(2, -1, 2) = 2(5\mathbf{i} + 2\mathbf{j} + 5\mathbf{k});$$

$$\|\nabla f\| = \sqrt{216} = 6\sqrt{6}$$

35. Let $f(x, y) = ax + by - c$, so $f(x, y) = 0$

$$\nabla f = a\mathbf{i} + b\mathbf{j}$$

$$\mathbf{u} = \pm\frac{a\mathbf{i} + b\mathbf{j}}{\sqrt{a^2 + b^2}}$$

37. Let $f(x, y) = \dfrac{x^2}{a^2} + \dfrac{y^2}{b^2} - 1$, so $f(x, y) = 0$

$$\nabla f = \frac{2x}{a^2}\mathbf{i} + \frac{2y}{b^2}\mathbf{j}$$

$$= 2a^{-2}b^{-2}(b^2 x\mathbf{i} + a^2 y\mathbf{j})$$

$$\mathbf{u} = \pm\frac{b^2 x_0\mathbf{i} + a^2 y_0\mathbf{j}}{\sqrt{b^4 x_0^2 + a^4 y_0^2}}$$

39. $\mathbf{u} = \cos\dfrac{\pi}{6}\mathbf{i} + \sin\dfrac{\pi}{6}\mathbf{j}$

$$= \tfrac{1}{2}(\sqrt{3}\,\mathbf{i} + \mathbf{j})$$

$$f(x, y) = x^2 + y^2$$

$$f_x = 2x$$

$$f_y = 2y$$

$$\nabla f = \nabla(x^2 + y^2)$$

$$= 2(x\mathbf{i} + y\mathbf{j})$$

$$\nabla f(1, 1) = 2(\mathbf{i} + \mathbf{j})$$

$$D_u f = \sqrt{3} + 1$$

41. $f(x, y) = e^{x^2 y^2}$

$$f_x = 2xy^2 e^{x^2 y^2}$$

$$f_y = 2x^2 y e^{x^2 y^2}$$

$$\nabla f = e^{x^2 y^2}(2xy^2 \mathbf{i} + 2x^2 y \mathbf{j})$$

$$\nabla f(1, -1) = 2e(\mathbf{i} - \mathbf{j});$$

$$\mathbf{V} = \langle 2 - 1, 3 - (-1) \rangle$$

$$= \langle 1, 4 \rangle$$

$$\mathbf{u} = \left\langle \frac{1}{\sqrt{17}}, \frac{4}{\sqrt{17}} \right\rangle$$

$$D_u f = -2e \left(\frac{3}{\sqrt{17}} \right)$$

$$= -\frac{6e}{\sqrt{17}}$$

43. $f(x, y, z) = xyz$

$$f_x = yz$$

$$f_y = xz$$

$$f_z = xy$$

$$\nabla f = \nabla xyz$$

$$= yz\mathbf{i} + xz\mathbf{j} + xy\mathbf{k}$$

$$\nabla f(1, -1, 2) = -2\mathbf{i} + 2\mathbf{j} - \mathbf{k}$$

$$\mathbf{v} \times \mathbf{w} = \begin{bmatrix} \mathbf{i} & \mathbf{j} & \mathbf{k} \\ 1 & -2 & 3 \\ 2 & 1 & -1 \end{bmatrix}$$

$$= -\mathbf{i} + 7\mathbf{j} + 5\mathbf{k}$$

$$\| \mathbf{v} \times \mathbf{w} \| = \sqrt{(-1)^2 + 7^2 + 5^2}$$

$$= \sqrt{75}$$

$$\mathbf{u} = \frac{1}{\sqrt{75}}(-\mathbf{i} + 7\mathbf{j} + 5\mathbf{k})$$

$$D_u f = \frac{2 + 14 - 5}{5\sqrt{3}}$$

$$= \frac{11\sqrt{3}}{15}$$

45. $T(x, y, z) = xy + yz + xz$

$$T_x = y + z$$

$$T_y = x + z$$

$$T_z = y + x$$

$$\nabla T = (y + z)\mathbf{i} + (x + z)\mathbf{j} + (x + y)\mathbf{k}$$

$$\nabla T_0 = \nabla T(1, 1, 1)$$

$$= 2(\mathbf{i} + \mathbf{j} + \mathbf{k})$$

The maximum rate of temperature change is

$$\|\nabla \mathbf{T}_0\| = 2\sqrt{3} \text{ in the direction of}$$

$$\mathbf{u} = \frac{\sqrt{3}}{3}(\mathbf{i} + \mathbf{j} + \mathbf{k})$$

47. $z = 1 - 3x^2 - \frac{5}{2}y^2$

$$z_x = -6x$$

$$z_y = -5y$$

$$\nabla z = -6x\mathbf{i} - 5y\mathbf{j}$$

$$\nabla z_0(\tfrac{1}{4}, -\tfrac{1}{2}) = -\tfrac{3}{2}\mathbf{i} + \tfrac{5}{2}\mathbf{j}$$

For the most rapid decrease, she should head

in the direction

$$-\nabla z_0 = \tfrac{3}{2}\mathbf{i} - \tfrac{5}{2}\mathbf{j}$$

Practically speaking, she should head in the

direction the water is running since streams

run perpendicular to the contours.

49. To find the directional derivative of f in the

direction of $\mathbf{u}$ we need f_x and f_y, which we

can find with a system of equations.

$$\begin{cases} (f_x\mathbf{i} + f_y\mathbf{j}) \cdot \left(\dfrac{3\mathbf{i} - 4\mathbf{j}}{5}\right) = 8 \\[2mm] (f_x\mathbf{i} + f_y\mathbf{j}) \cdot \left(\dfrac{12\mathbf{i} + 5\mathbf{j}}{13}\right) = 1 \end{cases}$$

$$\begin{cases} 3f_x - 4f_y = 40 \\ 12f_x + 5f_y = 13 \end{cases}$$

Solving simultaneously: $f_x = 4$, $f_y = -7$

Now for $\mathbf{v} = 3\mathbf{i} - 5\mathbf{j}$:

$$(f_x\mathbf{i} + f_y\mathbf{j}) \cdot \left(\frac{3\mathbf{i} - 5\mathbf{j}}{\sqrt{34}}\right) = (4\mathbf{i} - 7\mathbf{j}) \cdot \left(\frac{3\mathbf{i} - 5\mathbf{j}}{\sqrt{34}}\right)$$

$$= \frac{12 + 35}{\sqrt{34}} \approx 8.06$$

51. Let $\nabla f_0 = f_x\mathbf{i} + f_y\mathbf{j}$ be the gradient of f at

$P_0(1, 2)$. The unit vector from P_0 toward

$Q(3, -4)$ is $\mathbf{u} = \dfrac{1}{\sqrt{10}}(\mathbf{i} - 3\mathbf{j})$. Since the

maximal directional derivative points in the

direction of ∇f_0 and has magnitude 50, we

want to have

$$f_x\mathbf{i} + f_y\mathbf{j} = 50\mathbf{u}$$

$$= \frac{50}{\sqrt{10}}(\mathbf{i} - 3\mathbf{j})$$

Thus,

$$\nabla f(1, 2) = \frac{50(\mathbf{i} - 3\mathbf{j})}{\sqrt{10}}$$

$$= 5\sqrt{10}(\mathbf{i} - 3\mathbf{j})$$

53. a. Write $\dfrac{x^2}{a^2} + \dfrac{y^2}{b^2} + \dfrac{z^2}{c^2} = 1$ as $F(x, y, z) = 1$,

where $F(x, y, z) = \dfrac{x^2}{a^2} + \dfrac{y^2}{b^2} + \dfrac{z^2}{c^2}$. Then,

$$F_x = \frac{2x}{a^2}$$

$$F_y = \frac{2y}{b^2},$$

and

$$F_z = \frac{2z}{c^2}$$

so the tangent plane at $P_0(x_0, y_0, z_0)$ has

the equation

$$\frac{2x_0}{a^2}(x - x_0) + \frac{2y_0}{b^2}(y - y_0) + \frac{2z_0}{c^2}(z - z_0) = 0$$

$$\frac{x_0 x}{a^2} - \frac{x_0^2}{a^2} + \frac{y_0 y}{b^2} - \frac{y_0^2}{b^2} + \frac{z_0 z}{c^2} - \frac{z_0^2}{c^2} = 0$$

$$\frac{x_0 x}{a^2} + \frac{y_0 y}{b^2} + \frac{z_0 z}{c^2} = 1$$

b. As in part **a**, let

$$F(x,\ y,\ z) = \frac{x^2}{a^2} + \frac{y^2}{b^2} - \frac{z^2}{c^2}$$

and obtain

$$F_x = \frac{2x}{a^2},$$
$$F_y = \frac{2y}{b^2},$$

and

$$F_z = -\frac{2z}{c^2}$$

Thus, the tangent plane at $P_0(x_0,\ y_0,\ z_0)$

has the equation

$$\frac{2x_0}{a^2}(x - x_0) + \frac{2y_0}{b^2}(y - y_0) - \frac{2z_0}{c^2}(z - z_0) = 0$$

$$\frac{x_0 x}{a^2} - \frac{x_0^2}{a^2} + \frac{y_0 y}{b^2} - \frac{y_0^2}{b^2} - \frac{z_0 z}{c^2} + \frac{z_0^2}{c^2} = 0$$

$$\frac{x_0 x}{a^2} + \frac{y_0 y}{b^2} - \frac{z_0 z}{c^2} = 1$$

c. Write $\frac{z}{c} = \frac{x^2}{a^2} + \frac{y^2}{b^2}$ as $F(x,\ y,\ z) = 0$ where

$$F(x,\ y,\ z) = \frac{x^2}{a^2} + \frac{y^2}{b^2} - \frac{z}{c}. \text{ Then,}$$

$$F_x = \frac{2x}{a^2},$$
$$F_y = \frac{2y}{b^2},$$

and

$$F_z = \frac{-1}{c}$$

so the tangent plane at $P_0(x_0,\ y_0,\ z_0)$ has

the equation

$$\frac{2x_0}{a^2}(x - x_0) + \frac{2y_0}{b^2}(y - y_0) - \frac{1}{c}(z - z_0) = 0$$

$$\frac{2x_0 x}{a^2} - \frac{2x_0^2}{a^2} + \frac{2y_0 y}{b^2} - \frac{2y_0^2}{b^2} - \frac{z}{c} + \frac{z_0}{c} = 0$$

$$\frac{2x_0 x}{a^2} + \frac{2y_0 y}{b^2} - \frac{z}{c} = \frac{z_0}{c}$$

55. a. Note that $r_1 + r_2 = C$ (a constant) is a

level curve of the function $f = r_1 + r_2$.

By the normal property of the gradient,

we know that $\nabla(r_1 + r_2)$ is a normal to

$r_1 + r_2 = C$. So, $\mathbf{T} \cdot \nabla(r_1 + r_2) = 0$

b. From part **a**,

$$\mathbf{T} \cdot \nabla(r_1 + r_2) = 0$$

Let $\mathbf{R}_1 = \mathbf{PF}_1$ and $\mathbf{R}_2 = \mathbf{PF}_2$ be the

vectors from P to the two foci, so that

$$r_1 = \|\mathbf{R}_1\| \text{ and } r_2 = \|\mathbf{R}_2\|$$

By direct computation, it can be shown

that

$$\nabla r_1 = \frac{\mathbf{R}_1}{r_1} \text{ and } \nabla r_2 = \frac{\mathbf{R}_2}{r_2}$$

and by substituting into the vector

equation $\mathbf{T} \cdot \nabla(r_1 + r_2) = 0$, we have

$$\mathbf{T} \cdot \nabla r_1 = -\mathbf{T} \cdot \nabla r_2$$

$$\mathbf{T} \cdot \left(\frac{\mathbf{R}_1}{r_1}\right) = -\mathbf{T} \cdot \left(\frac{\mathbf{R}_2}{r_2}\right)$$

or

$$\frac{\|\mathbf{T}\|\|\mathbf{R_1}\|}{r_1}\cos(\pi - \theta_1) = -\frac{\|\mathbf{T}\|\|\mathbf{R_2}\|}{r_2}\cos\theta_2$$

so that

$$\cos(\pi - \theta_1) = -\cos\theta_2$$

$$\cos\theta_1 = \cos\theta_2$$

$$\theta_1 = \theta_2$$

57. **a.** $\nabla c = 0$ because $\frac{\partial}{\partial x}c = 0$

b. $\nabla(f + g) = \frac{\partial}{\partial x}(f + g)\mathbf{i} + \frac{\partial}{\partial y}(f + g)\mathbf{j}$

$$= (f_x + g_x)\mathbf{i} + (f_y + g_y)\mathbf{j}$$

$$= (f_x\mathbf{i} + f_y\mathbf{j}) + (g_x\mathbf{i} + g_y\mathbf{j})$$

$$= \nabla f + \nabla g$$

c. $\nabla\left(\frac{f}{g}\right) = \frac{\partial}{\partial x}\frac{f}{g}\mathbf{i} + \frac{\partial}{\partial y}\frac{f}{g}\mathbf{j}$

$$= \frac{gf_x - fg_x}{g^2}\mathbf{i} + \frac{gf_y - fg_y}{g^2}\mathbf{j}$$

$$= \frac{g(f_x\mathbf{i} + f_y\mathbf{j}) - f(g_x\mathbf{i} + g_y\mathbf{j})}{g^2}$$

$$= \frac{g\nabla f - f\nabla g}{g^2}$$

d. $\nabla(fg) = \frac{\partial}{\partial x}(fg)\mathbf{i} + \frac{\partial}{\partial y}(fg)\mathbf{j}$

$$= (fg_x + f_xg)\mathbf{i} + (fg_y + f_yg)\mathbf{j}$$

$$= f(g_x\mathbf{i} + g_y\mathbf{j}) + g(f_x\mathbf{i} + f_y\mathbf{j})$$

$$= f\nabla g + g\nabla f$$

59. General formula:

$$\nabla f = f_x\mathbf{i} + f_y\mathbf{j}$$

$$D_uf = f_x\cos\theta + f_y\sin\theta$$

In particular, for $f = xy^2e^{x-2y}$,

$$f_x = y^2e^{x-2y} + xy^2e^{x-2y}$$

$$f_y = 2xye^{x-2y} - 2xy^2e^{x-2y}$$

$$\nabla f = e^{x-2y}[(y^2 + xy^2)\mathbf{i} + (2xy - 2xy^2)\mathbf{j}]$$

$$\nabla f(-1, 3) = 12e^{-7}\mathbf{j}$$

$$D_uf = 12e^{-7}\sin\frac{\pi}{6}$$

$$= 6e^{-7}$$

61. True; since $\nabla f = f_x\mathbf{i} + f_y\mathbf{j} = \mathbf{0}$, it follows that $f_x(x, y) = f_y(x, y) = 0$ throughout the disk, so $f(x, y)$ must be a constant function.

63. Let $\mathbf{R} = x\mathbf{i} + y\mathbf{j} + z\mathbf{k}$

and let

$$r = \|\mathbf{R}\|$$

$$= \sqrt{x^2 + y^2 + z^2}$$

a. $\nabla r = \frac{x\mathbf{i} + y\mathbf{j} + z\mathbf{k}}{\sqrt{x^2 + y^2 + z^2}}$

$$= \frac{\mathbf{R}}{\|\mathbf{R}\|}$$

b. $\nabla r^n = \nabla(x^2 + y^2 + z^2)^{n/2}$

$$= \frac{n}{2}(r^2)^{n/2-1}(2x\mathbf{i} + 2y\mathbf{j} + 2z\mathbf{k})$$

$$= nr^{n-2}(x\mathbf{i} + y\mathbf{j} + z\mathbf{k})$$

$$= nr^{n-2}\,\mathbf{R}$$

11.7 Extrema of Functions of Two Variables, Pages 758-761

SURVIVAL HINT: *Note that, as in Section 4.1, the critical points are only* **candidates** *for extrema. If you are walking in the x-direction along a level trail on a hillside, $\partial f/\partial x = 0$ but $\partial f/\partial y$ may go to your right and down slope to your left. The point is not an extreme. If you are at a saddle point,* **both** *partials are zero and the candidate is still not an extreme. The candidate must be tested using the discriminant.*

1. A critical point of a function f defined on an

   open set S is a point (x_0, y_0) in S where either

   one of the following is true:

   (1) $f_x(x_0, y_0) = f_y(x_0, y_0) = 0$

   (2) At least one of $f_x(x_0, y_0)$ or $f_y(x_0, y_0)$

   does not exist.

3. Find all critical points of f in the interior of S.

   Find all boundary critical points. Evaluate f

   at the interior and boundary critical points.

   The largest of these is the absolute maximum

   on S and the smallest is the absolute minimum.

5. $f(x, y) = (x - 2)^2 + (y - 3)^4$

   $$f_x = 2(x - 2)$$

   $$f_y = 4(y - 3)^3$$

   $f_x = f_y = 0$ when $x = 2$, $y = 3$

   $$f_{xx} = 2$$

   $$f_{xy} = 0$$

   $$f_{yy} = 12(y - 3)^2$$

x	y	f_{xx}	f_{xy}	f_{yy}	D	Classify
2	3	2	0	0	0	inconclusive

   Note that f is a sum of squares, which implies

   that a minimum occurs at $(2, 3)$.

7. $f(x, y) = (1 + x^2 + y^2)e^{1 - x^2 - y^2}$

   $$f_x = e^{1 - x^2 - y^2}[(-2x)(1 + x^2 + y^2) + 2x]$$

   $$= -2x(x^2 + y^2)e^{1 - x^2 - y^2}$$

   $$f_y = e^{1 - x^2 - y^2}[(-2y)(1 + x^2 + y^2) + 2y]$$

   $$= -2y(x^2 + y^2)e^{1 - x^2 - y^2}$$

   $f_x = f_y = 0$ only when $x = y = 0$

   $$f_{xx} = 2(2x^4 + 2x^2y^2 - y^2 - 3x^2)e^{1 - x^2 - y^2}$$

   $$f_{xy} = 4xy(x^2 + y^2 - 1)e^{1 - x^2 - y^2}$$

   $$f_{yy} = 2(2y^4 + 2x^2y^2 - x^2 - 3y^2)e^{1 - x^2 - y^2}$$

x	y	f_{xx}	f_{xy}	f_{yy}	D	Classify
0	0	0	0	0	0	inconclusive

Examination of the graph of $f(x, y)$ suggests that a maximum occurs at $(0, 0)$.

9. $f(x, y) = x^2 + xy + y^2$

$$f_x = 2x + y$$

$$f_y = x + 2y$$

$f_x = f_y = 0$ only when $x = 0$, $y = 0$

$$f_{xx} = 2$$

$$f_{xy} = 1$$

$$f_{yy} = 2$$

x	y	f_{xx}	f_{xy}	f_{yy}	D	Classify
0	0	2	1	2	+	rel min

11. $f(x, y) = -x^3 + 9x - 4y^2$

$$f_x = -3x^2 + 9$$

$$f_y = -8y$$

$f_x = f_y = 0$ when $x = \pm\sqrt{3}$, $y = 0$

$$f_{xx} = -6x$$

$$f_{yy} = -8$$

$$f_{xy} = 0$$

x	y	f_{xx}	f_{xy}	f_{yy}	D	Classify
$\sqrt{3}$	0	$-6\sqrt{3}$	0	-8	+	rel max
$-\sqrt{3}$	0	$6\sqrt{3}$	0	-8	$-$	saddle point

13. $f(x, y) = (x^2 + 2y^2)e^{1 - x^2 - y^2}$

$$f_x = -2x(x^2 + 2y^2 - 1)e^{1 - x^2 - y^2}$$

$$f_y = -2y(x^2 + 2y^2 - 2)e^{1 - x^2 - y^2}$$

$f_x = f_y = 0$ when $x = 0$, $y = 0$ or when

$x = \pm 1$, $y = 0$; $x = 0$, $y = \pm 1$.

Note that $x^2 + 2y^2 = 1$, $x^2 + 2y^2 = 2$ is

impossible.

$$f_{xx} = 2(2x^4 + 4x^2y^2 - 5x^2 - 2y^2 + 1)e^{1 - x^2 - y^2}$$

$$f_{yy} = 2(2x^2y^2 - x^2 + 4y^4 - 10y^2 + 2)e^{1 - x^2 - y^2}$$

$$f_{xy} = 4xy(x^2 + 2y^2 - 3)e^{1 - x^2 - y^2}$$

x	y	f_{xx}	f_{xy}	f_{yy}	D	Classify
0	0	$2e$	0	$4e$	+	rel min
1	0	-4	0	2	$-$	saddle point
-1	0	-4	0	2	$-$	saddle point
0	1	-2	0	-8	+	rel max
0	-1	-2	0	-8	+	rel max

15. $f(x, y) = x^{-1} + y^{-1} + 2xy$

$$f_x = -\frac{1}{x^2} + 2y$$

$$f_y = -\frac{1}{y^2} + 2x$$

$f_x = f_y = 0$ when $y = \frac{1}{2x^2}$ and $x = \frac{1}{2y^2}$

Solving, we obtain $x = y$ and

$$x = \frac{1}{2}(2x^2)^2$$

$$= 2x^4$$

so $\left(\dfrac{1}{\sqrt[3]{2}}, \dfrac{1}{\sqrt[3]{2}}\right)$ is the critical point, since

$x \neq 0.$

$$f_{xx} = \frac{2}{x^3}$$

$$f_{yy} = \frac{2}{y^3}$$

$$f_{xy} = 2$$

x	y	f_{xx}	f_{xy}	f_{yy}	D	Classify
$2^{-1/3}$	$2^{-1/3}$	4	2	4	+	rel min

17. $f(x, y) = x^3 + y^3 + 3x^2 - 18y^2 + 81y + 5$

$$f_x = 3x^2 + 6x$$
$$= 3x(x + 2)$$
$$f_y = 3y^2 - 36y + 81$$
$$= 3(y - 3)(y - 9)$$

$f_x = f_y = 0$ when $x = 0, -2$ and $y = 3, 9$;

Critical points are $(0, 3)$, $(0, 9)$, $(-2, 3)$,

$(-2, 9)$.

$$f_{xx} = 6x + 6$$
$$f_{xy} = 0$$
$$f_{yy} = 6y - 36$$

x	y	f_{xx}	f_{xy}	f_{yy}	D	Classify
0	3	6	0	−18	−	saddle point
0	9	6	0	18	+	rel min
−2	9	−6	0	18	−	saddle point
−2	3	−6	0	−18	+	rel max

19. $f(x, y) = x^2 + y^2 - 6xy + 9x + 5y + 2$

$$f_x = 2x - 6y + 9$$
$$f_y = 2y - 6x + 5$$

$f_x = f_y = 0$ when $x = \frac{3}{2}$, $y = 2$

$$f_{xx} = 2$$
$$f_{yy} = 2$$
$$f_{xy} = -6$$

x	y	f_{xx}	f_{xy}	f_{yy}	D	Classify
$\frac{3}{2}$	2	2	−6	2	−	saddle point

21. $f(x, y) = x^2 + y^3 + \dfrac{768}{x + y}$ $\qquad x \neq -y$

$$f_x = 2x - \frac{768}{(x + y)^2}$$
$$f_y = 3y^2 - \frac{768}{(x + y)^2}$$

$f_x = f_y = 0$ when

$$2x(x + y)^2 = 3y^2(x + y)^2 = 768$$

Since $x \neq -y$, we have $2x = 3y^2$ and

$$2x\left(x \pm \sqrt{\frac{2x}{3}}\right)^2 = 768$$
$$x = 6, 8.99$$

so $y = 2$ and $y \approx -2.45$

$$f_{xx} = 2 + \frac{1,536}{(x + y)^3}$$
$$f_{xy} = \frac{1,536}{(x + y)^3}$$

$$f_{yy} = 6y + \frac{1{,}536}{(x+y)^3}$$

x	y	f_{xx}	f_{xy}	f_{yy}	D	Classify
6	2	5	3	15	+	rel min
8.99	-2.45	7.49	5.49	-9.21	$-$	saddle point

23. $f(x, y) = 3x^2 + 12x + 8y^3 - 12y^2 + 7$

$$f_x = 6x + 12$$

$$f_y = 24y^2 - 24y$$

$f_x = f_y = 0$ when $x = -2$, $y = 0$ or 1

Critical points are $(-2, 0)$ and $(-2, 1)$.

$$f_{xx} = 6$$

$$f_{xy} = 0$$

$$f_{yy} = 48y - 24$$

x	y	f_{xx}	f_{xy}	f_{yy}	D	Classify
-2	0	6	0	-24	$-$	saddle point
-2	1	6	0	24	+	rel min

25. $f(x, y) = xy - 2x - 5y;$

$$f_x = y - 2$$

$$f_y = x - 5$$

$f_x = f_y = 0$ only at $(5, 2)$

The boundary consists of three lines:

On $y = 0$, $x = t$ for $0 \le t \le 7$, we have

$$F_1(t) = f(t, 0) = 0 - 2t - 0$$

$$F_1{}'(t) = -2 \ne 0$$

On $x = 7$, $y = t$ for $0 \le t \le 7$

$$F_2(t) = f(7, t) = 7t - 14 - 5t$$

$$F_2{}'(t) = 2 \ne 0$$

On $x = t$, $y = t$ for $0 \le t \le 7$

$$F_3(t) = f(t, t) = t^2 - 7t$$

$$F_3{}'(t) = 2t - 7 \text{ boundary critical point}$$

$$\left(\tfrac{7}{2}, \tfrac{7}{2}\right) \text{ where } t = \tfrac{7}{2}$$

Thus, we have an interior critical point $(5, 2)$, a boundary critical point $\left(\tfrac{7}{2}, \tfrac{7}{2}\right)$, and three boundary endpoints $(0, 0)$, $(7, 0)$, and $(7, 7)$.

x	y	$f(x, y)$	
5	2	-10	
7/2	7/2	-12.25	
0	0	0	max
7	0	-14	min
7	7	0	max

The largest value of f on S is 0 and the smallest is -14.

SURVIVAL HINT: *In $\mathbb{R}^2$ the extreme value theorem requires a continuous function on a closed interval. Note that the $\mathbb{R}^3$ extension of the extreme value theorem requires that f be continuous on a closed and bounded set S. In $\mathbb{R}^2$ the curve may have corners, cups, or vertical tangents, as long as it is continuous. Likewise in $\mathbb{R}^3$ the surface may have creases, pointed peaks, and vertical tangents if it is continuous. It is* *sometimes difficult to verify the continuity of a surface. Most of your examples will be sums or products of continuous functions, which are continuous.*

27. $f(x, y) = 2\sin x + 5\cos y$

$$f_x = 2\cos x$$

$$f_y = -5\sin y$$

$f_x = f_y = 0$ when $x = \frac{\pi}{2}$, $y = 0$, π

Critical points are $(\frac{\pi}{2}, 0)$, $(\frac{\pi}{2}, \pi)$.

On $x = t$, $y = 0$ for $0 \le t \le 2$;

$$F_1(t) = f(t, 0)$$

$$= 2\sin t + 5$$

$F_1'(t) = 2\cos t = 0$ when $t = \frac{\pi}{2}$; point $(\frac{\pi}{2}, 0)$

On $x = 2$, $y = t$ for $0 \le t \le 5$;

$$F_2(t) = f(2, t)$$

$$= 2\sin 2 + 5\cos t$$

$F_2'(t) = -5\sin t = 0$ when $t = 0$, π;

Critical points are $(2, 0)$, $(2, \pi)$.

On $y = 5$, $x = t$ for $0 \le t \le 2$;

$$F_3(t) = f(t, 5)$$

$$= 2\sin t + 5\cos 5$$

$F_3'(t) = 2\cos t = 0$ when $t = \frac{\pi}{2}$;

Critical point $(\frac{\pi}{2}, 5)$.

On $x = 0$, $y = t$ for $0 \le t \le 5$;

$$F_4(t) = f(0, t)$$

$$= 5\cos t$$

$F_4'(t) = -5\sin t = 0$ when $t = 0$, π;

Critical points $(0, 0)$, $(0, \pi)$.

There is one interior critical point, $(\frac{\pi}{2}, \pi)$;

four boundary critical points $(\frac{\pi}{2}, 0)$, $(2, \pi)$,

$(\frac{\pi}{2}, 5)$, $(0, \pi)$; and four boundary endpoints

$(0, 0)$, $(2, 0)$, $(2, 5)$, $(0, 5)$.

x	y	$f(x, y)$	
$\pi/2$	0	7	max
2	π	-3.18	
$\pi/2$	5	3.42	
0	π	-5	min
0	0	5	
2	0	6.82	
2	5	3.24	
0	5	1.42	
$\pi/2$	π	-3	

The largest value of f on S is 7 and the smallest is -5.

29. $f(x, y) = x^2 + xy + y^2$

$$f_x = 2x + y$$

$$f_y = x + 2y$$

$f_x = f_y = 0$ only at $(0, 0)$

On the boundary $x^2 + y^2 = 1$, let

$x = \cos t$, $y = \sin t$ for $0 \le t \le 2\pi$.

Then,

$\quad F(t) = f(\cos t, \sin t)$

$\qquad = 1 + \cos t \sin t$

$F'(t) = -\sin^2 t + \cos^2 t = \cos 2t = 0$ when

$\quad t = \frac{\pi}{4};$ point $\left(\frac{\sqrt{2}}{2}, \frac{\sqrt{2}}{2} \right)$

$\quad t = \frac{3\pi}{4};$ point $\left(-\frac{\sqrt{2}}{2}, \frac{\sqrt{2}}{2} \right)$

$\quad t = \frac{5\pi}{4};$ point $\left(-\frac{\sqrt{2}}{2}, -\frac{\sqrt{2}}{2} \right)$

$\quad t = \frac{7\pi}{4};$ point $\left(\frac{\sqrt{2}}{2}, -\frac{\sqrt{2}}{2} \right)$

$f(0, 0) = 0$; this is a minimum since a

minimum must occur somewhere and this is

the only possibility.

$f\left(\frac{\sqrt{2}}{2}, \frac{\sqrt{2}}{2} \right) = f\left(-\frac{\sqrt{2}}{2}, -\frac{\sqrt{2}}{2} \right) = \frac{3}{2}$ max;

$f\left(-\frac{\sqrt{2}}{2}, \frac{\sqrt{2}}{2} \right) = f\left(\frac{\sqrt{2}}{2}, -\frac{\sqrt{2}}{2} \right) = \frac{1}{2}$

31. $m = \dfrac{5(25) - (1)(5)}{5(15) - 1^2} = \dfrac{60}{37} \approx 1.62$

$\quad b = \dfrac{15(5) - (1)(25)}{5(15) - 1^2} = \dfrac{25}{37} \approx 0.68$

$\quad y = 1.62x + 0.68$

33. $m = \dfrac{5(155.68) - (28.75)(27.14)}{5(184.47) - (28.75)^2} \approx -0.02$

$\quad b = \dfrac{184.47(27.14) - (28.75)(155.68)}{95.79} \approx 5.54$

$\quad y = -0.02x + 5.54$

35. Minimize $S = x^2 + y^2 + z^2$ subject to

$y^2 = 4 + xz$. We have,

$S(x, z) = x^2 + (4 + xz) + z^2$

$\qquad S_x = 2x + z$

$\qquad S_z = x + 2z$

$S_x = S_z = 0$ when $z = -2x = -\frac{1}{2}x$; $x = 0$,

$z = 0$, $y = \pm 2$.

The closest points are $(0, 2, 0)$ and $(0, -2, 0)$.

37. Let x, y, and z be the dimensions of the box.

The volume is $V = xyz$ or $z = V/(xy)$, and the

surface area is $S = xy + 2xz + 2yz$. We wish

to minimize

$S = xy + 2xz + 2yz$

$\quad = xy + \dfrac{2xV}{xy} + \dfrac{2yV}{xy}$

$\quad = xy + \dfrac{2V}{y} + \dfrac{2V}{x}$

We now find

$\qquad S_x = y - \dfrac{2V}{x^2}$

$\qquad S_y = x - \dfrac{2V}{y^2}$

Then $S_x = S_y = 0$ when $x = y = \sqrt[3]{2V}$, and

$z = \sqrt[3]{0.25V}$; $S_{xx} = \frac{4V}{x^3}$; $S_{xy} = 1$; $S_{yy} = \frac{4V}{y^3}$;

$D > 0$, so the dimensions for the minimum

construction are $x = \sqrt[3]{2V}$, $y = \sqrt[3]{2V}$,

$z = \sqrt[3]{0.25V}$.

39. If x, y, and z are the numbers we wish to

maximize $P = xyz$ subject to $x + y + z = 54$.

Since $z = 54 - x - y$, we have

$$P = xy(54 - x - y)$$
$$= 54xy - x^2y - xy^2$$

and

$$P_x = 54y - 2xy - y^2$$
$$P_y = 54x - x^2 - 2xy$$

Since $x > 0$, $y > 0$, it follows that $P_x = P_y = 0$

when $x = y = 18$. Since

$$D = P_{xx}P_{yy} - P_{xy}^2$$
$$= (-2y)(-2x) - (54 - 2x - 2y)^2$$

We have $P_{xx}(18, 18) < 0$ and $D(18, 18) > 0$, so

a relative maximum occurs at the critical point

$(18, 18)$. Thus, the product is maximized when

$x = 18$

$y = 18$

$z = 54 - 18 - 18 = 18$.

41. First, find the critical points for $T(x, y)$:

$$T_x = 4x - y = 0$$
$$T_y = -x + 2y - 2 = 0$$

The only critical point is $(\frac{2}{7}, \frac{8}{7})$. There are

three boundary lines:

I. $x = -1$

$$T_1 = 2 + y + y^2 - 2y + 1$$
$$= y^2 - y + 3$$
$$T_1' = 2y - 1$$
$$T_1' = 0 \text{ when } y = \tfrac{1}{2}; \text{ critical point: } (-1, \tfrac{1}{2})$$

II. $y = 2$

$$T_2 = 2x^2 - 2x + 4 - 4 + 1$$
$$= 2x^2 - 2x + 1$$
$$T_2' = 4x - 2$$
$$T_2' = 0 \text{ when } x = \tfrac{1}{2}; \text{ critical point: } (\tfrac{1}{2}, 2)$$

III. Line through $(-1, -2)$ and $(3, 2)$

$$\frac{y - 2}{x - 3} = \frac{4}{4} \text{ or } y = x - 1$$
$$T_3 = 2x^2 - x(x-1) + (x-1)^2 - 2(x-1) + 1$$
$$= 2x^2 - 3x + 4$$

$$T_3' = 4x - 3$$

$$T_3' = 0 \text{ when } x = \tfrac{3}{4}; \text{ critical point: } (\tfrac{3}{4}, -\tfrac{1}{4})$$

Evaluation:

	(x, y)	$T(x, y)$
interior critical point:	$(2/7, 8/7)$	$-1/7$
boundary critical points:	$(-1, 1/2)$	$11/4$
	$(1/2, 2)$	$1/2$
	$(3/4, -1/4)$	$23/8$
boundary end points:	$(-1, 2)$	5
	$(-1, -2)$	9
	$(3, 2)$	13

We see the minimum is $-1/7$ and the maximum is 13. The temperature is the greatest (13 °C) at (3, 2) and is least $(-1/7$ °C) at $(2/7, 8/7)$.

43. $R(x, y) = -x^2 - 2y^2 + 2xy + 8x + 5y$

$$R_x = -2x + 2y + 8 = 0$$

$$R_y = -4y + 2x + 5 = 0$$

Solving simultaneously, $x = \tfrac{21}{2}$, $y = \tfrac{13}{2}$.

$$R_{xx} = -2$$

$$R_{xy} = 2$$

$$R_{yy} = -4$$

$D > 0$, so the revenue is maximized at $(\tfrac{21}{2}, \tfrac{13}{2})$.

45. The profit for each bottle of California water is $x - 2$, and for each bottle of New York water is $y - 2$. The profit is

$$P(x, y) = (x - 2)(40 - 50x + 40y)$$
$$+ (y - 2)(20 + 60x - 70y)$$
$$= -50x^2 + 100xy - 70y^2 + 20x + 80y - 120$$

$$P_x = -100x + 100y + 20$$

$$P_y = 100x - 140y + 80$$

Solving simultaneously, $x = 2.7$, $y = 2.5$

$$P_{xx} = -100$$

$$P_{xy} = 100$$

$$P_{yy} = -140$$

Then $D > 0$, $P_{xx} < 0$, so the profit is maximized at (2.7, 2.5). The owner should charge $2.70 for California water and $2.50 for New York water.

47. Domestic profit $= x(60 - 0.2x + 0.05y) - 10x$
$$= x(50 - 0.2x + 0.05y)$$

Foreign profit $= y(50 - 0.1y + 0.05x) - 10y$
$$= y(40 - 0.1y + 0.05x)$$

$$P(x, y) = x(50 - 0.2x + 0.05y)$$
$$+ y(40 - 0.1y + 0.05x)$$

$$= -0.2x^2 + 0.1xy - 0.1y^2 + 50x + 40y$$

$$P_x = -0.4x + 0.1y + 50 = 0$$

$$P_y = 0.1x - 0.2y + 40 = 0$$

Solving simultaneously, $x = 200$, $y = 300$

$$P_{xx} = -0.4$$

$$P_{xy} = 0.1$$

$$P_{yy} = -0.2$$

$D > 0$ and $P_{xx} < 0$, so $(200, 300)$ is a

maximum. That is, 200 machines should be

supplied to the domestic market and 300 to the

foreign market.

49. $k \approx m \approx 1.35$ (use technology)

SURVIVAL HINT: *When doing a least squares regression, it is essential that you carefully organize and label your data.*

51. We use a graphing calculator to help us answer the questions of this problem.

a.

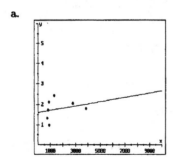

b. $y = 0.0001064x + 1.5965$, where y is the wine consumption and x is the DJIA.

c. When $x = 6,442$, the predicted wine consumption is 2.28 gal/person. This is higher than the actual consumption in 1997. Just because there seems to be correlation of certain data points, do not assume there is any inherent dependence.

d. Using $x = 10,790$, the predicted wine-consumption is 2.74 gal/person; check on the web to see if this is reasonable. For example, you might check with a source like the *Wine Institute* at

www.wineinstitute.org

53. a. Using a graphing calculator, we find the linear fit is

$$W = 3.42674 + 0.573164X$$

so m turns out to be a disappointing 0.573 vs the theoretical value of 2/3.

b. The poor agreement of m with the expected value is largely due to the exceptional performance of the 60 kg lifter.

55. $f(x, y) = 4x^2 e^y - 2x^4 - e^{4y}$

$$f_x = 8xe^y - 8x^3$$

$$f_y = 4x^2 e^y - 4e^{4y}$$

Then, $f_x = 0$ and $f_y = 0$ when

$$8xe^y - 8x^3 = 0 \quad 4x^2 e^y - 4e^{4y} = 0$$

$$8x(e^y - x^2) = 0 \quad 4e^y(x^2 - e^{3y}) = 0$$

$$e^y = x^2 \qquad\qquad e^{3y} = x^2$$

Note: in the first equation, if $x = 0$, then

$f_y \neq 0$, and in the second equation $e^y = 0$ is

impossible. Thus, $e^y = e^{3y}$, so $y = 0$, $x = \pm 1$.

The critical points are $(1, 0)$ and $(-1, 0)$. The

discriminant is

$$D = f_{xx}f_{yy} - f_{xy}^2$$

$$= (8e^y - 24x^2)(4x^2 e^y - 16e^{4y}) - (8xe^y)^2$$

and $D > 0$ with $f_{xx} < 0$ at both $(1, 0)$ and

$(-1, 0)$, so they both correspond to relative

maxima.

57. Tom travels $\sqrt{x^2 + 1.2^2}$ miles, Dick travels

$\sqrt{y^2 + 2.5^2}$ miles, and Mary travels

$4.3 - x - y$ miles. Thus, the total time of

travel is

$$T = \frac{\sqrt{x^2 + 1.2^2}}{2} + \frac{\sqrt{y^2 + 2.5^2}}{4} + \frac{4.3 - x - y}{6}$$

Then,

$$T_x = \frac{1}{2}\frac{x}{\sqrt{x^2 + 1.2^2}} - \frac{1}{6} = 0$$

$$T_y = \frac{1}{4}\frac{y}{\sqrt{y^2 + 2.5^2}} - \frac{1}{6} = 0$$

which has the unique solution $x \approx 0.4243$ and

$y \approx 2.2361$.

Boundary cases

If $x = 0$, then the time of travel is

$$T_1 = \frac{1.2}{2} + \frac{\sqrt{y^2 + 2.5^2}}{4} + \frac{4.3 - y}{6}$$

$$T_1' = \frac{1}{4}\frac{y}{\sqrt{y^2 + 2.5^2}} - \frac{1}{6} = 0$$

when $y \approx 2.2361$

We find $T_1 \approx 1.7825$

If $y = 0$, then the time of travel is

$$T_2 = \frac{\sqrt{x^2 + 1.2^2}}{2} + \frac{2.5}{4} + \frac{4.3 - x}{6}$$

$$T_2' = \frac{1}{2}\frac{x}{\sqrt{x^2 + 1.2^2}} - \frac{1}{6} = 0$$

when $x \approx 0.4243$

$T_2 \approx 1.9074$

If Tom trudges directly to B and Dick rows

from B to A, then the time of travel is

$$T_3 = \frac{1.2}{2} + \frac{2.5}{4} + \frac{4.3}{6} \approx 1.9417$$

Finally, if Tom trudges in a direct line toward

F, and Dick rows in a direct line toward F,

then

$$T_4: \frac{x}{1.2} = \frac{4.3}{3.7} \text{ or } x \approx 1.3946 \text{ and}$$

$$\frac{y}{2.5} = \frac{4.3}{3.7} \text{ or } y \approx 2.9054$$

We now evaluate the times:

Time	x	y	Time (in hours)
T	0.4243	2.2361	1.7482
T_1	0	2.2361	1.7825
T_2	0.4243	0	1.9074
T_3	0	0	1.9417
T_4	1.3946	2.9054	1.8781

The minimum time of travel occurs when Dick

waits 0.4243 miles from the line AS and Mary

waits $4.3 - 0.4243 - 2.2361 = 1.6396$ miles

from the finish line.

59. a. $z = f(x, y)$ has a minimum of 0 at the

origin because as (x, y) moves toward the

origin, the values of f drop toward 0.

b. $z = f(x, y)$ has a saddle point at the

origin. In the first and third quadrants, f

approaches 0 from above, while in the

second and fourth quadrants f approaches 0

from below as $(x, y) \to (0, 0)$. In the plane

$y = x$, the cross section is parabola-like

with a low point at $(0, 0)$. In the plane

$y = -x$, the cross section is parabola-like

with a high point at $(0, 0)$.

61. $F(m, b) = \sum\limits_{k=1}^{n} [y_k - (mx_k + b)]^2$

$F_m(m, b) = 2 \sum\limits_{k=1}^{n} [y_k - (mx_k + b)](-x_k)$

$F_b(m, b) = 2 \sum\limits_{k=1}^{n} [y_k - (mx_k + b)](-1)$

Then $F_m = F_b = 0$ when

$$\left(\sum_{k=1}^{n} x_k^2 \right) m + \left(\sum_{k=1}^{n} x_k \right) b = \sum_{k=1}^{n} x_k y_k$$

$$\left(\sum_{k=1}^{n} x_k \right) m + \left(\sum_{k=1}^{n} 1 \right) b = \sum_{k=1}^{n} y_k$$

or equivalently

$Am + Bb = D$

$Bm + nb = C$

where

$$A = \sum_{k=1}^{n} x_k^2, \ B = \sum_{k=1}^{n} x_k, \ C = \sum_{k=1}^{n} y_k,$$

and $D = \sum\limits_{k=1}^{n} x_k y_k$. Solving this system, we have

$$m = \frac{Dn - BC}{An - B^2} \qquad b = \frac{AC - BD}{An - B^2}$$

Thus,

$$m = \frac{n \sum_{k=1}^{n} x_k y_k - \left(\sum_{k=1}^{n} x_k \right) \left(\sum_{k=1}^{n} y_k \right)}{n \sum_{k=1}^{n} x_k^2 - \left(\sum_{k=1}^{n} x_k \right)^2} \text{ and}$$

$$b = \frac{\left(\sum_{k=1}^{n} x_k^2 \right) \left(\sum_{k=1}^{n} y_k \right) - \left(\sum_{k=1}^{n} x_k \right) \left(\sum_{k=1}^{n} x_k y_k \right)}{n \sum_{k=1}^{n} x_k^2 - \left(\sum_{k=1}^{n} x_k \right)^2}$$

11.8 Lagrange Multipliers, Pages 767-770

SURVIVAL HINT: *The method of Lagrange multipliers requires the solution of a system of equations. There is no single set of steps to follow in solving a system of equations. Substitution, addition, matrices, or some cleverness are usually required. Often one equation can be solved for λ, and that value substituted into the other equations involving λ. Solve the resulting equation with the constraint equation to find x and y. These values can then be used to evaluate λ, if necessary.*

1. $f(x, y) = xy; \quad g(x, y) = 2x + 2y$

$f_x = y, \, f_y = x, \, g_x = 2, \, g_y = 2$

Solve the system

$$\begin{cases} y = 2\lambda \\ x = 2\lambda \\ 2x + 2y = 5 \end{cases}$$

to find $x = y = \frac{5}{4}$

$f(\frac{5}{4}, \frac{5}{4}) = \frac{25}{16}$ is the constrained maximum.

3. $f(x, y) = 16 - x^2 - y^2; \, g(x, y) = x + 2y$

$f_x = -2x, \, f_y = -2y, \, g_x = 1, \, g_y = 2$

Solve the system

$$\begin{cases} -2x = \lambda \\ -2y = 2\lambda \\ x + 2y = 6 \end{cases}$$

to find $x = \frac{6}{5}, \, y = \frac{12}{5}$

$f(\frac{6}{5}, \frac{12}{5}) = \frac{44}{5}$ is the constrained maximum.

5. $f(x, y) = x^2 + y^2; \, g(x, y) = xy$

$f_x = 2x, \, f_y = 2y, \, g_x = y, \, g_y = x$

Solve the system

$$\begin{cases} 2x = \lambda y \\ 2y = \lambda x \\ xy = 1 \end{cases}$$

to find $x = y = 1$ and $x = y = -1$

$f(\pm 1, \pm 1) = 2$ is the constrained minimum.

7. $f(x, y) = x^2 - y^2; \, g(x, y) = x^2 + y^2$

$f_x = 2x, \, f_y = -2y, \, g_x = 2x, \, g_y = 2y$

Solve the system

$$\begin{cases} 2x = 2\lambda x \\ -2y = 2\lambda y \\ x^2 + y^2 = 4 \end{cases}$$

to find $x = 0$, $y = \pm 2$ or $x = \pm 2$, $y = 0$

$f(0, \pm 2) = -4$ is the constrained minimum.

$f(\pm 2, 0) = 4$ is a constrained maximum.

9. $f(x, y) = \cos x + \cos y$; $g(x, y) = y - x$

$f_x = -\sin x$, $f_y = -\sin y$, $g_x = -1$, $g_y = 1$

Solve the system

$$\begin{cases} -\sin x = -\lambda \\ -\sin y = \lambda \\ y = x + \frac{\pi}{4} \end{cases}$$

to find $x = \dfrac{(8n - 1)\pi}{8}$, $y = \dfrac{(8n + 1)\pi}{8}$

If $n = 1$, $f(\frac{7\pi}{8}, \frac{9\pi}{8}) \approx -1.8478$;

If $n = 2$, $f(\frac{15\pi}{8}, \frac{17\pi}{8}) \approx 1.8478$;

The constrained maximum is approximately

1.8478 for $f(-\frac{\pi}{8} + n\pi, \frac{\pi}{8} + n\pi)$ with n even.

11. $f(x, y) = \ln(xy^2)$; $g(x, y) = 2x^2 + 3y^2$

$f_x = \frac{1}{x}$; $f_y = \frac{2}{y}$; $g_x = 4x$; $g_y = 6y$

Solve the system

$$\begin{cases} \frac{1}{x} = (4x)\lambda \\ \frac{2}{y} = (6y)\lambda \\ 2x^2 + 3y^2 = 8 \end{cases}$$

to find $x = \sqrt{\frac{4}{3}}$, $(x = -\sqrt{\frac{4}{3}}$ is not in the

domain), $y = \pm \frac{4}{3}$;

$f(\sqrt{\frac{4}{3}}, \pm \frac{4}{3}) = \ln\left[\sqrt{\frac{4}{3}} \left(\frac{4}{3}\right)^2\right] = \frac{5}{2}\ln\frac{4}{3}$

The constrained maximum is approximately

0.72.

13. $f(x, y, z) = x^2 + y^2 + z^2$;

$g(x, y, z) = x - 2y + 3z$

$f_x = 2x$; $f_y = 2y$; $f_z = 2z$,

$g_x = 1$; $g_y = -2$; $g_z = 3$.

Solve the system

$$\begin{cases} 2x = \lambda \\ 2y = -2\lambda \\ 2z = 3\lambda \\ x - 2y + 3z = 4 \end{cases}$$

to find $\lambda = \frac{4}{7}$ and then $x = \frac{2}{7}$, $y = -\frac{4}{7}$, $z = \frac{6}{7}$

$f(\frac{2}{7}, -\frac{4}{7}, \frac{6}{7}) = \frac{8}{7}$ is the constrained minimum.

15. $f(x, y, z) = 2x^2 + 4y^2 + z^2$

$g(x, y, z) = 4x - 8y + 2z$

$f_x = 4x$, $f_y = 8y$, $f_z = 2z$, $g_x = 4$,

$g_y = -8$, $g_z = 2$

Solve the system

$$\begin{cases} 4x = 4\lambda \\ 8y = -8\lambda \\ 2z = 2\lambda \\ 4x - 8y + 2z = 10 \end{cases}$$

to find $\lambda = x = -y = z$ and then

$x = \frac{5}{7},\ y = -\frac{5}{7},\ z = \frac{5}{7}$

$f(\frac{5}{7}, -\frac{5}{7}, \frac{5}{7}) = \frac{25}{7}$ is the constrained minimum.

By using negative values for any two variables in the constraint equation, the third variable can be made arbitrarily large, thus f can be made arbitrarily large and does not have a maximum.

17. $f(x, y, z) = x - y + z;$

$g(x, y, z) = x^2 + y^2 + z^2$

$f_x = 1,\ f_y = -1,\ f_z = 1,$

$g_x = 2x,\ g_y = 2y,\ g_z = 2z$

Solve the system

$$\begin{cases} 1 = 2\lambda x \\ -1 = 2\lambda y \\ 1 = 2\lambda z \\ x^2 + y^2 + z^2 = 100 \end{cases}$$

to find $x = z = \pm\dfrac{10}{\sqrt{3}},\ y = \mp\dfrac{10}{\sqrt{3}}$

$f(\dfrac{10}{\sqrt{3}}, -\dfrac{10}{\sqrt{3}}, \dfrac{10}{\sqrt{3}}) = \dfrac{30}{\sqrt{3}}$

$$= 10\sqrt{3}$$

$$\approx 17.3$$

is the constrained maximum and

$f\left(-\dfrac{10}{\sqrt{3}}, \dfrac{10}{\sqrt{3}}, -\dfrac{10}{\sqrt{3}}\right) = -\dfrac{30}{\sqrt{3}}$

$$= -10\sqrt{3}$$

$$\approx -17.3$$

is the constrained minimum.

19. Minimize the square of the distance to obtain

$f(x, y, z) = x^2 + y^2 + z^2$, subject to

$g(x, y, z) = Ax + By + Cz = D$

$f_x = 2x,\ f_y = 2y,\ f_z = 2z,$

$g_x = A,\ g_y = B,\ g_z = C$

Solve the system

$$\begin{cases} 2x = A\lambda \\ 2y = B\lambda \\ 2z = C\lambda \\ Ax + By + Cz = D \end{cases}$$

to find $\lambda = \dfrac{2D}{A^2 + B^2 + C^2}$ and then

$$x = \frac{AD}{A^2 + B^2 + C^2}$$

$$y = \frac{BD}{A^2 + B^2 + C^2}$$

$$z = \frac{CD}{A^2 + B^2 + C^2}$$

Let $H = A^2 + B^2 + C^2$; the minimum distance is

$$S = \left[f\left(\frac{AD}{H}, \frac{BD}{H}, \frac{CD}{H}\right)\right]^{1/2}$$

$$= \frac{|D|}{\sqrt{A^2 + B^2 + C^2}}$$

21. Minimize the square of the distance to obtain

$$f(x, y, z) = x^2 + y^2 + z^2; \text{ subject to}$$

$$g(x, y, z) = 2x + y + z = 1$$

$$f_x = 2x, \ f_y = 2y, \ f_z = 2z,$$

$$g_x = 2, \ g_y = 1, \ g_z = 1$$

Solve the system

$$\begin{cases} 2x = 2\lambda \\ 2y = \lambda \\ 2z = \lambda \\ 2x + y + z = 1 \end{cases}$$

to find $x = \frac{1}{3}$, $y = \frac{1}{6}$, $z = \frac{1}{6}$

The nearest point is $(\frac{1}{3}, \frac{1}{6}, \frac{1}{6})$, and the minimum distance is

$$S = \left[f(\tfrac{1}{3}, \tfrac{1}{6}, \tfrac{1}{6}) \right]^{1/2}$$

$$= \frac{1}{\sqrt{6}}$$

$$\approx 0.4082$$

23. $f(x, y, z) = xy^2 z$

$$g(x, y, z) = x + y + z = 12$$

$$f_x = y^2 z, \ f_y = 2xyz, \ f_z = xy^2,$$

$$g_x = 1, \ g_y = 1, \ g_z = 1$$

Solve the system

$$\begin{cases} y^2 z = \lambda \\ xy^2 = \lambda \\ 2xyz = \lambda \\ x + y + z = 12 \end{cases}$$

to find $x = z = 3$, $y = 6$

The largest product is $f(3, 6, 3) = 324$

25. Minimize $T(x, y, z) = 100 - xy - xz - yz$

subject to $g(x, y, z) = 10$, where

$$g(x, y, z) = x + y + z;$$

$$T_x = -y - z$$

$$T_y = -x - z$$

$$T_z = -x - y$$

$$g_x = 1, \ g_y = 1, \ g_z = 1$$

$$\begin{cases} -y - z = \lambda \\ -x - z = \lambda \\ -x - y = \lambda \\ x + y + z = 10 \end{cases}$$

to obtain $x = y = z = -\lambda/2$, and then find

$$x = y = z = \frac{10}{3}$$

The lowest temperature is $T(\frac{10}{3}, \frac{10}{3}, \frac{10}{3}) = \frac{200}{3}$.

27. Let x, y denote the sides of the field. We wish to maximize $A = xy$ subject to

$$F(x, y) = 2x + 2y = 320$$

$$A_x = y, \ A_y = x,$$

$$F_x = 2, \ F_y = 2.$$

Solve the system

$$\begin{cases} y = 2\lambda \\ x = 2\lambda \\ 2x + 2y = 320 \end{cases}$$

to find $x = y = 80$ and the maximum value of

A is $(80)(80) = 6{,}400 \text{ yd}^2$

29. Let x and y be the radius and height of the

cylinder. Minimize the cost

$$f(x, y) = 2(2\pi x^2) + 2\pi xy$$

subject to

$$g(x, y) = 4\pi \text{ where } g(x, y) = \pi x^2 y$$

$$f_x = 8\pi x + 2\pi y$$

$$f_y = 2\pi x,$$

$$g_x = 2\pi xy, \ g_y = \pi x^2$$

Solve the system

$$\begin{cases} 8\pi x + 2\pi y = 2\lambda\pi xy \\ 2\pi x = \lambda\pi x^2 \\ \pi x^2 y = 4\pi \end{cases}$$

to obtain $y = 4x$, and then find the radius

$x = 1$ in. and the height $y = 4$ in.

31. Maximize $f(x, y) = 50x^{1/2}y^{3/2}$, subject to

$g(x, y) = 8$ where $g(x, y) = x + y$

$$f_x = 25x^{-1/2}y^{3/2}$$

$$f_y = 75x^{1/2}y^{1/2}$$

$$g_x = 1, \ g_y = 1$$

Solve the system

$$\begin{cases} 25x^{-1/2}y^{3/2} = \lambda \\ 75x^{1/2}y^{1/2} = \lambda \\ x + y = 8 \end{cases}$$

to find $x = 2$, $y = 6$. \$2,000 to development

and \$6,000 to promotion gives the maximum

sales of $f(2, 6) = 50\sqrt{2}(6\sqrt{6}) = 600\sqrt{3}$; this

is about 1,039 units.

33. Let s be the length of the living

space and y the depth.

The height of the

equilateral triangular

face is

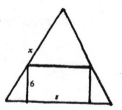

$$h = \frac{\sqrt{3}}{2}x$$

so by similar triangles,

$$\frac{\frac{\sqrt{3}}{2}x - 6}{\frac{\sqrt{3}}{2}x} = \frac{\frac{s}{2}}{\frac{x}{2}}$$

$$s = x - 4\sqrt{3}$$

The volume of the livable space is

$$V = 6(x - 4\sqrt{3})y = 6xy - 24\sqrt{3}\,y$$

and the surface area of the building is

$$S = 2\left[\frac{1}{2}\left(\frac{\sqrt{3}}{2}x^2\right)\right] + 2xy$$

$$= \frac{\sqrt{3}}{2}x^2 + 2xy$$

The problem is to maximize

$V = 6(x - 4\sqrt{3})y$ subject to

$$S = \frac{\sqrt{3}}{2}x^2 + 2xy = 500$$

We have, $V_x = 6y$, $V_y = 6x - 24\sqrt{3}$,

$S_x = \sqrt{3}\,x + 2y$, $S_y = 2x$, so we must solve

the system

$$\begin{cases} 6y = \lambda(\sqrt{3}\,x + 2y) \\ 6x - 24\sqrt{3} = \lambda(2x) \\ \frac{\sqrt{3}}{2}x^2 + 2xy = 500 \end{cases}$$

We find that $\lambda = \dfrac{6y}{\sqrt{3}\,x + 2y} = \dfrac{6x - 24\sqrt{3}}{2x}$

so that $y = \dfrac{x^2}{8} - \dfrac{\sqrt{3}}{2}x$, and then

$$\frac{\sqrt{3}}{2}x^2 + 2x\left(\frac{x^2}{8} - \frac{\sqrt{3}}{2}x\right) = 500$$

$$x \approx 13.87$$

and $y \approx 12.04$

35. Minimize $F = \dfrac{8m}{k^2}E = \dfrac{1}{x^2} + \dfrac{1}{y^2} + \dfrac{1}{z^2}$

subject to $V = xyz = C$. We have

$$F_x = \frac{-2}{x^3};\ F_y = \frac{-2}{y^3};\ F_z = \frac{-2}{z^3}$$

$$V_x = yz;\ V_y = xz;\ V_z = xy$$

We must solve the system of equations (for

nonnegative x, y, and z):

$$\begin{cases} \dfrac{-2}{x^3} = \lambda(yz) \\ \dfrac{-2}{y^3} = \lambda(xz) \\ \dfrac{-2}{z^3} = \lambda(xy) \\ xyz = C \end{cases}$$

Solving, we obtain $x = y = z = \sqrt[3]{C}$.

37. We must solve the system of equations:

$$\begin{cases} 2x = 20\lambda \\ 4y = 12\lambda \\ 20x + 12y = 100 \end{cases}$$

Solving, we obtain $\lambda = 100/236$ and then

$x \approx 4.24$, $y \approx 1.27$. In this case, the farmer

should apply 4.24 acre-ft of water and 1.27 lb

of fertilizer to maximize the yield.

39. Maximize $F = A^2 = s(s - a)(s - b)(s - c)$

Since $s = \frac{1}{2}(a + b + c)$, we substitute this into

the formula to find

$$F = \tfrac{1}{2}(a + b + c)\tfrac{1}{2}(b + c - a)\tfrac{1}{2}(a + c - b)\tfrac{1}{2}(a + b - c)$$

$$= \tfrac{1}{16}[2a^2b^2 + 2a^2c^2 + 2b^2c^2 - a^4 - b^4 - c^4]$$

subject to $P = a + b + c = P_0$. We must

solve the system of equations:

$$\begin{cases} F_a = \frac{1}{16}[-4a^3 + 4ab^2 + 4ac^2] = \lambda(1) \\ F_b = \frac{1}{16}[-4b^3 + 4ba^2 + 4bc^2] = \lambda(1) \\ F_c = \frac{1}{16}[-4c^3 + 4ca^2 + 4cb^2] = \lambda(1) \\ a + b + c = P_0 \end{cases}$$

Solving, we obtain $a = b = c = \frac{1}{3}P_0$, so the

triangle with maximum area is equilateral.

41. $f(x, y, z) = x^2 + y^2 + z^2$;

$g(x, y, z) = x + y = 4$

$h(x, y, z) = y + z = 6$

$f_x = 2x, f_y = 2y, f_z = 2z$

$g_x = 1, g_y = 1, g_z = 0$

$h_x = 0, h_y = 1, h_z = 1$

Solve the system

$$\begin{cases} 2x = \lambda \\ 2y = \lambda + \mu \\ 2z = \mu \\ x + y = 4 \\ y + z = 6 \end{cases}$$

to find $x = 2/3$, $y = 10/3$, $z = 8/3$

The minimum is $f(\frac{2}{3}, \frac{10}{3}, \frac{8}{3}) = \frac{56}{3}$

43. $f(x, y, z) = xy + xz$

$g(x, y, z) = 2x + 3z = 5$

$h(x, y, z) = xy = 4$

$f_x = y + z, f_y = x, f_z = x$

$g_x = 2, g_y = 0, g_z = 3$

$h_x = y, h_y = x, h_z = 0$

Solve the system

$$\begin{cases} y + z = 2\lambda + \mu y \\ x = \mu x \\ x = 3\lambda \\ 2x + 3z = 5 \\ xy = 4 \end{cases}$$

to find $\mu = 1$, $\lambda = \frac{5}{12}$, and then

$x = \frac{5}{4}$, $y = \frac{16}{5}$, $z = \frac{5}{6}$

The maximum is $f(\frac{5}{4}, \frac{16}{5}, \frac{5}{6}) = \frac{121}{24}$

45. PROFIT = REVENUE − COST

$$P(x, y) = \left(\frac{320y}{y + 2} + \frac{160x}{x + 4}\right)150$$

$$- \left(\frac{320y}{y + 2} + \frac{160x}{x + 4}\right)50 - 1{,}000(x + y)$$

a. $g(x, y) = x + y - 8$

$$P_x = \frac{100(160)(4)}{(x + 4)^2} - 1{,}000$$

$$P_y = \frac{100(320)(2)}{(y + 2)^2} - 1{,}000$$

$g_x = 1, g_y = 1$

Since $P_x = P_y = \lambda$,

$$\frac{100(160)(4)}{(x+4)^2} - 1,000$$

$$= \frac{100(320)(2)}{(y+2)^2} - 1,000$$

or $x + 4 = \pm(y + 2)$

Reject the negative solution as leading to

negative spending. Substituting

$y = x + 2$ in the constraint equation

$x + y = 8$ leads to $x = 3$ thousand dollars

for development and $y = 5$ thousand

dollars for promotion.

b. $\lambda = P_y = \dfrac{64,000}{49} - 1,000 \approx 306.122$

(for each \$1,000). Since the change in

this promotion/development is \$100, the

corresponding increase in profit is \$30.61.

Remember that the Lagrange multiplier is

the change in maximum profit for a unit

(one thousand dollars) change in the

constraint. The actual increase in profit

is \$29.68.

c. To maximize the profit when unlimited

funds are available maximize $P(x, y)$

without constraints. To do this, find the

critical points by setting $P_x = 0$ and

$P_y = 0$, that is

P_x: $\quad \dfrac{64}{(x+4)^2} - 1 = 0$

$\qquad\qquad (x+4)^2 = 64$

$\qquad\qquad\qquad x = 4 \quad$ and

P_y: $\quad \dfrac{64}{(y+2)^2} - 1 = 0$

$\qquad\qquad (y+2)^2 = 64$

$\qquad\qquad\qquad y = 6$

Thus, \$4,000 should be spent on

development and \$6,000 should be spent

on promotion to maximize profit.

d. If there were a restriction on the amount

spent on development and promotion,

then constraints would be

$$g(x, y) = x + y = k$$

for some positive constant k. The

corresponding Lagrange equations would

be

$$\frac{64}{(x+4)^2} - 1 = \lambda$$

$$\frac{64}{(y+2)^2} - 1 = \lambda$$

and $x + y = k$. To obtain the answer in

part **c**, eliminate λ. Beginning with the Lagrange equations from part **c**, set $\lambda = 0$ to obtain $64/(x+4)^2 - 1 = 0$ or $x = 4$, and $64/(y+2)^2 - 1 = 0$ or $y = 6$, just as we found in part c.

47.
$$P = 2pq + (2p + 2q)(1 - p - q)$$
$$= -2p^2 - 2pq - 2q^2 + 2p + 2q$$
$$P_p = -4p - 2q + 2$$
$$P_q = -2p - 4q + 2$$

To maximize P, we want $P_p = P_q = 0$, so

$$\begin{cases} -4p - 2q + 2 = 0 \\ -2p - 4q + 2 = 0 \end{cases}$$

Solving this system of equations, we obtain $p = q = \frac{1}{3}$, and thus $r = 1 - p - q = \frac{1}{3}$. To form the discriminant, we find

$$P_{pp} = -4 \qquad P_{qq} = -4 \qquad P_{pq} = -2$$

so $D = (-4)(-4) - (-2)^2 > 0$ and since $P_{pp} = -4 < 0$, it follows that the critical point $(\frac{1}{3}, \frac{1}{3})$ corresponds to a maximum. Finally,

$$P_{max} = P(\tfrac{1}{3}, \tfrac{1}{3}, \tfrac{1}{3})$$
$$= 2(\tfrac{1}{3})(\tfrac{1}{3}) + 2(\tfrac{1}{3})(\tfrac{1}{3}) + 2(\tfrac{1}{3})(\tfrac{1}{3})$$
$$= \tfrac{2}{3}$$

49. Let x, y, z be the length, width, and height of one-eighth of the rectangular box, respectively.

$$f(x, y, z) = xyz$$
$$g(x, y, z) = \frac{x^2}{a^2} + \frac{y^2}{b^2} + \frac{z^2}{c^2} = 1$$
$$f_x = yz, f_y = xz, f_z = xy$$
$$g_x = \frac{2x}{a^2}, g_y = \frac{2y}{b^2}, g_z = \frac{2z}{c^2}$$

Solve the system of equations:

$$yz = \frac{2x}{a^2}\lambda, \quad xz = \frac{2y}{b^2}\lambda, \quad xy = \frac{2z}{c^2}\lambda$$

Solving simultaneously,

$$x = \frac{a}{\sqrt{3}}, \ y = \frac{b}{\sqrt{3}}, \ z = \frac{c}{\sqrt{3}}$$

The maximum volume is

$$8f(\frac{a}{\sqrt{3}}, \frac{b}{\sqrt{3}}, \frac{c}{\sqrt{3}}) = \frac{8abc}{3\sqrt{3}} \text{ cubic units}$$

51. The goal is to minimize cost $C = px + qy$ subject to the fixed production function $Q(x, y) = Q_0$. Since $C_x = p$ and $C_y = q$, the three Lagrange equations are

$$\begin{cases} p = \lambda Q_x \\ q = \lambda Q_y \\ Q(x, y) = Q_0 \end{cases}$$

Solving this system leads to

$$\frac{Q_x}{p} = \frac{Q_y}{q}$$

53. Maximize $Q(x, y) = cx^\alpha y^\beta$ subject to

$C(x, y) = px + qy = k.$ We have

$Q_x = c\alpha x^{\alpha - 1} y^\beta;\ Q_y = c\beta x^\alpha y^{\beta - 1};$

$C_x = p;\ C_y = q.$

We must solve the system of equations

$$\begin{cases} c\alpha x^{\alpha - 1} y^\beta = \lambda(p) \\ c\beta x^\alpha y^{\beta - 1} = \lambda(q) \\ px + qy = k \end{cases}$$

a. Solving simultaneously,

$$x = \frac{k\alpha}{p(\alpha + \beta)} = \frac{k\alpha}{p} \text{ and } y = \frac{k\beta}{q}$$

b. If we drop the condition $\alpha + \beta = 1$, the maximum occurs at

$$x = \frac{k\alpha}{p(\alpha + \beta)},\ y = \frac{k\beta}{q(\alpha + \beta)}$$

If k is increased by 1 unit, the maximum output increases by

$$\frac{dQ}{dk} = \lambda = \frac{c\alpha \left[\dfrac{k\alpha}{p(\alpha + \beta)}\right]^{\alpha - 1} \left[\dfrac{k\beta}{q(\alpha + \beta)}\right]^\beta}{p}$$

$$= c\left(\frac{k}{\alpha + \beta}\right)^{\alpha + \beta - 1} \left(\frac{\alpha}{p}\right)^\alpha \left(\frac{\beta}{q}\right)^\beta$$

55. Answers to this research problem will vary.

CHAPTER 11 REVIEW

Proficiency Examination, Pages 770-771

SURVIVAL HINT: *To help you review the concepts of this chapter, **handwrite** the answers to each of these questions onto your own paper.*

1. A function of two variables is a rule that assigns to each ordered pair (x, y) in a set D a unique number $f(x, y)$.

2. The set D in the answer to Problem 1 is called the domain of the function, and the corresponding values of $f(x, y)$ constitute the range of f.

3. When the plane $z = C$ intersects the surface $z = f(x, y)$, the result is the space curve with the equation $f(x, y) = C$. Such an intersection is called the trace of the graph of f in the plane $z = C$. The set of points (x, y) in the xy-plane that satisfy $f(x, y) = C$ is called the level curve of f at C, and an entire family of level curves (or contour curves) is generated as C varies over the range of f.

4. The notation

$$\lim_{(x, y) \to (x_0, y_0)} f(x, y) = L$$

means that the functional values $f(x, y)$ can be made arbitrarily close to L by choosing a point (x, y) sufficiently close (but not equal) to the point (x_0, y_0). In other words, given some $\epsilon > 0$, we wish to find a $\delta > 0$ so that for any point (x, y) in the punctured disk of radius δ centered at (x_0, y_0), the functional value $f(x, y)$ lies between $L - \epsilon$ and $L + \epsilon$.

5. Suppose $\lim\limits_{(x, y)\to(x_0, y_0)} f(x, y) = L$ and

$$\lim\limits_{(x, y)\to(x_0, y_0)} g(x, y) = M$$

Then, for a constant a,

a. $\lim\limits_{(x, y)\to(x_0, y_0)} [af](x, y) = aL$

b. $\lim\limits_{(x, y)\to(x_0, y_0)} [f + g](x, y)$

$$= \left[\lim\limits_{(x, y)\to(x_0, y_0)} f(x, y) \right]$$

$$+ \left[\lim\limits_{(x, y)\to(x_0, y_0)} g(x, y) \right]$$

$$= L + M$$

c. $\lim\limits_{(x, y)\to(x_0, y_0)} [fg](x, y)$

$$= \left[\lim\limits_{(x, y)\to(x_0, y_0)} f(x, y) \right]$$

$$\times \left[\lim\limits_{(x, y)\to(x_0, y_0)} g(x, y) \right]$$

$$= LM$$

d. $\lim\limits_{(x, y)\to(x_0, y_0)} \left[\dfrac{f}{g}\right](x, y)$

$$= \dfrac{\lim\limits_{(x, y)\to(x_0, y_0)} f(x, y)}{\lim\limits_{(x, y)\to(x_0, y_0)} g(x, y)}$$

$$= \dfrac{L}{M} \quad (M \neq 0)$$

6. The function $f(x, y)$ is continuous at the point (x_0, y_0) if and only if

1. $f(x_0, y_0)$ is defined;

2. $\lim\limits_{(x,y)\to(x_0,y_0)} f(x, y)$ exists;

3. $\lim\limits_{(x, y)\to(x_0, y_0)} f(x, y) = f(x_0, y_0)$.

Also, f is continuous on a set S in its domain if it is continuous at each point in S.

7. If $z = f(x, y)$, then the (first) partial derivatives of f with respect to x and y are the functions f_x and f_y, respectively, defined by

$$f_x(x, y) = \lim\limits_{\Delta x \to 0} \frac{f(x + \Delta x, y) - f(x, y)}{\Delta x}$$

$$f_y(x, y) = \lim\limits_{\Delta y \to 0} \frac{f(x, y + \Delta y) - f(x, y)}{\Delta y}$$

provided the limits exist.

8. The line tangent at $P_0(x_0, y_0, z_0)$ to the trace of $z = f(x, y)$ in the plane $y = y_0$ has slope $f_x(x_0, y_0)$. Likewise, the line tangent at P_0 to

the trace of $z = f(x, y)$ in the plane

$x = x_0$ has slope $f_y(x_0, y_0)$.

9. $z = f(x, y)$; $\dfrac{\partial^2 f}{\partial x^2}, \dfrac{\partial^2 f}{\partial y^2}, \dfrac{\partial^2 f}{\partial x \partial y}, \dfrac{\partial^2 f}{\partial y \partial x}$

or $f_{xx}, f_{yy}, f_{yx}, f_{xy}$

10. Let $z = f(x, y)$; $\Delta z = \dfrac{\partial f}{\partial x}\Delta x + \dfrac{\partial f}{\partial y}\Delta y$

where $\Delta x = dx$, $\Delta y = dy$, and

$\Delta z = f_x \Delta x + f_y \Delta y$.

11. Suppose $f(x, y)$ is defined at each point in a circular disk that is centered at (x_0, y_0) and contains the point $(x_0 + \Delta x, y_0 + \Delta y)$.

Then f is said to be differentiable at (x_0, y_0) if the increment of f can be expressed as

$\Delta f = f_x(x_0, y_0)\Delta x + f_y(x_0, y_0)\Delta y + \epsilon_1 \Delta x + \epsilon_2 \Delta y$

where $\epsilon_1 \to 0$ and $\epsilon_2 \to 0$ as both $\Delta x \to 0$ and $\Delta y \to 0$ (and $\epsilon_1 = \epsilon_2 = 0$ when $\Delta x = \Delta y = 0$).

Also, $f(x, y)$ is said to be differentiable on the region R of the plane if f is differentiable at each point in R.

12. If $f(x, y)$ and its partial derivatives f_x and f_y are defined in an open region R containing the point $P(x_0, y_0)$ and f_x and f_y are continuous

at P, then

$\Delta f = f(x_0 + \Delta x, y_0 + \Delta y) - f(x_0, y_0)$

$\approx f_x(x_0, y_0)\Delta x + f_y(x_0, y_0)\Delta y$

so that

$f(x_0 + \Delta x, y_0 + \Delta y)$

$\approx f(x_0, y_0) + f_x(x_0, y_0)\Delta x + f_y(x_0, y_0)\Delta y$

13. If $z = f(x, y)$ and Δx and Δy are increments of x and y, respectively, and if we let $dx = \Delta x$ and $dy = \Delta y$ be differentials for x and y, respectively, then the total differential of $f(x, y)$ is

$df = \dfrac{\partial f}{\partial x}\, dx + \dfrac{\partial f}{\partial y}\, dy = f_x(x, y)\, dx + f_y(x, y)\, dy$

14. Let $f(x, y)$ be a differentiable function of x and y, and let $x = x(t)$ and $y = y(t)$ be differentiable functions of t. Then $z = f(x, y)$ is a differentiable function of t, and

$\dfrac{dz}{dt} = \dfrac{\partial z}{\partial x}\dfrac{dx}{dt} + \dfrac{\partial z}{\partial y}\dfrac{dy}{dt}$

15. Suppose $z = f(x, y)$ is differentiable at (x, y) and that the partial derivatives of $x = x(u, v)$ and $y = y(u, v)$ exist at (u, v). Then the composite function $z = f[x(u, v), y(u, v)]$ is differentiable at (u, v) with

$$\frac{\partial z}{\partial u} = \frac{\partial z}{\partial x}\frac{\partial x}{\partial u} + \frac{\partial z}{\partial y}\frac{\partial y}{\partial u} \text{ and } \frac{\partial z}{\partial v} = \frac{\partial z}{\partial x}\frac{\partial x}{\partial v} + \frac{\partial z}{\partial y}\frac{\partial y}{\partial v}$$

is

$$D_{\mathbf{u}}f(x,\ y) = \nabla f \cdot \mathbf{u}$$

16. Let f be a function of two variables, and let

$\mathbf{u} = u_1\mathbf{i} + u_2\mathbf{j}$ be a unit vector. The

directional derivative of f at $P_0(x_0,\ y_0)$ in

the direction of $\mathbf{u}$ is given by

$D_{\mathbf{u}}f(x_0,\ y_0)$

$= \lim_{h \to 0} \dfrac{f(x_0 + hu_1,\ y_0 + hu_2) - f(x_0,\ y_0)}{h}$

provided the limit exists.

17. Let f be a differentiable function at $(x,\ y)$ and

let $f(x,\ y)$ have partial derivatives $f_x(x,\ y)$

and $f_y(x,\ y)$. Then the gradient of f,

denoted by ∇f, is the vector given by

$$\nabla f(x,\ y) = f_x(x,\ y)\mathbf{i} + f_y(x,\ y)\mathbf{j}$$

18. Let f and g be differentiable functions. Then

a. $\nabla c = \mathbf{0}$ for any constant c

b. $\nabla(af + bg) = a\nabla f + b\nabla g$

c. $\nabla(fg) = f\nabla g + g\nabla f$

d. $\nabla\left(\dfrac{f}{g}\right) = \dfrac{g\nabla f - f\nabla g}{g^2}$ $g \neq 0$

e. $\nabla(f^n) = nf^{n-1}\nabla f$

19. If f is a differentiable function of x and y, then

the directional derivative at the point

$P_0(x_0,\ y_0)$ in the direction of the unit vector $\mathbf{u}$

20. Suppose f is differentiable and let ∇f_0 denote

the gradient at P_0. Then if $\nabla f_0 \neq \mathbf{0}$:

(1) The largest value of the directional

derivative of $D_{\mathbf{u}}f$ is $\|\nabla f_0\|$ and occurs

when the unit vector $\mathbf{u}$ points in the

direction of ∇f_0.

(2) The smallest value of $D_{\mathbf{u}}f$ is $-\|\nabla f_0\|$ and

occurs when $\mathbf{u}$ points in the direction of

$-\nabla f_0$.

21. Suppose the function f is differentiable at the

point P_0 and that the gradient at P_0 satisfies

$\nabla f_0 \neq \mathbf{0}$. Then ∇f_0 is orthogonal to the

level surface $f(x,\ y,\ z) = K$ at P_0.

22. Suppose the surface S has a nonzero normal

vector $\mathbf{N}$ at the point P_0. Then the line

through P_0 parallel to $\mathbf{N}$ is called the normal

line to S at P_0, and the plane through P_0

with normal vector $\mathbf{N}$ is the tangent plane to

S at P_0.

23. The function $f(x,\ y)$ is said to have an

absolute maximum at (x_0, y_0) if

$f(x_0, y_0) \geq f(x, y)$ for all (x, y) in the

domain D of f. Similarly, f has an absolute

minimum at (x_0, y_0) if $f(x_0, y_0) \leq f(x, y)$

for all (x, y) in D. Collectively, absolute

maxima and minima are called absolute

extrema.

24. Let f be a function defined at (x_0, y_0). Then

$f(x_0, y_0)$ is a relative maximum if

$f(x, y) \leq f(x_0, y_0)$ for all (x, y) in an open

disk containing (x_0, y_0). $f(x_0, y_0)$ is a relative

minimum if $f(x, y) \geq f(x_0, y_0)$ for all (x, y)

in an open disk containing (x_0, y_0).

Collectively, relative maxima and minima are

called relative extrema.

25. A critical point of a function f defined on an

open set S is a point (x_0, y_0) in S where either

one of the following is true:

(1) $f_x(x_0, y_0) = f_y(x_0, y_0) = 0$.

(2) $f_x(x_0, y_0)$ or $f_y(x_0, y_0)$ does not exist (one

or both).

26. Let $f(x, y)$ have a critical point at $P_0(x_0, y_0)$

and assume that f has continuous partial

derivatives in a disk centered at (x_0, y_0).

Let

$D = f_{xx}(x_0, y_0)f_{yy}(x_0, y_0) - [f_{xy}(x_0, y_0)]^2$

Then, a relative maximum occurs at P_0 if

$D > 0$ and $f_{xx}(x_0, y_0) < 0$

A relative minimum occurs at P_0 if

$D > 0$ and $f_{xx}(x_0, y_0) > 0$

A saddle point occurs at P_0 if $D < 0$.

If $D = 0$, then the test is inconclusive.

27. A function of two variables $f(x, y)$ assumes

an absolute extremum on any closed, bounded

set S in the plane where it is continuous.

Moreover, all absolute extrema must occur

either on the boundary of S or at critical

points in the interior of S.

28. Given a set of data points (x_k, y_k), a line

$y = mx + b$, called a regression line, is

obtained by minimizing the sum of squares of

distances $y_k - (mx_k + b)$.

29. Assume that f and g have continuous first

partial derivatives and that f has an

extremum at $P_0(x_0, y_0)$ on the smooth

constraint curve $g(x, y) = c$. If

$\nabla g(x_0, y_0) \neq 0$, there is a number λ such that

$\nabla f(x_0, y_0) = \lambda \nabla g(x_0, y_0)$.

30. Suppose f and g satisfy the hypotheses of

Lagrange's theorem, and suppose that $f(x, y)$

has an extremum (minimum and/or a

maximum) subject to the constraint

$g(x, y) = c$. Then to find the extreme values,

proceed as follows:

1. Simultaneously solve the following three

equations:

$f_x(x, y) = \lambda g_x(x, y)$

$f_y(x, y) = \lambda g_y(x, y)$

$g(x, y) = c$

2. Evaluate f at all points found in Step 1.

The largest of these values is the

maximum value of f and the smallest of

these values is the minimum value of f.

31. $f(x, y) = \sin^{-1} xy$

Recall $\dfrac{d}{dx} \sin^{-1} u = \dfrac{1}{\sqrt{1 - u^2}} \dfrac{du}{dx}$;

$f_x = \dfrac{y}{\sqrt{1 - x^2 y^2}}$

$f_y = \dfrac{x}{\sqrt{1 - x^2 y^2}}$

$f_{xy} = \dfrac{\sqrt{1 - x^2 y^2}\,(1) - \dfrac{y(-2x^2 y)}{2\sqrt{1 - x^2 y^2}}}{1 - x^2 y^2}$

$= \dfrac{1 - x^2 y^2 + x^2 y^2}{(1 - x^2 y^2)^{3/2}}$

$= \dfrac{1}{(1 - x^2 y^2)^{3/2}}$

$f_{yx} = \dfrac{\sqrt{1 - x^2 y^2}\,(1) - \dfrac{x(-2xy^2)}{2\sqrt{1 - x^2 y^2}}}{1 - x^2 y^2}$

$= \dfrac{1 - x^2 y^2 + x^2 y^2}{(1 - x^2 y^2)^{3/2}}$

$= \dfrac{1}{(1 - x^2 y^2)^{3/2}}$

32. $\dfrac{dw}{dt} = \dfrac{\partial w}{\partial x}\dfrac{dx}{dt} + \dfrac{\partial w}{\partial y}\dfrac{dy}{dt} + \dfrac{\partial w}{\partial z}\dfrac{dz}{dt}$

$= 2xy(t \cos t + \sin t)$

$\qquad + (x^2 + 2yz)(-t \sin t + \cos t) + y^2(2)$

If $t = \pi$, then $x = 0$, $y = -\pi$, $z = 2\pi$ and

$\dfrac{dw}{dt} = 0 - 2(-\pi)(2\pi) + 2\pi^2 = 6\pi^2$

33. $f(x, y, z) = xy + yz + xz$ at $(1, 2, -1)$

a. $\nabla f = (y + z)\mathbf{i} + (x + z)\mathbf{j} + (y + x)\mathbf{k}$

$\nabla f = \mathbf{i} + 3\mathbf{k}$ at $P_0(1, 2 - 1)$

b. $\mathbf{u} = \dfrac{P_0Q}{\|P_0Q\|}$

$= \dfrac{-2\mathbf{i} - \mathbf{j}}{\sqrt{5}}$

$D_{\mathbf{u}}(f) = \nabla f \cdot \mathbf{u}$

$= \dfrac{-2}{\sqrt{5}}$

$= \dfrac{-2\sqrt{5}}{5}$

c. The directional derivative has its greatest

value in the direction of the gradient,

$\mathbf{u} = \dfrac{\mathbf{i} + 3\mathbf{k}}{\sqrt{10}}$. The magnitude is

$\|\nabla f\| = \sqrt{10}$.

34. $\displaystyle\lim_{(x, y)\to(0, 0)} f(x, y)$ along the line $y = x$ is

$\displaystyle\lim_{x\to 0} \dfrac{x^3}{x^3 + x^3} = \dfrac{1}{2}$.

The limit does not equal $f(0, 0)$ so the

function is not continuous at $(0, 0)$.

35. $f(x, y) = \ln \dfrac{y}{x} = \ln y - \ln x$

$f_x = -\dfrac{1}{x}, \; f_y = \dfrac{1}{y}$,

$f_{yy} = -\dfrac{1}{y^2}$,

$f_{xy} = 0$

36. $f(x, y, z) = x^2 y + y^2 z + z^2 x$

$f_x = 2xy + z^2$

$f_y = x^2 + 2yz$

$f_z = y^2 + 2zx$

$f_x + f_y + f_z = 2xy + z^2 + x^2 + 2yz + y^2 + 2zx$

$= (x + y + z)^2$

37. $f(x, y) = x^4 + 2x^2 y^2 + y^4$

$\nabla f = (4x^3 + 4xy^2)\mathbf{i} + (4x^2 y + 4y^3)\mathbf{j}$

$\nabla f_0 = \nabla f(2, - 2) = 64(\mathbf{i} - \mathbf{j})$

A unit vector in the direction of $\dfrac{2\pi}{3}$ is

$\mathbf{u} = -\dfrac{1}{2}\mathbf{i} + \dfrac{\sqrt{3}}{2}\mathbf{j}$

$D_{\mathbf{u}}(f) = \nabla f_0 \cdot \mathbf{u}$

$= -32 - 32\sqrt{3}$

≈ -87.4

38. $f(x, y) = 12xy - 2x^2 - y^4$

$f_x = 12y - 4x$

$f_y = 12x - 4y^3$

$f_x = f_y = 0$ when $x = 3y$ and $36y = 4y^3$;

critical points are $(0, 0)$, $(9, 3)$ and

$(-9, -3)$.

$f_{xx} = -4$

$f_{xy} = 12$

$f_{yy} = -12y^2$

x	y	f_{xx}	f_{xy}	f_{yy}	D	Classify
0	0	-4	12	0	$-$	saddle point
9	3	-4	12	-108	$+$	rel max
-9	-3	-4	12	-108	$+$	rel max

39. $f(x, y) = x^2 + 2y^2 + 2x + 3$ subject to

$g(x, y) = x^2 + y^2 = 4$

$$\begin{cases} 2x + 2 = 2x\lambda \\ 4y = 2y\lambda \\ x^2 + y^2 = 4 \end{cases}$$

From the second equation: $\lambda = 2$ or $y = 0$.

If $\lambda = 2$, $2x + 2 = 4x$, $x = 1$, and from the

constraint equation $y = \pm\sqrt{3}$. If $y = 0$ the

constraint equation gives $x = \pm 2$. The set of

candidates: $(1, \sqrt{3})$, $(1, -\sqrt{3})$, $(2, 0)$, $(-2, 0)$.

$f(1, \sqrt{3}) = 12$; $f(1, -\sqrt{3}) = 12$; $f(2, 0) = 11$;

$f(-2, 0) = 3$. $f(x, y)$ has a maximum of 12

at $(1, \pm\sqrt{3})$ and a minimum of 3 at $(-2, 0)$.

40. $f(x, y) = x^2 - 4y^2 + 3x + 6y$

$f_x = 2x + 3$; $f_y = -8y + 6$

$f_x = f_y = 0$ if $x = -\frac{3}{2}$, $y = \frac{3}{4}$

For the boundary:

On $y = 0$, $-2 \le x \le 2$, $f_1 = x^2 + 3x$

$\qquad f_1' = 2x + 3 = 0$ when $x = -\frac{3}{2}$;

point $\left(-\frac{3}{2}, 0\right)$

On $x = 2$, $0 \le y \le 1$, $f_2 = -4y^2 + 6y + 10$

$\qquad f_2' = -8y + 6 = 0$ when $y = \frac{3}{4}$;

point $\left(2, \frac{3}{4}\right)$

On $y = 1$, $-2 \le x \le 2$, $f_3 = x^2 + 3x + 2$

$\qquad f_3' = 2x + 3 = 0$ when $x = -\frac{3}{2}$;

point $\left(-\frac{3}{2}, 1\right)$

On $x = -2$, $0 \le y \le 1$, $f_4 = -4y^2 + 6y - 2$;

$\qquad f_4' = -8y + 6 = 0$ when $y = \frac{3}{4}$;

point $\left(-2, \frac{3}{4}\right)$

The candidate points are $\left(-\frac{3}{2}, \frac{3}{4}\right)$, $\left(-\frac{3}{2}, 0\right)$,

$\left(2, \frac{3}{4}\right)$, $\left(-\frac{3}{2}, 1\right)$, $\left(-2, \frac{3}{4}\right)$, and the corner

points $(-2, 0)$, $(-2, 1)$, $(2, 0)$, $(2, 1)$.

x	y	$f(x, y)$	
$-3/2$	$3/4$	0	
$-3/2$	0	$-9/4$	min
2	$3/4$	$49/4$	max
$-3/2$	1	$-1/4$	
-2	$3/4$	$1/4$	
-2	0	-2	
-2	1	0	
2	0	10	
2	1	12	

The largest value of f is $49/4$ at $\left(2, \frac{3}{4}\right)$ and the

smallest is $-9/4$ at $(-3/2, 0)$.

CHAPTER 12

Multiple Integration

12.1 Double Integration Over Rectangular Regions, Pages 785-786

SURVIVAL HINT: *Be careful to use the x boundaries with dx and y boundaries with dy, for whatever iteration you choose.*

$$\int_{x_1}^{x_2} \int_{y_1}^{y_2} f(x,\ y)\ dy\ dx \quad \text{or} \quad \int_{y_1}^{y_2} \int_{x_1}^{x_2} f(x,\ y)\ dx\ dy$$

1.
$$\int_0^2 \int_0^1 (x^2 + xy + y^2)\ dy\ dx$$

$$= \int_0^2 \left[x^2 y + \frac{xy^2}{2} + \frac{y^3}{3} \right]\Bigg|_0^1\ dx$$

$$= \int_0^2 \left(x^2 + \frac{x}{2} + \frac{1}{3} \right)\ dx$$

$$= \frac{13}{3}$$

3.
$$\int_1^{e^2} \int_1^2 \left[\frac{1}{x} + \frac{1}{y} \right]\ dy\ dx = \int_1^{e^2} \left[\frac{y}{x} + \ln y \right]\Bigg|_1^2\ dx$$

$$= \int_1^{e^2} (\ln 2 + x^{-1})\ dx$$

$$= (e^2 - 1)(\ln 2) + 2$$

5.
$$\int_3^4 \int_1^2 \frac{x}{x - y}\ dy\ dx$$

$$= \int_3^4 - x \ln |x - y| \big|_1^2\ dx$$

$$= \int_3^4 [x \ln(x - 1) - x \ln(x - 2)]\ dx$$

$$= \frac{15}{2} \ln 3 - 10 \ln 2 + \frac{1}{2}$$

SURVIVAL HINT: *Problems 7-12 are concerned with the idea of evaluating a double integral by relating it to a volume.*

7.
$$\int\int_R 4\ dA = \overset{\text{height}}{\overset{\downarrow}{4}} \underbrace{(2)(4)}_{\text{area of base}}$$

$$= 32$$

9.
$$\int\int_R (4 - y)\ dA = \underbrace{\frac{1}{2}(4)(4)}_{\text{area of cross section}} \overset{\text{height}}{\overbrace{(3)}}$$

$$= 24$$

11.
$$\iint\limits_R \frac{y}{2}\, dA = \underbrace{\tfrac{1}{2}(4)(2)}_{\text{area of cross section}} \overbrace{(6)}^{\text{height}}$$

$$= 24$$

13.
$$\iint\limits_R x^2 y\, dA = \int_1^2 \int_0^1 x^2 y\, dy\, dx$$

$$= \int_1^2 \frac{x^2 y^2}{2}\Big|_0^1 \, dx$$

$$= \int_1^2 \frac{x^2}{2}\, dx$$

$$= \frac{x^3}{6}\Big|_1^2$$

$$= \frac{7}{6}$$

15.
$$\iint\limits_R 2x e^y\, dA = \int_{-1}^0 \int_0^{\ln 2} 2x e^y\, dy\, dx$$

$$= \int_{-1}^0 2x(e^{\ln 2} - 1)\, dx$$

$$= -1$$

17.
$$\iint\limits_R \frac{2xy\, dA}{x^2 + 1} = \int_0^1 \int_1^3 \frac{2xy}{x^2 + 1}\, dy\, dx$$

$$= 4\int_0^1 \frac{2x}{x^2 + 1}\, dx$$

$$= 4 \ln 2$$

19.
$$\iint\limits_R \sin(x + y)\, dA$$

$$= \int_0^{\pi/4} \int_0^{\pi/2} \sin(x + y)\, dy\, dx$$

$$= -\int_0^{\pi/4} [\cos(x + \tfrac{\pi}{2}) - \cos x]\, dx$$

$$= 1$$

21.
$$\iint\limits_R (2x + 3y)\, dA = \int_0^1 \int_0^2 (2x + 3y)\, dy\, dx$$

$$= \int_0^1 (4x + 6)\, dx$$

$$= 8$$

23.
$$\int_1^2 \int_1^2 \left(\frac{x}{y} + \frac{y}{x}\right) dy\, dx = \int_1^2 \left[x \ln y + \frac{1}{2}\frac{y^2}{x}\right]\Big|_1^2 \, dx$$

$$= \left[\tfrac{3}{2}\ln x + \tfrac{1}{2}(\ln 2)x^2\right]\Big|_1^2$$

$$= 3 \ln 2$$

25.
$$\int_0^1 \int_0^4 x^{1/2} y^{1/2}\, dy\, dx = \frac{2}{3}\int_0^1 x^{1/2}(8)\, dx$$

$$= \frac{32}{9}$$

27. $\displaystyle\int_0^1 \int_0^1 \frac{xy}{\sqrt{x^2 + y^2 + 1}}\, dy\, dx$

$$\boxed{\text{Let } u = x^2 + y^2 + 1;\ du = 2y\, dy}$$

$$= \int_0^1 x\sqrt{x^2 + y^2 + 1}\,\Big|_0^1\ dx$$

$$= \int_0^1 x\left[\sqrt{x^2 + 2} - \sqrt{x^2 + 1}\right] dx$$

$$= \tfrac{1}{3}\left[(x^2 + 2)^{3/2} - (x^2 + 1)^{3/2}\right]\Big|_0^1$$

$$= \sqrt{3} - \tfrac{4}{3}\sqrt{2} + \tfrac{1}{3}$$

29. $\displaystyle\int_0^1 \int_0^1 (x+y)^5\, dy\, dx = \tfrac{1}{6}\int_0^1 \left[(x+1)^6 - x^6\right] dx$

$$= 3$$

31. $\displaystyle\int_0^{\pi/2} \int_0^{\pi/2} (x\cos y + y\sin x)\, dy\, dx$

$$= \int_0^{\pi/2} \left[x\sin y + \tfrac{1}{2}y^2\sin x\right]\Big|_0^{\pi/2}\ dx$$

$$= \int_0^{\pi/2} \left(x + \frac{\pi^2}{8}\sin x\right) dx$$

$$= \left[\frac{x^2}{2} - \frac{\pi^2}{8}\cos x\right]\Big|_0^{\pi/2}$$

$$= \frac{\pi^2}{4}$$

33. A double integral $\displaystyle\int\int_R f(x, y)\, dA$ is often evaluated as an iterated integral

$$\int\left[\int f(x, y)\, dy\right] dx \text{ in which } y\text{-integration is}$$

performed first with x held constant, and then x-integration is performed. Likewise, the iterated integral $\displaystyle\int\left[\int f(x, y)\, dx\right] dy$ involves x-integration, then y-integration. Fubini's theorem says that the two iterated integrals are equal:

$$\int\left[\int f(x, y)\, dy\right] dx = \int\left[\int f(x, y)\, dx\right] dy$$

$$= \int\int_R f(x, y)\, dA$$

35. Answers vary; use Riemann sums to find

$$M = \int\int_R \delta(x, y)\, dA$$

37. $\displaystyle\int_0^2 \int_0^1 x(1 - x^2)^{1/2}e^{3y}\, dx\, dy = \tfrac{1}{3}\int_0^2 e^{3y}\, dy$

$$= \frac{e^6}{9} - \frac{1}{9}$$

39. $\displaystyle\int_1^3 \int_1^2 \frac{xy}{x^2+y^2}\,dy\,dx$

$$= \frac{1}{2}\int_1^3 \left[x\ln(x^2+4) - x\ln(x^2+1) \right] dx$$

$$= \frac{1}{4}\Big[(x^2+4)\ln(x^2+4) - (x^2+1)\ln(x^2+1) - 3\Big]\Big|_1^3$$

$$= \frac{1}{4}[13\ln 13 - 15\ln 5 - 8\ln 2]$$

$$\approx 0.91$$

41. The graph of the equation $z = f(x, y)$ is a paraboloid opening downward with vertex at $z = 4$ and intercepts at $(\pm 2, 0, 0)$ and $(0, \pm 2, 0)$. The integral represents the volume above the unit square in the first quadrant. The minimum value for z is $f(1, 1) = 2$. Since $z \geq 2$ over the given region R, the value of the integral will be greater than the volume of the "box" with unit base and height 2.

43. Approximations may vary depending on the the evaluation of z_k over each cell of the grid.

$$A \approx (0.25)^2 \sum_{k=1}^{16} z_k$$

$$\approx (0.0625)(23)$$

$$\approx 1.44$$

45. $\displaystyle\int\int_R \frac{\partial}{\partial y}\left[\frac{\partial f(x, y)}{\partial x}\right] dA$

$$= \int_{x_1}^{x_2} \int_{y_1}^{y_2} \frac{\partial}{\partial y}\left[\frac{\partial f(x, y)}{\partial x}\right] dy\,dx$$

$$= \int_{x_1}^{x_2} \frac{\partial}{\partial x} f(x, y)\Big|_{y=y_1}^{y=y_2} dx$$

$$= \int_{x_1}^{x_2} \left[\frac{\partial f(x, y_2)}{\partial x} - \frac{\partial f(x, y_1)}{\partial x}\right] dx$$

$$= f(x, y_2) - f(x, y_1)\Big|_{x=x_1}^{x=x_2}$$

$$= f(x_2, y_2) - f(x_1, y_2) - f(x_2, y_1) + f(x_1, y_1)$$

47. $\displaystyle\int_0^1 \int_0^1 \frac{y - x}{(x + y)^3}\,dx\,dy$ *Use partial fractions*

$$= \int_0^1 \int_0^1 \left[\frac{2y}{(x+y)^3} - \frac{1}{(x+y)^2}\right] dx\,dy$$

$$= \int_0^1 \left[2y\int_0^1 (x+y)^{-3}\,dx - \int_0^1 (x+y)^{-2}dx\right] dy$$

$$= \int_0^1 \left[2y(x+y)^{-2}\left(-\tfrac{1}{2}\right) + (x+y)^{-1} \right] \Big|_{x=0}^{x=1} dy$$

$$= \int_0^1 \frac{x}{(x+y)^2} \Big|_0^1 \ dy$$

$$= \int_0^1 (y+1)^{-2} \ dy$$

$$= -(y+1)^{-1} \Big|_0^1$$

$$= -\tfrac{1}{2} + 1$$

$$= \tfrac{1}{2}$$

$$\int_0^1 \int_0^1 \frac{y-x}{(x+y)^3} \ dy \ dx \qquad \textit{Use partial fractions}$$

$$= \int_0^1 \int_0^1 \left[\frac{2y}{(x+y)^3} - \frac{1}{(x+y)^2} \right] dy \ dx$$

$$= \int_0^1 \left[\frac{-x-2y}{(x+y)^2} + (x+y)^{-1} \right] \Big|_0^1 dx$$

$$\boxed{\text{Formula 49 where } u = y, \ b = x, \text{ and } a = 1}$$

$$= \int_0^1 \left[2\left(\frac{-1}{y+x} + \frac{x}{2(y+x)^2} \right) + (x+y)^{-1} \right] \Big|_0^1 dx$$

$$= \int_0^1 \frac{-1}{(x+1)^2} \ dx$$

$$= (x+1)^{-1} \Big|_0^1$$

$$= \tfrac{1}{2} - 1$$

$$= -\tfrac{1}{2}$$

Fubini's theorem does not apply because the integrand is not continuous at $(0, 0)$.

12.2 Double Integration Over Nonrectangular Regions, Pages 793-795

SURVIVAL HINT: *Draw a sketch of the region of integration and decide on vertical or horizontal strips. For vertical strips you must have numerical values for the x limits and either numerical or $y = g(x)$ expressions for the y-limits:*

$$\int_{x_1}^{x_2} \int_{y_1 = g_1(x)}^{y_2 = g_2(x)} f(x, y) \ dy \ dx$$

For horizontal strips, you must have numerical values for the y limits and either numerical or $x = h(y)$ expressions for the x-limits:

$$\int_{y_1}^{y_2} \int_{x_1 = h_1(x)}^{x_2 = h_2(x)} f(x, y) \ dx \ dy$$

1. If $f(x, y) \geq 0$ on R, then the volume of the region under $z = f(x, y)$ and above R is given by

$$V = \int_R \int f(x, y) \ dA.$$

SURVIVAL HINT: *Your might want to find out from your instructor if you should refer to your procedure as Type I and Type II, or if you should simply use the words vertically simple (vertical strips) or horizontally simple (horizontal strips), respectively.*

$$\int_0^1 \int_{-x^2}^{x^2} dy\ dx = \int_0^1 2x^2\ dx$$

$$= \frac{2}{3}$$

3. Vertically simple:

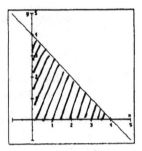

$$\int_0^4 \int_0^{4-x} xy\ dy\ dx = \int_0^4 \frac{x}{2}(4-x)^2\ dx$$

$$= \left[4x^2 - \frac{4}{3}x^3 + \frac{x^4}{8}\right]\Bigg|_0^4$$

$$= \frac{32}{3}$$

5. Vertically simple:

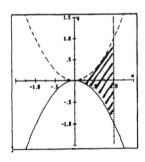

7. Horizontally simple:

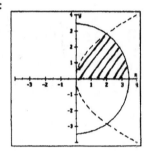

$$\int_0^{2\sqrt{3}} \int_{y^2/4}^{\sqrt{12-y^2}} dx\ dy$$

$$= \int_0^{2\sqrt{3}} \left(\sqrt{12-y^2} - \frac{y^2}{4}\right)dy$$

$$= \left(\frac{y}{2}\sqrt{12-y^2} + 6\sin^{-1}\frac{y}{2\sqrt{3}} - \frac{y^3}{12}\right)\Bigg|_0^{2\sqrt{3}}$$

$$\boxed{\text{Formula 23}}$$

$$= 3\pi - 2\sqrt{3}$$

$$\approx 5.9607$$

9. Vertically simple:

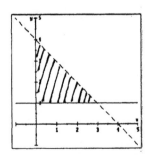

$$\int_0^3 \int_1^{4-x} (x+y) \, dy \, dx = \int_0^3 -\tfrac{1}{2}(x^2+2x-15) \, dx$$

$$= \frac{27}{2}$$

$$= \frac{1}{6} \sin^3 2$$

$$\approx 0.1253$$

15. Horizontally simple:

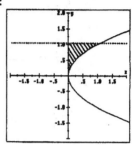

$$\int_0^{\pi/3} \int_0^{y^2} \tfrac{1}{y} \sin \tfrac{x}{y} \, dx \, dy = \int_0^{\pi/3} \left[-\cos \tfrac{x}{y} \right]\Big|_0^{y^2} dy$$

$$= \int_0^{\pi/3} (1 - \cos y) \, dy$$

$$= \frac{\pi}{3} - \frac{\sqrt{3}}{2}$$

$$\approx 0.1812$$

11. Vertically simple:

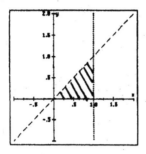

$$\int_0^1 \int_0^x (x^2 + 2y^2) \, dy \, dx = \int_0^1 \left(x^3 + \frac{2x^3}{3}\right) \, dx$$

$$= \frac{5}{12}$$

17. Horizontally simple:

$$\int_0^1 \int_0^{y^3} e^{x/y} \, dx \, dy = \int_0^1 y e^{x/y} \Big|_0^{y^3} dy$$

$$= \int_0^1 y(e^{y^2} - 1) \, dy$$

$$= \left[\tfrac{1}{2} e^{y^2} - \frac{y^2}{2} \right]\Big|_0^1$$

13. Vertically simple:

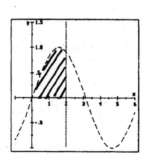

$$\int_0^2 \int_0^{\sin x} y \cos x \, dy \, dx = \tfrac{1}{2} \int_0^2 \sin^2 x \cos x \, dx$$

$$= \tfrac{1}{2}(e - 1) - \tfrac{1}{2}$$

$$= \tfrac{1}{2}e - 1$$

19. Vertically simple (change order):

$$\int_0^1 \int_{\tan^{-1}y}^{\pi/4} \sec x \, dx \, dy = \int_0^{\pi/4} \int_0^{\tan x} \sec x \, dy \, dx$$

$$= \int_0^{\pi/4} \sec x \, [y]\Big|_0^{\tan x} \, dx$$

$$= \int_0^{\pi/4} \sec x \tan x \, dx$$

$$= [\sec x]\Big|_0^{\pi/4}$$

$$= \sqrt{2} - 1$$

21. Vertically simple:

$$\iint_D (x + y) \, dA = \int_0^1 \int_x^1 (x + y) \, dy \, dx$$

$$= \int_0^1 \left(x + \tfrac{1}{2} - \tfrac{3x^2}{2} \right) dx$$

$$= \tfrac{1}{2}$$

23. Horizontally simple:

$$\iint_D y \, dA = \int_0^1 \int_{y^2}^{2-y} y \, dx \, dy$$

$$= \int_0^1 (2y - y^2 - y^3) \, dy$$

$$= \tfrac{5}{12}$$

25. Vertically simple:

$$\iint_D 2x \, dA = \int_0^1 \int_0^x 2x \, dy \, dx + \int_1^2 \int_0^{1/x^2} 2x \, dy \, dx$$

$$= \int_0^1 2x^2 \, dx + \int_0^2 \frac{2}{x} \, dx$$

$$= \tfrac{2}{3} + 2\ln 2$$

$$\approx 2.0530$$

27. Horizontally simple:

$$\iint_D 12x^2 e^{y^2} \, dA = \int_0^1 \int_y^{y^{1/3}} 12x^2 e^{y^2} \, dx \, dy$$

$$= 12 \int_0^1 e^{y^2} \int_y^{y^{1/3}} x^2 \, dx \, dy$$

$$= 4 \int_0^1 (y - y^3) e^{y^2} \, dy$$

$$= 4 \int_0^1 y e^{y^2} \, dy - 4 \int_0^1 y^3 e^{y^2} \, dy$$

$$\boxed{\text{Let } u = y^2; \; du = 2y \, dy}$$

$$= 2 \int_0^1 e^u \, du - 2 \int_0^1 u e^u \, du$$

$$= \left[2e^u - 2(u e^u - e^u) \right]\Big|_0^1$$

$$= 2e - 4$$

$$\approx 1.4366$$

29. We reverse the order:

$$\iint_D \frac{\sin x}{x} \, dx \, dy = \int_0^\pi \int_y^\pi \frac{\sin x}{x} \, dx \, dy$$

$$= \int_0^\pi \int_0^x \frac{\sin x}{x} \, dy \, dx$$

$$= \int_0^\pi \frac{\sin x}{x}(x) \, dx$$

$$= \left[-\cos x \right]\Big|_0^\pi$$

$$= 2$$

31. The parabola $y = x^2$ and the line $y = 2x + 3$ intersect when

$$x^2 = 2x + 3$$

$$x^2 - 2x - 3 = 0$$

$$(x - 3)(x + 1) = 0$$

$$x = -1, 3$$

That is, at $(-1, 1)$ and $(3, 9)$. It is vertically simple.

$$\iint_D x \, dy \, dx = \int_{-1}^3 \int_{x^2}^{2x+3} x \, dy \, dx$$

$$= \int_{-1}^3 xy \Big|_{x^2}^{2x+3} dx$$

$$= \int_{-1}^3 x[2x + 3 - x^2] \, dx$$

$$= \left[\tfrac{2}{3}x^3 + \tfrac{3}{2}x^2 - \tfrac{1}{4}x^4 \right]\Big|_{-1}^3$$

$$= \frac{32}{3}$$

33.

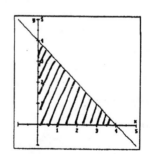

a. $$\int_0^4 \int_0^{4-x} xy \, dy \, dx = \frac{1}{2}\int_0^4 (x^3 - 8x^2 + 16x) \, dx$$

$$= \frac{32}{3}$$

b. $\displaystyle\int_0^4 \int_0^{4-y} xy \; dx \; dy = \frac{1}{2}\int_0^4 (y^3 - 8y^2 + 16y)\; dy$

$$= \frac{32}{3}$$

a. $\displaystyle\int_0^1 \int_{-x^2}^{x^2} dy \; dx = 2\int_0^1 x^2 \; dx$

$$= \frac{2}{3}$$

35.

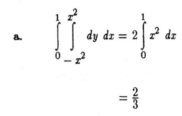

a. $\displaystyle\int_0^1 \int_x^{2x} e^{y-x} \; dy \; dx = \int_0^1 (e^x - 1)\; dx$

$$= e - 2$$

b. $\displaystyle\int_0^1 \int_{y/2}^{y} e^{y-x} \; dx \; dy + \int_1^2 \int_{y/2}^{1} e^{y-x} \; dx \; dy$

$$= -\int_0^1 (1 - e^{y/2})\; dy - \int_1^2 (e^{y-1} - e^{y/2})\; dy$$

$$= e - 2$$

37.

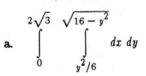

b. $\displaystyle\int_0^1 \int_{\sqrt{y}}^{1} dx \; dy + \int_{-1}^{0} \int_{\sqrt{-y}}^{1} dx \; dy$

$$= 2\int_0^1 \int_{\sqrt{y}}^{1} dx \; dy$$

$$= 2\int_0^1 (1 - \sqrt{y})\; dy$$

$$= \frac{2}{3}$$

39.

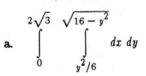

a. $\displaystyle\int_0^{2\sqrt{3}} \int_{y^2/6}^{\sqrt{16-y^2}} dx \; dy$

$$= \int_0^{2\sqrt{3}} \left(\sqrt{16 - y^2} - \frac{y^2}{6} \right) dy$$

$$= \int_0^{2\sqrt{3}} \sqrt{16 - y^2}\, dy - \frac{1}{6}\int_0^{2\sqrt{3}} y^2\, dy$$

$$\boxed{\text{Formula 231}}$$

$$= \left[8 \sin^{-1} \frac{y}{4} + \frac{y}{2}\sqrt{16 - y^2} - \frac{y^3}{18} \right]\Bigg|_0^{2\sqrt{3}}$$

$$= \frac{8\pi}{3} + \frac{2\sqrt{3}}{3}$$

$$\approx 9.5323$$

b. $$\int_0^2 \int_0^{\sqrt{6x}} dy\, dx + \int_2^4 \int_0^{\sqrt{16-x^2}} dy\, dx$$

$$= \int_0^2 \sqrt{6x}\, dx + \int_2^4 \sqrt{16 - x^2}\, dx$$

$$\boxed{\text{Formula 231}}$$

$$= \frac{2\sqrt{6}}{3} x^{3/2}\Bigg|_0^2 + \left[\frac{x\sqrt{16 - x^2}}{2} + 8\sin^{-1}\frac{x}{4} \right]\Bigg|_2^4$$

$$= \frac{8\sqrt{3}}{3} + \left(-2\sqrt{3} + \frac{8\pi}{3} \right)$$

$$= \frac{8\pi}{3} + \frac{2\sqrt{3}}{3}$$

$$\approx 9.5323$$

41.

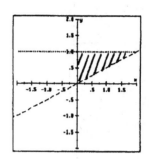

$$\int_0^2 \int_{x/2}^1 f(x,\, y)\, dy\, dx$$

43.

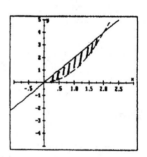

$$\int_0^2 \int_{x^2}^{2x} f(x,\, y)\, dy\, dx$$

45.

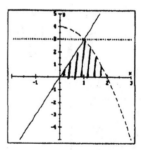

The curves intersect when

$$\frac{y}{3} = \sqrt{4 - y}$$

$$y^2 + 9y - 36 = 0$$

$$(y - 3)(y + 12) = 0$$

Intersection: $(1, 3)$

$$\int_0^1 \int_0^{3x} f(x, y) \, dy \, dx + \int_1^2 \int_0^{4-x^2} f(x, y) \, dy \, dx$$

47.

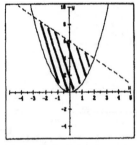

The curves intersect when

$$x^2 = 6 - x$$

$$x^2 + x - 6 = 0$$

$$(x+3)(x-2) = 0$$

Intersections: $(2, 4)$ and $(-3, 9)$

$$\int_0^4 \int_{-\sqrt{y}}^{\sqrt{y}} f(x, y) \, dx \, dy + \int_4^9 \int_{-\sqrt{y}}^{6-y} f(x, y) \, dx \, dy$$

49.

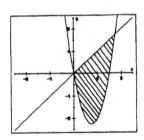

The line $y = x$ intersects the parabola

$y = x^2 - 6x$ when

$$x = x^2 - 6x$$

$$x^2 - 7x = 0$$

$$x = 0, 7$$

that is, at $(0, 0)$ and $(7, 7)$. Also

$$x^2 - 6x = y$$

$$x = 3 \pm \sqrt{9 + y}$$

$$\int_{-9}^0 \int_{3-\sqrt{9+y}}^{3+\sqrt{9+y}} f(x, y) \, dx \, dy + \int_0^7 \int_y^{3+\sqrt{9+y}} f(x, y) \, dx \, dy$$

51. $V = \displaystyle\int_0^{7/3} \int_0^{(7-3x)/2} (7 - 3x - 2y) \, dy \, dx$

53. Using symmetry of $x^2 + y^2 = 3$, and

$x^2 + y^2 + z^2 = 7$, we have

$$V = 8 \int_0^{\sqrt{3}} \int_0^{\sqrt{3 - x^2}} \sqrt{7 - x^2 - y^2} \, dy \, dx$$

55. The projection of the ellipsoid on the xy-plane

is the ellipse

$$\frac{x^2}{a^2} + \frac{y^2}{b^2} = 1.$$

Using symmetry of the ellipsoid, we have

$$V = 8 \int_0^a \int_0^{(b/a)\sqrt{a^2 - x^2}} c\sqrt{1 - \frac{x^2}{a^2} - \frac{y^2}{b^2}} \; dy \; dx$$

57. The volume removed from the sphere

$$x^2 + y^2 + z^2 = 2$$

by the square cylindrical hole centered at the origin is

$$V_r = 8 \int_0^1 \int_0^1 \sqrt{2 - x^2 - y^2} \; dy \; dx$$

We have used the symmetry of the sphere and square to simplify the integral. Thus, the volume that remains is

$$V = \tfrac{4}{3}\pi(\sqrt{2})^3 - 8 \int_0^1 \int_0^1 \sqrt{2 - x^2 - y^2} \; dy \; dx$$

59.

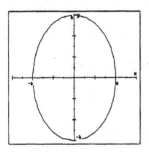

We find the limits of integration:

$$\frac{x^2}{a^2} + \frac{y^2}{b^2} = 1$$

$$y = \tfrac{b}{a}\sqrt{a^2 - x^2}$$

Using vertical strips and symmetry:

$$A = 4 \int_0^a \int_0^{(b/a)\sqrt{a^2 - x^2}} dy \; dx$$

Using horizontal strips:

$$4 \int_0^b \int_0^{(a/b)\sqrt{b^2 - y^2}} dx \; dy$$

$$= \frac{4a}{b} \int_0^b \sqrt{b^2 - y^2} \, dy$$

$$= \frac{4a}{b} \left[\frac{b^2}{2} \sin^{-1}\frac{y}{b} + \frac{y}{2}\sqrt{b^2 - y^2} \right]\Big|_0^b$$

$$= \frac{4a}{b} \left[\frac{\pi b^2}{4} \right]$$

$$= \pi a b$$

61.

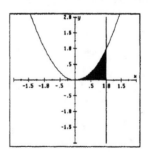

$$\int_0^1 \int_0^{x^2} 4x \; dy \; dx = 4 \int_0^1 x^3 \; dx$$

$$= 1$$

63.

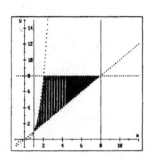

$$\int_{1}^{8} \int_{y/3}^{y} f(x, y) \, dx \, dy$$

65.

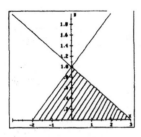

The lines intersect at $(-2, 0)$, $(3, 0)$, and $(0, 1)$.

$$\int_{0}^{1} \int_{2y-2}^{3-3y} (x^2 - xy - 1) \, dx \, dy$$

$$= \int_{0}^{1} \left[\frac{x^3}{3} - \frac{x^2 y}{2} - x \right] \Bigg|_{2(y-1)}^{3(1-y)} dy$$

$$= \int_{0}^{1} [9(1 - y)^3 - \tfrac{9}{2}(1 - y)^2 y - 3(1 - y)$$

$$- \tfrac{8}{3}(y - 1)^3 + 2(y - 1)^2 y + 2(y - 1)] \, dy$$

$$= \int_{0}^{1} [-\tfrac{85}{6}(y - 1)^3 - \tfrac{5}{2}(y - 1)^2 y + 5(y - 1)] \, dy$$

$$= \frac{5}{24}$$

67. $\displaystyle\int_{0}^{3} \int_{0}^{(1/3)\sqrt{36 - 4x^2}} (3x + y) \, dy \, dx$

$$= \int_{0}^{3} [2x\sqrt{9 - x^2} - \tfrac{2}{9}(x^2 - 9)] \, dx$$

$$\boxed{\text{Let } u = 9 - x^2; \; du = -2x \, dx}$$

$$= -\tfrac{2}{3} u^{3/2} \Big|_{9}^{0} + \left[-\tfrac{2}{27} x^3 + 2x \right] \Big|_{0}^{3}$$

$$= 22$$

69. The integral is the volume of a cylinder of radius 3 and height 3 minus the volume of a cone of radius 3 and height 3:

$$V = \pi r^2 h - \tfrac{1}{3}\pi r^2 h$$

$$= 3\pi(3)^2 - \tfrac{\pi}{3}(3^2)(3)$$

$$= 18\pi$$

71. The integral is the volume of a cylinder of radius 4 and height 4 minus the volume of a cone of radius 4 and height 4:

$$V = \pi r^2 h - \tfrac{1}{3}\pi r^2 h$$

$$= \tfrac{2\pi}{3}(4^2)(4)$$

$$= \tfrac{128\pi}{3}$$

73. Let $f(x, y) = e^{y \sin x}$. Then, since

$-1 \le \sin x \le 1$, we have

$$mA \le \int\int_D f(x, y)\, dA \le MA$$

$$\int_0^1 \int_{y-1}^{2-2y} e^{-y}\, dx\, dy \le \int\int_D e^{y \sin x}\, dx\, dy$$

$$\le \int_0^1 \int_{y-1}^{2-2y} e^y\, dx\, dy$$

$$3e^{-1} \le \int\int_D e^{y \sin x}\, dx\, dy \le 3(e - 2)$$

$$1.10 \le \int\int_D e^{y \sin x}\, dx\, dy \le 2.15$$

The value of the integral is between 1.10 and

2.15. (The actual value is approximately

1.63.)

12.3 Double Integrals in Polar Coordinates, Pages 801-804

SURVIVAL HINT: *Since polar equations are often periodic* [e.g. $f(0) = f(2\pi)$], *it is sometimes a good idea to make use of any symmetry in the graph. Find the area of the smallest symmetric piece and multiply by the number of such pieces.*

1.

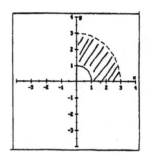

$$\int_0^{\pi/2} \int_1^3 r e^{-r^2}\, dr\, d\theta = -\tfrac{1}{2} \int_0^{\pi/2} (e^{-9} - e^{-1})\, d\theta$$

$$= -\tfrac{\pi}{4}(e^{-9} - e^{-1})$$

$$= \tfrac{\pi}{4}\left(\tfrac{1}{e} - \tfrac{1}{e^9}\right)$$

$$\approx 0.2888$$

3.

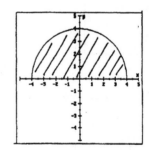

$$\int_0^{\pi} \int_0^4 r^2 \sin^2\theta\, dr\, d\theta = \tfrac{4^3}{3} \int_0^{\pi} \sin^2\theta\, d\theta$$

$$= \tfrac{32\pi}{3}$$

$$\approx 33.5103$$

5.

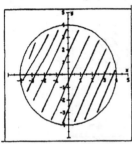

$$\int_0^{2\pi} \int_0^4 2r^2 \cos \theta \, dr \, d\theta = \frac{128}{3} \int_0^{2\pi} \cos \theta \, d\theta$$

$$= 0$$

7.

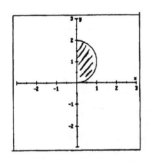

$$\int_0^{\pi/2} \int_0^{2 \sin \theta} dr \, d\theta = \int_0^{\pi/2} 2 \sin \theta \, d\theta$$

$$= 2$$

9. $\int_0^{2\pi} \int_0^4 r \, dr \, d\theta = 8 \int_0^{2\pi} d\theta$

$$= 16\pi$$

$$\approx 50.2655$$

11. $2 \int_0^{\pi} \int_0^{2(1-\cos \theta)} r \, dr \, d\theta$

$$= 4 \int_0^{\pi} \left[1 - 2 \cos \theta + \frac{1 + \cos 2\theta}{2} \right] d\theta$$

$$= 6\pi$$

13. $6 \int_0^{\pi/6} \int_0^{4 \cos 3\theta} r \, dr \, d\theta = 48 \int_0^{\pi/6} \cos^2 3\theta \, d\theta$

$$= \left[4 \sin 6\theta + 24\theta \right] \Big|_0^{\pi/6}$$

$$= 4\pi$$

15. $8 \int_0^{\pi/4} \int_0^{\cos 2\theta} r \, dr \, d\theta = 4 \int_0^{\pi/4} \cos^2 2\theta \, d\theta$

$$= \left[\frac{1}{2} \sin 4\theta + 2\theta \right] \Big|_0^{\pi/4}$$

$$= \frac{\pi}{2}$$

17. $r = 1$ and $r = 2 \sin \theta$; the circles intersect when

$$1 = 2 \sin \theta$$

$$\sin \theta = \frac{1}{2}$$

$$\theta = \sin^{-1} \frac{1}{2}$$

$$= \frac{\pi}{6}$$

Half the area is spanned by $\frac{\pi}{6} \leq \theta \leq \frac{\pi}{2}$

and $1 \le r \le 2 \sin \theta$, so we have:

$$2 \int_{\pi/6}^{\pi/2} \int_{1}^{2\sin\theta} r \, dr \, d\theta = \int_{\pi/6}^{\pi/2} (4 \sin^2\theta - 1) \, d\theta$$

$$= [\theta - 2\sin\theta \cos\theta] \Big|_{\pi/6}^{\pi/2}$$

$$= \tfrac{1}{6}(2\pi + 3\sqrt{3})$$

$$= \tfrac{\pi}{3} + \frac{\sqrt{3}}{2}$$

$$\approx 1.913$$

19. $r = 1$ and $r = 1 + \cos \theta$; use symmetry about

the polar axis and the equation of the quarter

circle.

$$2[\tfrac{1}{4}\pi(1)^2] + 2 \int_{\pi/2}^{\pi} \int_{0}^{1+\cos\theta} r \, dr \, d\theta$$

$$= \tfrac{\pi}{2} + 2 \int_{\pi/2}^{\pi} \left(\tfrac{1}{2} \cos^2\theta + \cos\theta + \tfrac{1}{2} \right) d\theta$$

$$= \tfrac{\pi}{2} + 2\left[\tfrac{1}{4} \cos\theta \sin\theta + \sin\theta + \tfrac{3}{4}\theta \right] \Big|_{\pi/12}^{\pi}$$

$$= \tfrac{\pi}{2} + \frac{3\pi - 8}{4}$$

$$= \tfrac{5\pi}{4} - 2$$

$$\approx 1.9270$$

21. $r = 3 \cos \theta$ and $r = 1 + \cos \theta$

The curves intersect when

$$3 \cos \theta = 1 + \cos \theta$$

$$\cos \theta = \tfrac{1}{2}$$

$$\theta = \cos^{-1} \tfrac{1}{2} = \tfrac{\pi}{3}$$

$$2 \int_{0}^{\pi/3} \int_{1+\cos\theta}^{3\cos\theta} r \, dr \, d\theta$$

$$= \int_{0}^{\pi/3} [9 \cos^2\theta - (1 + \cos\theta)^2] \, d\theta$$

$$= \int_{0}^{\pi/3} \left(8 \frac{\cos 2\theta + 1}{2} - 2 \cos\theta - 1 \right) d\theta$$

$$= \int_{0}^{\pi/3} (3 + 4 \cos 2\theta - 2 \cos\theta) \, d\theta$$

$$= \pi + 2 \sin \tfrac{2\pi}{3} - 2 \sin \tfrac{\pi}{3}$$

$$= \pi$$

23. The bottom half of the loop is scanned out as

θ varies from $2\pi/3$ to π. Thus, the area is

$$A = 2 \int_{2\pi/3}^{\pi} \int_{0}^{1 + 2\cos\theta} r \, dr \, d\theta$$

$$= 2 \int_{2\pi/3}^{\pi} \tfrac{1}{2}[1 + 2 \cos\theta]^2 \, d\theta$$

$$= \int\limits_{2\pi/3}^{\pi} (1 + 4 \cos\theta + 4 \cos^2\theta)\, d\theta$$

$$= \left[\theta + 4\sin\theta + 2\theta + \sin 2\theta\right]\Big|_{\pi/3}^{\pi}$$

$$= \pi - \frac{4\sqrt{3}}{2} + \frac{\sqrt{3}}{2}$$

$$= \pi - \frac{3\sqrt{3}}{2}$$

$$\approx 0.5435$$

SURVIVAL HINT: *One of the most common errors when converting from Cartesian to polar integration is forgetting the factor of "r." That is,*

$$dy\, dx = dA = r\, dr\, d\theta$$

25. $f(x, y) = y^2$; since $y = r\sin\theta$, $y^2 = r^2\sin^2\theta$

$$\iint\limits_{D} y^2\, dA = \int\limits_{0}^{2\pi}\int\limits_{0}^{a} r^2\sin^2\theta\, r\, dr\, d\theta$$

$$= \frac{a^4}{4}\int\limits_{0}^{2\pi}\frac{1 - \cos 2\theta}{2}\, d\theta$$

$$= \frac{a^4\pi}{4}$$

27. $f(x, y) = \dfrac{1}{a^2 + x^2 + y^2} = \dfrac{1}{a^2 + r^2}$

$$\iint\limits_{D}\frac{1}{a^2 + x^2 + y^2}\, dA = 4\int\limits_{0}^{\pi/2}\int\limits_{0}^{a}\frac{r}{a^2 + r^2}\, dr\, d\theta$$

$$= 2\int\limits_{0}^{\pi/2}\ln(a^2 + r^2)\Big|_{0}^{a}\, d\theta$$

$$= \int\limits_{0}^{\pi/2} 2\ln 2\, d\theta$$

$$= \pi\ln 2$$

29. $f(x, y) = \dfrac{1}{a + \sqrt{x^2 + y^2}} = \dfrac{1}{a + r}$

$$\iint\limits_{D}\frac{1}{a + \sqrt{x^2 + y^2}}\, dA$$

$$= \int\limits_{0}^{2\pi}\int\limits_{0}^{a}\frac{1}{a + r}\, r\, dr\, d\theta$$

$$= \int\limits_{0}^{2\pi}\int\limits_{0}^{a}\left(1 - \frac{a}{r + a}\right)\, dr\, d\theta$$

$$= \int\limits_{0}^{2\pi}(a - a\ln 2a + a\ln a)\, d\theta$$

$$= 2\pi a(1 + \ln\tfrac{1}{2}) \text{ or } 2\pi a(1 - \ln 2)$$

31. $\displaystyle\int\!\!\int_D y\,dA = \int_0^{2\pi}\!\!\int_0^2 r^2 \sin\theta\,dr\,d\theta$

$$= \int_0^{2\pi} \frac{8}{3}\sin\theta\,d\theta$$

$$= 0$$

33. $\displaystyle\int\!\!\int_D e^{x^2+y^2}\,dA = 4\int_0^{\pi/2}\!\!\int_0^3 re^{r^2}\,dr\,d\theta$

$$= 4\int_0^{\pi/2} \tfrac{1}{2}(e^9 - 1)\,d\theta$$

$$= \pi(e^9 - 1)$$

$$\approx 25{,}453$$

35. $\displaystyle\int\!\!\int_D \ln(x^2 + y^2 + 2)\,dA$

$$= \int_0^{\pi/2}\!\!\int_0^2 r\ln(r^2+2)\,dr\,d\theta$$

$$= \int_0^{\pi/2} \tfrac{1}{2}\Big[(r^2+2)\ln(r^2+2) - (r^2+2)\Big]\Big|_0^2\,d\theta$$

$$= \int_0^{\pi/2} (3\ln 6 - \ln 2 - 2)\,d\theta$$

$$= \tfrac{3}{2}\pi\ln 3 + \pi\ln 2 - \pi$$

$$\approx 4.2131$$

37. $\displaystyle V = \int_0^{2\pi}\!\!\int_0^2 (4 - r^2)r\,dr\,d\theta$

$$= \int_0^{2\pi}\Big[2r^2 - \frac{r^4}{4}\Big]\Big|_0^2\,d\theta$$

$$= \int_0^{2\pi} 4\,d\theta$$

$$= 8\pi$$

39. $\displaystyle\int_0^3\!\!\int_0^{\sqrt{9-x^2}} x\,dy\,dx = \int_0^{\pi/2}\!\!\int_0^3 (r\cos\theta)(r\,dr\,d\theta)$

$$= \int_0^{\pi/2} \cos\theta\Big[\frac{r^3}{3}\Big]\Big|_0^3\,d\theta$$

$$= 9\int_0^{\pi/2}\cos\theta\,d\theta$$

$$= 9[\sin\theta]\Big|_0^{\pi/2}$$

$$= 9$$

41. $\displaystyle\int_0^2\!\!\int_0^{\sqrt{4-y^2}} e^{x^2+y^2}\,dx\,dy = \int_0^{\pi/2}\!\!\int_0^2 e^{r^2} r\,dr\,d\theta$

$$= \int_0^{\pi/2} \Big[\tfrac{1}{2}e^{r^2}\Big]\Big|_0^2\,d\theta$$

$$= \int_0^{\pi/2} \Big[\tfrac{1}{2}e^4 - \tfrac{1}{2}e^0\Big]\,d\theta$$

$$= \left(\tfrac{1}{2}e^4 - \tfrac{1}{2}\right)\left(\tfrac{\pi}{2}\right)$$

$$= \tfrac{\pi}{4}(e^4 - 1)$$

$$\approx 42.096$$

$$= \int_0^{\pi/2} \left[-\sqrt{9-4} + \sqrt{9}\right] d\theta$$

$$= (3 - \sqrt{5})\left(\tfrac{\pi}{2}\right)$$

$$\approx 1.200$$

43. $\displaystyle \int_0^4 \int_0^{\sqrt{4y-y^2}} \frac{1}{\sqrt{x^2+y^2}}\, dx\, dy$

$$= \int_0^{\pi/2} \int_0^{4\sin\theta} \tfrac{1}{r}\, r\, dr\, d\theta$$

$$= \int_0^{\pi/2} 4\sin\theta\, d\theta$$

$$= \left[-4\cos\theta\right]\Big|_0^{\pi/2}$$

$$= 0 - (-4)$$

$$= 4$$

47. $\displaystyle \int_0^2 \int_0^{\sqrt{2x-x^2}} \frac{x-y}{x^2+y^2}\, dy\, dx$

$$= \int_0^{\pi/2} \int_0^{2\cos\theta} \frac{r\cos\theta - r\sin\theta}{r^2}\, r\, dr\, d\theta$$

$$= \int_0^{\pi/2} [\cos\theta - \sin\theta](2\cos\theta)\, d\theta$$

$$= \left[\tfrac{1}{2}\sin 2\theta + \tfrac{1}{2}\cos 2\theta + \theta\right]\Big|_0^{\pi/2}$$

$$= \frac{\pi - 2}{2}$$

$$\approx 0.5708$$

45. $\displaystyle \int_0^2 \int_0^{\sqrt{4-y^2}} \frac{1}{\sqrt{9-x^2-y^2}}\, dx\, dy$

$$= \int_0^{\pi/2} \int_0^2 \frac{1}{\sqrt{9-r^2}}\, r\, dr\, d\theta$$

$$= \int_0^{\pi/2} -(9-r^2)^{1/2}\Big|_0^2\, d\theta$$

49.

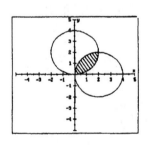

The curves intersect at

$\theta = \pi/4$ and at the pole.

$$\int_0^{\pi/4} \int_0^{4\sin\theta} r^3 \sin\theta \cos\theta \; dr \; d\theta$$

$$+ \int_{\pi/4}^{\pi/2} \int_0^{4\cos\theta} r^3 \sin\theta \cos\theta \; dr \; d\theta$$

$$= 4^3 \left[\int_0^{\pi/4} \sin^5\theta \cos\theta \; d\theta + \int_{\pi/4}^{\pi/2} \cos^5\theta \sin\theta \; d\theta \right]$$

$$= 4^3 \left(\left[\frac{\sin^6\theta}{6}\right]\Big|_0^{\pi/4} - \left[\frac{\cos^6\theta}{6}\right]\Big|_{\pi/4}^{\pi/2} \right)$$

$$= \frac{4^3}{6}\left(\left[\frac{1}{\sqrt{2}}\right]^6 + \left[\frac{1}{\sqrt{2}}\right]^6 \right)$$

$$= \frac{4^3}{6\cdot 4}$$

$$= \frac{8}{3}$$

51. The curves intersect at $\theta = \pi/4$. Making use of the symmetry we will integrate the $\sin\theta$ curve from 0 to $\pi/4$ to get half of the desired area.

$$A = 2 \int_0^{\pi/4} \int_0^{a\sin\theta} r \; dr \; d\theta$$

$$= 2 \int_0^{\pi/4} \frac{a^2 \sin^2\theta}{2} \; d\theta$$

$$= a^2 \int_0^{\pi/4} \frac{1 - \cos 2\theta}{2} \; d\theta$$

$$= \frac{a^2}{8}(\pi - 2)$$

53. $$V = \int_0^{2\pi} \int_3^5 \sqrt{25 - r^2} \; r \; dr \; d\theta$$

$$= \int_0^{2\pi} -\frac{1}{3}(25 - r^2)^{3/2}\Big|_3^5 \; d\theta$$

$$= \int_0^{2\pi} \frac{64}{3} \; d\theta$$

$$= \frac{128\pi}{3}$$

$$\approx 134.0413$$

55. The circle $x^2 + y^2 = 2x$ in polar form is $r = 2\cos\theta$. Half the solid is above the xy-plane and half is below.

$$V = \int\int_D z \; dA$$

$$= 4 \int_0^{\pi/2} \int_0^{2\cos\theta} \sqrt{4 - r^2} \; r \; dr \; d\theta$$

$$= 4 \int_0^{\pi/2} -\frac{8}{3}(\sin^3\theta - 1) \; d\theta$$

$$= -\frac{32}{3}\left[-\cos\theta + \frac{\cos^3\theta}{3} - \theta \right]\Big|_0^{\pi/2}$$

$$= \frac{16}{9}(3\pi - 4)$$

$$\approx 9.6440$$

57. The circle $x^2 + y^2 = x$ in polar form is

$r = \cos \theta$.

$$V = \int\int_D z \, dA$$

$$= 4 \int_0^{\pi/2} \int_0^{\cos \theta} (1 - r^2) \, r \, dr \, d\theta$$

$$= 4 \int_0^{\pi/2} \left(\frac{\cos^2\theta}{2} - \frac{\cos^4\theta}{4} \right) d\theta$$

$$= \frac{5\pi}{16}$$

$$\approx 0.9817$$

59. The curves intersect at $\theta = \frac{\pi}{4}$. For $0 \le \theta \le \frac{\pi}{4}$, we use the curve $r = 2\sin\theta$, and for $\frac{\pi}{4} \le \theta \le \frac{\pi}{2}$ we use the curve $r = 2\cos\theta$. Thus,

$$\int\int_D x \, dA = \int\int_D r \cos \theta (r \, dr \, d\theta)$$

$$= \int_0^{\pi/4} \int_0^{2 \sin \theta} r^2 \cos \theta \, dr \, d\theta$$

$$+ \int_{\pi/4}^{\pi/2} \int_0^{2 \cos \theta} r^2 \cos \theta \, dr \, d\theta$$

$$= \int_0^{\pi/4} \frac{1}{3}(2 \sin \theta)^3 \cos \theta \, d\theta$$

$$+ \int_{\pi/4}^{\pi/2} \frac{1}{3}(2 \cos \theta)^3 \cos \theta \, d\theta$$

$$\boxed{\text{Formula 320}}$$

$$= \frac{8}{3} \frac{\sin^4\theta}{4} \Big|_0^{\pi/4}$$

$$+ \frac{8}{3} \left[\frac{3\theta}{8} + \frac{\sin 2\theta}{4} + \frac{\sin 4\theta}{32} \right] \Big|_{\pi/4}^{\pi/2}$$

$$= \frac{1}{6} + \frac{(3\pi - 8)}{12}$$

$$= \frac{\pi}{4} - \frac{1}{2}$$

$$\approx 0.2854$$

61. The first solution can't be correct because $\cos \theta$ is negative for $\pi/2 < \theta \le \pi$. The second solution is correct.

63. The intersection of the plane $z = R - a$ and the sphere $r^2 + z^2 = R^2$ is

$$r^2 = R^2 - (R - a)^2$$

$$= a(2R - a)$$

Thus, the projected region is the disk

$D: r^2 \leq a(2R - a)$. The volume is given by

$$V = \int\int_D z\, dA$$

$$= \int_0^{2\pi} \int_0^{\sqrt{a(2R-a)}} [\sqrt{R^2 - r^2} - (R - a)]\, r\, dr\, d\theta$$

$$= 2\pi\left[-\frac{(R^2 - r^2)^{3/2}}{3} - (R - a)\frac{r^2}{2}\right]\Bigg|_0^{\sqrt{a(2R-a)}}$$

$$= \frac{2\pi}{3}\left[R^3 - (R - a)^3 - \frac{3a(R-a)(2R-a)}{2}\right]$$

$$= \frac{1}{3}\pi a^2(3R - a)$$

The volume of the cap is half the volume of the

hemisphere when

$$\tfrac{1}{3}\pi a^2(3R - a) = \tfrac{1}{2}(\tfrac{2}{3}\pi R^3)$$

$$a \approx 0.6527R \quad \text{(using technology)}$$

12.4 Surface Area, Pages 810–811

SURVIVAL HINT: *Do not try to evaluate any but the simplest of the radical expressions "in your head." You will make fewer errors if you take the time to write down the values f_x, f_y, square them, add 1, and then take the square root.*

1. $z = f(x, y) = 2 - \frac{1}{2}x - \frac{1}{4}y$

$$f_x = -\tfrac{1}{2};\, f_y = -\tfrac{1}{4}$$

$$\sqrt{f_x^{\,2} + f_y^{\,2} + 1} = \sqrt{\tfrac{1}{4} + \tfrac{1}{16} + 1}$$

$$= \frac{\sqrt{21}}{4}$$

$$S = \int_0^4 \int_0^{-2x+8} \frac{\sqrt{21}}{4}\, dy\, dx$$

$$= \frac{\sqrt{21}}{4}\int_0^4 (-2x+8)\, dx$$

$$= \frac{16\sqrt{21}}{4}$$

$$= 4\sqrt{21}$$

$$\approx 18.3303$$

3. $z = f(x, y) = x^2 + y^2$

$$f_x = 2x$$

$$f_y = 2y$$

$$\sqrt{f_x^{\,2} + f_y^{\,2} + 1} = \sqrt{4x^2 + 4y^2 + 1}$$

$$= \sqrt{4r^2 + 1}.$$

The projected region is $x^2 + y^2 \leq 1$; so

$$r \leq 1.$$

$$S = \int_0^{2\pi} \int_0^1 \sqrt{4r^2 + 1}\; r\, dr\, d\theta$$

$$= \frac{1}{8}\int_0^{2\pi} \int_0^1 \sqrt{4r^2 + 1}\;(8r\, dr)\, d\theta$$

$$= \frac{1}{12}\int_0^{2\pi} (5\sqrt{5} - 1)\, d\theta$$

$$= \frac{\pi}{6}(5\sqrt{5} - 1)$$

$$\approx 5.3304$$

5. $z = f(x, y) = \frac{1}{2}(12 - 3x - 6y)$

$$f_x = -\frac{3}{2}$$

$$f_y = -3$$

$$\sqrt{f_x^2 + f_y^2 + 1} = \sqrt{\tfrac{9}{4} + 9 + 1}$$

$$= \frac{7}{2}$$

$$S = \int_0^1 \int_0^x \frac{7}{2}\, dy\, dx$$

$$= \frac{7}{2}\int_0^1 x\, dx$$

$$= \frac{7}{4}$$

7. $z = f(x, y) = 9 - x^2$

$$f_x = -2x$$

$$f_y = 0$$

$$\sqrt{f_x^2 + f_y^2 + 1} = \sqrt{4x^2 + 1}$$

$$S = \int_0^2 \int_0^2 \sqrt{4x^2 + 1}\; dy\, dx$$

$$= \int_0^2 \sqrt{4x^2 + 1}(2\, dx)$$

$$= \left(x\sqrt{4x^2 + 1} + \tfrac{1}{2}\ln|2x + \sqrt{4x^2 + 1}|\right)\Big|_0^2$$

$$\boxed{\text{Formula 168}}$$

$$= 2\sqrt{17} + \tfrac{1}{2}\ln(4 + \sqrt{17})$$

$$\approx 9.2936$$

9. $z = f(x, y) = x^2$

$$f_x = 2x$$

$$f_y = 0$$

$$\sqrt{f_x^2 + f_y^2 + 1} = \sqrt{4x^2 + 1}$$

$$S = \int_0^4 \int_0^4 \sqrt{4x^2 + 1}\; dy\, dx$$

$$= 4\int_0^4 \sqrt{1 + 4x^2}\; dx$$

$$\boxed{\text{Formula 168}}$$

$$= 8\sqrt{65} + \ln(8 + \sqrt{65})$$

$$\approx 67.2745$$

SURVIVAL HINT: *In any expression where you find a factor $x^2 + y^2$ you should consider changing to polar coordinates.*

11. $z = f(x, y) = x^2 + y^2$

$$f_x = 2x$$

$$f_y = 2y$$

The projected region on the xy-plane is the disk $x^2 + y^2 \leq 1$. Converting to polar coordinates, we have

$$\sqrt{f_x^2 + f_y^2 + 1} = \sqrt{4x^2 + 4y^2 + 1}$$

$$= \sqrt{4r^2 + 1}$$

$$S = 4 \int_0^{\pi/2} \int_0^1 r\sqrt{4r^2 + 1} \ dr \ d\theta$$

$$= 4 \int_0^{\pi/2} \frac{5\sqrt{5} - 1}{12} \ d\theta$$

$$= \frac{\pi}{6}(5^{3/2} - 1)$$

$$\approx 5.3304$$

13. $z = f(x, y) = \sqrt{4 - x^2}$

$$f_x = \frac{-x}{\sqrt{4 - x^2}}$$

$$f_y = 0$$

$$\sqrt{f_x^2 + f_y^2 + 1} = \sqrt{\frac{x^2 + 4 - x^2}{4 - x^2}}$$

$$= \frac{2}{\sqrt{4 - x^2}}$$

The projected region is the square $0 \leq x \leq 2$, $0 \leq y \leq 2$. Thus,

$$S = \int_0^2 \int_0^2 \frac{2}{\sqrt{4 - x^2}} \ dy \ dx$$

$$= \int_0^2 \frac{4}{\sqrt{4 - x^2}} \ dx$$

$$= 4 \sin^{-1} \frac{x}{2}\Big|_0^2$$

$$= 2\pi$$

15. The intersection of the two surfaces (which is also the projection onto the z-plane) is found by eliminating the z variable: $x^2 + y^2 = 4$

$$z = f(x, y) = \sqrt{8 - x^2 - y^2};$$

$$f_x = \frac{-x}{\sqrt{8 - x^2 - y^2}}$$

$$f_y = \frac{-y}{\sqrt{8 - x^2 - y^2}}$$

$$\sqrt{f_x^2 + f_y^2 + 1} = \sqrt{\frac{x^2 + y^2 + 8 - x^2 - y^2}{8 - x^2 - y^2}}$$

$$= \frac{\sqrt{8}}{\sqrt{8 - r^2}}$$

Since half the surface lies above the xy-plane, we have

$$S = 2 \int_0^{2\pi} \int_0^2 \frac{2\sqrt{2}r}{\sqrt{8-r^2}}\, dr\, d\theta$$

$$= 2 \int_0^{2\pi} (8 - 4\sqrt{2})\, d\theta$$

$$= 16\pi(2 - \sqrt{2})$$

$$\approx 29.4448$$

17. $z = f(x, y) = x^2 + y$

$$f_x = 2x$$

$$f_y = 1$$

$$\sqrt{f_x^2 + f_y^2 + 1} = \sqrt{4x^2 + 2}$$

$$S = \int_0^2 \int_0^5 \sqrt{4x^2 + 2}\, dy\, dx$$

$$= 5\sqrt{2} \int_0^2 \sqrt{1 + 2x^2}\, dx$$

Formula 168

$$= 15\sqrt{2} + \tfrac{5}{2} \ln(2\sqrt{2} + 3)$$

$$\approx 25.6201$$

19. $z = f(x, y) = 4 - x^2 - y^2$

$$f_x = -2x$$

$$f_y = -2y$$

$$\sqrt{f_x^2 + f_y^2 + 1} = \sqrt{4x^2 + 4y^2 + 1}$$

$$= \sqrt{4r^2 + 1}$$

The projected region is $x^2 + y^2 \leq 4$ or $r \leq 2$.

$$S = \int_0^{2\pi} \int_0^2 \sqrt{4r^2 + 1}\; r\, dr\, d\theta$$

$$= \tfrac{1}{8} \int_0^{2\pi} \int_0^2 \sqrt{4r^2 + 1}\,(8r\, dr)\, d\theta$$

Let $u = 4r^2 + 1;\; du = 8r\, dr$

$$= \tfrac{1}{12} \int_0^{2\pi} (17\sqrt{17} - 1)\, d\theta$$

$$= \tfrac{\pi}{6} (17\sqrt{17} - 1)$$

$$\approx 36.1769$$

21. Let dA be a rectangular element of surface area. Project onto the xy-plane, which means trace out the shadow in the xy-plane when rays of light parallel to the z-axis hit the surface and then they shade (darken) a portion of the coordinate plane.

23. $z = f(x, y) = \frac{1}{C}(D - Ax - By)$

$$f_x = -\frac{A}{C}$$

$$f_y = -\frac{B}{C}$$

$$\sqrt{f_x^2 + f_y^2 + 1} = \sqrt{\frac{A^2}{C^2} + \frac{B^2}{C^2} + 1}$$

$$= \frac{\sqrt{A^2 + B^2 + C^2}}{C}$$

$$S = \int_0^{D/A} \int_0^{-\frac{A}{B}x + \frac{D}{B}} \frac{\sqrt{A^2 + B^2 + C^2}}{C} \, dy \, dx$$

$$= \frac{\sqrt{A^2 + B^2 + C^2}}{C} \int_0^{D/A} \left(-\frac{A}{B}x + \frac{D}{B} \right) dx$$

$$= \frac{D^2}{2ABC} \sqrt{A^2 + B^2 + C^2}$$

25. $z = f(x, y) = a - x - y$

$$f_x = -1$$

$$f_y = -1$$

$$\sqrt{f_x^2 + f_y^2 + 1} = \sqrt{3}$$

$$S = \int_0^{2\pi} \int_{a/2}^a \sqrt{3} \, r \, dr \, d\theta$$

$$= \int_0^{2\pi} \frac{3\sqrt{3}a^2}{8} \, d\theta$$

$$= \frac{3\sqrt{3}\pi a^2}{4}$$

27. $z = f(x, y) = \sqrt{x^2 + y^2};$

$$f_x = \frac{x}{\sqrt{x^2 + y^2}}$$

$$f_y = \frac{y}{\sqrt{x^2 + y^2}}$$

$$\sqrt{f_x^2 + f_y^2 + 1} = \sqrt{\frac{x^2 + y^2}{x^2 + y^2} + 1}$$

$$= \sqrt{2}$$

$$S = \sqrt{2} \int_0^{2\pi} \int_0^h r \, dr \, d\theta$$

$$= \frac{\sqrt{2}}{2} \int_0^{2\pi} h^2 \, d\theta$$

$$= \sqrt{2}\pi h^2$$

29. Solving the equations $x^2 + y^2 + z^2 = 9z$ for z, we obtain

$$z^2 - 9z + x^2 + y^2 = 0$$

$$z = \frac{9 + \sqrt{81 - 4(x^2 + y^2)}}{2}$$

(reject the negative root since it corresponds to the lower part of the sphere). Then,

$$z_x = \frac{-2x}{\sqrt{81 - 4(x^2 + y^2)}}$$

$$= \frac{-2r\cos\theta}{\sqrt{81 - 4r^2}}$$

$$z_y = \frac{-2y}{\sqrt{81 - 4(x^2 + y^2)}}$$

$$= \frac{-2r\sin\theta}{\sqrt{81 - 4r^2}}$$

$$\sqrt{z_x^2 + z_y^2 + 1} = \sqrt{\frac{4r^2\cos^2\theta + 4r^2\sin^2\theta}{81 - 4r^2} + 1}$$

$$= \frac{9}{\sqrt{81 - 4r^2}}$$

The projected region in the xy-plane is the intersection of $z^2 + r^2 = 9z$ with $r^2 = 4z$. Solving, we obtain $z = 0, 5$. (Reject $z = 0$, since $r = 0$ is a point.) The projected region is $r^2 = 4(5) = 20$, $r = \sqrt{20}$. Thus,

$$S = \int_0^{2\pi} \int_0^{\sqrt{20}} \frac{9}{\sqrt{81 - 4r^2}} \, r \, dr \, d\theta$$

$$= -\frac{9}{8} \int_0^{2\pi} \int_0^{\sqrt{20}} \frac{(-8r) \, dr}{\sqrt{81 - 4r^2}} \, d\theta$$

$$\boxed{\text{Let } u = 81 - 4r^2;\ du = -8r\, dr}$$

$$= -\frac{9}{4} \int_0^{2\pi} (1 - 9) \, d\theta$$

$$= 18(2\pi)$$

$$= 36\pi$$

31. Solve $z = e^{-x}\sin y$ for x:

$$z = e^{-x}\sin y$$

$$\frac{z}{\sin y} = e^{-x}$$

$$e^x = \frac{\sin y}{z}$$

$$x = \ln(\sin y) - \ln z$$

$$x_y = \frac{\cos y}{\sin y} = \cot y$$

$$x_z = \frac{-1}{z}$$

$$\sqrt{x_y^2 + x_z^2 + 1} = \sqrt{\cot^2 y + \frac{1}{z^2} + 1}$$

$$= \sqrt{\csc^2 y + z^{-2}}$$

$$S = \int_0^1 \int_0^y \sqrt{\csc^2 y + z^{-2}} \, dz \, dy$$

33. $z = f(x, y) = \cos(x^2 + y^2)$

$$f_x = -2x \sin(x^2 + y^2)$$

$$f_y = -2y \sin(x^2 + y^2)$$

$$\sqrt{f_x^2 + f_y^2 + 1} = \sqrt{(4x^2 + 4y^2)\sin^2(x^2 + y^2) + 1}$$

$$S = \int_0^{2\pi} \int_0^{\sqrt{\pi/2}} \sqrt{4r^2\sin^2 r^2 + 1} \, r \, dr \, d\theta$$

35. $z = f(x, y) = x^2 + 5xy + y^2$

The curves intersect when

$$xy = 5$$

$$x(6 - x) = 5$$

$$x^2 - 6x + 5 = 0$$

$$(x - 1)(x - 5) = 0$$

We see that the curves intersect when $x = 1$ and $x = 5$.

$$f_x = 2x + 5y$$

$$f_y = 2y + 5x$$

$$\sqrt{f_x^2 + f_y^2 + 1} = \sqrt{(2x + 5y)^2 + (2y + 5x)^2 + 1}$$

$$S = \int_1^5 \int_{5/x}^{6-x} \sqrt{(2x + 5y)^2 + (2y + 5x)^2 + 1} \; dy \; dx$$

37. $\mathbf{R}(u, v) = (2u \sin v)\mathbf{i} + (2u \cos v)\mathbf{j} + u^2\mathbf{k}$

$$\mathbf{R}_u = (2 \sin v)\mathbf{i} + (2 \cos v)\mathbf{j} + 2u\mathbf{k}$$

$$\mathbf{R}_v = (2u \cos v)\mathbf{i} + (-2u \sin v)\mathbf{j}$$

$$\mathbf{R}_u \times \mathbf{R}_v = \begin{vmatrix} \mathbf{i} & \mathbf{j} & \mathbf{k} \\ 2 \sin v & 2 \cos v & 2u \\ 2u \cos v & -2u \sin v & 0 \end{vmatrix}$$

$$= 4u^2 \sin v \, \mathbf{i} + 4u^2 \cos v \, \mathbf{j}$$

$$+ (-4u \sin^2 v - 4u \cos^2 v) \, \mathbf{k}$$

$$= 4u^2 \sin v \, \mathbf{i} + 4u^2 \cos v \, \mathbf{j} - 4u \, \mathbf{k}$$

$$\|\mathbf{R}_u \times \mathbf{R}_v\| = \sqrt{16u^4 \sin^2 v + 16u^4 \cos^2 v + 16u^2}$$

$$= \sqrt{16u^4 + 16u^2}$$

$$= 4|u|\sqrt{u^2 + 1}$$

39. $\mathbf{R}(u, v) = u\mathbf{i} + v^2\mathbf{j} + u^3\mathbf{k}$

$$\mathbf{R}_u = \mathbf{i} + 3u^2\mathbf{k}$$

$$\mathbf{R}_v = 2v\mathbf{j}$$

$$\mathbf{R}_u \times \mathbf{R}_v = \begin{vmatrix} \mathbf{i} & \mathbf{j} & \mathbf{k} \\ 1 & 0 & 3u^2 \\ 0 & 2v & 0 \end{vmatrix}$$

$$= -6u^2 v\mathbf{i} + 2v\mathbf{k}$$

$$\|\mathbf{R}_u \times \mathbf{R}_v\| = 2|v|\sqrt{9u^4 + 1}$$

41. $\mathbf{R}(u, v) = uv\mathbf{i} + (u - v)\mathbf{j} + (u + v)\mathbf{k}$

$$\mathbf{R}_u = v\mathbf{i} + \mathbf{j} + \mathbf{k}$$

$$\mathbf{R}_v = u\mathbf{i} - \mathbf{j} + \mathbf{k}$$

$$\mathbf{R}_u \times \mathbf{R}_v = \begin{vmatrix} \mathbf{i} & \mathbf{j} & \mathbf{k} \\ v & 1 & 1 \\ u & -1 & 1 \end{vmatrix}$$

$$= 2\mathbf{i} + (u - v)\mathbf{j} - (u + v)\mathbf{k}$$

$$\|\mathbf{R}_u \times \mathbf{R}_v\| = \sqrt{4 + (u - v)^2 + (-u - v)^2}$$

$$= \sqrt{4 + 2u^2 + 2v^2}$$

$$S = \int_0^{2\pi} \int_0^1 \sqrt{4 + 2r^2} \ r \ dr \ d\theta$$

$$= \frac{1}{4} \int_0^{2\pi} \int_0^1 \sqrt{4 + 2r^2} \ (4r \, dr) \ d\theta$$

$$= \frac{1}{6} \int_0^{2\pi} (6\sqrt{6} - 8) \ d\theta$$

$$= \frac{2\pi}{3}(3\sqrt{6} - 4)$$

$$\approx 7.01$$

43. **a.** $\mathbf{R}(u, v) = (u \sin v)\mathbf{i} + (u \cos v)\mathbf{j} + v\mathbf{k}$

$\mathbf{R}_u = (\sin v)\mathbf{i} + (\cos v)\mathbf{j},$

$\mathbf{R}_v = (u \cos v)\mathbf{i} + (-u \sin v)\mathbf{j} + \mathbf{k}$

$$\mathbf{R}_u \times \mathbf{R}_v = \begin{vmatrix} \mathbf{i} & \mathbf{j} & \mathbf{k} \\ \sin v & \cos v & 0 \\ u \cos v & -u \sin v & 1 \end{vmatrix}$$

$$= (\cos v)\mathbf{i} - (\sin v)\mathbf{j} - u\mathbf{k}$$

b. $\|\mathbf{R}_u \times \mathbf{R}_v\| = \sqrt{1 + u^2}$

$$S = \int_0^b \int_0^a \sqrt{1 + u^2} \ du \ dv$$

$$= \int_0^b \left[\frac{\ln\left| u + \sqrt{1 + u^2} \right|}{2} + \frac{u\sqrt{1 + u^2}}{2} \right]\Bigg|_0^a \ dv$$

$$= \frac{b}{2}[\ln(a + \sqrt{1 + a^2}) + a\sqrt{1 + a^2}]$$

45. $z = f(x, y) = \sqrt{a^2 - x^2};$

$$f_x = \frac{-x}{\sqrt{a^2 - x^2}}$$

$$f_y = 0$$

$$\sqrt{f_x^2 + f_y^2 + 1} = \sqrt{\frac{x^2 + a^2 - x^2}{a^2 - x^2}}$$

$$= \frac{a}{\sqrt{a^2 - x^2}}$$

$$S = 4 \int_0^a \int_0^h \frac{a \ dy \ dx}{\sqrt{a^2 - x^2}}$$

$$= 4 \int_0^a \frac{ha}{\sqrt{a^2 - x^2}} \ dx$$

$$= 4ah \sin^{-1} \frac{x}{a} \Big|_0^a$$

$$= 4ah\frac{\pi}{2}$$

$$= 2\pi ah$$

47. Since $F(x, y, z) = 0$, we have

$$\frac{\partial F}{\partial x} + \frac{\partial F}{\partial z}\frac{\partial z}{\partial x} = 0$$

and

$$\frac{\partial F}{\partial y} + \frac{\partial F}{\partial z}\frac{\partial z}{\partial y} = 0$$

so that

$$\frac{\partial z}{\partial x} = \frac{-F_x}{F_z}$$

$$\frac{\partial z}{\partial y} = \frac{-F_y}{F_z}$$

and

$$z_x{}^2 + z_y{}^2 + 1 = \left(\frac{-F_x}{F_z}\right)^2 + \left(\frac{-F_y}{F_z}\right)^2 + 1$$

$$= \frac{F_x{}^2 + F_y{}^2 + F_z{}^2}{F_z{}^2}$$

Thus,

$$A = \int\int_D \frac{\sqrt{F_x{}^2 + F_y{}^2 + F_z{}^2}}{|F_z|} \, dA_{xy}$$

To find the surface area of a sphere, let

$$f(x, y, z) = x^2 + y^2 + z^2 - R^2$$

$$f_x = 2x$$

$$f_y = 2y$$

$$f_z = 2z$$

$$\frac{\sqrt{f_x{}^2 + f_y{}^2 + f_z{}^2}}{|f_z|} = \frac{2R}{2|z|}$$

$$= \frac{R}{\sqrt{R^2 - x^2 - y^2}}$$

$$S = \int_0^R \int_0^{\sqrt{R^2 - x^2}} \frac{R}{\sqrt{R^2 - x^2 - y^2}} \, dy \, dx$$

$$= 8R \int_0^{\pi/2} \int_0^R \frac{r}{\sqrt{R^2 - r^2}} \, dr \, d\theta$$

$$= 8R \int_0^{\pi/2} R \, d\theta$$

$$= 4\pi R^2$$

12.5 Triple Integrals, Pages 820–822

SURVIVAL HINT: *It takes considerable skill and practice to determine which of the six possible permutations of dV will be "easiest" for a given problem. The most common situation is two functions of the form $z = f(x, y)$, in which case you would use $dz \, dA$, where the region A is determined by the intersection of the two functions. For other iterations, be certain that the limits for the first integration are either constants or functions of the remaining two variables, the limits of the second integration are constants or functions of the remaining variable, and the limits for the final integration are constants.*

1. Fubini's theorem says that a triple integral in x, y, and z can be evaluated as an iterated integral in any of the six possible orders of integration: xyz, xzy, yxz, yzx, zxy, zyx.

3. $\displaystyle\int_1^4 \int_{-2}^3 \int_2^5 dx \, dy \, dz = (3)(5)(3) = 45$

5. $\displaystyle\int_1^2 \int_0^1 \int_{-1}^2 8x^2yz^3 \, dx \, dy \, dz = 45$

7. $\displaystyle\int_{0}^{2}\int_{0}^{x}\int_{0}^{x+y} xyz \; dz \; dy \; dx$

$\displaystyle = \frac{1}{2}\int_{0}^{2}\int_{0}^{x} xy(x+y)^2 \; dy \; dz$

$\displaystyle = \frac{1}{2}\int_{0}^{2}\left(\frac{x^5}{2} + 2\frac{x^5}{3} + \frac{x^5}{4}\right) dx$

$\displaystyle = \frac{1}{24}\int_{0}^{2} (6+8+3)x^5 \; dx$

$\displaystyle = \frac{68}{9}$

9. $\displaystyle\int_{-1}^{2}\int_{0}^{\pi}\int_{1}^{4} yz \cos xy \; dz \; dx \; dy$

$\displaystyle = \frac{15}{2}\int_{-1}^{2}\int_{0}^{\pi} y \cos xy \; dx \; dy$

$\displaystyle = \frac{15}{2}\int_{-1}^{2} \sin \pi y \; dy$

$\displaystyle = -\frac{15}{\pi}$

$\displaystyle \approx -4.7746$

11. $\displaystyle\int_{0}^{1}\int_{0}^{y}\int_{0}^{\ln y} e^{z+2x} \; dz \; dx \; dy$

$\displaystyle = \int_{0}^{1}\int_{0}^{y} e^{2x}(y-1) \; dx \; dy$

$\displaystyle = \frac{1}{2}\int_{0}^{1} (e^{2y}-1)(y-1) \; dy$

$\displaystyle = \frac{1}{8}(5 - e^2)$

$\displaystyle \approx -0.2986$

13. $\displaystyle\int_{1}^{4}\int_{-1}^{2z}\int_{0}^{\sqrt{3}\,x} \frac{x-y}{x^2+y^2} \; dy \; dx \; dz$

$\displaystyle = \int_{1}^{4}\int_{-1}^{2z} \frac{1}{2}\left[2\tan^{-1}\frac{y}{x} - \ln(x^2+y^2)\right]\Big|_{0}^{\sqrt{3}\,x} dx \; dz$

$\displaystyle = \int_{0}^{4}\int_{-1}^{2z} \left[\frac{\pi}{3} - \ln 2\right] dx \; dz$

$\displaystyle = \left(\frac{\pi}{3} - \ln 2\right)\int_{1}^{4} (2z+1) \; dz$

$\displaystyle = 18\left(\frac{\pi}{3} - \ln 2\right)$

$\displaystyle \approx 6.3729$

15. $\displaystyle\iiint_D (x^2 y + y^2 z)\, dV$

$\displaystyle = \int_1^3 \int_{-1}^1 \int_2^4 (x^2 y + y^2 z)\, dz\, dy\, dx$

$\displaystyle = \int_1^3 \int_{-1}^1 (2x^2 y + 6y^2)\, dy\, dx$

$\displaystyle = \int_1^3 4\, dx$

$= 8$

17. $\displaystyle\iiint_D xyz\, dV$

$\displaystyle = \int_0^1 \int_0^{1-x} \int_0^{1-x-y} xyz\, dz\, dy\, dx$

$\displaystyle = \frac{1}{2} \int_0^1 \int_0^{1-x} xy(1 - x - y)^2\, dy\, dx$

$\displaystyle = \frac{1}{2} \int_0^1 \int_0^{1-x} \left[x(x-1)^2 y - 2x(x-1)y^2 + xy^3 \right] dy\, dx$

$\displaystyle = \frac{1}{2} \int_0^1 x(x - 1)^4 \left(\tfrac{1}{2} - \tfrac{2}{3} + \tfrac{1}{4} \right) dx$

$\displaystyle = \frac{1}{24} \int_0^1 x(x - 1)^4\, dx$

$\displaystyle = \frac{1}{24} \int_0^1 (x^5 - 4x^4 + 6x^3 - 4x^2 + x)\, dx$

$\displaystyle = \frac{1}{720}$

19. $\displaystyle\iiint_D xyz\, dV$

$\displaystyle = \int_0^1 \int_0^{\sqrt{1-y^2}} \int_{-\sqrt{1-y^2-z^2}}^{\sqrt{1-y^2-z^2}} xyz\, dx\, dz\, dy$

$\displaystyle = \int_0^1 \int_0^{\sqrt{1-y^2}} \frac{yzx^2}{2} \Bigg|_{-\sqrt{1-y^2-z^2}}^{\sqrt{1-x^2-z^2}}$

$= 0$

21. $\displaystyle\iiint_D e^z\, dV$

$\displaystyle = \int_0^1 \int_0^x \int_0^{x+y} e^z\, dz\, dy\, dx$

$\displaystyle = \int_0^1 \int_0^x (e^{x+y} - 1)\, dy\, dx$

$\displaystyle = \int_0^1 (e^{2x} - x - e^x)\, dx$

$\displaystyle = \frac{e^2}{2} - e$

≈ 0.9762

23. $\displaystyle\int_0^1\int_0^{-x+1}\int_0^{1-x-y} dz\,dy\,dx$

$$= \int_0^1\int_0^{-x+1}(1-x-y)\,dy\,dx$$

$$= \int_0^1\left(y - xy - \frac{y^2}{2}\right)\bigg|_0^{1-x}\,dx$$

$$= \int_0^1\left[1 - x - x(1-x) - \frac{(1-x)^2}{2}\right]dx$$

$$= \int_0^1\left[(1-x)^2 - \frac{(1-x)^2}{2}\right]dx$$

$$= \int_0^1\frac{(1-x)^2}{2}\,dx$$

$$= \frac{1}{6}$$

This result is easily verified with the formula for the volume of a pyramid.

25. $\displaystyle 8\int_1^2\int_2^A\int_3^B dz\,dy\,dx$ where

$$A = 2+\sqrt{1-(x-1)^2}\text{ and}$$

$$B = 3 + \sqrt{1-(x-1)^2 -(y-2)^2}$$

$$= 8\int_1^2\int_2^A\sqrt{1-(x-1)^2 -(y-2)^2}\,dy\,dx$$

$$= 8\int_1^2\frac{\pi}{4}(2x - x^2)\,dx$$

$$= \frac{4\pi}{3}$$

This result is easily verified with the formula for the volume of a sphere.

27. The intersection of the parabolic cylinder and the elliptic paraboloid gives the region of integration in the xy-plane:

$$4 - y^2 = x^2 + 3y^2$$

$$x^2 + 4y^2 = 4$$

$$V = 4\int_0^1\int_0^{2\sqrt{1-y^2}}\int_{x^2+3y^2}^{4-y^2} dz\,dx\,dy$$

$$= 4\int_0^1\int_0^{2\sqrt{1-y^2}}(4 - x^2 - 4y^2)\,dx\,dy$$

$$= 4\int_0^1\frac{16}{3}(1 - y^2)^{3/2}\,dy$$

$$= 4\pi$$

$$\approx 12.5664$$

29. The intersection of the two surfaces gives the region of integration in the xy-plane:

$$6 - x^2 - y^2 = 2x^2 + y^2$$

$$3x^2 + 2y^2 = 6$$

$$V = 4 \int_0^{\sqrt{2}} \int_0^{\sqrt{(6-3x^2)/2}} \int_{2x^2+y^2}^{6-x^2-y^2} dz\ dy\ dx$$

$$= 4 \int_0^{\sqrt{2}} \int_0^{\sqrt{(6-3x^2)/2}} [6 - x^2 - y^2 - (2x^2 + y^2)]\ dy\ dx$$

$$= 4 \int_0^{\sqrt{2}} \left[6y - 3x^2 y - \frac{2y^3}{3} \right] \Bigg|_0^{\sqrt{(6-3x^2)/2}} dx$$

$$= \int_0^{\sqrt{2}} [4\sqrt{6}(2 - x^2)^{3/2}\ dx$$

$$= 3\sqrt{6}\,\pi$$

$$\approx 23.0859$$

31. $\quad V = \int_{-1}^{1} \int_{-\sqrt{1-x^2}}^{\sqrt{1-x^2}} \int_{-\sqrt{1-z^2}}^{\sqrt{1-z^2}} dy\ dz\ dx$

$$= 8 \int_0^1 \int_0^{\sqrt{1-x^2}} \int_0^{\sqrt{1-z^2}} dy\ dz\ dx$$

$$= 8 \int_0^1 \int_0^{\sqrt{1-x^2}} \sqrt{1-z^2}\ dz\ dx$$

$$= 8 \int_0^1 \left[\frac{\sin^{-1} z}{2} + \frac{z}{2}\sqrt{1 - z^2} \right] \Bigg|_0^{\sqrt{1-x^2}} dx$$

$$= 4 \int_0^1 \left[\sin^{-1}\sqrt{1-x^2} + \sqrt{1-x^2}\sqrt{1-(1-x^2)} \right] dx$$

$$= 4 \int_0^1 \left[\sin^{-1}\sqrt{1-x^2} + x\sqrt{1-x^2} \right] dx$$

$$= 4(1) + 4\left(\tfrac{1}{3}\right)$$

$$= \frac{16}{3}$$

33. $\quad \displaystyle\int_0^1 \int_y^1 \int_0^y f(x,\ y,\ z)\ dz\ dx\ dy$

35. $\quad \displaystyle\int_0^2 \int_0^{\sqrt{4-z^2}} \int_0^{\sqrt{4-y^2-z^2}} f(x,\ y,\ z)\ dx\ dy\ dz$

37. $\quad \displaystyle\int_0^1 \int_{\sqrt[3]{x}}^1 \int_0^{1-y} f(x,\ y,\ z)\ dz\ dy\ dx$

39. Let $A = 3\sqrt{1 - \dfrac{x^2}{4}}$ and

$$B = 4\sqrt{1 - \frac{x^2}{4} - \frac{y^2}{9}}$$

$$V = 8 \int_0^2 \int_0^A \int_0^B dz\ dy\ dx$$

$$= 32 \int_0^2 \int_0^A \sqrt{1 - \frac{x^2}{4} - \frac{y^2}{9}} \, dy \, dx$$

$$x^2 + y^2 = 4$$

Let D be the part of the disk $x^2 + y^2 \leq 4$ in the first quadrant.

$$= 96 \int_0^2 \left\{ \frac{1 - \frac{x^2}{4}}{2} \sin^{-1} \frac{y}{3\sqrt{1 - \frac{x^2}{4}}} \right.$$

$$V = 4 \int\int_D \int_{3x^2 + 2y^2}^{16 - x^2 - 2y^2} dz \, dy \, dx$$

$$\left. + \frac{y}{6} \sqrt{1 - \frac{x^2}{4} - \frac{y^2}{9}} \right\} \Bigg|_0^{3\sqrt{1 - x^2/4}} dx$$

$$= 4 \int\int_D - 4(x^2 + y^2 - 4) \, dy \, dx$$

$$= 96 \int_0^2 \left\{ \frac{4 - x^2}{8} \sin^{-1}(1) \right.$$

$$\left. + \frac{1}{2} \sqrt{1 - \frac{x^2}{4}} \sqrt{1 - \frac{x^2}{4} - \frac{9\left(1 - \frac{x^2}{4}\right)}{9}} \right\} dx$$

$$= 4 \int_0^{\pi/2} \int_0^2 4r(4 - r^2) \, dr \, d\theta$$

$$= 96 \int_0^2 \left[\left(\frac{4 - x^2}{8} \right) \frac{\pi}{2} + 0 \right] dx$$

$$= 4 \int_0^{\pi/2} 16 \, d\theta$$

$$= 24\pi \int_0^2 \left(1 - \frac{x^2}{4} \right) dx$$

$$= 32\pi$$

$$= 24\pi \left(\frac{4}{3} \right)$$

$$= 32\pi$$

This result can be verified by using the formula for the volume of an ellipsoid

$$V = \frac{4}{3}\pi abc = \frac{4}{3}\pi(2)(3)(4) = 32\pi$$

43. $\displaystyle 4 \int_0^3 \int_0^{(1/3)\sqrt{9 - x^2}} \int_0^{x^2/9 + y^2} dz \, dy \, dx$

$$= 4 \int_0^3 \int_0^{(1/3)\sqrt{9 - x^2}} \left(\frac{x^2}{9} + y^2 \right) dy \, dx$$

41. The projected region of integration is found by intersecting the two surfaces

$$16 - x^2 - 2y^2 = 3x^2 + 2y^2$$

$$= 4 \int_0^3 \left[\frac{x^2}{27} \sqrt{9 - x^2} + \frac{1}{81} (9 - x^2)^{3/2} \right] dx$$

$$\boxed{\text{Formulas 233 and 245}}$$

$$= \frac{4}{27} \int_0^3 x^2 \sqrt{9 - x^2} \, dx + \frac{4}{81} \int_0^3 (9 - x^2)^{3/2} \, dx$$

$$= \frac{3\pi}{4} + \frac{3\pi}{4}$$

$$= \frac{3\pi}{2}$$

45. $\qquad V = 8 \int_0^{\pi/2} \int_0^R \int_0^{\sqrt{R^2 - r^2}} dz \; r \; dr \; d\theta$

$$= 8 \int_0^{\pi/2} \int_0^R r \sqrt{R^2 - r^2} \, dr \, d\theta$$

$$= \frac{8}{3} \int_0^{\pi/2} R^3 \, d\theta = \frac{4}{3} \pi R^3$$

47. $\quad z = c \left(1 - \frac{x^2}{a^2} - \frac{y^2}{b^2} \right)^{1/2}$

$$y = b \left(1 - \frac{x^2}{a^2} \right)^{1/2}$$

$$V = 8 \int_0^a \int_0^{b\sqrt{1 - x^2/a^2}} \int_0^{c\sqrt{1 - x^2/a^2 - y^2/b^2}} dz \; dy \; dx$$

$$= 8c \int_0^a \int_0^{b\sqrt{1 - x^2/a^2}} \left(1 - \frac{x^2}{a^2} - \frac{y^2}{b^2} \right)^{1/2} dy \; dx$$

$$= 4cb \int_0^a \left\{ \left(1 - \frac{x^2}{a^2} \right) \sin^{-1} \frac{y}{b\sqrt{1 - \frac{x^2}{a^2}}} \right.$$

$$\left. + \frac{y}{b} \sqrt{1 - \frac{x^2}{a^2} - \frac{y^2}{b^2}} \right\} \Bigg|_0^{b/\sqrt{1 - x^2/a^2}}$$

$$= 4bc \int_0^a \left[\left(1 - \frac{x^2}{a^2} \right) \frac{\pi}{2} \right] dx$$

$$= \frac{4\pi abc}{3}$$

49. It is generally true;

$$\int_a^b \int_c^d \int_r^s f(x) g(y) h(z) \; dz \; dy \; dx = \int_a^b f(x) \int_c^d g(y)$$

$$\int_r^s h(z) \; dz \; dy \; dx = \left[\int_a^b f(x) \; dx \right] \left[\int_c^d g(y) \; dy \right] \left[\int_r^s h(z) \; dz \right]$$

51. Using the results of Problem 50,

$$\int \int_D \int \sin(\pi - z)^3 \; dz \; dy \; dx$$

$$= \int_0^\pi \int_0^y \int_0^x \sin(\pi - z)^3 \; dz \; dx \; dy$$

$$= \frac{1}{2} \int_0^\pi (\pi - t)^2 \sin(\pi - t)^3 \; dt$$

$$= \frac{\cos(\pi - t)^3}{6} \Bigg|_0^\pi$$

$$= \frac{1 - \cos \pi^3}{6}$$

53. $\displaystyle\int_{1}^{2} \int_{-1}^{1} \int_{0}^{2} \int_{0}^{1} xyz^2 w^2 \; dx \; dy \; dz \; dw$

$$= \int_{1}^{2} \int_{-1}^{1} \int_{0}^{2} \frac{w^2 y z^2}{2} \; dy \; dz \; dw$$

$$= \int_{1}^{2} \int_{-1}^{1} w^2 z^2 \; dz \; dw$$

$$= \int_{1}^{2} \frac{2w^2}{3} \; dw$$

$$= \frac{14}{9}$$

12.6 Mass, Moments, and Probability Density Functions, Pages 829–833

SURVIVAL HINT: *When finding the center of mass for a lamina, you probably thought of the moment about a particular axis as a rotational force. The moment about the y-axis is the same as if all of the mass is at the centroid: $M_y = m\overline{x}$. The algebraic extension of this concept to volumes is exactly the same. The moment about the yz-plane is the same as if all of the mass is at the centroid: $M_{yz} = m\overline{x}$. However, the geometric visualization of "rotation" is impossible. A lamina can be rotated about a line by moving in $\mathbb{R}^3$. To "rotate" our volume about a plane we have to move into $\mathbb{R}^4$. Fortunately, the algebra is not restricted by our $\mathbb{R}^3$ world.*

1. Find the total mass. Compute the first moment with respect to the x-axis and the first moment with respect to the y-axis. Divide the first moments by the mass. The resulting coordinates are those of the center of mass.

3. A probability density function measures the probability that a continuous random variable X lies between two numbers on a number line ($\mathbb{R}$). A joint probability density function is the generalization to $\mathbb{R}^2$; that is, the probability that an ordered pair of continuous random variables (X, Y) lies within a particular region on a plane. Specifically, a joint probability density function for two random variables X and Y is a function $f(x, y)$ of two variables such that the probability that the point (X, Y) is in a region D satisfies

$$P[(x, y) \text{ is in } D] = \int_{D} \int f(x, y) \; dA$$

5. $\displaystyle m = \int_{0}^{3} \int_{0}^{4} 5 \; dy \; dx = 60$

$$M_x = \int\limits_0^3 \int\limits_0^4 5\ y\ dy\ dx$$

$$= \frac{5}{2} \int\limits_0^3 16\ dx$$

$$= 120$$

$$M_y = \int\limits_0^3 \int\limits_0^4 5\ x\ dy\ dx$$

$$= 20 \int\limits_0^3 x\ dx$$

$$= 90$$

$$(\overline{x}, \overline{y}) = \left(\frac{90}{60}, \frac{120}{60}\right) = \left(\frac{3}{2}, 2\right)$$

7. $$m = 2 \int\limits_0^2 \int\limits_{x^2}^{2x} dy\ dx$$

$$= 2 \int\limits_0^2 (2x - x^2)\ dx$$

$$= \frac{8}{3}$$

$$M_x = 2 \int\limits_0^2 \int\limits_{x^2}^{2x} y\ dy\ dx$$

$$= \int\limits_0^2 (4x^2 - x^4)\ dx$$

$$= \frac{64}{15}$$

$$M_y = 2 \int\limits_0^2 \int\limits_{x^2}^{2x} x\ dy\ dx$$

$$= 2 \int\limits_0^2 x(2x - x^2)\ dx$$

$$= \frac{8}{3}$$

$$(\overline{x}, \overline{y}) = \left(\frac{8(3)}{3(8)}, \frac{64(3)}{15(8)}\right) = \left(1, \frac{8}{5}\right)$$

9. $y = 2 - 3x^2$ intersects $3x + 2y = 1$ when

$$6x^2 - 3x - 3 = 0$$

$$3(x - 1)(2x + 1) = 0$$

$$x = 1,\ -\frac{1}{2}$$

$$m = \int\limits_{-1/2}^1 \int\limits_{(1-3x)/2}^{2-3x^2} dy\ dx$$

$$= \int\limits_{-1/2}^1 \left(-3x^2 + \frac{3}{2}x + \frac{3}{2}\right)\ dx$$

$$= \frac{27}{16}$$

$$M_x = \int\limits_{-1/2}^1 \int\limits_{(1-3x)/2}^{2-3x^2} y\ dy\ dx$$

$$= \frac{1}{2} \int_{-1/2}^{1} [(2-3x^2)^2 - \tfrac{1}{4}(1-3x)^2] \, dx$$

$$= \frac{27}{20}$$

$$M_y = \int_{-1/2}^{1} \int_{(1-3x)/2}^{2-3x^2} x \, dy \, dx$$

$$= \int_{-1/2}^{1} x[2 - 3x^2 - \tfrac{1}{2}(1-3x)] \, dx$$

$$= \frac{27}{64}$$

$$(\bar{x}, \bar{y}) = \left(\frac{27(16)}{64(27)}, \frac{27(16)}{20(27)}\right) = \left(\frac{1}{4}, \frac{4}{5}\right)$$

11.
$$m = 4\int_{0}^{4}\int_{0}^{-x+4}\int_{0}^{4-x-y} dz \, dy \, dx$$

$$= \frac{128}{3}$$

$$M_y = \int_{0}^{4}\int_{0}^{-x+4}\int_{0}^{4-x-y} 4x \, dz \, dy \, dx$$

$$= \frac{128}{3}$$

$$\bar{x} = \frac{M_y}{m} = 1$$

By symmetry $(\bar{x}, \bar{y}, \bar{z}) = (1, 1, 1)$

13.
$$m = 2\int_{0}^{3}\int_{0}^{\sqrt{9-x^2}} (x^2 + y^2) \, dy \, dx$$

$$= 2\int_{0}^{\pi/2}\int_{0}^{3} r^2 \, r \, dr \, d\theta$$

$$= 2\int_{0}^{\pi/2} \frac{81}{4} \, dr$$

$$= \frac{81\pi}{4}$$

$\bar{x} = 0$ (by symmetry)

$$M_x = 2\int_{0}^{3}\int_{0}^{\sqrt{9-x^2}} y(x^2 + y^2) \, dy \, dx$$

$$= 2\int_{0}^{3} [\tfrac{1}{2}x^2(9 - x^2) + \tfrac{1}{4}(81 - 18x^2 + x^4)] \, dx$$

$$= \frac{486}{5}$$

$$\bar{y} = \frac{486(4)}{5(81\pi)} = \frac{24}{5\pi}$$

$$(\bar{x}, \bar{y}) = \left(0, \frac{24}{5\pi}\right)$$

15.
$$m = \int_{0}^{5}\int_{6y/5}^{(60-6y)/5} 7x \, dx \, dy$$

$$= \int_{0}^{5} \frac{-504}{5}(y - 5) \, dy$$

$$= 1{,}260$$

$$= \frac{(\ln 2)^2}{2}$$

$$M_x = \int_0^6 \int_0^{5x/6} 7xy \; dy \; dx + \int_6^{12} \int_0^{-5x/6+10} 7xy \; dy \; dx$$

$$M_x = \int_1^2 \int_0^{\ln x} \frac{y}{x} \; dy \; dx$$

$$= \int_0^6 \frac{7x}{2}\left(\frac{5x}{6}\right)^2 dx + \int_6^{12} \frac{7x}{2}\left(-\frac{5x}{6} + 10\right)^2 \; dx$$

$$= \frac{1}{2} \int_1^2 \frac{(\ln x)^2}{x} \; dx$$

$$= \int_0^6 \frac{175}{72} x^3 \, dx + \int_6^{12} \left(\frac{175}{72} x^3 - \frac{175}{3} x^2 + 350x\right) dx$$

$$= \frac{(\ln 2)^3}{6}$$

$$= \frac{1{,}575}{2} + \frac{2{,}625}{2}$$

$$= 2{,}100$$

$$M_y = \int_1^2 \int_0^{\ln x} dy \; dx$$

$$M_y = 7 \int_0^5 \int_{6y/5}^{-(6/5)y+12} x^2 \; dx \; dy$$

$$= \int_1^2 \ln x \; dx$$

$$= \frac{7(6^3)}{3} \int_0^5 \left(-\frac{2y^3}{5^3} + \frac{6y^2}{5^2} - \frac{12y}{5} + 8\right) dy$$

$$= 2 \ln 2 - 1$$

$$= 8{,}820$$

$$(\overline{x}, \overline{y}) = \left(\frac{(2 \ln 2 - 1)(2)}{(\ln 2)^2}, \frac{(\ln 2)^3(2)}{6(\ln 2)^2}\right)$$

$$(\overline{x}, \overline{y}) = \left(\frac{8{,}820}{1{,}260}, \frac{2{,}100}{1{,}260}\right) = \left(7, \frac{5}{3}\right)$$

$$\approx (1.608, 0.231)$$

19. **a.** By symmetry $\overline{x} = 0$. $\delta = kr$.

17. $$m = \int_1^2 \int_0^{\ln x} x^{-1} \; dy \; dx$$

$$m = k \int_0^\pi \int_0^a (r) \; r \; dr \; d\theta$$

$$= \int_1^2 \frac{\ln x}{x} \; dx$$

$$= \frac{ka^3}{3} \int_0^\pi d\theta \; = \; \frac{k\pi a^3}{3}$$

$$M_x = k \int_0^\pi \int_0^a (r)(r \sin \theta) \; r \; dr \; d\theta$$

$$= \frac{ka^4}{4} \int_0^\pi \sin \theta \; d\theta$$

$$= \frac{ka^4}{4}(1 + 1)$$

$$= \frac{ka^4}{2}$$

$$\bar{y} = \frac{\frac{ka^4}{2}}{\frac{k\pi a^3}{3}}$$

$$= \frac{3a}{2\pi}$$

The centroid is at $\left(0, \frac{3a}{2\pi} \right)$

b. $\delta = k\theta$ (use $\delta = \theta$ as the k will cancel

in the quotient).

$$m = \int_0^\pi \int_0^a (\theta) \; r \; dr \; d\theta$$

$$= \frac{a^2}{2} \int_0^\pi \theta \; d\theta$$

$$= \frac{\pi^2 a^2}{4}$$

$$M_x = \int_0^\pi \int_0^a (\theta)(r \sin \theta) \; r \; dr \; d\theta$$

$$= \frac{a^3}{3} \int_0^\pi \theta \sin \theta \; d\theta$$

$$= \frac{\pi a^3}{3}$$

$$M_y = \int_0^\pi \int_0^a (\theta)(r \cos \theta) \; r \; dr \; d\theta$$

$$= \frac{a^3}{3} \int_0^\pi \theta \cos \theta \; d\theta$$

$$= -\frac{2a^3}{3}$$

$$(\bar{x}, \bar{y}) = \left(\frac{-\frac{2a^3}{3}}{\frac{\pi^2 a^2}{4}}, \frac{\frac{\pi a^3}{3}}{\frac{\pi^2 a^2}{4}} \right)$$

$$= \left(-\frac{8a}{3\pi^2}, \frac{4a}{3\pi} \right)$$

21.

$$m = \int_1^{e^2} \int_0^{\ln x} dy \; dx$$

$$= \int_1^{e^2} \ln x \; dx$$

$$= \left[x \ln x - x \right] \Big|_1^{e^2}$$

$$= e^2 + 1$$

$$M_x = \int_1^{e^2} \int_0^{\ln x} y \; dy \; dx$$

$$= \frac{1}{2} \int_1^{e^2} (\ln x)^2 \, dx$$

$$= \frac{1}{2} \left[x \ln^2 x - 2 \int_1^{e^2} \ln x \, dx \right]$$

$$= e^2 - 1$$

$$M_y = \int_1^{e^2} \int_0^{\ln x} x \, dy \, dx$$

$$= \int_1^{e^2} x \ln x \, dx$$

$$= \left[\frac{1}{2} x^2 \ln x - \frac{1}{4} x^2 \right] \Big|_1^{e^2}$$

$$= e^4 - \frac{e^4}{4} + \frac{1}{4}$$

$$= \frac{1}{4} (3e^4 + 1)$$

$$(\overline{x}, \overline{y}) = \left(\frac{3e^4 + 1}{4(e^2 + 1)}, \frac{e^2 - 1}{e^2 + 1} \right)$$

$$\approx (4.9110, 0.7616)$$

23. $\quad I_z = 4 \int_0^1 \int_0^1 (x^2 + y^2) \, x^2 y^2 \, dy \, dx$

$$= 4 \int_0^1 \int_0^1 (x^4 y^2 + x^2 y^4) \, dy \, dx$$

$$= \frac{4}{15} \int_0^1 (5x^4 + 3x^2) \, dx$$

$$= \frac{8}{15}$$

$$\approx 0.5333$$

25. $\quad m = \int_0^{\pi/2} \int_0^{\sqrt{2\sin 2\theta}} r \, dr \, d\theta$

$$= \int_0^{\pi/2} \sin 2\theta \, d\theta$$

$$= 1$$

$$M_y = \int_0^{\pi/2} \int_0^{\sqrt{2\sin 2\theta}} (r \cos \theta) \, r \, dr \, d\theta$$

$$= \int_0^{\pi/2} \frac{2\sqrt{2}}{3} \cos \theta (\sin 2\theta)^{3/2} \, d\theta$$

$$\approx 0.5554$$

(numerical computer approximation)

$\overline{x} \approx 0.5554, \overline{y} \approx 0.5554$ (by symmetry);

$(\overline{x}, \overline{y}) \approx (0.56, 0.56)$

27. $\quad m = \int_0^a \int_0^{bx/a} kx \, dy \, dx$

$$= \frac{kb}{a} \int_0^a x^2 \, dx$$

$$= \frac{a^2 bk}{3}$$

$$M_x = \int_0^a \int_0^{bx/a} kxy \, dy \, dx$$

$$= \frac{kb^2}{2a^2} \int_0^a x^3 \, dx$$

$$= \frac{ka^2 b^2}{8}$$

$$M_y = \int_0^a \int_0^{bx/a} kx^2 \, dy \, dx$$

$$= \frac{kb}{a} \int_0^a x^3 \, dx$$

$$= \frac{kba^3}{4}$$

$$(\overline{x}, \overline{y}) = \left(\frac{3ka^3 b}{4a^2 bk}, \frac{3ka^2 b^2}{8a^2 bk} \right)$$

$$= \left(\frac{3a}{4}, \frac{3b}{8} \right)$$

29. The distance from the point (x, y) to the line $x = a/2$ through the center of the lamina is $s = |x - a/2|$. Thus, the moment of inertia is

$$I_L = \int_{-\pi/2}^{\pi/2} \int_0^{a \cos \theta} (r \cos \theta - \tfrac{a}{2})^2 \, r \, dr \, d\theta$$

$$= \frac{a^4 \pi}{64}$$

31. If $\delta = 1$ then $m = A = \pi ab$. Using symmetry:

$$I_x = 4 \int_0^a \int_0^{\frac{b}{a}\sqrt{a^2 - x^2}} y^2 \, dy \, dx$$

$$= \frac{4}{3} \int_0^a \left(\frac{b}{a}\sqrt{a^2 - x^2} \right)^3 \, dx$$

$$= \frac{4b^3}{3a^3} \int_0^a (a^2 - x^2)^{3/2} \, dx$$

$$\boxed{\text{Formula 245}}$$

$$= \frac{b^3}{3a^3} \left(\frac{3a^4 \pi}{4} \right)$$

$$= \frac{ab^3 \pi}{4}$$

Substituting $m = \pi ab$:

$$I_x = \frac{mb^2}{4}$$

33. $$m = \frac{1}{8}\left(\frac{4}{3}\pi a^3 \right) = \frac{\pi a^3}{6}$$

$$M_{yz} = \int_0^{\pi/2} \int_0^a \int_0^{\sqrt{a^2 - r^2}} (r \cos \theta) \, r \, dz \, dr \, d\theta$$

$$= \int_0^{\pi/2} \int_0^a r^2 \cos\theta \sqrt{a^2 - r^2} \, dr \, d\theta$$

The probability is roughly 8%.

$$= \int_0^{\pi/2} \frac{\pi}{16} a^4 \cos\theta \, d\theta$$

$$= \frac{\pi a^4}{16}$$

$$\bar{x} = \frac{6\pi a^4}{16\pi a^3} = \frac{3a}{8}$$

By symmetry,

$$\bar{x} = \bar{y} = \bar{z} = \frac{3a}{8}$$

The centroid is $\left(\frac{3a}{8}, \frac{3a}{8}, \frac{3a}{8} \right)$.

35. $P(X + Y \le 1) = \displaystyle\int_0^1 \int_0^{1-x} xe^{-x}e^{-y} \, dy \, dx$

$$= -\int_0^1 (xe^{-1} - xe^{-x}) \, dx$$

$$= -\left(xe^{-x} + e^{-x} + \frac{x}{e} x \right) \Big|_0^1$$

$$= 1 - 2e^{-1} - \frac{1}{2}e^{-1}$$

$$= 1 - \frac{5}{2}e^{-1}$$

$$\approx 0.0803$$

37. $P(X + Y \le 8) = \dfrac{1}{8} \displaystyle\int_0^8 \int_0^{8-x} e^{-x/2} e^{-y/4} \, dy \, dx$

$$= -\frac{1}{2} \int_0^8 e^{-x/2}(e^{-2+x/4} - 1) \, dx$$

$$= (e^4 - 2e^2 + 1)e^{-4}$$

$$\approx 0.7476$$

The probability is roughly 75%.

39. $A = \displaystyle\int_0^3 \int_{2x/3}^{7-5x/3} y \, dy \, dx$

$$= \frac{63}{2}$$

$$A_1 = \int_0^3 \int_{2x/3}^{7-5x/3} xy \, dy \, dx$$

$$= \frac{231}{8}$$

$$A_2 = \int_0^3 \int_{2x/3}^{7-5x/3} y^2 \, dy \, dx$$

$$= \frac{469}{4}$$

$$\bar{x} = \frac{A_1}{A} = \frac{11}{12} \approx 0.9167$$

$$\bar{y} = \frac{A_2}{A} = \frac{67}{18} \approx 3.7222$$

41. $$A = \int_0^1 \int_{x^2}^1 dy \, dx$$

$$= \int_0^1 (1 - x^2) \, dx$$

$$= \frac{2}{3}$$

The average value is:

$$\frac{1}{A} \int_0^1 \int_{x^2}^1 e^x y^{-1/2} \, dy \, dx = \frac{3}{2} \int_0^1 (2e^x - 2xe^x) \, dx$$

$$= \frac{3}{2}(4e^x - 2xe^x)\Big|_0^1$$

$$= \frac{3}{2}(2e - 4)$$

$$= 3(e - 2)$$

$$\approx 2.1548$$

43. Since the sphere is symmetric in each of the eight octants, and xyz is positive in four ocatants and negative in four octants, the average value is 0.

45. $$m = \frac{ab}{2}$$

$$I_z = \int_0^a \int_0^{b - bx/a} (x^2 + y^2) \, dy \, dx$$

$$= \int_0^a \left[-\frac{bx^3}{a} - \frac{b^3 x^3}{3a^3} + \frac{b^3 x^2}{a^2} + bx^2 - \frac{b^3 x}{a} + \frac{b^3}{3} \right] dx$$

$$= \frac{ab(a^2 + b^2)}{12}$$

$$d^2 = \frac{2ab(a^2 + b^2)}{12ab}$$

$$d = \sqrt{\frac{a^2 + b^2}{6}}$$

47. $$m = \int_1^2 \int_1^{x^2} x^2 y \, dy \, dx$$

$$= \frac{1}{2} \int_1^2 x^2(x^4 - 1) \, dx$$

$$= \frac{332}{42}$$

$$I_x = \int_1^2 \int_1^{x^2} x^2 y \, y^2 \, dy \, dx$$

$$= \frac{1}{4} \int_1^2 x^2(x^8 - 1) \, dx$$

$$= \frac{1,516}{33}$$

$$d^2 = \frac{1,516(42)}{33(332)}$$

$$d = \sqrt{\frac{5,306}{913}}$$

$$\approx 2.4107$$

49. a. Fix x at $x = x_0$. Then,

$$\frac{\partial C}{\partial t}(x_0, t) = \frac{C_0}{\sqrt{k\pi t}} \frac{x_0{}^2}{4kt^2} e^{-x_0{}^2/(4kt)}$$

$$- \frac{C_0}{2\sqrt{k\pi}\, t^{3/2}} e^{-x_0{}^2/(4kt)}$$

This is 0 when

$$\frac{x_0{}^2}{4kt_m{}^2} - \frac{1}{2t_m} = 0$$

Thus,

$$t_m = \frac{x_0{}^2}{2k}, \text{ so } 2kt_m = x_0{}^2$$

and

$$C_m(x_0) = \frac{C_0}{\sqrt{k\pi\left(\frac{x_0{}^2}{2k}\right)}} \exp(\frac{-x_0{}^2}{2x_0{}^2})$$

$$= \sqrt{\frac{2}{\pi}}\, C_0 \frac{e^{-1/2}}{x_0}$$

b. We want x_m to satisfy

$$\sqrt{\frac{2}{\pi}}\, C_0 \frac{e^{-1/2}}{x_m} \geq 0.25\, C_0$$

$$4\sqrt{\frac{2}{\pi}}\, e^{-1/2} \geq x_m$$

$$x_m \leq 1.9358$$

The danger zone is approximately 1.9 miles.

c.
$$A = \int\limits_0^{x_m} \int\limits_0^{t_m} dt\; dx$$

$$\approx \int\limits_0^{1.9358} \int\limits_0^{x^2/(2k)} dt\; dx$$

$$= \int\limits_0^{1.9358} \frac{x^2}{2k}\, dx$$

$$= \frac{1}{2k}\left[\frac{x^3}{3}\right]\Bigg|_0^{1.9358}$$

$$\approx \frac{1.21}{k}$$

The average value is:

$$\frac{1}{A} \int\limits_0^{1.9358} \int\limits_0^{x^2/(2k)} \frac{C_0}{\sqrt{k\pi t}} \exp(\frac{-x^2}{4kt})\, dt\; dx$$

d. Answers vary. For example, we may say that the danger period at point x on the riverbank is the period of time $t_1 < t < t_2$ when the concentration exceeds $0.25\, C_0$.

51. a. $dF = \delta\, dV$; $d = h - z$;

$dW = \delta(h - z)\ dV$, so

$$W = \delta \iiint (h - z)\ dV$$

b. The volume is considered concentrated at

$(0, 0, 0)$; $d = h$, $F = \delta V$; $W = \delta h V$

c. $W = \delta(15)\pi(36)(10) = 5,400\pi\delta$

$$W = 4\delta \int\limits_{0}^{6} \int\limits_{0}^{\sqrt{36 - x^2}} \int\limits_{-5}^{5} (15 - z)\ dz\ dy\ dx$$

$$= 4\delta \int\limits_{0}^{6} \int\limits_{0}^{\sqrt{36 - x^2}} 150\ dy\ dx$$

$$= 5,400\pi\delta$$

53. Set up a coordinate system with the center of

mass of the lamina at the origin and its

length along the x-axis. The moment of

inertia about the z-axis is

$$I_z = 4 \int\limits_{0}^{\ell/2} \int\limits_{0}^{h/2} (x^2 + y^2)\ dy\ dx$$

$$= \int\limits_{0}^{\ell/2} \frac{h(12x^2 + h^2)}{6}\ dx$$

$$= \frac{h\ell(h^2 + \ell^2)}{12}$$

55. $S = 2\pi a(2\pi b) = 4\pi^2 ab$

Compare with Problem 46, Section 12.4:

$\mathbf{R}_u = -(a + b\cos v)\sin u\ \mathbf{i} + (a + b\cos v)\cos u\ \mathbf{j}$

$\mathbf{R}_v = -b\sin v\cos u\ \mathbf{i} - b\sin v\sin u\ \mathbf{j} + b\cos v\ \mathbf{k}$

$\mathbf{R}_u \times \mathbf{R}_v$

$$= \begin{vmatrix} \mathbf{i} & \mathbf{j} & \mathbf{k} \\ -(a+b\cos v)\sin u & (a+b\cos v)\cos u & 0 \\ -b\sin v\cos u & -b\sin v\sin u & b\cos v \end{vmatrix}$$

$= (b^2\cos^2 v + ab\cos v)(\cos u)\mathbf{i}$

$\quad + (b^2\cos^2 v + ab\cos v)(\sin u)\mathbf{j}$

$\quad + (b^2\sin v\cos v + ab\sin v)\mathbf{k}$

$\|\mathbf{R}_u \times \mathbf{R}_v\| = \left| ab + b^2\cos v \right|$

$$S = \int\limits_{0}^{2\pi} \int\limits_{0}^{2\pi} \left| ab + b^2\cos v \right|\ du\ dv$$

$$= \int\limits_{0}^{2\pi} 2\pi(ab + b^2\cos v)\ dv$$

$$= 4\pi^2 ab$$

They are the same.

12.7 Cylindrical and Spherical Coordinates, Pages 840-843

SURVIVAL HINT: *The best way to convert from one coordinate system to another is to understand the derivations and "visualize" the graph. Lacking that skill, you will find it necessary to memorize the transformation equations.*

1. Cylindrical coordinates are best used in problems with axial symmetry (that is, with cylinders), spherical coordinates in problems with radial symmetry (that is, with spheres), and rectangular coordinates in all other cases.

3. a. $(4, \pi/2, \sqrt{3})$

    b. $(\sqrt{19}, \pi/2, \cos^{-1}(\sqrt{3}/\sqrt{19}))$

5. a. $(\sqrt{5}, \tan^{-1}2, 3)$

    b. $(\sqrt{14}, \tan^{-1}2, \cos^{-1}(3/\sqrt{14}))$

7. a. $(-3/2, 3\sqrt{3}/2, -3)$

    b. $(\sqrt{18}, 2\pi/3, \cos^{-1}(-3/\sqrt{18}))$

        $= (3\sqrt{2}, \frac{2\pi}{3}, \frac{3\pi}{4})$

9. a. $(\sqrt{2}, \sqrt{2}, \pi)$

    b. $(\sqrt{\pi^2+4}, \pi/4, \cos^{-1}(\pi/\sqrt{\pi^2+4}))$

11. a. $(3/2, \sqrt{3}/2, -1)$ b. $(\sqrt{3}, \frac{\pi}{6}, -1)$

13. a. $(\sin 3 \cos 2, \sin 3 \sin 2, \cos 3)$

    b. $(\sin 3, 2, \cos 3)$

SURVIVAL HINT: *Cylindrical coordinates are dz with polar coordinates for the dA. Do not forget the "r."*
That is,

$dz\ dy\ dx = dV = dz\ r\ dr\ d\theta$

On the other hand, do not forget that when you find

$x^2 + y^2 + z^2$

in a function you should consider changing to spherical coordinates.

15. $z = r^2 \cos 2\theta$

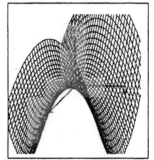

17. $r = \dfrac{6z}{\sqrt{4 - 13\cos^2\theta}}$ or

    $9r^2\cos^2\theta - 4r^2\sin^2\theta + 36z^2 = 0$

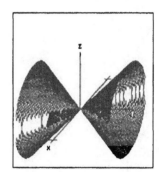

19. $\phi = \frac{\pi}{4}$

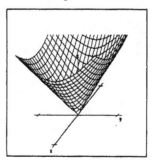

21. $\rho = \dfrac{4\cot\phi\csc\phi}{3 - 2\cos^2\theta}$ or

    $4\rho\cos\phi = \rho^2\sin^2\phi(\cos^2\theta + 3\sin^2\theta)$

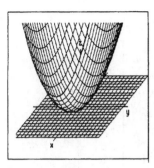

23. $z = 2xy$

25. $z = x^2 - y^2$

27. $xz = 1$

29.
$$\int_0^\pi \int_0^2 \int_0^{\sqrt{4-r^2}} r \sin \theta \; dz \; dr \; d\theta$$

$$= \int_0^\pi \int_0^2 (\sin \theta) \sqrt{4 - r^2} \; r \; dr \; d\theta$$

$$= \frac{8}{3} \int_0^\pi \sin \theta \; d\theta$$

$$= \frac{16}{3}$$

31.
$$\int_0^{\pi/2} \int_0^{2\pi} \int_0^2 \cos \phi \sin \phi \; d\rho \; d\theta \; d\phi$$

$$= 2 \int_0^{\pi/2} \int_0^{2\pi} \cos \phi \sin \phi \; d\theta \; d\phi$$

$$= 4\pi \int_0^{\pi/2} \cos \phi \sin \phi \; d\phi$$

$$= 2\pi$$

33.
$$\int_0^{2\pi} \int_0^4 \int_0^1 zr \; dz \; dr \; d\theta$$

$$= \frac{1}{2} \int_0^{2\pi} \int_0^4 r \; dr \; d\theta$$

$$= 4 \int_0^{2\pi} d\theta$$

$$= 8\pi$$

35. $\displaystyle\int_0^{\pi/2}\int_0^{\cos\theta}\int_0^{1-r^2} r\sin\theta\;dz\;dr\;d\theta$

$$=\int_0^{\pi/2}\int_0^{\cos\theta} r(1-r^2)\sin\theta\;dr\;d\theta$$

$$=\int_0^{\pi/2}\frac{\cos^2\theta\,\sin\theta}{2}\;d\theta-\int_0^{\pi/2}\frac{\cos^4\theta\,\sin\theta}{4}\;d\theta$$

$$=\left[-\frac{\cos^3\theta}{6}+\frac{\cos^5\theta}{20}\right]\Bigg|_0^{\pi/2}$$

$$=\frac{1}{6}-\frac{1}{20}$$

$$=\frac{7}{60}$$

37. $\dfrac{x^2}{a^2}+\dfrac{y^2}{b^2}=R^2\sin^2\phi$

and

$$\frac{z^2}{c^2}=R^2\cos^2\phi$$

so

$$\frac{x^2}{a^2}+\frac{y^2}{b^2}+\frac{z^2}{c^2}=R^2$$

39. $\displaystyle\iint_R\int (x^2+y^2)^2\;dx\;dy\;dz$

There are four octants where $z>0$.

41. $\displaystyle 4\int_0^{1/\pi}\int_0^{\pi/2}\int_0^a r^4\,r\;dr\;d\theta\;dz=\frac{4a^6}{6}\int_0^{1/\pi}\int_0^{\pi/2}d\theta\;dz$

$$=\frac{4a^6}{6}\left(\frac{\pi}{2}\right)\left(\frac{1}{\pi}\right)$$

$$=\frac{a^6}{3}$$

41. $\displaystyle m=\int_0^{2\pi}\int_0^9\int_r^9 r\;dz\;dr\;d\theta$

$$=\int_0^{2\pi}\int_0^9 (9r-r^2)\;dr\;d\theta$$

$$=\frac{243}{2}\int_0^{2\pi}d\theta$$

$$=243\pi$$

$\overline{x}=\overline{y}=0$ (by symmetry)

$$M_{xy}=\int_0^{2\pi}\int_0^9\int_r^9 rz\;dz\;dr\;d\theta$$

$$=\frac{1}{2}\int_0^{2\pi}\int_0^9 r(9^2-r^2)\;dr\;d\theta$$

$$=\frac{6{,}561}{8}\int_0^{2\pi}d\theta$$

$$=\frac{6{,}561\pi}{4}$$

$$\bar{z} = \frac{6,561\pi}{4(243\pi)} = \frac{27}{4}$$

Thus, the centroid is $(0, 0, \frac{27}{4})$

43. We use cylindrical coordinates.

a. $m = \displaystyle\int_0^{2\pi}\int_0^3\int_0^{\sqrt{9-r^2}} (r^2 \sin\theta \cos\theta + z)r\, dz\, dr\, d\theta$

b. $\bar{x} = \dfrac{1}{m}\displaystyle\int_0^{2\pi}\int_0^3\int_0^{\sqrt{9-r^2}} r\cos\theta(r^2\sin\theta\cos\theta + z)r\, dz\, dr\, d\theta$

c. $I_z = \displaystyle\int_0^{2\pi}\int_0^3\int_0^{\sqrt{9-r^2}} r^2(r^2\sin\theta\cos\theta + z)r\, dz\, dr\, d\theta$

45. The volume of the sphere is $V = \frac{4}{3}\pi(2)^3 = \frac{32}{3}\pi$.

In spherical coordinates, we have

$$x + y + z = \rho\sin\phi\cos\theta + \rho\sin\phi\sin\theta + \rho\cos\phi$$

$$= \rho[\sin\phi(\cos\theta + \sin\theta) + \cos\phi]$$

so the average value is

$$\frac{1}{V}\int_0^{2\pi}\int_0^{\pi}\int_0^a \rho[\sin\phi(\cos\theta + \sin\theta)$$

$$+ \cos\phi]\rho^2\sin\phi\, d\rho\, d\phi\, d\theta$$

$$= \frac{1}{V}\int_0^{2\pi}\int_0^{\pi} [\sin^2\phi(\cos\theta + \sin\theta)$$

$$+ \cos\phi\sin\phi]\frac{a^4}{4}\, d\phi\, d\theta = 0$$

We could have expected this answer because

$f(x, y, z) = x + y + z$ satisfies

$f(-x, -y, -z) = -f(x, y, z).$

47. $m = \displaystyle\int_0^{2\pi}\int_0^{\pi}\int_0^{2\sin\phi} \rho(\rho^2\sin\phi)\, d\rho\, d\phi\, d\theta$

$$= \int_0^{2\pi}\int_0^{\pi} \sin\phi\left[\frac{(2\sin\phi)^4}{4}\right] d\phi\, d\theta$$

$$= 4\int_0^{2\pi}\left[\frac{-\cos^5\phi}{5} + \frac{2\cos^3\phi}{3} - \cos\phi\right]\Bigg|_0^{\pi} d\theta$$

$$= \int_0^{2\pi}\frac{64}{15}\, d\theta$$

$$= \frac{128\pi}{15}$$

$$\approx 26.8083$$

49. $\displaystyle\iiint_R (x^2 + y^2 + z^2)\, dx\, dy\, dz$

$$= 8\int_0^{\pi/2}\int_0^{\pi/2}\int_0^{\sqrt{2}} \rho^4\sin\phi\, d\rho\, d\theta\, d\phi$$

$$= \frac{32\sqrt{2}}{5}\int_0^{\pi/2}\int_0^{\pi/2} \sin\phi\, d\theta\, d\phi$$

$$= \frac{16\sqrt{2}\pi}{5} \int_0^{\pi/2} \sin\phi \, d\phi$$

$$= \frac{16\pi\sqrt{2}}{5}$$

$$\approx 14.2172$$

51. $$\iiint_S \frac{dx \, dy \, dz}{\sqrt{x^2 + y^2 + z^2}}$$

$$= 8 \int_0^{\pi/2} \int_0^{\pi/2} \int_0^{\sqrt{3}} \rho \sin\phi \, d\rho \, d\theta \, d\phi$$

$$= 12 \int_0^{\pi/2} \int_0^{\pi/2} \sin\phi \, d\theta \, d\phi$$

$$= 6\pi \int_0^{\pi/2} \sin\phi \, d\phi$$

$$= 6\pi$$

$$\approx 18.8496$$

53. The projected region is $4 = x^2 + y^2$

$$V = \int_0^{2\pi} \int_0^1 \int_0^{4-r^2} dz \, r \, dr \, d\theta$$

$$= \int_0^{2\pi} \int_0^1 (4 - r^2) \, r \, dr \, d\theta$$

$$= \int_0^{2\pi} \frac{7}{4} \, d\theta$$

$$= \frac{7\pi}{2}$$

$$\approx 10.9956$$

55. $$V = 2 \int_0^{\pi/2} \int_0^{2\sin\theta} \int_0^{4-r^2} r \, dz \, dr \, d\theta$$

$$= 2 \int_0^{\pi/2} \int_0^{2\sin\theta} r(4 - r^2) \, dr \, d\theta$$

$$= 2 \int_0^{\pi/2} (8\sin^2\theta - 4\sin^4\theta) \, d\theta$$

$$= 2[4\theta - 4\sin\theta \cos\theta]\Big|_0^{\pi/2}$$

$$- 8\left[\frac{3\theta}{8} - \frac{\sin 2\theta}{4} + \frac{\sin 4\theta}{32}\right]\Big|_0^{\pi/2}$$

$$= \frac{5\pi}{2}$$

$$\approx 7.8540$$

57. The Spy will begin to drown when the water level reaches 3 ft above the floor (the xy plane); that is when $z = 3$.

$$\rho \cos\phi = 3$$

$$4(1 + \cos\phi) \cos\phi = 3$$

$$4\cos^2\phi + 4\cos\phi - 3 = 0$$

$$(2\cos\phi - 1)(2\cos\phi + 3) = 0$$

$$\cos \phi = \tfrac{1}{2}, -\tfrac{3}{2}$$

so in this application $\phi = \frac{\pi}{3}$ (reject $-\frac{3}{2}$). The

critical water level is $z = 3 = \rho \cos \phi$ or

$\rho = 3 \sec \phi$. We need to compute the

amount of water in the cave when the depth

is 3 ft and use the rate of 25 cu ft/min to

determine the time necessary for it to reach

that level. In spherical coordinates the

volume can be found as the sum of

two integrals:

$$V = \int_0^{2\pi} \int_0^{\pi/3} \int_0^{3\sec\phi} \rho^2 \sin\phi \; d\rho \; d\phi \; d\theta$$

$$+ \int_0^{2\pi} \int_{\pi/3}^{\pi/2} \int_0^{4(1+\cos\phi)} \rho^2 \sin\phi \; d\rho \; d\phi \; d\theta$$

$$= \int_0^{2\pi} \int_0^{\pi/3} \frac{(3 \sec\phi)^3}{3} \sin\phi \; d\phi \; d\theta$$

$$+ \int_0^{2\pi} \int_{\pi/3}^{\pi/2} \frac{[4(1+\cos\phi)]^3}{3} \sin\phi \; d\phi \; d\theta$$

$$= \tfrac{9}{2}(3)(2\pi) - \tfrac{64}{3}\left(\tfrac{1}{4} - \tfrac{81}{64}\right)(2\pi)$$

$$= \frac{211\pi}{3}$$

≈ 221 cu ft of water

At the incoming rate of 25 cu ft/min, it takes

about 8.84 min for the water to reach his

nose. So he drowns, you say — nonsense!

Any spy worth his salt can hold his breath for

a little more than a minute. He frees his

hands, stands (to buy more time), hops to the

door, pulls up the lever (to stop the water),

and opens the door. As the water drains from

the room, he unties his feet, and prepares to

pursue the Flamer.

59. **a.** Force $= \dfrac{GmM}{R^2} = \dfrac{Gm\delta(4\pi a^3)}{3R^2}$

b. With $R = 4$ and $a = 3$, we obtain

Force $= \dfrac{Gm\delta(9\pi)}{4}$

c. With a rectangular mass m, we got a

poor approximation using the center of

mass when the separating distance was

small. The approximation improved as

the distance increased. With the sphere

we always get perfect agreement.

Apparently the symmetry of the sphere

play the key role. Either a computer with

symbolic integration capability or determined work by hand will show that for the sphere the center of mass method gives the exact result.

61. $\quad I = I_x + I_y + I_z$

$$= \iint\limits_S \int (y^2 + z^2)\, dV + \iint\limits_S \int (x^2 + z^2)\, dV$$

$$+ \iint\limits_S \int (x^2 + y^2)\, dV$$

$$= \iint\limits_S \int (2x^2 + 2y^2 + 2z^2)\, dV$$

$$= 2 \int\limits_0^\pi \int\limits_0^{2\pi} \int\limits_0^1 \rho^2\, \rho^2 \sin\phi\, d\rho\, d\theta\, d\phi$$

$$= 2 \int\limits_0^{2\pi} \int\limits_0^\pi \int\limits_0^1 \rho^4 \sin\phi\, d\rho\, d\phi\, d\theta$$

$$= \frac{2}{5} \int\limits_0^{2\pi} \int\limits_0^\pi \sin\phi\, d\phi\, d\theta$$

$$= \frac{4}{5} \int\limits_0^{2\pi} d\theta$$

$$= \frac{8\pi}{5}$$

63. **a.** $\displaystyle V = \int\limits_0^\pi \int\limits_0^{2\pi} \int\limits_0^{1 + 0.2\sin 4\theta \sin 3\phi} \rho^2 \sin\phi\, d\rho\, d\theta\, d\phi$

$$= \int\limits_0^\pi \int\limits_0^{2\pi} \frac{\rho^3}{3} \sin\phi \bigg|_0^{1 + 0.2\sin 4\theta \sin 3\phi}\, d\theta\, d\phi$$

$$= \int\limits_0^\pi \int\limits_0^{2\pi} \bigg\{ \frac{\sin^3 4\theta\, \sin^3 3\phi \sin\phi}{375}$$

$$+ \frac{\sin^2 4\theta \sin^2 3\phi \sin\phi}{25} + \frac{\sin 4\theta\, \sin 3\phi\, \sin\phi}{5}$$

$$+ \frac{\sin\phi}{3} \bigg\}\, d\theta\, d\phi$$

$$= \int\limits_0^\pi \bigg\{ \bigg[-\sin^3 3\phi\, \sin\phi \bigg(\frac{\sin 4\theta\, \sin 8\theta}{9000} + \frac{\cos 4\theta}{2250} \bigg) \bigg]$$

$$+ \bigg[\sin^2 3\phi \sin\phi \bigg(\frac{\theta}{50} - \frac{\sin 8\theta}{400} \bigg) \bigg]$$

$$- \bigg[\sin 3\phi\, \sin\phi \bigg(\frac{\cos 4\theta}{20} \bigg) \bigg] + \frac{\sin\phi}{3}\, \theta \bigg\} \bigg|_0^{2\pi}\, d\phi$$

$$= \int\limits_0^\pi \bigg(\frac{\pi}{25} \sin^2 3\phi\, \sin\phi + \frac{2\pi}{3} \sin\phi \bigg)\, d\phi$$

$$= \bigg[\frac{\pi \cos 7\phi}{700} - \frac{\pi \cos 5\phi}{500} - \frac{103\pi \cos\phi}{150} \bigg] \bigg|_0^\pi$$

$$= \frac{3{,}608\pi}{2{,}625}$$

$$\approx 4.31804$$

b. Using a calculator, we find an approximate value of 4.31445.

12.8 Jacobians: Change of Variables, Pages 850-852

1. $x = u - v, \; y = u + v$

$$\frac{\partial(x, y)}{\partial(u, v)} = \begin{vmatrix} \dfrac{\partial x}{\partial u} & \dfrac{\partial x}{\partial v} \\ \dfrac{\partial y}{\partial u} & \dfrac{\partial y}{\partial v} \end{vmatrix}$$

$$= \begin{vmatrix} 1 & -1 \\ 1 & 1 \end{vmatrix}$$

$$= 2$$

3. $x = u^2, \; y = u + v$

$$\frac{\partial(x, y)}{\partial(u, v)} = \begin{vmatrix} \dfrac{\partial x}{\partial u} & \dfrac{\partial x}{\partial v} \\ \dfrac{\partial y}{\partial u} & \dfrac{\partial y}{\partial v} \end{vmatrix}$$

$$= \begin{vmatrix} 2u & 0 \\ 1 & 1 \end{vmatrix}$$

$$= 2u$$

5. $x = e^{u+v}, y = e^{u-v}$

$$\frac{\partial(x, y)}{\partial(u, v)} = \begin{vmatrix} \dfrac{\partial x}{\partial u} & \dfrac{\partial x}{\partial v} \\ \dfrac{\partial y}{\partial u} & \dfrac{\partial y}{\partial v} \end{vmatrix}$$

$$= \begin{vmatrix} e^{u+v} & e^{u+v} \\ e^{u-v} & -e^{u-v} \end{vmatrix}$$

$$= -2e^{u+v}e^{u-v} = -2e^{2u}$$

7. $x = e^u \sin v, \; y = e^u \cos v$

$$\frac{\partial(x, y)}{\partial(u, v)} = \begin{vmatrix} \dfrac{\partial x}{\partial u} & \dfrac{\partial x}{\partial v} \\ \dfrac{\partial y}{\partial u} & \dfrac{\partial y}{\partial v} \end{vmatrix}$$

$$= \begin{vmatrix} e^u \sin v & e^u \cos v \\ e^u \cos v & -e^u \sin v \end{vmatrix}$$

$$= -e^{2u}\sin^2 v - e^{2u}\cos^2 v$$

$$= -e^{2u}$$

9. $x = u + v - w, \; y = 2u - v + 3w,$

$z = -u + 2v - w$

$$\frac{\partial(x, y, z)}{\partial(u, v, w)} = \begin{vmatrix} 1 & 1 & -1 \\ 2 & -1 & 3 \\ -1 & 2 & -1 \end{vmatrix}$$

$$= -9$$

11. $x = u \cos v, \; y = u \sin v, \; z = we^{uv}$

$$\frac{\partial(x, y, z)}{\partial(u, v, w)} = \begin{vmatrix} \cos v & -u \sin v & 0 \\ \sin v & u \cos v & 0 \\ vwe^{uv} & uwe^{uv} & e^{uv} \end{vmatrix}$$

$$= u(\cos^2 v)e^{uv} + u(\sin^2 v)e^{uv}$$

$$= ue^{uv}$$

13. $u = 2x - 3y, \; v = x + 4y$

$$\frac{\partial(u, v)}{\partial(x, y)} = \begin{vmatrix} 2 & -3 \\ 1 & 4 \end{vmatrix}$$

$$= 11$$

so

$$\frac{\partial(x, y)}{\partial(u, v)} = \frac{1}{11}$$

15. $u = ye^{-x}, \; v = e^x; \; y = uv, \; x = \ln v$

$$\frac{\partial(x, y)}{\partial(u, v)} = \begin{vmatrix} 0 & \frac{1}{v} \\ v & u \end{vmatrix}$$

$$= -1$$

17. $u = x^2 - y^2, \; v = x^2 + y^2$

$$\frac{\partial(u, v)}{\partial(x, y)} = \begin{vmatrix} 2x & -2y \\ 2x & 2y \end{vmatrix}$$

$$= 4xy + 4xy = 8xy$$

$$\frac{\partial(x, y)}{\partial(u, v)} = \frac{1}{8xy}$$

19. $A(0, 5) \to (5, -5);$
$B(6, 5) \to (11, 1);$
$C(6, 0) \to (6, 6);$
$O(0, 0) \to (0, 0)$

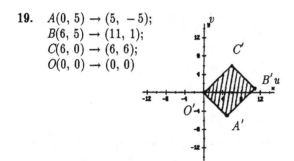

21. The boundary lines of the figure are $y = 0$,

$y = 4$, $y = 2x$, and $y = 2x - 10$. Thus, the

transformed boundaries are $v = 0$ and

$y = 4$: $u = x^2 - 16$, $v = 8x$ so $u = \left(\frac{v}{8}\right)^2 - 16$.

$y = 2x$: $u = -3x^2, \; v = 4x^2$ so $v = -\frac{4}{3}u$

$y = 2x - 10$: $u = -3x^2 + 40x - 100$,

$v = 4x^2 - 20x$, so (in parametric form)

$u = -3t^2 + 40t - 100, \; v = 4t^2 - 20t.$

The vertices of the figure are transformed as

follows:

$$A(5, 0) \to (25, 0)$$

$$B(7, 4) \to (33, 56);$$

$$C(2, 4) \to (-12, 16)$$

$$O(0, 0) \to (0, 0)$$

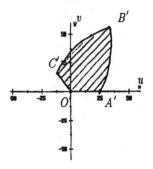

23. $x = u - uv, \; y = uv$

$$\frac{\partial(x, y)}{\partial(u, v)} = \begin{vmatrix} \frac{\partial x}{\partial u} & \frac{\partial x}{\partial v} \\ \frac{\partial y}{\partial u} & \frac{\partial y}{\partial v} \end{vmatrix}$$

$$= \begin{vmatrix} 1 - v & -u \\ v & u \end{vmatrix}$$

$$= u - uv + uv = u$$

$$dx\ dy\ =\ u\ du\ dv$$

25. Let $u = xy$, $v = \frac{y}{x}$. Then the transformed region has boundary lines $u = 1$, $u = 4$, and $v = 1$, $v = 4$. Thus, $x = \sqrt{\frac{u}{v}}$, $y = \sqrt{uv}$ and

$$\frac{\partial(x, y)}{\partial(u, v)} = \begin{vmatrix} \frac{\partial x}{\partial u} & \frac{\partial x}{\partial v} \\ \frac{\partial y}{\partial u} & \frac{\partial y}{\partial v} \end{vmatrix}$$

$$= \begin{vmatrix} \frac{1}{2\sqrt{uv}} & \frac{-1}{2v}\sqrt{\frac{u}{v}} \\ \frac{1}{2}\sqrt{\frac{v}{u}} & \frac{1}{2}\sqrt{\frac{u}{v}} \end{vmatrix}$$

$$= \frac{1}{2v}$$

The area is

$$A = \int_1^4 \int_1^4 \frac{1}{2v}\ dv\ du$$

$$= \int_1^4 \frac{1}{2} \ln|v| \Big|_1^4\ du$$

$$= \int_1^4 \frac{1}{2} \ln 4\ du$$

$$= 3(\ln 2)$$

SURVIVAL HINT: *For Problems 27-30, we have* $u = x - y$, $v = x + y$, *so* $x = \frac{u + v}{2}$, $y = \frac{v - u}{2}$. *The boundary lines* $x = 0$, $y = 0$, $x + y = 1$ *become* $-v = u$, $v = u$, $v = 1$, *respectively. The Jacobian of the transformation is*

$$\frac{\partial(x, y)}{\partial(u, v)} = \begin{vmatrix} \frac{\partial x}{\partial u} & \frac{\partial x}{\partial v} \\ \frac{\partial y}{\partial u} & \frac{\partial y}{\partial v} \end{vmatrix} = \begin{vmatrix} \frac{1}{2} & \frac{1}{2} \\ -\frac{1}{2} & \frac{1}{2} \end{vmatrix} = \frac{1}{2}$$

$$dy\ dx = \frac{1}{2}\ du\ dv.$$

27. $\displaystyle\int_D\!\!\int \left(\frac{x - y}{x + y}\right)^5\ dy\ dx = \int_0^1 \int_{-v}^{v} \frac{u^5}{v^5} \frac{1}{2}\ du\ dv$

$$= \frac{1}{2}\int_0^1 v^{-5} \frac{u^6}{6}\Big|_{-v}^{v}\ dv$$

$$= \frac{1}{12}\int_0^1 (v - v)\ dv$$

$$= 0$$

29. $\displaystyle\int_D\!\!\int (x - y)^5(x + y)^3\ dy\ dx$

$$= \frac{1}{2}\int_0^1 \int_{-v}^{v} u^5 v^3\ du\ dv$$

$$= \frac{1}{12}\int_0^1 (v^9 - v^9)\ dv$$

$$= 0$$

31. Find the Jacobian, map the region, then integrate.

SURVIVAL HINT: *For Problems 33-38, we have* $u = \frac{2x + y}{5}$, $v = \frac{x - 2y}{5}$ *so* $y = u - 2v$ *and* $x = 2u + v$

$$\frac{\partial(x, y)}{\partial(u, v)} = \begin{vmatrix} \frac{\partial x}{\partial u} & \frac{\partial x}{\partial v} \\ \frac{\partial y}{\partial u} & \frac{\partial y}{\partial v} \end{vmatrix} = \begin{vmatrix} 2 & 1 \\ 1 & -2 \end{vmatrix} = -5$$

$A(0, 0) \to (0, 0), \quad B(1, -2) \to (0, 1),$

$C(3, -1) \to (1, 1), \quad D(2, 1) \to (1, 0).$

R is the unit square in the uv-plane.

33. $\displaystyle\int_D \int \left(\frac{2x + y}{x - 2y + 5}\right)^2 dy\, dx$

$$= \int_0^1 \int_0^1 \left(\frac{u}{v+1}\right)^2 5 \ du\, dv$$

$$= \frac{5}{3} \int_0^1 (v + 1)^{-2} dv$$

$$= \frac{5}{6}$$

35. $\displaystyle\int_D \int (2x + y)^2 (x - 2y)\, dy\, dx$

$$= \int_0^1 \int_0^1 (5u)^2 (5v)\ 5\ du\, dv$$

$$= \frac{5^4}{3} \int_0^1 v\, dv$$

$$= \frac{625}{6}$$

37. $\displaystyle\int_D \int (2x + y)\tan^{-1}(x - 2y)\, dy\, dx$

$$= \int_0^1 \int_0^1 (5u)\tan^{-1}(5v)\ 5\ dv\, du$$

$$= \int_0^1 \left[25uv\tan^{-1}(5v) - \frac{5u\ln(25v^2 + 1)}{2}\right]\Bigg|_0^1 du$$

$$= \int_0^1 \left[25u\tan^{-1}5 - \frac{5u\ln 26}{2}\right] du$$

$$= \left[25\tan^{-1}5 - \frac{5}{2}\ln 26\right]\frac{u^2}{2}\Bigg|_0^1$$

$$= \frac{25}{2}\tan^{-1}5 - \frac{5}{4}\ln 26$$

$$\approx 13.0949$$

39. $u = 2x + y, \ v = 2y - x$ or

$$x = \frac{2u - v}{5}, \ y = \frac{u + 2v}{5}$$

$$\text{so } J = \begin{vmatrix} \frac{2}{5} & -\frac{1}{5} \\ \frac{1}{5} & \frac{2}{5} \end{vmatrix} = \frac{1}{5}$$

The trapezoidal region R is bounded by the lines

$x = 0, \ y = 0, \ y + 2x = 2,$ and $y + 2x = 8.$ The

transformed region R' is bounded by

$v = 2u, \ v = -\frac{1}{2}u, \ u = 2,$ and $u = 8.$

$$\int_R \int e^{(2y - x)/(y+2x)}\, dA = \int_2^8 \int_{-u/2}^{2u} \frac{1}{5} e^{v/u}\, dv\, du$$

$$= \int_2^8 \frac{1}{5} u e^{v/u}\Big|_{-u/2}^{2u} du$$

$$= \int_2^8 \frac{1}{5}(e^2 - e^{-1/2})\, u\, du$$

$$= \tfrac{1}{5}(e^2 - e^{-1/2})\tfrac{u^2}{2}\Big|_2^8$$

$$= 6(e^2 - e^{-1/2})$$

$$\approx 40.6952$$

41. By looking at the function we see a suitable

transformation can be obtained when

$a = b = s = 1$ and $r = -1$.

$u = x + y$, $v = -x + y$; so $y = \tfrac{1}{2}(u + v)$,

$x = \tfrac{1}{2}(u - v)$

$$\frac{\partial(x,\,y)}{\partial(u,\,v)} = \begin{vmatrix} \dfrac{\partial x}{\partial u} & \dfrac{\partial x}{\partial v} \\[2mm] \dfrac{\partial y}{\partial u} & \dfrac{\partial y}{\partial v} \end{vmatrix}$$

$$= \begin{vmatrix} \tfrac{1}{2} & -\tfrac{1}{2} \\[2mm] \tfrac{1}{2} & \tfrac{1}{2} \end{vmatrix}$$

$$= \tfrac{1}{2}$$

$A(0,0) \to (0,0)$, $B(1,1) \to (2,0)$,

$C(0,2) \to (2,2)$, $D(-1,1) \to (0,2)$.

$dy\,dx = \tfrac{1}{2}\,du\,dv$

$$\int_R \int \Big(\frac{x+y}{2}\Big)^2 e^{(y-x)/2}\,dy\,dx$$

$$= \int_0^2 \int_0^2 \Big(\frac{u}{2}\Big)^2 e^{v/2}\Big(\frac{1}{2}\Big)\,dv\,du$$

$$= \tfrac{1}{4}(e - 1)\int_0^2 u^2\,du$$

$$= \tfrac{2}{3}(e - 1)$$

$$\approx 1.1455$$

43. $x = ar\cos\theta$, $y = br\sin\theta$. Since

$\dfrac{x^2}{a^2} + \dfrac{y^2}{b^2} = r^2$, the transformed region is

$r \le 1$, $0 \le \theta \le \tfrac{\pi}{2}$.

$$\frac{\partial(x,\,y)}{\partial(r,\,\theta)} = \begin{vmatrix} \dfrac{\partial x}{\partial r} & \dfrac{\partial x}{\partial \theta} \\[2mm] \dfrac{\partial y}{\partial r} & \dfrac{\partial y}{\partial \theta} \end{vmatrix}$$

$$= \begin{vmatrix} a\cos\theta & -ar\sin\theta \\ b\sin\theta & br\cos\theta \end{vmatrix}$$

$$= abr$$

$dy\,dx = abr\,dr\,d\theta$

$$\int_{D^*} \int \exp\Big(-\frac{x^2}{a^2} - \frac{y^2}{b^2}\Big)\,dy\,dx = ab\int_0^{\pi/2}\int_0^1 re^{-r^2}\,dr\,d\theta$$

$$= \frac{ab}{2}\int_0^{\pi/2}(1 - e^{-1})\,d\theta$$

$$= \frac{ab\pi}{4}(1 - e^{-1})$$

45. Let $\begin{cases} u = \dfrac{x}{\sqrt{5}} \\[2mm] v = \dfrac{y}{2} \end{cases}$ Then,

$$\frac{\partial(x,\,y)}{\partial(u,\,v)} = \begin{vmatrix} \sqrt{5} & 0 \\ 0 & 2 \end{vmatrix}$$

$$= 2\sqrt{5}$$

and

$$\int\!\!\int_{D^*} e^{-(4x^2 + 5y^2)}\, dA$$

$$= \int_{-1}^{1} \int_{-\sqrt{1-v^2}}^{\sqrt{1-v^2}} e^{-4(\sqrt{5}\,u)^2 - 5(2v)^2}(2\sqrt{5})\, du\; dv$$

$$= 8\sqrt{5} \int_{0}^{1} \int_{0}^{\sqrt{1-v^2}} e^{-20(u^2 + v^2)}\, du\; dv$$

$$= 8\sqrt{5} \int_{0}^{2\pi} \int_{0}^{1} e^{-20r^2}\, r\, dr\, d\theta$$

$$= 8\sqrt{5} \int_{0}^{2\pi} \left[\frac{e^{-20r^2}}{-40}\right]\Bigg|_{0}^{1}\, d\theta$$

$$= \frac{8\sqrt{5}}{40}(1 - e^{-20})(2\pi)$$

$$\approx 2.80992$$

47. Let $u = xy^3$, $v = \dfrac{y}{x}$;

$$\frac{u}{v^3} = \frac{xy^3}{\left(\frac{y}{x}\right)^3} = x^4,$$

so $x = \sqrt[4]{\dfrac{u}{v^3}}$; $y = vx$ so $y = \sqrt[4]{uv}$

We have

$$\frac{\partial(u,\,v)}{\partial(x,\,y)} = \begin{vmatrix} y^3 & 3xy^2 \\ \dfrac{-y}{x^2} & \dfrac{1}{x} \end{vmatrix}$$

$$= \frac{y^3}{x} + \frac{3xy^3}{x^2}$$

$$= \frac{4y^3}{x}$$

So

$$\frac{\partial(x,\,y)}{\partial(u,\,v)} = \frac{x}{4y^3}$$

$$= \frac{1}{4}\frac{\sqrt[4]{\dfrac{u}{v^3}}}{\left(\sqrt[4]{uv}\right)^3}$$

$$= \frac{1}{4\sqrt{uv^3}}$$

The area of D^* is

$$A = \int_{1}^{2} \int_{3}^{6} \frac{1}{4\sqrt{uv^3}}\, du\; dv$$

$$= \frac{(3\sqrt{2} - 4)\sqrt{3}}{2}$$

$$\approx 0.2101$$

and

$$\bar{x} = \frac{1}{A} \int\!\!\int_{D^*} x \, dA$$

$$= \frac{1}{A} \int_1^2 \int_3^6 \sqrt[4]{\frac{u}{v^3}}\left(\frac{1}{4\sqrt{uv^3}}\right) du \, dv$$

$$= \frac{1}{A} \int_1^2 \int_3^6 \frac{1}{4}\, u^{-1/4} v^{-9/4} \, du \, dv$$

$$\approx \frac{0.2402}{A}$$

$$\approx 1.1430$$

$$\bar{y} = \frac{1}{A} \int\!\!\int_{D^*} y \, dA$$

$$= \frac{1}{A} \int_1^2 \int_3^6 \sqrt[4]{uv}\left(\frac{1}{4\sqrt{uv^3}}\right) du \, dv$$

$$= \frac{1}{A} \int_1^2 \int_3^6 \frac{1}{4}\, u^{-1/4} v^{-5/4} \, du \, dv$$

$$\approx \frac{0.3297}{A}$$

$$\approx 1.5690$$

Thus, the centroid is located at

(1.1430, 1.5690) in the xy-plane.

49. $x = au$, $y = bv$, $z = cw$.

$$\frac{\partial(x, y, z)}{\partial(u, v, w)} = \begin{vmatrix} a & 0 & 0 \\ 0 & b & 0 \\ 0 & 0 & c \end{vmatrix}$$

$$= abc$$

$$dy \, dx \, dz = abc \, du \, dv \, dw$$

$$\frac{x^2}{a^2} + \frac{y^2}{b^2} + \frac{z^2}{c^2} = u^2 + v^2 + w^2$$

Thus,

VOLUME OF THE ELLIPSOID $= abc$(VOLUME OF SPHERE)

$$abc(\tfrac{4}{3}\pi(1)) = \tfrac{4}{3}\pi\, abc$$

51. Let $u = x - y$, $v = x + y$ so that $x = \dfrac{u + v}{2}$,

$y = \dfrac{-u + v}{2}$ and

$$\frac{\partial(x, y)}{\partial(u, v)} = \begin{vmatrix} \frac{1}{2} & \frac{1}{2} \\ -\frac{1}{2} & \frac{1}{2} \end{vmatrix}$$

$$= \frac{1}{2}$$

The given region R is bounded by the lines

$x - 3y = 1$, $x + y = 1$, and $x = 4$ which

transform into $2u - v = 1$, $v = 1$, and

$u + v = 8$.

$$\int\!\!\int_R \ln\!\left(\frac{x - y}{x + y}\right) dy \, dx$$

$$= \frac{1}{2} \int_1^5 \int_{(v+1)/2}^{8-v} \ln(\tfrac{u}{v}) \, du \, dv$$

$$= \frac{1}{2} \int_{10}^{5} \left[u \ln \frac{u}{v} - u \right] \Big|_{(v+1)/2}^{8-v} \, dv$$

$$= \frac{1}{4} \int_{1}^{5} \left[-(v+1) \ln \frac{v+1}{2v} \right.$$

$$+ 2(8 - v) \ln \frac{8-v}{v} + 3(v - 5) \Big] \, dv$$

$$= \frac{1}{4} (49 \ln 7 - \frac{75}{2} \ln 5 - 27 \ln 3 + 6)$$

$$\approx 2.8333$$

53. We find $A = \frac{1}{5}$, $B = \frac{6}{5}$, so the transformed region

is $\frac{1}{5} u^2 + \frac{6}{5} v^2 = 1$. Since $u = x + 2y$,

$v = 2x - y$, we have $x = \frac{u+2v}{5}$, $y = \frac{2u-v}{5}$

and $J = \begin{vmatrix} \frac{1}{5} & \frac{2}{5} \\ \frac{2}{5} & -\frac{1}{5} \end{vmatrix} = -\frac{1}{5}$

Thus, the area is

$$A = \int \int_{R} dy \, dx$$

$$= 4 \int_{0}^{\sqrt{5/6}} \int_{0}^{\sqrt{5-6v^2}} \frac{1}{5} \, du \, dv$$

$$= \frac{\pi \sqrt{6}}{6}$$

$$\approx 1.2825$$

55. Let $u = x + y$, $v = 2y$, $x = u - v/2$, $y = v/2$.

The boundary $x + y = 0$ becomes $u = 0$;

$x + y = 2$ becomes $u = 2$; $y = 0$ becomes

$v = 0$; and $y = 1$ becomes $v = 2$.

$$J = \frac{\partial(x, y)}{\partial(u, v)}$$

$$= \begin{vmatrix} 1 & -\frac{1}{2} \\ 0 & \frac{1}{2} \end{vmatrix}$$

$$= \frac{1}{2}$$

$$dy \, dx = \frac{1}{2} \, du \, dv$$

$$\int_{R} \int f(x + y) \, dy \, dx = \frac{1}{2} \int_{0}^{2} \int_{0}^{2} f(u) \, dv \, du$$

$$= \int_{0}^{2} f(u) \, du$$

$$= \int_{0}^{2} f(t) \, dt$$

57. For the circumference, solve the equation

$$\frac{x^2}{a^2} + \frac{y^2}{b^2} = 1$$

for y to obtain $y = b \sqrt{1 - \frac{x^2}{a^2}}$. Then

$$\frac{dy}{dx} = \frac{-bx}{a\sqrt{a^2 - x^2}}$$

and the circumference, C, of the ellipse is given

by the improper integral

$$C = 4 \int_0^a \sqrt{1 + \left(\frac{-bx}{a\sqrt{a^2 - x^2}}\right)^2} \, dx$$

$$= 4 \int_0^a \frac{1}{a\sqrt{a^2 - x^2}} \sqrt{a^4 + (b^2 - a^2)x^2} \, dx$$

CHAPTER 12 REVIEW

Proficiency Examination, Pages 852-853

SURVIVAL HINT: *To help you review the concepts of this chapter, **handwrite** the answers to each of these questions onto your own paper.*

1. If f is defined on a closed, bounded region R in the xy-plane, then the double integral of f over R is defined by

$$\int_R \int f(x, y) \, dA = \lim_{\|P\| \to 0} \sum_{k=1}^N f(\overset{*}{x}_k, \overset{*}{y}_k) \Delta A_k$$

provided this limit exists. If the limit exists, we say that f is integrable over R.

2. If $f(x, y)$ is continuous over the rectangle R: $a \le x \le b$, $c \le y \le d$, then the double integral

$$\int_R \int f(x, y) \, dA$$

may be evaluated by either iterated integral; that is,

$$\int_R \int f(x, y) \, dA = \int_c^d \int_a^b f(x, y) \, dx \, dy$$

$$= \int_a^b \int_c^d f(x, y) \, dy \, dx$$

3. A type I region contains points (x, y) such that for each fixed x between constants a and b, y varies from $g_1(x)$ to $g_2(x)$, where g_1 and g_2 are continuous functions. This is vertically simple.

$$\int_D \int f(x, y) \, dA = \int_a^b \int_{g_1(x)}^{g_2(x)} f(x, y) \, dy \, dx$$

whenever both integrals exist.

4. A type II region contains points (x, y) such that for each fixed y between constants c and d, x varies from $h_1(y)$ to $h_2(y)$, where h_1 and h_2 are continuous functions. This is horizontally simple.

$$\int_D \int f(x, y) \, dA = \int_c^d \int_{h_1(y)}^{h_2(y)} f(x, y) \, dx \, dy$$

whenever both integrals exist.

5. The area of the region D in the xy-plane is given by $A = \int_D \int \, dA$

6. If f is continuous and $f(x, y) \geq 0$ on the region D, the volume of the solid under the surface $z = f(x, y)$ above the region D is given by $V = \displaystyle\int\int_D f(x, y)\, dA$

7. **a.** Linearity rule: for constants a and b,

$$\int\int_D [af(x, y) + bg(x, y)]\, dA$$

$$= a\int\int_D f(x, y)\, dA + b\int\int_D g(x, y)\, dA$$

b. Dominance rule: If $f(x, y) \geq g(x, y)$ throughout a region D, then

$$\int\int_D f(x, y)\, dA \geq \int\int_D g(x, y)\, dA$$

c. Subdivision rule: If the region of integration D can be subdivided into two subregions D_1 and D_2, then

$$\int\int_D f(x, y)\, dA$$

$$= \int\int_{D_1} f(x, y)\, dA + \int\int_{D_2} f(x, y)\, dA$$

8. If f is continuous in the polar region D such that for each fixed θ between α and β, r varies between $h_1(\theta)$ and $h_2(\theta)$, then

$$\int\int_D f(r, \theta)\, dA = \int_\alpha^\beta \int_{h_1(\theta)}^{h_2(\theta)} f(r, \theta)\; r\, dr\, d\theta$$

9. Assume that the function $f(x, y)$ has continuous partial derivatives f_x and f_y in a region R of the xy-plane. Then the portion of the surface $z = f(x, y)$ that lies over R has surface area

$$S = \int\int_R \sqrt{[f_x(x, y)]^2 + [f_y(x, y)]^2 + 1}\; dA$$

10. Let D be a region in the xy-plane on which x, y, z and their partial derivatives with respect to u and v are continuous. Also, let S be a surface defined by a vector function

$$\mathbf{R}(u, v) = x(u, v)\mathbf{i} + y(u, v)\mathbf{j} + z(u, v)\mathbf{k}$$

Then the surface area is defined by

$$S = \int\int_D \| \mathbf{R}_u(u, v) \times \mathbf{R}_v(u, v) \|\, du\, dv$$

11. If $f(x, y, z)$ is continuous over a rectangular solid R: $a \leq x \leq b, c \leq y \leq d$, $r \leq z \leq s$, then the triple integral may be evaluated by the iterated integral

$$\int\int\int_R f(x, y, z)\, dV = \int_r^s \int_c^d \int_a^b f(x, y, z)\, dx\, dy\, dz$$

The iterated integration can be performed in any order (with appropriate adjustments) to the limits of integration: $dx\ dy\ dz,\ dx\ dz\ dy,$ $dz\ dx\ dy,\ dy\ dx\ dz,\ dy\ dz\ dx,\ dz\ dy\ dx.$

12. If V is the volume of the solid region S, then

$$V = \int \int_S \int dV$$

13. If δ is a continuous density function on the lamina corresponding to a plane region R, then the mass m of the lamina is given by

$$m = \int_R \int \delta(x,\ y)\ dA$$

14. If δ is a continuous density function on a lamina corresponding to a plane region R, then the moments of mass with respect to the x-axis is

$$M_x = \int_R \int y\ \delta(x,\ y)\ dA$$

15. If m is the mass of the lamina, the center of mass is $(\overline{x},\ \overline{y})$, where

$$\overline{x} = \frac{M_y}{m} \quad \text{and} \quad \overline{y} = \frac{M_x}{m}$$

If the density δ is constant, the point $(\overline{x},\ \overline{y})$ is called the centroid of the region.

16. The moments of inertia of a lamina of variable density δ about the x- and y-axes, respectively, are

$$I_x = \int_R \int y^2 \delta(x,\ y)\ dA$$

and $$I_y = \int_R \int x^2 \delta(x,\ y)\ dA$$

17. A joint probability density function for the random variables X and Y is a continuous, nonnegative function $f(x,\ y)$ such that

$$P[(X,\ Y)\ \text{in}\ R] = \int_R \int f(x,\ y)\ dy\ dx$$

and $$\int_{-\infty}^{\infty} \int_{-\infty}^{\infty} f(x,\ y)\ dy\ dx = 1$$

where $P[(X,\ Y)\ \text{in}\ R]$ denotes the probability that $(X,\ Y)$ is in the region R in the xy-plane.

18. Rectangular to cylindrical:

$$r = \sqrt{x^2 + y^2};\ \tan\theta = \frac{y}{x};\ z = z$$

Rectangular to spherical:

$$\rho = \sqrt{x^2 + y^2 + z^2};\ \tan\theta = \frac{y}{x}$$

$$\phi = \cos^{-1}\left(\frac{z}{\sqrt{x^2 + y^2 + z^2}}\right)$$

Cylindrical to rectangular:

$x = r \cos \theta; \; y = r \sin \theta; \; z = z$

Cylindrical to spherical:

$\rho = \sqrt{r^2 + z^2}; \; \theta = \theta; \; \phi = \cos^{-1}\left(\dfrac{z}{\sqrt{r^2 + z^2}}\right)$

Spherical to rectangular:

$x = \rho \sin \phi \cos \theta; \; y = \rho \sin \phi \sin \theta; \; z = \rho \cos \phi$

Spherical to cylindrical:

$r = \rho \sin \phi; \; \theta = \theta; \; z = \rho \cos \phi$

19. Let f be a continuous function on the bounded, solid region S. Then the triple integral of f over S is given by:

a. $\displaystyle\iiint_S f(r, \theta, z) \; r \; dz \; dr \; d\theta$

in cylindrical coordinates

b. $\displaystyle\iiint_S f(\rho, \theta, \phi) \; \rho^2 \sin \phi \; d\rho \; d\theta \; d\phi$

in spherical coordinates

20. $\left|\dfrac{\partial(x, y)}{\partial(u, v)}\right| = \begin{vmatrix} \dfrac{\partial x}{\partial u} & \dfrac{\partial x}{\partial v} \\[2mm] \dfrac{\partial y}{\partial u} & \dfrac{\partial y}{\partial v} \end{vmatrix}$

$= \dfrac{\partial x}{\partial u}\dfrac{\partial y}{\partial v} - \dfrac{\partial y}{\partial u}\dfrac{\partial x}{\partial v}$

21. Let f be a continuous function on a region D,

and let T be a one-to-one transformation that maps the region D^* in the uv-plane onto a region D in the xy-plane under the change of variable $x = x(u, v)$, $y = y(u, v)$ where g and h are continuously differentiable on D^*. Then

$$\iint_D f(x, y) \; dy \; dx$$

$$= \iint_{D^*} f[x(u, v), y(u, v)] \, |J(u, v)| \; du \; dv$$

22. $\displaystyle\int_0^{\pi/3} \int_0^{\sin y} e^{-x} \cos y \; dx \; dy$

$= \displaystyle\int_0^{\pi/3} \left((e^{-\sin y})(-\cos y) + \cos y\right) dy$

$\boxed{\text{Let } u = -\sin y; \; dv = -\cos y \; dy}$

$= \displaystyle\int_0^{-\sqrt{3}/2} e^u \; du + \int_0^{\pi/3} \cos y \; dy$

$= e^{-\sqrt{3}/2} + \dfrac{\sqrt{3}}{2} - 1$

≈ 0.2866

23. $\displaystyle\int_{-1}^{1}\int_{0}^{z}\int_{y}^{y-z}(x+y-z)\,dx\,dy\,dz$

$\displaystyle=\int_{-1}^{1}\int_{0}^{z}\left(\frac{(y-z)^2}{2}+(y-z)y-(y-z)z\right.$

$\displaystyle\qquad\qquad\left.-\frac{y^2}{2}-y^2+yz\right)dy\,dz$

$\displaystyle=\int_{-1}^{1}\int_{0}^{z}\left(-2yz+\frac{3z^2}{2}\right)dy\,dz$

$\displaystyle=\int_{-1}^{1}\left[\frac{3}{2}yz^2-y^2z\right]\Big|_{0}^{z}dz$

$\displaystyle=\frac{1}{2}\int_{-1}^{1}z^3\,dz$

$=0$

24. $\displaystyle 2\int_{0}^{3}\int_{0}^{9-x^2}dy\,dx=2\int_{0}^{3}(9-x^2)\,dx$

$\displaystyle\qquad\qquad\qquad=36$

25. $\displaystyle A=\int_{0}^{\pi/2}\int_{0}^{1}\cos r^2\,r\,dr\,d\theta$

$\displaystyle=\frac{1}{2}\int_{0}^{\pi/2}\int_{0}^{1}(\cos r^2)(2r\,dr)\,d\theta$

$\displaystyle=\frac{1}{2}\int_{0}^{\pi/2}\sin 1\,d\theta$

$\displaystyle=\frac{\pi}{4}\sin 1$

≈ 0.6609

26. $z=c\left(1-\frac{x}{a}-\frac{y}{b}\right)$; $y=b\left(1-\frac{x}{a}\right)$ and the projected region on the xy-plane is the triangle bounded by $x=0$, $y=0$, and

$y=b\left(1-\frac{x}{a}\right)$

$\displaystyle V=\int_{0}^{a}\int_{0}^{b(1-x/a)}\int_{0}^{c(1-x/a-y/b)}dz\,dy\,dx$

$\displaystyle=-\frac{c}{ab}\int_{0}^{a}\int_{0}^{b(1-x/a)}(bx+ay-ab)\,dy\,dx$

$\displaystyle=\frac{bc}{2a^2}\int_{0}^{a}(x-a)^2\,dx$

$\displaystyle=\frac{abc}{6}$

27. The appliance fails during the first year if both components fail in that time; that is, if $(X,\,Y)$ lies in the square $0\le x\le 1$, $0\le y\le 1$. The probability of this occurring is

$P[0\le x\le 1,\,0\le y\le 1]$

$\displaystyle=\int_{0}^{1}\int_{0}^{1}\frac{1}{4}e^{-x/2}e^{-y/2}\,dy\,dx$

$\displaystyle=\int_{0}^{1}\frac{1}{4}e^{-x/2}[-2e^{-y/2}]\Big|_{0}^{1}\,dx$

$$= \left(\frac{1 - e^{-1/2}}{2} \right) \int_0^1 e^{-x/2} \, dx$$

$$= (1 - e^{-1/2})^2$$

$$\approx 0.1548$$

Thus, the probability of product failure is about 15%.

28. First find the intersection of the plane and the paraboloid:

$$x^2 + 2y^2 = 4x$$

$$(x - 2)^2 + 2y^2 = 4$$

$$\frac{(x - 2)^2}{4} + \frac{y^2}{2} = 1$$

This is a translated ellipse centered at $(2, 0)$ with x intercepts of 0 and 4.

$$V = \int_0^4 \int_0^{\sqrt{4x - x^2}/\sqrt{2}} \int_{x^2 + 2y^2}^{4x} dz \, dy \, dx$$

$$= \int_0^4 \int_0^{\sqrt{4x - x^2}/\sqrt{2}} (4x - x^2 - 2y^2) \, dy \, dx$$

$$= \int_0^4 \left\{ \frac{(4x - x^2)^{3/2}}{\sqrt{2}} - \frac{2}{3} \left(\frac{(4x - x^2)^{3/2}}{2\sqrt{2}} \right) \right\} \, dx$$

$$= \frac{2}{3\sqrt{2}} \int_0^4 (4x - x^2)^{3/2} \, dx$$

$$= \frac{\sqrt{2}}{3} \int_0^4 \left(2^2 - (x - 2)^2 \right)^{3/2} \, dx$$

$$\boxed{\text{Formula 245}}$$

$$= \frac{\sqrt{2}}{3} \left(\frac{(x - 2)(4x - x^2)^{3/2}}{4} \right.$$

$$+ \frac{12(x - 2)(4x - x^2)^{1/2}}{8}$$

$$+ \left. \frac{3}{8}(16) \sin^{-1} \frac{x - 2}{2} \right) \Big|_0^4$$

$$= \frac{\sqrt{2}}{3} \left(0 + 0 + 6(\tfrac{\pi}{2}) - 6\left(-\tfrac{\pi}{2} \right) \right)$$

$$= \frac{\sqrt{2}}{3}(6\pi)$$

$$= 2\pi\sqrt{2}$$

$$\approx 8.8858$$

29. The density is $\delta = r$ and the projected region is $x^2 + y^2 = 4$. In cylindrical coordinates,

$$m = \int_0^{2\pi} \int_0^2 \int_{r^2}^4 (r) \, r \, dz \, dr \, d\theta$$

$$= \int_0^{2\pi} \int_0^2 r^2(4 - r^2) \, dr \, d\theta$$

$$= \int_0^{2\pi} \left[\frac{4}{3} r^3 - \frac{1}{5} r^5 \right] \Big|_0^2 \, d\theta \qquad\qquad = \frac{1}{3}(e^2 + \frac{8}{e} - 3)$$

$$\approx 2.4440$$

$$= \int_0^{2\pi} \left(\frac{32}{3} - \frac{32}{5} \right) d\theta$$

$$= \frac{64}{15} \int_0^{2\pi} d\theta$$

$$= \frac{128\pi}{15}$$

$$\approx 26.8083$$

30. Let $u = x + y$ and $v = x - 2y$, so that

$x = \frac{1}{3}(2u + v)$, $y = \frac{1}{3}(u - v)$.

$$\frac{\partial(x,y)}{\partial(u,v)} = \begin{vmatrix} \frac{2}{3} & \frac{1}{3} \\ \frac{1}{3} & -\frac{1}{3} \end{vmatrix}$$

$$= -\frac{1}{3}$$

The region R is bounded by the lines $y = 0$,

$y = 2 - x$, $y = x$ which transform into the

lines $u = v$, $u = 2$, and $u = -2v$.

$$\int\limits_R \int (x + y)e^{x - 2y} \, dy \, dx$$

$$= \int_0^2 \int_{-u/2}^{u} u e^v \Big| -\frac{1}{3} \Big| \, dv \, du$$

$$= \frac{1}{3} \int_0^2 \left(u e^u - u e^{-u/2} \right) du$$

CHAPTER 13

Vector Analysis

13.1 Properties of a Vector Field: Divergence and Curl, Pages 866-867

Answers to Problems 1-8 may vary.

1. Answers vary; the operator

$$\nabla = \frac{\partial}{\partial x}\mathbf{i} + \frac{\partial}{\partial y}\mathbf{j} + \frac{\partial}{\partial z}\mathbf{k}$$

is called the del operator and is applicable to

a scalar function only:

$$\nabla f = f_x\mathbf{i} + f_y\mathbf{j} + f_z\mathbf{k}$$

The del operator is used to compute

divergence by

$$\text{div } \mathbf{F} = \nabla \cdot \mathbf{F}$$

This is applicable to vector functions only and

generates a scalar. It is also used to compute

curl by

$$\text{curl } \mathbf{F} = \nabla \times \mathbf{F}$$

This is applicable to vector functions only and

generates a vector.

Finally, we note that $\nabla^2 f = \nabla \cdot \nabla f$.

SURVIVAL HINT: *Remember that div F is always a scalar function and curl F is always a vector function.*

3. $\mathbf{F}(x, y, z) = x^2\mathbf{i} + xy\mathbf{j} + z^3\mathbf{k}$

$\text{div } \mathbf{F} = \nabla \cdot \mathbf{F}$

$$= \left(\frac{\partial}{\partial x}\mathbf{i} + \frac{\partial}{\partial y}\mathbf{j} + \frac{\partial}{\partial z}\mathbf{k}\right) \cdot (x^2\mathbf{i} + xy\mathbf{j} + z^3\mathbf{k})$$

$$= 2x + x + 3z^2$$

$$= 3x + 3z^2$$

$$\text{curl } \mathbf{F} = \nabla \times \mathbf{F} = \begin{vmatrix} \mathbf{i} & \mathbf{j} & \mathbf{k} \\ \frac{\partial}{\partial x} & \frac{\partial}{\partial y} & \frac{\partial}{\partial z} \\ x^2 & xy & z^3 \end{vmatrix}$$

$$= 0\mathbf{i} - 0\mathbf{j} + (y - 0)\mathbf{k} = y\mathbf{k}$$

5. $\mathbf{F}(x, y, z) = 2y\mathbf{j}$

$\text{div } \mathbf{F} = \nabla \cdot \mathbf{F}$

$$= \left(\frac{\partial}{\partial x}\mathbf{i} + \frac{\partial}{\partial y}\mathbf{j} + \frac{\partial}{\partial z}\mathbf{k}\right) \cdot (2y\mathbf{j})$$

$$= 2$$

$\text{curl } \mathbf{F} = \nabla \times \mathbf{F}$

$$= \begin{vmatrix} \mathbf{i} & \mathbf{j} & \mathbf{k} \\ \frac{\partial}{\partial x} & \frac{\partial}{\partial y} & \frac{\partial}{\partial z} \\ 0 & 2y & 0 \end{vmatrix}$$

$$= 0$$

7. $\mathbf{F}(x, y, z) = \mathbf{i} + \mathbf{j} + \mathbf{k}$

 div $\mathbf{F} = \nabla \cdot \mathbf{F}$

 $$= \left(\frac{\partial}{\partial x}\mathbf{i} + \frac{\partial}{\partial y}\mathbf{j} + \frac{\partial}{\partial z}\mathbf{k}\right) \cdot (\mathbf{i} + \mathbf{j} + \mathbf{k})$$

 $$= 0$$

 curl $\mathbf{F} = \nabla \times \mathbf{F}$

 $$= \begin{vmatrix} \mathbf{i} & \mathbf{j} & \mathbf{k} \\ \frac{\partial}{\partial x} & \frac{\partial}{\partial y} & \frac{\partial}{\partial z} \\ 1 & 1 & 1 \end{vmatrix}$$

 $$= \mathbf{0}$$

 Note: $\mathbf{F}$ is a constant vector, so the result is

 the same regardless of the given point.

9. $\mathbf{F}(x, y, z) = xyz\mathbf{i} + y\mathbf{j} + x\mathbf{k}$

 div $\mathbf{F} = \nabla \cdot \mathbf{F}$

 $$= \left(\frac{\partial}{\partial x}\mathbf{i} + \frac{\partial}{\partial y}\mathbf{j} + \frac{\partial}{\partial z}\mathbf{k}\right) \cdot (xyz\mathbf{i} + y\mathbf{j} + x\mathbf{k})$$

 $$= yz + 1$$

 curl $\mathbf{F} = \nabla \times \mathbf{F}$

 $$= \begin{vmatrix} \mathbf{i} & \mathbf{j} & \mathbf{k} \\ \frac{\partial}{\partial x} & \frac{\partial}{\partial y} & \frac{\partial}{\partial z} \\ xyz & y & x \end{vmatrix}$$

 $$= 0\mathbf{i} - (1 - xy)\mathbf{j} - xz\mathbf{k}$$

 At $(1, 2, 3)$,

div $\mathbf{F} = 2(3) + 1$

 $$= 7$$

 curl $\mathbf{F} = -(1 - 1(2))\mathbf{j} - (1)(3)\mathbf{k}$

 $$= \mathbf{j} - 3\mathbf{k}$$

11. $\mathbf{F}(x, y, z) = e^{-xy}\mathbf{i} + e^{xz}\mathbf{j} + e^{yz}\mathbf{k}$

 div $\mathbf{F} = \nabla \cdot \mathbf{F}$

 $$= \left(\frac{\partial}{\partial x}\mathbf{i} + \frac{\partial}{\partial y}\mathbf{j} + \frac{\partial}{\partial z}\mathbf{k}\right) \cdot (e^{-xy}\mathbf{i} + e^{xz}\mathbf{j} + e^{yz}\mathbf{k})$$

 $$= -ye^{-xy} + ye^{yz}$$

 curl $\mathbf{F} = \nabla \times \mathbf{F}$

 $$= \begin{vmatrix} \mathbf{i} & \mathbf{j} & \mathbf{k} \\ \frac{\partial}{\partial x} & \frac{\partial}{\partial y} & \frac{\partial}{\partial z} \\ e^{-xy} & e^{xz} & e^{yz} \end{vmatrix}$$

 $$= (ze^{yz} - xe^{xz})\mathbf{i} - 0\mathbf{j} + (ze^{xz} + xe^{-xy})\mathbf{k}$$

 At $(3, 2, 0)$,

 div $\mathbf{F} = -2e^{-6} + 2e^0$

 $$= 2 - 2e^{-6}$$

 curl $\mathbf{F} = (0 - 3e^0)\mathbf{i} + (0 + 3e^{-6})\mathbf{k}$

 $$= -3\mathbf{i} + 3e^{-6}\mathbf{k}$$

13. $\mathbf{F}(x, y) = (\sin x)\mathbf{i} + (\cos y)\mathbf{j}$

 div $\mathbf{F} = \nabla \cdot \mathbf{F}$

 $$= \left(\frac{\partial}{\partial x}\mathbf{i} + \frac{\partial}{\partial y}\mathbf{j}\right) \cdot [(\sin x)\mathbf{i} + (\cos y)\mathbf{j}]$$

 $$= \cos x - \sin y$$

$$\text{curl } \mathbf{F} = \nabla \times \mathbf{F}$$

$$= \begin{vmatrix} \mathbf{i} & \mathbf{j} & \mathbf{k} \\ \dfrac{\partial}{\partial x} & \dfrac{\partial}{\partial y} & \dfrac{\partial}{\partial z} \\ \sin x & \cos y & 0 \end{vmatrix}$$

$$= \mathbf{0}$$

15. $\mathbf{F}(x, y) = x\mathbf{i} - y\mathbf{j}$

div $\mathbf{F} = \nabla \cdot \mathbf{F}$

$$= \left(\dfrac{\partial}{\partial x}\mathbf{i} + \dfrac{\partial}{\partial y}\mathbf{j} \right) \cdot (x\mathbf{i} - y\mathbf{j})$$

$$= 0$$

curl $\mathbf{F} = \nabla \times \mathbf{F}$

$$= \begin{vmatrix} \mathbf{i} & \mathbf{j} & \mathbf{k} \\ \dfrac{\partial}{\partial x} & \dfrac{\partial}{\partial y} & \dfrac{\partial}{\partial z} \\ x & -y & 0 \end{vmatrix}$$

$$= \mathbf{0}$$

17. $\mathbf{F}(x, y) = \dfrac{x}{\sqrt{x^2 + y^2}}\,\mathbf{i} + \dfrac{y}{\sqrt{x^2 + y^2}}\,\mathbf{j}$

div $\mathbf{F} = \nabla \cdot \mathbf{F}$

$$= \left(\dfrac{\partial}{\partial x}\mathbf{i} + \dfrac{\partial}{\partial y}\mathbf{j} \right) \cdot \left(\dfrac{x}{\sqrt{x^2 + y^2}}\,\mathbf{i} + \dfrac{y}{\sqrt{x^2 + y^2}}\,\mathbf{j} \right)$$

$$= \dfrac{\sqrt{x^2 + y^2} - \dfrac{x^2}{\sqrt{x^2 + y^2}}}{x^2 + y^2}$$

$$+ \dfrac{\sqrt{x^2 + y^2} - \dfrac{y^2}{\sqrt{x^2 + y^2}}}{x^2 + y^2}$$

$$= \dfrac{1}{\sqrt{x^2 + y^2}}$$

curl $\mathbf{F} = \nabla \times \mathbf{F}$

$$= \begin{vmatrix} \mathbf{i} & \mathbf{j} & \mathbf{k} \\ \dfrac{\partial}{\partial x} & \dfrac{\partial}{\partial y} & \dfrac{\partial}{\partial z} \\ \dfrac{x}{\sqrt{x^2 + y^2}} & \dfrac{y}{\sqrt{x^2 + y^2}} & 0 \end{vmatrix}$$

$$= \mathbf{0}$$

19. $\mathbf{F}(x, y, z) = ax\mathbf{i} + by\mathbf{j} + c\mathbf{k}$

div $\mathbf{F} = \nabla \cdot \mathbf{F}$

$$= \left(\dfrac{\partial}{\partial x}\mathbf{i} + \dfrac{\partial}{\partial y}\mathbf{j} + \dfrac{\partial}{\partial z}\mathbf{k} \right) \cdot (ax\mathbf{i} + by\mathbf{j} + c\mathbf{k})$$

$$= a + b$$

curl $\mathbf{F} = \nabla \times \mathbf{F}$

$$= \begin{vmatrix} \mathbf{i} & \mathbf{j} & \mathbf{k} \\ \dfrac{\partial}{\partial x} & \dfrac{\partial}{\partial y} & \dfrac{\partial}{\partial z} \\ ax & by & c \end{vmatrix}$$

$$= \mathbf{0}$$

21. $\mathbf{F}(x, y, z) = x^2\mathbf{i} + y^2\mathbf{j} + z^2\mathbf{k}$

$\text{div } \mathbf{F} = \nabla \cdot \mathbf{F}$

$\quad = \left(\dfrac{\partial}{\partial x}\mathbf{i} + \dfrac{\partial}{\partial y}\mathbf{j} + \dfrac{\partial}{\partial z}\mathbf{k}\right) \cdot (x^2\mathbf{i} + y^2\mathbf{j} + z^2\mathbf{k})$

$\quad = 2x + 2y + 2z$

$\quad = 2(x + y + z)$

$\text{curl } \mathbf{F} = \begin{vmatrix} \mathbf{i} & \mathbf{j} & \mathbf{k} \\ \dfrac{\partial}{\partial x} & \dfrac{\partial}{\partial y} & \dfrac{\partial}{\partial z} \\ x^2 & y^2 & z^2 \end{vmatrix}$

$\quad = \mathbf{0}$

23. $\mathbf{F}(x, y, z) = xy\mathbf{i} + yz\mathbf{j} + xz\mathbf{k}$

$\text{div } \mathbf{F} = \nabla \cdot \mathbf{F}$

$\quad = \left(\dfrac{\partial}{\partial x}\mathbf{i} + \dfrac{\partial}{\partial y}\mathbf{j} + \dfrac{\partial}{\partial z}\mathbf{k}\right) \cdot (xy\mathbf{i} + yz\mathbf{j} + xz\mathbf{k})$

$\quad = x + y + z$

$\text{curl } \mathbf{F} = \begin{vmatrix} \mathbf{i} & \mathbf{j} & \mathbf{k} \\ \dfrac{\partial}{\partial x} & \dfrac{\partial}{\partial y} & \dfrac{\partial}{\partial z} \\ xy & yz & xz \end{vmatrix}$

$\quad = -y\mathbf{i} - z\mathbf{j} - x\mathbf{k}$

25. $\mathbf{F}(x, y, z) = xyz\mathbf{i} + x^2y^2z^2\mathbf{j} + y^2z^3\mathbf{k}$

$\text{div } \mathbf{F} = \nabla \cdot \mathbf{F}$

$\quad = \left(\dfrac{\partial}{\partial x}\mathbf{i} + \dfrac{\partial}{\partial y}\mathbf{j} + \dfrac{\partial}{\partial z}\mathbf{k}\right) \cdot (xyz\mathbf{i} + x^2y^2z^2\mathbf{j} + y^2z^3\mathbf{k})$

$\quad = yz + 2x^2yz^2 + 3y^2z^2$

$\text{curl } \mathbf{F} = \begin{vmatrix} \mathbf{i} & \mathbf{j} & \mathbf{k} \\ \dfrac{\partial}{\partial x} & \dfrac{\partial}{\partial y} & \dfrac{\partial}{\partial z} \\ xyz & x^2y^2z^2 & y^2z^3 \end{vmatrix}$

$\quad = (2yz^3 - 2x^2y^2z)\mathbf{i} + xy\mathbf{j} + (2xy^2z^2 - xz)\mathbf{k}$

27. $\mathbf{F}(x, y, z) = (z^2e^{-x})\mathbf{i} + (y^3\ln z)\mathbf{j} + (xe^{-y})\mathbf{k}$

$\text{div } \mathbf{F} = \nabla \cdot \mathbf{F}$

$\quad = \left(\dfrac{\partial}{\partial x}\mathbf{i} + \dfrac{\partial}{\partial y}\mathbf{j} + \dfrac{\partial}{\partial z}\mathbf{k}\right) \cdot \left(z^2e^{-x}\mathbf{i}\right.$

$\qquad \left. + y^3\ln z\mathbf{j} + xe^{-y}\mathbf{k}\right)$

$\quad = -z^2e^{-x} + 3y^2\ln z$

$\text{curl } \mathbf{F} = \begin{vmatrix} \mathbf{i} & \mathbf{j} & \mathbf{k} \\ \dfrac{\partial}{\partial x} & \dfrac{\partial}{\partial y} & \dfrac{\partial}{\partial z} \\ z^2e^{-x} & y^3\ln z & xe^{-y} \end{vmatrix}$

$\quad = \left(\dfrac{-y^3}{z} - xe^{-y}\right)\mathbf{i} + (2ze^{-x} - e^{-y})\mathbf{j}$

29. $u(x, y, z) = e^{-x}(\cos y - \sin y)$

$u_x = -e^{-x}(\cos y - \sin y)$

$u_{xx} = e^{-x}(\cos y - \sin y)$

$u_y = e^{-x}(-\sin y - \cos y)$

$u_{yy} = e^{-x}(-\cos y + \sin y)$

$u_{xx} + u_{yy} = 0$

u is harmonic

31. $w(x, y, z) = (x^2 + y^2 + z^2)^{-1/2}$

$$w_x = \frac{-x}{(x^2 + y^2 + z^2)^{3/2}}$$

$$w_{xx} = -\frac{(x^2 + y^2 + z^2)^{3/2} - 3x^2(x^2 + y^2 + z^2)^{1/2}}{(x^2 + y^2 + z^2)^3}$$

$$= \frac{2x^2 - y^2 - z^2}{(x^2 + y^2 + z^2)^{5/2}}$$

Similarly,

$$w_{yy} = \frac{2y^2 - x^2 - z^2}{(x^2 + y^2 + z^2)^{5/2}};$$

$$w_{zz} = \frac{2z^2 - x^2 - y^2}{(x^2 + y^2 + z^2)^{5/2}}$$

$$w_{xx} + w_{yy} + w_{zz} = 0$$

w is harmonic.

33. $\mathbf{B} = y^2 z \mathbf{i} + xz^3 \mathbf{j} + y^2 x^2 \mathbf{k}$

$$\text{div } \mathbf{B} = \frac{\partial(y^2 z)}{\partial x} + \frac{\partial(xz^3)}{\partial y} + \frac{\partial(y^2 x^2)}{\partial z}$$

$$= 0 + 0 + 0$$

$$= 0$$

Thus, $\mathbf{B}$ is incompressible.

35. $f(x, y, z) = x^2 y z^3$

$$\mathbf{F} = \nabla f$$

$$= 2xyz^3 \mathbf{i} + x^2 z^3 \mathbf{j} + 3x^2 y z^2 \mathbf{k}$$

$$\text{div } \mathbf{F} = 2yz^3 + 6x^2 yz$$

37. $\mathbf{F}(x, y, z) = xy\mathbf{i} + yz\mathbf{j} + z^2\mathbf{k}$

$\mathbf{G}(x, y, z) = x\mathbf{i} + y\mathbf{j} - z\mathbf{k}$

$$\mathbf{F} \times \mathbf{G} = \begin{vmatrix} \mathbf{i} & \mathbf{j} & \mathbf{k} \\ xy & yz & z^2 \\ x & y & -z \end{vmatrix}$$

$$= (-2yz^2)\mathbf{i} + (xyz + xz^2)\mathbf{j} + (xy^2 - xyz)\mathbf{k}$$

$$\text{curl}(\mathbf{F} \times \mathbf{G}) = \begin{vmatrix} \mathbf{i} & \mathbf{j} & \mathbf{k} \\ \dfrac{\partial}{\partial x} & \dfrac{\partial}{\partial y} & \dfrac{\partial}{\partial z} \\ -2yz^2 & xyz+xz^2 & xy^2-xyz \end{vmatrix}$$

$$= (2xy - xz - xy - 2xz)\mathbf{i}$$

$$- (y^2 - yz + 4yz)\mathbf{j} + (yz + z^2 + 2z^2)\mathbf{k}$$

$$= (xy - 3xz)\mathbf{i} - (y^2 + 3yz)\mathbf{j} + (yz + 3z^2)\mathbf{k}$$

39. $\mathbf{F}(x, y, z) = xy\mathbf{i} + yz\mathbf{j} + z^2\mathbf{k}$

$\mathbf{G}(x, y, z) = x\mathbf{i} + y\mathbf{j} - z\mathbf{k}$

$$\mathbf{F} \times \mathbf{G} = \begin{vmatrix} \mathbf{i} & \mathbf{j} & \mathbf{k} \\ xy & yz & z^2 \\ x & y & -z \end{vmatrix}$$

$$= (-2yz^2)\mathbf{i} + (xyz + xz^2)\mathbf{j} + (xy^2 - xyz)\mathbf{k}$$

$$\text{div}(\mathbf{F} \times \mathbf{G})$$

$$= \frac{\partial(-2yz^2)}{\partial x} + \frac{\partial(xyz + xz^2)}{\partial y} + \frac{\partial(xy^2 - xyz)}{\partial z}$$

$$= xz - xy$$

41. Given $\mathbf{R} = x\mathbf{i} + y\mathbf{j} + z\mathbf{k}$; let

$$\mathbf{A} = a\mathbf{i} + b\mathbf{j} + c\mathbf{k}$$

$$\mathbf{A} \times \mathbf{R} = \begin{vmatrix} \mathbf{i} & \mathbf{j} & \mathbf{k} \\ a & b & c \\ x & y & z \end{vmatrix}$$

$$= (bz - cy)\mathbf{i} - (az - cx)\mathbf{j} + (ay - bx)\mathbf{k}$$

$$\text{curl}(\mathbf{A} \times \mathbf{R}) = \begin{vmatrix} \mathbf{i} & \mathbf{j} & \mathbf{k} \\ \dfrac{\partial}{\partial x} & \dfrac{\partial}{\partial y} & \dfrac{\partial}{\partial z} \\ bz - cy & cx - az & ay - bx \end{vmatrix}$$

$$= 2a\mathbf{i} + 2b\mathbf{j} + 2c\mathbf{k}$$

$$= 2\mathbf{A}$$

43. Let $\omega = a\mathbf{i} + b\mathbf{j} + c\mathbf{k}$; then

a. $\mathbf{V} = \omega \times \mathbf{R} = \begin{vmatrix} \mathbf{i} & \mathbf{j} & \mathbf{k} \\ a & b & c \\ x & y & z \end{vmatrix}$

$$= (bz - cy)\mathbf{i} + (cx - az)\mathbf{j} + (ay - bx)\mathbf{k}$$

b. $\text{div } \mathbf{V} = 0 + 0 + 0 = 0$

$$\text{curl } \mathbf{V} = \begin{vmatrix} \mathbf{i} & \mathbf{j} & \mathbf{k} \\ \dfrac{\partial}{\partial x} & \dfrac{\partial}{\partial y} & \dfrac{\partial}{\partial z} \\ bz - cy & cx - az & ay - bx \end{vmatrix}$$

$$= 2a\mathbf{i} + 2b\mathbf{j} + 2c\mathbf{k}$$

$$= 2\omega$$

45. Let $\mathbf{F} = f_1\mathbf{i} + f_2\mathbf{j} + f_3\mathbf{k}$ and
$\mathbf{G} = g_1\mathbf{i} + g_2\mathbf{j} + g_3\mathbf{k}$

$$\mathbf{F} \times \mathbf{G} = \begin{vmatrix} \mathbf{i} & \mathbf{j} & \mathbf{k} \\ f_1 & f_2 & f_3 \\ g_1 & g_2 & g_3 \end{vmatrix}$$

$$= (f_2 g_3 - f_3 g_2)\mathbf{i} + (f_3 g_1 - f_1 g_3)\mathbf{j}$$
$$+ (f_1 g_2 - f_2 g_1)\mathbf{k}$$

$$\text{div}(\mathbf{F} \times \mathbf{G}) = (f_2 g_3 - f_3 g_2)_x + (f_3 g_1 - f_1 g_3)_y$$
$$+ (f_1 g_2 - f_2 g_1)_z$$

I. $(\text{div } \mathbf{F})(\text{div } \mathbf{G})$

$$= [(f_1)_x + (f_2)_y + (f_3)_z][(g_1)_x + (g_2)_y + (g_3)_z]$$

$$\neq \text{div}(\mathbf{F} \times \mathbf{G})$$

II. $(\text{curl } \mathbf{F}) \cdot \mathbf{G} - \mathbf{F} \cdot (\text{curl } \mathbf{G})$

$$= \langle (f_3)_y - (f_2)_z, (f_1)_z - (f_3)_x,$$
$$(f_2)_x - (f_1)_y \rangle \cdot \langle g_1, g_2, g_3 \rangle$$
$$- \langle f_1, f_2, f_3 \rangle \cdot \langle (g_3)_y - (g_2)_z,$$
$$(g_1)_z - (g_3)_x, (g_2)_x - (g_1)_y \rangle$$

$$= g_1[(f_3)_y - (f_2)_z] + g_2[(f_1)_z - (f_3)_x]$$
$$+ g_3[(f_2)_x - (f_1)_y] - f_1[(g_3)_y - (g_2)_z]$$
$$- f_2[(g_1)_z - (g_3)_x] - f_3[(g_2)_x - (g_1)_y]$$

$$= (f_2 g_3 - f_3 g_2)_x + (f_3 g_1 - f_1 g_3)_y$$
$$+ (f_1 g_2 - f_2 g_1)_z$$

$$= \text{div}(\mathbf{F} \times \mathbf{G})$$

III. $\mathbf{F}(\text{div } \mathbf{G}) + (\text{div } \mathbf{F})\mathbf{G}$ is a vector, so it can't possibility equal $\text{div}(\mathbf{F} \times \mathbf{G})$.

IV. Note that since

$$(\text{curl } \mathbf{F}) \cdot \mathbf{G} - \mathbf{F} \cdot (\text{curl } \mathbf{G}) = \text{div}(\mathbf{F} \times \mathbf{G})$$

we cannot also have

$$(\text{curl } \mathbf{F}) \cdot \mathbf{G} + \mathbf{F} \cdot (\text{curl } \mathbf{G}) = \text{div}(\mathbf{F} \times \mathbf{G})$$

unless

$$2\mathbf{F} \cdot (\text{curl } \mathbf{G}) = 0$$

which is not generally true, as the following

counterexample illustrates:

$$\mathbf{F} = z\mathbf{k}, \ \mathbf{G} = x\mathbf{j},$$

so

$$\mathbf{F} \cdot (\text{curl } \mathbf{G}) = z\mathbf{k} \cdot \mathbf{k}$$

$$= z$$

47. div $\mathbf{F}$ + div $\mathbf{G}$

$$= [(f_1)_x + (f_2)_y + (f_3)_z] + [(g_1)_x + (g_2)_y + (g_3)_z]$$

$$= [(f_1)_x + (g_1)_x] + [(f_2)_y + (g_2)_y] + [(f_3)_z + (g_3)_z]$$

$$= \text{div}(\mathbf{F} + \mathbf{G})$$

49. $\quad \text{curl}(c\mathbf{F}) = \begin{vmatrix} \mathbf{i} & \mathbf{j} & \mathbf{k} \\ \dfrac{\partial}{\partial x} & \dfrac{\partial}{\partial y} & \dfrac{\partial}{\partial z} \\ cf_1 & cf_2 & cf_3 \end{vmatrix}$

$$= c \begin{vmatrix} \mathbf{i} & \mathbf{j} & \mathbf{k} \\ \dfrac{\partial}{\partial x} & \dfrac{\partial}{\partial y} & \dfrac{\partial}{\partial z} \\ f_1 & f_2 & f_3 \end{vmatrix}$$

$$= c \ \text{curl } \mathbf{F}$$

51. Let $\mathbf{F} = M\mathbf{i} + N\mathbf{j} + P\mathbf{k}$

$$\text{div}(f\mathbf{F}) = (fM)_x + (fN)_y + (fP)_x$$

$$= f(M_x + N_y + P_z) + Mf_x + Nf_y + Pf_z$$

$$= f \ \text{div } \mathbf{F} + [f_x\mathbf{i} + f_y\mathbf{j} + f_z\mathbf{k}] \cdot [M\mathbf{i} + N\mathbf{j} + P\mathbf{k}]$$

$$= f \ \text{div } \mathbf{F} + \nabla f \cdot \mathbf{F}$$

53. Apply the formula in Problem 51, with

$\mathbf{F} = \nabla g$. Then

$$\text{div}(f\nabla g) = f \text{div} \nabla g + \nabla f \cdot \nabla g$$

55. Let $\mathbf{F} = M\mathbf{i} + N\mathbf{j} + P\mathbf{k}$

$$\text{div}(\text{curl } \mathbf{F}) = \text{div} \begin{vmatrix} \mathbf{i} & \mathbf{j} & \mathbf{k} \\ \dfrac{\partial}{\partial x} & \dfrac{\partial}{\partial y} & \dfrac{\partial}{\partial z} \\ M & N & P \end{vmatrix}$$

$$= \text{div}[(P_y - N_z)\mathbf{i} + (M_z - P_x)\mathbf{j} + (N_x - M_y)\mathbf{k}]$$

$$= P_{yx} - N_{zy} + M_{zy} - P_{xy} + N_{xz} - M_{yz}$$

$$= (P_{yx} - P_{xy}) + (M_{zy} - M_{yz}) + (N_{xz} - N_{zx})$$

$$= 0$$

57. Let $\mathbf{F} = \dfrac{1}{r^3}(x\mathbf{i} + y\mathbf{j} + z\mathbf{k})$

div $\mathbf{F}$

$$= \frac{-(2x^2 - y^2 - z^2) - (2y^2 - x^2 - z^2) - (2z^2 - x^2 - y^2)}{(x^2 + y^2 + z^2)^{5/2}}$$

$$= 0$$

59. $\operatorname{div}(r\mathbf{R}) = \frac{\partial}{\partial x}(rx) + \frac{\partial}{\partial y}(ry) + \frac{\partial}{\partial z}(rz)$

$$= (r + \tfrac{x^2}{r}) + (r + \tfrac{y^2}{r}) + (r + \tfrac{z^2}{r})$$

$$= \frac{3r^2 + (x^2 + y^2 + z^2)}{r}$$

$$= \frac{3r^2 + r^2}{r}$$

$$= 4r$$

61. $\operatorname{div} \nabla(fg) = \nabla \cdot (f\nabla g + g\nabla f)$

$$= f(\nabla \cdot \nabla g) + \nabla f \cdot \nabla g + \nabla g \cdot \nabla f + g(\nabla \cdot \nabla f)$$

$$= f \operatorname{div}(\nabla g) + 2\nabla f \cdot \nabla g + g \operatorname{div}(\nabla f)$$

13.2 Line Integrals, Pages 875-877

1. Let $f(x, y, z)$ be a function of t where

$x = x(t)$, $y = y(t)$, $z = z(t)$ over $a \le t \le b$ is

given. Then

$$\int_C f \, ds$$

$$= \int_a^b f[x(t),\, y(t),\, z(t)] \sqrt{[x'(t)]^2 + [y'(t)]^2 + [z'(t)]^2} \, dt$$

and

$$\int_C f \, dx = \int_a^b f[x(t),\, y(t),\, z(t)] \, x'(t) \, dt$$

3. $ds = \sqrt{[x'(t)]^2 + [y'(t)]^2} \, dt$

$$= \sqrt{(3t^{1/2})^2 + (3)^2} \, dt$$

$$= \sqrt{9t + 9} \, dt$$

$$\int_C \frac{1}{3 + y} \, ds = \int_0^1 \frac{1}{3 + 3t} \sqrt{9t + 9} \, dt$$

$$= 2\sqrt{t + 1} \Big|_0^1$$

$$= 2\sqrt{2} - 2$$

5. $ds = \sqrt{[x'(t)]^2 + [y'(t)]^2} \, dt$

$$= \sqrt{4 + 16t^6} \, dt$$

$$\int_C \frac{y^2}{x^3} \, ds = \int_1^2 \frac{(t^4)^2}{(2t)^3} \sqrt{4 + 16t^6} \, dt$$

$$= \frac{1}{4} \int_1^2 t^5 \sqrt{1 + 4t^6} \, dt$$

$$= \left[\frac{(1 + 4t^6)^{3/2}}{144} \right]\Big|_1^2$$

$$= \frac{1}{144}(257^{3/2} - 5^{3/2})$$

$$\approx 28.534$$

7. $x = t$, $y = 4t^2$, $-1 \le t \le 0$

$$\int_C (-y \, dx + x \, dy) = \int_{-1}^0 \left[-(4t^2) \, dt + t(8t \, dt) \right]$$

$$= \int_{-1}^0 \left[-4t^2 + 8t^2 \right] dt$$

$$= \frac{4}{3}$$

9. $x = t$, $y = \dfrac{2t - 1}{4}$, $4 \leq t \leq 8$

$$\int_C (x\ dy\ -\ y\ dx) = \int_4^8 \left[t(\tfrac{1}{2}\ dt)\ -\ \left(\frac{2t - 1}{4} \right) dt \right]$$

$$= \frac{1}{4} \int_4^8 dt$$

$$= 1$$

SURVIVAL HINT: *If C is a piecewise smooth curve, then the line integral needs to be computed for each smooth segment.*

11. C needs to be considered in two pieces:

Let $x = t$, then $y = -2t$ on $-1 \leq t \leq 0$

and $y = 2t$ on $0 \leq t \leq 1$

On $[-1, 0]$:

$[(x + y)^2 dx - (x - y)^2 dy]$

$= (-t)^2 dt - (3t)^2 (-2dt)$

$= 19 t^2 dt$

On $[0, 1]$:

$[(x + y)^2 dx - (x - y)^2 dy]$

$= (3t)^2 dt - (-t)^2 (2dt)$

$= 7 t^2 dt$

$\displaystyle\int_C [(x + y)^2 dx - (x - y)^2 dy]$

$= \displaystyle\int_{-1}^0 19 t^2 dt + \int_0^1 7 t^2 dt$

$= \dfrac{19}{3} + \dfrac{7}{3}$

$= \dfrac{26}{3}$

13. a. Let $x = \cos\theta$ and $y = \sin\theta$ on $0 \leq \theta \leq \dfrac{\pi}{2}$

$(x^2 + y^2)\ dx + 2xy\ dy$

$= (1)(-\sin\theta\ d\theta) + 2\cos\theta \sin\theta (\cos\theta)\ d\theta$

$\displaystyle\int_C [(x^2 + y^2)\ dx + 2xy\ dy]$

$= \displaystyle\int_0^{\pi/2} (-\sin\theta + 2\sin\theta\cos^2\theta)\ d\theta$

$= -\dfrac{1}{3}$

b. Let $x = 1 - t$ and $y = t$ on $0 \leq t \leq 1$

$(x^2 + y^2)\ dx + 2xy\ dy$

$= [(1 - t)^2 + t^2](-dt) + 2(1 - t)(t)\ dt$

$= (-4t^2 + 4t - 1)\ dt$

$\displaystyle\int_C [(x^2 + y^2)\ dx + 2xy\ dy]$

$= \displaystyle\int_0^1 (-4t^2 + 4t - 1)\ dt$

$= -\dfrac{1}{3}$

15. We must use two regions:

On $[0, 2]$: $x = t$, $y = 0$

$[x^2 y\ dx + (x^2 - y^2)\ dy] = 0$

On $[2, 4]$ $x = 2$, $y = t$

$$[x^2 y\, dx + (x^2 - y^2)\, dy] = (4 - t^2)\, dt$$

$$\int_C [x^2 y\, dx + (x^2 - y^2)\, dy]$$

$$= \int_0^4 (4 - t^2)\, dt$$

$$= -\frac{16}{3}$$

17. C must be divided into two smooth pieces.

On the first piece: $x = 1 - t$, $y = 2t^2$,

$0 \le t \le 1$

$$-y^2\, dx + x^2\, dy$$

$$= -(2t^2)^2(-dt) + (1 - t)^2(4t\, dt)$$

On the second piece: $x = -t$ and $y = 2$,

$0 \le t \le 1$

$$-y^2\, dx + x^2\, dy$$

$$= -4(-dt) + t^2(0)$$

$$\int_C (-y^2\, dx + x^2\, dy)$$

$$= \int_0^1 [4t^4 + 4t(1 - t)^2]\, dt + \int_0^1 4\, dt$$

$$= \frac{77}{15}$$

19. On the first piece $x = t$, $y = t^2$ for $0 \le t \le 1$.

$$x^2 y\, dx - xy\, dy = t^4\, dt - t(t^2)(2t\, dt)$$

$$= -t^4\, dt$$

On the second piece $x = 1 - t$, $y = 1 - t$ for

$0 \le t \le 1$.

$$x^2 y\, dx - xy\, dy$$

$$= (1 - t)^3(-dt) - (1 - t)^2(-dt)$$

$$= (t^3 - 2t^2 + t)\, dt$$

$$\oint_C (x^2 y\, dx - xy\, dy)$$

$$= \int_0^1 (-t^4)\, dt + \int_0^1 (t^3 - 2t^2 + t)\, dt$$

$$= -\frac{7}{60}$$

21. $\mathbf{F} \cdot d\mathbf{R} = [(10t + t)\mathbf{i} + 2t\mathbf{j}] \cdot (2\mathbf{i} + \mathbf{j})\, dt$

$$= 24t\, dt$$

$$\int_0^1 24t\, dt = 12$$

23. $\mathbf{F} = \left\langle \dfrac{x}{\sqrt{x^2 + y^2}}, \dfrac{-y}{\sqrt{x^2 + y^2}} \right\rangle = \langle \cos t, -\sin t \rangle$

$\mathbf{R} = (a \cos t)\mathbf{i} + (a \sin t)\mathbf{j}$, for $0 \le t \le \frac{\pi}{2}$

$d\mathbf{R} = (-a \sin t\, dt)\mathbf{i} + (a \cos t\, dt)\mathbf{j}$,

$$\mathbf{F} \cdot d\mathbf{R} = -a \sin t \cos t - a \sin t \cos t$$

$$= -2a \sin t \cos t$$

$$\int_C \mathbf{F} \cdot d\mathbf{R} = -2a \int_0^{\pi/2} \cos t \sin t \, dt$$

$$= -a$$

25. **a.** $\int_C (x \, dx + y \, dy + z \, dz)$

$$= \int_0^{\pi/2} [\cos t(-\sin t) + \sin t(\cos t) + t] \, dt$$

$$= \frac{\pi^2}{8}$$

b. $C: x = 1 - t, \ y = t, \ z = \frac{\pi}{2}t, \ 0 \leq t \leq 1$

$$\int_C (x \, dx + y \, dy + z \, dz)$$

$$= \int_0^1 [(1 - t)(-1) + t + (\tfrac{\pi}{2} t)(\tfrac{\pi}{2})] \, dt$$

$$= \frac{\pi^2}{8}$$

27. **a.** For the arc: $x = t^2, \ y = t, \ z = 0, \ 0 \leq t \leq 1$

$$5xy \, dx + 10yz \, dy + z \, dz = 10t^4 \, dt$$

For the line: $x = 1, \ y = 1, \ z = t, \ 0 \leq t \leq 1$

$$5xy \, dx + 10yz \, dy + z \, dz = t \, dt$$

$$\int_C (5xy \, dx + 10yz \, dy + z \, dz)$$

$$= \int_0^1 10t^4 \, dt + \int_0^1 t \, dt = \frac{5}{2}$$

b. $x = y = z = t$ for $0 \leq t \leq 1$

$$\int_C (5xy \, dx + 10yz \, dy + z \, dz)$$

$$= \int_0^1 (5t^2 \, dt + 10t^2 \, dt + t \, dt)$$

$$= \int_0^1 (15t^2 + t) \, dt$$

$$= \frac{11}{2}$$

29. $C: \quad x^2 + 4y^2 - 8y + 3 = 0$

$$x^2 + 4(y^2 - 2y + 1) = -3 + 4$$

$$x^2 + 4(y - 1)^2 = 1$$

Let $x = \cos t, \ y = \frac{1}{2} \sin t + 1, \ z = 0$;

$$0 \leq t < 2\pi$$

$$\mathbf{F} = x\mathbf{i} + xy\,\mathbf{j} + x^2yz\mathbf{k}$$

$$= (\cos t)\mathbf{i} + (\cos t)\left(\tfrac{1}{2}\sin t + 1\right)\mathbf{j}$$

$$+ (\cos t)^2\left(\tfrac{1}{2}\sin t + 1\right)(0)\mathbf{k}$$

$$\mathbf{R} = (\cos t)\mathbf{i} + \left(\tfrac{1}{2}\sin t + 1\right)\mathbf{j} + 0\mathbf{k}$$

$$d\mathbf{R} = \left[(-\sin t)\mathbf{i} + \left(\tfrac{1}{2} \cos t \right)\mathbf{j} \right] dt$$

$$\oint_C \mathbf{F} \cdot d\mathbf{R} = \int_0^{2\pi} \left[-\cos t \sin t + \tfrac{1}{4} \cos^2 t \sin t + \tfrac{1}{2} \cos^2 t \right] dt$$

$$= \tfrac{\pi}{2}$$

31. $x = t,\ y = t^2,\ z = \tfrac{2}{3}t^3$ for $0 \le t \le 1$

$$ds = \sqrt{(1)^2 + (2t)^2 + (2t^2)^2}\ dt$$

$$= \sqrt{1 + 4t^2 + 4t^4}\ dt$$

$$= (2t^2 + 1)\ dt$$

$$\int_C 2xy^2 z\ ds = \int_0^1 2t(t^2)^2 \left(\tfrac{2}{3}t^3 \right)(2t^2 + 1)\ dt$$

$$= \int_0^1 \tfrac{4}{3}t^8 (2t^2 + 1)\ dt$$

$$= \tfrac{116}{297}$$

33. $\mathbf{F} = y^2\mathbf{i} + x^2\mathbf{j} - (x + z)\mathbf{k}$

C_1: $x = y = t,\ z = 0;\ 0 \le t \le 1;\ \mathbf{R} = t\mathbf{i} + t\mathbf{j}$

$$\mathbf{F} = t^2\mathbf{i} + t^2\mathbf{j} - t\mathbf{k};\ d\mathbf{R} = (\mathbf{i} + \mathbf{j})\ dt$$

C_2: $x = 1,\ y = 1 - t,\ z = 0;\ 0 \le t \le 1$

$$\mathbf{R} = \mathbf{i} + (1 - t)\mathbf{j}$$

$$\mathbf{F} = (1 - t)^2\mathbf{i} + \mathbf{j} - \mathbf{k};\ d\mathbf{R} = -\mathbf{j}\ dt$$

C_3: $x = 1 - t,\ y = 0,\ z = 0;\ 0 \le t \le 1$

$$\mathbf{R} = (1 - t)\mathbf{i}$$

$$d\mathbf{R} = -\mathbf{i}\ dt$$

$$\mathbf{F} = (1 - t^2)\mathbf{j} + (t - 1)\mathbf{k}$$

$$\int_C \mathbf{F} \cdot d\mathbf{R} = \int_0^1 2t^2\ dt - \int_0^1 dt + \int_0^1 0\ dt$$

$$= -\tfrac{1}{3}$$

35. $\mathbf{F} = -x\mathbf{i} + 2\mathbf{j}$

C_1: $(0, 0)$ to $(0, 1)$: $x = 0,\ y = t,\ 0 \le t \le 1$

$$\mathbf{R} = t\mathbf{j}$$

$$d\mathbf{R} = \mathbf{j}\ dt$$

$$\mathbf{F} = 2\mathbf{j}$$

C_2: $(0, 1)$ to $(2, 1)$; $x = t,\ y = 1,\ 0 \le t \le 2$

$$\mathbf{R} = t\mathbf{i} + \mathbf{j}$$

$$d\mathbf{R} = \mathbf{i}\ dt$$

$$\mathbf{F} = -t\mathbf{i} + 2\mathbf{j}$$

C_3: $(2, 1)$ to $(1, 0)$; $x = 2 - t,\ y = 1 - t,\ 0 \le t \le 1$

$$\mathbf{R} = (2 - t)\mathbf{i} + (1 - t)\mathbf{j}$$

$$d\mathbf{R} = (-\mathbf{i} - \mathbf{j})\ dt$$

$$\mathbf{F} = -(2 - t)\mathbf{i} + 2\mathbf{j}$$

C_4: $(1, 0)$ to $(0, 0)$, $x = 1 - t,\ y = 0,\ 0 \le t \le 1$

$$\mathbf{R} = (1 - t)\mathbf{i}$$

$$d\mathbf{R} = -\mathbf{i}\, dt$$

$$\mathbf{F} = -(1 - t)\mathbf{i} + 2\mathbf{j}$$

$$\oint_C \mathbf{F} \cdot \mathbf{T}\, ds = \oint_C \mathbf{F} \cdot d\mathbf{R}$$

$$= \int_0^1 2\, dt + \int_0^2 -t\, dt$$

$$+ \int_0^1 (2 - t - 2)\, dt + \int_0^1 (1 - t)\, dt$$

$$= 2 - 2 - \tfrac{1}{2} + \tfrac{1}{2}$$

$$= 0$$

37. $\dfrac{dx}{dt} = 2 \cos t(-\sin t)$

$$\dfrac{dy}{dt} = 2 \sin t \cos t$$

$$ds = \sqrt{4 \cos^2 t \sin^2 t + 4 \sin^2 t \cos^2 t}\, dt$$

$$= 2\sqrt{2}\, |\sin t \cos t|\, dt$$

$$\int_C (x + y)\, ds = 2\sqrt{2} \int_{-\pi/4}^0 (\cos^2 t + \sin^2 t)\, |\sin t \cos t|\, dt$$

$$= -2\sqrt{2} \int_{-\pi/4}^0 \sin t \cos t\, dt$$

$$= \dfrac{\sqrt{2}}{2}$$

$$\approx 0.7071$$

39. For $0 \le t \le 2\pi$, $x = \cos t$, $y = \sin t$;

$$dx = -\sin t\, dt$$

$$dy = \cos t\, dt$$

$$\dfrac{x\, dy - y\, dx}{x^2 + y^2} = \dfrac{\cos^2 t + \sin^2 t}{\cos^2 t + \sin^2 t}\, dt$$

$$= 1\, dt$$

$$\oint_C \dfrac{x\, dy - y\, dx}{x^2 + y^2} = \int_0^{2\pi} 1\, dt$$

$$= 2\pi$$

41. $\mathbf{F} = a\mathbf{i} + \mathbf{j}$

$$x = t, \quad y = at$$

$$\mathbf{R}(t) = t\mathbf{i} + at\mathbf{j}$$

$$dW = \mathbf{F} \cdot d\mathbf{R}$$

$$= (a\mathbf{i} + \mathbf{j}) \cdot (\mathbf{i} + a\mathbf{j})\, dt$$

$$= 2a\, dt$$

$$W = 2a \int_a^0 dt$$

$$= -2a^2$$

43. $\mathbf{F} = (x^2 + y^2)\mathbf{i} + (x + y)\mathbf{j}$

$$C_1: \mathbf{R} = \cos t\,\mathbf{i} + \sin t\,\mathbf{j}; \quad 0 \le t \le \pi$$

$$C_2: \mathbf{R} = t\mathbf{i}; \quad -1 \le t \le 1$$

$$\int_0^\pi [(\cos^2 t + \sin^2 t)(-\sin t) + (\cos t + \sin t)(\cos t)]\, dt$$

$$+ \int_{-1}^{1} t^2 \, dt$$

$$= -2 + \frac{\pi}{2} + \frac{2}{3}$$

$$= \frac{\pi}{2} - \frac{4}{3}$$

$$\approx 0.2375$$

45. $\mathbf{F} = (y^2 - z^2)\mathbf{i} + 2yz\mathbf{j} - x^2\mathbf{k}$

$$W = \int_C \mathbf{F} \cdot d\mathbf{R}$$

$$= \int_0^1 [(t^2)^2 - (t^3)^2 + 2(t^2)(t^3)(2t) - t^2(3t^2)] \, dt$$

$$= \int_0^1 (t^4 - t^6 + 4t^6 - 3t^4) \, dt$$

$$= \frac{1}{35}$$

47. $\mathbf{F} = x\mathbf{i} + y\mathbf{j} + (xz - y)\mathbf{k}$

$\mathbf{R} = 2t\mathbf{i} + t\mathbf{j} + 2t\mathbf{k};\ 0 \le t \le 1$

$d\mathbf{R} = (2\mathbf{i} + \mathbf{j} + 2\mathbf{k}) \, dt$

$\mathbf{F} = 2t\mathbf{i} + t\mathbf{j} + (4t^2 - t)\mathbf{k}$

$$W = \int_C \mathbf{F} \cdot d\mathbf{R}$$

$$= \int_0^1 [(2t)(2) + t + 2(4t^2 - t)] \, dt$$

$$= \int_0^1 (8t^2 + 3t) \, dt$$

$$= \frac{25}{6}$$

49. $ds = \sqrt{(\sqrt{2} \cos t)^2 + (-\sin t)^2 + (-\sin t)^2} \, dt$

$$= \sqrt{2} \, dt$$

$$m = \int_C \delta \, ds$$

$$= \int_C xyz \, ds$$

$$= \int_0^{\pi} (\sqrt{2} \sin t)(\cos t)(\cos t)(\sqrt{2} \, dt)$$

$$= \frac{4}{3}$$

$$\overline{x} = \frac{1}{m} \int_C x(\delta \, ds)$$

$$= \frac{1}{m} \int_C x^2 yz \, ds$$

$$= \frac{1}{m} \int_0^{\pi} (\sqrt{2} \sin t)^2 (\cos t)(\cos t)(\sqrt{2} \, dt)$$

$$= \frac{1}{m} \int_0^{\pi} 2\sqrt{2} \sin^2 t \cos^2 t \, dt$$

$$= \frac{1}{m} \left(\frac{\sqrt{2}\pi}{4m} \right)$$

$$= \frac{3\sqrt{2}\pi}{16}$$

$$\overline{y} = \frac{1}{m} \int_C y(\delta \, ds)$$

$$= \frac{1}{m} \int_C xy^2 z \, ds$$

$$= \frac{1}{m} \int_0^\pi (\sqrt{2} \sin t)(\cos^2 t)(\cos t)(\sqrt{2} \, dt)$$

$$= 0$$

$$\overline{z} = \frac{1}{m} \int_C z(\delta \, ds)$$

$$= \frac{1}{m} \int_C xyz^2 \, ds$$

$$= \frac{1}{m} \int_0^\pi (\sqrt{2} \sin t)(\cos t)(\cos^2 t)(\sqrt{2} \, dt)$$

$$= 0$$

The center of mass is $\left(\frac{3\sqrt{2}}{16}\pi, 0, 0 \right)$.

51. $x = 2t, \ y = t^2, \ z = \frac{1}{3}t^3$ for $0 \le t \le 2$

$$ds = \sqrt{(2)^2 + (2t)^2 + (t^2)^2} \, dt$$

$$= \sqrt{(t^2 + 2)^2} \, dt$$

$$= (t^2 + 2) \, dt$$

$$m = \int_C ds$$

$$= \int_0^2 (t^2 + 2) \, dt$$

$$= \frac{20}{3}$$

$$\overline{x} = \frac{1}{m} \int_C x \, ds$$

$$= \frac{1}{m} \int_0^2 (2t)(t^2 + 2) \, dt$$

$$= \frac{12}{5}$$

$$\overline{y} = \frac{1}{m} \int_C y \, ds$$

$$= \frac{1}{m} \int_0^2 t^2(t^2 + 2) \, dt$$

$$= \frac{44}{25}$$

$$\overline{z} = \frac{1}{m} \int_C z \, ds$$

$$= \frac{1}{m} \int_0^2 \frac{1}{3}t^3(t^2 + 2) \, dt$$

$$= \frac{3}{20}\left(\frac{56}{9} \right)$$

$$= \frac{14}{15}$$

The centroid is $\left(\frac{12}{5}, \frac{44}{25}, \frac{14}{15} \right)$.

55. Let t measure the rotation in radians. The

laborer climbs 50 ft in $5(2\pi) = 10\pi$ radians, so z increases at the rate of $50/(10\pi) = 5/\pi$ ft/radians. The helical path is

$$\mathbf{R}(t) = \langle 10 \cos t,\ 10 \sin t,\ \tfrac{5t}{\pi} \rangle$$

and the force exerted by the laborer and the sand is

$$\mathbf{F} = \langle 0,\ 0,\ 180 + 40 \rangle$$

The work done is

$$W = \int_C \mathbf{F} \cdot d\mathbf{R}$$

$$= \int_0^{10\pi} (180 + 40) \left(\frac{5}{\pi} \right) dt$$

$$= 11{,}000 \text{ ft-lb}$$

55. $\mathbf{R} = 5{,}000(\cos t\mathbf{i} + \sin t\mathbf{j})$;

$$d\mathbf{R} = 5{,}000(-\sin t\mathbf{i} + \cos t\mathbf{j})\ dt$$

$$\mathbf{F} = 5{,}000(5{,}000 \cos t\mathbf{i} + 5{,}000 \sin t\mathbf{j})$$

$$W = \oint_C \mathbf{F} \cdot d\mathbf{R}$$

$$= \int_0^{2\pi} (5{,}000)^3 (-\cos t \sin t + \sin t \cos t)\ dt$$

$$= 0$$

57. Counterexample; $f(x) = \sin x$ and C: $x = t$,

$y = 0$ for $0 \le t \le 2\pi$.

$$\int_C f(x)\ ds = \int_0^{2\pi} \sin t \sqrt{1 + 0}\ dt$$

$$= 0$$

but $f(x) \ne 0$ on C. Note that f is not the zero function, but does not need to be.

13.3 The Fundamental Theorem and Path Independence, Pages 884-887

SURVIVAL HINT: *The fundamental theorem of calculus provides a simple method for the evaluation of a definite integral. Likewise, the fundamental theorem for line integrals provides a simple method for the evaluation of a line integral. Just as the hypotheses of the fundamental theorem of calculus requires a continuous f', the hypotheses of the fundamental theorem for line integrals require a conservative vector field. If this hypothesis is met, we have independence of path, and the value of the line integral is the difference in the scalar potentials at the ending and beginning points. So the essence of a problem becomes verifying that $\mathbf{F}$ is conservative, and finding f.*

1. $\mathbf{F}$ is conservative if $\mathbf{F} = \nabla f$ where f is continuously differentiable.

3. Answers vary; the line integral $\int_C \mathbf{F} \cdot d\mathbf{R}$ is *independent of path* in a region D if for any two points P and Q in D the line integral along every piecewise smooth curve in D from

P to Q has the same value. The following ways can be used to determine whether a given line integral is path independent.

(1) **F** is conservative on D; that is, $\mathbf{F} = \nabla f$ for some function f defined on D

(2) $\displaystyle\oint_C \mathbf{F} \cdot d\mathbf{R} = 0$ for every piecewise smooth closed curve in D.

5. $\mathbf{F} = 2xy^3\mathbf{i} + 3x^2y^2\mathbf{j}$

Since

$$\frac{\partial}{\partial y}(2xy^3) = \frac{\partial}{\partial x}(3y^2x^2)$$

$$= 6xy^2$$

the field is conservative. Also,

$$\frac{\partial f}{\partial x} = 2xy^3$$

$f(x, y) = x^2y^3 + c(y)$, so

$$\frac{\partial f}{\partial y} = 3x^2y^2 + c'(y)$$

$$= 3x^2y^2$$

$c'(y) = 0$, so $c(y) = K$. If we pick $K = 0$,

then $f(x, y) = x^2y^3$.

7. $\mathbf{F} = (-y + e^x\sin y)\mathbf{i} + (x + 2)e^x\cos y\,\mathbf{j}$

$$\frac{\partial}{\partial y}(-y + e^x\sin y) = -1 + e^x\cos y$$

$$\frac{\partial}{\partial x}(x + 2)e^x\cos y = e^x\cos y(x + 3)$$

These are not the same, so by the cross-partials test, the field is not conservative.

9. $\mathbf{F} = (e^{2x}\sin y)\mathbf{i} + (e^{2x}\cos y)\mathbf{j}$

$$\frac{\partial}{\partial y}(e^{2x}\sin y) = e^{2x}\cos y$$

$$\frac{\partial}{\partial x}(e^{2x}\cos y) = 2e^{2x}\cos y$$

These are not the same, so by the cross-partials test, the field is not conservative.

11. $\displaystyle\int_C [(3x + 2y)\,dx - (2x + 3y)\,dy]$

$$\frac{\partial}{\partial y}(3x + 2y) = 2$$

$$\frac{\partial}{\partial x}(-2x - 3y) = -2$$

These are not the same, so by the cross-partials test, the field is not conservative.

a. $x = \cos t,\ y = \sin t;\ 0 \le t \le \pi$

$$\int_0^\pi [(3\cos t + 2\sin t)(-\sin t)$$

$$- (2\cos t + 3\sin t)(\cos t)]\,dt$$

$$= \int_0^\pi (-6\cos t \sin t - 2)\,dt = -2\pi$$

b. C_1: $x = 1 - t$, $y = t$; $0 \le t \le 1$

C_2: $x = -t$, $y = 1 - t$; $0 \le t \le 1$

$$\int_0^1 [3(1-t) + 2t](-dt) - \int_0^1 [2(1-t) + 3t] \, dt$$

$$+ \int_0^1 [-3t + 2(1-t)](-dt)$$

$$- \int_0^1 [2(-t) + 3(1-t)](-dt)$$

$$= -4$$

c. The circular part is the same as in part **a**,

and the line segment is

C_3: $x = t - 1$, $y = 0$, $0 \le t \le 2$

$$\oint_C \mathbf{F} \cdot d\mathbf{R} = -2\pi + \int_0^2 3(t-1) \, dt$$

$$= -2\pi + 0$$

$$= -2\pi$$

13. $\int_C [2xy \, dx + x^2 \, dy]$

$\dfrac{\partial u}{\partial y} = 2x$

$\dfrac{\partial v}{\partial x} = 2x$

$u_y = v_x$

so the line integral is path independent.

a. The integral is 0 because the path is

closed.

b. C: $x = t$, $y = 2t^2$; $0 \le t \le 2$

$$\int_0^2 8t^3 \, dt = 32$$

c. Because of path independence, it is the

same as **b**, namely 32.

SURVIVAL HINT: *Note the answers to parts **b** and **c** must be the same since the line integral is path independent.*

15. $\mathbf{F}(x, y) = 2xy\mathbf{i} + x^2\mathbf{j}$

$\dfrac{\partial u}{\partial y} = 2x$

$\dfrac{\partial v}{\partial x} = 2x$

F is conservative.

$f(x, y) = x^2 y + c(y)$

$\dfrac{\partial f}{\partial y} = x^2 + c'(y)$

$= x^2$

$c'(y) = 0$, so $c(y) = 0$

$f(x, y) = x^2 y$

$$\int_A^B \mathbf{F} \cdot d\mathbf{R} = f(1, 1) - f(0, 0)$$

$$= 1 - 0$$

$$= 1$$

17. $\mathbf{F}(x, y) = (2x - y)\mathbf{i} + (y^2 - x)\mathbf{j}$

$\dfrac{\partial u}{\partial y} = -1$

$\dfrac{\partial v}{\partial x} = -1$

$\mathbf{F}$ is conservative

$f(x, y) = x^2 - yx + c(y)$

$\dfrac{\partial f}{\partial y} = -x + c'(y) = -x + y^2$

$c'(y) = y^2$, so $c(y) = \dfrac{y^3}{3}$

$f(x, y) = x^2 - xy + \frac{1}{3}y^3$

$\displaystyle\int_C \mathbf{F} \cdot d\mathbf{R} = f(1, 1) - f(0, 0)$

$\qquad = \dfrac{1}{3} - 0$

$\qquad = \dfrac{1}{3}$

19. $\mathbf{F}(x, y) = \dfrac{(y + 1)\mathbf{i} - x\mathbf{j}}{(y + 1)^2}$

$\dfrac{\partial}{\partial y}\left[\dfrac{y + 1}{(y+1)^2}\right] = \dfrac{-1}{(y+1)^2}$

$\dfrac{\partial}{\partial x}\left[\dfrac{-x}{(y+1)^2}\right] = \dfrac{-1}{(y+1)^2};$

$\mathbf{F}$ is conservative;

$\dfrac{\partial f}{\partial x} = \dfrac{1}{y + 1}$

$f(x, y) = \dfrac{x}{y + 1} + c(y)$

$\dfrac{\partial f}{\partial y} = \dfrac{-x}{(y+1)^2} + c'(y)$

$\qquad = \dfrac{-x}{(y + 1)^2}$

so $c'(y) = 0$.

For $c(y) = 0$; $f(x, y) = \dfrac{x}{y + 1}$

$\displaystyle\int_C \mathbf{F} \cdot d\mathbf{R} = f(1, 1) - f(0, 0)$

$\qquad = \dfrac{1}{2} - 0$

$\qquad = \dfrac{1}{2}$

SURVIVAL HINT: *In Problems 20-25, the directions asks for a scalar potential function, and for this reason, we choose to pick the constant to be 0, as shown in Example 3 of this section.*

21. $\mathbf{F}(x, y, z) = yze^{xy}\mathbf{i} + xze^{xy}\mathbf{j} + e^{xy}\mathbf{k}$

$\text{curl } \mathbf{F} = \begin{vmatrix} \mathbf{i} & \mathbf{j} & \mathbf{k} \\ \dfrac{\partial}{\partial x} & \dfrac{\partial}{\partial y} & \dfrac{\partial}{\partial z} \\ yze^{xy} & xze^{xy} & e^{xy} \end{vmatrix}$

$= (xe^{xy} - xe^{xy})\mathbf{i} - (ye^{xy} - ye^{xy})\mathbf{j}$

$\qquad + (ze^{xy} + xyze^{xy} - ze^{xy} - xyze^{xy})\mathbf{k}$

$= \mathbf{0}$

$\mathbf{F}$ is conservative from Theorem 13.5.

$\dfrac{\partial f}{\partial x} = yze^{xy}$

$$f = ze^{xy} + c(y, z)$$

$$\frac{\partial f}{\partial y} = xze^{xy} + \frac{\partial c}{\partial y}$$

$$= xze^{xy}$$

$$\frac{\partial c}{\partial y} = 0, \ c = c_1(z)$$

$$f = ze^{xy} + c_1(z)$$

$$\frac{\partial f}{\partial z} = e^{xy} + c_1'(z)$$

$$= e^{xy}$$

$$c_1' = 0, \ c_1 = 0$$

$$f(x, y, z) = ze^{xy}$$

23. $\mathbf{F}(x, y, z) = (x^2 + y^2 + z^2)(x\mathbf{i} + y\mathbf{j} + z\mathbf{k})$

$$\text{curl } \mathbf{F} = \begin{vmatrix} \mathbf{i} & \mathbf{j} & \mathbf{k} \\ \frac{\partial}{\partial x} & \frac{\partial}{\partial y} & \frac{\partial}{\partial z} \\ rx & ry & rz \end{vmatrix}$$

$$\text{where } r = x^2 + y^2 + z^2$$

$$= (2yz - 2yz)\mathbf{i} - (2xz - 2xz)\mathbf{j} + (2xy - 2xy)\mathbf{k}$$

$$= \mathbf{0}$$

$\mathbf{F}$ is conservative from Theorem 13.5.

$$\frac{\partial f}{\partial x} = x(x^2 + y^2 + z^2)$$

$$f = \tfrac{1}{4}(x^2 + y^2 + z^2)^2 + c(y, z)$$

$$\frac{\partial f}{\partial y} = y(x^2 + y^2 + z^2) + \frac{\partial c}{\partial y}$$

$$= y(x^2 + y^2 + z^2)$$

$$\frac{\partial c}{\partial y} = 0, \ c = c_1(z)$$

$$f = \tfrac{1}{4}(x^2 + y^2 + z^2)^2 + c_1(z)$$

$$\frac{\partial f}{\partial z} = z(x^2 + y^2 + z^2) + c_1'(z)$$

$$= z(x^2 + y^2 + z^2)$$

$$c_1' = 0, \ c_1 = 0$$

$$f(x, y, z) = \tfrac{1}{4}(x^2 + y^2 + z^2)^2$$

25. $\mathbf{F}(x, y, z) = (xy^2 + yz)\mathbf{i} + (x^2y + xz + 3y^2z)\mathbf{j}$
$$+ (xy + y^3)\mathbf{k}$$

$$\text{curl } \mathbf{F} = \begin{vmatrix} \mathbf{i} & \mathbf{j} & \mathbf{k} \\ \frac{\partial}{\partial x} & \frac{\partial}{\partial y} & \frac{\partial}{\partial z} \\ xy^2 + yz & x^2y + xz + 3y^2z & xy + y^3 \end{vmatrix}$$

$$= (x + 3y^2 - x - 3y^2)\mathbf{i} - (y - y)\mathbf{j}$$
$$+ (2xy + z - 2xy - z)\mathbf{k}$$

$$= \mathbf{0}$$

$\mathbf{F}$ is conservative from Theorem 13.5.

$$\frac{\partial f}{\partial x} = xy^2 + yz$$

$$f = \tfrac{1}{2}x^2y^2 + xyz + c(y, z)$$

$$\frac{\partial f}{\partial y} = x^2y + xz + \frac{\partial c}{\partial y}$$

$$= x^2y + xz + 3y^2z$$

$$\frac{\partial c}{\partial y} = 3y^2 z,$$

$$c = y^3 z + c_1(z);$$

$$f = \tfrac{1}{2}x^2 y^2 + xyz + y^3 z + c_1(z)$$

$$\frac{\partial f}{\partial z} = xy + y^3 + c_1{}'(z)$$

$$= xy + y^3$$

$$c_1{}' = 0, \; c_1 = 0$$

$$f(x, y, z) = \tfrac{1}{2}x^2 y^2 + xyz + y^3 z$$

27. $\mathbf{F} = \langle \sin z, \; -z \sin y, \; x \cos z + \cos y \rangle$

$$\text{curl } \mathbf{F} = \begin{vmatrix} \mathbf{i} & \mathbf{j} & \mathbf{k} \\ \frac{\partial}{\partial x} & \frac{\partial}{\partial y} & \frac{\partial}{\partial z} \\ \sin z & -z\sin y & x\cos z + \cos y \end{vmatrix}$$

$= \mathbf{0}$; thus, $\mathbf{F}$ is conservative.

$$\frac{\partial f}{\partial x} = \sin z$$

$$f = x \sin z + c(y, z)$$

$$\frac{\partial f}{\partial y} = -z \sin y$$

$$= \frac{\partial}{\partial y}(x \sin z + c)$$

$$= \frac{\partial c}{\partial y}$$

$$\frac{\partial c}{\partial y} = -z \sin y, \text{ so } c = z \cos y + c_1(z)$$

$$f = x \sin z + z \cos y + c_1(z)$$

$$\frac{\partial f}{\partial z} = x \cos z + \cos y$$

$$= \frac{\partial}{\partial z}[x \sin z + z \cos y + c_1(z)]$$

$$= x \cos z + \cos y + c_1{}'(z)$$

$$c_1{}'(z) = 0, \text{ so } c_1(z) = 0$$

$$f = x \sin z + z \cos y$$

$$\int_C \mathbf{F} \cdot d\mathbf{R} = f(0, -1, 1) - f(1, 0, -1)$$

$$= [0 + (1)\cos(-1)]$$

$$\quad - [(1)\sin(-1) + (-1)(1)]$$

$$= \cos(-1) - \sin(-1) + 1$$

$$= \cos 1 + \sin 1 + 1$$

29. $\mathbf{F} = \left\langle \dfrac{y}{1 + x^2} + \tan^{-1} z, \; \tan^{-1} x, \; \dfrac{x}{1 + z^2} \right\rangle$

curl $\mathbf{F} = 0$, so $\mathbf{F}$ is conservative.

$$\frac{\partial f}{\partial x} = \frac{y}{1 + x^2} + \tan^{-1} z$$

$$f = y \tan^{-1} x + x \tan^{-1} z + c(y, z)$$

$$\frac{\partial f}{\partial y} = \tan^{-1} x$$

$$= \frac{\partial}{\partial y}(y \tan^{-1} x + x \tan^{-1} z + c)$$

$$= \tan^{-1} x + \frac{\partial c}{\partial y},$$

$$\text{so } \frac{\partial c}{\partial y} = 0 \text{ and } c = c_1(z)$$

$$f = y \tan^{-1} x + x \tan^{-1} z + c_1(z)$$

$$\frac{\partial f}{\partial z} = \frac{x}{1 + z^2}$$

$$= \frac{\partial}{\partial z}[y \tan^{-1} x + x \tan^{-1} z + c_1(z)]$$

$$= \frac{x}{1 + z^2} + c_1{}'(z), \text{ so}$$

$$c_1'(z) = 0, \text{ and } c_1 = 0$$

$$f = y \tan^{-1}x + x \tan^{-1}z$$

$$\int_C \mathbf{F} \cdot d\mathbf{R} = f(0, -1, 1) - f(1, 0, -1)$$

$$= [-\tan^{-1}(0) + 0]$$

$$- [0 + (1)\tan^{-1}(-1)]$$

$$= -\tan^{-1}(-1) = \frac{\pi}{4}$$

31. $\frac{\partial u}{\partial y} = 2y; \frac{\partial v}{\partial x} = 2y;$ **F** is conservative

$$\frac{\partial f}{\partial x} = 3x^2 + 2x + y^2;$$

$$f(x, y) = x^3 + x^2 + xy^2 + c(y)$$

$$\frac{\partial f}{\partial y} = 2xy + c'(y)$$

$$= 2xy + y^3$$

$$c'(y) = y^3, \text{ so } c(y) = \tfrac{1}{4}y^4$$

$$f(x, y) = x^3 + x^2 + xy^2 + \frac{y^4}{4}$$

$$\int_C [(3x^2 + 2x + y^2)dx + (2xy + y^3)dy]$$

$$= \int_C \mathbf{F} \cdot d\mathbf{R}$$

$$= f(1, 1) - f(0, 0)$$

$$= \frac{13}{4} - 0$$

$$= \frac{13}{4}$$

33. $u = y - x^2$ and $y = x + y^2$

$$\frac{\partial u}{\partial y} = 1$$

$$\frac{\partial v}{\partial x} = 1$$

F is conservative.

$$\frac{\partial f}{\partial x} = y - x^2$$

$$f(x, y) = xy - \frac{x^3}{3} + c(y)$$

$$\frac{\partial f}{\partial y} = x + c'(y)$$

$$= x + y^2;$$

$$c'(y) = y^2, \text{ so } c(y) = \frac{y^3}{3}$$

$$f(x, y) = xy - \frac{x^3}{3} + \frac{y^3}{3}$$

$$\int_C [(y - x^2)dx + (x + y^2)dy]$$

$$= f(0, 3) - f(-1, -1)$$

$$= 9 - 1$$

$$= 8$$

35. $u = \sin y$ and $v = 3 + x \cos y$

$$\frac{\partial u}{\partial y} = \cos y$$

$$\frac{\partial v}{\partial x} = \cos y$$

F is conservative.

$$\frac{\partial f}{\partial x} = \sin y$$

$$f(x, y) = x \sin y + c(y)$$

$$\frac{\partial f}{\partial y} = x \cos y + c'(y)$$

$$= 3 + x \cos y$$

$c'(y) = 3$, so $c(y) = 3y$

$$f(x, y) = x \sin y + 3y$$

When $t = 0$, $(x, y) = (0, 0)$

when $t = 1$, $(x, y) = (-2, \pi/2)$

$$\int_C [(\sin y)\,dx + (3 + x \cos y)\,dy]$$

$$= f(-2, \pi/2) - f(0, 0)$$

$$= \tfrac{1}{2}(3\pi - 4)$$

$$\approx 2.7124$$

37. $u = y$ and $v = x$

$$\frac{\partial u}{\partial y} = 1$$

$$\frac{\partial v}{\partial x} = 1$$

$\mathbf{F} = y\mathbf{i} + x\mathbf{j}$; $\mathbf{F}$ is conservative.

$$\frac{\partial f}{\partial x} = y$$

$$f(x, y) = xy + c(y)$$

$$\frac{\partial f}{\partial y} = x + c'(y)$$

$$= x$$

$c'(y) = 0$, so $c(y) = 0$

$$f(x, y) = xy$$

$$\int_C (y\mathbf{i} + x\mathbf{j}) \cdot d\mathbf{R} = f(2, 4) - f(0, 0)$$

$$= 8$$

39. $u = 2y$ and $v = 2x$

$$\frac{\partial u}{\partial y} = 2$$

$$\frac{\partial v}{\partial x} = 2$$

$$\mathbf{F} = 2y\mathbf{i} + 2x\mathbf{j}$$

$\mathbf{F}$ is conservative.

$$\frac{\partial f}{\partial x} = 2y$$

$$f(x, y) = 2xy + c(y)$$

$$\frac{\partial f}{\partial y} = 2x + c'(y)$$

$c'(y) = 0$, so $c(y) = 0$

$$f(x, y) = 2xy$$

$$\int_C (2y\,dx + 2x\,dy) = f(4, 4) - f(0, 0)$$

$$= 32$$

41. $u = \tan^{-1} \dfrac{y}{x} - \dfrac{xy}{x^2 + y^2}$

$$v = \frac{x^2}{x^2 + y^2} + e^{-y}(1 - y)$$

$$\frac{\partial}{\partial x}\left[\frac{x^2}{x^2 + y^2} + e^{-y}(1 - y) \right]$$

$$= \frac{2x(x^2 + y^2) - 2x^3}{(x^2 + y^2)^2}$$

$$= \frac{2xy^2}{(x^2 + y^2)^2};$$

$$\frac{\partial}{\partial y}\left[\tan^{-1}\frac{y}{x} - \frac{xy}{x^2 + y^2}\right]$$

$$= \frac{1/x}{1 + y^2/x^2} - \frac{x^3 - xy^2}{(x^2 + y^2)^2}$$

$$= \frac{2xy^2}{(x^2 + y^2)^2}$$

F is conservative.

$$\frac{\partial f}{\partial y} = \frac{x^2}{x^2 + y^2} + e^{-y}(1 - y)$$

$$= \frac{1}{1 + \left(\frac{y}{x}\right)^2} + e^{-y} - ye^{-y}$$

$$f(x, y) = x\tan^{-1}\frac{y}{x} - e^{-y} + e^{-y}(y+1) + e(x)$$

$$= x\tan^{-1}\frac{y}{x} + ye^{-y} + c(x)$$

$$\frac{\partial f}{\partial x} = x\frac{-\frac{y}{x^2}}{1 + \left(\frac{y}{x}\right)^2} + \tan^{-1}\left(\frac{y}{x}\right) + c'(x)$$

$$= -\frac{xy}{x^2 + y^2} + \tan^{-1}\left(\frac{y}{x}\right) + c'(x)$$

$$= \tan^{-1}\left(\frac{y}{x}\right) - \frac{xy}{x^2 + y^2}$$

$c'(x) = 0$, so $c(x) = 0$

$$f(x, y) = x\tan^{-1}\frac{y}{x} + ye^{-y}$$

$$\int_C\left\{\left[\tan^{-1}\frac{y}{x} - \frac{xy}{x^2 + y^2}\right]dx\right.$$

$$\left. + \left[\frac{x^2}{x^2 + y^2} + e^{-y}(1 - y)\right]dy\right\}$$

$$= f(-1, 2) - f(1, 1)$$

$$= -\tan^{-1}(-2) + 2e^{-2} - \left(\frac{\pi}{4} + e^{-1}\right)$$

$$\approx 0.2245$$

43. We want $\mathbf{G} = \langle g(x)(x^4 + y^4), -g(x)(xy^3)\rangle$ to be conservative, so

$$\frac{\partial}{\partial y}[g(x)(x^4 + y^4)] = \frac{\partial}{\partial x}[-g(x)(xy^3)]$$

$$g(x)(4y^3) = -g'(x)(xy^3) - g(x)y^3$$

$$\frac{g'(x)}{g(x)} = \frac{-5}{x}$$

$$\ln g(x) = -5\ln x + C_1$$

We assume $g(x) > 0$, without loss of generality.

$$g(x) = Cx^{-5}$$

for any constant $C \neq 0$.

45. a. $\mathbf{F} = -KmM\left\langle\frac{x}{r^3}, \frac{y}{r^3}, \frac{z}{r^3}\right\rangle$

$$\text{where } r = \sqrt{x^2 + y^2 + z^2}$$

$$\frac{\partial f}{\partial x} = \frac{-KmMx}{(x^2 + y^2 + z^2)^{3/2}}$$

$$f = \frac{KmM}{\sqrt{x^2 + y^2 + z^2}} + c_1(y, z)$$

$$\frac{\partial f}{\partial y} = \frac{-KmMy}{(x^2 + y^2 + z^2)^{3/2}} + \frac{\partial c_1}{\partial y}$$

$$= \frac{-KmMy}{(x^2 + y^2 + z^2)^{3/2}}$$

$\frac{\partial c_1}{\partial y} = 0$, use $c_1(y, z) = c_2(z)$

$$f = \frac{KmM}{\sqrt{x^2 + y^2 + z^2}} + c_2(z)$$

$$\frac{\partial f}{\partial z} = \frac{-KmMz}{(x^2 + y^2 + z^2)^{3/2}} + c_2{}'(z)$$

$$= \frac{-KmMz}{(x^2 + y^2 + z^2)^{3/2}}$$

$c_2{}'(z) = 0$; use $c_2(z) = 0$

$$f(x, y, z) = \frac{KmM}{\sqrt{x^2 + y^2 + z^2}}$$

$$= kmM(\tfrac{1}{r})$$

Note: Physicists sometimes use $\phi = -\frac{kmM}{r}$
as the gravitational potential.

b. $$W = \int_P^Q \mathbf{F} \cdot d\mathbf{R}$$

$$= f(a_2, b_2, c_2) - f(a_1, b_1, c_1)$$

$$= \frac{kmM}{\sqrt{a_2{}^2 + b_2{}^2 + c_2{}^2}} - \frac{kmM}{\sqrt{a_1{}^2 + b_1{}^2 + c_1{}^2}}$$

47. The bucket moves in the circle C described by

$$\mathbf{R}(t) = (3 \cos t)\mathbf{i} + (3 \sin t)\mathbf{j}; \ 0 \le t \le 2\pi$$

The bucket and water have mass $m = \frac{30}{32}$ and

moves with speed $v = 2\pi(3) = 6\pi$ ft/s.

The magnitude of the force $\mathbf{F}$ is

$$\|\mathbf{F}\| = \frac{mv^2}{r} = \frac{30}{32}\left(\frac{1}{3}\right)(6\pi)^2 = \frac{45}{4}\pi^2$$

and since $\mathbf{F}$ always points toward the center

of the circle, we have

$$\mathbf{F} = \frac{45}{4}\pi^2(-\cos t\,\mathbf{i} - \sin t\,\mathbf{j})$$

Thus, the work performed is

$$W = \oint_C \mathbf{F} \cdot d\mathbf{R}$$

$$= \int_0^{2\pi} \frac{45}{4}\pi^2[(-\cos t)(-3\sin t)$$

$$+ (-\sin t)(3\cos t)]\, dt = 0$$

Alternatively, note that $\mathbf{F}$ is conservative,

so $\oint_C \mathbf{F} \cdot d\mathbf{R} = 0$ around the closed curve C.

49. a. Since $\frac{\partial}{\partial y}(-a) = 0 = \frac{\partial}{\partial x}(-b)$, the wind

force $\mathbf{F}$ is conservative, so the work

integral $W = \int_C \mathbf{F} \cdot d\mathbf{R}$ is independent of

the path C. The line segment C_1 between

$A(0, 0)$ and $B(2, 1)$ can be parameterized

as C_1: $x = 2t$, $y = t$; $0 \le t \le 1$.

$$W = \int_C \mathbf{F} \cdot d\mathbf{R}$$

$$= \int_0^1 [-a(2) - b(1)]\, dt$$

$$= -(2a + b)$$

b. The force $\mathbf{F}_1 = \langle -a, -ae^{-y} \rangle$ is also conservative since

$$\frac{\partial}{\partial y}(-a) = 0 = \frac{\partial}{\partial x}(-ae^{-y})$$

The work, computed along the line segment C_1 given in part **a** is

$$W_1 = \int_{C_1} \mathbf{F}_1 \cdot d\mathbf{R}$$

$$= \int_0^1 [-a(2) - ae^{-t}(1)]\, dt$$

$$= a(e^{-1} - 3)$$

c. The force $\mathbf{F}_2 = \langle -a, -ae^{-y+x/9} \rangle$ is not conservative since

$$\frac{\partial}{\partial y}(-a) = 0$$

but

$$\frac{\partial}{\partial x}(-ae^{-y+x/9}) = \frac{-a}{9}e^{-y+x/9}$$

The work performed varies with the path chosen. For instance, let C_1 be the linear path used in parts **a** and **b** and let C_2 be the path along the x-axis from $(0, 0)$ to $(2, 0)$, then along $x = 2$ from $(2, 0)$ to $(2, 1)$. Then

$$W_2 = \int_{C_1} \mathbf{F}_2 \cdot d\mathbf{R}$$

$$= \int_0^1 -a\, dx - ae^{-y+x/9}\, dy$$

$$= \int_0^1 [-a(2) - ae^{-t+2t/9}]\, dt$$

$$= \int_0^1 -a[2 + e^{-7t/9}]\, dt$$

$$= -a\left[2t - \frac{9}{7}e^{-7t/9}\right]\Big|_0^1$$

$$= -a\left[\left(2 - \frac{9}{7}e^{-7/9}\right) - \left(0 - \frac{9}{7}\right)\right]$$

$$= \frac{a}{7}(9e^{-7/9} - 23)$$

and

$$W_3 = \int_{C_2} \mathbf{F}_2 \cdot d\mathbf{R}$$

$$= \int_0^1 -a(2)\, dt + \int_0^1 -ae^{-t+2/9}\, dt$$

$$= -ae^{-7/9}(2e^{7/9} + e - 1)$$

51. If $\mathbf{F}$ is conservative, then

$$\mathbf{F} = \nabla\phi = \langle \phi_x, \phi_y, \phi_z \rangle$$

so

$$M = \phi_x,\ N = \phi_y,\ P = \phi_z$$

Thus, using the equality of mixed partials, we have

$$\frac{\partial M}{\partial y} = \phi_{xy} = \phi_{yx} = \frac{\partial N}{\partial x}$$

$$\frac{\partial N}{\partial z} = \phi_{yz} = \phi_{zy} = \frac{\partial P}{\partial y}$$

$$\frac{\partial P}{\partial x} = \phi_{zx} = \phi_{xz} = \frac{\partial M}{\partial z}$$

Thus, $\frac{\partial P}{\partial y} = \frac{\partial N}{\partial z}$; $\frac{\partial M}{\partial z} = \frac{\partial P}{\partial x}$; $\frac{\partial N}{\partial x} = \frac{\partial M}{\partial y}$

53. Following the same procedure as in Problem 45, we can show that $\mathbf{F} = \mathbf{R}/r^3$ is conservative in D with scalar potential function $f = -1/r$. The work done in moving the object from distance r_1 to r_2 along a curve C is given by

$$W = \int_C \mathbf{F} \cdot d\mathbf{R} = \int_C \nabla\left(\frac{-1}{r}\right) \cdot d\mathbf{R}$$

$$= \left(\frac{-1}{r_2}\right) - \left(\frac{-1}{r_1}\right)$$

$$= \frac{1}{r_1} - \frac{1}{r_2}$$

55. a. Let $Q_1(x_1, y_1)$ be any point on the same horizontal line as $Q(x_0, y)$.

$$f(x, y) = \int_{(a, b)}^{(x, y)} \mathbf{F} \cdot d\mathbf{R}$$

$$= \int_{C_1} \mathbf{F} \cdot d\mathbf{R} + \int_{C_2} \mathbf{F} \cdot d\mathbf{R}$$

$$= \int_{(a, b)}^{(x_1, y)} \mathbf{F} \cdot d\mathbf{R} + \int_{C_2} \mathbf{F} \cdot d\mathbf{R}$$

Since the first integral only depends on y (x_1 is constant), we have

$$\frac{\partial f}{\partial x} = 0 + \frac{\partial}{\partial x}\int_{C_2} [M(x, y)\, dx + N(x, y)\, dy]$$

$$= \frac{\partial}{\partial x}\int_{C_2} M(x, y)\, dx$$

since $\int_{C_2} N(x, y)\, dy = 0$ along the horizontal line segment $\overline{Q_1 Q}$ (because $dy = 0$). Thus, we have

$$\frac{\partial f}{\partial x} = \frac{\partial}{\partial x}\int_{x_1}^{x} M(u, y)\, du = M(x, y)$$

by the second fundamental theorem of calculus.

b. Let $Q_2(x, y_1)$ be any point on the same vertical line as $Q(x, y)$, and let C_3 be a path in D joining $P(a, b)$ to Q_2 and C_4 be the vertical line segment from Q_2 to Q. Then,

$$f(x, y) = \int_{(a, b)}^{(x, y_1)} \mathbf{F} \cdot d\mathbf{R} + \int_{C_4} \mathbf{F} \cdot d\mathbf{R}$$

and

$$\frac{\partial f}{\partial y} = \frac{\partial}{\partial y} \int_{C_4} \mathbf{F} \cdot d\mathbf{R}$$

$$= \frac{\partial}{\partial y} \int_{y_1}^{y} N(x, u) \, du = N(x, y)$$

13.4 Green's Theorem, Page 896-898

SURVIVAL HINT: *When using Green's theorem to find area, take the time to write out the values of dx and dy before substituting (except in the simplest of problems), and do not forget the factor of $\frac{1}{2}$.*

1. $$\oint_C (y^2 \, dx + x^2 \, dy) = \int_0^1 \int_0^1 (2x - 2y) \, dy \, dx$$

$$= \int_0^1 (2x - 1) \, dx$$

$$= 0$$

Alternatively, on C:

$C_1: y = 0;\ C_2: x = 1;\ C_3: y = 1;\ C_4: x = 0$

$$\int_0^1 0 \, dx + \int_0^1 dy + \int_1^0 dx + \int_1^0 0 \, dy = 0$$

3. Since the orientation is clockwise,

$$\oint_C \mathbf{F} \cdot d\mathbf{R} = -\int_0^1 \int_y^{2-y} (0 - 3) \, dx \, dy$$

$$= 3 \int_0^1 (2 - 2y) \, dy$$

$= 3$

Alternatively, we can parameterize C:

$C_1: x = t,\ y = t;\ 0 \le t \le 1$

$C_2: x = 1 + t,\ y = 1 - t;\ 0 \le t \le 1$

$C_3: x = 2 - t,\ y = 0;\ 0 \le t \le 2$

$$\oint_C \mathbf{F} \cdot d\mathbf{R} = \int_0^1 [2t^2 + 3t - 3t^2] \, dt$$

$$+ \int_0^1 [2(1 + t)^2 + 3(1 - t) - 3(1 - t)^2(-1)] \, dt$$

$$+ \int_0^2 [2(2 - t)^2(-1)] \, dt$$

$$= \frac{7}{6} + \frac{43}{6} - \frac{16}{3}$$

$= 3$

5. The orientation is clockwise, so

$$\oint_C 4xy \, dx = -\int\int_D (-4x) \, dA$$

$$= 4 \int_0^{2\pi} \int_0^1 r \cos \theta \, (r \, dr \, d\theta)$$

$= 0$

Alternatively, on C: $x = \sin \theta,\ y = \cos \theta,$

$dx = \cos \theta \, d\theta,\ dy = -\sin \theta \, d\theta$

$$4 \int_0^{2\pi} \cos^2\theta \sin \theta \, d\theta = 0$$

7.
$$\oint_C (2y\,dx - x\,dy) = \int\int_D (-1-2)\,dA$$

$$= -3\left[\tfrac{1}{2}\pi(2)^2\right]$$

$$= -6\pi$$

9.
$$\oint_C (x\sin x\,dx - e^{y^2}\,dy) = \int\int_D 0\,dA$$

$$= 0$$

11.
$$\oint_C [(x - y^2)\,dx + 2xy\,dy]$$

$$= \int\int_D (2y + 2y)\,dA$$

$$= \int_0^2\int_0^2 4y\,dy\,dx$$

$$= \int_0^2 8\,dx$$

$$= 16$$

13. Since the orientation is clockwise,

$$\oint_C [\sin x \cos y\,dx + \cos x \sin y\,dy]$$

$$= -\int\int_D \left[\frac{\partial}{\partial x}(\cos x \sin y) - \frac{\partial}{\partial y}(\sin x \cos y)\right] dA$$

$$= -\int_0^2\int_0^2 [-\sin x \sin y + \sin x \sin y]\,dy\,dx$$

$$= 0$$

15.
$$W = \oint_C \mathbf{F} \cdot d\mathbf{R}$$

$$= \oint_C [(3y - 4x)\mathbf{i} + (4x - y)\mathbf{j}] \cdot [dx\,\mathbf{i} + dy\,\mathbf{j}]$$

$$= \int\int_D \left[\frac{\partial}{\partial x}(4x - y) - \frac{\partial}{\partial y}(3y - 4x)\right] dA$$

$$= \int\int_D (4 - 3)\,dA = 2(1)\pi = 2\pi$$

since the semimajor and semiminor axes of

the ellipse have length 2 and 1, respectively.

17. Parameterize the circle by $x = 2\cos t$,

$$y = 2\sin t,\ 0 \le t \le 2\pi$$

$$A = \tfrac{1}{2}\oint_C (-y\,dx + x\,dy)$$

$$= \tfrac{1}{2}\int_0^{2\pi} [(-2\sin t)(-2\sin t) + (2\cos t)(2\cos t)]\,dt$$

$$= \tfrac{1}{2}\int_0^{2\pi} 4\,dt = 4\pi$$

Check: $A = \pi r^2 = \pi(2)^2 = 4\pi$.

19. C_1: $x = t,\ y = 0;\ 0 \le t \le 4$

C_2: $x = 4 - t,\ y = t;\ 0 \le t \le 3$

C_3: $x = 1 - t,\ y = 3;\ 0 \le t \le 1$

C_4: $x = 0,\ y = 3 - t;\ 0 \le t \le 3$

$$A = \frac{1}{2} \oint_C (-y\ dx + x\ dy)$$

$$= \frac{1}{2} \int_0^4 0\ dt + \frac{1}{2} \int_0^3 [-t(-1) + (4-t)]\ dt$$

$$+ \frac{1}{2} \int_0^1 (-3)(-1)\ dt + \frac{1}{2} \int_0^3 0\ dt$$

$$= \frac{15}{2}$$

Check: $A = \frac{1}{2}(b_1 + b_2)h = \frac{1}{2}(1+4)(3) = \frac{15}{2}$

21. $\oint_C (x^2 y\ dx - y^2 x\ dy) = \int\int_D (-y^2 - x^2)\ dA$

$$= -\int\int_D r^2\ dA$$

$$= -\int_0^\pi \int_0^a r^3\ dr\ d\theta$$

$$= -\frac{\pi a^4}{4}$$

23. $I = \oint_C [(5 - xy - y^2)\ dx - (2xy - x^2)\ dy]$

$$= \int\int_D (-2y + 2x + x + 2y)\ dA$$

$$= 3 \int\int_D x\ dA$$

$$= 3M_y$$

Solving,

$$M_y = \frac{1}{3} I$$

The square has area $A = 1$, so

$$\bar{x} = \frac{M_y}{A} = \frac{\frac{1}{3} I}{1},$$

Thus, $I = 3\bar{x}$

25. $\oint_C x^2\ dy = \int\int_D (2x - 0)\ dA$

$$= 2 \int\int_D x\ dA$$

$$= 2\bar{x} A$$

$$\oint_C y^2\ dx = \int\int_D (0 - 2y)\ dA$$

$$= -2 \int\int_D y\ dA$$

$$= -2\bar{y} A$$

Thus,

$$\bar{x} = \frac{1}{2A} \oint_C x^2\ dy$$

and

$$\bar{y} = -\frac{1}{2A} \oint_C y^2\ dy.$$

27. $\oint_C \left[\left(\frac{-y}{x^2} + \frac{1}{x} \right) dx + \frac{1}{x}\ dy \right] = \int\int_D \left[-\frac{1}{x^2} + \frac{1}{x^2} \right] dA$

$$= 0$$

29.

Let C_1 be a circle centered at the origin with radius R so small that all of C_1 is contained within the given curve C. Assume C_1 is oriented clockwise, and let D be the region between C_1 and C. Then, according to Green's theorem for doubly-connected regions,

$$\oint_C \frac{x\,dx + y\,dy}{x^2 + y^2} + \oint_{C_1} \frac{x\,dx + y\,dy}{x^2 + y^2}$$

$$= \int\int_D \left[\frac{\partial}{\partial x}\left(\frac{y}{x^2 + y^2}\right) - \frac{\partial}{\partial y}\left(\frac{x}{x^2 + y^2}\right) \right] dA$$

$$= \int\int_D \left[\frac{-2xy}{(x^2 + y^2)^2} - \frac{-2xy}{(x^2 + y^2)^2} \right] dA$$

$$= 0$$

To evaluate the line integral about C_1, use the parameterization

$$C_1: x = \sin\theta,\ y = \cos\theta;\ 0 \le \theta \le 2\pi$$

(Remember, C_1 is oriented clockwise.) Thus,

$$\oint_C \frac{x\,dx + y\,dy}{x^2 + y^2}$$

$$= -\oint_{C_1} \frac{x\,dx + y\,dy}{x^2 + y^2}$$

$$= -\int_0^{2\pi} \frac{(\sin\theta)(\cos\theta) + (\cos\theta)(-\sin\theta)}{\sin^2\theta + \cos^2\theta}\, d\theta$$

$$= 0$$

31.
$$\oint_C \frac{-(y+2)\,dx + (x-1)\,dy}{(x-1)^2 + (y+2)^2}$$

$$= \int\int_D \frac{-[(x-1)^2 - (y+2)^2] - [(y+2)^2 - (x-1)^2]}{[(x-1)^2 + (y+2)^2]^2}\, dA$$

$$= 0$$

33. $\nabla f = (2xy - 2y)\mathbf{i} + (x^2 - 2x + 2y)\mathbf{j}$

Since $\mathbf{N} = \dfrac{dy}{ds}\mathbf{i} - \dfrac{dx}{ds}\mathbf{j}$ is a unit normal to the curve.

$$\oint_C \frac{\partial f}{\partial n}\, ds = \oint_C (\nabla f \cdot \mathbf{N})\, ds$$

$$= \oint_C [(2xy - 2y)\,dy - (x^2 - 2x + 2y)\,dx]$$

$$= \int_0^1 \int_0^1 (2y + 2)\, dy\, dx$$

$$= 3$$

35.
$$\oint_C [(x - 3y)\,dx + (2x - y^2)]\,dy$$

$$= \int\int_D (2 + 3)\, dA$$

$$= 5A$$

37. If $\dfrac{\partial u}{\partial y} = \dfrac{\partial v}{\partial x}$, then for any closed curve C, we

have

$$\oint_C \mathbf{F} \cdot d\mathbf{R} = \oint_C [u \, dx + v \, dy]$$

$$= \int\!\!\int_D \left(\frac{\partial v}{\partial x} - \frac{\partial u}{\partial y} \right) dA$$

$$= 0$$

where D is the interior of C. Thus, $\displaystyle\oint_C \mathbf{F} \cdot d\mathbf{R}$

is independent of path and $\mathbf{F}$ is conservative.

Conversely, if $\mathbf{F}$ is conservative, then

$\displaystyle\oint_C \mathbf{F} \cdot d\mathbf{R} = 0$ for any closed curve C in D.

Thus,

$$0 = \oint_C [u \, dx + v \, dy]$$

$$= \int\!\!\int_D \left(\frac{\partial v}{\partial x} - \frac{\partial u}{\partial y} \right) dA$$

which is true for every closed curve C only if

$$\frac{\partial v}{\partial x} = \frac{\partial u}{\partial y}$$

39. a. $\displaystyle\oint_C f \frac{\partial g}{\partial n} \, ds = \oint_C f \nabla g \cdot \mathbf{N} \, ds$

$$= \oint_C f(g_x \mathbf{i} + g_y \mathbf{j}) \cdot (dy \, \mathbf{i} - dx \, \mathbf{j})$$

$$= \oint_C [f g_x \, dy - f g_y \, dx]$$

$$= \int\!\!\int_D \left[\frac{\partial}{\partial x}(f g_x) + \frac{\partial}{\partial y}(f g_y) \right] dA$$

$$= \int\!\!\int_D [f_x g_x + f g_{xx} + f_y g_y + f g_{yy}] \, dA$$

$$= \int\!\!\int_D [f(g_{xx} + g_{yy}) + f_x g_x + f_y g_y] \, dA$$

$$= \int\!\!\int_D [f \nabla^2 g + \nabla f \cdot \nabla g] \, dA$$

b. $\displaystyle\oint_C \left(f \frac{\partial g}{\partial n} - g \frac{\partial f}{\partial n} \right) ds$

$$= \oint_C (f \nabla g \cdot \mathbf{N} - g \nabla f \cdot \mathbf{N}) \, ds$$

$$= \oint_C [f(g_x \mathbf{i} + g_y \mathbf{j}) - g(f_x \mathbf{i} + f_y \mathbf{j})] \cdot (dy \mathbf{i} - dx \mathbf{j})$$

$$= \oint_C [(f g_x - g f_x) \, dy + (f g_y - g f_y)(- dx)]$$

$$= \int\!\!\int_D \left[\frac{\partial}{\partial x}(f g_x - g f_x) + \frac{\partial}{\partial y}(f g_y - g f_y) \right] dA$$

$$= \int\!\!\int_D [f g_{xx} + g_x f_x - g f_{xx} - f_x g_x + f g_{yy}$$
$$+ f_y g_y - g f_{yy} - g_y f_y] \, dA$$

$$= \int\!\!\int_D [f(g_{xx} + g_{yy}) - g(f_{xx} + f_{yy})] \, dA$$

$$= \int\int_D [f\nabla^2 g - g\nabla^2 f] \, dA$$

41. Introduce a "cut" line C_4 joining C_1 and C_2 and another C_5 joining C_2 to C_3. Then if

$$\mathbf{F} = M(x, y)\mathbf{i} + N(x, y)\mathbf{j}$$

is continuously differentiable in the shaded region D, the conditions of Green's theorem are satisfied by the curve C formed by traveling counterclockwise around C_1, to the left along C_4, clockwise around the bottom of C_2, to the left along C_5, clockwise around C_3, to the right along C_5, clockwise around the top of C_2, the back to C_1 along C_4. Then,

$$\int_C (M \, dx + N \, dy) = \int\int_D \left[\frac{\partial N}{\partial x} - \frac{\partial M}{\partial y}\right] dA$$

$$= \int_{C_1} (M \, dx + N \, dy) + \int_{C_2} (M \, dx + N \, dy)$$

$$+ \int_{C_3} (M \, dx + N \, dy)$$

If F is conservative in D and C_2 and C_3 are traversed counterclockwise instead of clockwise, then

$$0 = \int_{C_1} - \int_{C_2} - \int_{C_3}$$

so that

$$\int_{C_1} = \int_{C_2} + \int_{C_3}$$

43. If R has two holes, there are three possibilities for

$$\oint_C \mathbf{F} \cdot d\mathbf{R}:$$

Case I: C surrounds neither hole

Since $\mathbf{F}$ is conservative, any curve C in R that does not "surround" the hole will C surround the hole will have

$$\oint_C \mathbf{F} \cdot d\mathbf{R} = 0$$

In this case, C surrounds neither hole, so

$$\oint_C \mathbf{F} \cdot d\mathbf{R} = 0$$

Case II: C surrounds just one hole.

There are four possibilities depending on which hole is surrounded and whether the orientation of C is clockwise or counterclockwise.

Case III: *C* **surrounds both holes.**

Again, There are two possibilities. Either *C* circles both holes with the same orientation or it circles one hole with one orientation and the other with the opposite orientation (a "figure- eight").

Thus, there are a total of seven possible values

$$\oint_C \mathbf{F} \cdot d\mathbf{R}$$

13.5 Surface Integrals, pages 906-907

SURVIVAL HINT: *If you think of a surface integral as a double integral over a curved surface, rather than a flat region of a plane, then the radical is the "slope" factor to transform R into S. Since there are several steps in finding the value of the radical, taking the time to write them out carefully will save time in the long run, as you will have fewer errors.*

1. $z = 2 - y;\ z_x = 0,\ z_y = -1$

$$dS = \sqrt{0^2 + (-1)^2 + 1}\ dA$$

$$= \sqrt{2}\ dy\ dx$$

$$\iint_S xy\ dS = \int_0^2 \int_0^2 xy\ \sqrt{2}\ dy\ dx$$

$$= 2\sqrt{2} \int_0^2 x\ dx$$

$$= 4\sqrt{2}$$

3. $z = 5;\ dS = dA$

$$\iint_S xy\ dS = \iint_R xy\ dA$$

$$= \int_0^{2\pi} \int_0^1 (r \cos\theta)(r \sin\theta)\ r\ dr\ d\theta$$

$$= 0$$

5. $z = 4 - x - 2y;\ z_x = -1,\ z_y = -2$

$$dS = \sqrt{(-1)^2 + (-2)^2 + 1}\ dA$$

$$= \sqrt{6}\ dA$$

$$\iint_S (x^2 + y^2)\,dS = \int_0^4 \int_0^2 (x^2 + y^2)\ \sqrt{6}\ dy\ dx$$

$$= \sqrt{6} \int_0^4 \left(2x^2 + \frac{8}{3}\right)\ dx$$

$$= \frac{160\sqrt{6}}{3}$$

7. $z = 4,\ z_x = 0,\ z_y = 0$

$$dS = \sqrt{0^2 + 0^2 + 1}\ dA$$

$$\int_0^{2\pi} \int_0^1 r^2(r\ dr\ d\theta) = \frac{\pi}{2}$$

For Problems 9-14, $z = \sqrt{4 - x^2 - y^2}$;

$$z_x{}^2 + z_y{}^2 + 1 = \left(\frac{-x}{z}\right)^2 + \left(\frac{-y}{z}\right)^2 + 1$$

$$= \frac{4}{z^2}$$

$$dS = \sqrt{z_x^2 + z_y^2 + 1}\ dA = \frac{2}{z}\ dA$$

R is the circular disk $x^2 + y^2 \leq 4$.

9. $\displaystyle\iint\limits_{S} z\ dS = \iint\limits_{R} z\left(\frac{2}{z}\right)dA$

$$= \int_0^{2\pi}\int_0^2 2r\ dr\ d\theta$$

$$= 8\pi$$

11. $\displaystyle\iint\limits_{S}(x - 2y)\,dS$

$$= \iint\limits_{R}(x - 2y)\left(\frac{2}{z}\right)dA$$

$$= \int_0^{2\pi}\int_0^2 (r\cos\theta - 2r\sin\theta)\frac{2}{\sqrt{4 - r^2}}\,r\ dr\ d\theta$$

$$= 2\pi\int_0^{2\pi}(\cos\theta - 2\sin\theta)\ d\theta$$

$$= 0$$

13. $\displaystyle\iint\limits_{S}(x^2 + y^2)z\ dS = \iint\limits_{R}(x^2 + y^2)z\left(\frac{2}{z}\right)dA$

$$= 2\int_0^{2\pi}\int_0^2 r^2(r\ dr\ d\theta)$$

$$= 16\pi$$

For Problems 15-18, $z = x^2 + y^2 = r^2$; $z_x = 2x$, $z_y = 2y$;

$$dS = \sqrt{(2x)^2 + (2y)^2 + 1}\ dA$$

$$= \sqrt{4r^2 + 1}\ dA$$

the projected region R is the disk $x^2 + y^2 \leq 4$.

15. $\displaystyle\iint\limits_{S}z\ dS = \int_0^{2\pi}\int_0^2 r^2\sqrt{4r^2 + 1}\ r\ dr\ d\theta$

$$= 4\pi\int_0^2 r^3\sqrt{r^2 + \left(\tfrac{1}{2}\right)^2}\ dr\ \boxed{\text{Formula 171}}$$

$$= 4\pi\left[\frac{(r^2 + \frac{1}{4})^{5/2}}{5} - \frac{\frac{1}{4}(r^2 + \frac{1}{4})^{3/2}}{3}\right]\Bigg|_0^2$$

$$= 4\pi\left[\left(\frac{4}{5} - \frac{1}{30}\right)\frac{17}{4}\sqrt{\frac{17}{4}} + \frac{1}{30}\left(\frac{1}{8}\right)\right]$$

$$= 4\pi\left(\frac{391\sqrt{17}}{240} + \frac{1}{240}\right)$$

$$= \frac{\pi}{60}\left(391\sqrt{17} + 1\right)$$

$$\approx 84.4635$$

17. $\displaystyle\iint\limits_{S}\sqrt{1 + 4z}\ dS = \int_0^{2\pi}\int_0^2\sqrt{1 + 4r^2}\ r\ dr\ d\theta$

$$= 2\pi\left(\frac{17\sqrt{17}}{12} - \frac{1}{12}\right)$$

$$= \frac{\pi}{6}(17\sqrt{17} - 1)$$

$$\approx 36.1769$$

19. $z = \sqrt{1 - x^2 - y^2}$

$$z_x = \frac{-x}{z}$$

$$z_y = \frac{-y}{z}$$

$$z_x{}^2 + z_y{}^2 + 1 = \frac{x^2}{z^2} + \frac{y^2}{z^2} + 1$$

$$= \frac{1}{z^2}$$

$dS = \frac{1}{z} dA;$

projected region R is $x^2 + y^2 \le 1$

$$\int\!\!\int_S (x^2 + y^2) dS$$

$$= \int_0^{2\pi}\!\!\int_0^1 \frac{r^2}{\sqrt{1 - r^2}}\, r\, dr\, d\theta$$

$$= \int_0^{2\pi}\!\!\int_0^1 \frac{r^3 dr}{\sqrt{1 - r^2}}\, d\theta$$

$$\boxed{\text{Formula 227}}$$

$$= \int_0^{2\pi} \left[\frac{(1 - r^2)^{3/2}}{3} - (1 - r^2)^{1/2} \right]\Bigg|_0^1 d\theta$$

$$= \int_0^{2\pi} \frac{2}{3}\, d\theta$$

$$= \frac{4\pi}{3}$$

21. $z = x + 1$

$$dS = \sqrt{1^2 + 0^2 + 1}\, dA$$

$$= \sqrt{2}\, r\, dr\, d\theta$$

The projected region R is the disk $x^2 + y^2 \le 1$

$$\int\!\!\int_R (r^2 + z^2)\, \sqrt{2}\, r\, dr\, d\theta$$

$$= \int_0^{2\pi}\!\!\int_0^1 [r^2 + (r\cos\theta + 1)^2]\, \sqrt{2}\, r\, dr\, d\theta$$

$$= \int_0^{2\pi}\!\!\int_0^1 [r^3 + r^3\cos^2\theta + 2r^2\cos\theta + r]\sqrt{2}\, dr\, d\theta$$

$$= \int_0^{2\pi} \frac{\sqrt{2}}{12}(3\cos^2\theta + 8\cos\theta + 9)\, d\theta$$

$$= \frac{7\pi\sqrt{2}}{4}$$

23. $z = 2 - 5x + 4y$

$$z_x = -5$$

$$z_y = 4$$

$$dS = \sqrt{(-5)^2 + 4^2 + 1}\, dA$$

$$= \sqrt{42}\, dy\, dx$$

$$\mathbf{N} = \frac{1}{\sqrt{42}}(5\mathbf{i} - 4\mathbf{j} + \mathbf{k})$$

$$\mathbf{F} = x\mathbf{i} + 2y\mathbf{j} - 3z\mathbf{k};$$

$$\mathbf{F} \cdot \mathbf{N} = \frac{1}{\sqrt{42}}(5x - 8y - 3z)$$

The projected region R is $0 \le x \le 1,\, 0 \le y \le 1$

$$\iint_S \mathbf{F} \cdot \mathbf{N} \, dS = \iint_R \frac{1}{\sqrt{42}}(5x - 8y - 3z)\sqrt{42} \, dy \, dx$$

$$= \int_0^1 \int_0^1 [5x - 8y - 3(2 - 5x + 4y)] \, dy \, dx$$

$$= \int_0^1 (20x - 16) \, dx = -6$$

25. $f(x,\, y,\, z) = x^2 + y^2 + z^2 = 1;$

$\nabla f = 2x\mathbf{i} + 2y\mathbf{j} + 2z\mathbf{k};$

$dS = \frac{1}{z} \, dA_{xy}$

$\mathbf{N} = x\mathbf{i} + y\mathbf{j} + z\mathbf{k}$

$\mathbf{F} \cdot \mathbf{N} = x^2 + y^2$

$$\iint_S \mathbf{F} \cdot \mathbf{N} \, dS = \int_0^{2\pi} \int_0^1 \frac{r^3 \, dr \, d\theta}{\sqrt{1 - r^2}}$$

$$= 2\pi \int_0^1 r^2 \frac{r \, dr}{\sqrt{1 - r^2}}$$

$$= \frac{4\pi}{3}$$

27. $z = y + 1$

$z_x = 0$

$z_y = 1;$

$$dS = \sqrt{0^2 + 1^2 + 1^2} \, dA$$

$$= \sqrt{2} \, dy \, dx$$

$f(x,\, y,\, z) = z - y - 1;$

$$\mathbf{N} = \frac{\nabla f}{\|\nabla f\|}$$

$$= \frac{-\mathbf{j} + \mathbf{k}}{\sqrt{2}}$$

$\mathbf{F} = x^2\mathbf{i} + y^2\mathbf{j} + z^2\mathbf{k}$

$$\mathbf{F} \cdot \mathbf{N} = \frac{-y^2 + z^2}{\sqrt{2}}$$

$$= \frac{2y + 1}{\sqrt{2}}$$

The projected region R is the disk $x^2 + y^2 \le 1$

$$\iint_S \mathbf{F} \cdot \mathbf{N} \, dS = \iint_R \frac{1}{\sqrt{2}}(2y + 1)(\sqrt{2} \, dy \, dx)$$

$$= \int_0^{2\pi} \int_0^1 (2r \sin \theta + 1) \, r \, dr \, d\theta$$

$$= \int_0^{2\pi} \left(\frac{2}{3} \sin \theta + \frac{1}{2}\right) d\theta$$

$$= \pi$$

29. The projected region on the xy-plane is the rectangle $R: -\sqrt{3} \le x \le \sqrt{3},\, 0 \le y \le 1$. Let $f = z + x^2 - 3$, so $f_x = 2x,\, f_y = 0$. Then

$$dS = \sqrt{f_x^2 + f_y^2 + 1} \, dA$$

$$= \sqrt{(2x)^2 + 0 + 1} \, dA$$

$$= \sqrt{4x^2 + 1} \, dA$$

Since $z = 3 - x^2$, we have

$$\int\int_S \mathbf{F} \cdot \mathbf{N} \, dS$$

$$= \int\int_S \langle y, z^2, -2z \rangle \cdot \langle f_x, f_y, 1 \rangle \, dS$$

$$= \int\int_R \langle y, z^2, -2z \rangle \cdot \langle 2x, 0, 1 \rangle \sqrt{4x^2 + 1} \, dA$$

$$= \int_{-\sqrt{3}}^{\sqrt{3}} \int_0^1 [2xy - 2(3 - x^2)]\sqrt{4x^2 + 1} \, dy \, dx$$

$$= \int_{-\sqrt{3}}^{\sqrt{3}} (2x^2 + x - 6)\sqrt{4x^2 + 1} \, dx$$

$$= \left\{ \tfrac{1}{48}(24x^3 + 16x^2 - 141x + 4)\sqrt{4x^2 + 1} \right.$$

$$\left. - \tfrac{49}{32} \ln \left| \sqrt{4x^2 + 1} + 2x \right| \right\} \Big|_{-\sqrt{3}}^{\sqrt{3}}$$

$$= \tfrac{49}{32} \ln(25 - 4\sqrt{39}) - \tfrac{23}{8}\sqrt{39}$$

$$\approx -23.944$$

31. $\quad \mathbf{R} = u^2\mathbf{i} + v\mathbf{j} + u\mathbf{k}$

$$\mathbf{R}_u = 2u\mathbf{i} + \mathbf{k}; \ \mathbf{R}_v = \mathbf{j}$$

$$\mathbf{R}_u \times \mathbf{R}_v = \begin{vmatrix} \mathbf{i} & \mathbf{j} & \mathbf{k} \\ 2u & 0 & 1 \\ 0 & 1 & 0 \end{vmatrix}$$

$$= -\mathbf{i} + 2u\mathbf{k}$$

$$\|\mathbf{R}_u \times \mathbf{R}_v\| = \sqrt{1 + 4u^2};$$

$$\int\int_S (x - y^2 + z) \, dS$$

$$= \int_0^1 \int_0^1 (u^2 - v^2 + u)\sqrt{1 + 4u^2} \, dv \, du$$

$$= \int_0^1 (u^2 - \tfrac{1}{3} + u)\sqrt{1 + 4u^2} \, du$$

$$= -\frac{19\ln(\sqrt{5} + 2)}{192} + \frac{17\sqrt{5}}{32} - \frac{1}{12}$$

$$\approx 0.9617$$

33. $\quad \mathbf{R} = u\mathbf{i} - u^2\mathbf{j} + v\mathbf{k}$

$$\mathbf{R}_u = \mathbf{i} - 2u\mathbf{j}; \ \mathbf{R}_v = \mathbf{k}$$

$$\mathbf{R}_u \times \mathbf{R}_v = \begin{bmatrix} \mathbf{i} & \mathbf{j} & \mathbf{k} \\ 1 & -2u & 0 \\ 0 & 0 & 1 \end{bmatrix}$$

$$= -2u\mathbf{i} - \mathbf{j}$$

$$\|\mathbf{R}_u \times \mathbf{R}_v\| = \sqrt{1 + 4u^2};$$

$$\int\int_S (x^2 + y - z)\, dS$$

$$= \int_0^2 \int_0^1 [u^2 - u^2 - v]\sqrt{1 + 4u^2})\, dv\, du$$

$$= -\frac{1}{2}\int_0^2 \sqrt{1 + 4u^2}\, du$$

$$= -\frac{1}{8}[4\sqrt{17} + \ln(4 + \sqrt{17})]$$

$$\approx -2.3234$$

35. $\mathbf{R} = \langle \sin u \cos v,\ \sin u \sin v,\ \cos u \rangle$

$\mathbf{R}_u = \langle \cos u \cos v,\ \cos u \sin v,\ -\sin u \rangle$

$\mathbf{R}_v = \langle -\sin u \sin v,\ \sin u \cos v,\ 0 \rangle$

$$\mathbf{R}_u \times \mathbf{R}_v = \begin{bmatrix} \mathbf{i} & \mathbf{j} & \mathbf{k} \\ \cos u \cos v & \cos u \sin v & -\sin u \\ -\sin u \sin v & \sin u \cos v & 0 \end{bmatrix}$$

$$= (\sin^2 u \cos v)\mathbf{i} - (0 - \sin^2 u \sin v)\mathbf{j}$$

$$+ (\sin u \cos u)\mathbf{k}$$

$\mathbf{F} = \langle x^2,\ 0,\ z \rangle$

$$= \langle \sin^2 u \cos^2 v,\ 0,\ \cos u \rangle$$

$\mathbf{F} \cdot (\mathbf{R}_u \times \mathbf{R}_v) = (\sin^2 u \cos^2 v)(\sin^2 u \cos v)$

$$+ \cos u(\sin u \cos u)$$

$$= \sin^4 u \cos^3 v + \sin u \cos^2 u$$

$$\int\int_S \mathbf{F} \cdot \mathbf{N}\, dS = \int\int_D \mathbf{F} \cdot (\mathbf{R}_u \times \mathbf{R}_v)\, du\, dv$$

$$= \int_0^\pi \int_0^{2\pi} (\sin^4 u \cos^3 v + \sin u \cos^2 u)\, dv\, du$$

$$= \int_0^\pi 2\pi \sin u \cos^2 u\, du$$

$$= \frac{4\pi}{3}$$

37. $z = 4 - x - 2y$

$z_x = -1,$

$z_y = -2$

$$dS = \sqrt{(-1)^2 + (-2)^2 + 1}\, dA$$

$$= \sqrt{6}\, dy\, dx$$

The projected region R is

$x + 2y \le 4;\ x \ge 0,\ y \ge 0$

$$m = \int\int_R x(\sqrt{6}\, dy\, dx)$$

$$= \sqrt{6}\int_0^4 \int_0^{(4-x)/2} x\, dy\, dx$$

$$= \sqrt{6}\int_0^4 \frac{x(4 - x)}{2}\, dx$$

$$= \frac{16\sqrt{6}}{3}$$

$$\approx 13.0639$$

39. $z = x^2 + y^2$

$$z_x = 2x$$

$$z_y = 2y$$

$$dS = \sqrt{(2x)^2 + (2y)^2 + 1} \; dA$$

$$= \sqrt{1 + 4r^2} \; r \; dr \; d\theta$$

The projected region R is the disk

$x^2 + y^2 \leq 1$; $\delta = z = r^2$

$$m = \int\int_R z \; dS$$

$$= \int_0^{2\pi} \int_0^1 r^2(4r^2 + 1)^{1/2} \; r \; dr \; d\theta$$

$$= \int_0^{2\pi} \left(\frac{5\sqrt{5}}{24} + \frac{1}{120} \right) d\theta$$

$$= \frac{\pi}{60}(25\sqrt{5} + 1)$$

$$\approx 2.9794$$

41. $z = \sqrt{5 - r^2}$

$$dS = \sqrt{\frac{r^2}{z^2} + 1} \; dA$$

$$= \frac{\sqrt{5}}{z} \; r \; dr \; d\theta$$

The plane $z = 1$ intersects $r^2 + z^2 = 5$ where

$r^2 = 4$, so the projected region R is the disk $r^2 \leq 4$;

$$m = \int\int_R \theta^2 \; dS$$

$$= \int_0^{2\pi} \int_0^2 \theta^2 \frac{\sqrt{5}}{\sqrt{5 - r^2}} \; r \; dr \; d\theta$$

$$= \int_0^{2\pi} \theta^2(5 - \sqrt{5}) \; d\theta$$

$$= \frac{8}{3}\pi^3(5 - \sqrt{5})$$

$$\approx 228.5313$$

43. $\rho\mathbf{V} = \rho\langle xy, \; yz, \; xz\rangle$

Let $G = x^2 + y^2 + z - 9$

$$\nabla G = \langle 2x, \; 2y, \; 1\rangle$$

The rate of fluid flow is:

$$\int\int_S \rho\mathbf{V} \cdot \mathbf{N} \; dS$$

$$\int\int_S \rho\mathbf{V} \cdot \nabla G \; dA$$

$$= \int\int_{x^2 + y^2 \leq 9} \rho\langle xy, \; yz, \; xz\rangle \cdot \langle 2x, \; 2y, \; 1\rangle \; dA$$

$$= \int\int_{x^2 + y^2 \leq 9} \rho[2x^2y + 2y^2z + xz] \; dA$$

$$= \rho \iint\limits_{x^2 + y^2 \leq 9} [2x^2 y + (2y^2 + x)(9 - x^2 - y^2)] \, dA$$

$$\text{Since } z = 9 - x^2 - y^2$$

$$= \rho \int_0^{2\pi} \int_0^3 [2r^2\cos^2\theta(r\sin\theta)$$

$$+ (2r^2\sin^2\theta + r\cos\theta)(9 - r^2)] \, r \, dr \, d\theta$$

$$= \rho \int_0^{2\pi} \frac{81}{10} \big[12\cos^2\theta\sin\theta + 4\cos\theta + 15\sin^2\theta\big] \, d\theta$$

$$= \left(\frac{243\pi}{2}\right)\rho$$

45. a. $z = \sqrt{a^2 - x^2 - y^2}$

$$z_x = \tfrac{1}{2}(a^2 - x^2 - y^2)^{-1/2}(-2x)$$

$$= \frac{-x}{z}$$

$$z_y = \tfrac{1}{2}(a^2 - x^2 - y^2)^{-1/2}(-2y)$$

$$= \frac{-y}{z}$$

$$dS = \sqrt{\left(\frac{-x}{z}\right)^2 + \left(\frac{-y}{z}\right)^2 + 1} \, dA$$

$$= \frac{a}{\sqrt{a^2 - r^2}} \, r \, dr \, d\theta$$

The sphere and cone intersect where

$2(x^2 + y^2) = a^2$, so the projected region R

is the disk $r \leq a/\sqrt{2}$;

$$\delta = x^2 y^2 z$$

$$= (r\cos\theta)^2(r\sin\theta)^2\sqrt{a^2 - r^2}$$

$$m = \int_R \int x^2 y^2 z \, dS$$

$$= \int_0^{2\pi} \int_0^{a/\sqrt{2}} r^4\cos^2\theta\sin^2\theta\sqrt{a^2 - r^2}\left(\frac{a}{\sqrt{a^2 - r^2}}\right) r \, dr \, d\theta$$

$$= \int_0^{2\pi} \int_0^{a/\sqrt{2}} ar^5\cos^2\theta\sin^2\theta \, dr \, d\theta$$

$$= \frac{a^7}{48} \int_0^{2\pi} \cos^2\theta\sin^2\theta \, d\theta$$

$$= \frac{\pi a^7}{192}$$

b. $z = \sqrt{a^2 - x^2 - y^2}$

$$z_x = \frac{-x}{z}$$

$$z_y = \frac{-y}{z}$$

$$dS = \frac{ar \, dr \, d\theta}{\sqrt{a^2 - r^2}}$$

The cone with vertex angle ϕ has the equation

$z = a\cot\phi$ and intersects the sphere where

$r^2 + r^2\cot^2\phi = a^2$, so the projected region R

is $r^2(1 + \cot^2\phi_0) \leq a^2$ or $r \leq a\sin\phi_0$.

$$m = \int_0^{2\pi} \int_0^{a\sin\phi_0} \frac{ar \, dr \, d\theta}{\sqrt{a^2 - r^2}}$$

$$= 2\pi a(a - a\sqrt{1 - \sin^2\phi_0})$$

$$= 2\pi a^2(1 - \cos\phi_0)$$

47. $z = \sqrt{a^2 - x^2 - y^2}$

$z_x = \frac{-x}{z}$

$z_y = \frac{-y}{z}$

$dS = \sqrt{\left(\frac{x}{z}\right)^2 + \left(\frac{y}{z}\right)^2 + 1} \; dA$

$\quad = \frac{ar \; dr \; d\theta}{\sqrt{a^2 - r^2}}$

$m = 2 \int_0^{2\pi} \int_0^a a(a^2 - r^2)^{-1/2} \, r \, dr \, d\theta$

$\quad = 4\pi a^2$

$I_z = 2a \int_0^{2\pi} \int_0^a r^3(a^2 - r^2)^{-1/2} \, dr \, d\theta$

$\quad = \frac{16a^4}{3} \int_0^{\pi/2} d\theta$

$\quad = \frac{8\pi a^4}{3}$

$\quad = \frac{2}{3} m a^2$

13.6 Stokes' Theorem, Pages 914-916

SURVIVAL HINT: *Green's theorem relates a double integral over a flat region of a plane to a line integral over the boundary of the region. Stokes' theorem relates a double integral over a curved surface in $\mathbb{R}^3$ to the line integral over its boundary.*

As with any theorem, Stokes' theorem has hypotheses that must be verified before the conclusion can be applied. In each case, verify that the orientation of C is compatible with the orientation on S.

1. Evaluating the line integral $\oint_C \mathbf{F} \cdot d\mathbf{R}$ where C is the curve $x = 3\cos\theta$, $y = 3\sin\theta$, $z = 0; \; 0 \le \theta \le 2\pi$.

$\oint_C \mathbf{F} \cdot d\mathbf{R} = \int_C (z \, dx + 2x \, dy + 3y \, dz)$

$\quad = \int_0^{2\pi} 0 + 2 \int_0^{2\pi} 9 \cos^2\theta \; d\theta + 3 \int_0^{2\pi} 0$

$\quad = 18\pi$

Evaluating the integral $\iint_S (\text{curl } \mathbf{F} \cdot \mathbf{N}) \, dS$

$\text{curl } \mathbf{F} = \begin{vmatrix} \mathbf{i} & \mathbf{j} & \mathbf{k} \\ \frac{\partial}{\partial x} & \frac{\partial}{\partial y} & \frac{\partial}{\partial z} \\ z & 2x & 3y \end{vmatrix}$

$\quad = 3\mathbf{i} + \mathbf{j} + 2\mathbf{k}$

$f(x, \, y, \, z) = x^2 + y^2 + z^2 = 9$

$\nabla f = \langle 2x, \, 2y, \, 2z \rangle$

$\mathbf{N} = \frac{1}{3} \langle x, \, y, \, z \rangle$

$z = \sqrt{9 - x^2 - y^2}$

$dS = \sqrt{\left(\frac{-x}{z}\right)^2 + \left(\frac{-y}{z}\right)^2 + 1} \; dA$

$\quad = \frac{3}{\sqrt{9 - r^2}} \, r \, dr \, d\theta$

The projected region D is the disk

$$x^2 + y^2 \le 9.$$

$$\iint\limits_S (\text{curl } \mathbf{F} \cdot \mathbf{N}) \, dS$$

$$= \iint\limits_S \frac{1}{3}(3x + y + 2z) \, dS$$

$$= \frac{1}{3}\int_0^{2\pi}\int_0^3 (3r\cos\theta + r\sin\theta + 2\sqrt{9 - r^2})\frac{3r \, dr}{\sqrt{9 - r^2}} \, d\theta$$

$$= \int_0^{2\pi}\int_0^3 \frac{3r^2 \cos\theta \, dr \, d\theta}{\sqrt{9 - r^2}} + \int_0^{2\pi}\int_0^3 \frac{r^2 \sin\theta \, dr \, d\theta}{\sqrt{9 - r^2}}$$

$$+ \int_0^{2\pi}\int_0^3 2r \, dr \, d\theta$$

$$= 0 + 0 + 18\pi$$

$$= 18\pi$$

3. Evaluating the line integral $\oint\limits_C \mathbf{F} \cdot d\mathbf{R}$

C_1: $x + 2y = 3$; C_2: $2y + z = 3$; C_3: $x + z = 3$.

Parameterizing all three with $0 \le t \le \frac{3}{2}$:

C_1: $x = 3 - 2t$, $y = t$, $z = 0$,

$$\mathbf{R} = (3 - 2t)\mathbf{i} + t\mathbf{j},$$

$$d\mathbf{R} = (-2\mathbf{i} + \mathbf{j}) \, dt,$$

$$\mathbf{F} \cdot d\mathbf{R} = [-2(x + 2z) + (y - x)] \, dt$$

$$= [-3x + y - 4z] \, dt$$

$$= (7t - 9) \, dt$$

$$I_1 = \int_0^{3/2} (7t - 9) \, dt$$

$$= -\frac{45}{8}$$

C_2: $x = 0$, $y = \frac{3}{2} - t$, $z = 2t$,

$$\mathbf{R} = (\tfrac{3}{2} - t)\mathbf{j} + 2t\mathbf{k},$$

$$d\mathbf{R} = (-\mathbf{j} + 2\mathbf{k}) \, dt,$$

$$\mathbf{F} \cdot d\mathbf{R} = [-(y - x) + 2(z - y)] \, dt$$

$$= [x - 3y + 2z] \, dt$$

$$= [7t - \tfrac{9}{2}] \, dt$$

$$I_2 = \int_0^{3/2} \left(7t - \tfrac{9}{2}\right) dt$$

$$= \frac{9}{8}$$

C_3: $x = 2t$, $y = 0$, $z = 3 - 2t$,

$$\mathbf{R} = 2t\mathbf{i} + (3 - 2t)\mathbf{k},$$

$$d\mathbf{R} = (2\mathbf{i} - 2\mathbf{k}) \, dt,$$

$$\mathbf{F} \cdot d\mathbf{R} = [2(x + 2z) - 2(z - y)] \, dt$$

$$= [2x + 2y + 2z] \, dt$$

$$= 6 \, dt$$

$$I_3 = \int_0^{3/2} 6 \, dt$$

$$= 9$$

$$\oint_C \mathbf{F} \cdot d\mathbf{R} = I_1 + I_2 + I_3$$

$$= -\frac{45}{8} + \frac{9}{8} + 9$$

$$= \frac{9}{2}$$

Evaluating the integral $\iint_S (\text{curl } \mathbf{F} \cdot \mathbf{N}) \, dS$

$$\text{curl } \mathbf{F} = \begin{vmatrix} \mathbf{i} & \mathbf{j} & \mathbf{k} \\ \dfrac{\partial}{\partial x} & \dfrac{\partial}{\partial y} & \dfrac{\partial}{\partial z} \\ x+2z & y-x & z-y \end{vmatrix}$$

$$= -\mathbf{i} + 2\mathbf{j} - \mathbf{k}$$

$$\mathbf{N} = \frac{1}{\sqrt{6}}(\mathbf{i} + 2\mathbf{j} + \mathbf{k})$$

$$dS = \sqrt{6} \, dy \, dx$$

$$\iint_S (\text{curl } \mathbf{F} \cdot \mathbf{N}) \, dS = \iint_R \frac{-1 + 4 - 1}{\sqrt{6}} \sqrt{6} \, dy \, dx$$

$$= 2(\text{the area of the } \triangle)$$

$$= 2\left(\frac{9}{4}\right)$$

$$= \frac{9}{2}$$

5. Evaluating the line integral $\oint_C \mathbf{F} \cdot d\mathbf{R}$

C: $x = 2 \cos \theta$, $y = 2 \sin \theta$, $z = 0$; $0 \le \theta \le 2\pi$

$$\oint_C \mathbf{F} \cdot d\mathbf{R} = \oint_C (2y \, dx - 6z \, dy + 3x \, dz)$$

$$= \int_0^{2\pi} 2(2 \sin \theta)(-2 \sin \theta) \, d\theta$$

$$= -8\pi$$

Evaluating the integral $\iint_S (\text{curl } \mathbf{F} \cdot \mathbf{N}) \, dS$

$$\text{curl } \mathbf{F} = \begin{vmatrix} \mathbf{i} & \mathbf{j} & \mathbf{k} \\ \dfrac{\partial}{\partial x} & \dfrac{\partial}{\partial y} & \dfrac{\partial}{\partial z} \\ 2y & -6z & 3x \end{vmatrix}$$

$$= 6\mathbf{i} - 3\mathbf{j} - 2\mathbf{k}$$

$$f(x, y, z) = x^2 + y^2 + z$$

$$= 4$$

$$\nabla f = 2x\mathbf{i} + 2y\mathbf{j} + \mathbf{k}$$

$$\mathbf{N} = \frac{2x\mathbf{i} + 2y\mathbf{j} + \mathbf{k}}{\sqrt{4(x^2 + y^2) + 1}}$$

$$z = 4 - x^2 - y^2$$

$$z_x = -2x$$

$$z_y = -2y$$

$$dS = \sqrt{4x^2 + 4y^2 + 1} \, dA$$

$$(\text{curl } \mathbf{F} \cdot \mathbf{N}) \, dS = \frac{12x - 6y - 2}{\sqrt{4(x^2 + y^2) + 1}} \sqrt{4(x^2 + y^2) + 1} \, dA$$

$$= (12x - 6y - 2) \, dA$$

$$\iint\limits_{S} (\text{curl } \mathbf{F} \cdot \mathbf{N}) \; dS$$

$$= \int\limits_{0}^{2\pi} \int\limits_{0}^{2} (12r\cos\theta - 6r\sin\theta - 2)r \; dr \; d\theta$$

$$= \int\limits_{0}^{2\pi} (32\cos\theta - 16\sin\theta - 4) \; d\theta = -8\pi$$

7. $\mathbf{F} = z\mathbf{i} + x\mathbf{j} + y\mathbf{k}$

$$\text{curl } \mathbf{F} = \begin{vmatrix} \mathbf{i} & \mathbf{j} & \mathbf{k} \\ \frac{\partial}{\partial x} & \frac{\partial}{\partial y} & \frac{\partial}{\partial z} \\ z & x & y \end{vmatrix}$$

$$= \mathbf{i} + \mathbf{j} + \mathbf{k}$$

The triangle is the portion of the plane

$2x + y + 3z = 6$ in the first octant, and since

the orientation is clockwise, the normal is

$$\mathbf{N} = \frac{-2\mathbf{i} - \mathbf{j} - 3\mathbf{k}}{\sqrt{14}}$$

so curl $\mathbf{F} \cdot \mathbf{N} = \frac{-6}{\sqrt{14}}$. Since

$$z = \tfrac{1}{3}(6 - 2x - y)$$

we have $z_x = -\frac{2}{3}$, $z_y = -\frac{1}{3}$, and

$$dS = \sqrt{\left(-\tfrac{2}{3}\right)^2 + \left(-\tfrac{1}{3}\right)^2 + 1} \; dA$$

$$= \frac{\sqrt{14}}{3} \; dy \; dx$$

The projected region D is $2x + y \le 6$ for

$x \ge 0$, $y \ge 0$. By Stokes' theorem

$$\oint\limits_{C} \mathbf{F} \cdot d\mathbf{R} = \iint\limits_{S} (\text{curl } \mathbf{F} \cdot \mathbf{N}) \; dS$$

$$= \int\limits_{0}^{3} \int\limits_{0}^{6-2x} \left(-\frac{6}{\sqrt{14}}\right)\left(\frac{\sqrt{14}}{3}\right) dy \; dx$$

$$= \int\limits_{0}^{3} -2(6 - 2x) \; dx$$

$$= -18$$

9. $\mathbf{F} = 2xy^2z\mathbf{i} + 2x^2yz\mathbf{j} + (x^2y^2 - 2z)\mathbf{k};$

$$\text{curl } \mathbf{F} = \begin{vmatrix} \mathbf{i} & \mathbf{j} & \mathbf{k} \\ \frac{\partial}{\partial x} & \frac{\partial}{\partial y} & \frac{\partial}{\partial z} \\ 2xy^2z & 2x^2yz & x^2y^2 - 2z \end{vmatrix}$$

$$= \mathbf{0}$$

$$\oint\limits_{C} [2xy^2z \; dx + 2x^2yz \; dy + (x^2y^2 - 2z) dz]$$

$$= \iint\limits_{S} (\text{curl } \mathbf{F} \cdot \mathbf{N}) \; dS \quad \text{(Stokes' theorem)}$$

$$= 0$$

since curl $\mathbf{F} = \mathbf{0}$.

11. $\mathbf{F} = y\mathbf{i} + z\mathbf{j} + x\mathbf{k}$

$$\text{curl } \mathbf{F} = \begin{vmatrix} \mathbf{i} & \mathbf{j} & \mathbf{k} \\ \frac{\partial}{\partial x} & \frac{\partial}{\partial y} & \frac{\partial}{\partial z} \\ y & z & x \end{vmatrix}$$

$$= -(\mathbf{i} + \mathbf{j} + \mathbf{k})$$

Take S to be the boundary plane $x + y = 2$.

Since the curve is traversed clockwise, as

viewed from above.

$$N = \frac{-(i + j)}{\sqrt{2}}$$

and

$$\text{curl } \mathbf{F} \cdot \mathbf{N} = \frac{2}{\sqrt{2}}.$$

Projecting onto the xz-plane, we have

$y = 2 - x$, so $y_x = -1$, $y_z = 0$, and

$dS = \sqrt{2}\, dA$. The plane and the sphere

intersect when

$$x^2 + (2 - x)^2 + z^2 = 4$$

so the projected region D is the interior of the

ellipse

$$(x - 1)^2 + \tfrac{1}{2} z^2 = 1$$

By Stokes' theorem

$$\oint_C \mathbf{F} \cdot d\mathbf{R} = \int\int_S (\text{curl } \mathbf{F} \cdot \mathbf{N})\, dS$$

$$= \int\int_D \left(\frac{2}{\sqrt{2}}\right)\left(\sqrt{2}\right) dA$$

$$= 2 \int\int_D dA$$

$$= 2\,(\text{AREA OF ELLIPSE})$$

$$= 2\left(\pi(1)\sqrt{2}\right)$$

$$= 2\sqrt{2}\,\pi$$

Note: We used the formula for the area of an

ellipse with semimajor axis of length $a = \sqrt{2}$

and semiminor axis of length $b = 1$:

$$\pi ab = \pi(1)\sqrt{2}.$$

13. $\mathbf{F} = 3y\mathbf{i} + 2z\mathbf{j} - 5x\mathbf{k}$;

$$\text{curl } \mathbf{F} = \begin{vmatrix} \mathbf{i} & \mathbf{j} & \mathbf{k} \\ \frac{\partial}{\partial x} & \frac{\partial}{\partial y} & \frac{\partial}{\partial z} \\ 3y & 2z & -5x \end{vmatrix}$$

$$= -2\mathbf{i} + 5\mathbf{j} - 3\mathbf{k}$$

Take S to be the xy-plane, so $\mathbf{N} = \mathbf{k}$ and

curl $\mathbf{F} \cdot \mathbf{N} = -3$. We have $dS = dA$ and the

projected region D is the disk $x^2 + y^2 \leq 1$.

By Stokes' theorem

$$\oint_C \mathbf{F} \cdot d\mathbf{R} = \int\int_S (\text{curl } \mathbf{F} \cdot \mathbf{N})\, dS$$

$$= \int\int_D (-3)\, dA$$

$$= -3(\text{AREA OF } D)$$

$$= -3\pi(1)^2$$

$$= -3\pi$$

15. $\mathbf{F} = (y^2 + z^2)\mathbf{i} + (x^2 + y^2)\mathbf{j} + (x^2 + y^2)\mathbf{k};$

$$\text{curl } \mathbf{F} = \begin{vmatrix} \mathbf{i} & \mathbf{j} & \mathbf{k} \\ \dfrac{\partial}{\partial x} & \dfrac{\partial}{\partial y} & \dfrac{\partial}{\partial z} \\ y^2 + z^2 & x^2 + y^2 & x^2 + y^2 \end{vmatrix}$$

$$= (2y)\mathbf{i} - (2x - 2z)\mathbf{j} + (2x - 2y)\mathbf{k}$$

The plane containing the triangle is

$x + y + z = 1$, and the projected region in the

xy-plane is the triangle D bounded by the x-

axis, the y-axis, and the line $x + y = 1$. If

$G = x + y + z - 1$, we have

$\nabla G = \mathbf{i} + \mathbf{j} + \mathbf{k}$, and by Stokes' theorem

$$\oint_C \mathbf{F} \cdot d\mathbf{R} = \int\int_S (\text{curl } \mathbf{F} \cdot \mathbf{N}) \, dS$$

$$= \int\int_D (\text{curl } \mathbf{F} \cdot \nabla G) \, dA$$

$$= \int\int_D \langle 2y, \, 2z - 2x, \, 2x - 2y \rangle \cdot \langle 1, \, 1, \, 1 \rangle \, dA$$

$$= \int_0^1 \int_0^{1-x} [2y + 2z - 2x + 2x - 2y] \, dy \, dx$$

$$= \int_0^1 \int_0^{1-x} 2(1 - x - y) \, dy \, dx$$

$$= \int_0^1 (x - 1)^2 \, dx$$

$$= \frac{1}{3}$$

17. The boundary curve C is a square in the xy-

plane ($z = 0$):

C_1: $x = t$, $y = 0$ for $0 \le t \le 1$

C_2: $x = 1$, $y = t$ for $0 \le t \le 1$

C_3: $x = 1 - t$, $y = 1$ for $0 \le t \le 1$

C_4: $x = 0$, $y = 1 - t$ for $0 \le t \le 1$

$$\int\int_S (\text{curl } \mathbf{F} \cdot \mathbf{N}) \, dS$$

$$= \oint_C (xy \, dx - z \, dy)$$

$$= \int_0^1 0 \, dt + \int_0^1 0 \, dt + \int_0^1 (1 - t)(-1) \, dt$$

$$\qquad + \int_0^1 0 \, dt$$

$$= -\frac{1}{2}$$

19. The paraboloid and the plane intersect where

$x^2 + y^2 = y$ or $r = \sin \theta$ (in polar form).

Thus, the curve C can be parameterized as:

$$x = (\sin \theta) \cos \theta = \tfrac{1}{2} \sin 2\theta$$

$$y = z = \sin \theta \, \sin \theta = \tfrac{1}{2}(1 - \cos 2\theta), \, 0 \le \theta \le \pi$$

$$\iint_S (\text{curl } \mathbf{F} \cdot \mathbf{N}) \ dS$$

$$= \oint_C (xy \ dx + x^2 \ dy + z^2 \ dz)$$

$$= \int_0^\pi [\tfrac{1}{4} \sin 2\theta (1 - \cos 2\theta)(\cos 2\theta)$$

$$+ \tfrac{1}{4} \sin^2 2\theta (\sin 2\theta)$$

$$+ \tfrac{1}{4}(1 - \cos 2\theta)^2 (\sin 2\theta)] \ d\theta$$

$$= \int_0^\pi [1 - \cos 2\theta + \sin^2 2\theta] \tfrac{1}{4} \sin 2\theta \ d\theta$$

$$= 0$$

21. The curve C is the circle $x^2 + y^2 = 4$ in the xy-plane:

$$x = 2 \cos t, \ y = 2 \sin t, \ z = 0; \ 0 \le t \le 2\pi$$

$$\iint_S (\text{curl } \mathbf{F} \cdot \mathbf{N}) \ dS$$

$$= \oint_C (4y \ dx + z \ dy + 2y \ dz)$$

$$= \int_0^{2\pi} [4(2 \sin t)(-2 \sin t) + 0 + 0] \ dt$$

$$= -16\pi$$

23. $\quad \text{curl } \mathbf{F} = \begin{vmatrix} \mathbf{i} & \mathbf{j} & \mathbf{k} \\ \dfrac{\partial}{\partial x} & \dfrac{\partial}{\partial y} & \dfrac{\partial}{\partial z} \\ 6x^2 e^{yz} & 2x^3 ze^{yz} & 2x^3 ye^{yz} \end{vmatrix}$

$$= (2x^3 e^{yz} + 2x^3 yze^{yz} - 2x^3 e^{yz} - 2x^3 zye^{yz})\mathbf{i}$$

$$- (6x^2 ye^{yz} - 6x^2 ye^{yz})\mathbf{j}$$

$$+ (6x^2 ze^{yz} - 6x^2 ze^{yz})\mathbf{k}$$

$$= \mathbf{0}$$

Thus, $\displaystyle\iint_S (\text{curl } \mathbf{F} \cdot \mathbf{N}) \ dS = 0.$

25. Let S be the part of the plane $x + y + z = 1$ that lies in the first octant.

$$\mathbf{F} = (1 + y)z\mathbf{i} + (1 + z)x\mathbf{j} + (1 + x)y\mathbf{k}$$

$$f(x, y, z) = x + y + z - 1$$

$$\mathbf{N} = \frac{1}{\sqrt{3}}(\mathbf{i} + \mathbf{j} + \mathbf{k})$$

$$dS = \sqrt{3} \ dA$$

$$\text{curl } \mathbf{F} = \begin{vmatrix} \mathbf{i} & \mathbf{j} & \mathbf{k} \\ \dfrac{\partial}{\partial x} & \dfrac{\partial}{\partial y} & \dfrac{\partial}{\partial z} \\ (1+y)z & (1+z)x & (1+x)y \end{vmatrix}$$

$$= \mathbf{i} + \mathbf{j} + \mathbf{k}$$

$$(\text{curl } \mathbf{F} \cdot \mathbf{N}) \ dS = 3 \ dA$$

$$\oint_C \mathbf{F} \cdot d\mathbf{R} = \iint_D (\text{curl } \mathbf{F} \cdot \mathbf{N}) \ dS$$

$$= \iint_R 3 \ dA$$

$$= \frac{3}{2} \quad \textit{since the projection in the } xy\text{-}$$

plane is a triangle with vertices $(0, 0)$, $(0, 1)$, $(1, 0)$.

27. $\mathbf{V} = x\mathbf{i} + (z - x)\mathbf{j} + y\mathbf{k}$

$$\text{curl } \mathbf{V} = \begin{vmatrix} \mathbf{i} & \mathbf{j} & \mathbf{k} \\ \frac{\partial}{\partial x} & \frac{\partial}{\partial y} & \frac{\partial}{\partial z} \\ x & z-x & y \end{vmatrix}$$

$$= -\mathbf{k}$$

Take S to be the hemisphere

$z = \sqrt{1 - x^2 - y^2}$, so $\mathbf{N} = x\mathbf{i} + y\mathbf{j} + z\mathbf{k}$

curl $\mathbf{V} \cdot \mathbf{N} = -z$, and

$$dS = \sqrt{\left(\frac{-x}{z}\right)^2 + \left(\frac{-y}{z}\right)^2 + 1}\; dA$$

$$= \frac{1}{z}\, dA.$$

The projected region D is the disk $x^2 + y^2 \le y$

or $r \le \sin\theta$ (in polar form), for $0 \le \theta \le \pi$.

By Stokes' theorem

$$\oint_C \mathbf{V} \cdot d\mathbf{R} = \int\int_S (\text{curl } \mathbf{V} \cdot \mathbf{N})\, dS$$

$$= \int\int_D (-z)\left(\tfrac{1}{z}\, dA\right)$$

$$= - (\text{AREA OF } D)$$

$$= -\pi(\tfrac{1}{2})^2$$

$$= -\frac{\pi}{4}$$

29. $\mathbf{V} = (e^{x^2} + z)\mathbf{i} + (x + \sin y^3)\mathbf{j} + [y + \ln(\tan^{-1}z)]\mathbf{k}$

$$\text{curl } \mathbf{V} = \begin{vmatrix} \mathbf{i} & \mathbf{j} & \mathbf{k} \\ \frac{\partial}{\partial x} & \frac{\partial}{\partial y} & \frac{\partial}{\partial z} \\ e^{x^2} + z & x + \sin y^3 & y + \ln(\tan^{-1}z) \end{vmatrix}$$

$$= \mathbf{i} + \mathbf{j} + \mathbf{k}$$

The sphere and the cone intersect where

$$x^2 + y^2 + (x^2 + y^2) = 1$$

or, equivalently, $z = \frac{\sqrt{2}}{2}$. Take S to be the

plane $z = \frac{\sqrt{2}}{2}$, so $\mathbf{N} = \mathbf{k}$, curl $\mathbf{V} \cdot \mathbf{N} = 1$, and

$dS = dA$. The projected region D is the disk

$x^2 + y^2 \le \frac{1}{2}$. By Stokes' theorem

$$\oint_C \mathbf{V} \cdot d\mathbf{R} = \int\int_S (\text{curl } \mathbf{V} \cdot \mathbf{N})\, dS$$

$$= \int\int_D 1\, dA$$

$$= \text{AREA OF DISK}$$

$$= \pi\left(\frac{\sqrt{2}}{2}\right)^2$$

$$= \frac{\pi}{2}$$

31. $\mathbf{F} = y^2\mathbf{i} + xy\mathbf{j} + xz\mathbf{k}$,

C: $z = 0$, $x = \cos\theta$, $y = \sin\theta$; $0 \le \theta \le 2\pi$

$$\int\int_S (\text{curl } \mathbf{F} \cdot \mathbf{N}) \; dS = \oint_C (y^2 \, dx + xy \; dy + xz \; dz)$$

$$= \int_0^{2\pi} [\sin^2\theta(-\sin\theta) + (\cos\theta\sin\theta)\cos\theta] \; d\theta$$

$$= 0$$

33.
$$\text{curl } \mathbf{F} = \begin{vmatrix} \mathbf{i} & \mathbf{j} & \mathbf{k} \\ \frac{\partial}{\partial x} & \frac{\partial}{\partial y} & \frac{\partial}{\partial z} \\ 4z & -3x & 2y \end{vmatrix}$$

$$= 2\mathbf{i} + 4\mathbf{j} - 3\mathbf{k}$$

Let C be a positively-oriented curve bounding the region (surface) D in the plane

$$5x + 3y + 2z = 4$$

If

$$G = 5x + 3y + 2z - 4$$

then $\nabla G = \langle 5, 3, 2 \rangle$ and by Stokes' theorem

$$\oint_C \mathbf{F} \cdot d\mathbf{R} = \int\int_S (\text{curl } \mathbf{F} \cdot \mathbf{N}) \; dS$$

$$= \int\int_D (\text{curl } \mathbf{F} \cdot \nabla G) \sqrt{38} \; dA$$

$$= \int\int_D \langle 2, 4, -3 \rangle \cdot \langle 5, 3, 2 \rangle \sqrt{38} \; dA$$

$$= \int\int_D 16\sqrt{38} \; dA$$

$$= 16\sqrt{38} \,(\text{AREA OF D})$$

Thus, $\oint_C \mathbf{F} \cdot d\mathbf{R}$ depends on only the area bounded by C, and not on the curve C itself, so

$$\oint_{C_1} = \oint_{C_2}$$

as claimed.

35. Let C be the curve of intersection of the top half of the ellipsoid with the xy-plane, oriented counterclockwise and let $-C$ denote the same curve oriented clockwise. However, in relation to the outer normal of the bottom half of the ellipsoid, $-C$ has positive orientation. Thus, by Stokes' theorem

$$\int\int_S (\text{curl } \mathbf{F} \cdot \mathbf{N}) \; dS = \oint_C \mathbf{F} \cdot d\mathbf{R} + \oint_{-C} \mathbf{F} \cdot d\mathbf{R}$$

$$= \oint_C \mathbf{F} \cdot d\mathbf{R} - \oint_C \mathbf{F} \cdot d\mathbf{R}$$

$$= 0$$

The same procedure applies whenever S is a closed surface.

37. $\mu I = \int\int_S \mu \, \mathbf{J} \cdot \mathbf{N} \; dS$

$$= \int\!\!\int_S (\text{curl } \mathbf{B} \cdot \mathbf{N}) \ dS, \quad \text{since } \mu J = \text{curl } B$$

$$= \oint_C \mathbf{B} \cdot d\mathbf{R} \qquad \text{by Stokes' theorem}$$

39. $\text{curl}(f \nabla g + g \nabla f)$

$$= \begin{vmatrix} \mathbf{i} & \mathbf{j} & \mathbf{k} \\ \frac{\partial}{\partial x} & \frac{\partial}{\partial y} & \frac{\partial}{\partial z} \\ fg_x + gf_x & fg_y + gf_y & fg_z + gf_z \end{vmatrix}$$

$$= \mathbf{0}$$

By Stokes' theorem,

$$\oint_C (f \nabla g + g\nabla f) \cdot d\mathbf{R}$$

$$= \int\!\!\int_S \text{curl}(f \nabla g + g\nabla f) \cdot \mathbf{N} \ dS$$

$$= 0$$

13.7 The Divergence Theorem, Pages 922-924

SURVIVAL HINT: *Green's theorem, which found a relationship between a double integral over a region in a plane and the line integral of its boundary, is extended to the divergence theorem. This gives the relationship between the triple integral over a portion of space, D, and the surface integral over the boundary, S, of that region in $\mathbb{R}^3$.*

1. Evaluating the surface integral $\int\!\!\int_S \mathbf{F} \cdot \mathbf{N} \ dS$

For the top portion:

$$\mathbf{F} = xz\mathbf{i} + y^2\mathbf{j} + 2z\mathbf{k}$$

$$z = \sqrt{4 - x^2 - y^2}$$

$$z_x = -\frac{x}{z}$$

$$z_y = -\frac{y}{z}$$

$$dS = \frac{2}{z} \ dA$$

(upward normal): $\mathbf{N} = \frac{x}{2}\mathbf{i} + \frac{y}{2}\mathbf{j} + \frac{z}{2}\mathbf{k}$

$$\int\!\!\int_{S_T} \mathbf{F} \cdot \mathbf{N} \ dS$$

$$= \int\!\!\int_R \left(x^2 + \frac{y^3}{z} + 2z \right) dA$$

$$= \int\!\!\int_R \left(x^2 + \frac{y^3}{\sqrt{4 - x^2 - y^2}} + 2\sqrt{4 - x^2 - y^2} \right) dy \ dx$$

$$= \int_0^{2\pi}\!\!\int_0^2 \left(r^3 \cos^2\theta + \frac{r^4\sin^3\theta}{\sqrt{4 - r^2}} + 2r\sqrt{4 - r^2} \right) dr \ d\theta$$

$$= \int_0^{2\pi} \left(4 \cos^2\theta + 3\pi \sin^3\theta + \frac{16}{3} \right) d\theta$$

$$= \frac{44\pi}{3}$$

For the bottom portion:

$$\mathbf{F} = xz\mathbf{i} + y^2\mathbf{j} + 2z\mathbf{k}$$

$$z = -\sqrt{4 - x^2 - y^2}$$

$$z_x = \frac{x}{\sqrt{4 - x^2 - y^2}}$$

$$z_y = \frac{y}{\sqrt{4 - x^2 - y^2}}$$

$$dS = -\frac{2}{z}\, dA$$

(downward normal):

$$\mathbf{N} = \frac{x}{2}\mathbf{i} + \frac{y}{2}\mathbf{j} + \frac{z}{2}\mathbf{k}$$

$$\iint_{S_B} \mathbf{F} \cdot \mathbf{N}\, dS$$

$$= \iint_R \left(-x^2 + \frac{y^3}{\sqrt{4 - x^2 - y^2}} + 2\sqrt{4 - x^2 - y^2} \right) dA$$

$$= \int_0^{2\pi} \int_0^2 \left[-r^3\cos^2\theta + \frac{r^4\sin^3\theta}{\sqrt{4 - r^2}} + 2r\sqrt{4 - r^2} \right] dr\, d\theta$$

$$= \frac{20\pi}{3}$$

Thus,

$$\iint_S \mathbf{F} \cdot \mathbf{N}\, dS = \frac{44\pi}{3} + \frac{20\pi}{3} = \frac{64\pi}{3}$$

Evaluating the integral:

$$\iiint_D \operatorname{div} \mathbf{F}\, dV$$

$$= \iiint_D (z + 2y + 2)\, dV$$

$$= \int_0^\pi \int_0^{2\pi} \int_0^2 \left(\rho\cos\phi + 2\rho\sin\phi\sin\theta + 2 \right)\rho^2\sin\phi \, d\rho\, d\theta\, d\phi$$

$$= \int_0^\pi \int_0^{2\pi} \left(4\,\sin\phi\cos\phi + 8\sin^2\phi\sin\theta + \frac{16}{3}\sin\phi \right) d\theta\, d\phi$$

$$= \int_0^\pi \left(8\pi\,\sin\phi\,\cos\phi + \frac{32\pi}{3}\,\sin\phi \right) d\phi$$

$$= \frac{64\pi}{3}$$

3. Evaluating the surface integral $\displaystyle\iint_S \mathbf{F} \cdot \mathbf{N}\, dS$

$$\mathbf{F} = 2y^2\mathbf{j}$$

$$f(x,\, y,\, z) = x + 4y + z - 8$$

$$= 0$$

$$\nabla f = \mathbf{i} + 4\mathbf{j} + \mathbf{k}$$

$$\mathbf{N} = \frac{\mathbf{i} + 4\mathbf{j} + \mathbf{k}}{3\sqrt{2}}$$

$$\mathbf{F} \cdot \mathbf{N} = \frac{8y^2}{3\sqrt{2}}$$

$$dS = 3\sqrt{2}\, dA$$

$$\iint_S \mathbf{F} \cdot \mathbf{N}\, dS = \int_0^2 \int_0^{8-4y} \frac{8y^2}{3\sqrt{2}}\, 3\sqrt{2}\, dx\, dy$$

$$= 8\int_0^2 y^2(8 - 4y)\, dy$$

$$= \frac{128}{3}$$

Since $\operatorname{div} \mathbf{F} = 4y$, we have

$$\iiint_D \operatorname{div} \mathbf{F}\, dV = \iiint_D 4y\, dV$$

$$= 4\int_0^2 \int_0^{8-4y} \int_0^{8-x-4y} y\, dz\, dx\, dy$$

$$= 4\int_0^2 \int_0^{8-4y} (8y - xy - 4y^2)\, dx\, dy$$

$$= 4 \int_0^2 \left(32y - 32y^2 + 8y^3 \right) \, dy$$

$$= 32 \int_0^2 \left(4y - 4y^2 + y^3 \right) \, dy$$

$$= \frac{128}{3}$$

5. $\text{div } \mathbf{F} = 3$;

$$\int\int_S \mathbf{F} \cdot \mathbf{N} \, dS = \int\int\int_D \text{div } \mathbf{F} \, dV$$

$$= 3V$$

$$= 3$$

7. $\mathbf{F} = (\cos yz)\mathbf{i} + e^{xz}\mathbf{j} + 3z^2\mathbf{k}$

$\text{div } \mathbf{F} = 6z$

$$\int\int_S \mathbf{F} \cdot \mathbf{N} \, dS = \int\int\int_D \text{div } \mathbf{F} \, dV$$

$$= 6 \int_0^{2\pi} \int_0^2 \int_0^{\sqrt{4-r^2}} r \, z \, dz \, dr \, d\theta$$

$$= 3 \int_0^{2\pi} \int_0^2 (4 - r^2) \, r \, dr \, d\theta$$

$$= 3 \int_0^{2\pi} (8 - 4) \, d\theta$$

$$= 24\pi$$

9. Let S_C denote the surface of the closed cube.

Then $S_C = S \cup S_m$, where S_m is the surface of the missing face. For S_C,

$$\mathbf{F} = (x^2 + y^2 - z^2)\mathbf{i} + (x^2 y)\mathbf{j} + (3z)\mathbf{k}$$

$\text{div } \mathbf{F} = 2x + x^2 + 3$

$$\int\int_{S_C} \mathbf{F} \cdot \mathbf{N} \, dS = \int\int\int_D \text{div } \mathbf{F} \, dV$$

$$= \int_0^1 \int_0^1 \int_0^1 (2x + x^2 + 3) \, dz \, dy \, dx$$

$$= \frac{13}{3}$$

On S_m, $z = 0$ and $\mathbf{N} = -\mathbf{k}$, so $\mathbf{F} \cdot \mathbf{N} = 0$ and

$$\int\int_S \mathbf{F} \cdot \mathbf{N} \, dS = \int\int_{S_C} \mathbf{F} \cdot \mathbf{N} \, dS - \int\int_{S_m} \mathbf{F} \cdot \mathbf{N} \, dS$$

$$= \frac{13}{3} - 0$$

$$= \frac{13}{3}$$

11. Let S_C denote the closed surface consisting of the paraboloid S and the top disk S_d. Then

$$\mathbf{F} = x\mathbf{i} + y\mathbf{j} + z\mathbf{k}$$

$\text{div } \mathbf{F} = 1 + 1 + 1 = 3$,

$$\int\int_{S_C} \mathbf{F} \cdot \mathbf{N} \, dS = \int\int\int_D \text{div } \mathbf{F} \, dV$$

$$= \int_0^{2\pi} \int_0^3 \int_{r^2}^9 3 \, r \, dz \, dr \, d\theta$$

$$= 3 \int_0^{2\pi} \int_0^3 (9 - r^2) \, r \, dr \, d\theta$$

$$= 6\pi \int_0^3 (9r - r^3) \, dr$$

$$= \frac{243\pi}{2}$$

For the disk $x^2 + y^2 \leq 9$, $z = 9$ and $\mathbf{N} = \mathbf{k}$,

so $\mathbf{F} \cdot \mathbf{N} = z = 9$; $dS = dA$ and

$$\iint_{S_d} \mathbf{F} \cdot \mathbf{N} \, dS = \iint_{S_d} 9 \, dA$$

$$= 9(\text{AREA OF DISK})$$

$$= 9\pi(3)^2$$

$$= 81\pi$$

Thus,

$$\iint_S \mathbf{F} \cdot \mathbf{N} \, dS = \iint_{S_C} \mathbf{F} \cdot \mathbf{N} \, dS - \iint_{S_d} \mathbf{F} \cdot \mathbf{N} \, dS$$

$$= \frac{243\pi}{2} - 81\pi$$

$$= \frac{81\pi}{2}$$

13. $\mathbf{F} = x^2\mathbf{i} + y^2\mathbf{j} + z^2\mathbf{k}$

div $\mathbf{F} = 2(x + y + z)$;

$$\iint_S \mathbf{F} \cdot \mathbf{N} \, dS$$

$$= \iiint_D \text{div } \mathbf{F} \, dV$$

$$= \iiint_D 2(x + y + z) \, dV$$

$$= \int_0^{2\pi} \int_0^2 \int_{-\sqrt{4-r^2}}^{\sqrt{4-r^2}} 2[r\cos\theta + r\sin\theta + z] \, r \, dz \, dr \, d\theta$$

$$= \int_0^{2\pi} \int_0^2 4(\cos\theta + \sin\theta) \, r^2\sqrt{4 - r^2} \, dr \, d\theta$$

$$= \int_0^{2\pi} 4\pi(\cos\theta + \sin\theta) \, d\theta$$

$$= 0$$

15. $\mathbf{F} = x\mathbf{i} + y\mathbf{j} + (z^2 - 1)\mathbf{k}$

div $\mathbf{F} = 1 + 1 + 2z$

$$= 2(z + 1)$$

$$\iint_S \mathbf{F} \cdot \mathbf{N} \, dS = \iiint_D \text{div } \mathbf{F} \, dV$$

$$= 2 \int_0^{2\pi} \int_0^2 \int_0^1 (z + 1) \, dz \, r \, dr \, d\theta$$

$$= 3 \int_0^{2\pi} \int_0^2 r \, dr \, d\theta$$

$$= 6 \int_0^{2\pi} d\theta$$

$$= 6(2\pi)$$

$$= 12\pi$$

17. $\mathbf{F} = \langle xy^2, \ yz^2, \ x^2y \rangle$

$\text{div } \mathbf{F} = y^2 + z^2 = (\rho \sin \phi \sin \theta)^2 + (\rho \cos \phi)^2$

$= \rho^2 [\sin^2 \phi \sin^2 \theta + \cos^2 \phi]$

By the divergence theorem

$$\iint\limits_{S} \mathbf{F} \cdot \mathbf{N} \ dS = \iiint\limits_{R} \text{div } \mathbf{F} \ dV$$

where R is the solid region bounded by S.

Thus,

$$\iint\limits_{S} \mathbf{F} \cdot \mathbf{N} \ dS$$

$$= \iiint\limits_{R} \text{div } \mathbf{F} \ dV$$

$$= \int_{0}^{2\pi} \int_{0}^{\pi/4} \int_{0}^{2} \rho^2 [\sin^2 \phi \sin^2 \theta + \cos^2 \phi] \rho^2 \sin\phi \ d\rho \ d\phi \ d\theta$$

$$= \int_{0}^{2\pi} \int_{0}^{\pi/4} \int_{0}^{2} [\sin^3 \phi \sin^2 \theta + \cos^2 \phi \sin \phi] \rho^4 \ d\rho \ d\phi \ d\theta$$

$$= \int_{0}^{2\pi} \int_{0}^{\pi/4} [\sin^3 \phi \sin^2 \theta + \cos^2 \phi \sin \phi] \left[\frac{\rho^5}{5}\right]\Bigg|_{0}^{2} \ d\phi \ d\theta$$

$$= \frac{32}{5} \int_{0}^{2\pi} \left[\left(-\frac{1}{3}\sin^2\phi - \frac{2}{3}\right)\cos\phi \ \sin^2\theta - \frac{1}{3}\cos^3\phi\right]\Bigg|_{0}^{\pi/4} \ d\theta$$

$$= \frac{32}{5} \int_{0}^{2\pi} \left[\left(\frac{2}{3} - \frac{5\sqrt{2}}{12}\right)\sin^2\theta - \frac{2\sqrt{2}}{24} + \frac{1}{3}\right] \ d\theta$$

$$= \frac{32}{5}\left(\frac{16 - 7\sqrt{2}}{12}\right)\pi$$

$$= \frac{8(16 - 7\sqrt{2})}{15}\pi$$

Even though the region R is described in spherical coordinates, it is easier to use cylindrical coordinates because $y^2 + z^2$ is complicated when expanded in spherical form. In cylindrical coordinates, the sphere $\rho = 2$ becomes $r^2 + z^2 = 4$, the cone $\phi = \frac{\pi}{4}$ becomes $z = r$, and the projected region on the xy-plane is the disk $r \leq \sqrt{2}$. Thus, we have the (somewhat easier) alternate calculation:

$$\iint\limits_{S} \mathbf{F} \cdot \mathbf{N} \ dS$$

$$= \iiint\limits_{R} \text{div } \mathbf{F} \ dV$$

$$= \int_0^{2\pi} \int_0^{\sqrt{2}} \int_r^{\sqrt{4-r^2}} (r^2 \sin^2\theta + z^2) r \, dz \, dr \, d\theta$$

$$= \int_0^{2\pi} \int_0^{\sqrt{2}} \left[(r^3 \sin^2\theta) z + \frac{r}{3} z^3 \right] \Big|_r^{\sqrt{4-r^2}} \, dr \, d\theta$$

$$= \int_0^{2\pi} \int_0^{\sqrt{2}} \left\{ r^3 \sin^2\theta (\sqrt{4-r^2} - r) + \frac{r}{3} \left[(4-r^2)^{3/2} - r^3 \right] \right\} d\theta$$

$$= \int_0^{2\pi} -\frac{8}{15} \left[(5\sqrt{2} - 8)\sin^2\theta + \sqrt{2} - 4 \right] d\theta$$

$$= \frac{8\pi}{15}(16 - 7\sqrt{2})$$

19. $\operatorname{div} \mathbf{F} = 3x^2 + 3y^2 + 3a^2 = 3(r^2 + a^2)$

$$\iint_S \mathbf{F} \cdot \mathbf{N} \, dS = \iiint_D \operatorname{div} \mathbf{F} \, dV$$

$$= \int_0^{2\pi} \int_0^a \int_0^1 3(r^2 + a^2) \, r \, dz \, dr \, d\theta$$

$$= 3 \int_0^{2\pi} \int_0^a (r^3 + a^2 r) \, dr \, d\theta$$

$$= 3 \int_0^{2\pi} \frac{3a^4}{4} \, d\theta$$

$$= \frac{9\pi a^4}{2}$$

21. $\mathbf{R} = x\mathbf{i} + y\mathbf{j} + z\mathbf{k}$

$\|\mathbf{R}\|\mathbf{R} = a(x\mathbf{i} + y\mathbf{j} + z\mathbf{k})$

$\operatorname{div}(\|\mathbf{R}\|\mathbf{R}) = a + a + a$

$$= 3a$$

$$\iint_S \|\mathbf{R}\|\mathbf{R} \cdot \mathbf{N} \, dS = \iiint_D 3a \, dV$$

$$= 3a\left(\frac{4}{3}\pi a^3\right)$$

$$= 4\pi a^4$$

23. Applying the divergence theorem with

$\mathbf{F} = f\nabla f$, we obtain

$$\iint_S (f\nabla f) \cdot \mathbf{N} \, dS = \iiint_G \operatorname{div}(f\nabla f) \, dV$$

Therefore, since

$$\operatorname{div}(f\nabla f) = f \operatorname{div} \nabla f + \nabla f \cdot \nabla f$$

$$= f \operatorname{div} \nabla f + \|\nabla f\|^2$$

it follows that

$$\iint_S (f\nabla f) \cdot \mathbf{N} \, dS = \iiint_G \|\nabla f\|^2 \, dV$$

if $f \operatorname{div} \nabla f = 0$; that is, $\operatorname{div} \nabla f = 0$.

25. a. $\displaystyle \iint_S \frac{\partial u}{\partial n} \, dS = \iint_S \nabla u \cdot \mathbf{N} \, dS$

$$= \iiint_D \operatorname{div}(\nabla u) \, dV$$

$$= \iiint_D \nabla \cdot \nabla u \, dV$$

$$= \int\int\int_D \nabla^2 u \, dV$$

b. $\operatorname{div}(u\nabla v) = \nabla \cdot (u\nabla v)$

$$= u\nabla^2 v + \nabla u \cdot \nabla v$$

$$= (x + y + z)(3) + \langle 1, 1, 1\rangle \cdot \langle x, y, z\rangle$$

$$= 4(x + y + z)$$

$$\int\int_S (u\nabla v)\cdot \mathbf{N}\, dS = \int\int\int_D \operatorname{div}\,(u\nabla v)\, dV$$

$$= \int_0^1\int_0^1\int_0^1 4(x + y + z)\, dz\, dy\, dx$$

$$= 6$$

27. $\displaystyle \int\int_S \frac{\partial g}{\partial n}\, dS = \int\int_S \nabla g \cdot \mathbf{N}\, dS$

$$= \int\int\int_D \nabla \cdot (\nabla g)\, dV$$

divergence theorem

$$= \int\int\int_D \nabla^2 g \, dV$$

$$= 0 \quad \textit{g is harmonic}$$

29. If K is a variable,

$$\operatorname{div}(K\nabla T) = \nabla \cdot (K\nabla T)$$

$$= K\nabla^2 T + \nabla K \cdot \nabla T$$

Thus, if $\mathbf{F} = -K\nabla T$ is the velocity of heat

flow,

$$\int\int_S \mathbf{F}\cdot \mathbf{N}\, dlS = \int\int\int_D \operatorname{div}(-K\nabla T)\, dV$$

$$= \int\int\int_D (-1)(K\nabla^2 T + \nabla K \cdot \nabla T)\, dV$$

$$= \int\int\int_D (-1)\left(\sigma\delta\frac{\partial T}{\partial t}\right)\, dV$$

Shrinking the body to a single point, we

obtain

$$K\nabla^2 T + \nabla K \cdot \nabla T = \sigma\rho\,\frac{\partial T}{\partial t}$$

31. Suppose the surface S encloses the origin.

Then by Gauss' law (part **b** of Problem 30),

we have

$$\int\int_S \mathbf{D}\cdot \mathbf{N}\, dS = \int\int_S (\epsilon\mathbf{E})\cdot \mathbf{N}\, dS$$

$$= \frac{q}{\epsilon}(\epsilon)$$

$$= q$$

$$= \int\int\int_V Q\, dV$$

Let S' be a sphere centered at the origin that

is entirely contained within S, and let S'' be

the surface of the region inside S but outside

S'(as in the proof of part **b** of Problem 30).

Then the divergence theorem applies to S'', and we have

$$\iiint_V Q \, dV = \iint_{S''} \mathbf{D} \cdot \mathbf{N} \, dS$$

$$= \iiint_{\text{interior of } S} \text{div } \mathbf{D} \, dV$$

If we take S' to be smaller and smaller ($\rho \to 0$, where ρ is the radius of S'), then in the limit, we have

$$\iiint_V Q \, dV = \iiint_V \text{div } \mathbf{D} \, dV$$

Finally, since this equation holds for *every* region contained within a surface that encloses the origin, it follows that the integrands must be equal; that is,

$$Q = \text{div } \mathbf{D}$$

CHAPTER 13 REVIEW

Proficiency Examination, Pages 925-926

SURVIVAL HINT: *To help you review the concepts of this chapter, **handwrite** the answers to each of these questions onto your own paper.*

1. A vector field is a collection S of points in space together with a rule that assigns to each point (x, y, z) in S exactly one vector $\mathbf{V}(x, y, z)$.

2. The divergence of a vector field
    $$\mathbf{V}(x, y, z) = u(x,y,z)\mathbf{i} + v(x,y,z)\mathbf{j} + w(x,y,z)\mathbf{k}$$
    is denoted by div $\mathbf{V}$ and is defined by
    $$\text{div } \mathbf{V} = \frac{\partial u}{\partial x}(x, y, z) + \frac{\partial v}{\partial y}(x, y, z) + \frac{\partial w}{\partial z}(x, y, z)$$

3. The curl of a vector field
    $$\mathbf{V}(x, y, z) = u(x,y,z)\mathbf{i} + v(x,y,z)\mathbf{j} + w(x,y,z)\mathbf{k}$$
    is denoted by curl $\mathbf{V}$ and is defined by
    $$\text{curl } \mathbf{V} = \left(\frac{\partial w}{\partial y} - \frac{\partial v}{\partial z}\right)\mathbf{i} + \left(\frac{\partial u}{\partial z} - \frac{\partial w}{\partial x}\right)\mathbf{j} + \left(\frac{\partial v}{\partial x} - \frac{\partial u}{\partial y}\right)\mathbf{k}$$

4. The del operator is defined by
    $$\nabla = \frac{\partial}{\partial x}\mathbf{i} + \frac{\partial}{\partial y}\mathbf{j} + \frac{\partial}{\partial z}\mathbf{k}$$

5. The Laplacian of f is
    $$\nabla^2 f = \nabla \cdot \nabla f$$
    $$= \frac{\partial^2 f}{\partial x^2} + \frac{\partial^2 f}{\partial y^2} + \frac{\partial^2 f}{\partial z^2}$$
    $$= f_{xx} + f_{yy} + f_{zz}$$

    The equation $\nabla^2 f = 0$ is called Laplace's equation.

6. A line integral involves taking a limit of a Riemann sum formed by parameterizing with

respect to a curve in space. An "ordinary" Riemann integral is formed by parameterizing with respect to the x-axis.

7. Let
$$\mathbf{F}(x, y, z) = u(x, y, z)\mathbf{i} + v(x, y, z)\mathbf{j} + w(x, y, z)\mathbf{k}$$
be a vector field, and let C be the curve with parametric representation
$$\mathbf{R}(t) = x(t)\mathbf{i} + y(t)\mathbf{j} + z(t)\mathbf{k} \quad \text{for } a \le t \le b$$
Using $d\mathbf{R} = dx\,\mathbf{i} + dy\,\mathbf{j} + dz\,\mathbf{k}$, we denote the line integral of $\mathbf{F}$ over C by $\int_C \mathbf{F} \cdot d\mathbf{R}$ and define it by
$$\int_C \mathbf{F} \cdot d\mathbf{R} = \int_C (u\,dx + v\,dy + w\,dz)$$
$$= \int_a^b \left\{ u[x(t),\ y(t),\ z(t)]\,\frac{dx}{dt} + v[x(t),\ y(t),\ z(t)]\,\frac{dy}{dt} \right.$$
$$\left. + w[x(t),\ y(t),\ z(t)]\,\frac{dz}{dt} \right\} dt$$

8. Let $\mathbf{F}$ be a continuous force field over a domain D. Then the **work** W done by $\mathbf{F}$ as an object moves along a smooth curve C in D is given by the line integral
$$W = \int_C \mathbf{F} \cdot d\mathbf{R}$$

9. Let f be continuous on a smooth curve C. If C is defined by
$$\mathbf{R}(t) = x(t)\mathbf{i} + y(t)\mathbf{j} + z(t)\mathbf{k}, \text{ where } a \le t \le b,$$
then $\int_C f(x, y, z)\,ds$
$$= \int_a^b f[x(t),\ y(t), z(t)]\sqrt{[x'(t)]^2 + [y'(t)]^2 + [z'(t)]^2}\ dt$$

10. Let $\mathbf{F}$ be a conservative vector field on the region D and let f be a scalar potential function for $\mathbf{F}$; that is, $\nabla f = \mathbf{F}$. Then, if C is any piecewise smooth curve lying entirely within D, with initial point P and terminal point Q, we have
$$\int_C \mathbf{F} \cdot d\mathbf{R} = f(Q) - f(P)$$
Thus, the line integral $\int_C \mathbf{F} \cdot d\mathbf{R}$ is independent of path in D.

11. A vector field $\mathbf{F}$ is said to be conservative in a region D if it can be represented in D as the gradient of a continuously differentiable function f, which is then called a scalar potential of $\mathbf{F}$. That is, $\mathbf{F} = \nabla f$ for (x, y) in D.

12. If $\mathbf{F}$ is a conservative vector field and $\mathbf{F} = \nabla f$,

then f is a scalar potential for **F**.

13. A Jordan curve is a closed curve with no self intersections.

14. Let D be a simply connected region with a positively oriented piecewise smooth boundary C. Then if the vector field

$$\mathbf{F}(x,\ y) = M(x,\ y)\mathbf{i} + N(x,\ y)\mathbf{j}$$

is continuously differentiable on D, we have

$$\oint_C (M\ dx + N\ dy) = \int\int_D \left(\frac{\partial N}{\partial x} - \frac{\partial M}{\partial y} \right) dA$$

15. $$A = \frac{1}{2} \oint_C (-y\ dx + x\ dy)$$

$$= - \oint_C y\ dx$$

$$= \oint_C x\ dy$$

16. The normal derivative of f is denoted by $\partial f / \partial n$ and is the directional derivative of f in the direction of the normal **N** pointing to the exterior of the domain of f.

$$\frac{\partial f}{\partial n} = \nabla f \cdot \mathbf{N}$$

where **N** is the outer unit normal.

17. Let S be a surface defined by $z = f(x,\ y)$ and R_{xy} its projection on the xy-plane. If f, f_x, and f_y are continuous in R_{xy} and g is continuous on S, then the surface integral of g over S is

$$\int\int_S g(x,y,z)\,dS$$

$$= \int\int_R g(x,y,f(x,y))\sqrt{[f_x(x,y)]^2 + [f_y(x,y)]^2 + 1}\ dA$$

Similar definitions hold for projections on yz and xz planes.

18. If a surface S is defined parametrically by the vector function

$$\mathbf{R}(u,\ v) = x(u,\ v)\mathbf{i} + y(u,\ v)\mathbf{j} + z(u,\ v)\mathbf{k}$$

and $f(x,\ y,\ z)$ is continuous on D, the surface integral of f over D is given by

$$\int\int_S f(x,\ y,\ z)\,dS = \int\int_D f(\mathbf{R})\,\|\mathbf{R}_u \times \mathbf{R}_v\|\,du\,dv$$

19. The flux integral of a vector field **F** across a surface S is given by

$$\int\int_S \mathbf{F} \cdot \mathbf{N}\ dS$$

20. Let S be an oriented surface with unit normal vector $\mathbf{N}$, and assume that S is bounded by a closed, piecewise smooth curve C whose orientation is compatible with that of S. If $\mathbf{F}$ is a vector field that is continuously differentiable on S, then

$$\oint_C \mathbf{F} \cdot d\mathbf{R} = \int_S \int (\text{curl } \mathbf{F} \cdot \mathbf{N}) \, dS$$

21. If $\mathbf{F}$ and curl $\mathbf{F}$ are continuous in a simply connected region D, then $\mathbf{F}$ is conservative in D if and only if curl $\mathbf{F} = \mathbf{0}$

22. Let D be a region in space bounded by a smooth, orientable closed surface S. If $\mathbf{F}$ is a continuous vector field whose components have continuous partial derivatives in D, then

$$\int_S \int \mathbf{F} \cdot \mathbf{N} \, dS = \int \int_D \int \text{div } \mathbf{F} \, dV$$

where $\mathbf{N}$ is an outward unit normal to the surface S.

23. $\mathbf{F} = yz\mathbf{i} + xz\mathbf{j} + xy\mathbf{k} = M\mathbf{i} + N\mathbf{j} + P\mathbf{k}$

$$\text{curl } \mathbf{F} = \begin{vmatrix} \mathbf{i} & \mathbf{j} & \mathbf{k} \\ \frac{\partial}{\partial x} & \frac{\partial}{\partial y} & \frac{\partial}{\partial z} \\ yz & xz & xy \end{vmatrix}$$

$$= (x - x)\mathbf{i} - (y - y)\mathbf{j} + (z - z)\mathbf{k}$$

$$= \mathbf{0}$$

so $\mathbf{F}$ is conservative.

$\frac{\partial f}{\partial x} = yz$, so $f = xyz + a(y, z)$ and

$\frac{\partial f}{\partial y} = xz + \frac{\partial a}{\partial y} = xz$ so $\frac{\partial a}{\partial y} = 0$ and,

$a(y, z) = b(z)$; $f = xyz + b(z)$, but

$\frac{\partial f}{\partial z} = xy + \frac{\partial b}{\partial z} = xy$, so $\frac{\partial b}{\partial z} = 0$,

$b(z) = C$ and $f = xyz$

(where we take $C = 0$)

24. $\mathbf{F} = x^2 y\mathbf{i} - e^{yz}\mathbf{j} + \frac{x}{2}\mathbf{k}$

div $\mathbf{F} = 2xy - ze^{yz}$

$$\text{curl } \mathbf{F} = \begin{vmatrix} \mathbf{i} & \mathbf{j} & \mathbf{k} \\ \frac{\partial}{\partial x} & \frac{\partial}{\partial y} & \frac{\partial}{\partial z} \\ x^2 y & -e^{yz} & \frac{x}{2} \end{vmatrix}$$

$$= ye^{yz}\mathbf{i} - \frac{1}{2}\mathbf{j} - x^2\mathbf{k}$$

25. $\oint_C \mathbf{F} \cdot d\mathbf{R} = \int_T \int \left[\frac{\partial}{\partial x}(3y^2) - \frac{\partial}{\partial y}(2x + y) \right] dA$

$$= \int_T \int (-1) \, dA$$

$$= -(\text{AREA OF } T)$$

$$= -\tfrac{1}{2}(2)(2)$$

$$= -2$$

26. By Stokes' theorem

$$\oint_C \mathbf{F} \cdot d\mathbf{R} = \int\int_S (\text{curl } \mathbf{F} \cdot \mathbf{N})\, dS$$

$$\text{curl } \mathbf{F} = \begin{vmatrix} \mathbf{i} & \mathbf{j} & \mathbf{k} \\ \frac{\partial}{\partial x} & \frac{\partial}{\partial y} & \frac{\partial}{\partial z} \\ 2y & z & y \end{vmatrix}$$

$$= (1-1)\mathbf{i} - 2\mathbf{k}$$

$$= -2\mathbf{k}$$

The intersection of the sphere and the plane:

$$x^2 + y^2 + (x+2)^2 = 4(x+2)$$

$$2x^2 + y^2 = 4$$

$$\frac{x^2}{2} + \frac{y^2}{4} = 1$$

$$g(x, y, z) = z - x - 2 = 0$$

$$\mathbf{N} = \frac{-\mathbf{i} + \mathbf{k}}{\sqrt{2}}$$

$$\text{curl } \mathbf{F} \cdot \mathbf{N} = -\frac{2}{\sqrt{2}}$$

$$dS = \sqrt{2}\, dy\, dx$$

$$\int\int_S (\text{curl } \mathbf{F} \cdot \mathbf{N})\, dS$$

$$= -\int\int_S \sqrt{2}\sqrt{2}\, dy\, dx$$

$$= -2(\text{the area of the ellipse, } \pi ab)$$

$$= -2\pi(\sqrt{2})(2) = -4\pi\sqrt{2}$$

27. By the divergence theorem

$$\int\int_S \mathbf{F} \cdot \mathbf{N}\, dS = \int\int\int_D \text{div } \mathbf{F}\, dV$$

$$\mathbf{F} = x^2\mathbf{i} + (y+z)\mathbf{j} - 2z\mathbf{k};$$

$$\text{div } \mathbf{F} = 2x + 1 - 2 = 2x - 1$$

$$\int\int\int_D \text{div } \mathbf{F}\, dV = \int\int\int_D (2x-1)\, dV$$

$$= \int_0^1\int_0^1\int_0^1 (2x-1)\, dz\, dy\, dx$$

$$= \int_0^1 (2x-1)\, dx$$

$$= 0$$

28. $$\mathbf{F} \cdot d\mathbf{R} = \frac{x\, dx}{(x^2+y^2)^2} + \frac{y\, dy}{(x^2+y^2)^2}$$

$$= M\, dx + N\, dy$$

$$\frac{\partial N}{\partial x} = \frac{-2y(2x)}{(x^2+y^2)^3}$$

$$= \frac{-4xy}{(x^2+y^2)^3}$$

$$\frac{\partial M}{\partial y} = \frac{-2x(2y)}{(x^2+y^2)^3}$$

$$= \frac{-4xy}{(x^2+y^2)^3}$$

$$\frac{\partial N}{\partial x} = \frac{\partial M}{\partial y} \text{ so } \mathbf{F} \text{ is conservative and}$$

independent of path. Since C is a closed

path the value of the line integral is 0.

29. Since

$$\text{curl } \mathbf{R} = \begin{vmatrix} \mathbf{i} & \mathbf{j} & \mathbf{k} \\ \dfrac{\partial}{\partial x} & \dfrac{\partial}{\partial y} & \dfrac{\partial}{\partial z} \\ x & y & z \end{vmatrix}$$

$$= \mathbf{0}$$

it follows that $\mathbf{R}$ and hence $\mathbf{F} = m\omega^2 \mathbf{R}$ are

conservative. Ignore for the moment

the factor $m\omega^2$.

$$\frac{\partial f}{\partial x} = x$$

so

$$f = \frac{x^2}{2} + c(y, z)$$

and

$$\frac{\partial f}{\partial y} = \frac{\partial c}{\partial y} = y$$

$$c = \frac{y^2}{2} + c_1(z)$$

$$f = \frac{x^2}{2} + \frac{y^2}{2} + c_1(z)$$

but

$$\frac{\partial f}{\partial z} = \frac{\partial c_1}{\partial z} = z$$

so

$$c_1 = \frac{z^2}{2} + C$$

$$f = \frac{x^2}{2} + \frac{y^2}{2} + \frac{z^2}{2}$$

Inserting the scalar factor (and taking $C = 0$):

$$\phi = \frac{m\omega^2}{2} f$$

$$= \frac{m\omega^2}{2}(x^2 + y^2 + z^2)$$

is a scalar potential function for $\mathbf{F}$.

30. Since $\mathbf{F}$ is conservative with scalar potential

$\phi = \dfrac{m\omega^2}{2}(x^2 + y^2 + z^2)$, we have

$$W = \int_C \mathbf{F} \cdot d\mathbf{R}$$

$$= \phi(-3, 0, 2) - \phi(3, 0, 2)$$

$$= \frac{m\omega^2}{2}\left[(-3)^2 + 0^2 + 2^2\right] - \frac{m\omega^2}{2}[3^2 + 0^2 + 2^2]$$

$$= 0$$

This result could have been anticipated as z is

constant and $\mathbf{R}$ is symmetric about the y-axis.

Cumulative Review for Chapters 11-13, Pages 933-934

SURVIVAL HINT: *This is the last of four cumulative reviews included in the text. You can use one or more of these reviews when studying for final examinations. If you do not have the time to actually do all of these problems, try looking at each one to see if you recall the concept involved and how to proceed with the solution. If you are confident about your ability to solve the problem, do not spend the time. If you feel a little uncertain about the problem, refer to the* appropriate section, review the concepts, look in your old homework for a similar problem, and then see if you can work it. Be more concerned about understanding the concept than about obtaining a correct answer. Do not spend a lot of your time looking for algebraic or arithmetic errors, but rather focus on important ideas.

1. *Answers vary.* Calculus is the study of dynamic processes (rather than static). It is the study of infinitesimals, the behavior of functions at or near a point. There are three fundamental ideas of calculus; the notion of limits, derivatives (the limit of difference quotients), and integrals (the sum of infinitesimal quantities). Calculus is also the study of transformations of reference frames (coordinate systems), and motion with respect to reference frames (vector analysis).

3. Multivariable calculus involves limits, derivatives, and integrals of functions of more than one independent variable.

5. $f(x, y) = x^2 e^{y/x}$

$$f_x = e^{y/x}(2x - y)$$

$$f_y = x e^{y/x}$$

$$f_{xy} = e^{y/x}(1 - y/x)$$

7. $f(x, y) = y \sin^2 x + \cos xy$

$$f_x = 2y \sin x \cos x - y \sin xy$$

$$f_y = \sin^2 x - x \sin xy$$

$$f_{xy} = \sin 2x - xy \cos xy - \sin xy$$

9. $f(x, y) = \dfrac{x^2 + y^2}{x - y}$

$$f_x = \frac{(x - y)(2x) - (x^2 + y^2)}{(x - y)^2}$$

$$= \frac{x^2 - 2xy - y^2}{(x - y)^2}$$

$$f_y = \frac{(x - y)(2y) + (x^2 + y^2)}{(x - y)^2}$$

$$= \frac{x^2 + 2xy - y^2}{(x - y)^2}$$

$$f_{xy} = \frac{(x - y)^2(-2x - 2y) + (x^2 - 2xy - y^2)(2)(x - y)}{(x - y)^4}$$

$$= \frac{-4xy}{(x - y)^3}$$

11. $$\int_0^4 \int_0^{\sqrt{x}} 3x^5 \, dy \, dx = \int_0^4 3x^{11/2} \, dx$$

$$= \frac{49,152}{13}$$

13. $$\int_0^{15\pi} \int_0^{\pi} \int_0^{\sin \phi} \rho^3 \sin \phi \, d\rho \, d\theta \, d\phi$$

$$= \frac{1}{4} \int_0^{15\pi} \int_0^{\pi} \sin^5\phi \ d\theta \ d\phi$$

$$= \frac{\pi}{4} \int_0^{15\pi} \sin^5\phi \ d\phi$$

$$= \frac{4\pi}{15}$$

15. $\displaystyle\int\int_R ye^{xy} \ dA = \int_0^1 \int_0^2 ye^{xy} \ dy \ dx$

$$= \int_0^2 \int_0^1 ye^{xy} \ dx \ dy$$

$$= \int_0^2 (e^y - 1) \ dy$$

$$= e^2 - 3$$

17. $\displaystyle\int\int_R x \sin xy \ dA = \int_0^{\pi} \int_0^1 x \sin xy \ dy \ dx$

$$= - \int_0^{\pi} (\cos x - 1) \ dx$$

$$= \pi$$

19. $\displaystyle\int_C (5 \ xy \ dx + 10 \ yz \ dy + z \ dz)$

$$= \int_0^1 [5(t^2)(t)(2t) + 10(t)(2t^3) + (2t^3)(6t^2)] \ dt$$

$$= \int_0^1 (30t^4 + 12t^5) \ dt = 8$$

21. $\dfrac{ds}{dt} = \sqrt{\left(\dfrac{dx}{dt}\right)^2 + \left(\dfrac{dy}{dt}\right)^2}$

$$= \sqrt{[3(\cos^2 t)(-\sin t)]^2 + [3(\sin^2 t)(\cos t)]^2}$$

$$= 3|\sin t \cos t| \sqrt{\cos^2 t + \sin^2 t}$$

$$= 3|\sin t \cos t|$$

$$\int_C (x^3 + y^3) \ ds$$

$$= \int_0^{2\pi} (\cos^9 t + \sin^9 t)3|\sin t \cos t| \ dt$$

$$= 3 \int_0^{\pi/2} (\cos^9 t + \sin^9 t) \sin t \cos t \ dt$$

$$+ 3 \int_{\pi/2}^{\pi} (\cos^9 t + \sin^9 t)(-\sin t \cos t) \ dt$$

$$+ 3 \int_{\pi}^{3\pi/2} (\cos^9 t + \sin^9 t) \sin t \cos t \ dt$$

$$+ 3 \int_{3\pi/2}^{2\pi} (\cos^9 t + \sin^9 t)(-\sin t \cos t) \ dt$$

$$= \frac{6}{11} + 0 - \frac{6}{11} + 0$$

$$= 0$$

23. Minimize $S = 2(xy + xz + yz)$ subject to

$$V_0 = xyz;$$

$$\begin{cases} 2y + 2z = \lambda(yz) \\ 2x + 2z = \lambda(xz) \\ 2x + 2y = \lambda(xy) \\ xyz = V_0 \end{cases}$$

Solving this system simultaneously, we find

$$x = y = z = \sqrt[3]{V_0}$$

25. $T(x, y, z) = \frac{1}{10}(x^2 + y + z^3);$

$$\nabla T = \frac{1}{10}[2xi + j + 3z^2 k]$$

$$\nabla T_0 = \nabla T(-2, 9, 1)$$

$$= \frac{1}{10}(-4i + j + 3k)$$

a. $P_0 Q = 3i - 12j + 4k$

$$u = \frac{P_0 Q}{\|P_0 Q\|} = \frac{3i - 12j + 4k}{\sqrt{9 + 144 + 16}}$$
$$= \frac{1}{13}(3i - 12j + 4k)$$

$$D_u(T) = \nabla T_0 \cdot u$$
$$= -\frac{12}{10(13)}$$
$$= -\frac{6}{65}$$

b. It will move in the direction of

$$\nabla T_0 = \frac{1}{10}(-4i + j + 3k)$$

c. The maximal rate is $\|\nabla T_0\| = \frac{1}{10}\sqrt{26}$

27. $z = x^2 + y^2; z_x = 2x, z_y = 2y$

$$dS = \sqrt{(2x)^2 + (2y)^2 + 1}\ dA$$

$$= \sqrt{4r^2 + 1}\ r\ dr\ d\theta$$

The projected region is the disk $x^2 + y^2 \le 16$

$(r \le 4)$

$$S = \int_0^{2\pi} \int_0^4 r(4r^2 + 1)^{1/2}\ dr\ d\theta$$

$$= \frac{\pi}{6}(65\sqrt{65} - 1)$$

29. $F = \langle x - 2y, y - 2x \rangle$

$$u(x, y) = x - 2y; \frac{\partial u}{\partial y} = -2;$$

$$v(x, y) = y - 2x; \frac{\partial v}{\partial x} = -2,$$

so F is conservative.

$$f_x = x - 2y$$

$$f = \frac{x^2}{2} - 2xy + a(y)$$

$$f_y = -2x + a'(y)$$

$$= -2x + y$$

$$a(y) = \frac{y^2}{2}$$

$$f(x, y) = \frac{x^2}{2} + \frac{y^2}{2} - 2xy$$

$$W = \int_C F \cdot dR$$

$$= f(0, 1) - f(1, 0)$$

$$= \frac{1}{2} - \frac{1}{2} = 0$$

CHAPTER 14

Introduction to Differential Equations

14.1 First-Order Differential Equations, Pages 944-946

SURVIVAL HINT: *In this chapter you will encounter solutions to differential equations such as*

$$y = e^{\ln|\cos x|^{-1} + C}$$

The simplification steps (which are not usually shown) are:

$$y = e^{\ln|\cos x|^{-1} + C}$$

$$y = e^{\ln|\cos x|^{-1}} e^C$$

$$y = e^C(|\cos x|^{-1}) \quad since \ e^{\ln M} = M$$

$$y = B\sec x \quad since \ e^C \ is \ a \ constant, \ |B|$$

Do you see how the constant changed from C to B? You will need to carry out these steps frequently.

1. $xy \, dx = (x - 5) \, dy$

$$\int \frac{x}{x - 5} \, dx = \int y^{-1} \, dy$$

$$\int \left(1 + \frac{5}{x - 5}\right) dx = \int y^{-1} \, dy$$

$$x + 5\ln|x - 5| = \ln|y| + C_1$$

$$\ln|y| = x + \ln|x - 5|^5 + C$$

$$y = e^{x + \ln|x - 5|^5 + C}$$

SURVIVAL HINT: *In this problem we left the answer as a function of x. However, many differential equation problems have a form which is not easily solvable for y. For this reason, it is often acceptable to leave your answer in a form which is not technically simplified. For this problem, it would probably be acceptable to leave your answer as*

$$\ln|y| = x + \ln|x - 5|^5 + C$$

SURVIVAL HINT: *The letters D.E. are often used as an abbreviation for* **differential equation.**

3. $(e^{2x} + 9) \dfrac{dy}{dx} = y$

$$\int y^{-1} \, dy = \int \frac{dx}{e^{2x} + 9}$$

$$\ln|y| = -\frac{1}{18}[\ln(e^{2x} + 9) - 2x] + C$$

5. $9 \, dx - x\sqrt{x^2 - 9} \, dy = 0$

$$\int dy = \int \frac{9 \, dx}{x\sqrt{x^2 - 9}}$$

$$y = 3\sec^{-1}\left|\frac{x}{3}\right| + C$$

7. $\dfrac{dy}{dx} + 2xy = 4x$ passing through $(0, 0)$

$$p(x) = 2x, \ q(x) = 4x, \ I(x) = e^{\int 2x \, dx} = e^{x^2}$$

$$y = \frac{1}{e^{x^2}}\left[\int e^{x^2}(4x)\ dx + C\right]$$

$$= e^{-x^2}\left[2e^{x^2} + C\right]$$

$$= 2 + Ce^{-x^2}$$

Particular value: $0 = 2 + Ce^0;\ C = -2$

Particular solution: $y = 2(1 - e^{-x^2})$

9. $\frac{dy}{dx} + y = \cos x$ passing through $(0, 0)$

$p(x) = 1,\ q(x) = \cos x,\ I(x) = e^{\int dx} = e^x$

$$y = \frac{1}{e^x}\left[\int e^x\cos x\ dx + C\right]$$

$$= e^{-x}(\tfrac{1}{2}e^x \cos x + \tfrac{1}{2}e^x\sin x + C)$$

$$= \tfrac{1}{2}\cos x + \tfrac{1}{2}\sin x + Ce^{-x}$$

Particular value: $0 = \frac{1}{2}(1 + 0) + C;\ C = -\frac{1}{2}$

Particular solution:

$$y = \tfrac{1}{2}(\cos x + \sin x) - \tfrac{1}{2}e^{-x}$$

11. $x\frac{dy}{dx} - 2y = x^3$ passing through $(2, -1)$

$$\frac{dy}{dx} - \frac{2}{x}y = x^2$$

$p(x) = -\frac{2}{x},\ q(x) = x^2,\ I(x) = e^{\int(-2/x)\ dx} = x^{-2}$

$$y = x^2\left[\int x^{-2} \cdot x^2\ dx + C\right]$$

$$= x^3 + Cx^2$$

Particular value: $-1 = 8 + 4C;\ C = -\frac{9}{4}$

Particular solution: $y = x^3 - \frac{9}{4}x^2$

13. $(3x - y)\ dx + (x + 3y)\ dy = 0$

Show the equation is homogeneous.

$$\frac{dy}{dx} = \frac{-(3x - y)}{x + 3y} \cdot \frac{1/x}{1/x} = \frac{-(3 - \frac{y}{x})}{1 + 3(\frac{y}{x})}$$

Let $f(v) = \frac{-3 + v}{1 + 3v}$ where $v = \frac{y}{x}$

$$\frac{dv}{\frac{-3 + v}{1 + 3v} - v} = \frac{dx}{x}$$

$$\int \frac{1 + 3v}{-3v^2 - 3}\ dv = \int \frac{dx}{x}$$

$$-\tfrac{1}{2}\ln(v^2 + 1) - \tfrac{1}{3}\tan^{-1}v = \ln|x| + C$$

$$-\tfrac{1}{2}\ln\left(\frac{y^2}{x^2} + 1\right) - \ln|x| - \tfrac{1}{3}\tan^{-1}\frac{y}{x} = C$$

$$-\ln\sqrt{x^2 + y^2} - \tfrac{1}{3}\tan^{-1}\frac{y}{x} = C$$

15. $(3x - y)\ dx + (x - 3y)\ dy = 0$

Show the equation is homogeneous.

$$\frac{dy}{dx} = \frac{-(3x - y)}{x - 3y}$$

$$= \frac{-(3 - \frac{y}{x})}{1 - 3(\frac{y}{x})}$$

Let $v = \frac{y}{x},\ v + x\frac{dv}{dx} = \frac{-(3 - v)}{1 - 3v}$

$$\int \frac{1 - 3v}{3v^2 - 3} \, dv = \int \frac{dx}{x}$$

$$-\tfrac{1}{3}\ln\left|(v + 1)^2(v - 1)\right| = \ln|x| + C_1$$

$$(v + 1)^2(v - 1) = Cx^{-3}$$

$$\left(\tfrac{y}{x} + 1\right)^2\left(\tfrac{y}{x} - 1\right) = Cx^{-3}$$

$$(y + x)^2(y - x) = C$$

17. $(-6y^2 + 3xy + 2x^2) \, dx + x^2 \, dy = 0$

Show the equation is homogeneous.

$$\frac{dy}{dx} = \frac{6y^2 - 3xy - 2x^2}{x^2}$$

$$= 6\left(\tfrac{y}{x}\right)^2 - 3\left(\tfrac{y}{x}\right) - 2$$

Let $f(v) = 6v^2 - 3v - 2$ where $v = \tfrac{y}{x}$

$$\frac{dv}{6v^2 - 3v - 2 - v} = \frac{dx}{x}$$

$$\int \frac{dv}{6v^2 - 4v - 2} = \int \frac{dx}{x}$$

$$-\tfrac{1}{8}\ln\left|\frac{3v + 1}{v - 1}\right| = \ln|x| + C_1$$

$$\tfrac{1}{8}\ln\left|\frac{v - 1}{3v + 1}\right| = \ln|x| + C_1$$

$$\frac{v - 1}{3v + 1} = Cx^8$$

$$\frac{\tfrac{y}{x} - 1}{3\left(\tfrac{y}{x}\right) + 1} = Cx^8$$

$$\frac{y - x}{3y + x} = Cx^8$$

19. $(3x^2y + \tan y) \, dx + (x^3 + x \sec^2 y) \, dy = 0$

$$M(x, y) = 3x^2y + \tan y \qquad \frac{\partial M}{\partial y} = 3x^2 + \sec^2 y$$

$$N(x, y) = x^3 + x\sec^2 y \qquad \frac{\partial N}{\partial x} = 3x^2 + \sec^2 y$$

Since $\frac{\partial M}{\partial y} = \frac{\partial N}{\partial x}$, the D.E. is exact.

$$f(x, y) = \int (3x^2y + \tan y) \, dx$$

$$= x^3y + x\tan y + u(y)$$

$$\frac{\partial f}{\partial y} = x^3 + x\sec^2 y + u'(y)$$

Compare this with $N(x, y)$ to see $u'(y) = 0$, so $u(y) = C_1$. Thus,

$$x^3y + x\tan y = C \quad .$$

21. $\left[\frac{1}{1 + x^2} + \frac{2x}{x^2 + y^2}\right] dx + \left[\frac{2y}{x^2 + y^2} - e^{-y}\right] dy$

$$= 0$$

$$M(x, y) = \frac{1}{1 + x^2} + \frac{2x}{x^2 + y^2} \qquad \frac{\partial M}{\partial y} = \frac{-4xy}{(x^2 + y^2)^2}$$

$$N(x, y) = \frac{2y}{x^2 + y^2} - e^{-y} \qquad \frac{\partial N}{\partial x} = \frac{-4xy}{(x^2 + y^2)^2}$$

Since $\frac{\partial M}{\partial y} = \frac{\partial N}{\partial x}$, the D.E. is exact.

$$f(x, y) = \int \left(\frac{1}{1 + x^2} + \frac{2x}{x^2 + y^2}\right) dx$$

$$= \tan^{-1}x + \ln(x^2 + y^2) + u(y)$$

$$\frac{\partial f}{\partial y} = \frac{2y}{x^2 + y^2} + u'(y)$$

Compare with $N(x, y)$ to see $u'(y) = -e^{-y}$

Integrate to find $u(y) = e^{-y} + C_1$. Thus,

$$\tan^{-1}x + \ln(x^2 + y^2) + e^{-y} = C$$

23. $[2x \cos 2y - 3y(1 - 2x)]\, dx$

$$- [2x^2\sin 2y + 3(2 + x - x^2)]\, dy = 0$$

$M(x, y) = 2x \cos 2y - 3y(1 - 2x)$

$N(x, y) = -2x^2\sin 2y - 3(2 + x - x^2)$

$$\frac{\partial M}{\partial y} = -4x \sin 2y - 3(1 - 2x)$$

$$\frac{\partial N}{\partial x} = -4x \sin 2y - 3(1 - 2x)$$

Since $\frac{\partial M}{\partial y} = \frac{\partial N}{\partial x}$, the D.E. is exact.

$$f(x, y) = \int [2x \cos 2y - 3y(1 - 2x)]\, dx$$

$$= x^2\cos 2y - 3xy + 3x^2y + u(y)$$

$$\frac{\partial f}{\partial y} = -2x^2\sin 2y - 3x + 3x^2 + u'(y)$$

Compare with $N(x, y)$ to see that $u'(y) = -6$

Integrate to find $u(y) = -6y + C_1$. Thus,

$$x^2\cos 2y - 3xy + 3x^2y - 6y = C$$

25. **a.** $\frac{dy}{dx} - y = x$ is a first-order D.E.

$$I(x) = e^{\int -dx} = e^{-x}$$

$$y = e^x\left[\int xe^{-x}\, dx + C\right]$$

$$= e^x[-xe^{-x} - e^{-x} + C]$$

$$= -x - 1 + Ce^x$$

At $(1, 2)$: $2 = -1 - 1 + Ce$; $C = 4e^{-1}$

Particular solution: $y = -x - 1 + 4e^{x-1}$

b.

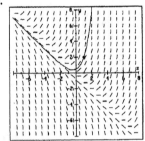

c.

n	x_n	y_n	$f(x_n, y_n)$	$y_{n+1} = y_n + 0.2f(x_n, y_n)$
0	1	2	3	2.60
1	1.2	2.60	3.80	3.36
2	1.4	3.36	4.76	4.31
3	1.6	4.31	5.91	5.49
4	1.8	5.49	7.29	6.95
5	2.0	6.95	8.95	8.74

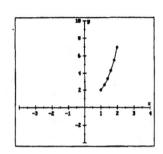

27. $\dfrac{dy}{dx} = \dfrac{x + y}{y - x} = f(x, y); \ h = 0.1$

n	x_n	y_n	$f(x_n, y_n)$	$y_{n+1} = y_n + 0.1 f(x_n, y_n)$
0	0	1	1	1.10
1	0.1	1.10	1.20	1.22
2	0.2	1.22	1.39	1.36
3	0.3	1.36	1.57	1.52
4	0.4	1.52	1.72	1.69
5	0.5	1.69	1.84	1.87

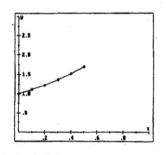

29. $\dfrac{dy}{dx} = \dfrac{5x - 3xy}{1 + x^2} = f(x, y); \ h = 0.1$

n	x_n	y_n	$f(x_n, y_n)$	$y_{n+1} = y_n + 0.1 f(x_n, y_n)$
0	0	0	0	0
1	0.1	0	0.50	0.05
2	0.2	0.05	0.93	0.14
3	0.3	0.14	1.26	0.27
4	0.4	0.27	1.45	0.41
5	0.5	0.41	1.50	0.56

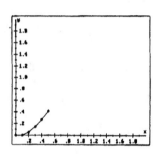

31. $y \, dx + (y - x) \, dy = 0$

$g(x, y) = y^n$

$M = y; \ N = y - x$

We want

$$\frac{\partial}{\partial y}[y^n y] = \frac{\partial}{\partial x}[y^n(y - x)]$$

$$(n + 1)y^n = -y^n$$

$$n + 1 = -1$$

$$n = -2$$

Thus, the equation is

$$y^{-2}y \, dx + y^{-2}(y - x) \, dy = 0$$

$$\frac{\partial f}{\partial x} = y^{-1}; \ f = \frac{x}{y} + u(y)$$

$$\frac{\partial f}{\partial y} = \frac{-x}{y^2} + u'(y) = \frac{-x}{y^2} + y^{-1}$$

$$u'(y) = y^{-1}$$

$$u(y) = \ln|y| + C_1$$

Solution: $\frac{x}{y} + \ln|y| = C$

33. $(2xy^2 + 3x^2y - y^3) \, dx$

$$+ \ (2x^2y + x^3 - 3xy^2) \, dy = 0$$

$$M(x, y) = 2xy^2 + 3x^2y - y^3$$

$$N(x, y) = 2x^2y + x^3 - 3xy^2$$

$$\frac{\partial M}{\partial y} = 4xy + 3x^2 - 3y^2$$

$$\frac{\partial N}{\partial x} = 4xy + 3x^2 - 3y^2$$

Since $\frac{\partial M}{\partial y} = \frac{\partial N}{\partial x}$, the D.E. is exact.

$$f(x, y) = \int (2xy^2 + 3x^2y - y^3) \, dx$$

$$= x^2y^2 + x^3y - xy^3 + u(y)$$

$$\frac{\partial f}{\partial y} = 2x^2y + x^3 - 3xy^2 + u'(y)$$

Compare with $N(x, y)$ to see that $u'(y) = 0$ so

$u(y) = C_1$. Thus,

$$x^2y^2 + x^3y - xy^3 = C$$

35. $\left(\frac{2x}{y} - \frac{y^2}{x^2}\right) dx + \left(\frac{2y}{x} - \frac{x^2}{y^2} + 3\right) dy = 0$

$$M(x, y) = \frac{2x}{y} - \frac{y^2}{x^2} \qquad \frac{\partial M}{\partial y} = -\frac{2x}{y^2} - \frac{2y}{x^2}$$

$$N(x, y) = \frac{2y}{x} - \frac{x^2}{y^2} + 3 \qquad \frac{\partial N}{\partial x} = -\frac{2y}{x^2} - \frac{2x}{y^2}$$

Since $\frac{\partial M}{\partial y} = \frac{\partial N}{\partial x}$, the D.E. is exact.

$$f(x, y) = \int \left(\frac{2x}{y} - \frac{y^2}{x^2}\right) dx = \frac{x^2}{y} + \frac{y^2}{x} + u(y)$$

$$\frac{\partial f}{\partial y} = -\frac{x^2}{y^2} + \frac{2y}{x} + u'(y)$$

Compare with $N(x, y)$ to see that $u'(y) = 3$ so

that $u(y) = 3y + C_1$. Thus,

$$\frac{x^2}{y} + \frac{y^2}{x} + 3y = C$$

37. $e^{y-x}\sin x \, dx - \csc x \, dy = 0$ is separable.

$$\int e^{-y} \, dy = \int e^{-x} \frac{\sin x}{\csc x} \, dx$$

$$= \int e^{-x}\sin^2 x \, dx$$

$$= \frac{-e^{-x}\sin^2 x}{5} - \frac{2e^{-x}\sin x(x \cos x)}{5}$$

$$-\frac{2}{5}e^{-x} + C_1$$

Thus,

$$-e^{-y} = -\frac{2e^{-x}}{5}\sin x\cos x - \frac{e^{-x}}{5}\sin^2 x - \frac{2}{5}e^{-x} + C_1$$

$$5e^x = e^y(2\sin x \cos x + \sin^2 x + 2) + Ce^{x+y}$$

39. $(3x^2 - y\sin xy)\, dx - (x\sin xy)\, dy = 0$

$$M(x,\, y) = 3x^2 - y\sin xy$$

$$N(x,\, y) = -x\sin xy$$

$$\frac{\partial M}{\partial y} = -xy\cos xy - \sin xy$$

$$\frac{\partial N}{\partial x} = -\sin xy - xy\cos xy$$

Since $\dfrac{\partial M}{\partial y} = \dfrac{\partial N}{\partial x}$, the D.E. is exact.

$$f(x,\, y) = \int (3x^2 - y\sin xy)\, dx$$

$$= x^3 + \cos xy + u(y)$$

$$\frac{\partial f}{\partial y} = -x\sin xy + u'(y)$$

Compare with $N(x,\, y)$ to see that $u'(y) = 0$,

so $u(y) = C_1$; thus,

$$x^3 + \cos xy = C$$

41. $(y - \sin^2 x)\, dx + (\sin x)\, dy = 0$

This is a first-order linear D.E.

$$\frac{dy}{dx} + \frac{y}{\sin x} = \sin x$$

$$p(x) = \csc x,\ q(x) = \sin x,$$

$$I(x) = e^{\int \csc x\, dx}$$

$$= e^{\ln[1/(\csc x + \cot x)]} = \frac{1}{\csc x + \cot x}$$

$$y = (\csc x + \cot x)\left[\int \frac{\sin x}{\csc x + \cot x}\, dx + C_1\right]$$

$$\int \frac{\sin x}{\csc x + \cot x}\, dx = \int \frac{\sin^2 x}{1 + \cos x}\, dx$$

$$= \int \frac{(1 - \cos^2 x)}{1 + \cos x}\, dx$$

$$= \int (1 - \cos x)\, dx$$

$$= x - \sin x + C_2$$

Thus,

$$y = (\csc x + \cot x)(x - \sin x + C)$$

43. The given D.E. is homogeneous.

$$\frac{dy}{dx} = \frac{y}{x} - \sqrt{1 + \left(\frac{y}{x}\right)^2}$$

Let $f(v) = v - \sqrt{1 + v^2}$ where $v = \dfrac{y}{x}$

$$\int \frac{dv}{v - \sqrt{1 + v^2} - v} = \int \frac{dx}{x}$$

$$-\int \frac{dv}{\sqrt{1 + v^2}} = \int \frac{dx}{x}$$

$$-\ln\left|\sqrt{v^2 + 1} + v\right| = \ln|x| + C_1$$

$$-\ln\left|\sqrt{\frac{y^2}{x^2} + 1} + \frac{y}{x}\right| = \ln|x| + C_1$$

$$-\ln\left|\sqrt{x^2 + y^2} + y\right| + \ln|x| - \ln|x| = C_1$$

$$\ln\left|\sqrt{x^2 + y^2} + y\right| = C$$

Alternate form: $\sqrt{x^2 + y^2} + y = B$

(where $B = e^C$)

45. The given D.E. is homogeneous.

$$\left(x \sin^2 \frac{y}{x} - y\right) dx + x\, dy = 0$$

$$\frac{dy}{dx} = \frac{y}{x} - \sin^2 \frac{y}{x}$$

Let $f(v) = v - \sin^2 v$ where $v = \frac{y}{x}$

$$\frac{dv}{v - \sin^2 v - v} = \frac{dx}{x}$$

$$-\int \frac{dv}{\sin^2 v} = \int \frac{dx}{x}$$

$$\cot v = \ln|x| + C$$

$$\cot \frac{y}{x} = \ln|x| + C$$

If $x = \frac{4}{\pi}$ when $y = 1$, we have

$$\cot \frac{\pi}{4} = \ln\left(\frac{4}{\pi}\right) + C \text{ so that } C = 1 - \ln \frac{4}{\pi}$$

$$\cot \frac{y}{x} = \ln|x| + 1 - \ln \frac{4}{\pi}$$

$$\cot \frac{y}{x} = \ln \frac{\pi|x|}{4} + 1$$

47. The D.E. is first-order linear.

$$x \frac{dy}{dx} - 3y = x^3$$

$$\frac{dy}{dx} - \frac{3}{x}y = x^2$$

$$p(x) = -\frac{3}{x}, \; q(x) = x^2,$$

$$I(x) = e^{\int (-3/x)\, dx} = e^{-3\ln|x|} = x^{-3}$$

$$y = x^3 \left[\int x^{-3}x^2 \, dx + C \right]$$

$$= x^3 \ln|x| + Cx^3$$

If $x = 1$, then $y = 1$, so $C = 1$, so

$$y = x^3 \ln|x| + x^3$$

49. $[\sin(x^2 + y) + 2x^2\cos(x^2 + y)]\, dx$

$$+ [x\cos(x^2 + y)]\, dy = 0$$

$$M(x, y) = \sin(x^2 + y) + 2x^2\cos(x^2 + y)$$

$$N(x, y) = x \cos(x^2 + y)$$

$$\frac{\partial M}{\partial y} = \cos(x^2 + y) - 2x^2\sin(x^2 + y)$$

$$\frac{\partial N}{\partial x} = \cos(x^2 + y) - 2x^2\sin(x^2 + y)$$

Since $\frac{\partial M}{\partial y} = \frac{\partial N}{\partial x}$, the D.E. is exact.

$$f(x, y) = \int x\cos(x^2 + y)\, dy$$

$$= x\sin(x^2 + y) + u(x)$$

$$\frac{\partial f}{\partial x} = \sin(x^2 + y) + 2x^2\cos(x^2 + y) + u'(x)$$

Compare with $M(x, y)$ to see that $u'(x) = 0$ so

$u(x) = C$. Also, since $x = y = 0$, $C = 0$.

Thus,

$$x\sin(x^2 + y) = 0$$

51. $y\,e^x dy = (y^2 + 2y + 2)\,dx$ is separable.

$$\int \frac{y\,dy}{y^2 + 2y + 2} = \int e^{-x}\,dx$$

$$\tfrac{1}{2}\ln\left| y^2 + 2y + 2 \right| - \tan^{-1}(y + 1) = -e^{-x} + C$$

If $x = 0$, $y = -1$, so $0 - 0 = -1 + C$ or $C = 1$

Thus,

$$\tfrac{1}{2}\ln\left(y^2 + 2y + 2 \right) - \tan^{-1}(y + 1) + e^{-x} = 1$$

53.   If $\alpha \neq \beta$, Example 2 shows that

$$Q(t) = \frac{\alpha\beta[e^{(\alpha - \beta)kt} - 1]}{\alpha e^{(\alpha - \beta)kt} - \beta}$$

If $\alpha > \beta$,

$$\lim_{t \to +\infty} Q(t) = \frac{\alpha\beta(\alpha - \beta)ke^{(\alpha - \beta)kt}}{\alpha(\alpha - \beta)ke^{(\alpha - \beta)kt}} = \beta$$

by l'Hôpital's rule

If $\alpha < \beta$, then the exponent of e is negative

and

$$\lim_{t \to +\infty} Q(t) = \frac{\alpha\beta[0 - 1]}{\alpha(0) - \beta} = \alpha$$

If $\alpha = \beta$,

$$\frac{dQ}{dt} = k(\alpha - Q)^2$$

$$\int \frac{dQ}{(\alpha - Q)^2} = \int k\,dt$$

$$\frac{1}{\alpha - Q} = kt + C$$

$$\alpha - Q = \frac{1}{kt + C}$$

$$Q(t) = \frac{-1}{kt + C} + \alpha$$

$Q(0) = 0$, so $C = \tfrac{1}{\alpha}$, and we see that

$$Q(t) = \frac{-1}{kt + \tfrac{1}{\alpha}} + \alpha = \frac{\alpha^2 kt}{\alpha kt + 1}$$

In "the long run,"

$$\lim_{t \to +\infty} Q(t) = \lim_{t \to \infty} \left[\frac{\alpha^2 k}{\alpha k} \right] = \alpha$$

by l'Hôpital's rule

55.   a. $$\frac{dP}{dt} = P(k - \ell P) - h$$

$$\int \frac{dP}{P^2 - \tfrac{k}{\ell}P + \tfrac{h}{\ell}} = \int -\ell\,dt$$

Since $h < \tfrac{k^2}{4\ell}$, we have

$$P^2 - \tfrac{k}{\ell}P + \tfrac{h}{\ell} = (P - r_1)(P - r_2)$$

where $r_1 = \dfrac{k + \sqrt{k^2 - 4h\ell}}{2\ell}$

and $r_2 = \dfrac{k - \sqrt{k^2 - 4h\ell}}{2\ell}$

Let $D = \sqrt{k^2 - 4h\ell}$, so that $r_1 - r_2 = \dfrac{D}{\ell}$

$$\int \frac{dP}{P^2 - \frac{k}{\ell}P + \frac{h}{\ell}} = \int \frac{dP}{(P - r_1)(P - r_2)}$$

$$= \int \frac{\ell}{D}\left[\frac{1}{P - r_1} - \frac{1}{P - r_2}\right] dP$$

$$= \frac{\ell}{D} \ln\left|\frac{P - r_1}{P - r_2}\right| + C_1$$

Since $\int -\ell\, dt = -\ell t + C_2$

we now have a solution to the given D.E.

$$\frac{\ell}{D} \ln\left|\frac{P - r_1}{P - r_2}\right| = -\ell t + C$$

$$P(t) = \frac{r_1 - r_2 Be^{-Dt}}{1 - Be^{-Dt}}$$

Since $P = P_0$ when $t = 0$, we have

$$P_0 = \frac{r_1 - r_2 B}{1 - B}$$

$$B = \frac{P_0 - r_1}{P_0 - r_2}$$

so

$$P(t) = \frac{r_1(P_0 - r_2) - r_2(P_0 - r_1)e^{-Dt}}{(P_0 - r_2) - (P_0 - r_1)e^{-Dt}}$$

b. $\lim\limits_{t \to +\infty} P(t) = r_1$

57. a. Reverse the roles of x and y: The

integrating factor is

$$I(y) = e^{\int R(y)\, dy}$$

and the general solution has the form

$$x = \frac{1}{I(y)}\left[\int I(y)\, S(y)\, dy + C\right]$$

b. $y\, dx - 2x\, dy = y^4 e^{-y}\, dy$

$$\frac{dx}{dy} - \frac{2x}{y} = y^3 e^{-y}$$

$$I(y) = e^{\int -2/y\, dy} = y^{-2}$$

$$x = y^2\left[\int y^{-2}(y^3 e^{-y})\, dy + C\right]$$

$$= -y^2[(y + 1)e^{-y} + C]$$

59. Since units are in yards, the Flamer runs at 4 yd/s and Blohardt pursues at 5 yd/s. Suppose the Spy runs at v yd/s and after t seconds is at the point $S(x, y)$. Since he runs toward the Flamer at all times, the tangent line to his path at S passes through the Flamer's position $(0, 4t)$ at time t. This means that

$$y' = \frac{4t - y}{-x}; \quad t = \frac{y - xy'}{4}$$

The distance the Spy runs in t seconds is vt and is also equal to the arc length of his path from (x, y) to $(100, 0)$. Thus, we have

$$v\left(\frac{y - xy'}{4}\right) = \int_x^{100} \sqrt{1 + (y')^2}\, dx$$

Differentiating both sides of this equation with respect to x, we obtain

$$\frac{v}{4}(-xy'') = -\sqrt{1+(y')^2}$$

To solve this differential equation, let $p = y'$. Then

$$xp' = \frac{4}{v}\sqrt{1+p^2}$$

$$\int \frac{dp}{\sqrt{1+p^2}} = \frac{4}{v}\int \frac{dx}{x}$$

$$\ln\left|p + \sqrt{1+p^2}\right| = \frac{4}{v}\ln|x| + C$$

Since $p = 0$ when $x = 100$, the Spy starts from rest at $(100,0)$, we have

$$\ln 1 = \frac{4}{v}\ln 100 + C_1$$

$$C_1 = -\frac{4}{v}\ln 100$$

so that

$$\ln\left|p + \sqrt{1+p^2}\right| = \frac{4}{v}[\ln x - \ln 100]$$

$$p + \sqrt{1+p^2} = \left(\frac{x}{100}\right)^{4/v}$$

Solving for p, we find

$$p = \frac{1}{2}\left[\left(\frac{x}{100}\right)^{4/v} - \left(\frac{x}{100}\right)^{-4/v}\right]$$

Since $\frac{dy}{dx} = p$, we now integrate with respect to x to obtain

$$y = \int \frac{1}{2}\left[\left(\frac{x}{100}\right)^{4/v} - \left(\frac{x}{100}\right)^{-4/v}\right]dx$$

$$= 50\left[\frac{v}{v+4}\left(\frac{x}{100}\right)^{1+4/v} - \frac{v}{v-4}\left(\frac{x}{100}\right)^{1-4/v}\right] + C_2$$

Since $y = 0$ when $x = 100$,

$$0 = 50\left[\frac{v}{v+4}(1) - \frac{v}{v-4}(1)\right] + C_2$$

$$C_2 = 50\left[\frac{8v}{v^2-16}\right]$$

and

$$y = 50\left\{\frac{v}{v+4}\left(\frac{x}{100}\right)^{1+4/v}\right.$$

$$\left. - \frac{v}{v-4}\left(\frac{x}{100}\right)^{1-4/v}\right\} + \frac{400v}{v^2-16}$$

Blohardt catches the Flamer when

$$5T = 4T + 80$$

$$T = 80$$

When $T = 80$ seconds, both are $4(80) = 320$ yards above the Flamer's starting point, so they are at $(0, 320)$. The Spy reaches the y-axis when $x = 0$, at the point $\left(0, \frac{400v}{v^2-16}\right)$, which means that he and Blohardt catch up to the Flamer at the same time if

$$\frac{400v}{v^2-16} = 320$$

$$v \approx 4.674 \text{ yd/s} \approx 14 \text{ ft/s}$$

Just as the Spy starts to run, Blohardt

fires a warning shot, and the Flamer surrend-
ers. The Spy quickly dives behind a big rock
and watches as his erstwhile superior is led
away to an unknown fate. "It likely was a
trap," he rationalizes. "Besides, he did try to
drown me, and that Death Force was pretty
nasty, too."

A little smile curls on his lips as he
makes one last calculation. "Who do you
suppose will replace him as Chief of Spies?"
he muses.

61. a. $z = \frac{1}{y - u}$ so that $y = u + \frac{1}{z}$

$$\frac{dy}{dx} = \frac{du}{dx} - \frac{1}{z^2}\frac{dz}{dx}$$

and by substituting into the Riccati

equation, we obtain

$$\frac{du}{dx} - \frac{1}{z^2}\frac{dz}{dx} = P(x)(u + \tfrac{1}{z})^2 + Q(x)(u + \tfrac{1}{z}) + R(x)$$

or

$$\frac{du}{dx} - \frac{1}{z^2}\frac{dz}{dx} = P(x)u^2 + Q(x)u + R(x)$$
$$+ \left[P(x)\frac{1}{z^2} + 2P(x)u\frac{1}{z} + Q(x)\frac{1}{z} \right]$$

But u is assumed to satisfy the Riccati

equation, so

$$\frac{du}{dx} = P(x)u^2 + Q(x)u + R(x)$$

and

$$\frac{du}{dx} - \frac{1}{z^2}\frac{dz}{dx} = \frac{du}{dx} + \left[P(x)\frac{1}{z^2} + 2P(x)u\frac{1}{z} + Q(x)\frac{1}{z} \right]$$

Multiply by $-z^2$:

$$\frac{dz}{dx} = -P(x) - 2P(x)uz - Q(x)z$$

or

$$\frac{dz}{dx} + [2P(x)u + Q(x)]z = -P(x)$$

which is a first-order linear equation, as

required.

b. The first-order linear equation in part **a**
can be solved to obtain

$$z = \frac{C}{w(x)} + \frac{1}{w(x)}\int -P(x)w(x)\ dx$$

where

$$w(x) = \exp\left(\int [2P(x)u(x) + Q(x)]\ dx \right)$$

and then

$$y = u(x) + \frac{1}{z(x)}$$

is a general solution of the Riccati

equation.

c. $\frac{dy}{dx} = \frac{1}{x^2}y^2 + \frac{2}{x}y - 2$ implies

$P(x) = \frac{1}{x^2}$, $Q(x) = \frac{2}{x}$, and $R(x) = -2$

To find a particular solution of the form

$u = Ax$, note that $du/dx = A$ and by

substituting, we obtain

$$A = \frac{1}{x^2}(A^2 x^2) + \frac{2}{x}(Ax) - 2$$

$$A = A^2 + 2A - 2$$

$$A^2 + A - 2 = 0$$

$$A = 1, -2$$

For $A = 1$, the particular solution is

$u = x$ and $y = u + 1/z = x + 1/z$.

Substituting into the equation

$$\frac{dy}{dx} = \frac{1}{x^2}y^2 + \frac{2}{x}y - 2$$

we obtain

$$1 - \frac{1}{z^2}\frac{dz}{dx} = \frac{1}{x^2}\left(x + \frac{1}{z}\right)^2 + \frac{2}{x}\left(x + \frac{1}{z}\right) - 2$$

$$= 1 + \frac{4}{xz} + \frac{1}{x^2 z^2}$$

$$\frac{dz}{dx} = -\frac{4z}{x} - \frac{1}{x^2}$$

$$\frac{dz}{dx} + \frac{4z}{x} = -\frac{1}{x^2}$$

This is a first-order linear equation with

the integrating factor

$$I(x) = e^{\int (4/x)\, dx} = e^{4\ln x} = x^4$$

so

$$z = \frac{1}{x^4}\left[\int x^4\left(-\frac{1}{x^2}\right)dx + C\right]$$

$$= \frac{1}{x^4}\left[-\frac{x^3}{3} + C\right]$$

$$= -\frac{1}{3x} + \frac{C}{x^4}$$

Thus, the general solution to the given

Riccati equation is

$$y = x + \frac{1}{z}$$

$$= x + \frac{1}{\frac{C}{x^4} - \frac{1}{3x}}$$

$$= x + \frac{1}{\frac{3C - x^3}{3x^4}}$$

$$= \frac{3Cx + 2x^4}{3C - x^3}$$

14.2 Second-Order Homogeneous Linear Differential Equations, Pages 955-957

1. $y'' + y' = 0$

$$r^2 + r = 0$$

$$r(r + 1) = 0$$

$$r = 0, -1$$

Particular solutions: $y_1 = 1$, $y_2 = e^{-x}$

General solution: $y = C_1 + C_2 e^{-x}$

3. $y'' + 6y' + 5y = 0$

$r^2 + 6r + 5 = 0$

$(r + 5)(r + 1) = 0$

$r = -5, -1$

Particular solutions: $y_1 = e^{-5x}$, $y_2 = e^{-x}$

General solution: $y = C_1 e^{-5x} + C_2 e^{-x}$

5. $y'' - y' - 6y = 0$

$r^2 - r - 6 = 0$

$(r - 3)(r + 2) = 0$

$r = 3, -2$

Particular solutions: $y_1 = e^{3x}$, $y_2 = e^{-2x}$

General solution: $y = C_1 e^{3x} + C_2 e^{-2x}$

7. $2y'' - 5y' - 3y = 0$

$2r^2 - 5r - 3 = 0$

$(2r + 1)(r - 3) = 0$

$r = -\frac{1}{2}, 3$

Particular solutions: $y_1 = e^{(-1/2)x}$, $y_2 = e^{3x}$

General solution: $y = C_1 e^{(-1/2)x} + C_2 e^{3x}$

9. $y'' - y = 0$

$r^2 - 1 = 0$

$(r - 1)(r + 1) = 0$

$r = 1, -1$

Particular solutions: $y_1 = e^x$, $y_2 = e^{-x}$

General solution: $y = C_1 e^x + C_2 e^{-x}$

11. $y'' + 11y = 0$

$r^2 + 11 = 0$

$r = \pm\sqrt{11}\, i$

$\alpha = 0, \beta = \sqrt{11}$

General solution: $y = C_1 \cos\sqrt{11}\, x + C_2 \sin\sqrt{11}\, x$

13. $7y'' + 3y' + 5y = 0$

$7r^2 + 3r + 5 = 0$

$r = \dfrac{-3 \pm \sqrt{9 - 4(7)(5)}}{2(7)}$

$= -\dfrac{3}{14} \pm \dfrac{\sqrt{131}}{14} i$

$\alpha = -\dfrac{3}{14}, \beta = \dfrac{\sqrt{131}}{14}$

General solution:

$$y = e^{(-3/14)x}\left[C_1 \cos\left(\frac{\sqrt{131}}{14} x\right) + C_2 \sin\left(\frac{\sqrt{131}}{14} x\right)\right]$$

15. $y''' + y'' = 0$

$r^3 + r^2 = 0$

$r^2(r + 1) = 0$

$r = 0 \text{ (mult. 2)}, -1$

$y = C_1 + C_2 x + C_3 e^{-x}$

17. $y^{(4)} + y''' + 2y'' = 0$

$$r^4 + r^3 + 2r^2 = 0$$

$$r^2(r^2 + r + 2) = 0$$

$$r = 0 \text{ (mult. 2)}, \ r = -\tfrac{1}{2} \pm \tfrac{\sqrt{7}}{2}i$$

$$y = C_1 + C_2 x$$

$$+ \, e^{-(1/2)x}\left[C_3 \cos\!\left(\frac{\sqrt{7}}{2}x\right) + C_4 \sin\!\left(\frac{\sqrt{7}}{2}x\right) \right]$$

19. $y''' + 2y'' - 5y' - 6y = 0$

$$r^3 + 2r^2 - 5r - 6 = 0$$

$$(r + 1)(r - 2)(r + 3) = 0$$

$$r = 2, \ -3, \ -1$$

$$y = C_1 e^{2x} + C_2 e^{-3x} + C_3 e^{-x}$$

21. $y'' - 10y' + 25y = 0$

$$r^2 - 10r + 25 = 0$$

$$(r - 5)^2 = 0$$

$$r = 5 \ \text{(mult. 2)}$$

$$y = C_1 e^{5x} + C_2 x e^{5x}$$

$$y(0) = 1, \text{ so } 1 = C_1(1) + 0; \ C_1 = 1$$

$$y' = 5e^{5x} + C_2(e^{5x} + 5xe^{5x})$$

$$y'(0) = -1, \text{ so } -1 = 5(1) + C_2(1 + 0);$$

$$C_2 = -6$$

$$y = e^{5x} - 6xe^{5x} \ \text{ or } \ e^{5x}(1 - 6x)$$

23. $y'' - 12y' + 11y = 0$

$$r^2 - 12r + 11 = 0$$

$$(r - 11)(r - 1) = 0$$

$$r = 11, \ 1$$

$$y = C_1 e^x + C_2 e^{11x}$$

$$y(0) = 3, \text{ so } 3 = C_1 + C_2$$

$$y' = C_1 e^x + 11 C_2 e^{11x}$$

$$y'(0) = 11, \text{ so } 11 = C_1 + 11 C_2$$

$$\text{Solve} \begin{cases} C_1 + C_2 = 3 \\ C_1 + 11 C_2 = 11 \end{cases}$$

to find $C_1 = \tfrac{11}{5}, \ C_2 = \tfrac{4}{5}$

$$y = \tfrac{11}{5} e^x + \tfrac{4}{5} e^{11x}$$

25. $y''' + 10y'' + 25y' = 0$

$$r^3 + 10r^2 + 25r = 0$$

$$r = 0, \ -5 \ \text{(mult. 2)}$$

$$y = C_1 + C_2 e^{-5x} + C_3 x e^{-5x}$$

$$y(0) = 3, \text{ so } 3 = C_1 + C_2 \cdot 1$$

$$y'(0) = 2, \text{ so } 2 = -5C_2 \cdot 1 + C_3(1 - 0)$$

$$y''(0) = -1, \text{ so}$$

$$-1 = 25C_2 \cdot 1 + C_3(-10 \cdot 1 + 0)$$

$$\text{Solve} \begin{cases} C_1 + C_2 = 3 \\ -5C_2 + C_3 = 2 \\ 25C_2 - 10C_3 = -1 \end{cases}$$

to find $C_1 = \frac{94}{25}$, $C_2 = -\frac{19}{25}$, $C_3 = -\frac{9}{5}$

$y = \frac{94}{25} - \frac{19}{25}e^{-5x} - \frac{9}{5}xe^{-5x}$

27. $W(e^{-2x}, e^{3x}) = \begin{vmatrix} e^{-2x} & e^{3x} \\ -2e^{-2x} & 3e^{3x} \end{vmatrix}$

$= 3e^x + 2e^x = 5e^x \neq 0$

29. $W(e^{-x}\cos x, \ e^{-x}\sin x)$

$= \begin{vmatrix} e^{-x}\cos x & e^{-x}\sin x \\ e^{-x}(-\cos x - \sin x) & e^{-x}(\cos x - \sin x) \end{vmatrix}$

$= (e^{-x}\cos x)e^{-x}(\cos x - \sin x)$

$\qquad - e^{-x}(-e^{-x})\sin x(\sin x + \cos x)$

$= e^{-2x} \neq 0$

31. $y_1 = e^{ax}$; assume $y_2 = ve^{ax}$

$y_2' = v'e^{ax} + ave^{ax}$;

$y_2'' = v''e^{ax} + 2av'e^{ax} + a^2ve^{ax}$

Substitute into the given D.E.:

$(v'' + 2av' + a^2v)e^{ax} - 2a(v' + av)e^{ax} + bve^{ax} = 0$

$(a^2 - 2a^2 + b)ve^{ax} = 0$

Hence,

$\qquad v'' = 0$ since $a^2 = b$;

$\qquad v = x$ is one solution to $v'' = 0$

so $y_2 = xe^{ax}$

$W(ae^{ax}, xe^{ax}) = \begin{vmatrix} e^{ax} & xe^{ax} \\ ae^{ax} & (ax+1)e^{ax} \end{vmatrix}$

$= e^{2ax} \neq 0$

Thus, e^{ax} and xe^{ax} are linearly independent.

33. $2y'' - y' - 6y = 0$

$y_1 = e^{2x}$

$y_2 = ve^{2x}$

$y_2' = v'e^{2x} + 2ve^{2x}$

$y_2'' = v''e^{2x} + 4v'e^{2x} + 4ve^{2x}$

$2(v'' + 4v' + 4v)e^{2x} - (v' + 2v)e^{2x}$

$\qquad - 6ve^{2x} = 0;$

$2v'' + 7v' = 0$

$\frac{v''}{v'} = -\frac{7}{2}$

$\ln|v'| = -\frac{7}{2}x$

$v' = e^{-7x/2}$

$v = -\frac{2}{7}e^{-7x/2}$

$y_2 = \left(-\frac{2}{7}e^{-7x/2}\right)e^{2x}$

$= -\frac{2}{7}e^{-3x/2}$

$W(e^{2x}, -\frac{2}{7}e^{-3x/2}) = \begin{vmatrix} e^{2x} & -\frac{2}{7}e^{-3x/2} \\ 2e^{2x} & \frac{3}{7}e^{-3x/2} \end{vmatrix}$

$= \frac{3}{7}e^{x/2} + \frac{4}{7}e^{x/2}$

$= e^{x/2} \neq 0$

General solution: $y = C_1 e^{2x} + C_2 e^{-3x/2}$

35. $x^2 y'' + xy' - 4y = 0$

$$y_1 = x^{-2}$$

$$y_2 = vy_1 = vx^{-2}$$

$$y_2' = v'x^{-2} - 2vx^{-3}$$

$$y_2'' = v''x^{-2} - 4v'x^{-3} + 6vx^{-4}$$

$$x^2(v''x^{-2} - 4v'x^{-3} + 6vx^{-4})$$

$$+ x(v'x^{-2} - 2vx^{-3}) - 4vx^{-2} = 0$$

$$v'' - 3v'x^{-1} = 0$$

$$\frac{v''}{v'} = \frac{3}{x}$$

$$\ln|v'| = 3 \ln|x|$$

$$v' = x^3$$

$$v = \tfrac{1}{4}x^4$$

$$y_2 = vx^{-2} = \tfrac{1}{4}x^2$$

$$W(x^{-2}, \tfrac{1}{4}x^2) = \begin{vmatrix} x^{-2} & \tfrac{1}{4}x^2 \\ -2x^{-3} & \tfrac{1}{2}x \end{vmatrix}$$

$$= x^{-1} \neq 0$$

General solution: $y = C_1 x^{-2} + C_2 x^2$

37. $(1-x)^2 y'' - (1-x)y' - y = 0$; $y_1 = 1 - x$

$$y_2 = vy_1 = v(1 - x)$$

$$y_2' = v'(1 - x) - v$$

$$y_2'' = v''(1 - x) - 2v'$$

$$(1 - x)^2[v''(1 - x) - 2v']$$

$$- (1 - x)[v'(1 - x) - v] - v(1 - x) = 0$$

$$(1 - x)^3 v'' - 3(1 - x)^2 v' = 0$$

$$\frac{v''}{v'} = \frac{3}{1 - x}$$

$$\ln|v'| = -3\ln|1 - x|$$

$$v' = (1 - x)^{-3}$$

$$v = \tfrac{1}{2}(1 - x)^{-2}$$

$$y_2 = vy_1 = \tfrac{1}{2}(1 - x)^{-1}$$

$$W(1 - x, \tfrac{1}{2}(1 - x)^{-1})$$

$$= \begin{vmatrix} 1 - x & \tfrac{1}{2}(1 - x)^{-1} \\ -1 & \tfrac{1}{2}(1 - x)^{-2} \end{vmatrix}$$

$$= (1 - x)^{-1} \neq 0$$

General solution:

$$y = C_1(1 - x) + C_2(1 - x)^{-1}$$

For Problems 38-43, the spring constant is

$$k = \frac{16 \text{ lb}}{(8/12) \text{ ft}} = 24 \text{ lb/ft}$$

and the mass is

$$m = \frac{16 \text{ lb}}{32 \text{ ft/sec}^2} = \tfrac{1}{2}\text{slug}$$

The governing equation has the form

$$\tfrac{1}{2}y'' + cy' + 24y = 0$$

39. $\frac{1}{2}y'' = -24y$ with $y_0 = \frac{10}{12} = \frac{5}{6}$ ft,

$y_0' = v_0 = -6$

Using y and y' found in Problem 38,

$y(0) = \frac{5}{6}$, so $C_1 = \frac{5}{6}$

$y'(0) = -6$ so $-6 = 4\sqrt{3}\, C_2$ or

$C_2 = \frac{-6}{4\sqrt{3}} = -\frac{1}{2}\sqrt{3}$

Solution: $y = \frac{5}{6}\cos(4\sqrt{3}\,t) - \frac{1}{2}\sqrt{3}\sin(4\sqrt{3}\,t)$

41. $\frac{1}{2}y'' = -24y$ with $y_0 = -1$ ft, $y_0' = v_0 = 0$

Using y and y' found in Problem 38,

$y(0) = -1$, so $C_1 = -1$

$y'(0) = 0$ so $C_2 = 0$

Solution: $y = -\cos(4\sqrt{3}\,t)$

43. In this problem we have $c = 0.08$.

$0.5y'' + 0.08y' + 24y = 0$

The characteristic equation is:

$0.5r^2 + 0.08r + 24 = 0$

$r \approx -0.08 \pm 6.92774i$

$y = e^{-0.08t}\big[C_1\cos 6.93t + C_2\sin 6.93t\big]$

$y_0(0) = 1$, so $C_1 = 1$

$y' = -0.08e^{-0.08t}\big[C_1\cos 6.93t + C_2\sin 6.93t\big]$
$+ e^{-0.08t}\big[-6.93C_1\sin 6.93t + 6.93C_2\cos 6.93t\big]$

$y'(0) = 0$, so $C_2 \approx \frac{0.08}{6.93} \approx 0.0115$

Solution: $y = e^{-0.08t}\big[\cos 6.93t + 0.0115\sin 6.93t\big]$

45. $y''' - 6y'' + 12y' - 8y = 0$

The characteristic equation is:

$r^3 - 6r^2 + 12r - 8 = 0$

$(r-2)^3 = 0$

$r = 2$ (mult. 3)

$y = C_1e^{2x} + C_2xe^{2x} + C_3x^2e^{2x}$, so y_1, y_2, and

y_3 are particular solutions. To show they are

linearly independent, consider

$W(y_1, y_2, y_3)$

$= \begin{vmatrix} e^{2x} & xe^{2x} & x^2e^{2x} \\ 2e^{2x} & e^{2x}(1+2x) & 2xe^{2x}(1+x) \\ 4e^{2x} & 4e^{2x}(1+x) & 2e^{2x}(1+4x+2x^2) \end{vmatrix}$

$= 2e^{6x} \neq 0$

47. Answers vary.

49. $my'' + ky = 0$

The characteristic equation is:

$mr^2 + k = 0$, so $r = \pm\sqrt{\frac{k}{m}}\,i$ $(k > 0)$

$x(t) = C_1\cos\left(\sqrt{\frac{k}{m}}\,t\right) + C_2\sin\left(\sqrt{\frac{k}{m}}\,t\right)$

$= \cos\left(\sqrt{\frac{k}{m}}\,t + C\right)\sqrt{C_1^2 + C_2^2}$

where $C = \tan^{-1}\left(-\frac{C_2}{C_1}\right)$ because

$$\cos C = \frac{C_1}{\sqrt{C_1^2 + C_2^2}}, \ \sin C = \frac{-C_2}{\sqrt{C_1^2 + C_2^2}}$$

51. $ay'' + by' + cy = 0$

The characteristic equation is:

$$ar^2 + br + c = 0$$

$$r = \frac{-b \pm \sqrt{b^2 - 4ac}}{2a}$$

Let $\alpha = -\frac{b}{2a}$, $\beta = \frac{\sqrt{4ac - b^2}}{2a}$;

$y_1 = e^{\alpha x}\cos \beta x$

$y_1' = e^{\alpha x}(\alpha \cos \beta x - \beta \sin \beta x)$

$y_1'' = e^{\alpha x}[(\alpha^2 - \beta^2)\cos \beta x + (-2\alpha\beta)\sin \beta x]$

Substitute into the original D.E. equation:

$ay_1'' + by_1' + cy_1$

$= ae^{\alpha x}[(\alpha^2 - \beta^2)\cos \beta x + (-2\alpha\beta)\sin \beta x]$

$\qquad + be^{\alpha x}[\alpha \cos \beta x - \beta \sin \beta x] + ce^{\alpha x}\cos \beta x$

$= 0$

Thus, $e^{\alpha x}\cos \beta x$ satisfies the differential equation and a similar computation shows that $y_2 = e^{\beta x}\sin \beta x$ is also a solution.

Finally, show that y_1 and y_2 are linearly independent, we compute the Wronskan

$W(e^{\alpha x}\cos \beta x, \ e^{\alpha x}\sin \beta x)$

$= e^{2\alpha x}[\beta(\cos^2 \beta x + \sin^2 \beta x)]$

$= \beta e^{2\alpha x} \neq 0$

53. **a.** $m(t)s''(t) + m'(t)v_0 + m(t)g = 0$

The weight of the rocket and fuel at time t is $(w - rt)$ and its mass is

$$\frac{w - rt}{g} = m(t)$$

Thus,

$$s''(t) = -\frac{m'(t)}{m(t)}v_0 - g$$

$$= -\frac{-\frac{r}{g}}{\frac{w - rt}{g}}v_0 - g$$

$$= \frac{rv_0}{w - rt} - g$$

Integrating, we obtain

$$s'(t) = -v_0 \ln|w - rt| - gt + C_1$$

and since $s'(0) = 0$, we have

$$s'(0) = -v_0 \ln(w - 0) - g(0) + C_1,$$

so $C_1 = v_0 \ln w$

Thus,

$$s'(t) = -v_0 \ln(w - rt) - gt + v_0 \ln w$$

$$= -v_0 \ln\left(\frac{w - rt}{w}\right) - gt$$

b. Integrating a second time, we find

$$s(t) = -v_0\left[t\ln\left(\frac{w - rt}{w}\right) - t - \frac{w}{r}\ln(w - rt)\right]$$

$$-\frac{gt^2}{2} + C_2$$

Since $s(0) = 0$,

$$0 = -v_0[0 - 0 - \frac{w}{r}\ln w] - 0 + C_2$$

so $C_2 = -\frac{v_0}{r}w\ln w$

Thus,

$$s(t) = -v_0\left[t\ln\left(\frac{w-rt}{w}\right) - t - \frac{w}{r}\ln(w - rt)\right]$$
$$-\frac{gt^2}{2} - \frac{v_0}{r}w\ln w$$
$$= \frac{v_0(w - rt)}{r}\ln\left(\frac{w - rt}{w}\right) - \frac{1}{2}gt^2 + v_0t$$

c. The fuel is consumed when $rt = w_f$; that

is, when $t = w_f/r$.

d. At the time when $t = w_f/r$, the height is

$$s\left(\frac{w_f}{r}\right) = \frac{v_0\left[w - r\left(\frac{w_f}{r}\right)\right]}{r}\ln\left[\frac{w - r\left(\frac{w_f}{r}\right)}{w}\right]$$

$$- \frac{1}{2}g\left(\frac{w_f}{r}\right)^2 + v_0\left(\frac{w_f}{r}\right)$$

$$= \frac{v_0(w - w_f)}{r}\ln\left(\frac{w - w_f}{w}\right)$$

$$- \frac{1}{2}\frac{gw_f^2}{r^2} + \frac{v_0w_f}{r}$$

14.3 Second-Order Nonhomogeneous Linear Differential Equations, Pages 965-966

1. $y'' - 6y' = 0$; $\quad r^2 - 6r = 0$

$$r = 0, 6$$

$$y_h = C_1 + C_2e^{6x}; F(x) = e^{2x}$$

$$\overline{y}_p = Ae^{2x}$$

3. $y'' + 2y' + 2y = 0$

$$r^2 + 2r + 2 = 0$$

$$r = -1 \pm i$$

$$y_h = e^{-x}(C_1 \cos x + C_2 \sin x); F(x) = e^{-x}$$

$$\overline{y}_p = Ae^{-x}$$

5. $y'' + 2y' + 2y = 0$

$$y_h = e^{-x}(C_1 \cos x + C_2 \sin x);$$

See the solution to Problem 3.

$$F(x) = (\sin x)e^{-x}$$

$$\overline{y}_p = xe^{-x}(A \cos x + B \sin x)$$

7. $y'' + 4y' + 5y = 0$

$$r^2 + 4r + 5 = 0$$

$$r = -2 \pm i$$

$$y_h = e^{-2x}(C_1\cos x + C_2\sin x);$$

$$F(x) = e^{-2x}(x + \cos x)$$

$$\overline{y}_p = (A + Bx)e^{-2x} + xe^{-2x}(C \cos x + D \sin x)$$

For Problems 9-16, the characteristic equation is

$$r^2 + 6r + 9 = 0$$

$$(r + 3)^2 = 0$$

$$r = -3 \quad (\text{mult. 2})$$

$$y_h = (C_1 + C_2 x)e^{-3x}$$

9. $F(x) = 3x^3 - 5x;$

$$\overline{y}_p = A_3 x^3 + A_2 x^2 + A_1 x + A_0$$

11. $F(x) = x^3 \cos x;$

$$\overline{y}_p = (A_3 x^3 + A_2 x^2 + A_1 x + A_0)\cos x$$
$$+ (B_3 x^3 + B_2 x^2 + B_1 x + B_0)\sin x$$

13. $F(x) = e^{2x} + \cos 3x;$

$$\overline{y}_p = A_0 e^{2x} + B_0 \cos 3x + C_0 \sin 3x$$

15. $F(x) = 4x^3 - x^2 + 5 - 3e^{-x}$

$$\overline{y}_p = A_3 x^3 + A_2 x^2 + A_1 x + A_0 + B_0 e^{-x}$$

17. $y'' + y' = 0$

Characteristic equation: $\quad r^2 + r = 0$

$$r(r + 1) = 0$$

$$r = 0, -1$$

$$y_h = C_1 + C_2 e^{-x}$$

$F(x) = -3x^2 + 7$

$\overline{y}_p = A_2 x^3 + A_1 x^2 + A_0 x$

$\overline{y}_p{}' = 3A_2 x^2 + 2A_1 x + A_0$

$\overline{y}_p{}'' = 6A_2 x + 2A_1$

Substitute into the given D.E.:

$$(6A_2 x + 2A_1) + (3A_2 x^2 + 2A_1 x + A_0)$$

$$= -3x^2 + 7$$

This gives rise to the system of equations

$$\begin{cases} 3A_2 = -3 \\ 2A_1 + 6A_2 = 0 \quad \text{which has solution:} \\ A_0 + 2A_1 = 7 \quad A_0 = 1, \ A_1 = 3, \ A_2 = -1 \end{cases}$$

Thus,

$$\overline{y}_p = -x^3 + 3x^2 + x$$

Solution: $y = C_1 + C_2 e^{-x} - x^3 + 3x^2 + x$

19. $y'' + 8y' + 15y = 0$

Characteristic equation: $\quad r^2 + 8r + 15 = 0$

$$(r + 3)(r + 5) = 0$$

$$r = -3, -5$$

$$y_h = C_1 e^{-3x} + C_2 e^{-5x}$$

$F(x) = 3e^{2x}$

$\overline{y}_p = Ae^{2x}; \ \overline{y}_p{}' = 2Ae^{2x}; \ \overline{y}_p{}'' = 4Ae^{2x}$

Substitute into the given D.E.:

$$4Ae^{2x} + 8(2Ae^{2x}) + 15(Ae^{2x}) = 3e^{2x}$$

Solving, we obtain $A = \frac{3}{35}$

Solution: $y = C_1 e^{-3x} + C_2 e^{-5x} + \frac{3}{35}e^{2x}$

21. $y'' + 2y' + 2y = 0$

Characteristic equation: $\quad r^2 + 2r + 2 = 0$

$$r = -1 \pm i$$

$$y_h = e^{-x}[C_1\cos x + C_2 \sin x]$$

$F(x) = \cos x$

$\overline{y}_p = A\cos x + B\sin x$

$\overline{y}_p{}' = -A\sin x + B\cos x$

$\overline{y}_p{}'' = -A\cos x - B\sin x$

Substitute into the given D.E.:

$$[-A\cos x - B\sin x] + 2[-A\sin x + B\cos x]$$
$$+ 2[A\cos x + B\sin x] = \cos x$$

This gives rise to the system of equations

$\begin{cases} A + 2B = 1 \\ -2A + B = 0 \end{cases}$ which has solution:

$$A = \tfrac{1}{5},\ B = \tfrac{2}{5}$$

Solution:

$$y = e^{-x}(C_1\cos x + C_2 \sin x) + \tfrac{1}{5}\cos x + \tfrac{2}{5}\sin x$$

23. $7y'' + 6y' - y = 0$

Characteristic equation: $7r^2 + 6r - 1 = 0$

$$r = -1, \tfrac{1}{7}$$

$$y_h = C_1 e^{x/7} + C_2 e^{-x}$$

$F(x) = e^{-x}(x + 1)$

$\overline{y}_p = e^{-x}(Ax + B)x$

$\overline{y}_p{}' = [-Ax^2 + (2A - B)x + B]e^{-x}$

$\overline{y}_p{}'' = [Ax^2 + (B-4A)x + (2A-2B)]e^{-x}$

Substitute into the given D.E.:

$$7[Ax^2 + (B - 4A)x + (2A - 2B)]e^{-x}$$
$$+ 6[-Ax^2 + (2A - B)x + B]e^{-x}$$
$$- [Ax^2 + Bx]e^{-x} = e^{-x}(x + 1)$$

This gives rise to the system

$\begin{cases} -16A = 1 \\ 14A - 8B = 1 \end{cases}$ which has solution:

$$A = -\tfrac{1}{16},\ B = -\tfrac{15}{64}$$

Solution:

$$y = C_1 e^{x/7} + C_2 e^{-x} + e^{-x}\left(-\tfrac{1}{16}x^2 - \tfrac{15}{64}x\right)$$

25. $y'' - y' = 0$

Characteristic equation: $r^2 - r = 0$

$$r(r - 1) = 0$$

$$r = 0, 1$$

$$y_h = C_1 + C_2 e^x$$

$F(x) = x^3 - x + 5$

$\overline{y}_p = A_3 x^3 + A_2 x^2 + A_1 x + A_0$; since we have

$r = 0$ (mult. 1), we multiply by x:

$\overline{y}_p = A_3 x^4 + A_2 x^3 + A_1 x^2 + A_0 x$

$\overline{y}_p{}' = 4A_3 x^3 + 3A_2 x^2 + 2A_1 x + A_0$

$\overline{y}_p{}'' = 12A_3 x^2 + 6A_2 x + 2A_1$

Substitute into the given D.E.:

$$[12A_3 x^2 + 6A_2 x + 2A_1] - [4A_3 x^3 + 3A_2 x^2$$
$$+ 2A_1 x + A_0] = x^3 - x + 5$$

This gives rise to the system of equations

$$\begin{cases} -4A_3 = 1 \\ 12A_3 - 3A_2 = 0 \\ 6A_2 - 2A_1 = -1 \\ 2A_1 - A_0 = 5 \end{cases}$$

which has solution:

$$A_0 = -10, \; A_1 = -\tfrac{5}{2},$$
$$A_2 = -1, \; A_3 = -\tfrac{1}{4}$$

Solution:

$$y = C_1 + C_2 e^x - (\tfrac{1}{4}x^4 + x^3 + \tfrac{5}{2}x^2 + 10x)$$

27. $y'' + 2y' + y = 0$

Characteristic equation: $\quad r^2 + 2r + 1 = 0$

$$(r + 1)^2 = 0$$

$$r = -1 \text{ (mult. 2)}$$

$$y_h = C_1 e^{-x} + C_2 x e^{-x}$$

$$F(x) = (4 + x)e^{-x}$$

$$\overline{y}_p = e^{-x}(Ax + B)x^2$$

$$\overline{y}_p{}' = e^{-x}[-Ax^3 + (3A - B)x^2 + 2Bx]$$

$$\overline{y}_p{}'' = e^{-x}[Ax^3 - (6A - B)x^2$$
$$+ (6A - 4B)x + 2B]$$

Substitute into the given D.E.:

$$e^{-x}[Ax^3 - (6A - B)x^2 + (6A - 4B)x + 2B]$$

$$+ 2e^{-x}[-Ax^3 + (3A - B)x^2 + 2Bx]$$

$$+ e^{-x}(Ax^3 + Bx^2)$$

$$= (4 + x)e^{-x}$$

This gives rise to the system of equations

$$\begin{cases} 6A = 1 \\ 2B = 4 \end{cases}$$

which has solution:

$$A = \tfrac{1}{6}, \; B = 2$$

Solution:

$$y = C_1 e^{-x} + C_2 x e^{-x} + e^{-x}(\tfrac{1}{6}x^3 + 2x^2)$$

29. $y'' + y = 0$

Characteristic equation: $\quad r^2 + 1 = 0$

$$r = \pm i$$

$$y_1 = \cos x, \; y_2 = \sin x$$

$$F(x) = \tan x$$

$$\overline{y}_p = u y_1 + v y_2$$

$$\overline{y}_p = u \cos x + v \sin x$$

$$u' = \frac{-\sin x(\tan x)}{1}$$

$$= -\frac{\sin^2 x}{\cos x}$$

$$u = \int \frac{-\sin^2 x}{\cos x} \, dx$$

$$= -\ln|\sec x + \tan x| + \sin x$$

$$v' = \frac{\cos x(\tan x)}{1} = \sin x$$

$$v = \int \sin x \, dx = -\cos x$$

$$\overline{y}_p = (-\ln|\sec x + \tan x| + \sin x)\cos x$$

$$+ (-\cos x)\sin x$$

$$= -\cos x \ln|\sec x + \tan x|$$

Solution:

$$y = C_1\cos x + C_2\sin x - \cos x\ln|\sec x + \tan x|$$

31. $y'' - y' - 6y = 0$

Characteristic equation: $\quad r^2 - r - 6 = 0$

$$(r - 3)(r + 2) = 0$$

$$r = 3, \; -2$$

$$y_1 = e^{3x}, \; y_2 = e^{-2x}$$

$$F(x) = x^2 e^{2x}$$

$$\overline{y}_p = uy_1 + vy_2; \text{ Let } D = y_1 y_2{}' - y_2 y_1{}' = -5e^x$$

$$u' = \frac{-y_2 F(x)}{D}$$

$$= \frac{-e^{-2x}(x^2 e^{2x})}{-5e^x}$$

$$= \tfrac{1}{5}x^2 e^{-x}$$

$$u = \int \tfrac{1}{5}x^2 e^{-x}dx$$

$$= -\tfrac{1}{5}(x^2 + 2x + 2)e^{-x}$$

$$v' = \frac{y_1 F(x)}{D}$$

$$= \frac{e^{3x}(x^2 e^{2x})}{-5e^x}$$

$$= -\tfrac{1}{5}x^2 e^{4x}$$

$$v = \int -\tfrac{1}{5}x^2 e^{4x}dx$$

$$= -\tfrac{1}{160}(8x^2 - 4x + 1)e^{4x}$$

$$y_p = [-\tfrac{1}{5}(x^2 + 2x + 2)e^{-x}]e^{3x}$$

$$+ [-\tfrac{1}{160}(8x^2 - 4x + 1)e^{4x}]e^{-2x}$$

$$= -\tfrac{1}{32}(8x^2 + 12x + 13)e^{2x}$$

Solution:

$$y = C_1 e^{3x} + C_2 e^{-2x} - \tfrac{1}{32}(8x^2 + 12x + 13)e^{2x}$$

33. $y'' + 4y = 0$

Characteristic equation: $\quad r^2 + 4 = 0$

$$r = \pm 2i$$

$$y_1 = \cos 2x, \; y_2 = \sin 2x$$

$$F(x) = \sec 2x \tan 2x$$

Look for a particular solution of the form

$$\overline{y}_p = uy_1 + vy_2$$

$$= u\cos 2x + v\sin 2x$$

$$u' = \frac{(-\sin 2x)(\sec 2x \tan 2x)}{2}$$

$$= -\tfrac{1}{2}\tan^2 2x$$

$$u = \int -\tfrac{1}{2}\tan^2 2x \; dx$$

$$= \tfrac{x}{2} - \tfrac{1}{4}\tan 2x$$

$$v' = \frac{\cos 2x(\sec 2x \tan 2x)}{2}$$

$$= \tfrac{1}{2}\tan 2x$$

$$v = \int \tfrac{1}{2}\tan 2x\ dx = -\tfrac{1}{4}\ln|\cos 2x|$$

Solution:

$$y = C_1\cos 2x + C_2\sin 2x + \tfrac{1}{2}x\cos 2x$$

$$- \tfrac{1}{4}\sin 2x(\ln|\cos 2x|)$$

35. $y'' + 2y' + y = 0$

Characteristic equation: $\qquad r^2 + 2r + 1 = 0$

$$r = 1\ (\text{mult. }2)$$

$$y_1 = e^{-x},\ y_2 = xe^{-x}$$

$$F(x) = e^{-x}\ln x$$

Look for a particular solution of the form

$$\overline{y}_p = u\,e^{-x} + vxe^{-x}$$

$$u' = \frac{(-xe^{-x})e^{-x}(\ln x)}{e^{-x}(-x+1)e^{-x} - xe^{-x}(-e^{-x})}$$

$$= -x\ln x$$

$$u = \int -x\ln x\ dx$$

$$= \frac{x^2}{4} - \frac{x^2}{2}\ln x$$

$$v' = \frac{e^{-x}e^{-x}\ln x}{e^{-2x}}$$

$$= \ln x$$

$$v = \int \ln x\ dx$$

$$= x\ln x - x$$

$$\overline{y}_p = \left[-\frac{x^2}{4}(2\ln x - 1)\right]e^{-x} + [x\ln x - x]xe^{-x}$$

$$= \tfrac{1}{4}x^2(2\ln x - 3)e^{-x}$$

Solution:

$$y = C_1 e^{-x} + C_2 xe^{-x} + \tfrac{1}{4}x^2(2\ln x - 3)e^{-x}$$

37. $y'' - y' = 0$

Characteristic equation: $\qquad r^2 - r = 0$

$$r = 0,\ 1$$

$$y_1 = 1,\ y_2 = e^x$$

$$F(x) = \cos^2 x$$

Look for a particular solution of the form

$$\overline{y}_p = u + ve^x$$

$$u' = \frac{-e^x\cos^2 x}{(1)e^x - e^x(0)}$$

$$= -\cos^2 x$$

$$u = \int -\cos^2 x\ dx$$

$$= -\tfrac{1}{4}\sin 2x - \tfrac{1}{2}x$$

$$v' = \frac{(1)\cos^2 x}{e^x}$$

$$= e^{-x}\cos^2 x$$

$$v = \int e^{-x}\cos^2 x \, dx$$

$$= -\tfrac{1}{5}e^{-x}(\cos^2 x - \sin 2x + 2)$$

$$\overline{y}_p = (1)\left(-\tfrac{1}{4}\sin 2x - \tfrac{1}{2}x\right)$$

$$+ e^x[-\tfrac{1}{5}e^{-x}(\cos^2 x - \sin 2x + 2)]$$

$$= -\tfrac{1}{20}\sin 2x - \tfrac{1}{2}x - \tfrac{1}{5}\cos^2 x - \tfrac{2}{5}$$

Solution:

$$y = C_1 + C_2 e^x - \tfrac{1}{20}\sin 2x - \tfrac{1}{2}x - \tfrac{1}{5}\cos^2 x$$

Note: the term $-\tfrac{2}{5}$ in y_p is "absorbed" by the

arbitrary constant C_1.

39. From Problem 37,

$$y = C_1 + C_2 e^x + 2\left(-\tfrac{1}{20}\sin 2x - \tfrac{1}{2}x - \tfrac{1}{5}\cos^2 x\right)$$

$$y(0) = C_1 + C_2(1) + 0 + 0 - \tfrac{2}{5}$$

$$y' = C_2 e^x - \tfrac{1}{5}\cos 2x + \tfrac{2}{5}\sin 2x - 1$$

$$y'(0) = C_2(1) - \tfrac{1}{5}(1) + 0 - 1$$

Solve the system:

$$\begin{cases} C_1 + C_2 = \tfrac{2}{5} \\ C_2 = \tfrac{6}{5} \end{cases} \quad C_1 = -\tfrac{4}{5}, \ C_2 = \tfrac{6}{5}$$

Solution:

$$y = -\tfrac{4}{5} + \tfrac{6}{5}e^x - \tfrac{1}{10}\sin 2x - x - \tfrac{2}{5}\cos^2 x$$

41. $y'' + 9y = 0$

Characteristic equation: $\quad r^2 + 9 = 0$

$$r = \pm 3i$$

$$y_h = C_1\cos 3x + C_2\sin 3x$$

$$F(x) = 4e^{3x}$$

$$\overline{y}_p = Ae^{3x}; \ \overline{y}_p{}' = 3Ae^{3x}; \ \overline{y}_p{}'' = 9Ae^{3x}$$

Substitute into the given D.E.:

$$9Ae^{3x} + 9Ae^{3x} = 4e^{3x}, \text{ or } A = \tfrac{2}{9}$$

General solution:

$$y = C_1\cos 3x + C_2\sin 3x + \tfrac{2}{9}e^{3x}$$

$$y(0) = C_1 \cdot 1 + C_2 \cdot 0 + \tfrac{2}{9} = 0 \text{ or } C_1 = -\tfrac{2}{9}$$

$$y' = \tfrac{2}{3}\sin 3x + 3C_2\cos 3x + \tfrac{2}{3}e^{3x}$$

$$y'(0) = 0 + 3C_2 \cdot 1 + \tfrac{2}{3} = 2 \text{ or } C_2 = \tfrac{4}{9}$$

Solution: $\ y = -\tfrac{2}{9}\cos 3x + \tfrac{4}{9}\sin 3x + \tfrac{2}{9}e^{3x}$

43. $y'' + 9y = 0$

Characteristic equation: $\quad r^2 + 9 = 0$

$$r = \pm 3i$$

$$y_h = C_1\cos 3x + C_2\sin 3x$$

$$F(x) = x$$

$$\overline{y}_p = Ax + B; \ \overline{y}_p{}' = A; \ \overline{y}_p{}'' = 0$$

Substitute into the given D.E. equation:

$$0 + 9(Ax + B) = x$$

Solve:

$$\begin{cases} 9A = 1 \\ 9B = 0 \end{cases} \quad \text{which has solution:}$$

$$A = \tfrac{1}{9}, \ B = 0$$

Thus,

$$y = C_1\cos 3x + C_2\sin 3x + \tfrac{1}{9}x$$

$$y' = -3C_1\sin 3x + 3C_2\cos 3x + \tfrac{1}{9}$$

$$y'(0) = 4 = 3C_2 + \tfrac{1}{9}; \ C_1 = 0, \ C_2 = \tfrac{35}{27}$$

Solution: $y = \tfrac{35}{27}\sin 3x + \tfrac{1}{9}x$

45. $y'' + y = 0$

Characteristic equation: $\qquad r^2 + 1 = 0$

$$r = \pm i$$

$$y_1 = \cos x, \ y_2 = \sin x$$

$$F(x) = \cot x$$

$$\overline{y}_p = uy_1 + vy_2$$

$$u' = \frac{-\sin x(\cot x)}{1}$$

$$= -\cos x$$

$$u = \int -\cos x \, dx$$

$$= -\sin x$$

$$v' = \frac{\cos x(\cot x)}{1}$$

$$= \frac{\cos^2 x}{\sin x}$$

$$v = \int \frac{\cos^2 x}{\sin x} \, dx$$

$$= -\ln|\csc x + \cot x| + \cos x$$

$$\overline{y}_p = (-\sin x)\cos x$$

$$+ (-\ln|\csc x + \cot x| + \cos x)\sin x$$

$$= -\sin x \ln|\csc x + \cot x|$$

General solution:

$$y = C_1\cos x + C_2\sin x - \sin x \ln|\csc x + \cot x|$$

$$y(\tfrac{\pi}{2}) = 0 = C_1 \cdot 0 + C_2 \cdot 1 - 1 \cdot \ln|1 + 0| = C_2$$

$$y' = -C_1\sin x + C_2\cos x - \cos x \ln|\csc x + \cot x| + 1$$

$$y'(\tfrac{\pi}{2}) = 5 = C_1 \cdot (-1) + C_2 \cdot 0 + 0 + 1; \ C_1 = -4$$

Solution:

$$y = -4\cos x - \sin x \ln|\csc x + \cot x|$$

47. $y'' + 5y' + 6y = F(x)$

From Figure 14.12, we find

$$F(x) = \begin{cases} 2x & \text{for } 0 \leq x \leq 1 \\ 2 & \text{for } 1 < x < 3 \\ -2x + 8 & \text{for } 3 \leq x \leq 4 \end{cases}$$

Characteristic equation: $\qquad r^2 + 5r + 6 = 0$

$$r = -2, \ -3$$

$$y_h = C_1 e^{-2x} + C_2 e^{-3x}$$

If $0 \leq x \leq 1$,

$$\overline{y}_1 = A_1 x + A_2$$

$$\overline{y}_1{}' = A_1$$

$$\overline{y}_1{}'' = 0$$

Substitute into the given D.E. equation:

$$0 + 5A_1 + 6(A_1 x + A_2) = 2x$$

This gives rise to the system of equations:

$$\begin{cases} 6A_1 = 2 \\ 5A_1 + 6A_2 = 0 \end{cases}$$ which has solution: $A_1 = \frac{1}{3}, A_2 = -\frac{5}{18}$

$$\overline{y}_1 = \frac{1}{3}x - \frac{5}{18}$$

If $1 < x < 3$, $\overline{y}_2 = C$; $6C = 2$; $C = \frac{1}{3}$, so

$$\overline{y}_2 = \frac{1}{3}$$

If $3 \le x \le 4$,

$$\overline{y}_3 = B_1 x + B_2$$

$$\overline{y}_3{}' = B_1$$

$$\overline{y}_3{}'' = 0$$

Substitute into the given D.E.:

$$0 + 5B_1 + 6(B_1 x + B_2) = -2x + 8$$

This gives rise to the system of equations:

$$\begin{cases} 6B_1 = -2 \\ 5B_1 + 6B_2 = 8 \end{cases}$$ which has solution: $B_1 = -\frac{1}{3}, B_2 = \frac{29}{18}$

$$\overline{y}_3 = -\frac{1}{3}x + \frac{29}{18}$$

General solution:

$$y = C_1 e^{-2x} + C_2 e^{-3x} + G(x)$$

where

$$G(x) = \begin{cases} \frac{1}{3}x - \frac{5}{18} & \text{for } 0 \le x \le 1 \\ \frac{1}{3} & \text{for } 1 < x < 3 \\ -\frac{1}{3}x + \frac{29}{18} & \text{for } 3x \le x \le 4 \end{cases}$$

49. Substitute the given values into the governing equation from the solution to Problem 48:

$$L \frac{dI}{dt} + RI + \frac{Q}{C} = E$$

$$(1)I'' + 10I' + 9I = 9\cos t$$

Characteristic equation: $r^2 + 10r + 9 = 0$

$$r = -9, -1$$

$$I_h(t) = C_1 e^{-9t} + C_2 e^{-t}$$

$$\overline{y}_p = A_1 \cos t + A_2 \sin t$$

$$\overline{y}_p{}' = -A_1 \sin t + A_2 \cos t$$

$$\overline{y}_p{}'' = -A_1 \cos t - A_2 \sin t$$

Substitute into the given D.E.:

$$(-A_1 \cos t - A_2 \sin t) + 10(-A_1 \sin t$$

$$+ A_2 \cos t) + 9(A_1 \cos t + A_2 \sin t) = 9 \cos t$$

This gives rise to the system of equations

$$\begin{cases} 8A_1 + 10A_2 = 9 \\ -10A_1 + 8A_2 = 0 \end{cases}$$ which has solution: $A_1 = \frac{18}{41}, A_2 = \frac{45}{82}$

General solution:

$$I(t) = C_1 e^{-9t} + C_2 e^{-t} + \tfrac{18}{41}\cos t + \tfrac{45}{82}\sin t$$

$$I(0) = C_1(1) + C_2(1) + \tfrac{18}{41}(1) + 0 = 0$$

$$I'(t) = -9C_1 e^{-9t} - C_2 e^{-t} - \tfrac{18}{41}\sin t + \tfrac{45}{82}\cos t$$

$$I'(0) = -9C_1(1) - C_2(1) - 0 + \tfrac{45}{82} = 0$$

This gives rise to the system of equations

$$\begin{cases} C_1 + C_2 = -\tfrac{18}{41} \\ -9C_1 - C_2 = -\tfrac{45}{82} \end{cases}$$

which has solution: $C_1 = \tfrac{81}{656}$, $C_2 = -\tfrac{9}{16}$

Solution:

$$I(t) = \tfrac{81}{656} e^{-9t} - \tfrac{9}{16} e^{-t} + \tfrac{18}{41}\cos t + \tfrac{45}{82}\sin t$$

51. From Problem 49 where

$$\frac{dE}{dt} = 5\sin t + 5t\cos t$$

$$I'' + 10I' + 9I = 5\sin t + 5t\cos t$$

Characteristic equation: $\quad r^2 + 10r + 9 = 0$

$$r = -1, -9$$

$$I_h(t) = C_1 e^{-t} + C_2 e^{-9t}$$

$$\overline{y}_p = A_1\cos t + A_2\sin t + t(B_1\cos t + B_2\sin t)$$

$$\overline{y}_p{}' = -A_1\sin t + A_2\cos t + B_1\cos t$$

$$+ B_2\sin t + t(-B_1\sin t + B_2\cos t)$$

$$= (A_2 + B_1)\cos t + (B_2 - A_1)\sin t$$

$$+ t(B_2\cos t - B_1\sin t)$$

$$\overline{y}_p{}'' = (2B_2 - A_1)\cos t + (-A_2 - 2B_1)\sin t$$

$$+ t(-B_2\sin t - B_1\cos t)$$

Substitute into the given D.E.:

$$[(2B_2 - A_1)\cos t + (-A_2 - 2B_1)\sin t$$

$$+ t(-B_2\sin t - B_1\cos t)]$$

$$+ 10[(A_2 + B_1)\cos t + (B_2 - A_1)\sin t$$

$$+ t(B_2\cos t - B_1\sin t)]$$

$$+ 9[A_1\cos t + A_2\sin t + t(B_1\cos t + B_2\sin t)]$$

$$= 5\sin t + 5t\cos t$$

Thus, equating the coefficients,

$$A_1 = \frac{-500}{1{,}681} \approx -0.297,$$

$$A_2 = \frac{-225}{3{,}362} \approx -0.067$$

$$B_1 = \frac{10}{41} \approx 0.244,$$

$$B_2 = \frac{25}{82} \approx 0.305$$

$$I(t) \approx C_1 e^{-t} + C_2 e^{-9t} - 0.297\cos t$$

$$- 0.067\sin t + t(0.244\cos t + 0.305\sin t)$$

$$I(0) = C_1 + C_2 - 0.297 = 0$$

$$I'(t) = -C_1 e^{-t} - 9C_2 e^{-9t} + 0.177\cos t$$

$$+ 0.602\sin t + t(0.305\cos t - 0.244\sin t)$$

$$I'(0) = 0 = -C_1 - 9C_2 + 0.177$$

This gives rise to the system of equations

$$\begin{cases} C_1 + C_2 = 0.297 \\ C_1 + 9C_2 = 0.177 \end{cases}$$

which has solution: $C_1 \approx 0.312$, $C_2 \approx -0.015$

Solution:

$I(t) = 0.312\, e^{-t} - 0.015 e^{-9t} - 0.297 \cos t$

$\quad - 0.067 \sin t + t(0.244 \cos t + 0.305 \sin t)$

53. $(1 + x^2)y'' + 2xy' - 2y = 0$

Assume a solution of the form $y = x^n$. Then

$(1+x^2)[n(n-1)x^{n-2}] + 2x[nx^{n-1}] - 2x^n = 0$

is satisfied when $n = 1$, so $y_1 = x$ is one

solution. Assume $y_2 = wx$, so $y_2' = w'x + w$;

$y_2'' = w''x + 2w'$.

$(1+x^2)(w''x + 2w') + 2x(w'x + w) - 2wx = 0$

$(x + x^3)w'' + (4x^2 + 2)w' = 0$

$$\frac{w''}{w'} = \frac{-(4x^2 + 2)}{x + x^3}$$

$$\ln|w'| = \int \frac{-(4x^2+2)}{x + x^3}\, dx$$

$$= -\ln(x^4 + x^2)$$

$$w' = \frac{1}{x^4 + x^2}$$

$$w = -\tan^{-1}x - \frac{1}{x}$$

A second solution is

$$y_2 = x\left(-\tan^{-1}x - \frac{1}{x}\right) = -x\tan^{-1}x - 1$$

and y_1, y_2 are linearly independent since

$$W = y_1 y_2' - y_2 y_1' = \frac{1}{x^2 + 1} \neq 0$$

Next, find a particular solution $\overline{y}_p$ of

$$y'' + \frac{2x}{1+x^2}y' - \frac{2}{1+x^2}y = \frac{\tan^{-1}x}{1+x^2}$$

of the form $\overline{y}_p = uy_1 + wy_2$:

$$u' = \frac{-(-x\tan^{-1}x - 1)\left(\dfrac{\tan^{-1}x}{1+x^2}\right)}{\dfrac{1}{x^2+1}}$$

$$u = \int (x\tan^{-1}x + 1)\tan^{-1}x\, dx$$

$$= \tfrac{1}{2}(x^2 + 1)(\tan^{-1}x)^2$$

$$v' = \frac{x\left(\dfrac{\tan^{-1}x}{1+x^2}\right)}{\dfrac{1}{x^2+1}}$$

$$= x\tan^{-1}x$$

$$v = \int x\tan^{-1}x\, dx$$

$$= \left(\frac{x^2}{2} + \frac{1}{2}\right)\tan^{-1}x - \frac{x}{2}$$

$$\overline{y}_p = \left[\tfrac{1}{2}(x^2 + 1)(\tan^{-1}x)^2\right]x$$

$$+ \left[\left(\frac{x^2}{2} + \frac{1}{2}\right)\tan^{-1}x - \frac{x}{2}\right](-x\tan^{-1}x - 1)$$

$$= -\tfrac{1}{2}\tan^{-1}x + \tfrac{x}{2}$$

Thus, the general solution is

$$y = C_1 x + C_2(x\tan^{-1}x + 1) - \tfrac{1}{2}\tan^{-1}x$$

CHAPTER 14 REVIEW

Proficiency Examination, Page 966

SURVIVAL HINT: *To help you review the concepts of this chapter,* **handwrite** *the answers to each of these questions onto your own paper.*

1. A separable differential equation is one that can be written in the form

$$\frac{dy}{dx} = \frac{g(x)}{f(y)}$$

and then can be solved by separating the variables and integrating each side.

2. A differential equation of the form

$$M(x,\ y)\ dx + N(x,\ y)\ dy = 0$$

said to be a homogeneous differential equation if it can be written in the form

$$\frac{dy}{dx} = f(\tfrac{y}{x})$$

In other words, dy/dx is isolated on one side of the equation and the other side can be expressed as a function of y/x.

3. A first-order linear differential equation is one of the form

$$\frac{dy}{dx} + p(x)y = q(x)$$

Its general solution is given by

$$y = \frac{1}{I(x)}\left[\int I(x)\ q(x)\ dx + C\right]$$

where $I(x)$ is the integrating factor

$$I(x) = e^{\int p(x)\mathrm{d}x}$$

4. An exact differential equation is one that can be written in the general form

$$M(x,\ y)\,dx + N(x,\ y)\,dy = 0$$

where M and N are functions of x and y that satisfy the cross-derivative test

$$\frac{\partial M}{\partial y} = \frac{\partial N}{\partial x}$$

5. Euler's method is a procedure for approximating a solution of the initial value problem

$$\frac{dy}{dx} = f(x,\ y),\ y(x_0) = y_0$$

It depends on the fact that the portion of the solution curve near $(x_n,\ y_n)$ is close to the line

$$y = y_n + f(x_n,\ y_n)(x - x_n)$$

6. The functions $y_1,\ y_2,\ \ldots,\ y_n$ are said to be *linearly independent* if the equation

$$C_1 y_1 + C_2 y_2 + \ldots + C_n y_n = 0$$

has only the trivial solution

$$C_1 = C_2 = \ldots = C_n = 0$$

Otherwise the y_k's are *linearly dependent*.

7. The *Wronskian* $W(y_1, y_2, \ldots, y_n)$ of n functions $y_1, y_2, \ldots, y_n$ having $n - 1$ derivatives on an interval I is defined to be the determinant function $W(y_1, y_2, \ldots, y_n)$

$$= \begin{vmatrix} y_1 & y_2 & \cdots & y_n \\ y_1' & y_2' & \cdots & y_n' \\ \vdots & \vdots & & \vdots \\ y_1^{(n-1)} & y_2^{(n-1)} & \cdots & y_n^{(n-1)} \end{vmatrix}$$

The functions $y_1, \cdots, y_n$ are linearly independent if and only if $W \neq 0$ on I.

8. **a.** The *characteristic equation* of

$$ay'' + by' + cy = 0 \text{ is the equation}$$

$$ar^2 + br + c = 0.$$

b. If r_1 and r_2 are the roots of the characteristic equation $ar^2 + br + c = 0$, then the general solution of the homogeneous linear equation $ay'' + by' + cy = 0$ can be expressed in one of these forms:

$b^2 - 4ac > 0$: The general solution is

$$y = C_1 e^{r_1 x} + C_2 e^{r_2 x}$$

$b^2 - 4ac = 0$: The general solution is

$$y = C_1 e^{-bx/2} + C_2 x e^{-bx/2}$$

$$= (C_1 + C_2 x) e^{-bx/2}$$

$b^2 - 4ac < 0$: The general solution is

$$y = e^{-bx/2} \left\{ C_1 \cos \left(\frac{\sqrt{4ac - b^2}}{2a} x \right) \right. $$
$$\left. + C_2 \sin \left(\frac{\sqrt{4ac - b^2}}{2a} x \right) \right\}$$

9. Let y_p be a particular solution of the nonhomogeneous second-order linear equation $y'' + ay' + by = F(x)$. Let y_h be the general solution of the related homogeneous equation

$$y'' + ay' + by = 0.$$

Then the general solution of

$$y'' + ay' + by = F(x),$$

is given by the sum $y = y_h + y_p$.

10. To solve $y'' + ay' + by = F(x)$ when $F(x)$ is one of the following forms:

(1) $F(x) = P_n(x)$, a polynomial of degree n

(2) $F(x) = P_n(x) e^{kx}$

(3) $F(x) = e^{kx}[P_n(x) \cos\alpha x + Q_n(x) \sin\alpha x]$,

where $Q_n(x)$ is another polynomial of

degree n

Outline of the procedure:

1. The solution is of the form $y = y_h + y_p$,

   where y_h is the general solution and y_p is

   a particular solution.

2. Find y_h by solving the homogeneous

   equation $y'' + ay' + by = 0$

3. Find y_p by picking an appropriate trial

   solution $\overline{y}_p$:

   a. Form: $P_n(x) = c_n x^n + \ldots + c_1 x + c_0$

   Corresponding trial expression:

   $A_n x^n + \ldots + A_1 x + A_0$

   b. Form: $P_n(x)e^{kx}$

   Corresponding trial expression:

   $[A_n x^n + A_{n-1} x^{n-1} + \ldots + A_0]e^{kx}$

   c. Form: $e^{kx}[P_n(x) \cos\alpha x + Q_n(x) \sin\alpha x]$

   Corresponding trial expression:

   $e^{kx}[(A_n x^n + \ldots + A_0)\cos\alpha x$

   $\qquad + (B_n x^n + \ldots + B_0)\sin\alpha x]$

4. If no term in the trial expression $\overline{y}_p$

appears in the general homogeneous

solution y_h, the particular solution can be

found by substituting $\overline{y}_p$ into the

equation $y'' + ay' + by = F(x)$ and

solving for the undetermined coefficients.

5. If any term in the trial expression $\overline{y}_p$

   appears in y_h, multiply $\overline{y}_p$ by x^k, where k

   is the smallest integer such that no term

   in $x^k \overline{y}_p$ is in y_k. Then proceed as in Step

   4, using $x^k \overline{y}_p$ as the trial solution.

11. To find the general solution of

$$y'' + P(x)y' + Q(x)y = F(x)$$

1. Find the general solution,

$$y_h = C_1 y_1 + C_2 y_2$$

to the related homogeneous equation

$$y'' + Py' + Qy = 0.$$

2. Set $y_p = uy_1 + vy_2$ and substitute into

   the formulas:

$$u' = \frac{-y_2 F(x)}{y_1 y_2' - y_2 y_1'}$$

$$v' = \frac{y_1 F(x)}{y_1 y_2' - y_2 y_1'}$$

3. Integrate u' and v' to find u and v.

4. A particular solution is $y_p = uy_1 + vy_2$,

   and the general solution is $y = y_h + y_p$.

12. $$\frac{dy}{dx} = \sqrt{\frac{1 - y^2}{1 + x^2}}$$

$$\int \frac{dy}{\sqrt{1 - y^2}} = \int \frac{dx}{\sqrt{1 + x^2}}$$

$$\sin^{-1} y = \sinh^{-1} x + C$$

13. $$\frac{x}{y^2} dx - \frac{x^2}{y^3} dy = 0$$

$$\int x^{-1} dx = \int y^{-1} dy$$

$$\ln|x| = \ln|y| + C$$

$$y = Bx$$

14. $$\frac{dy}{dx} = \frac{2x + y}{3x} = \frac{2}{3} + \frac{y}{3x}$$

$$\frac{dy}{dx} - \frac{1}{3x} y = \frac{2}{3}$$

$$p(x) = -\frac{1}{3x}, \ q(x) = \frac{2}{3}$$

$$I(x) = e^{\int -1/(3x) dx} = x^{-1/3}$$

$$y = x^{1/3} \left[\int x^{-1/3} \left(\frac{2}{3} \right) dx + C \right]$$

$$= x^{1/3} \left[x^{2/3} + C \right]$$

$$= x + C \sqrt[3]{x}$$

15. $$xy \ dy = (x^2 - y^2) \ dx$$

$$\frac{dy}{dx} = \frac{x^2 - y^2}{xy}$$

$$= \frac{x}{y} - \frac{y}{x}$$

Let $f(v) = v^{-1} - v$ where $v = \frac{y}{x}$.

$$\int \frac{dv}{\frac{1}{v} - v - v} = \int \frac{dx}{x}$$

$$\int \frac{v \ dv}{1 - 2v^2} = \int \frac{dx}{x}$$

$$-\frac{1}{4} \ln|1 - 2v^2| = \ln|x| + C_1$$

$$1 - 2v^2 = Cx^{-4}$$

$$1 - 2\left(\frac{y}{x} \right)^2 = Cx^{-4}$$

$$x^2(x^2 - 2y^2) = C$$

16. $$x^2 \ dy - (x^2 + y^2) \ dx = 0$$

$$\frac{dy}{dx} = \frac{x^2 + y^2}{x^2} = 1 + \left(\frac{y}{x} \right)^2$$

Let $f(x) = 1 + v^2$ where $v = \frac{y}{x}$

$$\int \frac{dv}{1 + v^2 - v} = \int \frac{dx}{x}$$

$$\frac{2}{\sqrt{3}} \tan^{-1} \left[\frac{2}{\sqrt{3}} \left(\frac{y}{x} - \frac{1}{2} \right) \right] = \ln|x| + C$$

17. $$y'' + 2y' + 2y = \sin x$$

Characteristic equation: $\quad r^2 + 2r + 2 = 0$

$$r = -1 \pm i$$

$$y_h = e^{-x}(C_1\cos x + C_2\sin x)$$

$$\overline{y}_p = A\cos x + B\sin x$$

$$\overline{y}_p{}' = -A\sin x + B\cos x$$

$$\overline{y}_p{}'' = -A\cos x - B\sin x$$

Substitute into the given D.E.:

$$(-A\cos x - B\sin x) + 2(-A\sin x + B\cos x)$$

$$+ 2(A\cos x + B\sin x) = \sin x$$

This gives rise to the system of equations

$$\begin{cases} A + 2B = 0 \\ -2A + B = 1 \end{cases}$$ which has solution: $A = -\frac{2}{5}, B = \frac{1}{5}$

Solution:

$$y = e^{-x}(C_1\cos x + C_2\sin x) - \frac{2}{5}\cos x + \frac{1}{5}\sin x$$

18. $(3x^2 e^{-y} + y^{-2} + 2xy^{-3})\,dx$

$$+ (-x^3 e^{-y} - 2xy^{-3} - 3x^2 y^{-4})\,dy = 0$$

$$M(x, y) = 3x^2 e^{-y} + y^{-2} + 2xy^{-3}$$

$$N(x, y) = -x^3 e^{-y} - 2xy^{-3} - 3x^2 y^{-4}$$

$$\frac{\partial M}{\partial y} = -3x^2 e^{-y} - 2y^{-3} - 6xy^{-4}$$

$$\frac{\partial N}{\partial x} = -3x^2 e^{-y} - 2y^{-3} - 6xy^{-4}$$

Since $\frac{\partial M}{\partial y} = \frac{\partial N}{\partial x}$, the D.E. is exact.

$$f(x, y) = \int (3x^2 e^{-y} + y^{-2} + 2xy^{-3})\,dx$$

$$= x^3 e^{-y} + xy^{-2} + x^2 y^{-3} + u(y)$$

$$\frac{\partial f}{\partial y} = -x^3 e^{-y} - 2xy^{-3} - 3x^2 y^{-4} + u'(y)$$

Compare with $N(x, y)$ to see that

$u'(y) = 0$, so $u(y) = C$. Thus,

$$x^3 e^{-y} + xy^{-2} + x^2 y^{-3} = C$$

19. Governing equation:

$$mx'' + cx' + kx = 0$$

where m = mass, c = damping factor, and

k = spring constant. Also, $x(0) = \frac{4}{12} = \frac{1}{3}$ ft

and $x'(0) = 0$; that is, no initial velocity since

the spring is released. Given:

$m = \frac{8}{32} = \frac{1}{4}$ slug, $c = 0.8$, and $k = 30$.

$$0.25x'' + 0.8x' + 30x = 0$$

Characteristic equation:

$$0.25r^2 + 0.8r + 30 = 0$$

$$r \approx -1.60 \pm 10.84i$$

General solution:

$$x = e^{-1.60t}[C_1\cos 10.84t + C_2\sin 10.84t]$$

$$x(0) = \frac{1}{3} = (1)[C_1(1) + C_2(0)]; \ C_1 = \frac{1}{3}$$

$$x' = -1.6e^{-1.60t}[C_1\cos 10.84t + C_2\sin 10.84t]$$

$$+ e^{-1.60t}[-10.84C_1\sin 10.84t$$

$$+ 10.84 C_2 \cos 10.84t]$$

$$x'(0) = 0 = -1.60(1)[C_1(1) + C_2(0)]$$

$$+ (1)[0 + 10.84 C_2(1)]$$

Since $C_1 = \frac{1}{3}$, we see $C_2 \approx 0.05$

Solution:

$$x(t) = e^{-1.60t}[0.33 \cos 10.84t + 0.05 \sin 10.84t]$$

20. The governing differential equation is

$$0.1 \frac{dI}{dt} + 25I + \frac{Q}{(200)10^{-6}} = 50 \cos 100t$$

where Q is the charge and $dQ/dt = I$.

Converting this equation into an equation for

Q, we obtain

$$\frac{d^2Q}{dt^2} + 250 \frac{dQ}{dt} + 50,000Q = 500 \cos 100t$$

Characteristic equation:

$$r^2 + 250r + 50,000 = 0$$

$$r = -125 \pm 25\sqrt{55}\,i$$

$$Q_h(t) = e^{-125t}(C_1 \cos 25\sqrt{55}\,t + C_2 \sin 25\sqrt{55}\,t)$$

$$\overline{y}_p = A \cos 100t + B \sin 100t$$

$$\overline{y}_p{}' = 100(-A \sin 100t + B \cos 100t)$$

$$\overline{y}_p{}'' = 10,000(-A \cos 100t - B \sin 100t)$$

Substitute into the given D.E.:

$$10,000(-A \cos 100t - B \sin 100t)$$

$$+ 25,000(-A \sin 100\ t + B \cos 100t)$$

$$+ 50,000(A \cos 100t + B \sin 100t)$$

$$= 500 \cos 100t$$

This gives rise to the system of equations

$$\begin{cases} 40,000A + 25,000B = 500 \\ -25,000A + 40,000B = 0 \end{cases}$$

which has solution

$$A = \frac{4}{445} \approx 0.00899, \; B = \frac{1}{178} \approx 0.0056$$

so

$$Q(t) = e^{-125t}[C_1 \cos 25\sqrt{55}\,t + C_2 \sin 25\sqrt{55}\,t]$$

$$+ \frac{4}{445} \cos 100t + \frac{1}{178} \sin 100t$$

To find the current, we differentiate

$$I = \frac{dQ}{dt} \approx e^{-125t}[(-125C_1 + 185.4\,C_2)\cos 185.4t]$$

$$+ e^{-125t}(-185.4C_1 - 125C_2)\sin 185.4t]$$

$$+ 0.562 \cos 100t - 0.899 \sin 100\ t$$

Since $Q(0) = 0$ and $I(0) = 0$, we have

$$\begin{cases} C_1 + 0.00899 = 0 \\ -125C_1 + 185.4C_2 + 0.562 = 0 \end{cases}$$

which has solution $C_1 \approx -0.008992$,

$C_2 \approx -0.0091$, and the solution is

$$I(t) = e^{-125t}[-0.562 \cos 185.4t + 2.79 \sin 185.4t]$$

$$+ 0.562 \cos 100t - 0.899 \sin 100t$$

APPENDIX A

Introduction to the Theory of Limits

1. Prove $\lim_{x \to 1} (2x - 5) = -3$; we see

$f(x) = 2x - 5$ and $L = -3$. For any chosen value of ϵ we want the distance between $f(x)$ and L to be less than ϵ. In absolute value notation this is

$$\left| f(x) - L \right| < \epsilon$$

For our function and limit we need

$$\left| (2x - 5) - (-3) \right| < \epsilon$$
$$\left| 2x - 2 \right| < \epsilon$$
$$2 \left| x - 1 \right| < \epsilon$$

So the distance between x and 1 needs to be less than $\frac{\epsilon}{2}$. Therefore, let $\delta = \frac{\epsilon}{2}$.

3. $\lim_{x \to 1} (3x + 1) = 4 \neq 5$; the interval on the y-axis in the neighborhood corresponding to $x = 1$ can be arbitrarily restrictive.

$$0 < \left| x - 1 \right| < \delta$$

must apply and the given statement is false. The actual limit is 4.

5. Prove $\lim_{x \to 2} (x^2 + 2) = 6$; we see

$f(x) = x^2 + 2$ and $L = 6$. For any chosen value of ϵ we want the distance between $f(x)$ and L to be less than ϵ. In absolute value notation this is

$$\left| f(x) - L \right| < \epsilon$$

For our function and limit we need

$$\left| (x^2 + 2) - (6) \right| < \epsilon$$
$$\left| x^2 - 4 \right| < \epsilon$$
$$\left| x - 2 \right| \left| x + 2 \right| < \epsilon$$

As we approach 2, $\left| x + 2 \right|$ is less than 5.

If the distance between x and 2, written as $\left| x - 2 \right|$, is to be less than δ, we need

$\left| x - 2 \right| (5) < \epsilon = 5\delta$. Therefore, let

$\delta = $ minimum of $\{1, \frac{\epsilon}{5}\}$.

7. Prove $\lim_{x \to 2} (x + 3) = 5$; we see

$f(x) = x + 3$ and $L = 5$.

$$\begin{aligned} \left| f(x) - L \right| &= \left| (x + 3) - 5 \right| \\ &= \left| x - 2 \right| \\ &< \delta \end{aligned}$$

Choose $\delta = \epsilon$.

9. $\lim_{x \to 2} (3x + 7) = 1$; we see

$f(x) = 3x + 7$ and $L = 1$.

$$\begin{aligned} \left| f(x) - L \right| &= \left| (3x + 7) - 1 \right| \\ &= \left| 3x + 6 \right| \\ &= 3 \left| x + 2 \right| \\ &< 3\delta \end{aligned}$$

Choose $\delta = \frac{\epsilon}{3}$.

11. $\lim_{x \to 2} (x^2 + 2) = 6$; we see

$f(x) = x^2 + 2$ and $L = 6$.

$$\begin{aligned} \left| f(x) - L \right| &= \left| (x^2 + 2) - 6 \right| \\ &= \left| x^2 - 4 \right| \\ &= \left| x - 2 \right| \left| x + 2 \right| \end{aligned}$$

Choose $\delta = \min(1, \frac{\epsilon}{5})$, as $\left| x - 2 \right| < \delta$ and

$\left| x - 2 \right| < 1$ means

$$-1 < x < 3$$

or

$$1 < x + 2 < 5$$

so

$$\left| x - 2 \right| \left| x + 2 \right| < \delta(5)$$
$$= \epsilon$$

13. In order for $f(x)$ to be continuous at $x = 0$, $\lim_{x \to 0} f(x)$ must equal $f(0)$. To show that $\lim_{x \to 0} \sin \frac{1}{x} = 0$ we need to show that for any $\epsilon > 0$ there exists a $\delta > 0$ such that

$\left| f(x) - L \right| < \epsilon$ when $\left| x - 0 \right| < \delta$.

Arbitrarily letting $\epsilon = 0.5$ we need to find a δ-interval about 0 such that $\left| \sin \frac{1}{x} \right| < 0.5$.

However any interval about 0 contains a point $x = \frac{2}{\pi n}$ (with n odd) where $\sin \frac{1}{x}$

$= \sin \frac{\pi n}{2} = \pm 1$. Therefore, there does not exist a δ-interval about 0 such that

$\left| \sin \frac{1}{x} \right| < 0.5$, and $f(x)$ must be discontinuous at $x = 0$.

15. Given $\epsilon > 0$ there exists δ_1 and δ_2 such that

$\left| f(x) - L_1 \right| < \frac{\epsilon}{2}$ for $0 < \left| x - c \right| < \delta_1$

and

$\left| g(x) - L_2 \right| < \frac{\epsilon}{2}$ for $0 < \left| x - c \right| < \delta_2$.

Then

$$\left| f(x) \ - \ g(x) \ - \ (L_1 \ - \ L_2) \right|$$

$$= \left| [f(x) \ - \ L_1] \ - \ [g(x) \ - \ L_2] \right|$$

$$\leq \left| f(x) \ - \ L_1 \right| + \left| g(x) \ - \ L_2 \right|$$

$$< \frac{\epsilon}{2} + \frac{\epsilon}{2}$$

$$= \epsilon$$

if $0 < |x \ - \ c| \ < \ \min(\delta_1, \delta_2)$.

Thus $\lim_{x \to c} \left| f(x) \ - \ g(x) \right| = L_1 \ - \ L_2$.

17. Given $\epsilon > 0$,

$$\left| f(x) \ - \ 0 \right| < \sqrt{\epsilon} \text{ for } 0 < |x \ - \ c| < \delta_1$$

$$\left| g(x) \ - \ 0 \right| < \sqrt{\epsilon} \text{ for } 0 < |x \ - \ c| < \delta_2$$

So

$$\left| f(x)g(x) \ - \ 0 \right| = \left| f(x) \ - \ 0 \right| \left| g(x) \ - \ 0 \right|$$

$$< \ \sqrt{\epsilon} \ \sqrt{\epsilon}$$

$$= \epsilon$$

if $0 < |x \ - \ c| < \min(\delta_1, \delta_2)$. Thus,

$\lim_{x \to c} f(x)g(x) = 0$.

19. Let $h(x) = f(x) \ - \ g(x)$. Since $f(x) \geq g(x)$ throughout an open interval containing c, the limit limitation theorem guarantees that

$$\lim_{x \to c} h(x) = \lim_{x \to c} [f(x) \ - \ g(x)]$$

$$= \lim_{x \to c} f(x) \ - \ \lim_{x \to c} g(x) \ \geq \ 0 \text{ or}$$

$\lim_{x \to c} f(x) \ \geq \ \lim_{x \to c} g(x)$.

21. By hypothesis, $\lim_{x \to c} f(x) = L$, and by Problem 20, $\lim_{x \to c} \left| f(x) \right| = |L|$, or if $\epsilon = \frac{|L|}{2} > 0$ then there exists δ so that $\left| \left| f(x) \right| \ - |L| \right| < \frac{|L|}{2}$ for all δ such that $0 < |x \ - \ c| < \delta$. Thus,

$$-\frac{|L|}{2} < \left(\left| f(x) \right| - |L| \right) < \frac{|L|}{2} \text{ and}$$

$$\frac{|L|}{2} < \left| f(x) \right| < \frac{3|L|}{2}.$$

23. $\frac{1}{4}[(f + g)^2 \ - \ (f \ - \ g)^2]$

$$= \frac{1}{4}[f^2 + 2fg + g^2 \ - \ f^2 + 2fg \ - \ g^2]$$

$$= \frac{1}{4}[2fg + 2fg]$$

$$= fg$$

Thus,

$$\lim_{x \to c} fg = \lim_{x \to c} \frac{1}{4}[(f + g)^2 \ - \ (f \ - \ g)^2]$$

$$= \frac{1}{4} \lim_{x \to c} [(f + g)^2 \ - \ (f \ - \ g)^2]$$

by Problem 14. By Problem 15, it also equals

$$\frac{1}{4}\left[\lim_{x \to c}(f + g)^2 \ - \ \lim_{x \to c}(f \ - \ g)^2 \right]$$

by Problem 22 it becomes

$$\frac{1}{4}\left[(L + M)^2 \ - \ (L \ - \ M)^2 \right] = LM$$

25. **a.** Since f is continuous at L, we have $\lim_{w \to L} f(w) = f(L)$ which means that there exists a δ_1 such that $\left| f(w) \ - \ f(L) \right| < \epsilon_1$

whenever $|w - L| < \delta_1$.

b. Since $\lim_{x \to c} g(x) = L$, by hypothesis,

$$|g(x) - L| < \epsilon_2$$

whenever $0 < |x - c| < \delta_2$.

Let $w = g(x)$ in part **a**, and $\epsilon_2 = \delta_1$.

Then

$$|f[g(x)] - f(L)| < \epsilon_1$$

whenever $0 < |x - c| < \delta_2$.